Encyclopaedia of MODERN GENETICS ANALYSIS INTEGRATING GENES AND GENOMES

Encyclopaedia of MODERN GENETICS ANALYSIS INTEGRATING GENES AND GENOMES

Volume III: Fundamentals of Modern Statistical Genetics

Contributors

Jianliang Ni, Weidong Mao., and Tao He et al.

AURIS Reference

www.aurisreference.com

Encyclopaedia of Modern Genetics Analysis Integrating Genes and Genomes

Volume III: Fundamentals of Modern Statistical Genetics

Contributors: Jianliang Ni, Weidong Mao., and Tao He al.

Published by Auris Reference Limited

www.aurisreference.com

United Kingdom

Fundamentals of Modern Statistical Genetics

ISBN: 978-1-78154-772-4 (Set of 3 Volumes)

British Library Cataloguing in Publication Data
A CIP record for this book is available from the British Library

Printed in the United Kingdom

Exclusively distributed by CBS Publishers & Distributors Pvt. Ltd.

Sales & Distribution Rights only for India, Pakistan, Bangladesh, Sri Lanka, Nepal and Bhutan.This book is not to be sold outside these territories.

Contents

List of Abbreviations

AFLP	Amplified Fragment Length Polymorphism
BWT	Burrows-Wheeler Transform
CCA	Canonical Correlation Analysis
CTS	Cecum total score
CHDWB	Center for Health Discovery and Well Being
COAG	Clarification of Optimal Anticoagulation through Genetics
CA	Cluster Analysis
CNV	copy number variation
DSMB	Data Safety and Monitoring Board
ED	Euclidian Distance
EM	Expectation maximization
eQTL	Expression quantitative trait loci (
FDR	False discovery rate
FBAT	Family-based association tests
FWER	Family-wise error rate
GWAS	genome wide association study
GWA	genome-wide association
HWE	Hardy–Weinberg equilibrium
HMM	Hidden Markov model
ITAMs	immunoreceptor tyrosine-based activation motifs
IME	Individual mutation error
ISE	Individual switch error
INR	International Normalized Ratio
IQR	Interquartile range
KIRs	Killer cell immunoglobin-like receptors
LDA	Linear Discriminant Analysis
LD	Linkage disequilibrium
LMN	Lymphocytes, monocytes, neutrophils
MHC	Major histocompatibility complex
DSM	Manual of mental disorders
MLEs	`Maximum likelihood estimates
MANOVA	Multivariate Analysis of Variance
NGS	Next generation sequencing (
ORs	odds ratios
PLS	Partial Least Squares
PADI4	Peptidyl arginine deiminase-4
PCR	Polymerase chain reaction
PCA	Principal Component Analysis
QNM	Quantile normalization
QTL	Quantitative trait locus

RFLPs	Restriction Fragment Length Polymorphic
RIN	RNA Integrity Number,
RMSE	Root of mean squared error
SSR	Simple Sequence Repeats
SMRT	Single Molecule Real Time
SNPs	single nucleotide polymorphism
SBWR	Spleen/body weight ratio
SD	standard deviations
TDT	Transmission/disequilibrium test
WGCNA	Weighted Correlation Network Analysis
WTCCC	Wellcome Trust Case Control Consortium

List of Contributors

Tao He
Department of Statistics and Probability, Michigan State University, East Lansing, Michigan, United States of America

Jian Sa
Department of Statistics and Probability, Michigan State University, East Lansing, Michigan, United States of America
Division of Medical Statistics, School of Public Health, Shanxi Medical University, Taiyuan, Shanxi, China

Ping-Shou Zhong
Department of Statistics and Probability, Michigan State University, East Lansing, Michigan, United States of America

Yuehua Cui
Department of Statistics and Probability, Michigan State University, East Lansing, Michigan, United States of America
Division of Medical Statistics, School of Public Health, Shanxi Medical University, Taiyuan, Shanxi, China

Bernd Genser
Department of Epidemiology and Population Health, London School of Hygiene and Tropical Medicine, London, UK
Instituto de Saúde Coletiva, Federal University of Bahia, Salvador, Brazil

Philip J Cooper
Centre for Infection, St George's University of London, London, UK
Instituto de Microbiologia, Universidad San Francisco de Quito, Quito, Ecuador

Maria Yazdanbakhsh
Department of Parasitology, Leiden University Medical Center, Leiden, The Netherlands

Mauricio L Barreto
Instituto de Saúde Coletiva, Federal University of Bahia, Salvador, Brazil

Laura C Rodrigues
Instituto de Saúde Coletiva, Federal University of Bahia, Salvador, Brazil

Mogens Fenger

Clinical Biochemistry, Molecular Biology, and Genetics, KBA339, Hvidovre, Denmark

Yao Li
Center for Computational Biology, Beijing Forestry University, Beijing, People's Republic of China
Department of Statistics, West Virginia University, Morgantown, West Virginia, United States of America

Yunqian Guo
Center for Computational Biology, Beijing Forestry University, Beijing, People's Republic of China

Jianxin Wang
Center for Computational Biology, Beijing Forestry University, Beijing, People's Republic of China

Wei Hou
Department of Biostatistics, University of Florida, Gainesville, Florida, United States of America

Myron N. Chang
Department of Biostatistics, University of Florida, Gainesville, Florida, United States of America

Duanping Liao
Department of Public Health Sciences, Penn State College of Medicine, Hershey, Pennsylvania, United States of America

Rongling Wu
Center for Computational Biology, Beijing Forestry University, Beijing, People's Republic of China
Department of Public Health Sciences, Penn State College of Medicine, Hershey, Pennsylvania, United States of America

Miaoqing Shen
Boyce Thompson Institute for Plant Research, Ithaca, New York, United States of America
United States Department of Agriculture, Agricultural Research Service, RW Holley Center for Agriculture and Health, Ithaca, New York, United States of America

Corey D. Broeckling
Colorado State University, Proteomics and Metabolomics Facility, Fort Collins, Colorado, United States of America

Elly Yiyi Chu
United States Department of Agriculture, Agricultural Research Service, RW Holley Center for Agriculture and Health, Ithaca, New York, United States of America

Gregory Ziegler
United States Department of Agriculture, Agricultural Research Service, Plant Genetics Research Unit, St. Louis, Missouri, United States of America
Donald Danforth Plant Science Center, St. Louis, Missouri, United States of America

Ivan R. Baxter
United States Department of Agriculture, Agricultural Research Service, Plant Genetics Research Unit, St. Louis, Missouri, United States of America

Jessica E. Prenni
Colorado State University, Proteomics and Metabolomics Facility, Fort Collins, Colorado, United States of America

Owen A. Hoekenga
United States Department of Agriculture, Agricultural Research Service, RW Holley Center for Agriculture and Health, Ithaca, New York, United States of America

A. N. Diaz-Lacava
Institute for Medical Biometry, Informatics, and Epidemiology, University of Bonn, 53127 Bonn, Germany
Cologne Center for Genomics, University of Cologne, 50931 Cologne, Germany
DNA Analysis Unit, Official College of Pharmacists and Biochemists, C1184ABA Buenos Aires, Argentina

M. Walier
Institute for Medical Biometry, Informatics, and Epidemiology, University of Bonn, 53127 Bonn, Germany

D. Holler
Institute for Medical Biometry, Informatics, and Epidemiology, University of Bonn, 53127 Bonn, Germany

M. Steffens
Institute for Medical Biometry, Informatics, and Epidemiology, University of Bonn, 53127 Bonn, Germany

C. Gieger
Research Unit of Molecular Epidemiology, Helmholtz Zentrum München, German Research Center for Environmental Health, 85764 Neuherberg, Germany
Institute of Epidemiology II, Helmholtz Zentrum München, German Research Center for Environmental Health, 85764 Neuherberg, Germany

C. Furlanello
FBK, 38122 Trento, Italy

C. Lamina
Division of Genetic Epidemiology, Department of Medical Genetics, Molecular and Clinical Pharmacology, Medical University of Innsbruck, 6020 Innsbruck, Austria

H. E. Wichmann
Institute of Medical Informatics, Biometry and Epidemiology, Chair of Epidemiology, Ludwig-Maximilians-University, 81377 Munich, Germany
Institute of Epidemiology I, Helmholtz Zentrum München, German Research Center for Environmental Health, 85764 Neuherberg, Germany
Institute of Medical Statistics and Epidemiology, Technical University Munich, 81675 Munich, Germany

T. Becker
Institute for Medical Biometry, Informatics, and Epidemiology, University of Bonn, 53127 Bonn, Germany
German Center for Neurodegenerative Diseases (DZNE), 53127 Bonn, Germany

C. O. Aremu
Department of Crop Production and Soil Science Ladoke Akintola University of Technology, Ogbomoso, Oyo state Landmark University, Omu-Aran Nigeria

Alex Clarke
Imperial College London, Faculty of Medicine, Section of Molecular Genetics and Rheumatology, Hammersmith Hospital

Timothy J Vyse
Imperial College London, Faculty of Medicine, Section of Molecular Genetics and Rheumatology, Hammersmith Hospital

Benjamin French
Department of Biostatistics and Epidemiology, University of Pennsylvania School of Medicine, 423 Guardian Drive, Philadelphia, Pennsylvania 19104 USA

Jungnam Joo
Office of Biostatistics Research, National Heart, Lung and Blood Institute, 6701 Rockledge Drive MSC 7913, Bethesda, Maryland 20892 USA

Nancy L Geller
Office of Biostatistics Research, National Heart, Lung and Blood Institute, 6701 Rockledge Drive MSC 7913, Bethesda, Maryland 20892 USA

Stephen E Kimmel
Department of Biostatistics and Epidemiology, University of Pennsylvania School of Medicine, 423 Guardian Drive, Philadelphia, Pennsylvania 19104 USA

Yves Rosenberg
Atherothrombosis and Coronary Artery Disease Branch, National Heart, Lung and Blood Institute, 6701 Rockledge Drive MSC 7956, Bethesda, Maryland 20892 USA

Jeffrey L Anderson
JL Sorenson Heart and Lung Center, Intermountain Medical Center, 5121 S Cottonwood St, Murray, Utah 84107 USA

Brian F Gage
Division of General Medical Sciences, Washington University School of Medicine, 660 S Euclid Ave, St. Louis, Missouri 63110 USA

Julie A Johnson
Department of Pharmacotherapy and Translational Research, University of Florida College of Pharmacy, Box 100486, Gainesville, Florida 32610 USA.

Jonas H Ellenberg
Department of Biostatistics and Epidemiology, University of Pennsylvania School of Medicine, 423 Guardian Drive, Philadelphia, Pennsylvania 19104 USA

Michael Lässig
Institut für Theoretische Physik, Universität zu Köln

Heping Zhang
Yale School of Public Health, Yale University, 60 College Street, New Heaven, Connecticut 06520-8034, USA

Lin Li
Department of Biostatistics, Harvard School of Public Health, Boston, MA, USA

Michael Kabesch
Department of Pediatric Pneumology and Allergy, KUNO University Children's Hospital Regensburg, Regensburg, Germany

Emmanuelle Bouzigon
INSERM, Genetic Variation and Human Diseases Unit, U946, Paris, France
Sorbonne Paris Cité, Institut Universitaire d'Hématologie, Université Paris Diderot, Paris, France

Florence Demenais
INSERM, Genetic Variation and Human Diseases Unit, U946, Paris, France
Sorbonne Paris Cité, Institut Universitaire d'Hématologie, Université Paris Diderot, Paris, France

Martin Farrall
Wellcome Trust Centre for Human Genetics, Oxford, UK

Miriam F. Moffatt
Molecular Genetics and Genomics Section, National Heart and Lung Institute, Imperial College London, London, UK

Xihong Lin
Department of Biostatistics, Harvard School of Public Health, Boston, MA, USA

Liming Liang
Department of Biostatistics, Harvard School of Public Health, Boston, MA, USA
Department of Epidemiology, Harvard School of Public Health, Boston, MA, USA

Jihua Wu
Section on Statistical Genetics, Department of Biostatistics, University of Alabama at Birmingham, Birmingham, AL, USA

Guo-Bo Chen
Section on Statistical Genetics, Department of Biostatistics, University of Alabama at Birmingham, Birmingham, AL, USA
Queensland Brain Institute, The University of Queensland, St. Lucia, QLD, Australia

Degui Zhi
Section on Statistical Genetics, Department of Biostatistics, University of Alabama at Birmingham, Birmingham, AL, USA

Nianjun Liu
Section on Statistical Genetics, Department of Biostatistics, University of Alabama at Birmingham, Birmingham, AL, USA

Kui Zhang
Section on Statistical Genetics, Department of Biostatistics, University of Alabama at Birmingham, Birmingham, AL, USA

Debashis Ghosh
Department of Statistics and Public Health Sciences, Penn State University, 514A Wartik Building, University Park, PA 16802, USA

Zhaohui S. Qin
Department of Biostatistics and Bioinformatics, Rollins School of Public Health, Center for Comprehensive Informatics, Emory University, 1518 Clifton Rd., N.E., Atlanta, GA 30322, USA

Shaopu Qin
School of Biology, Georgia Institute of Technology, Atlanta, GA, USA

Jinhee Kim
School of Biology, Georgia Institute of Technology, Atlanta, GA, USA

Dalia Arafat
School of Biology, Georgia Institute of Technology, Atlanta, GA, USA

Greg Gibson
School of Biology, Georgia Institute of Technology, Atlanta, GA, USA

Preface

Encyclopaedia of Modern Genetics Analysis Integrating Genes and Genomes covers the statistical models and methods that are used to understand human genetics, following the historical and recent developments of human genetics. Statistical genetics is a scientific field concerned with the development and application of statistical methods for drawing inferences from genetic data. The term is most commonly used in the context of human genetics. Rapid advances in genomic technologies have galvanized the development of new techniques to interpret genetics and genomics data sets. Fundamentals of Modern Statistical Genetics emphasizes on the statistical models and methods that are used to understand genetics, following the historical and recent developments of genetics. First chapter presents an efficient statistical model for genome-wide estimating and testing the cytoplasmic effect, nuclear DNA imprinting effect as well as the interaction between them under reciprocal backcross and F2 designs derived from inbred lines. Second chapter provides an overview of the range of statistical methods that can be used to answer different immunological study questions. We discuss specific aspects of immunological studies and give examples of typical scientific questions related to immunological data. We review classical bivariate and multivariate statistical techniques (factor analysis, cluster analysis, discriminant analysis) and more advanced methods aimed to explore causal relationships and illustrate their application to immunological data. Third chapter aims at identifying genes in biochemical and physiological processes to reveal genetic causes of rare and common diseases. In fourth chapter, we propose a statistical design for detecting imprinted loci that control quantitative traits based on a random set of three-generation families from a natural population in humans. This design provides a pathway for characterizing the effects of imprinted genes on a complex trait or disease at different generations and testing transgenerational changes of imprinted effects. In fifth chapter, we describe the mass spectrometry based profiling of maize kernels, a model system for genomic studies and a cornerstone of the agroeconomy. Sixth chapter aims to investigate fine-scale patterns of genetic heterogeneity in modern humans from a geographic perspective, a genetic geostatistical approach framed within a geographic information system is presented. Seventh chapter reveals the underlying importance of genetic diversity and reviews useable statistical techniques for identifying and grouping genotypes for intraspecies crop improvement. In eighth chapter, we focus on rheumatoid arthritis, systemic lupus erythematosus and ankylosing spondylitis and describe some of the recently described genes that underlie these conditions and the extent to which they overlap. Ninth chapter reveals on statistical design of personalized medicine interventions. Chapter ten provides an introductory review on how genes interact to produce biological functions. In eleventh chapter, we present the challenges and methods from a statistical perspective and focus on genetic association studies. In last chapter, we propose using eQTL weights as prior information in SNP based association tests to improve test power while maintaining control of the family-wise error rate (FWER) or the false discovery rate (FDR).

Chapter 1

STATISTICAL DISSECTION OF CYTO-NUCLEAR EPISTASIS SUBJECT TO GENOMIC IMPRINTING IN LINE CROSSES

Tao He[1], Jian Sa[1, 2], Ping-Shou Zhong[1], Yuehua Cui[1, 2]

[1]Department of Statistics and Probability, Michigan State University, East Lansing, Michigan, United States of America

[2]Division of Medical Statistics, School of Public Health, Shanxi Medical University, Taiyuan, Shanxi, China

ABSTRACT

Cytoplasm contains important metabolism reaction organelles such as mitochondria and chloroplast (in plant). In particular, mitochondria contains special DNA information which can be passed to offsprings through maternal gametes, and has been confirmed to play a pivotal role in nuclear activities. Experimental evidences have documented the importance of cyto-nuclear interactions in affecting important biological traits. While studies have also pointed out the role of interaction between imprinting nuclear DNA and cytoplasm, no statistical method has been developed to efficiently model such effect and further quantify its effect size. In this work, we developed an efficient statistical model for genome-wide estimating and testing the cytoplasmic effect, nuclear DNA imprinting effect as well as the interaction between them under reciprocal backcross and F_2 designs derived from inbred lines. Parameters are estimated under maximum likelihood framework implemented with the EM algorithm. Extensive simulations show good performance in a variety of scenarios. The utility of the method is demonstrated by analyzing a published data set in an F_2 family derived from C3H/HeJBir and C57BL/6 J mouse strains. Important cyto-nuclear interactions were identified. Our approach provides a quantitative framework for identifying and estimating cyto-nuclear interactions subject to genomic imprinting involved in the genetic control of complex traits.

INTRODUCTION

One of the central foci in biological study is to unravel the genetic secrets of complex traits of agricultural, evolutional and biomedical importance. Quantitative trait locus (QTL) mapping has been the major tool for this purpose over decades [1]–[3]. In QTL mapping, the identified QTLs are chromosome segments harboring potential genetic variants that could give rise to phenotypical manifestation. Large successes have been witnessed in the past [4]. However, there are still many phenomena that could not be explained by Mendelian genetics, leading to the new exploration of research focus on epigenetics [5].

Genomic imprinting, one of the major epigenetic phenomena, plays key roles in controlling embryonic growth and development [6], [7]. Let subscript letter M and F denote the parental origin of two alleles in a diploid organism, then a locus with two alleles A and a is thought to be imprinted if two heterozygotes $A_M a_F$ and $a_M A_F$ have different expressions [8]. The malfunction of imprinted genes could potentially lead to abnormal characters such as cancers or other genetic disorders [9].

Genomic imprinting effect is considered as one type of parent-of-origin effect due to allelic effect with specific parental origin. In contrast to this, maternal effect or cytoplasmic effect is also considered as one type of parent-of-origin effect in which the offspring›s expression is influenced by maternal parent. For example, a mother›s genotype, even if not transmitted to her offspring, may influence in utero conditions and increase risk and/or interact with genetic predisposition for particular diseases among those offspring [10]. For cytoplasm, it contains a wide variety of organelles such as mitochondria and chloroplast (in plant). Almost all the reactions of cellular metabolism take place in such an environment. It has been demonstrated that cytoplasm plays a central role in coordinating the activities of nuclear genetic materials[11]–[15]. Thus, the identification of cyto-nuclear interaction could shed novel insights into the genetic and epigenetic control of phenotypic variation. A number of empirical studies have documented the significant contribution of cyto-nuclear interaction to phenotypic variation in organisms such as wheat, rice, mice, yeast and Drosophila [16]–[19].

On the other hand, the existence of such parent-of-origin effects may lead to incorrect interpretation of the (marginal) effects of particular genes when performing genetic mapping studies, unless such effects are appropriately accounted for in the analysis [20]. Statistical methods for dissecting genomic imprinting effect has been extensively studied in literature (e.g., [21]–[26]). Tang et al. [27] developed a model to evaluate cyto-nuclear interaction effect based on experimental crosses. However, how the two types of parent-of-origin effects, one in nuclear level and one in cytoplasmic level, interact to influence phenotypic variation is largely unknown due to the lack of proper statistical models.

In this work, we discuss potential scenarios of parent-of-origin effects, and present a statistical method to dissect the cyto-nuclear interaction effects subject to genomic imprinting. The developed framework is based on experimental crosses which can be realized through two different designs, reciprocal backcross and F_2 design. When an F_2 design is considered, sex-specific difference in recombination fractions, which is initially discovered in [28], [29] and later observed in many species, is incorporated into our model to distinguish the genetic differences between two reciprocal heterozygote. Such information can be found in literature, such as the female-to-male recombination ratio of 1.6:1 for human [30], 1.4:1 for pig [31], 1.4:1 for dog [32], 1.25:1 for mouse [33] on average across the whole genome. A genome-wide scan for the identification of iQTL mapping cyto-nuclear epistasis is performed based on the interval mapping theory, and parameters are estimated based on the framework of maximum likelihood method implemented with the EM algorithm [34]. Extensive simulations are conducted to evaluate the performance of our model under different scenarios, such as different sample sizes, different heritability levels, and different gene effects. The utility of the method is illustrated by applying it to a genome-wide scan of four traits in an F_2 family derived from two inbred mouse strains.

STATISTICAL METHODS

Genetic Designs

Consider a design initiated with two inbreed lines with two segregating alleles A and a. Let subscript letter M and F represent the parental origin of offspring alleles inherited from the mother and father, respectively. A complete dissection of the cyto-nuclear interaction subject to imprinting needs experimental designs that can distinguish the quantitative variation between two heterozygotes $A_M a_F$ and $a_M A_F$, and also against the cytoplasmic effect. For this purpose, a reciprocal backcross or F_2 design is proposed so that variations and interactions can be fully introduced.

Figure 1 illustrates the reciprocal backcross design. Let the maternal parent carrying genotype AA (denoted as P2) has a cytoplasmic effect in contrast to the maternal line carrying genotype aa (denoted as P1) as the reference line. In the diagram, individuals that carry the maternal effect coming from the maternal parent with genotype AA are denoted by gray squares. Four possible backcrosses can be initiated as illustrated in Fig. 1. As shown in the figure, any backcross offsprings coming from the middle two designs carry the cytoplasmic maternal effect derived from the AA genotype.

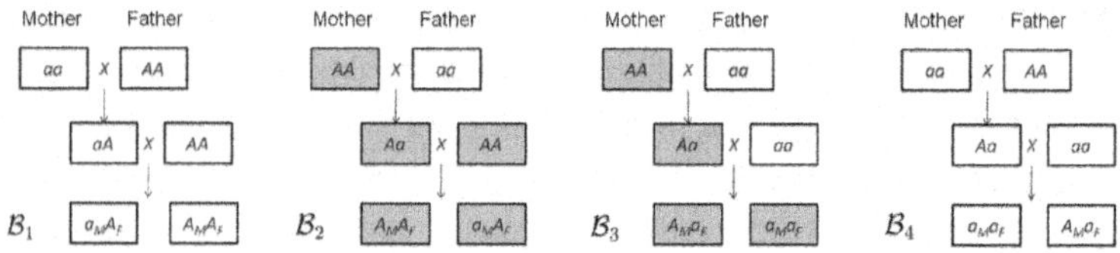

Figure 1: A reciprocal backcross design.

Figure 2 shows the F_2 design. Denote the left one as design S_1, and the right one as design S_2. The cross of two F_1's generates four possible allele-specific F_2 genotypes. Assuming there is a cytoplasmic effect, F_2 offsprings may show different phenotypes depending on the genotype of the maternal parental lines. For example, if maternal cytoplasmic effect exists, the offspring phenotypic value for AA genotype may be different depending on whether it comes from the S_1 or S_2 design. In the F_2 design, the two reciprocal heterozygotes A_Ma_F and a_MA_F cannot be distinguished in general. Sex-specific recombination difference in male and female needs to be considered in order to distinguish the two (Cui et al. 2006) [25].

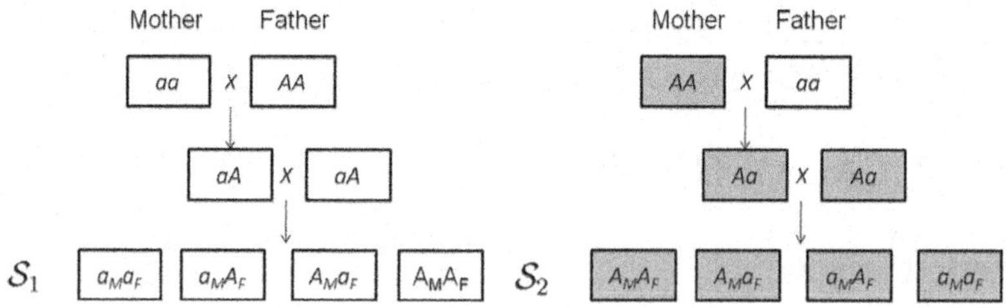

Figure 2: A reciprocal F_2 **design.**

Statistical Parameterization

For a particular cross, let y_j $(j=1,\cdots,n)$ denote the phenotypic value of interest. Following Tang et al. [27], the one-QTL genetic model can be expressed as,

$$y_j=\mu+cx_{1j}+ax_{2j}+dx_{3j}+ix_{4j}+i_{ca}x_{5j}+i_{cd}x_{6j}+i_{ci}x_{7j}+e_j \tag{1}$$

where μ is the overall mean; c is the cytoplasmic effect; a, d and i are the additive, dominance and imprinting effects of a QTL, respectively; and i_{ca}, i_{cd} and i_{ci} are the cytoplasm by additive, cytoplasm by dominance and cytoplasm by imprinting interactions, respectively; x_{1j} is an indicator variable, denoting $x_{1j}=1$ for the AA maternal cytoplasm and $x_{1j}=-1$ for the aa maternal cytoplasm; $x_{2j}, x_{3j}, \cdots, x_{7j}$ are other indicators of various effects describing the additive, dominance and the interaction effect between the cytoplasm and genetic variables.

For the S_1 design in the F_2 population initiated with cross $AA \times aa$, the mean genotypic values of four possible genotypes formed by different allelic combination from the two F_1parents can be expressed as,

$$\begin{cases} \mu_1 = \mu + c + a + i_{ca}, & \text{for } A_M A_F, \\ \mu_2 = \mu + c + d + i + i_{cd} + i_{ci}, & \text{for } A_M a_F, \\ \mu_3 = \mu + c + d - i + i_{cd} - i_{ci}, & \text{for } a_M A_F, \\ \mu_4 = \mu + c - a - i_{ca}, & \text{for } a_M a_F. \end{cases} \tag{2}$$

Similarly for the S_2 design initiated with cross $aa \times AA$, the genetic model can be expressed as,

$$\begin{cases} \mu_5 = \mu - c + a - i_{ca}, & \text{for } A_M A_F, \\ \mu_6 = \mu - c + d + i - i_{cd} - i_{ci}, & \text{for } A_M a_F, \\ \mu_7 = \mu - c + d - i - i_{cd} + i_{ci}, & \text{for } a_M A_F, \\ \mu_8 = \mu - c - a + i_{ca}, & \text{for } a_M a_F. \end{cases} \tag{3}$$

For the reciprocal backcross design which consists of B_1, B_2, B_3 and B_4, the indicator variables in Eq. (1) describing different QTL genotypes are defined in Table 1.

Table 1: QTL genotypes and corresponding genetic components under different backcross designs

Backcross design	BC QTL genotype	c	a	d	i	i_{ca}	i_{cd}	i_{ci}
B_1	$A_M A_F$	−1	1	0	0	−1	0	0
	$a_M A_F$	−1	0	1	−1	0	−1	1
B_2	$A_M A_F$	1	1	0	0	1	0	0
	$a_M A_F$	1	0	1	−1	0	1	−1
B_3	$A_M a_F$	1	0	1	1	0	1	1
	$a_M a_F$	1	−1	0	0	−1	0	0
B_4	$A_M a_F$	−1	0	1	1	0	−1	−1
	$a_M a_F$	−1	−1	0	0	1	0	0

doi:10.1371/journal.pone.0091702.t001

For simplicity, we will use matrix form to rewrite the models. Let us denote $\boldsymbol{\beta} = (\mu, c, a, d, i, i_{ca}, i_{cd}, i_{ci})^{\mathrm{T}}$, $\boldsymbol{\mu} = (\mu_1, \mu_2, \mu_3, \mu_4, \mu_5, \mu_6, \mu_7, \mu_8)^{\mathrm{T}}$. Then the relationship between the eight genetic means and the eight parameters can be written as

$$\boldsymbol{\mu} = \mathbf{D}\boldsymbol{\beta}, \tag{4}$$

Where

$$\mathbf{D}=\mathbf{D}^{BC}=\begin{pmatrix} 1 & -1 & 1 & 0 & 0 & -1 & 0 & 0 \\ 1 & -1 & 0 & 1 & -1 & 0 & -1 & 1 \\ 1 & 1 & 1 & 0 & 0 & 1 & 0 & 0 \\ 1 & 1 & 0 & 1 & -1 & 0 & 1 & -1 \\ 1 & 1 & 0 & 1 & 1 & 0 & 1 & 1 \\ 1 & 1 & -1 & 0 & 0 & -1 & 0 & 0 \\ 1 & -1 & 0 & 1 & 1 & 0 & -1 & -1 \\ 1 & -1 & -1 & 0 & 0 & 1 & 0 & 0 \end{pmatrix}, \quad (5)$$

for the reciprocal backcross design, and

$$\mathbf{D}=\mathbf{D}^{F_2}=\begin{pmatrix} 1 & 1 & 1 & 0 & 0 & 1 & 0 & 0 \\ 1 & 1 & 0 & 1 & 1 & 0 & 1 & 1 \\ 1 & 1 & 0 & 1 & -1 & 0 & 1 & -1 \\ 1 & 1 & -1 & 0 & 0 & -1 & 0 & 0 \\ 1 & -1 & 1 & 0 & 0 & -1 & 0 & 0 \\ 1 & -1 & 0 & 1 & 1 & 0 & -1 & -1 \\ 1 & -1 & 0 & 1 & -1 & 0 & -1 & 1 \\ 1 & -1 & -1 & 0 & 0 & 1 & 0 & 0 \end{pmatrix}, \quad (6)$$

for the F_2 design. For the purpose of illustration, the following estimation and inference is demonstrated through the F_2 design. The same procedure applies to BC design too.

The mixture Model and the Likelihood

Statistical methods for QTL interval mapping based on a mixture model traced back to the work by Lander and Botstein [1]. In the mixture model, each observation y is modeled as a weighted mixture of J (known and finite) components, and each component, which corresponds to a certain genotype category depending on the underlying genetic design, follows a certain distribution f_j with weight π_j. Conditional on the marker genotype **M** and unknown parameters $\boldsymbol{\phi}$ and η, the density of the observed y has the following expression

$$y \sim p(y|\mathrm{M},\boldsymbol{\phi},\eta)=\pi_1 f_1(y|\mathrm{M},\phi_1,\eta)+\ldots+\pi_J f_J(y|\mathrm{M},\phi_J,\eta), \quad (7)$$

where $\boldsymbol{\pi}=(\pi_1,\ldots,\pi_J)^T$ refers to the mixture proportions which are constrained nonnegative and $\sum_{j=1}^{J}\pi_j=1$; $\boldsymbol{\phi}=(\phi_1,\ldots,\phi_J)^T$ is a vector for the component-specific parameters, with ϕ_j being specific to jth component; and η consists of parameters (i.e., residual variance) that are common to all components.

For the F_2 design we described above, there are four genotypes at each locus $(A_M A_F, A_M a_F, a_M A_F, \text{and } a_M a_F)$. The genotype of the QTL is generally unobservable, but can be inferred by using the two flanking markers' information. Given the flanking marker genotypes of the ith individual, the conditional probabilities $\boldsymbol{\pi}_i = (\pi_{A_M A_F|i}, \pi_{A_M a_F|i}, \pi_{a_M A_F|i}, \pi_{a_M a_F|i})$ of the QTL genotype can be calculated. These conditional probabilities become the mixture proportions in the mixture model (7). Let us denote $\boldsymbol{\pi}_i = (\pi_{1|i}, \pi_{2|i}, \pi_{3|i}, \pi_{4|i})$ to simplify the notation. These conditional probabilities are expressed in terms of sex-specific recombination rates in order to distinguish the two reciprocal heterozygotes. Please refer to Cui et al. [25] for the conditional probabilities of QTL genotypes given marker genotypes in terms of sex-specific recombination fractions for an F_2 design.

Assume the total number of F_2 offsprings for design S_1 and S_2 are n_1 and n_2 respectively, and let $n = n_1 + n_2$. Phenotype data for a certain quantitative trait can be observed and recorded as a vector $y = (y_1, \ldots, y_n)$, where $\mathbf{y}^{(1)} = (y_1, \ldots, y_{n_1})$ are from design S_1 and $\mathbf{y}^{(2)} = (y_{n_1+1}, \ldots, y_n)$ are from design S_2. Marker information can be reorganized as matrix $\mathbf{M} = (\mathbf{M}_1; \mathbf{M}_2)^T$, where the jth $(j = 1, \ldots, n_1)$ row of $\mathbf{M}_1$ (the jth row of matrix M) contains all the marker information of the jth individual under design S_1 and the kth $(k = 1, \ldots, n_2)$ row of $\mathbf{M}_2$ (the $(n_1 + k)$th row of matrix M) contains all the marker information of the kth individual under design S_2. Based on the mixture model (7), and with the independence assumption, the joint likelihood function for the F_2 family with total n individuals, constructed by combining design S_1 and S_2 together, can be formulated as,

$$L(\Theta|\mathrm{M}, \mathrm{y}) = \prod_{i=1}^{n_1} [\pi_{1|i} f_1(y_i) + \pi_{2|i} f_2(y_i) + \pi_{3|i} f_3(y_i) + \pi_{4|i} f_4(y_i)]$$

$$\times \prod_{i=n_1+1}^{n} [\pi_{1|i} f_5(y_i) + \pi_{2|i} f_6(y_i) + \pi_{3|i} f_7(y_i) + \pi_{4|i} f_8(y_i)], \tag{8}$$

where the unknown vector Θ contains the QTL position, QTL effects and residual variance, and the density function f_j $(j = 1, \ldots, 8)$ is assumed to follow a normal distribution with mean μ_j and common variance σ^2. More specifically, parameter vector Θ can be divided into two subsets, Θ_l and Θ_g, where Θ_l describes the location of QTL and Θ_g contains all the genetic parameters, including QTL-effects vector $\boldsymbol{\beta} = (\mu, c, a, d, i, i_{ca}, i_{cd}, i_{ci})^T$ and residual variance σ^2 in our model.

Parameter Estimation

To estimate the unknown parameters $\Theta = (\Theta_l, \Theta_k)$, several algorithms could be implemented, such as Expectation-Maximization (EM), Newton Raphson and Fisher Scoring. Among all these methods, EM algorithm, which was initially developed by Dempster et al. [34], is most commonly used in QTL mapping study. In this paper, EM algorithm is applied to obtain the maximum likelihood estimates (MLEs). This procedure involves differentiating the log-likelihood function with respect to each unknown parameter, letting the derivatives equal to zero, and solving the log-likelihood equation for the corresponding parameter.

For the QTL position which is unknown in the model, we did not estimate parameters Θ_l directly. As commonly treated in QTL mapping studies, we applied a grid search approach to estimate the putative QTL position via scanning the entire linkage genome by 1 or 2 cM increment flanked by two markers and did a hypothesis testing at each putative position. A likelihood ratio or LOD profile plot can be generated to graphically display the LR or LOD test statistic for a putative QTL at each testing position. The genomic position which corresponds to a peak in the profile plot is the MLE of the QTL location, given that the peak passes the genome-wide threshold determined by the permutation tests detailed below. Bootstrap methods can be applied to assess the confidence interval of the estimated position [35].

Hypothesis Test

Testing the overall QTL effect on the quantitative trait is the first step toward a complete dissection of different genetic contributions to the trait. Once the MLEs of the parameters are obtained, the presence of QTL responsible for the variation of the quantitative phenotype can be tested by using the following hypotheses,

$$\begin{cases} H_0: & a = d = i = i_{ca} = i_{cd} = i_{ci} = 0 \\ H_1: & \text{not all equal zero} \end{cases}$$

The test statistic for testing the above hypotheses is calculated as the log-likelihood ratio test statistic (LR) of the full model (H_1) over the reduced model(H_0):

$$LR = -2\log[L(\tilde{\Omega}) - L(\hat{\Omega})] \tag{9}$$

where $\tilde{\Omega}$ and $\hat{\Omega}$ denote the MLEs of the unknown parameters under H_0 and H_1, respectively. Since a genome-wide scan involves multiple correlated tests, we use the permutation tests proposed by Churchill and Doerge [36] to find the threshold value.

If there is a QTL, a number of other hypothesis tests can also be performed to test the property of the detected QTL. To test the imprinting effect, we can simply formulate the hypothesis as $H_0 : i=0$ vs $H_1 : i \neq 0$ to assess the mean difference of the two reciprocal heterozygotes. To test the cytoplasmic maternal effect, the hypothesis can be stated as $H_0 : c=0$ vs $H_1 : c \neq 0$. The epistatic effects of all interaction terms can also be tested as

$$\begin{cases} H_0 : & i_{ca}=i_{cd}=i_{ci}=0 \\ H_1 : & \text{not all equal zero} \end{cases}$$

Similarly, additive and dominance effects can be tested as

$$\begin{cases} H_0 : & a=d=0 \\ H_1 : & \text{not all equal zero} \end{cases}$$

If specific interest is focused, for instance, the interaction of imprinting and maternal effect, the hypothesis can be formulated as $H_0 : i_{ci}=0$ vs $H_1 : i_{ci} \neq 0$. All the above tests can be done by applying the likelihood ratio test in which the test statistic asymptotically follows a χ^2 distribution with degrees of freedom equal to the difference of the parameters under the null and the alternative hypotheses. For example, when testing $H_0 : a=d=0$, the LR test statistics is compared with the χ_2^2 cutoff with 2 degrees of freedom.

RESULTS

Monte Carlo Simulation

Monte Carlo simulations were performed to investigate the statistical behavior of our model. We simulated an F_2 population, with one half of the population coming from design S_1 and the other half coming from design S_2. A genome with 100 cM long linkage group, composed of 6 equidistant markers, was constructed. The position of QTL was assumed to be located at 48 cM away from the first marker on the linkage group. The marker genotypes in the F_2 population were simulated by mimicking sex-specific recombination fractions in mice, i.e., $r_M=1.25r_P$[33]. A series of simulation study with different sample size ($n=400$ vs $n=800$) and different heritability levels ($H^2=0.1,0.25,0.4$) was conducted to examine the impacts of parameter spaces on parameter estimation and testing power. These simulation designs, which were aimed to give a better understanding of model performance under different situations, can provide biologists some empirical evidences to design their experiments.

In the simulation study, the residual variance was calculated under different heritability levels. For the F_2 design, the genetic variances of various terms can be calculated as follows:

$$\sigma_a^2 = a^2/2,$$

$$\sigma_d^2 = d^2/4,\ \sigma_i^2 = i^2/2,\ \sigma_c^2 = c^2,\ \sigma_G^2 = \sigma_a^2 + \sigma_d^2 + \sigma_i^2 + i_{ca}^2/2 + i_{cd}^2/2 + i_{ci}^2/2 + ci_{cd},$$

and the broad sense heritability level can be expressed as

$$H^2 = \sigma_G^2/(\sigma_G^2 + \sigma_c^2 + \sigma_e^2).$$

For given genetic parameters and the heritability level, the residual variance can be calculated as

$$\sigma_e^2 = (\sigma_Q^2 + \sigma_{cQ}^2)(1/H^2 - 1) - \sigma_c^2,$$

from which the phenotype data can be generated.

The MLEs of the QTL position and effect parameters, based on 200 simulation replicates under different heritability levels and sample sizes, are displayed in Table 2. The square root of mean squared error (RMSE) of parameter estimates are given in parenthesis to show the estimation accuracy. As we expected, the accuracy of parameter estimates increases as the sample size and heritability level increase. For instance, the RMSE of estimated QTL position decreases from 13.28 to 3.81, an 71% increase in accuracy when the sample size increases from 400 to 800 under a fixed heritability level 0.1. The other parameter estimates show the same pattern. If we increase the heritability level when the sample size is fixed, a clear reduction in RMSE can be observed. For example, with 400 samples, the RMSE of estimated QTL position decreases from 13.28 to 4.99, then to 2.77 as H^2 gradually increases from 0.1 to 0.4. From the decreasing RMSEs of the parameter estimates, we observed that simply increasing sample size is less efficient than increasing heritability level in order to increase the precision of parameter estimation. Since high heritability corresponds to small environmental variability [37], reducing environmental variation is of more practically important than just simply increasing sample size.

Table 2: The MLEs of the QTL position and effect parameters based on 200 simulation replicates under different heritabilities and sample sizes

H^2	n	**Position 48 cM**	$\mu=10$	$c=1$	$a=1$	$d=0.8$	$i=0.6$	$i_{ca}=0.6$	$i_{cd}=0.5$	$i_{ci}=0.4$	σ^2
0.1	400	47.93	10.00	0.99	0.96	0.78	-	0.59	0.49	-	3.679
		(13.28)	(0.33)	(0.35)	(0.35)	(0.55)	-	(0.32)	(0.53)	-	(0.22)
	800	47.35	10.01	1.02	0.98	0.77	-	0.60	0.45	-	3.73
		(3.81)	(0.23)	(0.21)	(0.24)	(0.39)	-	(0.22)	(0.33)	-	(0.15)
0.25	400	47.04	10.02	1.00	0.96	0.78	-	0.60	0.50	-	1.99
		(4.99)	(0.17)	(0.18)	(0.16)	(0.26)	-	(0.17)	(0.26)	-	(0.11)
	800	47.65	10.00	1.01	0.98	0.81	-	0.59	0.50	-	2.01
		(2.29)	(0.12)	(0.12)	(0.12)	(0.20)	-	(0.10)	(0.18)	-	(0.08)
0.4	400	48.02	9.99	1.00	0.99	0.81	-	0.60	0.50	-	1.23
		(2.77)	(0.10)	(0.11)	(0.09)	(0.16)	-	(0.10)	(0.17)	-	(0.06)
	800	47.95	9.99	0.99	0.99	0.81	-	0.60	0.51	-	1.24
		(1.78)	(0.07)	(0.07)	(0.08)	(0.12)	-	(0.07)	(0.11)	-	(0.04)

The squared roots of the mean squared errors (RMSE) of the MLEs are given in parentheses.
The locations of the QTL is described by the map distances (in cM) from the first marker of the linkage group. The hypothesized σ^2 value is 3.81 for $H^2=0.1$, 2.04 for $H^2=0.25$ and 1.26 for $H^2=0.4$.
doi:10.1371/journal.pone.0091702.t002

Note that we did not list the estimation of the imprinting effect i and cyto-imprinting interaction i_{ci}, which are not estimable under the F_2 design. The reason is that the imprinting direction cannot be inferred from the F_2 design [25]. However, we can still conduct hypothesis test to infer the imprinting effect as well as its interaction with cytoplasmic effect. To further investigate the testing performance of cytoplasmic and imprinting effects, we introduced two proportions, namely η_c and η_i, where $\eta_c=\sigma_c^2/\sigma_G^2$ and $\eta_i=\sigma_i^2/\sigma_G^2$. We can evaluate the test power under different cytoplasmic and imprinting effect sizes. Given all other genetic parameter values fixed (as shown in Table 2), simple algebra shows that the cytoplasmic effect c and imprinting effect i can be calculated for a given value of η_c or η_i, i.e.,

$$c=\frac{i_{cd}\eta_c+\sqrt{(i_{cd}\eta_c)^2+4\eta_c Q_1}}{2}, \text{ and } i=\sqrt{\frac{2\eta_i Q_2}{1-\eta_i}}$$

where

$$Q_1=\sigma_a^2+\sigma_d^2+\sigma_i^2+i_{ca}^2/2+i_{cd}^2/2+i_{ci}^2/2,$$

and

$$Q_2=\sigma_a^2+\sigma_d^2+i_{ca}^2/2+i_{cd}^2/2+i_{ci}^2/2+ci_{cd}.$$

Based on 1000 simulation runs under different heritabilities, sample sizes and variation proportions, the power of cytoplasmic effect test, imprinting effect test, interaction effects test and additive/dominance effects test are listed in Table 3. As we expect, the test power increases with the increasing of sample size and heritability level. For example, the cytoplasmic testing power increases from 0.663 to 0.905 as sample size increases from 400 to 800, a 36.5% increase in power for fixed $\eta_c=0.1$ and $H^2=0.25$. The same pattern is observed for the imprinting test. As the proportion of variance explained by the cytoplasmic and imprinting effect increases, the power increases accordingly. Noted that when both η_c or η_i, are zeros, the testing power corresponds to the type I error rate for the corresponding factor. From the table we can see that the size of imprinting test is well controlled under different sample sizes and heritability levels. For the cytoplasmic effect, the size is a little inflated under $n=400$, but it gets close to the nominal 5% level as sample size increases to 800, especially under large heritability (e.g., $H^2=0.4$). In sum, the simulation evidences show that the model performs reasonably well in both parameter estimation and testing.

Table 3: The power of four hypothesis tests based on 1000 samplings under different heritabilities, sample sizes and variation proportions

n	H^2	Power[1]			Power[2]			Power[3]	Power[4]
		$\eta_c=0\%$	$\eta_c=10\%$	$\eta_c=20\%$	$\eta_i=0\%$	$\eta_i=10\%$	$\eta_i=20\%$		
400	0.10	0.088	0.353	0.528	0.066	0.085	0.089	0.654	0.678
	0.25	0.081	0.663	0.928	0.051	0.263	0.432	0.990	0.992
	0.40	0.064	0.930	0.998	0.050	0.807	0.942	1.000	1.000
800	0.10	0.073	0.464	0.753	0.059	0.100	0.126	0.898	0.927
	0.25	0.066	0.905	0.998	0.047	0.473	0.713	1.000	1.000
	0.40	0.049	0.997	1.000	0.050	0.980	0.997	1.000	1.000

Power^k ($k=1,2,3,4$) refer to the powers for testing 1) $H_0: c=0$ vs $H_1: c\neq 0$; 2) $H_0: i=0$ vs $H_1: i\neq 0$; 3) $H_0: i_{ca}=i_{cd}=i_d=0$ vs H_1 : not all equal to 0; and 4) $H_0: a=d=0$ vs H_1 : not all equal to 0, respectively. For a given η_c, all other effect values are fixed as 0.8 except for c, which can be calculated in terms of η_c and other parameters. The hypothesized c value is 0 for $\eta_c=0$, 0.461 for $\eta_c=0.1$ and 0.679 for $\eta_c=0.2$. Similarly, the value of i, which depends on imprinting effect variation proportion η_i, is 0 for $\eta_i=0$, 0.680 for $\eta_i=0.1$ and 1.020 for $\eta_i=0.2$.
doi:10.1371/journal.pone.0091702.t003

A Case Study

We applied the model to a published F_2 cross data set based on design S_1 and design S_2aimed to find QTLs that contribute to variation in quantitative traits related to colitis severity in 1L-10-deficient mice [38]. The data contain 411 F_2 mice derived from inbred strains, where 203 mice are from design S_1 and 208 mice are from design S_2. Ninety-one markers were obtained with an average length of ~15 cM spanning acroos the 19 autosome chromosomes. For more information about the data, the readers are referred to the original paper [38].

It has been reported that on average the female chromosome is 25% longer in genetic distance between homologous loci than the male in mice [38]. The sex-specific recombination fraction, expressed as $r_M=1.25r_P$, was reconstructed based on the marker information (see Cui et al. for details [25]). The method was applied to four phenotypes, cecum total score (CTS), spleen/body weight ratio (SBWR), mesenteric lymph node(MLN)/body weight ratio(MBWR), and secretory IgA (SIgA) level, where cecum total score was graded by using colitis-related criteria, including severity, hyperplasia, ulceration and the percentage of area involved. Box-Cox transformation was applied to all traits before fitting the Gaussian-mixture model. The genome-wide LOD profile plots for the four phenotypes are shown in Figure 3, where the solid blue curves correspond to the LOD values and the dashed red lines correspond to the 5% genome-wide threshold values out of 1000 permutations. The LOD score is calculated as $\log_{10}$LR, where LR is obtained from equation (9) to test the null hypothesis: $H_0 : a = d = i = i_{ca} = i_{cd} = i_{ci} = 0$.

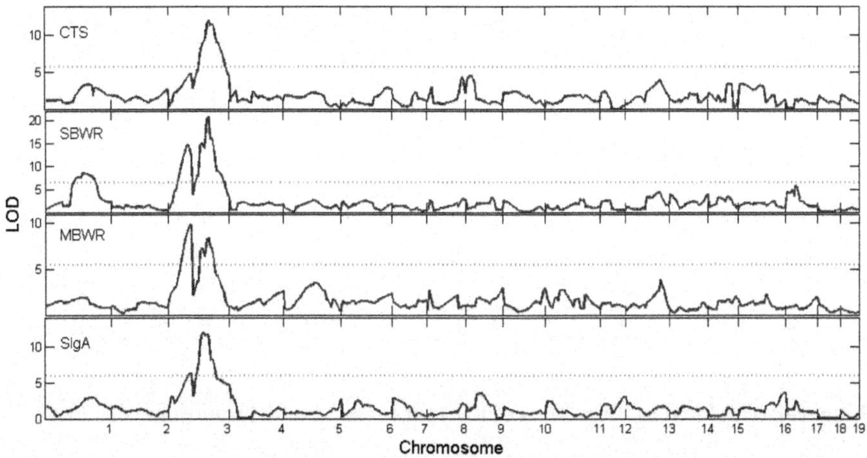

Figure 3: The LOD profiles of the four traits across the 19 chromosomes using the linkage map constructed from microsatellite markers [38]. The genomic positions corresponding to the peak of the curve are the MLEs of the QTL locations.

As shown in Figure 3, one QTL on chromosome 3 is detected for cecum total score trait, two QTLs on chromosomes 3 and one on chromosome 1 are detected for spleen/body weight ratio trait, three QTLs on chromosome 3 are detected for MLN/body weight ratio trait, and two QTLs located on chromosome 3 are detected for SIgA trait. The QTL located at 60.6 cM on chromosome 3 is common to three traits. The one located at 52.6 cM for SIgA trait is very close to it. It is highly possible that it is the same QTL that controls the four traits. Such a pleiotropic effect needs to be further evaluated. It should be mentioned that the QTL detected in the original paper for the four traits is located at 61.8 cM on chromosome 3 [38], which is 1.2 cM away from the one we found. Such a difference may arise because of the capitalization of sex-specific recombination rates and different models fitted.

In addition to the QTL identified in our analysis and the original paper, some other major QTLs which are not detected in Farmer et al. [14], such as those at 52.6 cM and 38.6 cM on chromosome 3, stand out in our model and therefore need further investigation. Almost all the QTLs on chromosome 3 are clustered together, whose local LOD profiles of four traits are shown in Figure 4.

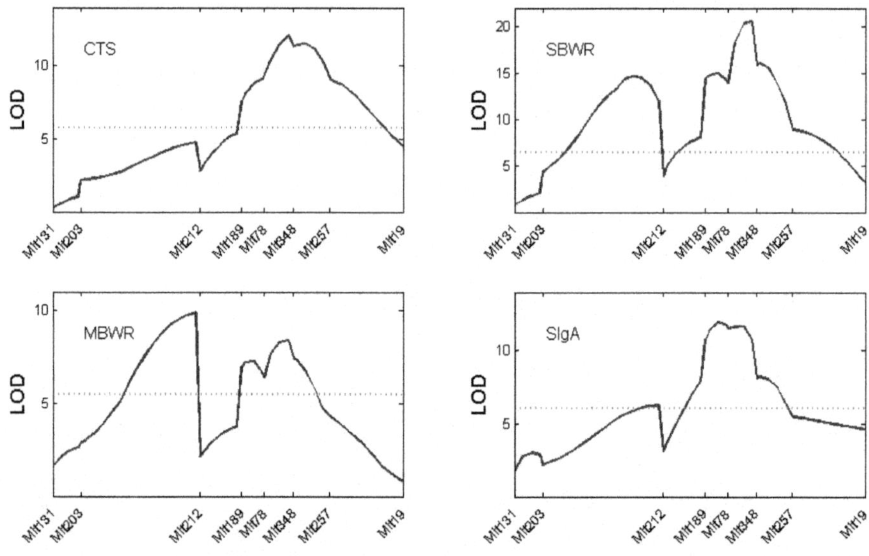

Figure 4: The local LOD profiles of the four traits across chromosomes 3.

The marker interval for each QTL is listed in Table 4, which also tabulates the p-values under four different tests for each estimated QTL using permutation tests. From the test results, it can be seen that most QTLs have strong additive and dominance genetic effect, except for the spleen/body weight ratio QTL located at D1Mit156+31.1 cM on chromosome 1. This QTL shows evidence of cytoplasmic, imprinting as well as cyto-nuclear interaction effects (p-values for the three tests are 0.0282, 0.0296 and 0.0073, respectively), but shows no sign of additive and dominance effect. In addition, the MLN/body weight ratio QTL located at D3Mit78+5.6 cM shows evidence of cytoplasmic effect. In summary, we identified one QTL on chromosome 1 with evidence of cyto-nuclear interaction effect and this QTL also shows evidence of cytoplasmic and imprinting effect. Further functional validation is needed to confirm the results.

Table 4: List of QTL positions, corresponding marker intervals and p-values under different tests for the four traits

Trait	Ch	QTL postion	Marker interval	p-value[1]	p-value[2]	p-value[3]	p-value[4]
CTS	3	60.6 cM	D3Mit78-D3Mit348 (D3Mit78+5.6 cM)	0.6680	0.3046	0.5005	$<10^{-11*}$
SBWR	3	60.6 cM	D3Mit78-D3Mit348 (D3Mit78+5.6 cM)	0.4322	0.6107	0.4502	$<10^{-16*}$
	3	32.6 cM	D3Mit203-D3Mit212 (D3Mit203+11.4 cM)	0.2003	0.9502	0.2700	$<10^{-13*}$
	1	63.9 cM	D3Mit156-D1Mit17 (D1Mit156+31.1 cM)	0.0282*	0.0296*	0.0073*	0.8064
MBWR	3	38.6 cM	D3Mit203-D3Mit212 (D3Mit203+27.4 cM)	0.3574	0.2651	0.1658	$<10^{-9*}$
	3	60.6 cM	D3Mit78-D3Mit348 (D3Mit78+5.6 cM)	0.0102*	0.9708	0.4831	$<10^{-8*}$
	3	52.6 cM	D3Mit189-D3Mit78 (D3Mit189+2.9 cM)	0.2120	0.6421	0.7095	$<10^{-8*}$
SIgA	3	52.6 cM	D3Mit189-D3Mit78 (D3Mit189+2.9 cM)	0.1016	0.6488	0.3195	$<10^{-9*}$
	3	38.6 cM	D3Mit203-D3Mit212 (D3Mit203+27.4 cM)	0.7174	0.4051	0.8020	$<10^{-7*}$

p-value[k] ($k=1,2,3,4$) refer to the p-values obtained with the likelihood ratio tests for testing 1) $H_0: c=0$ vs $H_1: c\neq 0$; 2) $H_0: i=0$ vs $H_1: i\neq 0$; 3) $H_0: i_{ca}=i_{cd}=i_{ci}=0$ vs H_1: not all equal to 0; and 4) $H_0: a=d=0$ vs H_1: not all equal to 0, respectively. Significant test results are indicated with the "*" sign.

doi:10.1371/journal.pone.0091702.t004

DISCUSSION

The cytoplasmic environment influences the expression of nuclear information in a very complicated way, which is still an unravel mystery to many organisms. For example, Burgess and Husband have demonstrated great maternal contributions to the fitness of mulberry hybrids [39]. While it is an important parent-of-origin effect affecting offspring fitness, genomic imprinting, another source of parent-of-origin effect can also lead to phenotypic variation. Increasing evidence from cytoplasmic substation and cell fusion experiments suggests that weakness of hybrids may connect with the interactions between cytoplasm and nuclear [40],[41], and the evidences about interaction between cytoplasm and imprinting have also been observed [42], [43]. As the source of genetic variation for many traits, these two types of parent-of-origin effects are often confounded, making it difficult to distinguish without proper statistical dissection. Although the role of cross-talk between the two sets of factors on phenotypic variation has been recognized, which genes are involved in the process and in what form they respond to the cytoplasmic changes are still unclear.

In this paper, we developed a statistical model to evaluate the cytoplasmic environment and nuclear gene interaction subject to imprinting effect within the framework of QTL interval mapping. The model that considers eight genetic factors which measure the degree of imprinting, cytoplasmic, additive, dominance effects as well as the interaction effects among them, provides a complete dissection of cyto-nuclear epistasis subject to imprinting effect. A number of hypothesis tests can be performed not only to assess major genetic effect(s) responsible for phenotypic variation, but also to find the statistical evidence for the existence of imprinting, cytoplasmic effect as well as the cyto-nuclear interactions. Simulation study showed relative good performance of the model under the F_2 design, in which parameters are estimated efficiently with modest heritability and sample size. Low heritability level ($H^2=0.1$) and small sample size ($n=400$) result in large mean squared error of parameter estimation. This result is valuable in practice as we need to be careful about the interpretation of genetic effects obtained in real data analysis when the proportion of variance explained by the QTL is small (i.e., low heritability). Although our model cannot estimate the imprinting effect (so the cyto-imprinting interaction effect) due to the nature of the F_2 design, existence test of imprinting (or cyto-imprinting interaction effect) can be achieved. Nevertheless, such imprinting estimation problem can be solved under the reciprocal backcross design illustrated in Figure 1.

In the real data analysis, one QTL located on chromosome 1 were found to have significant cytoplasmic, imprinting effect and cyto-nuclear interaction effects for spleen/body weight ratio. Other than that, no imprinting effect was found for all other QTLs, and only one on chromosome 3 that shows cytoplasmic effect for the MLN/body weigh ratio trait. It is worth mentioning that several QTL on chromosome 3 are detected by our model, but they are located relatively close to each other, as shown in Figure 3. For the sake of cautiousness, we reported all of them. However, these detected clustered QTLs may be due to the limitation of interval mapping, which can be overcome by fitting a composite interval mapping model as following,

$$y_j=\mu+cx_{1j}+ax_{2j}+dx_{3j}+ix_{4j}+i_{ca}x_{5j}+i_{cd}x_{6j}+i_{ci}x_{7j}$$

$$+\sum_{k=1}^{K}(a_kx_{2jk}+d_kx_{3jk}+i_kx_{4jk}+i_{ca_k}x_{5jk}+i_{cd_k}x_{6jk}+i_{ci_k}x_{7jk})+e_j$$

where $x_{2jk},\cdots,x_{7jk}$ are corresponding variables for the kth marker, assuming total Kmarkers are selected for controlling background genetic effect. Although more variables are introduced in the model, theoretically some dimension-reduction techniques such as LASSO, can be applied to implement the variable selection for each marker before fitting them into the final model [44]. The composite interval mapping is know for its improved resolution in QTL detection. Regardless of the potential limitations mentioned above, the integration of cyto-nuclear interactions into the QTL mapping framework provides a testable platform with feasible experimental design for biologists to test the existence of cytoplasmic and imprinting effects, as well as the interactions of interested. The proposed model will have important biological implications with potentials to lift a corner of the great veil of the genetic system.

AUTHOR CONTRIBUTIONS

Conceived and designed the experiments: YC. Analyzed the data: TH. Wrote the paper: TH YC. Participated in analysis: JS PSZ.

REFERENCES

1. Lander E, Bostein E (1989) Mapping Mendelian factors underlying quantitative traits using RFLP linkage maps. Genetics 121: 185–199.
2. Zeng Z (1994) Precision mapping of quantitative trait loci. Genetics 136: 228–235.
3. Wu R, Ma C, Gallo-Meagher M, Littell R, Casella G (2002) Statistical methods for dissectiong triploid endosperm traits using molecular markers: an autogamous model. Genetics 162: 875–892.
4. Mackay T (2001) Quantitative trait loci in Drosophila. Nat Rev Genet 2: 11–20. doi: 10.1038/nrg1206
5. Jirtle R, Skinner M (2007) Environmental epigenomics and disease susceptibility. Nat Rev Genet 8: 253–262. doi: 10.1038/nrg2045
6. Isles A, Holland A (2005) Imprinted genes and mother-offspring interactions. Early Hum Dev 81: 73–77. doi: 10.1016/j.earlhumdev.2004.10.006
7. Tycko B, Morison I (2002) Physiological functions of imprinted genes. J Cell Physiol 192: 245–258. doi: 10.1002/jcp.10129
8. Reik W, Walter J (2001) Genomic imprinting: parental influence on the genome. Nat Rev Genet 2: 21–32. doi: 10.1038/35047554
9. Falls J, Pulford D, Wylie A, Jirtle R (1999) Genomic imprinting: implications for human disease. Am J Pathol 154: 635–647. doi: 10.1016/s0002-9440(10)65309-6
10. Clerget-Darpoux F, Babron MC, Deschamps I, Hors J (1991) Complementation and maternal effect in insulin-dependent diabetes. Am J Hum Genet 49: 42–48.
11. Nanney D (1953) Nucleo-cytoplasmic interaction during conjugation in Tetrahymena. Biol Bull 105: 133–148. doi: 10.2307/1538562

12. Haga N (1995) Elucidation of nucleus-cytoplasm interaction: change in ability of the nucleus to express sexuality according to clonal age in Paramecium. J Cell Sci 108: 3671–3676.
13. Elfgang C, Rosorius O, Hofer L, Jaksche H, Hauber J, et al. (1999) Evidence for specific nucleocytoplasmic transport pathways used by leucine-rich nuclear export signals. Proc Natl Acad Sci USA 96: 6229–6234. doi: 10.1073/pnas.96.11.6229
14. Chinnery P (2003) Searching for nuclear-mitochondrial genes. Trends Genet 19: 60–62. doi: 10.1016/s0168-9525(02)00030-6
15. Rand D, Haney R, Fry A (2004) Cytonuclear coevolution: the genomics of cooperation. Trends Ecol Evol 19: 645–653. doi: 10.1016/j.tree.2004.10.003
16. Roubertoux P, Sluyter F, Carlier M, Marcet B, Maaroufveray F (2003) Mitochondrial DNA modifies cognition in interaction with the nuclear genome and age in mice. Nat Genet 35: 65–69. doi: 10.1038/ng1230
17. Tao D, Hu F, Yang J, Yang G, Yang Y, et al. (2004) Cytoplasm and cytoplasm-nucleus interactions affect afronomic traits in japonica rice. Euphytica 135: 129–134. doi: 10.1023/b:euph.0000009548.81467.73
18. Zeyl C, Andreson B, Weninck E (2005) Nuclear-mitochondrial epistasis for fitness in Saccharomyces cerevisiae. Evolution 59: 910–914. doi: 10.1554/04-487
19. Rand D, Fry A, Sheldahl L (2006) Nuclear-mitochondrial epistasis and Drosophila aging: introgression of Drosophila simulans mtDNA modifies longevity in D. melanogaster nuclear backgrounds. Genetics 172: 329–341. doi: 10.1534/genetics.105.046698
20. Li S, Lu Q, Fu W, Romero R, Cui Y (2009) A regularized regression approach for dissecting genetic conflicts that increase disease risk in pregnancy. Stat Appl Genet Mol Biol 8: 45. doi: 10.2202/1544-6115.1474
21. Hanson R, Kobes S, Lindsay R, Kmowler W (2001) Assessment of parent-of-origin effects in linkage analysis of quantitative traits. Am J Hum Genet 68: 951–962. doi: 10.1086/319508
22. de Koning DJ, Bovenhuis H, van Arendonk J (2002) On the detection of imprinted quantitative trait loci in experimental crosses of outbred species. Genetics 161: 931–938.
23. Cui Y (2007) A statistical framework for genome-wide scanning and testing imprinted quantitative trait loci. J Theo Biol 244: 115–126. doi: 10.1016/j.jtbi.2006.07.009

24. Cui Y, Cheverud J, Wu R (2007) A statistical model for dissecting genomic imprinting through genetic mapping. Genetica 130: 227–239. doi: 10.1007/s10709-006-9101-x
25. Cui Y, Lu Q, Cheverud J, Littell R, Wu R (2006) Model for mapping imprinted quantitative trait loci in an inbred F2 design. Genomics 87: 543–551. doi: 10.1016/j.ygeno.2005.11.021
26. Li G, Cui Y (2010) A general statistical framework for dissecting parent-of-origin effects underlying triploid endosperm traits in flowering plants. Ann Appl Stat 4: 1214–1233. doi: 10.1214/09-aoas323
27. Tang Z, Wang X, Hu Z, Yang Z, Xu C (2007) Genetic Dissection of Cytonuclear Epistasis in Line Crosses. Genetics 177: 669–672. doi: 10.1534/genetics.107.074302
28. Haldane J (1922) The part played by recurrent muation in evolution. Am Nat 67: 5–9.
29. Huxley J (1928) Sexual difference of linkage Grammarus chereuxi. J Genet 20: 145–156.
30. Dib C, Fauer S, Fizames C, Samson S, Drouot N, et al. (1996) A comprehensive genetiv map of the human genome based on 5,264 microsatellites. Nature 380: 152–154. doi: 10.1038/380152a0
31. Marklund L, Moller M, Hoyheim B, Davies W, Fredholm M, et al. (1996) A comprehensive linkage map of the pig based on a wild pig-Large White intercross. Anim Genet 27: 255–269. doi: 10.1111/j.1365-2052.1996.tb00487.x
32. Neff M, Broman K, Mellersh C, Ray K, Acland GM, et al. (1999) A second-generation genetic linkage map of the domestic dog, Canis familiaris. Genetics 151: 803–820.
33. Dietrich W, Miller J, Steen R, Merchant M, Damron-Boles D, et al. (1996) A comprehensive genetic map of the mouse genome. Nature 380: 149–152. doi: 10.1038/380149a0
34. Dempster A, Laird N, Rubin D (1977) Maximum likelihood from incomplete data via the EM algorthim. J Roy Stat Soc B 39: 1–38.
35. Manichaikul A, Dupuis J, Sen S, Broman K (2006) Poor Performance of Bootstrap Confidence Intervals for the Location of a Quantitative Trait Locus. Genetics 174. doi: 10.1534/genetics.106.061549
36. Churchill G, Doerge RW (1994) Empirical threshold values for quantitative trait mapping. Genetics 138: 963–971.
37. Lynch M, Walsh B (1998) Genetics and Analysis of Quantitative Traits. Sunderland, MA: Sinauer Associates, Inc.

38. Farmer M, Sundberg J, Bristol I, Churchill G, Li R, et al. (2001) A major quantitative trait locus on chromosome 3 controls colitis severity in 1L-10-deficient mice. PNAS 98: 13820–13825. doi: 10.1073/pnas.241258698

39. Burgess K, Husband B (2004) Maternal and paternal contributions to the fitness of hybrids between red and white mulberry. Am J Bot 91: 1802–1808. doi: 10.3732/ajb.91.11.1802

40. Levin D (2003) The cytoplasmic factor in plant speciation. Syst Bot 28: 5–11.

41. Rhode J, Cruzan M (2005) Contributions of heterosis and epistasis to hybrid fitness. Am Nat 166: E124–E139. doi: 10.1086/491798

42. Moore T, Haig D (1991) Genomic imprinting in mammalian development: a parental tug-of-war. Tren Genet 7: 45–49. doi: 10.1016/0168-9525(91)90230-n

43. Cardoso M, Leonhardt H (1999) DNA Methyltransferase Is Actively Retained in the Cytoplasm during Early Development. J Cell Bio 147: 25–32. doi: 10.1083/jcb.147.1.25

44. Li Z, Sillanpää M (2012) Overview of LASSO-related penalized regression methods for quantitative trait mapping and genomic selection. Theor Appl Genet 125: 419–435. doi: 10.1007/s00122-012-1892-9

Chapter 2

A GUIDE TO MODERN STATISTICAL ANALYSIS OF IMMUNOLOGICAL DATA

Bernd Genser[1, 2], Philip J Cooper[3, 4], Maria Yazdanbakhsh[5], Mauricio L Barreto[2] and Laura C Rodrigues[2]

[1]Department of Epidemiology and Population Health, London School of Hygiene and Tropical Medicine, London, UK

[2]Instituto de Saúde Coletiva, Federal University of Bahia, Salvador, Brazil

[3]Centre for Infection, St George's University of London, London, UK

[4]Instituto de Microbiologia, Universidad San Francisco de Quito, Quito, Ecuador

[5]Department of Parasitology, Leiden University Medical Center, Leiden, The Netherlands

ABSTRACT

Background

The number of subjects that can be recruited in immunological studies and the number of immunological parameters that can be measured has increased rapidly over the past decade and is likely to continue to expand. Large and complex immunological datasets can now be used to investigate complex scientific questions, but to make the most of the potential in such data and to get the right answers sophisticated statistical approaches are necessary. Such approaches are used in many other scientific disciplines, but immunological studies on the whole still use simple statistical techniques for data analysis.

Results

The paper provides an overview of the range of statistical methods that can be used to answer different immunological study questions. We discuss specific aspects of immunological studies and give examples of typical scientific questions related to immunological data. We review classical bivariate and multivariate statistical techniques (factor analysis, cluster analysis, discriminant analysis) and more advanced methods aimed to explore causal relationships (path analysis/structural equation modelling) and illustrate their application to immunological data. We show the main features of each method, the type of study question they can answer, the type of data they can be applied to, the assumptions required for each method and the software that can be used.

Conclusion

This paper will help the immunologist to choose the correct statistical approach for a particular research question.

BACKGROUND

The understanding of the importance of immunological mechanisms underlying human disease and the identification of associated immunological markers have grown enormously over the past ten years and the number of published immunological studies that investigate the relationships between human disease and cytokines and other immunological parameters has increased rapidly. Technical developments in sample processing and sophisticated immunological techniques permit the analysis of more immunological parameters in larger samples of human subjects, containing information that allows not only the measurement of simple associations between two parameters but also the exploration of the complex relationships between immunity, disease, environmental, social and genetic factors. The potential complexity of the possible relationships between large numbers of immunological parameters poses a special challenge for the applied immunologist: how to select the appropriate statistical techniques to extract the maximum relevant information from complex datasets and avoid spurious findings.

Immunologists tend to use simple statistical approaches even when multiple relationships between immunological parameters are expected, [1, 2] instead of multivariate statistical approaches that can analyse simultaneously multiple measurements on the same individual. Multivariate statistical analysis techniques are being widely applied in other scientific fields and numerous books and articles have been published that describe these techniques in detail. Unfortunately, this literature is not easily accessible to the applied immunologist without a detailed knowledge of statistics and few articles have been written demonstrating the application of statistical techniques to immunological data [3, 4].

This paper provides an overview of statistical analysis techniques that may be considered for the analysis of immunological data. We discuss specific aspects of immunological studies, give examples of typical scientific questions related to immunological data and present a statistical framework to help the immunologist to choose the correct statistical approach for a particular research questions. Although we have focused on cytokine data in the examples provided, the methods presented are applicable to most other immunological parameters.

SPECIFIC ASPECTS OF IMMUNOLOGICAL STUDIES RELEVANT FOR STATISTICAL ANALYSIS

In the following section we discuss specific aspects of immunological studies that are relevant for statistical analysis.

Structure of Immunological Data

Before analysing immunological data it is very important to examine the structure of the data because most statistical methods will only give the correct answer if the data has the characteristics required for the use of that method ("satisfy the data assumptions"). For example, common data assumptions are that the observations are approximately normally distributed or that the variances are similar across different subpopulations. Unfortunately, immunological data very frequently do not meet these assumptions and investigators are obliged to either apply data transformations (e.g. a logarithmic transformation to make skewed data approximately normally distributed), or to choose an alternative statistical techniques with less stringent data assumptions (e.g. using a non-parametric statistical approach that does not require the data to be normally distributed [5] instead of a parametric statistical technique that does). A further important aspect of immunological data is that different immunological parameters measured in the same study subject are frequently highly correlated ("multicollinearity"). Hence the application of statistical techniques that assume independence among the observations is often not valid and in such situations a method should be used that takes into account the fact that study variables may be the result of a common underlying biological mechanism. Examples of underlying biological mechanisms that can not be directly observed but will influence that value of more than one immunological variable are: "immune maturation," "down regulation" or "Th2 shift."

Complexity of the Relationships in Immunological Parameters

Immunological parameters are often involved in complex immunological mechanisms; and relationships between immunological parameters may be changeable. For example, a specific parameter (e.g. a cytokine) may have different effects in different cell populations, at different times and in the presence (or absence) of other immunological parameters. We often aim to explain the complete causal pathway from a non-immunological factor (e.g. exposure to an allergen) to an outcome (e.g. atopy or asthma). Clearly simple univariate statistical analysis would not be able to identify such inter-relationships among several study variables and the underlying immunological mechanisms that cannot be measured; multivariate statistical techniques are required that can examine multiple parameters simultaneously. A fundamental step to guide statistical analysis is to make the hypothesis explicit in a conceptual framework [6]. Conceptual frameworkspresent the proposed inter-relationships among the study variables and define any larger underlying immunological mechanisms assumed to influence their values. Conceptual frameworks should be detailed and explicit as they are used to guide the analysis.

Two further important aspects of immunological data that are not the focus of this paper but should be mentioned are:

Reproducibility of the Measurement of an Immunological Parameter

Reproducibility reflects how often we obtain the same result using the same laboratory test and sample. Some variation is expected for any measurement and statistical analysis must take into account the degree of variation. Although reproducibility of immunologic measurements is well defined for some immunological outcomes, particularly those used for diagnostic purposes, such as antibody levels associated with vaccine protection (e.g. levels >10 IU/mL for anti-HBS for vaccine response to hepatitis B vaccine), and for phenotypic characterization (CD4 counts or CD4/CD8 ratio for evaluation of immune status in HIV), reproducibility is not well defined for most immunological parameters. This is independent of the separate but important issue of repeatability between centres in multicentre studies that measure the same immunological parameters in different laboratories or even the measurement of the same parameter between different studies.

Multiple Testing

The problem of multiple testing is becoming increasingly relevant in immunological studies as the number of immunological parameters that can be measured increases and investigators conduct a large number of statistical tests on the same study data [7]. A specific concern in statistical analysis is to separate associations that occur by chance (because of «random variation» or «noise») from those reflecting true biological relationships («systematic variation,» often assumed to be a causal relationship). Most researchers use a statistical significance level («type I error,» for example P = 0.05) to decide whether the result of an analysis is likely to be due to chance. Conducting multiple hypothesis testing may result in substantial inflation of type I errors (depending on the degree of dependence between the tests). For example, if the value $p < 0.05$ is used, conducting twenty independent significance tests within a data set is likely to result in one comparison being significant just by chance. There are numerous multiple comparison procedures to adjust statistical analysis for type I error inflation, for example simultaneous test procedures, such as the approaches by Bonferroni, Tukey, Scheffé or Dunnet or more sophisticated step-wise procedures, such as the techniques by Newman-Keuls or Ryan. A good overview about the most important multiple comparison techniques is given by Toothaker [8].

RESEARCH OBJECTIVES OF IMMUNOLOGICAL STUDIES

In this section we list typical research questions from immunological studies.

Common objectives of immunological studies can be grouped into four overall categories:

i) Those that investigate patterns of associations between several immunological parameters, without assuming any causal relationship (and therefore not classifying study variables as dependent variables (i.e. outcomes) and independent variables (i.e. explanatory variables or covariates).

For example, typical research questions of such studies are:

- To assess the magnitude of the correlation between different cytokines or to quantify the balance between levels of cytokine expression. For example, the research question might be to measure the correlation between pro-inflammatory and anti-inflammatory cytokines (e.g. correlation between TNF-α and IL-10) or to quantify the «balance» between pro-inflammatory and anti-inflammatory cytokines (e.g. by calculating the ratio TNF-α/IL-10).
- To identify highly correlated cytokines and to place them into groups which reflect an unobserved underlying mechanism. For example, Th1-related immune responses such as IFN-γ and TNF-α may mediate an inflammatory disease. Depending on the question being investigated, it may be more appropriate to first use a statistical analysis approach to «reduce the data», i.e., to aggregate the correlated Th1 related cytokines to form a «summarising variable» that reflects the underlying immunological mechanism (e.g. «degree of Th1 immune response») and use that summary variable in the analysis rather than using all the variables with the original cytokine levels.

- To identify individuals with similar profiles of immunological parameters and to place them into groups (so called "clusters"). For example, patients with a clinical outcome might be defined as "atopics» or «non-atopics» based on the values of skin prick tests; or subjects with a specific infection may be classified into groups (eg, active, chronic, or past) defined by the overall elevation in antibodies (e.g. IgE, IgA, IgM or IgG subclasses). However, within the same group of patients, clusters with distinct or overlapping profiles might be distinguishable and subsequent analyses might show associations between distinct clusters and disease (or some other outcome).

ii) The second group of research objectives investigates causal relationships between one or more immunological parameters (e.g. different cytokines, or summary measures) and other study variables (e.g. an outcome such as asthma). To guide the investigation of causation it is important to have developed an a priori causal pathway model. This will allow the appropriate definition of variables, i.e. defining which variables are dependent variables (outcomes), intervening variables (mediating the effect) or independent variables (exposures, confounding factors and effect modifiers) and will determine the choice of statistical approach.

Possible research objectives for causality include:

- Identification of determinants of immunological profiles. The objective may be to compare the expression of cytokines between two or more groups defined by an exposure, e.g. people infected or not infected with helminths, or people vaccinated and non-vaccinated in a vaccine trial where the vaccine exposure is assumed to influence the levels of the immunological parameter to be measured. For example, the question may be to determine if BCG vaccination influences the levels of IFN-γ secreted by mononuclear cells stimulated in vitro with a mycobacterial antigen. The immunological parameter is the outcome or dependent variable.
- Identification of clinical consequences of immunological profiles, (immunological parameter as the risk factor) or, in other words to identify associations between an immunological parameter and clinical (or other) outcomes. For example, immunologists are interested in predicting the probability of a disease occurrence by measurement of cytokine levels. For example, whether elevated TNF-α levels are associated with active disease in rheumatoid arthritis? The immunological parameter is the risk factor (often called «exposure») or independent variable.

iii) The third group consists of more complex research questions that may include two or more of the objectives described above. Such questions may examine the role of cytokines in larger causal constructs, including more than one risk factor, intervening variables that mediate and modify an effect, and outcomes; and inter relationships between them. An example will be to investigate the causal inter relationships between early life infections, level of expression of pattern recognition receptors (such as Nods), activity of pro-inflammatory cytokines (IFN-γ and TNF-α) and the development of inflammatory bowel disease.

iv) The field of in silico immunology (computer analysis generally in conjunction with informatics or immuno-informatics) is a rapidly developing and expanding field and has been used to address several types of study questions, such as:

- The prediction of immunogenic sequences from microbial genomes to predict potential vaccine candidates [9].
- The prediction of protein sequences in therapeutic antibodies that may be associated with adverse reactions [10].
- Identification of regulatory molecules in the innate immune system [11].

These approaches are generally high-throughput analyses of large data sets (e.g. microbial genomes, human genome, etc) using available software (e.g. EpiMatrix) to either generate or test hypotheses and have been reviewed in detail elsewhere [12–15].

In silico statistical analyses use many of the multivariate statistical techniques discussed later in this review (e.g. cluster analysis). Because this is a highly specialised field for which there are many computational tools available [9], in silicoimmunology will not be discussed further in this review.

STATISTICAL METHODS FOR ANALYSIS OF IMMUNO-LOGICAL DATA

We conducted a systematic literature search in the database MEDLINE (1980–2005) to review statistical methods that have been previously applied to cytokine data. Because the objective was to get a crude overview rather than to reveal the exact number of papers published in this area we defined quite sensitive search criteria using the following key words:"cytokine$" or terms to identify specific cytokines (e.g. among others «IL$,» «interleukin$,» IF$, interferon$, TNF$, etc.) and common univariate and multivariate statistical techniques (e.g. among others "linear regression," «analysis of variance,»»cluster analysis,» «factor analysis» etc.).

Table 1 shows the results of our search. The most widely used methods found were simple statistical approaches that investigate the relationship between two variables (so called bivariate meth ods – also called univariate methods when variables are classified as dependent and independent variables). We frequently found standard methods to compare means of immunological parameters between independent groups (e.g. t-test, analysis of variance or their non-parametric equivalents), bivariate correlation analysis (Pearson's or Spearman's correlation coefficients) and univariate linear regression. By contrast, multivariate techniques (i.e. statistical approaches that consider three or more study variables simultaneously) were less frequently applied to cytokine data. Several studies used factor analysis (to identify groups of correlated immunological parameters) or cluster analysis (to identify groups of individuals with similar immunological profiles) or discrimination techniques such as logistic regression, discriminant analysis (to identify causes or consequences of immunological profiles). We also found a few examples of advanced modelling techniques (path analysis/structural equation modelling) that simultaneously model multiple relationships between the study variables.

Table 1: Results of a literature review conducted in the medical database MEDLINE (1980–2006) about statistical methods found in immunological studies investigating cytokine expressions

Statistical methods	**Number of references**
Univariate techniques	
Analysis of variance	2908
T-test	420
Mann-Whitney U-test	316
Wilcoxon/McNemar test	193
Univariate linear regression	163
Bivariate correlation analysis	157
Kruskal-Wallis H-test	95
Repeated measures analysis of variance	31
Friedman test	7
Non linear regression	5
Multivariate techniques	
Logistic regression	629
Cluster analysis	192
Multivariate analysis of variance	144

Multiple linear regression	91
Factor analysis/Principal components analysis	80
Analysis of covariance	56
Linear discriminant analysis	51
Partial correlation coefficient	24
Multinomial logistic regression	9
Multivariate analysis of covariance	7
Path analysis/Structural equation modelling	4

In the following section we provide an overview of statistical methods that can be considered for analysing immunological data that should help the applied immunologists without a detailed knowledge of statistics to select the appropriate statistical technique for each particular research question. The definition of which method is the most appropriate is strongly dependent on the research objective, the type of data collected, whether data assumptions are fulfilled and whether the sample size is sufficient. We begin with a short introduction to these topics.

Exploratory Data Analysis

An important first step in analysing immunological data is exploring and describing the data. Whatever the research question investigators should first explore the data in tables showing summary statistics (e.g. means, standard deviations, etc.) and apply graphical methods such as bar charts, histograms, Box and Whisker plots, or scatter plots. For multivariate data, scatter plot matrices are a powerful tool to examine associations among several immunological parameters [16]. A good overview of statistical methods for Exploratory Data Analysis is provided by Tukey [17].

Data Assumptions

The next thing investigators have to consider when selecting a statistical method is whether their data meet a number ofdata assumptions. The first assumption is the scale of measurement of the data, i.e. whether the data type is categorical (e.g. groups like male and female), ordinal (e.g. groups with a logical order, like order of birth) or continuous (also called metric [measured on a defined scale]). This restricts the statistical methods that can be used. When the investigator has to deal with continuous data the second assumption to test is whether the data follow a theoretical distribution (e.g. normal distribution). Distributional assumptions can be tested graphically by diagnostic plots or by applying a statistical test that compares the distribution of the data with a theoretical distribution. When the original data do not meet distributional assumptions required for a particular statistical technique (e.g. normal distribution to apply a t-test) a common approach is transforming the data to meet the data assumptions, for example to use the logarithm of the values [4]. However, immunological data frequently do not meet data assumptions even after applying different data transformations. In such cases it may be more appropriate to use an alternative statistical approach that requires fewer data assumptions (e.g. applying a non parametric Wilcoxon test instead of a t-test) [5]. Another approach for continuous variables that do not fulfil distributional assumptions is «categorising» the measurements by biological meaningful cut-off values (e.g. level >= 2.5: «positive», level < 2.5: «negative») or using centiles, and then apply a statistical test that is appropriate for categorical data. A less frequently applied strategy that allows for the use of a parametric approach even when data assumptions are violated is the «robust resampling variance estimator [18].» The concept behind this approach is to draw repeated random samples from the data, and then to estimate the parameter of interest (e.g. the mean) from each sample and finally to obtain an estimate of the variance by calculating the variance of the parameter across the samples. «Robust estimation» means the approach will provide valid estimations even when data assumptions (e.g. normal distribution) are violated [3]. For example, in a hypothetical immunological study IL-10 measurements have been obtained from a sample of 100 individuals. By drawing 500 random samples from these 100 individuals each including 50 subjects, a robust estimate of the variance of the parameter of interest (the population mean) can be calculated from the mean of each sample and from the variance of the mean (across the samples).

Sample Size Issues

A second major issue that immunologist frequently face and that also substantially affects the statistical approach to be used is to estimate the appropriate sample size for a statistical analysis. Whereas for univariate techniques (e.g. t-test, ANOVA etc.) sample size can be determined by conducting a power analysis [19], sample size formulae are available only for a couple of multivariate techniques. Cohen's Power analysis guide is a useful review for calculating sample sizes for common univariate and some multivariate techniques (analysis of covariance and multiple regression) [20]. Unfortunately, the theory underlying sample size estimation for other multivariate techniques is less developed. Recommendations for sample size for studies using factor analysis are that more subjects should be included than the number of unique correlations present in the correlation matrix [21, 22]. Authors suggest that for cluster analysis sample size calculations are dependent on how the investigator believes the study population is clustered. If some small clusters (say less than 10) are expected, then sufficient subjects will be required to sample at least 5 to 10 people in the smallest cluster [23]. There is a statistical framework and software available for sample size calculation for advanced techniques such as structural equation modelling and path analysis [24, 25].

SELECTING THE APPROPRIATE STATISTICAL METHOD FOR EACH PARTICULAR RESEARCH QUESTION

In this section we provide an overview of statistical methods that we consider useful for statistical analysis of immunological data. Our guide for data analysis was written with the "classical statistical approach" in mind, i.e. we provide statistical techniques to extract the maximum information from the present data. We do not discuss Bayesian statistical methods that also might be useful for application to immunological data seeking to combine a priori information with the information captured in the present data; a good overview is provided by Lee [26]. Moreover, statistical methods are important also for standardising immunological techniques and this is discussed elsewhere [27].

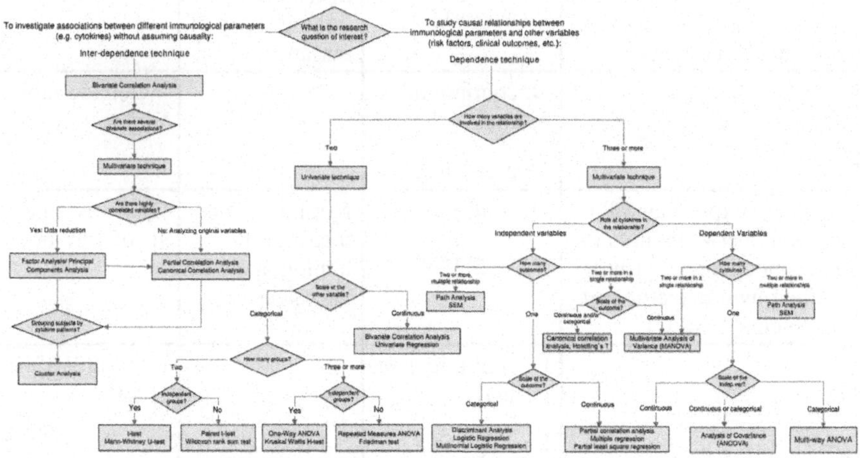

Figure 1: Selecting the appropriate statistical technique for analysis of immunological data.

We have prepared a summary table (Table 2) that lists the most important univariate and multivariate methods and links them to typical immunological research questions. In addition, we provide a flowchart that will help the immunologist to select the appropriate statistical analysis according to the research objective and the number and type of study variables to be analysed

(Figure 1). Most examples refer to cytokine data but the application to other continuous immunological data is straightforward.

Table 2: Selection of important statistical methods suitable for the analysis of immunological data

Example of research question	Type of data [D: dependent, I: independent]	Other data assumptions	Statistical method[1]
Univariate techniques			
Univariate group mean comparison techniques			
Compare expression of a cytokine between two independent groups (e.g. treatment vs. control)	D: continuous I: categorical	Normal distribution homogeneity of variances	t-test
	D: continuous or ordinal I: categorical		Mann Whitney-U test
Compare expression of a cytokine between two related groups (e.g. before and after treatment)	D: continuous I: categorical	Normal distribution, homogeneity of variances	Paired t-test
	D: continuous or ordinal I: categorical		Wilcoxon rank sum test
Compare expression of a cytokine between three or more independent groups defined by one factor (e.g. treatments A, B, C)	D: continuous I: categorical	Normal distribution, homogeneity of variances	One-way analysis of variance
	D: continuous or ordinal I: categorical		Kruskal Wallis – H test
Compare expression of a cytokine between three or more related groups (e.g. measurements 1, 2, and 3 weeks after treatment)	D: continuous I: categorical	Multivariate normal distribution, assumptions about covariance	Repeated measurements analysis of variance

	D: continuous or ordinal I: categorical		Friedman's ANOVA
Correlation and regression analysis			
Quantify association between two cytokines or a cytokine and another continuous variable	D: continuous I: continuous	Linear relationship, normality	Pearson correlation coefficient
	D: continuous or ordinal I: continuous or ordinal	Linear relationship	Spearman rank correlation coefficient
Predicting expression of a cytokine by a continuous independent variable	D: continuous I: continuous	Specified relationship (e.g. linearity for linear regression), normal distribution (for parametric regression)	Univariate regression
Multivariate techniques			
Multivariate correlation and regression techniques			
Quantify associations between two cytokines adjusted for the effect of a third continuous variable	All variables: continuous	Linear relationship, normality	Partial correlation coefficient
Predicting a continuous outcome (e.g. a cytokine) by several continuous or categorical independent variables	D: continuous I: continuous, ordinal or categorical	Specified relationship (e.g. linearity for linear regression), normal distribution for parametric regression, No multi-collinearity	Multiple regression
		Specified relationship, multicollinearity	Partial least squares regression

Quantifying the magnitude of correlation between two groups of continuous variables (e.g. Th1 and Th2 related cytokines)	All variables: continuous		Canonical correlation analysis
Multivariate group mean comparison procedures			
Compare cytokine expressions between three or more independent groups defined by two or more factors (e.g. treatment and gender)	D: continuous I: categorical	Normal distribution, homogeneity of variances	Multi-way analysis of variance (ANOVA)
Simultaneously compare expressions of two or more cytokines between three or more independent groups defined by two or more factors	D: continuous I: categorical	Multivariate normal distribution, homogeneity of covariance matrices	Multivariate analysis of variance (MANOVA)
Compare cytokine expressions between three or more related groups defined by two or more factors (e.g. measurements at different time points during a study and treatment)	D: continuous I: categorical	Multivariate normality, homogeneity of covariance matrices	Multi-way repeated measurements analysis of variance
Grouping set of correlated cytokines to summary variables ("principal components")	All variables: continuous	High degree of multicollinearity	Factor analysis/ Principal components analysis
Grouping subjects in homogenous subgroups according to similar expression levels of two or more cytokines	All variables: continuous	Low degree of multicollinearity	Cluster analysis
Classification procedures			
Explaining or predicting group membership of two or more independent groups by cytokine levels	D: categorical I: continuous	Multivariate normal distribution, equal covariance matrices, low degree of multicollinearity	Linear discriminant analysis

Explaining or predicting group membership of two independent groups by cytokine levels	D: categorical I: continuous, ordinal or categorical		Logistic regression
Explaining or predicting group membership of three or more groups by cytokine levels	D: categorical I: continuous, ordinal or categorical		Multinomial logistic regression
Advanced techniques for multiple relationships			
Modelling multiple relationships between several immunological parameters and one or more outcome variables	All variables: categorical, ordinal or continuous data	Conceptual framework specifying the multiple relationships among the study variables	Path analysis/ Structural equation modelling

[1] All univariate and multivariate statistical approaches listed above can be implemented in general purpose statistical packages, e.g. among others S-PLUS® (Insightful Corporation, Seattle, WA), SAS® (SAS Institute Cary, NC, USA), SPSS® (Chicago: SPSS Inc.) or STATA® (StataCorp.Stata Statistical Software. College Station, TX: StataCorp LP). Path analysis/structural equation modelling can be implemented in STATA andSPSS that provide the extensions modules GLLAMM and AMOS, respectively, as well as in several special purpose software packages, e.g. among others LISREL® (Scientific Software International, Inc, IL, USA) or MX® (MCV, Department of Psychiatry, Richmond, VA, USA).

Inter-dependence Techniques

The first group embraces so called inter-dependence techniques, i.e. statistical methods aimed to explore relationships between study variables without assuming any causal relationship. These techniques are appropriate when the investigator cannot define (or may not wish to define) which variable is the independent variable (cause, exposure) and which is the dependent variable (effect, outcome). For example, inter-dependence techniques might be very useful in immunological studies to examine relationships between different cytokines measured in the same individual.

Correlation Analysis

One commonly applied inter-dependence approach is bivariate correlation analysis that aims to assess the magnitude of the linear relationship between two continuous variables (e.g. two cytokines, or a cytokine and another continuous variable. For example, Hartel et al conducted an observational, cross-sectional study in children aged between 1 and 96 months and in adults to investigate age-related changes in cytokine production [2]. The association between cytokine levels and age was analyzed using non-parametric rank correlation coefficients.

If there are more than two immunological parameters of interest and one finds several significant bivariate correlations, amultivariate correlation analysis should be conducted to examine the degree of multicollinearity in the data, i.e. whether multiple relationships are present between three or more study variables. A simple but useful approach to examine associations between three variables is to conduct stratified bivariate correlation analysis across strata defined by levels of a third variable. For example, to examine the association between a Th1-related and a Th2-related cytokine could be examined after stratifying by low and high levels of expression of IL-10.

Data Reduction Techniques

Datasets with several highly correlated immunological parameters can be simplified using "data reduction techniques."These methods are especially appropriate when it is assumed that many variables reflect aspects of an underlying process which is not directly measured. The most common technique is Principal Component Analysis (PCA) [28], which is a special type of Factor Analysis [29]. The idea behind this approach is to create summary variables (called «principal components»), that capture most of the information of the original data. For example, the technique can be used to derive two principal components from several correlated cytokines. After using this technique it is essential to consider whether the components identified are biologically plausible (e.g. it would be important to observe that the classification of cytokines grouped into two groups is consistent with the findings of published literature).

A useful feature of factor analysis and especially PCA is that the weights (the «factor loadings») for each variable within the components can be interpreted as correlation measures between the observed variable and the underlying unobservable component. Data assumptions in factor analysis are more conceptual than statistical. From a statistical point of view, normality is only necessary if a statistical test is applied to measure the significance of the factors, but these tests are rarely used in practice. The more important conceptual assumptions are that some underlying structure does exist in the set of selected variables and that results in some degree of multicollinearity.

In immunological studies factor analysis or PCA could be applied to extract information on the Th1- or Th2-related immune response from a set of cytokines or the ratio of a Th1- and Th2-score could be calculated to quantify the degree of «Th1/Th2 bias [30, 31].» For example, Turner et al [31] derived summary variables (called «principal components») from 11 different cytokines that were believed to better reflect the underlying mechanism and that were used in further analyses [32] instead of the original parameters. Data reduction techniques should be used when the objective of the study is not to investigate the role of each parameter but the role of the underlying mechanism.

Cluster Analysis

Cluster Analysis (CA) is the appropriate statistical approach when the researcher seeks to group individuals (not variables) according to their values of study variables (e.g. cytokine levels) [23]. CA groups individuals so that subjects in the same cluster have similar profiles of the parameters being studied (i.e. a high «within-cluster homogeneity») and subjects from different clusters have quite different immunological profiles (i.e. a high «between-cluster heterogeneity»). To perform CA the researcher has to define the variables on which the clustering is to be based and the type of cluster algorithm.Agglomerative algorithms treat each observation as a cluster and group similar individuals into clusters, while divisive algorithms start with the whole study population as a single cluster and divide the population by identifying homogeneous subgroups. The most appropriate clustering approach for a particular dataset depends on the type of data collected and the research question. Agglomerative clustering is preferable when there are extreme values in the data (outliers).

CA has strong mathematical properties but not statistical foundations. Data assumptions (e.g. normality and linearity) that are important in other multivariate techniques are of little importance in cluster analysis. However, the researcher is encouraged to examine the degree of multicollinearity in the data because each variable is weighted and variables that are multi-collinear are implicitly more heavily weighted in the clustering algorithm. For example, a cluster solution derived from a dataset with five highly correlated Th1-related cytokines and two correlated Th2-cytokines would substantially overestimate the importance of the Th1-component in the clustering. This can be avoided by first applying a data reduction technique (e.g. PCA) to derive the «principal components» that quantify the magnitude of Th1/Th2-immune response and afterwards clustering the individuals with respect to these immunological components.

In immunological studies cluster analysis may be useful to identify groups of individuals with similar immunological patterns (e.g. cytokine or antibody levels) that reflect an unknown common underlying immunological mechanism. For example, Mutapi et al [16] sought to group people infected with S. mansoni into clusters defined by levels of parasite specific IgE, IgA, IgM and the IgG subclasses. The authors identified two clusters, a cluster with high levels of IgM and low levels of all other antibodies and a cluster with high IgM and IgG1 and medium IgG4 and low levels of all other antibodies. In further analyses the authors investigated whether epidemiological features of schistosomiasis were associated with cluster membership and whether treatment changed this.

Another useful technique to study associations among immunological parameters without assuming any causal relationship is Canonical Correlation Analysis (CCA) [33] an approach that aims to quantify the correlation between two predefined sets of variables. In our MEDLINE search we did not find any previous applications of CCA to immunological data but we suggest that this statistical approach might be very useful to quantify the magnitude of correlation between two different sets of immunological parameters. A hypothetical example is where the investigators were interested to quantify the correlation between Th1- and Th2-related cytokines in individuals with and without helminths infections, or in atopic or non-atopic study subjects.

Dependence Techniques

The second group of techniques are statistical dependence techniques that are appropriate when the study investigates causation, i.e. variables can be classified as independent (cause, exposure) and dependent variables (effect, outcome), based on concrete a priori hypotheses about the underlying biological mechanisms. For example, an immunological parameter might be considered as an independent variable when it is the proposed cause (e.g. an autoantibody in an autoimmune disease) or the dependent variable when it is considered to be an effect (or outcome [e.g. the production of interferon gamma produced by lymphocytes stimulated with mycobacterial antigen following BCG vaccination]), or an intervening variable or intermediate factor (mediating variable) within a complex causal chain (e.g. the cytokine IL-13 would be an intermediate factor in studies that examine the relationship between exposure to aeroallergens and the development of atopic asthma). For example, Black et al [34] studied the IFN-γ response to Mycobacterium tuberculosis (as the outcome variable) before and after receiving BCG vaccination (the independent variable). Cooper et al [35] studied the effect of cholera vaccine (the independent variable) on the IL-2 response to recombinant cholera toxin B (as the outcome). In both studies, the authors pre-specified the classification of study variables before conducting statistical analysis. A more complex hypothetical example is the investigation of the effect of the impact of the intensity of infection with helminths on cytokine expressions (e.g. using Th2-related cytokines as outcomes) in a study of the relationship between helminths infections and atopy, and when the investigators may wish to consider the same cytokines as determinants, or risk factors, for atopy within the same study.

The choice of which statistical dependence technique is most appropriate for a particular research question will depend on the study design, the number and «scaling» (i.e. continuous, ordinal or categorical) of the study variables and other data assumptions (see Figure 1).

Univariate Dependence Techniques

These methods are appropriate when there is only one dependent and one independent variable. Common techniques areunivariate group mean comparison procedures [36] (aimed to compare the levels of a continuous variable (e.g. cytokine expressions) between groups of individuals pre-defined by an exposure that is considered to cause the immunological profile (e.g. vaccinated or not). The number of groups and whether the groups are independent or related will determine which approach is the most appropriate for each situation (see overview in Table 2).

By contrast, univariate regression analysis [37] is the best approach when the investigator seeks to model the relationship between two continuous variables and is able to decide which one is the outcome variable. The most common approach isLinear Regression a technique with stringent data assumptions (linearity of the relationship and normality of the error distribution). Robust alternatives when data assumptions are violated are non-parametric or non-linear regression techniques. Examples of applications for regression analysis in immunological studies are predicting the levels of expression of a continuous outcome variable (e.g. a cytokine) by a continuous variable (e.g. age or another continuous immunological parameter) when a causal relationship can be assumed.

Multivariate dependence techniques are needed when there are three or more variables involved and at least one variable can be considered the dependent variable.

Classification Techniques

These methods are required when there are several independent variables (e.g. cytokines) and one categorical outcome (e.g. atopy). One classical approach is Linear Discriminant Analysis (LDA), a method that derives linear combinations of the independent variables (called discriminant functions) that best discriminate between the two outcome groups (defined on the basis of the independent variable) [38]. LDA requires continuous normally distributed independent variables. A flexible approach which can be used with non normally distributed data and involves categorical independent variables is Logistic Regression [39].

In immunological studies classification techniques can be very useful to identify immunological profiles that best discriminate two or more pre-defined groups of interest (e.g. atopy vs. non atopy). For example, Gama et al [40] used LDA aimed to identify an immunological marker based on six cytokines (IL-2, IL-4, IL-10, IL-12, IFN-γ and TNF-α) to discriminate between clinical and asymptomatic forms of visceral leishmaniasis. The authors found that TNF-α, IL-10 and IL-4 were highly correlated with the clinical form, while IL-2, IL-12 and IFN-γ were correlated with the asymptomatic form. In another study, logistic regression was used to explore the role of parasite induced IL10 in decreasing the frequency of atopy: Van Biggelaar et al [41] sought to predict positive skin prick tests to house dust mite in children by mite-specific IgE, total IgE, IL-5 and IL-10 to worm and used logistic regression to show that positive skin prick test was positively associated with mite specific IgE but negatively associated with IL-10; and the probability of a skin test positivity was a result of the interaction between level of mite IgE and worm IL10.

Multivariate Group Mean Comparison Techniques

These techniques such as Multivariate Analysis of Variance (MANOVA) or Multi-way Analysis of Variance (Multi-way ANOVA), are used to compare the distributions of one or more continuous variables between groups defined by one ("one-way") or more ("multi-way") factors of interest (see overview Table 2) [21]. In contrast to LDA, where the groups are assumed to define a categorical outcome (e.g. asthmatic vs. non asthmatic) and the independent variables (e.g. cytokines) are used to discriminate between groups, in (M)ANOVA the groups are defined by the investigator considering one or more independent variables (e.g. treatments, vaccination status). A very useful application of MANOVA in immunological studies is to simultaneously compare the levels of two or more cytokines (e.g. IFN-γ, TNF-α) between two groups (asthmatic vs. non asthmatic). By applying MANOVA instead of repeated application of ANOVA the investigator can avoid the problem of type I error inflation for the whole experiment.

Multivariate Analysis of Covariance (MANCOVA) is an extension of MANOVA that additionally allows to control for the effect of an other continuous variable to be controlled (e.g. a confounder) [21]. An application of MANCOVA in immunological studies could be to simultaneously compare the expression levels of different cytokines (e.g. IF-γ, TNF-α, etc.) across groups defined by one or more experimental factors (e.g. vaccinated or control) and adjusted for age.

Multiple Regression Techniques

Multiple Regression is appropriate when the research question is to predict a single continuous dependent variable by a set of continuous and/or categorical independent variables [37]. The standard approach frequently used is multiple linear regression, however there are alternatives (e.g. non-parametric, non-linear multiple regression) when data assumptions are not met [42]. Regression analysis could be applied in immunological studies to predict the expression of a cytokine by explanatory variables (e.g. a set of other cytokines or other immunological parameters) or to predict the values of a continuous outcome variable (e.g. intensity of parasitic infection) by the expressions of one or more cytokines. For example, Dodoo et al [43] used multiple linear regression to predict malaria-related outcomes (fever, hemoglobin concentration) by levels of different cytokines (IFN-γ, TNF-α, IL-12, IL-10, TGF-β).

Partial Least Squares (PLS) Regression [44] is an extension of multiple linear regression for constructing predictive models when the factors are many and highly collinear. The approach could be very useful for analysis of immunological data, e.g. when the objective is to predict an outcome by a large set of highly correlated immunological parameters. Technically, the approach is a combination between principal components analysis and multiple linear regression, i.e. it produces factor scores as linear combinations of the original predictor variables so that there is no correlation between the factor score variables used in the predictive regression model.

Advanced Techniques

Path Analysis and Structural Equation Modelling

All the multivariate statistical methods that have been mentioned above have one common limitation: although they may include many variables, they all assume the presence of one single relationship between them. However, in modern immunological studies investigators often assume multiple relationships among immunological parameters and other study variables so that a simple multivariate approach might not be sufficient to reflect the complexity of the underlying immunological process. For example, in an immuno-epidemiological study conducted to study risk factors for asthma and allergy, investigators could define a conceptual framework that assumes multiple relationships between risk factors (e.g. allergens, vaccines, early life infections), immunological profiles (e.g. cytokine expression levels) and the occurrence of outcomes (e.g. asthma, atopy) (Fig 2). To model such complex immunological processes it will be important to simultaneously model all these multiple associations. Path Analysis and Structural Equation Modelling (SEM) are techniques developed by geneticists [45, 46] and economists [47] that can handle multiple relationships among study variables simultaneously and have been frequently applied in other scientific fields (e.g. economics, social sciences) [48]. SEM is an extension of path analysis that allows also for so called "latent variables" (a conceptual term for unobserved variables, see Appendix). A structural equation model consists of two components: "a measurement model" that defines how the observed measurements (called indicators) are related to the unobserved latent variables and a "structural model" that defines the assumed relationships between the observed study variables and one or more latent variables.

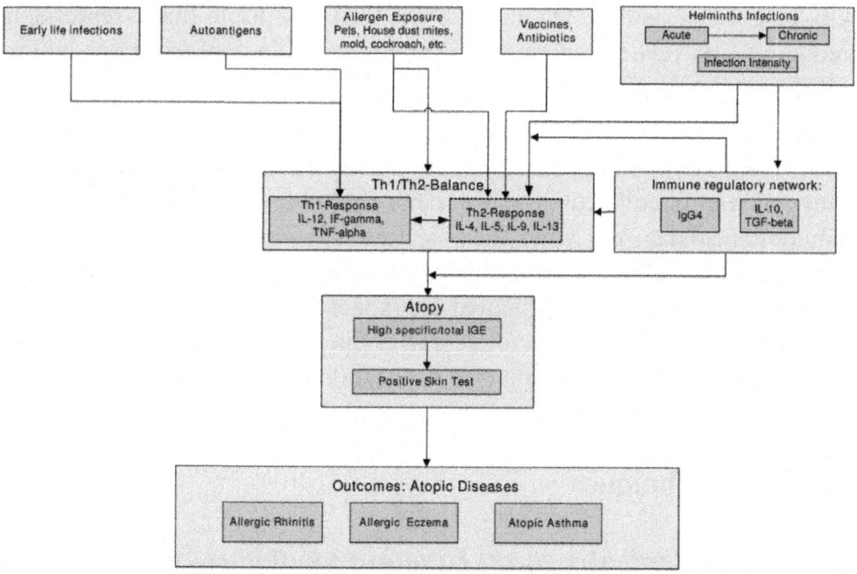

Figure 2: Conceptual framework that specifies multiple associations between potential risk factors, immunological parameters and outcomes (atopy and asthma).

The concept of latent variables and SEM is likely to be useful in immunological studies because immunologists frequently hypothesise that the measured immunological parameters are the result of unobservable underlying complex immunological processes. For example, imagine a hypothetical immunological study where different cytokines are measured to quantify two important immunological components. Figure 3 shows the path diagram of the study in which multiple relationships are assumed between the study variables. In this example we consider that the cytokines IL-12, IFN-γand TNF-α represent an unobservable latent variable «Th1-related immune response» and the cytokines IL-4, IL-9 and IL-13 represent the latent variable «Th2-related immune response». Further, the structural model assumes relationships between the two latent variables (c1) and the «effects» (c2, c3) of these on an outcome variable (e.g. atopy, asthma). In SEM the concept of the «measurement model» is similar to factor analysis in which a linear relationship between the (observed) indicator variables and the (unobserved) latent variable is assumed. For example, in Figure 3, the «indicator loadings i1–i6» reflect the magnitude of association of each cytokine with the latent variable. However, the distinction between the two analyses is that in factor analysis the principal components are extracted to maximize the degree of variance explained by a specified number of factors while in SEM the investigator has to define a priory a path diagram that specifies which variables are the indicators of the underlying latent variables and the correlations of the measurements with the latent variables are derived that best reflect the whole conceptual framework e.g. maximizing the correlation between all latent and observed variables.

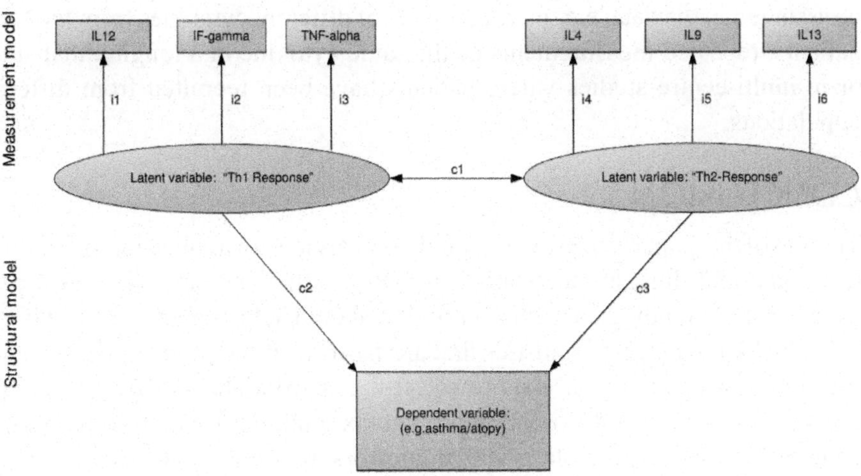

Figure 3: Example path diagram that specifies a structural equation model with two latent variables.

Application of these techniques to immunological data should allow the inference of complex immunological phenomena. For example, Chan et al [49] used SEM to study the role of obesity-associated dyslipidaemia, endothelial activation and two cytokines (IL-6 and TNF-α) in a complex causal chain of the metabolic syndrome. There are few applications of conceptual frameworks, path analysis and SEM to immunological data in the literature. However, because of increasing sample size and advancing knowledge about underlying complex immunological mechanisms, these approaches have a potentially important role in the analysis of data from modern immunological studies. The concept of latent variables in SEM will be especially useful because immunological mechanism can rarely be observed directly and must therefore be inferred through the measurement of different immunological indicators. Moreover, the unique feature of SEM that allows the analysis of multiple relationships between the study variables would be appropriate for the inference of causal chains of immunological processes.

Mixed Effects Models

The application of Random or Mixed Effects Models [50] for statistical analysis of clustered or longitudinal data is becoming a popular method in medicine. The basic idea of the approach is to adjust data analysis for an effect of the clustering by introducing a «random effect», i.e. an unobserved random variable that is specific to the clustering unit. In immunological studies the application of mixed effects models could be very useful when clustering must be assumed in the data, e.g. measurements of different cytokines from the same patients, repeated measurements of the same cytokine in a longitudinal study or in multi-centre studies where patients have been recruited from different populations.

CONCLUSION

The aim of this paper is to provide a modern overview for applied immunologists to explain and illustrate the statistical methods that can be employed for the analysis of immunological data. Our review should help immunologists without a detailed knowledge of statistics that are faced with the problem of statistical analysis of immunological data to select the appropriate statistical technique that will allow the valid extraction of the maximum information from the data collected. However, the statistical framework presented here should not be used as a substitute for an experienced biostatistician who should be involved from the beginning of the study for advice on study design, calculation of the sample size and planning of the statistical analysis. The systematic literature review illustrates the fact that most immunological studies still employ simple statistical approaches to immunological data even when multiple inter-relationships among several study variables are expected. We think it is important that more sophisticated statistical techniques are used for complex immunological data that will permit a better understanding of complex underlying immunological mechanism. Our focus has been on multivariate techniques that permit the analysis of multiple study variables simultaneously and hope that our examples from both real and hypothetical immunological studies will stimulate immunologists to make more use of these techniques. Moreover, bearing in mind the complexity of research questions addressed by modern immunological studies, we have introduced the idea of conceptual frameworks, latent variables and advanced statistical techniques (e.g. path analysis, SEM), providing a toolbox that should help the investigator to analyse multiple relationships among several study variables simultaneously. Finally, it should be pointed out that statistical techniques are a tool for the inference of underlying mechanisms and can never substitute for a priori hypotheses that are based on a sound knowledge of the scientific literature.

ACKNOWLEDGEMENTS

This research was funded by The Wellcome Trust, UK, HCPC Latin America Excellence Centre Program, Ref. 072405/Z/03/Z. PJC is supported also by the Wellcome Trust, grant no. 074679. BG is supported also by the Fundação de Amparo à Pesquisa do Estado da Bahia, Brazil (FAPESB), contract no. 1360/2006. The authors thank Dr Franca Hartgers and Dr Victoria Wright for their useful comments on the manuscript.

AUTHORS' CONTRIBUTIONS

BG and LCR had the idea for the paper, conducted the literature review, developed the guide for statistical analysis and wrote the manuscript. PJC and MY participated in writing the sections about research objectives of immunological studies and specific aspects of immunological data and evaluated the applicability of the proposed statistical framework for data from modern immunological studies. MLB participated in developing the statistical framework and helped to draft the manuscript. All authors have read and approved the final manuscript.

REFERENCES

1. Geiger SM, Massara CL, Bethony J, Soboslay PT, Correa-Oliveira R: Cellular responses and cytokine production in post-treatment hookworm patients from an endemic area in Brazil. Clin Exp Immunol. 2004, 136: 334-40. 10.1111/j.1365-2249.2004.02449.x.
2. Hartel C, Adam N, Strunk T, Temming P, Muller-Steinhardt M, Schultz C: Cytokine responses correlate differentially with age in infancy and early childhood. Clin Exp Immunol. 2005, 142: 446-53.

3. McGuinness D, Bennett S, Riley E: Statistical analysis of highly skewed immune response data. J Immunol Methods. 1997, 201: 99-114. 10.1016/S0022-1759(96)00216-5.
4. Bennett S, Riley EM: The statistical analysis of data from immunoepidemiological studies. J Immunol Methods. 1992, 146: 229-39. 10.1016/0022-1759(92)90232-I.
5. Hollander M, Wolfe DA: Nonparametric Statistical Methods. 1999, Wiley, 2
6. CG Victora, Huttly SR, Fuchs SC, Olinto MT: The role of conceptual frameworks in epidemiological analysis: a hierarchical approach. Int J Epidemiol. 1997, 26: 224-7. 10.1093/ije/26.1.224.
7. Hsu J: Multiple Comparisons. 1996, Chapman & Hall/CRC
8. Toothaker L: Multiple Comparison Procedures. 1992, London: SAGE University Paper
9. De Groot AS, Sbai H, Aubin CS, McMurry J, Martin W: Immuno-informatics: Mining genomes for vaccine components. Immunol Cell Biol. 2002, 80: 255-69. 10.1046/j.1440-1711.2002.01092.x.
10. Koren E, De Groot AS, Jawa V, Beck KD, Boone T, Rivera D, Li L, Mytych D, Koscec M, Weeraratne D, et al.,: Clinical validation of the "in silico" prediction of immunogenicity of a human recombinant therapeutic protein. Clin Immunol. 2007, 124: 26-32. 10.1016/j.clim.2007.03.544.
11. Gilchrist M, Thorsson V, Li B, Rust AG, Korb M, Kennedy K, Hai T, Bolouri H, Aderem A: Systems biology approaches identify ATF3 as a negative regulator of Toll-like receptor 4. Nature. 2006, 441: 173-8. 10.1038/nature04768.
12. Chakraborty AK, Dustin ML, Shaw AS: In silico models for cellular and molecular immunology: successes, promises and challenges. Nat Immunol. 2003, 4: 933-6. 10.1038/ni1003-933.
13. De Groot AS, Moise L: Prediction of immunogenicity for therapeutic proteins: state of the art. Curr Opin Drug Discov Devel. 2007, 10: 332-40.
14. De Groot AS, Moise L: New tools, new approaches and new ideas for vaccine development. Expert Rev Vaccines. 2007, 6: 125-7. 10.1586/14760584.6.2.125.
15. Hedeler C, Paton NW, Behnke JM, Bradley JE, Hamshere MG, Else KJ: A classification of tasks for the systematic study of immune response using functional genomics data. Parasitology. 2006, 132: 157-67. 10.1017/S0031182005008796.

16. Mutapi F, Mduluza T, Roddam AW: Cluster analysis of schistosome-specific antibody responses partitions the population into distinct epidemiological groups. Immunol Lett. 2005, 96: 231-40. 10.1016/j.imlet.2004.08.017.
17. Tukey JW: Exporatory Data Analysis. 1977, Boston: Addison-Wesley
18. Good PI: Resampling Methods: A Practical Guide to Data Analysis. 2005, Birkhauser
19. Chmura Kraemer H, Thiemann S: How many subjects? Statistical Power Analysis in Research. 1987, Sage Publications
20. Cohen J: Statistical Power Analysis for the Behavioral Sciences. 1988, Hillside, New Jersey: Lawrence Erlbaum Associates
21. Hair JF, Tatham RL, Anderson RE, Black W: Multivariate Data Analysis. 1998, Prentice Hall, 5
22. MacCallum RC, Widaman KF, Zhang S, Hong S: Sample size in factor analysis. Psychological Methods. 84-99.
23. Everitt BS, Landau S, Leese M: Cluster Analysis. 1993, London: Arnold
24. MacCallum RC, Browne MW, Sugawara HM: Power analysis and determination of sample size for covariance structure modeling. Psychological Methods. 1996, 130-149. 10.1037/1082-989X.1.2.130.
25. MacCallum RC, Hong S: Power analysis in covariance structure modeling using GFI and AGFI. Multivariate Behavioral Research. 1997, 193-210. 10.1207/s15327906mbr3202_5.
26. Lee PM: Bayesian Statistics: An Introduction. 2004, London: Arnold Publisher
27. Plikaytis BD, Carlone GM: Statistical considerations for vaccine immunogenicity trials. Part 1: Introduction and bioassay design and analysis. Vaccine. 2005, 23: 1596-605. 10.1016/j.vaccine.2004.06.046.
28. Jolliffe IT: Principal Component Analysis. 2002, Springer, 2
29. Gorsuch RL: Factor Analysis. 1983, Lawrence Erlbaum
30. Jackson JA, Turner JD, Kamal M, Wright V, Bickle Q, Else KJ, Ramsan M, Bradley JE: Gastrointestinal nematode infection is associated with variation in innate immune responsiveness. Microbes Infect. 2005
31. Turner JD, Faulkner H, Kamgno J, Cormont F, Van Snick J, Else KJ, Grencis RK, Behnke JM, Boussinesq M, Bradley JE: Th2 cytokines are associated with reduced worm burdens in a human intestinal helminth infection. J Infect Dis. 2003, 188: 1768-75. 10.1086/379370.
32. Jackson JA, Turner JD, Rentoul L, Faulkner H, Behnke JM, Hoyle M, Grencis RK, Else KJ, Kamgno J, Bradley JE, et al.,: Cytokine response

profiles predict species-specific infection patterns in human GI nematodes. Int J Parasitol. 2004, 34: 1237-44. 10.1016/j.ijpara.2004.07.009.

33. Thompson B: Canonical Correlation Analysis: Uses and Interpretation (Quantitative Applications in the Social Sciences). 1984, Sage Publications

34. Black GF, Weir RE, Floyd S, Bliss L, Warndorff DK, Crampin AC, Ngwira B, Sichali L, Nazareth B, Blackwell JM, et al.,: BCG-induced increase in interferon-gamma response to mycobacterial antigens and efficacy of BCG vaccination in Malawi and the UK: two randomised controlled studies. Lancet. 2002, 359: 1393-401. 10.1016/S0140-6736(02)08353-8.

35. Cooper PJ, Chico M, Sandoval C, Espinel I, Guevara A, Levine MM, Griffin GE, Nutman TB: Human infection with Ascaris lumbricoides is associated with suppression of the interleukin-2 response to recombinant cholera toxin B subunit following vaccination with the live oral cholera vaccine CVD 103-HgR. Infect Immun. 2001, 69: 1574-80. 10.1128/IAI.69.3.1574-1580.2001.

36. Montgomery DC: Design and Analysis of Experiments. 2000, Wiley, 5

37. Kleinbaum D, Kupper L, Muller KE, Nizam A: Applied Regression Analysis and Multivariable Methods. 1997, Duxbury Press

38. Huberty CJ: Applied Discriminant Analysis. 1994, Wiley

39. Hosmer DW, Lemeshow S: Applied Logistic Regression. 2000, Wiley, 2

40. Gama ME, Costa JM, Pereira JC, Gomes CM, Corbett CE: Serum cytokine profile in the subclinical form of visceral leishmaniasis. Braz J Med Biol Res. 2004, 37: 129-36. 10.1590/S0100-879X2004000100018.

41. van den Biggelaar AH, van Ree R, Rodrigues LC, Lell B, Deelder AM, Kremsner PG, Yazdanbakhsh M: Decreased atopy in children infected with Schistosoma haematobium: a role for parasite-induced interleukin-10. Lancet. 2000, 356: 1723-7. 10.1016/S0140-6736(00)03206-2.

42. Takezawa K: Introduction to nonparametric regression. 2005, Wiley

43. Dodoo D, Omer FM, Todd J, Akanmori BD, Koram KA, Riley EM: Absolute levels and ratios of proinflammatory and anti-inflammatory cytokine production in vitro predict clinical immunity to Plasmodium falciparum malaria. J Infect Dis. 2002, 185: 971-9. 10.1086/339408.

44. Geladi P, Kowalski B: Partial least squares regression: A tutorial. Analytica Chimica Acta. 1986, 185: 1-17. 10.1016/0003-2670(86)80028-9.

45. Wright S: Correlation and causation. Journal of Agricultural Research. 1921, 557-85.

46. MacCallum RC, Austin JT: Applications of structural equation modeling in psychological research. Annu Rev Psychol. 2000, 51: 201-26. 10.1146/annurev.psych.51.1.201.

47. Haavelmo T: The statistical implications of a system of simultaneous equations. Econometrica. 1943, 1-12. 10.2307/1905714.

48. Skrondal A, Rabe-Hesketh S: Generalized Latent Variable Modeling. Multilevel, Longitudinal, and Structural Equation Models. 2004, Chapman & Hall/CRC

49. Chan JC, Cheung JC, Stehouwer CD, Emeis JJ, Tong PC, Ko GT, Yudkin JS: The roles of obesity-associated dyslipidaemia, endothelial activation and cytokines in the Metabolic Syndrome – an analysis by structural equation modelling. Int J Obes Relat Metab Disord. 2002, 26: 994-1008. 10.1038/sj.ijo.0802017.

50. Brown H, Prescott R: Applied Mixed Models in Medicine. 1999, Chichester: John Wiley & Sons, Ltd

Chapter 3

NEXT GENERATION GENETICS

Mogens Fenger

Clinical Biochemistry, Molecular Biology, and Genetics, KBA339, Hvidovre, Denmark

ABSTRACT

One of the goals in genetic research aims at identifying genes in biochemical and physiological processes to reveal genetic causes of rare and common diseases. Previous obstacles such as costly genotyping or sequencing have been reduced with the chip-based genome wide association studies (GWAS), now culminating with the latest toy—next generation sequencing methodologies (NGS). Concomitantly, computer technologies have evolved to an increasing use of multicore processors and distributed computing on large networks or grids. Although the technologies are not perfect, we now have unprecedented opportunities to perform genetic studies not possible just 10 years ago. The hype about these new technologies have been large, but all the promises have however not been fulfilled entirely as hoped for. Maybe because the hype has been more about the technologies as such, and less about their intended use. Millions of genetic variations have been detected by GWAS and NGS, but only a few have been linked to diseases—with almost no practical clinical significance. A major reason for this apparent deadlock is the inadequacy of the models used, which are based on the traditional "Mendelian" approach, in which one gene is supposed to have a main effect on a trait or a disease. However, most genes claimed to be associated with a disease have small effects and only a tiny fraction of the genetic variance has been captured.

In this short notice it is argued, why this traditional approach should be supplemented or even replaced by modeling approaches in accordance with the complexity of biological systems, if we shall have any reasonable hope to understand the genetics behind any trait and bring genetics into practical use in medicine for common diseases (Costanzo et al., 2010; Ramanan et al., 2012).

EVOLUTION, FITNESS, AND EPISTASIS

Evolution of phyla is a complex process governed by genomic as well as environmental factors (Marshall, 2006). Much theoretic and practical work about evolution are based on theories of adaptive landscapes of fitness and natural selection, as advocated by Fisher in his geometrical model of adaption (Fisher, 1930;Martin and Lenormand, 2006; Chevin et al., 2010; Weinreich and Knies, 2013). In this model fitness is determined in a multidimensional landscape of phenotypes or traits, on which a selection pressure is imposed that limits the number of viable phenotypes. Although the space of theoretically phenotypes increases with the complexity of an organism, this may come with a cost of decreasing adaptability (Fisher, 1930; Orr, 2000; Martin and Lenormand, 2006; Borenstein and Krakauer, 2008). The Fisher model(s) is not explicitly rooted in genetic models but rather considers the phenotypic pleiotropic effect of mutations, i.e., particular genes and loci are not formulated in the model. In contrast, Wrights formulation of evolution (Wright, 1920) can be described as a multidimensional mutational or genetic landscape in which each dimension corresponds to a specific locus. In "modern" terms these ideas can be formulated as the occurrence of stabilizing selection acting on the increasing mutational load possibly involving pleiotropic behavior of a given mutation, that is a mutation may affect several endophenotypes (Weinreich et al., 2006; Masel and Siegal, 2009). Pleiotropicity also means that a selection pressure imposed on one endophenotype may not only constrict the number of viable genotypes but also inflict a collateral selection on other endophenotypes and genes (Gavrilets and De Jong, 1993; Snitkin and Segre, 2011). The latter may be regarded as "innocent" bystanders and the preserved genotypes may just be those that happen to be around at the moment of selection.

Fitness is a measure of the capability of survival and reproduction of a species as the result of integrated action of many subprocesses conditional on the imposed selection pressure. However, less fitness may not necessarily result in an entire loss of a phenotype or trait, but may prevail and in fact increase fitness if the selection pressure changes. This scenario is supported by the long known fact that a mutation may have major effect in one genotypic background, but may only have a minor influence in another and hence escape purging by selection (Nijhout and Paulsen, 1997; Kouyos et al., 2006). The fitness landscape (or any other trait landscape) may thus be roughed with several local optima. This is clearly obvious in the landscape of species, but is also present within a species (Marshall, 2006).

For long the question has been if a mutation impose a pure additive effect on fitness or if epistasis (the effect of mutations in a gene or regulatory structure on the effects of other genes) is the prevalent driving force in evolution. The effect of any mutation (genic as well as exgenic including possible changes in epigenetic processes) may be increased, buffered, or ameliorated in particular genetic backgrounds, while having negative (even lethal) effect in other genetic backgrounds. Buffering is the essence in evolutionary theory of canalization and organismal robustness, in which the phenotype appears robust to mutational perturbations. Mutations may accumulate and appear as cryptic or neutral variations as long as they are not selected against (Masel and Siegal, 2009). These cryptic genetic variations may be revealed if some genetic or environmental changes happen affecting the fitness and then contribute to evolvability (McGuigan and Sgró, 2009).

Canalization (or buffering) implies that the phenotypic mean tends to be preserved when a mutation occurs, but the cost is diminished variation of the phenotype, as new (deleterious) mutations are buffered leaving less degrees of freedom of variation compared to the pre-mutational genotype. Thus, a particular phenotype representing a local maximum in the phenotypic landscape, is generated by an ensemble of genotypes, each depicting a path or trajectory of the genetic network. Generally, the probability of a given genetic path being accessible to generate a phenotype decreases with the number of mutations. However, as the number of paths increases exponentially with the number of mutations a large and increasing number of paths may eventually generate a phenotype. This hypothesis has been confirmed empirically (Dowell et al., 2010; Franke et al., 2011). These and many other studies have firmly established epistasis as a primary driving force in evolution and as a fundamental principle in governing biological processes (Rice, 1998; Segre et al., 2005; Weinreich et al., 2005, 2013; Bershtein et al., 2006; Borenstein and Krakauer, 2008; Pavlicev et al., 2008; Chevin et al., 2010; Lunzer et al., 2010; Breen et al., 2012; Huang et al., 2012; Hemani et al., 2013; Weinreich and Knies, 2013).

The mechanism behind interactions and epistasis has been extensively studied and includes concepts as sign epistasis (Weinreich et al., 2005), reciprocal epistasis (Poelwijk et al., 2011), and the expansion of the protein universe (Povolotskaya and Kondrashov, 2010) to mention just a few outstanding contributions. The reader is referred to the cited work and to the vast literature appearing now.

THE GENETIC AND PHENOTYPIC SPACES

Complex species like humans are organized in interacting and interdependent functional units called organs or multicellular tissues. This extends the complexity of the genetics to another level. Despite the constrains this impose, the phenotypic space is vast.

Suppose that a diploid organism like *Homo sapiens* with 23 sets of chromosomes only harbor one mutation in each chromosome. The theoretically number of gametes emerging by random segregation amounts to approximately 8.4 million. If all gametes are viable then the number of possible zygotes will be more than $7*10^{13}$ or more than 11.000 fold the number human beings ever lived on planet Earth. Most probably a vast amount of the gametes or zygotes are not viable, but nevertheless, genetic variations so far discovered runs in the millions. This maps to as many phenotypes and hence, two human beings will never be genetically identical.

Similarly, in a physiological process like blood pressure, which are regulated by say 100 interacting genes, more than 10^{30} networks with exactly the same topology can be constructed if just one mutation is present in each gene. This would map to as many physiological states and dynamics. Adding to the number of genes their alternative spliced forms, the vast number of posttranslational modifications of proteins, non-protein regulatory elements (metabolites, small regulatory RNAs), epigenetic modifications, non-genic regulatory and genome-organizing structures, and not the least interactions and communications between cells in a multicellular organisms like humans, the combinatorial space of interactions and hence phenotypes is (almost) infinite.

POPULATION STRUCTURE AND GENETIC NETWORKS

Two basic aspects must be addressed in population genetics: (1) biological processes even in its most simple forms are blue-printed in the genome of interacting networks of genes; (2) the expression of the biological processes and phenotypes are conditional on the genetic variations and their inherent epistasis.

The genetic networks coding a trait can be mapped as a "continuum" reflecting the physiological states they define (see the Figure 1). Neighboring networks can be distinguished by variations in one or several genes or non-genetic regulatory structures, but may appear physiological similar as most genetic variations have small effects. The sensitivity of a network to external factors is encoded in the genome, and it is the variations in the process-specific genes and regulatory structures that determines the range of the response to an external perturbation. Identifying genetic networks are neither simple, nor transparent: functional networks are multipartite structures and are not secluded entities but rather interconnected with other networks (e.g., the glucose and the fat metabolisms are highly intermingled processes). Nevertheless, it may be possible to define a reasonable number of sub-ensembles of networks to be interpretable.

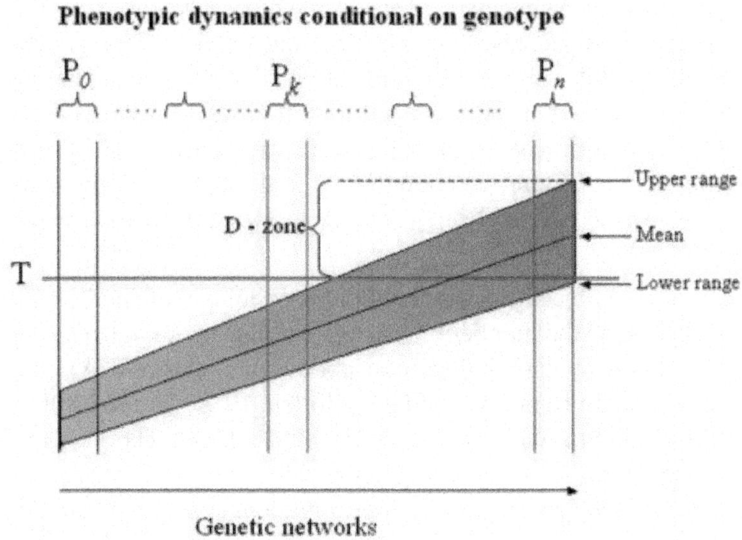

Figure 1: Risk profile in genetic networks. The population in this example is partitioned into more homogeneous subpopulations (P_i) by the LCA-SEM procedure indicated by the sets of two vertical lines. Each subpopulation is genetically defined by an ensemble of networks with exactly the same topology but differs in genetic variations. The networks in each subpopulation arise by successive mutational incidences that are balanced (buffered) to generate a phenotype similar for all subjects in the subpopulation (see also the text). In reality the subpopulations represent different local maxima in a miltidimensional phenotypic landscape, but are for illustrative purposes collapsed to flat, two-dimensional presentation. The range of the phenotype (e.g., diastolic blood pressure) depends on extra-genetic factors, but can never exceed the range defined by the genotype. Thus, some subjects (genotypes) will never exceed the threshold (T), while others will experience the extreme phenotype (e.g., diastolic hypertension) regardless of extra-genetic factors. The D(anger) - zone indicates the subjects or subpopulations which may be classified in this examples as diastolic hypertensive. However, this may depend on the circumstances under which the blood pressure is measured, i.e., subjects may be classified as normotensive although they have a massive propensity to develop hypertension. Dichotomizing the trait in a population is thus a dubious affaire and compromise most association studies to the point that information of the genetics of, in this case diastolic blood pressure, is entirely lost.

HETEROGENEITY

Population heterogeneity refers to the mixture of phenotypically homogeneous subpopulations, although in the extreme no subpopulation is truly homogenous as each subject harbors a unique genomewide genotype. A more or less well-defined phenotype thus comprise an ensemble of genotypes in the population.

The initial task is then to cluster subjects into more physiological homogeneous subpopulations to increase the accuracy and power of the genetic analysis (Fenger et al., 2008, 2011). The application of appropriate cluster algorithms is generally ill posed, as no universal formal criteria for the best clustering is available. Many of the well-known classification procedures implement some data-reduction or feature selection (Saeys et al., 2007), but any manipulations of the data space are likely to result in loss of information and should be avoided.

Allocating subjects to subpopulations is an art of modeling hidden or latent variables as the number of subpopulations is not known *a priori*. A way to resolve this is by applying the concept of latent class (LCA) in a structural equation modeling framework (SEM) (Bollen, 1989; Muthen, 2002; Skrondal and Rabe-Hesketh, 2007; Fenger et al., 2008, 2011). The idea of the LCA-SEM approach is to model a physiological process, and therefore the most appropriate study population would be a random selection of subjects as each subject provide information of the physiological process. Genetic structures and variations are not necessarily modeled directly, but are embedded as latent variables in the SEM structure and are mapped or reflected by the measured variables. Modeling in this framework addresses two pivotal issues in complex data: resolving the heterogeneity in the population, and simultaneously evaluating the data structure within the sub-populations. This approach outperforms most other classification methods in almost all aspects (Magidson and Vermunt, 2002).

An emerging line of methods implement ensembles of classification functions (Polikar, 2006). These approaches are particularly attractive when features in multi-source or distributed data sets are partly or completely disjoint, or if access to data in data set is limited to a subset of objects. Thus, the problem of missing data and hence reduced power may be circumvented to some extent and represent a potential alternative to imputing missing genetic data.

Undoubtedly, new and promising methods will merge, in particular as theoretical ideas mentioned below are integrated in future developments.

INHERITANCE: GENES OR INFORMATION?

Understanding and integrating the wealth of genetic data in a biological and medical context requires new approaches and techniques. Fortunately many new approaches are emerging increasingly embracing the nature of biological systems. In particularly the recent developments in network theory (Dorogovtsev and Mendes, 2002; Newman, 2009) including concepts of modularity (Newman and Girvan, 2004), stochastic block modeling (Karrer and Newman, 2011), statistical mechanics (Reichardt and Bornholdt, 2006;Ronhovde and Nussinov, 2009), and information theory (Anand and Bianconi, 2010) are promising.

All these methods comprise passing of information that have its corollary in genetics. Stabilizing selection may give rise to prevailing linkage disequilibrium of genes within and across chromosomes (Fenger et al., 2011). Such linkage disequilibrium arise as a consequence of preserving physiological processes regardless of the physical structure of the genome. Thus, inheritance is not a simple matter of passing on genetic material, but rather to combine information harbored in the genome into a viable organism. Information theoretic approaches in genetics may therefore be more promising (how abstract it may be) than traditional association methods, although transformation of these theories to biological structures may not be always straightforward.

IS VALIDATION IN GENETICS ACTUALLY POSSIBLE?

Validation is a standard requirements in genetic association studies today. However, validation of an association of a genetic variant to a trait or disease is often not or only sporadically obtained and for that reason a gene may be dismissed as disease related (Ioannidis, 2007; Shriner and Vaughan, 2011), or even to be pivotal in a physiological process (Fenger et al., 2011; Spijkers et al., 2011). It should hopefully be clear from the discussions above that validation should be expected to be an exception. A non-validated association may simply indicate a local optimum in the phenotypic landscape that happens to be detected because the genotypes in a population are permissive for expressing an apparent main effect. It has indeed been demonstrated that the lack of validation actually suggest more complex genetic structures governs a trait (Greene et al., 2009; Fenger et al., 2011), and including epistasis in the analysis may eventual confirm the importance of non-validated associations.

In the end the importance of genetic variations should be confirmed by cellular experiments. If possible, studies should be done in the cells where the genes in a network has its effect (which we often do not know). A gargantuan endeavor and at the moment maybe wild-fetched, but eventually any genetic variation has to be substantiated in a real biological context - not just as a statistical phenomenon.

CONFLICT OF INTEREST STATEMENT

The author declares that the research was conducted in the absence of any commercial or financial relationships that could be construed as a potential conflict of interest.

ACKNOWLEDGMENT

Elin R. Carlsson is thanked for helpful suggestions to this Opinion.

REFERENCES

1. Anand, K., and Bianconi, G. (2010). Gibbs entropy of network ensembles by cavity methods. *Phys. Rev. E Stat. Nonlin. Soft Matter Phys.* 82:011116. doi: 10.1103/PhysRevE.82.011116
2. Bershtein, S., Segal, M., Bekerman, R., Tokuriki, N., and Tawfik, D. S. (2006). Robustness-epistasis link shapes the fitness landscape of a randomly drifting protein. *Nature* 444, 929–932. doi: 10.1038/nature05385
3. Bollen, K. A. (1989). *Structural Equations with Latent Variables*. Hobroken, NJ: John Wiley and Sons.
4. Borenstein, E., and Krakauer, D. C. (2008). An end to endless forms: epistasis, phenotype distribution bias, and nonuniform evolution.*PLoS Comput. Biol.* 4:e1000202. doi: 10.1371/journal.pcbi.1000202
5. Breen, M. S., Kemena, C., Vlasov, P. K., Notredame, C., and Kondrashov, F. A. (2012). Epistasis as the primary factor in molecular evolution. *Nature* 490, 535–538. doi: 10.1038/nature11510
6. Chevin, L. M., Martin, G., and Lenormand, T. (2010). Fisher›s model and the genomics of adaption: restricted pleiotropy, heterogenous mutation, and parallel evolution. *Evolution* 64, 3213–3231. doi: 10.1111/j.1558-5646.2010.01058.x
7. Costanzo, M., Baryshnikova, A., Bellay, J., Kim, Y., Spear, E. D., Sevier, C. S., et al. (2010). The genetic landscape of a cell. *Science* 327, 425–431. doi: 10.1126/science.1180823
8. Dorogovtsev, S. N., and Mendes, J. F. (2002). Evolution of networks. *Adv. Phys.* 51, 1079–1187. doi: 10.1080/00018730110112519
9. Dowell, R. D., Ryan, O., Jansen, A., Cheung, D., Agarwala, S., Danford, T., et al. (2010). Genotype to phenotype: a complex problem.*Science* 328, 469. doi: 10.1126/science.1189015
10. Fenger, M., Linneberg, A., Jorgensen, T., Madsbad, S., Sobye, K., Eugen-Olsen, J., et al. (2011). Genetics of the ceramide/sphingosine-1-phosphate rheostat in blood pressure regulation and hypertension. *BMC Genet.* 12:44. doi: 10.1186/1471-2156-12-44
11. Fenger, M., Linneberg, A., Werge, T., and Jorgensen, T. (2008). Analysis of heterogeneity and epistasis in physiological mixed populations by combined structural equation modelling and latent class analysis. *BMC*

Genet. 9:43. doi: 10.1186/1471-2156-9-43

12. Fisher, R. A. (1930). *The Genetical Theory of Natural Selection*. London: Nature Publishing Group.
13. Franke, J., Klözer, A., de Visser, J. A., and Krug, J. (2011). Evolutionary accessibility of mutational pathways. *PLoS Comput. Biol.* 7:e1002134. doi: 10.1371/journal.pcbi.1002134
14. Gavrilets, S., and De Jong, G. (1993). Pleiotropic models of polygenic variation, stabilizing selection, and epistasis. *Genetics* 134, 609–625.
15. Greene, C. S., Penrod, N. M., Williams, S. M., and Moore, J. H. (2009). Failure to replicate a genetic association may provide important clues about genetic architecture. *PLoS ONE* 4:e5639. doi: 10.1371/journal.pone.0005639
16. Hemani, G., Knott, S., and Haley, C. (2013). An Evolutionary Perspective on epistasis and the missing heritability. *PLoS Genet.* 9:e1003295. doi: 10.1371/journal.pgen.1003295
17. Huang, W., Richards, S., Carbone, M. A., Zhu, D., Anholt, R. R. H., Ayroles, J. F., et al. (2012). Epistasis dominates the genetic architecture of Drosophila quantitative traits. *Proc. Natl. Acad. Sci. U.S.A.* 109, 15553–15559. doi: 10.1073/pnas.1213423109
18. Ioannidis, J. P. (2007). Non-replication and inconsistency in the genome-wide association setting. *Hum. Hered.* 64, 203–213. doi: 10.1159/000103512
19. Karrer, B., and Newman, M. E. (2011). Stochastic blockmodels and community structure in networks. *Phys. Rev. E Stat. Nonlin. Soft Matter Phys.* 83:016107. doi: 10.1103/PhysRevE.83.016107
20. Kouyos, R. D., Otto, S. P., and Bonhoeffer, S. (2006). Effect of varying epistasis on the evolution of recombination. *Genetics* 173, 589–597. doi: 10.1534/genetics.105.053108
21. Lunzer, M., Golding, G. B., and Dean, A. M. (2010). Pervasive cryptic epistasis in molecular evolution. *PLoS Genet.* 6:e1001162. doi: 10.1371/journal.pgen.1001162
22. Magidson, J., and Vermunt, J. K. (2002). Latent class models for clustering: a comparison with K-means. *Can. J. Market. Res.* 20, 37–44. Available online at: http://statisticalinnovations.com/technicalsupport/cjmr.pdf
23. Marshall, C. R. (2006). Explaining the Cambrian "Explosion" of animals. *Annu. Rev. Earth Planet. Sci.* 34, 355–384. doi: 10.1146/annurev.earth.33.031504.103001

24. Martin, G., and Lenormand, T. (2006). A general multivariate extension of Fisher's geometrical model and the distribution of mutation fitness effects across species. *Evolution* 60, 893–907. doi: 10.1111/j.0014-3820.2006.tb01169.x
25. Masel, J., and Siegal, M. L. (2009). Robustness: mechanisms and consequences. *Trends Genet.* 25, 395–403. doi: 10.1016/j.tig.2009.07.005
26. McGuigan, K., and Sgró C. M. (2009). Evolutionary consequences of cryptic genetic variation. *Trends Ecol. Evol.* 24, 305–311. doi: 10.1016/j.tree.2009.02.001
27. Muthen, B. O. (2002). Beyond SEM: general latent variabel modeling. *Behaviormetrika* 29, 81–117. doi: 10.2333/bhmk.29.81
28. Newman, M. E. J. (2009). The structure and functions of complex networks. *SIAM Rev.* 45, 167–256. doi: 10.1137/S003614450342480
29. Newman, M. E. J., and Girvan, M. (2004). Finding and evaluating community structure in networks. *Phys. Rev. E Stat. Nonlin. Soft Matter Phys.* 69:026113. doi: 10.1103/PhysRevE.69.026113
30. Nijhout, H. F., and Paulsen, S. M. (1997). Develomental models and polygenic characters. *Am. Nat.* 149, 394–405. doi: 10.1086/285996
31. Orr, H. A. (2000). Adaptation and the cost of complexity. *Evolution* 54, 13–20. doi: 10.1111/j.0014-3820.2000.tb00002.x
32. Pavlicev, M., Kenney-Hunt, J. P., Norgard, E. A., Roseman, C. C., Wolf, J. B., and Cheverud, J. M. (2008). Genetic variation in pleiotropy: differential epistasis as a source of variation in the allometric relationship between long bone lengths and body weight.*Evolution* 62, 199–213. doi: 10.1111/j.1558-5646.2007.00255.x
33. Poelwijk, F. J., T-ânase-Nicola, S., Kiviet, D. J., and Tans, S. J. (2011). Reciprocal sign epistasis is a necessary condition for multi-peaked fitness landscapes. *J. Theor. Biol.* 272, 141–144. doi: 10.1016/j.jtbi.2010.12.015
34. Polikar, R. (2006). Ensemble based systems in decision making. *IEEE Circuits Syst. Mag.* 21–45. doi: 10.1109/MCAS.2006.1688199
35. Povolotskaya, I. S., and Kondrashov, F. A. (2010). Sequence space and the ongoing expansion of the protein universe. *Nature* 465, 922–926. doi: 10.1038/nature09105
36. Ramanan, V. K., Shen, L., Moore, J. H., and Saykin, A. J. (2012). Pathway analysis of genomic data: concepts, methods, and prospects for future development. *Trends Genet.* 28, 323–332. doi: 10.1016/j.tig.2012.03.004

37. Reichardt, J., and Bornholdt, S. (2006). Statistical mechanics of community detection. *Phys. Rev. E Stat. Nonlin. Soft Matter Phys.* 74:016110. doi: 10.1103/PhysRevE.74.016110

38. Rice, S. H. (1998). The evolution of canalization and the breaking of von Baer's laws: modeling the evolution of development with epistasis. *Evolution* 52, 647–656. doi: 10.2307/2411260

39. Ronhovde, P., and Nussinov, Z. (2009). Multiresolution community detection for megascale networks by information-based replica correlations. *Phys. Rev. E Stat. Nonlin. Soft Matter Phys*. 80:016109. doi: 10.1103/PhysRevE.80.016109

40. Saeys, Y., Inza, I., and Larranaga, P. (2007). A review of feature selection techniques in bioinformatics. *Bioinformatics* 23, 2507–2517. doi: 10.1093/bioinformatics/btm344

41. Segre, D., Deluna, A., Church, G. M., and Kishony, R. (2005). Modular epistasis in yeast metabolism. *Nat. Genet*. 37, 77–83. doi: 10.1038/ng1489

42. Shriner, D., and Vaughan, L. K. (2011). A unified framework for multi-locus association analysis of both common and rare variants. *BMC Genomics* 12:89. doi: 10.1186/1471-2164-12-89

43. Skrondal, A., and Rabe-Hesketh, S. (2007). Latent variable modelling: a survey. *Scand. J. Stat*. 34, 712–745. doi: 10.1111/j.1467-9469.2007.00573.x

44. Snitkin, E. S., and Segre, D. (2011). Epistatic interaction maps relative to multiple metabolic phenotypes. *PLoS Genet*. 7:e1001294. doi: 10.1371/journal.pgen.1001294

45. Spijkers, L. J., van den Akker, R. F., Janssen, B. J., Debets, J. J., De Mey, J. G., Stroes, E. S., et al. (2011). Hypertension is associated with marked alterations in sphingolipid biology: a potential role for ceramide. *PLoS ONE* 6:e21817. doi: 10.1371/journal.pone.0021817

46. Weinreich, D. M., Delaney, N. F., Depristo, M. A., and Hartl, D. L. (2006). Darwinian evolution can follow only very few mutational paths to fitter proteins. *Science* 312, 111–114. doi: 10.1126/science.1123539

47. Weinreich, D. M., and Knies, J. L. (2013). Fishers›s geometric model of adaption meets the functional synthesis: data on pairwise epistasis fro fitness yields insights into the shape and size of phenotypic space. *Evolution* 67, 2957–2972. doi: 10.1111/evo.12156

48. Weinreich, D. M., Lan, Y., Wylie, C. S., and Heckendorn, R. B. (2013). Should evolutionary geneticists worry about higher-order epistasis? *Curr. Opin. Genet. Dev*. 23, 700–707. doi: 10.1016/j.gde.2013.10.007

49. Weinreich, D. M., Watson, R. A., and Chao, L. (2005). Perspective: sign epistasis and genetic constraint on evolutionary trajectories.*Evolution* 59, 1165–1174. doi: 10.1554/04-272

50. Wright, S. (1920). The relative importance of heredity and environment in determining the piebald pattern of guinea-pigs. *Proc. Natl. Acad. Sci. U.S.A.* 6, 320–332. doi: 10.1073/pnas.6.6.320

Chapter 4

A STATISTICAL DESIGN FOR TESTING TRANSGENERATIONAL GENOMIC IMPRINTING IN NATURAL HUMAN POPULATIONS

Yao Li[1,2], Yunqian Guo[1], Jianxin Wang[1], Wei Hou[3], Myron N. Chang[3], Duanping Liao[4], Rongling Wu[1,4]

[1]Center for Computational Biology, Beijing Forestry University, Beijing, People's Republic of China

[2] Department of Statistics, West Virginia University, Morgantown, West Virginia, United States of America

[3] Department of Biostatistics, University of Florida, Gainesville, Florida, United States of America

[4]Department of Public Health Sciences, Penn State College of Medicine, Hershey, Pennsylvania, United States of America

ABSTRACT

Genomic imprinting is a phenomenon in which the same allele is expressed differently, depending on its parental origin. Such a phenomenon, also called the parent-of-origin effect, has been recognized to play a pivotal role in embryological development and pathogenesis in many species. Here we propose a statistical design for detecting imprinted loci that control quantitative traits based on a random set of three-generation families from a natural population in humans. This design provides a pathway for characterizing the effects of imprinted genes on a complex trait or disease at different generations and testing transgenerational changes of imprinted effects. The design is integrated with population and cytogenetic principles of gene segregation and transmission from a previous generation to next. The implementation of the EM algorithm within the design framework leads to the estimation of genetic parameters that define imprinted effects. A simulation study is used to investigate the statistical properties of the model and validate its utilization. This new design, coupled with increasingly used genome-wide association studies, should have an immediate implication for studying the genetic architecture of complex traits in humans.

INTRODUCTION

Genomic imprinting arises from a gene when either the maternally or paternally derived copy of it is expressed while the other copy is silenced [1], [2]. Caused by epigenetic modifications such as DNA methylation established during gametogenesis and maintained throughout somatic development in the offspring, genetic imprinting has been shown to play a pivotal role in regulating the formation, development, function, and evolution of complex traits and diseases [3], [4], [5], [6], [7], [8], [9], [10]. While most studies of genetic imprinting focus on the epigenetic and molecular mechanisms of this phenomenon [7], [11], the number and distribution of imprinted genes and their epistatic interactions for quantitative traits are poorly understood, limiting the scope of our inference about the effects of imprinting genes on the diversity of biological traits or processes. Several authors have started to use genome-wide association and linkage studies to identify the regions of the genome that contain imprinted sequence variants and further understand the epigenetic variation of complex traits [12], [13],[14], [15].

In a series of recent studies, Cheverud, Wolf, and colleagues categorized genetic imprinting into different types based on the pattern of its expression, i.e., maternal expression, paternal expression, bipolar dominance, polar overdominance, and polar underdominance [14], [15]. With a three-generation F2 design, they identified these types of imprinted quantitative trait loci (iQTL) affecting body weight and growth in mice, displaying much more complex and diverse effect patterns than previously assumed. A different design based on reciprocal backcrosses was proposed to test and estimate the distribution of iQTL responsible for physiological traits related to endosperm development in maize [16]. By modeling identical-by-descent relationships in multiple related families of canines, Liu et al. [13] derived a random effect model based on linkage analysis to genome-wide scan for the existence of iQTL that affect canine hip dysplasia. In a recent study, Wang et al. [9] used reciprocal F2 designs to identify the additive and dominant effects of iQTLs and their interactions with imprinting effects for hyperoxic acute lung injury survival time in mice. These authors also explore the transgenerational inheritance of iQTLs.

While epigenetic marks resulting in genomic imprinting can be generally stable in an organism›s lifetime, they may undergo reprogramming, i.e., a faithful clearing of the epigenetic state established in the previous generation, in the new generation during gametogenesis and early embryogenesis [17], [18],

. However, a growing body of evidence since the early 1980s indicates that genes may escape such reprogramming and, thus, inherit their imprinting effects into next generations [20], [21], [22], [23], [24], [25]. Two fundamental questions will naturally arise from this discovery: how common are imprinted genes of this type and how strong is the evidence for their existence in humans and other organisms? If epigenetic changes through imprinted genes can be inherited across generations, this would significantly alter the way we think about the inheritance of phenotype [26], [27]. Such transgenerational epigenetic inheritance, i.e., modifications of the chromosomes that pass to the next generation through gametes, may be related with health and diseases with a mechanism for transmitting environmental exposure information that alters gene expression in the next generation(s) [28]. The identification of imprinted loci displaying transgenerational epigenetic inheritance will be greatly helpful for addressing the two questions mentioned above, in a quest to elucidate the detailed genetic architecture of complex traits and diseases.

The motivation of this study is to develop a novel strategy for identifying imprinted genes for a quantitative trait and understanding the transgenerational changes of their effects with a three-generation family design by sampling multiple unrelated nuclear families, each composed of the grandfather, grandmother, father, mother, and grandchildren, from a natural population. This transgenerational design contains information about how alleles at different loci co-transmit during meiosis from one generation to next and, thus, has been widely used for genetic linkage analysis [29], [30]. By tracing the inheritance of alleles at a gene(s) from a paternal or maternal parent, this design allows the characterization of parent-of-origin of alleles and provides a powerful way to estimate genetic imprinting effects. Because only genotypes can be observed, we formulate a mixture model to specify allelic configurations in terms of parental origins of the alleles. The EM algorithm is implemented to estimate the effects of imprinted genes and their changes across generations. A testing procedure is proposed to study the pattern of transgenerational epigenetic inheritance. The statistical behavior of the model is examined through simulation studies.

RESULTS

Simulation studies were performed to examine the statistical behavior of the model. A three-generation design is simulated which include a certain number of first-generation families sorted into 9 mating types according to the genotype frequencies. Assume that the allele frequencies of a gene are 0.6 and 0.4 in a natural population at Hardy-Weinberg equilibrium. Our simulation will focus on the investigation of the impacts of different sampling strategies and heritabilities on parameter estimation and model power. For a given sample size, two sampling strategies are simulated, (1) a large family number and small family size, and (2) a small family number and large family size.

The first strategy samples 200 unrelated grandfathers and 200 unrelated grandmothers, who marry to form 200 the first-generation families. Each first-generation family is assumed to have one son who, as the father, form a second-generation family with the mother from the natural population. There is one child for each second-generation family. This allocation results in a total of 1000 subjects. All members in the design are typed for the gene, but only the fathers and offspring of the third generation are phenotyped for a normally distributed trait. The second strategy samples 50 unrelated grandfathers and 50 unrelated grandmothers. In each first-generation family, 3 sons are simulated, forming 150 second-generation families in which 4 children are assumed. This strategy also results in 1000 subjects.

Different genetic effects of the gene, additive, dominant, and imprinting, are simulated for the second- and third-generations using the designed. Two different heritability levels, 0.1 and 0.4, are simulated for each generation, from which variances are determined. Table 1 tabulates the estimates of population and quantitative genetic parameters from the three-generation design. As expected, allele frequency can be very well estimated. The model provides reasonable estimation accuracy and precision for all genetic parameters under different sampling strategies, even for a modest heritability level. Under both strategies, the model has great power (0.85 or higher) to test the significance of individual genetic effects, additive, dominant, and imprinting, expressed in different generations. The model is also powerful to detect differences of genetic effects between two consecutive generations. More interesting, the difference of imprinting effect between different generations, i.e., transgenerational inheritance of genetic imprinting, can be discerned with power 0.80 using our statistical design.

Table 1: The maximum likelihood estimates (MLEs) of additive (a), dominant (d), and imprinting effects (i) of a functional SNP on a complex trait in parental (F) and offspring (O) generations under two different strategies

Genetic	True	Strategy 1		Strategy 2	
Parameter	**Value**	$H^2=0.1$	$H^2=0.4$	$H^2=0.1$	$H^2=0.4$
a_F	1.0	1.0198(0.0236)	1.0028(0.0112)	1.0387(0.0333)	0.9975(0.0125)
d_F	0.6	0.5702(0.0340)	0.6184(0.0144)	0.6272(0.0392)	0.6046(0.0185)
i_F	0.6	0.6024(0.0306)	0.5991(0.0093)	0.5897(0.0328)	0.6037(0.012)
a_O	1.0	0.9779(0.0429)	0.9889(0.0167)	1.0136(0.0243)	1.0082(0.0108)
d_O	1.5	1.5218(0.0462)	1.4397(0.0242)	1.5465(0.0389)	1.4962(0.0138)
i_O	1.0	0.9853(0.0393)	0.9885(0.0138)	1.0272(0.0271)	1.0212(0.0119)

The esimates are the means of MLEs obtained from 200 simulation replicates, with standard errors given in parentheses.
doi:10.1371/journal.pone.0016858.t001

One major aim of this study is to estimate the change of genetic effects over generation. Although our model has great power to detect the transgenerational change of genetic effects, its false positive rates should also be assessed. We conducted an additional simulation study to address this issue by simulating a SNP that has the same genetic effects between the two generations. The model detects a small proportion of simulation replicates ($<6\%$) which displays transgenerational differences in all types of genetic effects including additive, dominant, and imprinting. This suggests that the model has a small type I error rate for detecting the transgenerational difference of overall genetic effects. We particularly tested the type I error rate for the transgenerational difference of genetic imprinting, which is reasonably small ($<8\%$).

The haplotype model is also examined through simulation studies. We simulated two SNPs with a recombination fraction of $r = 0.05$ that are segregating in a human population. Of the four haplotypes, one is assumed to function as a risk haplotype. The remaining is collectively called the non-risk haplotype. The genetic values of composite diplotypes constituted by risk and non-risk haplotypes include the additive (a), dominant (d), and imprinting (i) genetic effects. We assume that some of these effects are different, and the others are the same between the parental and offspring generations. Combinations of different heritabilities between the two generations are simulated.

Table 2: Simulation results for transgeneration imprinting effects comparisons

				First Generation Parameters			Second Generation Parameters		
N	H^2		$r=0.05$	$a_F=1$	$d_F=0.5$	$i_F=0.4$	$a_O=1$	$d_O=1.5$	$i_O=0.6$
400	0.1	0.1	0.0561(0.0049)	0.9936(0.0185)	0.5228(0.0263)	0.3824(0.0195)	1.0182(0.0266)	1.4761(0.0405)	0.6592(0.0269)
	0.1	0.4	0.0450(0.0049)	0.9975(0.0162)	0.4690(0.0279)	0.4437(0.0184)	0.9860(0.0106)	1.4936(0.0174)	0.5937(0.0115)
	0.4	0.1	0.0639(0.0055)	1.0011(0.0082)	0.5032(0.0107)	0.4013(0.0082)	1.0210(0.0277)	1.5275(0.0383)	0.5631(0.0285)
	0.4	0.4	0.0681(0.0054)	1.0137(0.0075)	0.5072(0.0115)	0.3916(0.0083)	0.9939(0.0108)	0.4920(0.0144)	0.5975(0.0110)
800	0.1	0.1	0.0486(0.0044)	1.0003(0.0134)	0.4610(0.0183)	0.4162(0.0126)	1.0079(0.0194)	1.5219(0.0284)	0.6322(0.0175)
	0.1	0.4	0.0461(0.0040)	0.9956(0.0135)	0.4933(0.0180)	0.3796(0.0114)	1.0049(0.0075)	1.4978(0.0112)	0.5874(0.0072)
	0.4	0.1	0.0516(0.0041)	1.0047(0.0046)	0.5074(0.0077)	0.3915(0.0055)	0.9916(0.0083)	1.5027(0.0091)	0.6002(0.0073)
	0.4	0.4	0.0567(0.0036)	1.0011(0.0056)	0.5023(0.0080)	0.3976(0.0061)	0.9773(0.0077)	1.5069(0.0104)	0.5926(0.0083)
2000	0.1	0.1	0.0516(0.0032)	1.0109(0.0079)	0.4951(0.0116)	0.4059(0.0089)	1.0053(0.0122)	1.5095(0.0150)	0.5878(0.0119)
	0.1	0.4	0.0536(0.0029)	1.0078(0.0094)	0.5283(0.0107)	0.4038(0.0099)	1.0017(0.0042)	1.5011(0.0064)	0.5912(0.0053)
	0.4	0.1	0.0488(0.0027)	0.9996(0.0034)	0.5076(0.0044)	0.4064(0.0036)	0.9830(0.0115)	1.4997(0.0152)	0.6001(0.0138)
	0.4	0.4	0.0545(0.0028)	1.0009(0.0033)	0.5043(0.0047)	0.3986(0.0033)	0.9993(0.0050)	1.5012(0.0064)	0.5996(0.0048)

The genetic design scenarios are chosen as the combination of different heritabilities and sample sizes. They are: $H_1^2=0.1/0.4$, $H_2^2=0.1/0.4$, $n=400,800,2000$.
doi:10.1371/journal.pone.0016858.t002

Table 2 gives the results of simulation for different heritabilities and sample sizes (all subjects used). Overall, all parameters can be estimated reasonably well. As expected, the precision of parameter estimation increases with heritability and sample size. The additive genetic effects in both generations can well be estimated with a modest sample size (say 400) for a small heritability (0.1). More sample sizes (say 800) are needed to provide a good estimate for genetic imprinting effects for a small heritability. To well estimate dominant genetic effects, an even larger sample size (say 2000) is required for the same level of heritability.

DISCUSSION

The traditional view of quantitative trait expression analysis assumes that the maternally and paternally derived alleles of each gene are expressed simultaneously at a similar level. However, this view is violated by a growing body of evidence that alleles are expressed from only one of the two parental chromosomes [1], [2]. This so-called genetic imprinting or parent-of-origin effect has been thought to play a pivotal role in regulating the phenotypic variation of a complex trait [3], [4], [6], [8], [9], [12], [13], [14], [15]. With the discovery of more imprinting genes involved in trait control through molecular and bioinformatics approaches, we will be in a position to elucidate the genetic architecture of quantitative variation for various organisms including humans.

Recent evidence shows that epigenetic inheritance in humans may experience a transgenerational change. This would represent a significant shift in our current understanding of inheritance and disease aetiology. Despite the development of new technologies that are reducing the time and cost of genotyping by several orders of magnitude [31], [32], the understanding of the underlying genetic events will be challenging. In this article, we present a computational model for identifying the genomic imprinting effect of genes on quantitative phenotypes and transgenerational change of genomic imprinting using a multigenerational sampling design for human families. The model formulates a general framework for testing the difference of genetic effects between different generations. By including multiple SNPs, the model was extended to estimate genomic imprinting and its transgenerational change expressed at the haplotype level. Although several models have been developed to estimate genomic imprinting for binary disease traits [33], [34], our model is among the first for estimating genetic imprinting operational in regulating the variation of quantitative traits and is certainly the first of its kind that can discern the transgenerational change of genetic imprinting.

Although no real data were analyzed for the moment, this model presents a conceptual design by which new data can be collected according to the sampling strategy proposed and then analyzed by the computational algorithm derived. Based on computer simulation, the model should display convincing statistical properties in parameter estimation and test and can be applied to a practical data set. However, several issues need to be addressed when the model is attempted to solve broader genetic questions. First, the maternal effects that cause parent-of-origin effects of alleles may be confounded with imprinting effects [35], which should be separated by developing a proper design in order to better study the patterns of gene expression and evolutionary dynamics.

Second, this study assumes the unisex (sons) produced from the first-generation family. One can also assume daughters with no change of the model, allowing the test of genomic imprinting between mother and offspring. In fact, our model can involve both sexes so that in the second generation sex-specific genetic effects can be characterized. If the sexes in the third generation are considered, the model can be extended to study the transgenerational changes of gene-sex interactions. Third, it is possible that part of parental genotypes are missing in practice. To infer genomic imprinting using such data sets, a multi-hierachical mixture model can be derived to estimate the missing parental genotypes based on observed offspring genotypes. Fourth, although a basic premise of epigenetic processes was that, once established, these marks were maintained through rounds of mitotic cell division and stable for the life of the organism, several recent studies have shown that at some loci the epigenetic state can be altered by the environment [36]. The questions are how common are genes of this type and how strong is the evidence for their existence in humans? The development of our design and model will help to address these biological questions of fundamental importance in elucidating the genetic architecture of complex traits.

METHODS

Sampling Strategies

Suppose there is a natural human population at Hardy–Weinberg equilibrium (HWE) from which a panel of three-generation families, each composed of the grandfather, grandmother, father, mother, and grandchildren, are sampled. Each member in a family is typed for single nucleotide polymorphisms (SNPs) from the human genome. Consider a quantitative trait affected by a SNP with two alleles A in a frequency of p and a in a frequency of q, leading to three genotypes AA, Aa, and aa with the frequencies of p^2, $2pq$, and q^2, respectively. In the grandparent generation, these three genotypes are mating randomly to produce nine cross types. Given a cross type, the genotypes of sons or daughters can be inferred. Here we first assume one sex (say son) in the second generation, although both sexes can be considered. The sons from a family serve as the father to mate with the females as the mother derived from a natural population, with genotypes, AA, Aa, and aa, characterized by frequencies p^2, $2pq$, and q^2, respectively. Each of such second-generation families produces a certain number of grandchildren. The genotype frequencies in the third generation are derived according to Mendel's first law.

According to this design, the grandfathers and grandmothers are founders whose parents are unknown. Alleles of sons from a first-generation family can be traced directly or indirectly, but the females used to generate the second-generation family are the founders with the unknown origin of alleles. For this reason, we will measure the phenotype for sons from the first-generation families and grandchildren from the second-generation families. This design will allow us to characterize imprinting effects of a gene in the second- and third-generations.

Genetic Models

There are three genotypes AA, Aa, and aa, for a biallelic gene according to Mendelian segregation pattern. Considering the parent-of-origin of alleles, these genotypes are described by four configurations, $A|A$ (coded as 2), $A|a$ (coded as 1), $a|A$ (coded as $1'$), and $a|a$(coded as 0), where symbol $|$ is used to separate the maternally- (left) and paternally-derived alleles (right). The genotypic values of the four configurations in two different generations are defined as follows:

Configuration	Paternal	Offspring
$A\|A$	$\mu_2^F=\mu_F+a_F$	$\mu_2^O=\mu_O+a_O,$
$A\|a$	$\mu_1^F=\mu_F+d_F+i_F$	$\mu_1^O=\mu_O+d_O+i_O,$
$a\|A$	$\mu_{1'}^F=\mu_F+d_F-i_F$	$\mu_{1'}^O=\mu_O+d_O-i_O,$
$a\|a$	$\mu_0^F=\mu_F-a_F$	$\mu_0^O=\mu_O-a_O,$

(1)

where μ_F and μ_O are the overall means of the paternal and offspring generations, a_F, d_F, and i_F are the additive, dominant and imprinting genetic effects of the gene in the parental generation, and a_O, d_O, and i_O are the additive, dominant and imprinting genetic effects of the gene in the offspring generation.

The difference in the genetic architecture of a complex trait between two different generations is described as

$$\Delta_a=a_F-a_O, \tag{2}$$

$$\Delta_d=d_F-d_O, \tag{3}$$

$$\Delta_i=i_F-i_O. \tag{4}$$

By testing whether these differences are equal to zero jointly or individually, we can determine the transgenerational changes of the pattern of genetic control. If a significant imprinting effect is detected, we can test the type of genetic imprinting, i.e., parental or maternal dominance, by incorporating the imprinting models of Cheverud et al. [14].

Estimation

The grandfather and grandmother in the first generation from a natural population constitutes $3 \times 3 = 9$ mating types for three genotypes. For the jth first-generation mating type listed in Table $(j = 1, \ldots, 9)$, let N_j denote the family number of this mating type. Each first-generation family may have one or multiple sons who serve the father of the second generation. Those families in the second generation with the father derived from the jth first-generation mating type and the mother of a particular genotype from the natural population are summed together, denoted by N_{jl}^M, for mother genotype l (l=2 for AA, 1 for Aa, and 0 for aa). Thus, we have a total of $N_l^M = \sum_{j=1}^{9} N_{jl}^M$ second-generation mothers who carry genotype l.

It is not difficult to derive the maximum likelihood estimate of allele frequency from the three-generation family design as

$$p = \frac{4N_1 + 3(N_2 + N_4) + 2(N_3 + N_5 + N_7) + (N_6 + N_8) + 2N_2^M + N_1^M}{\sum_{j=1}^{9} 4N_j + 2(N_2^M + N_1^M + N_0^M)}$$

$$q = \frac{4N_8 + 3(N_5 + N_7) + 2(N_2 + N_4 + N_6) + (N_1 + N_3) + 2N_0^M + N_1^M}{\sum_{j=1}^{9} 4N_j + 2(N_2^M + N_1^M + N_0^M)}.$$

The male individuals from the first generation are typed for the marker, with four distinct configurations, $A|A$ (2), $A|a$ (1), $a|A$ (1′), and $a|a$ (0). Let n_{jk}^F denote the cumulative number of male individuals (as the father for the second generation) bearing configuration k $(k = 2, 1, 1', 0)$ from n_j first-generation families. In the third generation, only genotypes rather than configurations can be observed. We use N_{jkls}^O to denote the number of children who carry genotype s $(s = 2, 1, 0)$ from a second-generation family with father k (from the jth first-generation mating type) and mother l from a natural population. The phenotypic values measured are expressed as y_{jki}^F $(i = 1, \ldots, n_{jk}^F)$ for the second-generation fathers and y_{jklsi}^O $(i = 1, \ldots, N_{jkls}^O)$ for the third-generation children. Both y_{jki}^F and y_{jklsi}^O are assumed to follow a normal distribution with mean depending on genotypes and residual variances σ_F^2 and σ_O^2, respectively.

Since offspring genotypes depend on parental genotypes, the log-likelihood of paternal and offspring parameters given marker (**M**) and phenotypic (y) data from the three generations is decomposed into two components, one related to the paternal parameters and the second related to the offspring parameters given the paternal parameters, expressed as

$$L(\Omega_F,\Omega_O|y_F,y_O,\mathbf{M}_F^{\{1\}},\mathbf{M}_M^{\{1\}},\mathbf{M}_F^{\{2\}},\mathbf{M}_M^{\{2\}},\mathbf{M}_O^{\{3\}})$$

$$=L(\Omega_F|y_F,\mathbf{M}_F^{\{1\}},\mathbf{M}_M^{\{1\}},\mathbf{M}_F^{\{2\}})+$$
$$L(\Omega_O|y_O,\mathbf{M}_F^{\{1\}},\mathbf{M}_M^{\{1\}},\mathbf{M}_F^{\{2\}},\mathbf{M}_M^{\{2\}},\mathbf{M}_O^{\{3\}}), \quad (5)$$

where $\Omega_F=(\mu_F,a_F,d_F,i_F,\sigma_F^2)$ are the paternal parameters and $\Omega_O=(\mu_O,a_O,d_O,i_O,\sigma_O^2)$ are the offspring parameters. Maximizing joint likelihood (5) is equivalent to maximizing its two likelihood components independently. The estimates of parameters Ω_F that maximize the first component can be obtained with the EM algorithm. In the E step, the posterior probability with which the double heterozygote father of the second generation from the 5th first-generation mating type in Table has a particular configuration is calculated by

$$\Phi_{51i}^F=\frac{\frac{1}{2}f_1(y_{51i}^F)}{\frac{1}{2}f_1(y_{51i}^F)+\frac{1}{2}f_{1'}(y_{51'i}^F)} \text{ and } \Phi_{51'i}^F=\frac{\frac{1}{2}f_{1'}(y_{51'i}^F)}{\frac{1}{2}f_1(y_{51i}^F)+\frac{1}{2}f_{1'}(y_{51'i}^F)} \quad (6)$$

In the M step, the genotypic values of configurations and variance are calculated by

$$\mu_2^F=\frac{\sum_{i=1}^{N_{12}^F}y_{12i}^F+\sum_{i=1}^{N_{22}^F}y_{22i}^F+\sum_{i=1}^{N_{42}^F}y_{42i}^F+\sum_{i=1}^{N_{52}^F}y_{52i}^F}{N_{12}^F+N_{22}^F+N_{42}^F+N_{52}^F},$$

$$\mu_1^F=\frac{\sum_{i=1}^{N_{21}^F}y_{21i}^F+\sum_{i=1}^{N_{31}^F}y_{31i}^F+\sum_{i=1}^{N_{51}^F}\Phi_{51i}^F y_{51i}^F+\sum_{i=1}^{N_{61}^F}y_{61i}^F}{N_{21}^F+N_{31}^F+\sum_{i=1}^{N_{51}^F}\Phi_{51i}^F+N_{61}^F},$$

$$\mu_{1'}^F=\frac{\sum_{i=1}^{N_{41'}^F}y_{41pi}^F+\sum_{i=1}^{N_{51'}^F}\Phi_{51'i}^F y_{51'i}^F+\sum_{i=1}^{N_{71'}^F}y_{71'i}^F+\sum_{i=1}^{N_{81'}^F}y_{81'i}^F}{N_{41'}^F+\sum_{i=1}^{N_{51'}^F}\Phi_{51'i}^F+N_{71'}^F+N_{81'}^F},$$

$$\mu_0^F=\frac{\sum_{i=1}^{N_{50}^F}y_{50i}^F+\sum_{i=1}^{N_{60}^F}y_{60i}^F+\sum_{i=1}^{N_{80}^F}y_{80i}^F+\sum_{i=1}^{N_{90}^F}y_{90i}^F}{N_{50}^F+N_{60}^F+N_{80}^F+N_{90}^F},$$

$$\sigma_F^2 = \frac{1}{\sum_{j=1}^{9} N_j} \left\{ \sum_{i=1}^{N_{12}^F} (y_{12i}^F - \mu_2^F)^2 + \sum_{i=1}^{N_{22}^F} (y_{22i}^F - \mu_2^F)^2 \right.$$

$$+ \sum_{i=1}^{N_{21}^F} (y_{21i}^F - \mu_1^F)^2 + \sum_{i=1}^{N_{31}^F} (y_{31i}^F - \mu_1^F)^2$$

$$+ \sum_{i=1}^{N_{42}^F} (y_{42i}^F - \mu_2^F)^2 + \sum_{i=1}^{N_{41'}^F} [(y_{41'i}^F - \mu_{1'}^F)^2$$

$$+ \sum_{i=1}^{N_{52}^F} (y_{52i}^F - \mu_2^F)^2 + \sum_{i=1}^{N_{51}^F} (y_{51i}^F - \mu_1^F)^2$$

$$+ \sum_{i=1}^{N_{51'}^F} (y_{51'i}^F - \mu_{1'}^F)^2 + \sum_{i=1}^{N_{50}^F} (y_{50i}^F - \mu_0^F)^2$$

$$+ \sum_{i=1}^{N_{61}^F} (y_{61i}^F - \mu_1^F)^2 + \sum_{i=1}^{N_{60}^F} (y_{60i}^F - \mu_0^F)^2$$

$$+ \sum_{i=1}^{N_{71'}^F} (y_{71'i}^F - \mu_{1'}^F)^2 + \sum_{i=1}^{N_{81'}^F} (y_{81'i}^F - \mu_{1'}^F)^2$$

$$\left. + \sum_{i=1}^{N_{80}^F} (y_{80i}^F - \mu_0^F)^2 + \sum_{i=1}^{N_{90}^F} (y_{90i}^F - \mu_0^F)^2 \right\}. \quad (7)$$

The EM algorithm can also be implemented to estimate genetic parameters $\mathbf{\Omega}_O$ in the third generation that maximize the second component in (5). In the E step, the posterior probability with which the double heterozygote offspring of the third generation derived from the combination of two double heterozygote parents in the second generation has a particular configuration is calculated by

$$\Phi_{jkl1i}^O = \frac{\frac{1}{4} f_1(y_{jkli}^O)}{\frac{1}{4} f_1(y_{jkli}^O) + \frac{1}{4} f_{1'}(y_{jkli}^O)} \text{ and } \Phi_{jkl1'i}^O = \frac{\frac{1}{4} f_{1'}(y_{jkli}^O)}{\frac{1}{4} f_1(y_{jkli}^O) + \frac{1}{4} f_{1'}(y_{jkli}^O)} \quad (8)$$

In the M step, the genotypic values of configurations and variance are calculated by

$$\mu_2^O = \frac{\sum_{j=1}^9 \sum_{k=0}^2 \sum_{l=0}^2 \sum_{i=1}^{N_{jkl2}^O} y_{jkl2i}^O \xi_{jkl2i}}{\sum_{j=1}^9 \sum_{k=0}^2 \sum_{l=0}^2 N_{jkl2}^O}, k=0,1',1,2; l=0,1,2,$$

$$\mu_1^O = \frac{\sum_{j=1}^9 \sum_{k=0}^2 \sum_{l=0}^2 \sum_{i=1}^{N_{jkl}^O} \Phi_{jkl1i}^O y_{jkl1i}^O \xi_{jkl1i}}{\sum_{j=1}^9 \sum_{k=0}^2 \sum_{l=0}^2 \sum_{i=1}^{N_{jkl}^O} \Phi_{jkl1i}^O \xi_{jkl1i}}, k=0,1',1,2; l=0,1,2,$$

$$\mu_{1'}^O = \frac{\sum_{j=1}^9 \sum_{k=0}^2 \sum_{l=0}^2 \sum_{i=1}^{N_{jkl}^O} \Phi_{jkl1'i}^O y_{jkl1i}^O \xi_{jkl1i}}{\sum_{j=1}^9 \sum_{k=0}^2 \sum_{l=0}^2 \sum_{i=1}^{N_{jkl}^O} \Phi_{jkl1'i}^O \xi_{jkl1i}}, k=0,1',1,2; l=0,1,2,$$

$$\mu_0^O = \frac{\sum_{j=1}^9 \sum_{k=0}^2 \sum_{l=0}^2 \sum_{i=1}^{N_{jkl0}^O} y_{jkl0i}^O \xi_{jkl0i}}{\sum_{j=1}^9 \sum_{k=0}^2 \sum_{l=0}^2 N_{jkl0}^O}, k=0,1',1,2; l=0,1,2,$$

$$\sigma_O^2 = \frac{1}{\sum_{j=1}^9 \sum_{k=0}^2 \sum_{l=0}^2 (N_{jkl2}^O + N_{jkl1}^O + N_{jkl0}^O)}$$

$$\times \sum_{j=1}^{9} \sum_{k=0}^{2} \sum_{l=0}^{2} \left\{ \sum_{i=1}^{N_{jkl2}^O} \xi_{jkl2i} (y_{jkl2i}^O - \mu_2^O)^2 + \sum_{i=1}^{N_{jkl1}^O} \xi_{jkl1i} \left[\Phi_{jkl1i}^O (y_{jkl1i}^O - \mu_1^O)^2 \right. \right.$$

$$\left. \left. + \Phi_{jkl1'i}^O (y_{jkl1i}^O - \mu_{1'}^O)^2 \right] + \sum_{i=1}^{N_{jkl0}^O} \xi_{jkl0i} (y_{jkl0i}^O - \mu_0^O)^2 \right\}, k=0,1',1,2; l=0,1,2,$$

where ξ_{jkl2i}, ξ_{jkl1i}, and ξ_{jkl0i} are the indicator variables that are defined as 1 if offspring i in the third generation from the combination of father k from the jth first-generation mating type and mother l from the natural population has genotype AA, Aa, and aa, respectively, and 0 otherwise. The EM steps are iterated between equations (6) and (7) to obtain the MLEs of Ω_F and between equations (8) and (9) to obtain the MLEs of Ω_O.

Hypothesis Tests

It is imperative to know whether there exists a significant association between a specific SNP and a complex trait and how a significant SNP triggers an additive, dominant, or imprinting effect on the trait. To test for the overall significant association of SNP genotype and trait phenotype, we generate the following hypotheses:

$$H_0 : a_F = d_F = i_F = a_O = d_O = i_O = 0$$

H_1 : At least one of these equalities above does not hold.

The log-likelihood ratio under the null and alternative hypotheses is calculated. Since the null hypothesis contains a nuisance parameter, allele frequency, this log-likelihood ratio test statistic may have an unclear distribution. For this reason, the critical threshold for claiming the existence of a significant SNP is determined from permutation tests [37]. If our interest is in testing whether there is an additive, dominant, or imprinting effect, the null hypothesis should be $H_0 : a_F = a_O = 0$, $H_0 : d_F = d_O = 0$, and $H_0 : i_F = i_O = 0$, respectively. Because each of these null hypotheses is nested within its alternative, the log-likelihood ratio test statistic can be thought to asymptotically follow a χ^2-distribution for a large sample size.

The transgenerational changes of different genetic effects can also be tested. The null hypotheses used to test whether the additive, dominant, and imprinting effects display significant changes from one generation to next are expressed as $H_0 : \Delta_a = 0$, $H_0 : \Delta_d = 0$, and $H_0 : \Delta_i = 0$, respectively. These null hypotheses can be considered singly or jointly, in order to better study the transgenerational changes of the genetic architecture of a trait.

Haplotyping Model

Recent molecular surveys suggest that the human genome contains many discrete haplotype blocks that are sites of closely located SNPs [38], [39], [40]. Each block may have a few common haplotypes which account for a large proportion of chromosomal variation. Between adjacent blocks are there large regions, called hotspots, in which recombination events occur with high frequencies. Several algorithms have been developed to identify a minimal subset of SNPs, i.e., tagging SNPs, that can characterize the most common haplotypes [41]. The number and type of tagging SNPs within each haplotype block can be determined prior to association studies. In this section, we will derive a model for detecting the association between haplotypes constructed by alleles at a set of SNPs and complex traits.

For the simplicity of our description, consider two SNPs **A** (with two alleles A and a) and **B**(with two alleles B and b). They form four haplotypes AB, Ab, aB, and ab, of which one that is distinct from the rest three is defined as a risk haplotype W and all the others are defined as a non-risk haplotype w [42]. Risk and non-risk haplotypes from the maternal and paternal parents generate four composite diplotypes, $W|W$, $W|w$, $w|W$, and $w|w$, whose genotypic values are described by the additive (a), dominant (d), and imprinting genetic effects (i). Cheng et al. [43] and Wang et al. [44] proposed a two- and three-SNP model for estimating and testing genetic imprinting effects in a natural population, respectively. Wu et al.'s procedure[45] allows the choice of an optimal number and combination of risk haplotypes within a multiallelic model framework. Here, we adopted Cheng et al.›s two-SNP model to estimate haplotype imprinting genetic effects and their transgenerational change.

In this example, four haplotypes AB, Ab, aB, and ab have frequencies denoted as p_{11}, p_{10}, p_{01}, and p_{00}, respectively. The two SNPs yield nine joint genotypes, $AABB$ (coded as 1),$AABb$ (coded as 2), …, $aabb$ (coded as 9), which are actually observed. Each subject must bear one of these genotypes, and the parents in each family will be one of 9 × 9=81 possible genotype by genotype combinations. If each parent for a combination is homozygous for both SNPs, their offspring will have one genotype. As long as one parent is heterozygous for one SNP, the offspring will have two or more genotypes. However, only when both SNPs are heterozygous for at least one parent, the genotype frequencies of offspring will be determined by the recombination fraction between the markers (r). Tables show the structure and frequencies of mother by father genotype combinations under random mating and their offspring genotype frequencies in the second and third generation, respectively. For a double heterozygote $AaBb$, its observed genotype may be derived from two possible diplotypes, $AB|ab$ (with the relative proportion of $\phi=\frac{p_{11}p_{00}}{p_{11}p_{00}+p_{10}p_{01}}$) or $Ab|aB$) (with the relative proportion of $1-\phi=\frac{p_{10}p_{01}}{p_{11}p_{00}+p_{10}p_{01}}$). Each of these two diplotypes produce four haplotypes AB, Ab, aB, and ab, whose frequencies are expressed as

		Haplotype			
Diplotype	Proportion	AB	Ab	aB	ab
$AB\|ab$	ϕ	$\frac{1}{2}(1-r)$	$\frac{1}{2}r$	$\frac{1}{2}r$	$\frac{1}{2}(1-r)$
$Ab\|aB$	$1-\phi$	$\frac{1}{2}r$	$\frac{1}{2}(1-r)$	$\frac{1}{2}(1-r)$	$\frac{1}{2}r$

A similar likelihood (5) cane be formulated for haplotype models. A complicated EM algorithm is derived to estimate haplotype frequencies using the parental information. Let N_{ij} denote the observation of mating type between genotype i for one parent and genotype j for the second parent. In the E step, calculate the proportion of a diplotype for a heterozygous genotype for a particular mating design by

$$\psi_1 = \frac{(1-\phi)r}{\omega_1}$$

$$\psi_2 = \frac{\phi r}{\omega_2}$$

$$\psi_3 = \frac{(1-\phi)r+\phi r}{\omega_1+\omega_2}$$

$$\psi_4 = \frac{(1-\phi)^2r^2+\phi(1-\phi)r(1-r)}{\omega_1^2}$$

$$\psi_5 = \frac{\phi^2r^2+\phi(1-\phi)r(1-r)}{\omega_2^2}$$

$$\psi_6 = \frac{\phi(1-\phi)r^2+[\phi^2+(1-\phi)^2]r(1-r)}{2\psi_1\omega_2}$$

$$\psi_7 = \frac{\phi^2r^2+2\phi(1-\phi)r(1-r)+(1-\phi)^2r^2}{2(\omega_1^2+\phi_2^2)},$$

where $\omega_1 = (1-\phi)r+\phi(1-r), \omega_2 = \phi r+(1-\phi)(1-r)$.

In the M step, estimate the haplotype frequencies and recombination fraction by

$$\begin{aligned}
p_{11} = \{&4N_{11} + 3(N_{12} + N_{21} + N_{14} + N_{41}) + 2(N_{22} + N_{44} + N_{13} + N_{31} \\
&+ N_{16} + N_{61} + N_{17} + N_{71} + N_{18} + N_{81} + N_{19} + N_{91} + N_{24} + N_{42}) \\
&+ N_{23} + N_{32} + N_{26} + N_{62} + N_{27} + N_{72} + N_{28} + N_{82} + N_{29} + N_{92} \\
&+ N_{34} + N_{43} + N_{46} + N_{64} + N_{47} + N_{74} + N_{48} + N_{84} + N_{49} + N_{94} \\
&+ \phi[3(N_{15} + N_{51}) + 2(N_{25} + N_{52} + N_{45} + N_{54}) + N_{35} + N_{53} + N_{56} \\
&+ N_{65} + N_{57} + N_{75} + N_{58} + N_{85} + N_{59} + N_{95}] \\
&+ (1-\phi)[2(N_{15} + N_{51}) + N_{25} + N_{52} + N_{45} + N_{54}] + 2\phi N_{55}\} \\
&/(4 \sum_{i=1,j=1}^{9,9} N_{ij}),
\end{aligned}$$

$$\begin{aligned}
p_{10} = \{&4N_{33} + 3(N_{23} + N_{32} + N_{36} + N_{63}) + 2(N_{22} + N_{66} + N_{13} + N_{31} \\
&+ N_{26} + N_{62} + N_{34} + N_{43} + N_{37} + N_{73} + N_{38} + N_{83} + N_{39} + N_{93}) \\
&+ N_{12} + N_{21} + N_{16} + N_{61} + N_{24} + N_{42} + N_{27} + N_{72} + N_{28} + N_{82} \\
&+ N_{29} + N_{92} + N_{46} + N_{64} + N_{67} + N_{76} + N_{68} + N_{86} + N_{69} + N_{96} \\
&+ \phi[2(N_{35} + N_{53}) + N_{25} + N_{52} + N_{56} + N_{65}] \\
&+ (1-\phi)[3(N_{35} + N_{53}) + 2(N_{25} + N_{52} + N_{56} + N_{65}) + N_{15} + N_{51} \\
&+ N_{45} + N_{54} + N_{57} + N_{75} + N_{58} + N_{85} + N_{59} + N_{95}] + 2(1-\phi)N_{55}\} \\
&/(4 \sum_{i=1,j=1}^{9,9} N_{ij}),
\end{aligned}$$

$$\begin{aligned}
p_{01} = \{&4N_{77} + 3(N_{47} + N_{74} + N_{78} + N_{87}) + 2(N_{44} + N_{88} + N_{17} + N_{71} \\
&+ N_{27} + N_{72} + N_{37} + N_{73} + N_{67} + N_{76} + N_{48} + N_{84} + N_{79} + N_{97}) \\
&+ N_{41} + N_{18} + N_{81} + N_{24} + N_{42} + N_{28} + N_{82} + N_{34} + N_{43} + N_{38} \\
&+ N_{83} + N_{46} + N_{64} + N_{49} + N_{94} + N_{68} + N_{86} + N_{89} + N_{98} \\
&+ \phi[2(N_{57} + N_{75}) + N_{45} + N_{54} + N_{58} + N_{85}] \\
&+ (1-\phi)[3(N_{57} + N_{75}) + 2(N_{45} + N_{54} + N_{58} + N_{85}) + N_{15} + N_{51} \\
&+ N_{25} + N_{52} + N_{35} + N_{53} + N_{56} + N_{65} + N_{59} + N_{95}] + 2(1-\phi)N_{55}\} \\
&/(4 \sum_{i=1,j=1}^{9,9} N_{ij}),
\end{aligned}$$

$$
\begin{aligned}
p_{00} = \{&4N_{99} + 3(N_{69} + N_{96} + N_{89} + N_{98}) + 2(N_{66} + N_{88} + N_{19} + N_{91} \\
&+ N_{29} + N_{92} + N_{39} + N_{93} + N_{49} + N_{94} + N_{68} + N_{86} + N_{79} + N_{97}) \\
&+ N_{16} + N_{61} + N_{18} + N_{81} + N_{26} + N_{62} + N_{28} + N_{82} + N_{36} + N_{63} \\
&+ N_{38} + N_{83} + N_{46} + N_{64} + N_{48} + N_{84} + N_{67} + N_{76} + N_{78} + N_{87} \\
&+ \phi[3(N_{59} + N_{95}) + 2(N_{56} + N_{65} + N_{58} + N_{85}) + N_{15} + N_{51} + N_{25} \\
&+ N_{52} + N_{35} + N_{53} + N_{45} + N_{54} + N_{57} + N_{75}] \\
&+ (1-\phi)[2(N_{59} + N_{95}) + N_{56} + N_{65} + N_{58} + N_{85}] + 2\phi N_{55}\} \\
&/(4 \sum_{i=1,j=1}^{9,9} N_{ij}).
\end{aligned}
$$

$$
\begin{aligned}
r = \{&\psi_1(N_{51}^1 + N_{51}^5 + N_{52}^1 + N_{52}^6 + N_{53}^2 + N_{53}^6 + N_{54}^1 + N_{54}^8 + N_{56}^2 \\
&+ N_{56}^9 + N_{57}^4 + N_{57}^8 + N_{58}^4 + N_{58}^9 + N_{59}^5 + N_{59}^9 + N_{15}^1 + N_{15}^5 + N_{25}^1 \\
&+ N_{25}^6 + N_{35}^2 + N_{35}^6 + N_{45}^1 + N_{45}^8 + N_{65}^2 + N_{65}^9 + N_{75}^4 + N_{75}^8 + N_{85}^4 \\
&+ N_{85}^9 + N_{95}^5 + N_{95}^9) + \psi_2(N_{51}^2 + N_{51}^4 + N_{52}^3 + N_{52}^4 + N_{53}^3 + N_{53}^5 \\
&+ N_{54}^2 + N_{54}^7 + N_{56}^3 + N_{56}^8 + N_{57}^5 + N_{57}^7 + N_{58}^6 + N_{58}^7 + N_{59}^6 + N_{59}^8 \\
&+ N_{15}^2 + N_{15}^4 + N_{25}^3 + N_{25}^4 + N_{35}^3 + N_{35}^5 + N_{45}^2 + N_{45}^7 + N_{65}^3 + N_{65}^8 \\
&+ N_{75}^5 + N_{75}^7 + \psi_3(N_{52}^2 + N_{52}^5 + N_{54}^4 + N_{54}^5 + N_{56}^5 + N_{56}^6 + N_{58}^5 \\
&+ N_{58}^8 + N_{25}^2 + N_{25}^5 + N_{45}^4 + N_{45}^5 + N_{85}^6 + N_{85}^7 + N_{95}^6 + N_{95}^8) + N_{65}^6 \\
&+ N_{85}^5 + N_{85}^8) + \psi_4(N_{55}^1 + N_{55}^9) + N_{65}^5 + \psi_5(N_{55}^3 + N_{55}^7) \\
&+ \psi_6(N_{55}^2 + N_{55}^4 + N_{55}^4 + N_{55}^6 + N_{55}^8) + \psi_7 N_{55}^5\} \\
&/(\sum_{j=1}^{9} N_{5j} + \sum_{i=1}^{9} N_{i5} - N_{55})
\end{aligned}
$$

In the M step, the equations for estimating additive, dominant, imprinting effects expressed in paternal and offspring generations are also derived. The E and M steps are iterated until the estimates converge to a stable value. These stable values are the maximum likelihood estimates (MLEs) of parameters. The estimated haplotype frequencies and recombination fraction are embedded into a mixture model for estimating genotypic values and variances for different generations.

AUTHOR CONTRIBUTIONS

Conceived and designed the experiments: RW. Performed the experiments: JW DL. Analyzed the data: YL YG WH JW MNC. Contributed reagents/materials/ analysis tools: YL WH JW. Wrote the paper: RW YL.

REFERENCES

1. Reik W, Walter J (2001) Genomic imprinting: parental influence on the genome. Nat Rev Genet 2: 21–32.
2. Wilkins JF, Haig D (2003) What good is genomic imprinting: The function of parent-specific gene expression. Nat Rev Genet 4: 359–368.
3. Itier JM, Tremp G, Léonard JF, Multon MC, et al. (1998) Imprinted gene in postnatal growth role. Nature 393: 125–126.
4. Li LL, Keverne EB, Aparicio SA, Ishino F, et al. (1999) Regulation of maternal behaviour and offpring growth by paternally expressed Peg3. Science 284: 330–333.
5. Isles AR, Wilkinson LS (2000) Imprinted genes, cognition and behaviour. Trend Cogn Sci 4: 309–318.
6. Constancia M, Kelsey G, Reik W (2004) Resourceful imprinting. Nature 432: 53–57.
7. Wood AJ, Oakey RJ (2006) Genomic imprinting in mammals: Emerging themes and established theories. PLoS Genet 2(11): e147.
8. Wilkinson LS, Davies W, Isles AR (2007) Genomic imprinting effects on brain development and function. Nat Rev Neurosci 4: 1–19.
9. Wang CG, Wang Z, Luo JT, Li Q, et al. (2010) A model for transgenerational imprinting variation in complex traits. PLoS ONE 5(7): e11396.
10. Frost JM, Moore GE (2010) The importance of imprinting in the human placenta. PLoS Genet 6(7): e1001015.
11. Sha K (2008) A mechanistic view of genomic imprinting. Ann Rev Genom Hum Genet 9: 197–216.
12. De Koning DJ, Rattniek AP, Harlizius B, Arendonk JAM, et al. (2000) Genome-wide scan for body composition in pigs reveals important role of imprinting. Proc Natl Acad Sci U S A 97: 7947–7950.
13. Liu T, Todhunter RJ, Wu S, Hou W, et al. (2007) A random model for mapping imprinted quantitative trait loci in a structured pedigree: An implication for mapping canine hip dysplasia. Genomics 90: 276–284.

14. Cheverud JM, Hager R, Roseman C, Fawcett G, et al. (2008) Genomic imprinting effects on adult body composition in mice. Proc Natl Acad Sci U S A 105: 4253–4258.
15. Wolf JB, Cheverud JM, Roseman C, Hager R (2008) Genome-wide analysis reveals a complex pattern of genomic imprinting in mice. PLoS Genet 4: e1000091.
16. Li YC, Coelho CM, Liu T, Wu S, et al. (2007) A statistical strategy to estimate maternal-zygotic interactions and parent-of-origin effects of QTLs for seed development. PLoS ONE 3: e3131.
17. Morgan HD, Santos F, Green K, Dean W, et al. (2005) Epigenetic reprogramming in mammals. Hum Mol Genet 14: R47–R58.
18. Sasaki H, Matsui Y (2008) Epigenetic events in mammalian germ-cell development: reprogramming and beyond. Nat Rev Genet 9: 129–140.
19. Tal O, Kisdi E, Jablonka E (2010) Epigenetic contribution to covariance between relatives. Genetics 184: 1037–1050.
20. McGrath J, Solter D (1984) Inability of mouse blastomere nuclei transferred to enucleated zygotes to support development in vitro. Science 226: 1317–1319.
21. Surani MA, Barton SC, Norris ML (1984) Development of reconstituted mouse eggs suggests imprinting of the genome during gametogenesis. Nature 308: 548–550.
22. Morgan HD, Sutherland HG, Martin DI, Whitelaw E (1999) Epigenetic inheritance at the agouti locus in the mouse. Nat Genet 23: 314–318.
23. Cropley JE, Suter CM, Beckman KB, Martin DI (2006) Germ-line epigenetic modification of the murine Avy allele by nutritional supplementation. Proc Natl Acad Sci U S A 103: 17308–17312.
24. Skinner MK (2008) What is an epigenetic transgenerational phenotype? F3 or F2. Reprod Toxic 25: 2–6.
25. Dolinoy DC, Weidman JR, Waterland RA, Jirtle RL (2006) Maternal genistein alters coat color and protects Avy mouse offspring from obesity by modifying the fetal epigenome. Environ Health Perspect 114: 567–572.
26. Whitelaw NC, Whitelaw E (2008) Transgenerational epigenetic inheritance in health and disease. Curr Opin Genet Dev 18: 273–279.
27. Youngson NA, Whitelaw E (2008) Transgenerational epigenetic effects. Ann Rev Genom Hum Genet 9: 233–257.

28. Pembrey ME, Bygren LO, Kaati G, Edvinsson S, et al. (2006) Sex-specific, male-line transgenerational responses in humans. Europ J Hum Genet 14: 159–166.

29. Wu RL, Ma CX, Casella G (2007) Statistical Genetics of Quantitative Traits: Linkage, Map, and QTLs. New York: Springer.

30. Li Q, Wu RL (2009) A multilocus model for constructing a linkage disequilibrium map in human populations. Stat Appl Genet Mol Biol. 8, Iss. 1, Article 18.

31. Chan EY (2005) Advances in sequencing technology. Mutant Res 573: 13–40.

32. Beckmann JS, Estivill X, Antonarakis SE (2007) Copy number variants and genetic traits: closer to the resolution of phenotypic to genotypic variability. Nat Rev Genet 8: 639–646.

33. Weinberg CR, Wilcox AJ, Lie RT (1998) A log-linear approach to case-parent triad data: Assessing effects of disease genes that act directly or through maternal effects, and may be subject to parental imprinting. Am J Hum Genet 62: 969–978.

34. Cordell HJ, Barratt BJ, Clayton DG (2004) Case/pseudocontrol analysis in genetic association studies: a unified framework for detection of genotype and haplotype associations, gene-gene and gene-environment interactions and parent-of-origin effects. Genet Epid 26: 167–185.

35. Hager R, Cheverud JM, Wolf JB (2008) Maternal effects as the cause of parent-of-origin dependent effects that mimic genomic imprinting. Genetics 178: 755–1762.

36. Jirtle RL, Skinner MK (2007) Environmental epigenomics and disease susceptibility. Nat Rev Genet 8: 253–262.

37. Churchill GA, Doerge RW (1994) Empirical threshold values for quantitative triat mapping. Genetics 138: 963–971.

38. Dawson E, Abecasis GR, Bumpstead S, Chen Y, et al. (2002) A first-generation linkage disequilibrium map of human chromosome. Nature 418: 544–548.

39. Gabriel SB, Schaffer SF, Nguyen H, Moore JM, et al. (2002) The structure of haplotype blocks in the human genome. Science 296: 2225–2229.

40. Patil N, Berno AJ, Hinds DA, Barrett WA, et al. (2001) Blocks of limited haplotype diversity revealed by high-resolution scanning of human chromosome 21. Science 294: 1719–1723.

41. Zhang K, Deng M, Chen T, Waterman MS, Sun F (2002) A dynamic programming algorithm for haplotype block partitioning. Proc Natl Acad Sci U S A 99: 7335–7339.

42. Wu RL, Lin M (2008) Statistical and Computational Pharmacogenomics. London: Chapman & Hall/CRC.

43. Cheng Y, Berg A, Wu S, Li Y, Wu RL (2009) Computing genetic imprinting expressed by haplotypes. Method Mol Biol 573: 189–212.

44. Wang CG, Cheng Y, Liu T, Li Q, et al. (2008) A computational model for sex-specific genetic architecture of complex traits in humans. Mol Pain 4: 13.

45. Wu S, Yang J, Wang CG, Wu RL (2007) A general quantitative genetic model for haplotyping a complex trait in humans. Curr Genom 8: 343–350.

Chapter 5

LEVERAGING NON-TARGETED METABOLITE PROFILING VIA STATISTICAL GENOMICS

Miaoqing Shen[1,2], Corey D. Broeckling[3], Elly Yiyi Chu[2], Gregory Ziegler[4,5], Ivan R. Baxter[4], Jessica E. Prenni[3], Owen A. Hoekenga[2]

[1] Boyce Thompson Institute for Plant Research, Ithaca, New York, United States of America

[2] United States Department of Agriculture, Agricultural Research Service, RW Holley Center for Agriculture and Health, Ithaca, New York, United States of America

[3] Colorado State University, Proteomics and Metabolomics Facility, Fort Collins, Colorado, United States of America

[4] United States Department of Agriculture, Agricultural Research Service, Plant Genetics Research Unit, St. Louis, Missouri, United States of America

[5] Donald Danforth Plant Science Center, St. Louis, Missouri, United States of America

ABSTRACT

One of the challenges of systems biology is to integrate multiple sources of data in order to build a cohesive view of the system of study. Here we describe the mass spectrometry based profiling of maize kernels, a model system for genomic studies and a cornerstone of the agroeconomy. Using a network analysis, we can include 97.5% of the 8,710 features detected from 210 varieties into a single framework. More conservatively, 47.1% of compounds detected can be organized into a network with 48 distinct modules. Eigenvalues were calculated for each module and then used as inputs for genome-wide association studies. Nineteen modules returned significant results, illustrating the genetic control of biochemical networks within the maize kernel. Our approach leverages the correlations between the genome and metabolome to mutually enhance their annotation and thus enable biological interpretation. This method is applicable to any organism with sufficient bioinformatic resources.

INTRODUCTION

Assembling increasingly large datasets due to the enhanced efficiency of various phenotyping technologies (e.g. metabolomic and proteomic profiling or nucleic acid sequencing) permits increasingly comprehensive views of biological processes. However, the problem of analysis and visualization in systems biology has led some commenters to question how best to "drink from a fire hose" [1]. Statistical methodologies that are highly inclusive can help solve the dual problem of analysis and visualization [2], [3]. Here, we describe the use of weighted correlation network analysis (WGCNA) as a method to integrate mass spectrometry-based simultaneous of the maize kernel, isolated from a diverse panel of inbred maize varieties previously utilized for genome wide association studies of multiple traits [4], [5], [6], [7], [8]. We assert that the use of such a study panel allows us to leverage the genetic and genomic resources already available to enhance our annotation and analysis of mass spectrometry results. Likewise, this approach also improves the annotation of a genome by providing specific metabolites and chemistries to describe the roles of predicted proteins. This approach relies heavily on software written in the R programming language, which should enable wide adoption by the scientific community due to the lack of associated cost [9].

Our choice of study system was deliberate. Maize has incredible genetic and phenotypic diversity, providing an ideal resource for systems biology studies [10]. Variation in plant yield, composition, and morphological traits has been reported in multiple collections of diverse inbred varieties and related biparental mapping populations [4], [5], [6], [7], [8], [11]. Much has also been learned about the structural and genetic variation within the maize genome [12], [13]. The quality of maize grain is a key factor for breeders and other stakeholders, but the development of biomarkers to assist breeding and transgenic crop improvement remains challenging [14],[15], [16], [17]. Metabolomics, relying on the use of mass signals as markers, provides a rapid approach to characterize related varieties and enable the description of existing and novel quality traits [18], [19]. The aim of metabolomics is to provide a comprehensive and quantitative analysis of a vast number of components in a specific biological sample, and identify as many metabolites as possible [20], [21], [22].

Metabolomic analyses of plants can be especially challenging, as plants contain great chemical diversity especially in secondary metabolites [18], [23]. These secondary metabolites help keep plants' systems working properly, play roles in the response to genetic or environmental changes, and have powerful physiological effects in humans or animals [20]. Although mass spectrometry-based metabolomics enables the measurement of hundreds or thousands of compounds from a single complex sample, the plant metabolome is still poorly defined and the identification process for specific compounds remains challenging [24], [25], [26], [27], [28],[29]. However, metabolite profiling is not mutually exclusive of statistical genetic and genomics-based approaches [30]; the combination of systems biology strategies is mutually supportive and beneficial.

Genome-wide association studies (GWAS) consider nucleotide variation patterns, relative to population structure, to identify correlations between particular genomic regions and phenotypes. Often, susceptibility to particular diseases or metabolic syndromes is analyzed using GWAS, as this statistical genomics approach is useful in both humans and model systems [7], [31], [32]. GWAS have been applied to metabolomic datasets, to identify SNPs that may be causally linked to particular biochemical processes or pathways [14], [33], [34],[35], [36]. However, in each of these cases the unit of analysis has been individual metabolites, which does little to improve the efficiency of calculation or maximize the benefit of measuring hundreds or thousands of analytes.

Here, we describe non-targeted metabolite profiling of whole maize kernels. The study of grain quality was approached from the perspective of maize as a foodstuff, thus methanolic extracts were isolated from cooked kernels. Our choice of study panel gave us access to more than 1 million SNPs to support GWAS [12]. We applied WGCNA to organize our data into modules that contained multiple markers that also enabled the identification of networks under genetic regulation. This condensation step allowed us to reduce the complexity of the dataset, addressing the multiple testing problem that is endemic to systems biology, and increase computational efficiency. GWAS were applied to weighted averages for each module (hereafter, module eigenvalues), to identify SNPs associated with collections of biochemical markers. We suggest that the WGCNA procedure does not excessively smooth the data, as SNPs correlated with module eigenvalues were significantly correlated with specific compounds assigned to those modules. Finally, module eigenvalues were used in linear regression models to analyze traits that were recalcitrant to GWAS.

MATERIALS AND METHODS

Materials

HPLC-grade acetonitrile, methanol, and formic acid were purchased from Fisher (Pittsburgh, PA); UPLC HSS C18 column (1.8 μm, 2.1 mm×100 mm), sample vials, UPLC column test mix and leucine enkephalin were purchased from Waters (Milford, MA); and all other reagents were available through Sigma (St. Louis, MO), or as indicated.

Sample Preparation

A maize inbred diversity panel was grown in 2010 on the Musgrave Research Farm of Cornell University (Poplar Ridge, NY)(Flint-Garcia et al., 2005). Duplicated trials were grown using a randomized field design with regular check rows of the B73 accession; 210 of the 282 accessions produced sufficient grain for subsequent analysis, largely due to flowering time issues. Whole maize kernels (n=50) were covered with an equal volume of 18 megaohm water and autoclaved for 30 minutes to fully cook the grain. Samples were then freeze-dried and ground to a fine flour using a consumer-grade grain mill (KoMo Medium Mill, Pleasant Hill Grain, Hampton, NE). Ground, cooked samples were frozen at −20°C until extracted with a 1:1 mixture of water and methanol. After 10 min sonication, extracts were centrifuged for 10 min at 4000 rpm. The supernatant was filtered through 0.45 μm filter. Two independent biological replicates were obtained and analyzed, although only one is discussed here.

UPLC and Mass Spectrometer

Sample injections were performed with an ACQUITY UPLC system (Waters), equipped with a Waters Acquity UPLC HSS C18 column. The samples were injected by means of a 7.5 μL partial loop injection with 3 technical replicates by randomizing all injections. Mobile phase A consisted of 0.1% formic acid in water and mobile phase B contained 0.1% formic acid in acetonitrile. The following gradient was used: 4.5 min 2.4% B, 0.5 min 40% B, 3.5 min 64% B, and 3.5 min 97.6% B. Flow rate was 0.4 ml/min and column temperature was maintained at 40°C. The eluent from the column was delivered to a Xevo G2 TOF (Waters). The mass spectrometer operated in a positive mode using a samples cone voltage of 20 V and a capillary voltage of 2.5 kV with the temperature of source and desolvation at 120°C and 350°C and the flow rate of nitrogen desolvation gas at 850 L/h. Data were acquired in a centroid mode from 50 to1,200 m/z with scan time of 0.2 sec. MS data were collected at a collision energy of 6 V with alternative collection of MSE mode using a ramped collision energy of 20–40 V. Leucine enkephalin was used as the lock mass compound (m/z 556.2771 in positive) and infused at 10 μl/min with a concentration of 1 ng/μl. The lock mass was acquired in all injections of samples to ensure accuracy and reproducibility. The instrument was calibrated using sodium formate at a concentration of 5 mM with mass accuracy within 1 ppm.

Data Transfer and Statistics Approaches

A variety of statistical procedures were employed to analyze data using R (version 2.13.1) or JMP (version 9, SAS Institute, Cary NC). MarkerLynx (v4.1, Waters) was used to integrate and align MS data points and to convert them into exact mass and retention time signals. Principal component analysis (PCA) was performed using Pareto-scaled data on all detected features for initial charactering the separation of maize variables and checking repeatability for technical replicates. The MarkerLynx generated feature list was somewhat larger than that obtained using XCMS, but WGCNA produced highly similar outcomes from both datasets. To use XCMS to identify and annotate features, raw data files were converted to NetCDF format using the Waters DataBridge software [37]. Peak detection and alignment was performed on both the low and high collision energy channels (MS and MS^E) using XCMS software (version 1.22.1 [38]). Reconstruction of indiscriminant MS/MS spectra (idMS/MS) was performed as described [37], with exception that rather than utilizing CAMERA groupings, the grouping was centered around the retention time of the feature of interest, with a 2 second window on either side. Reconstructed spectra presented in this paper are supplied as File S1, an msp format spectral file suitable for viewing using NIST MS search program. A correlational filter was then used to find features that demonstrated similar quantitative patterns to the feature of interest. Reconstructed MS and idMS/MS spectra were exported as an '.msp' formatted spectral library using a custom R script. The library was batch searched against the MassBank database [39], and manually searched against the NIST (http://www.nist.gov/srd/nist1a.cfm) and Metlin databases [40]. Identification confidence scores were assigned as described [37]. For multiply charged peptides, the spectra were manually converted to.mgf format, and the precursor ion was manually interpreted based on the MS spectrum. idMS/MS spectra were searched against the NCBInr protein database using a taxonomy filter for maize (version 07/12/12) (43,920 sequence entries) using the Mascot database search engine (version 2.3). Search parameters were set as follows: monoisotopic mass, parent ion mass tolerance of 0.05 Da, fragment ion mass tolerance of 0.1 Da, no enzyme specificity, and variable modification of oxidation of Met.

The XCMS also fills empty cells based on retention time and mass specific signal, thus reducing the frequency of zero values in the dataset. The XCMS generated data matrix, with an intensity value for each feature and each sample, was used as input for the clustering analysis by WGCNA. Weighted correlation network analysis (WGCNA) was produced with the R package, creating unsigned networks where both positive and negative correlations could be clustered into a single module [2], [3]. WGCNA used autoscaled data in order to reduce the dominance of dynamic, high-concentration metabolites (Table S1). Module eigenvalues for each module were calculated for each of the 210 maize accessions, providing a condensed dataset of derived variables for subsequent genetic analysis. Cytoscape, the open source bioinformatics software, was used to illustrate metabolite networks [41].

Genome-wide Association Study (GWAS)

GAPIT, the genome association and prediction integrated tool (http://www.maizegenetics.net/gapit) was used to perform GWAS and genome prediction. Module eigenvalues from WGCNA were used as maize phenotypic traits in GWAS. The genotypic data are publicly available from panzea.org (http://www.panzea.org/lit/data_sets.html#genos). Previously, the maize diversity panel had been characterized using next-generation sequence analysis such that more than 1 million SNPs across the maize genome are available to characterize the genetic diversity [12]. A kinship matrix was calculated [42] and population structure was modeled as a fixed effect [43]. Significant SNPs (FDR corrected $p<0.001$ and $p<0.05$) were identified for each module eigenvalue with a requirement that the SNP be present with ≥5% allele frequency.

RESULTS

Non-Targeted Metabolite Profiling

Our long-term goal is to characterize phenotypic variation in maize grain quality and to identify the genetic and environmental factors that influence the metabolomic composition of this important staple food and model plant. This effort will provide information to better describe the existing food supply and also project what new grain quality traits may be achievable in the future using conventional plant breeding. Towards this goal, we chose to use mass spectrometry based non-targeted metabolite profiling of maize meal prepared from cooked, whole kernels (Figure 1). While it may be counterintuitive to treat samples in this way, our dataset represents a genetically diverse sample of a foodstuff that could be consumed by either humans or animals. This choice helps to define the range of normal and acceptable variation within a highly diverse crop plant [17]. More than 8,710 metabolomic features were detected from the whole kernel methanolic extracts (Table S1). Principal component analysis (PCA) gave an initial characterization of the profiling results. PCA explains about 22% of the variance with 2 PC's (Figure S1). The performance of PCA for this dataset is typical as the composition varies very widely across genetically distant accessions.

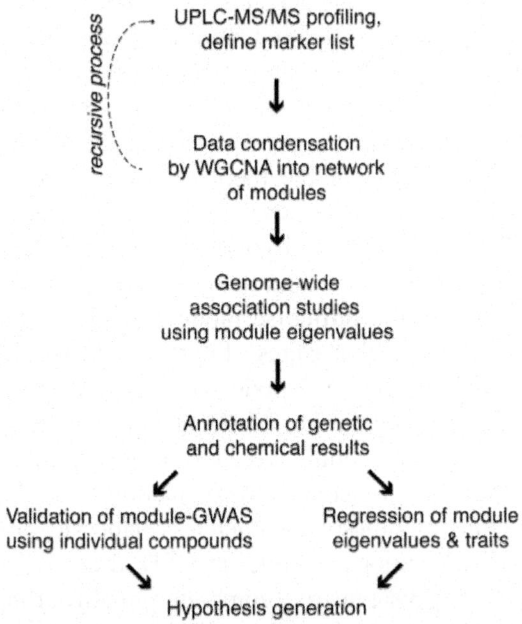

Figure 1: Genomics-assisted chemistry & chemistry-assisted genomics.

This flow chart describes the process by which statistical genetics and genomics can enable metabolite profiling to have greater power and impact.

Data Condensation by Weighted Correlation Network Analysis (WGCNA)

One of the endemic problems of systems biology is the multiple testing problem, wherein the number of variables measured dwarfs observations. One potential solution to this problem is to condense the dataset into a smaller number of distinct groups (hereafter, modules), normalizing the issue of observations and variables. WGCNA is an approach to display model network relationships, identifying co-regulated groups of features (hereafter, nodes) such as patterns of gene expression [2]. It can also be used to visualize metabolite networks and increase the comprehensiveness of non-targeted metabolomics [3]. WGCNA describes the relationships between all of the input variables, summarizing the correlation and connectivity of all nodes. The network can be more or less elaborate, depending on the rules set for inclusion into the network. A principal component is calculated for each module for each variety, summarizing the contribution of all nodes included into a particular module, which is referred to by a randomly assigned color. This principal component (hereafter, module eigenvalue) can be used for correlation tests or ANOVA.

For our dataset, 97.5% of the detected molecular features (nodes) could be included in a network with 56 defined modules (Table S2). The network was then pruned to require that the minimum connectivity between nodes exceed 4 standard deviations (SD) above the mean connectivity observed between all nodes. At this threshold, the network contained 48 modules and 4,102 nodes (47% of nodes, 3.1% of the theoretical connections; Figure 2). The network was redefined under even stricter terms, using a 6SD threshold (Table S2). As the modules were defined by the strength of the correlations among members, modules varied in size and membership according to the inclusion threshold. For example, the turquoise module in the initial description had 2,105 nodes and ~3.73 million edges (Table S2). At the 4SD threshold, the turquoise module reduced to 1,597 nodes with more than 0.62 million edges, while at the 6SD shrinking further to 635 nodes with 40,217 connections. Nodes within the turquoise module were also connected with members of the black module, which likewise contained connections to both the turquoise and purple modules. Other modules were much less elaborate; orange contained 81 nodes in its initial description, 63 nodes at 1 SD, dropping to 9 nodes and 56 connections at 4SD, and disappearing completely at 6SD (Table S2). At the 4SD threshold some modules broke into distinct clusters as connections that helped to define the original module, using the original definitions, dropped below the significance threshold (Figure 2). This facet of the WGCNA procedure represents both a strength and a weakness for the approach. Information can be applied to poorly connected members of a particular module using guilt by association on tightly connected central elements. However, as the module eigenvalues are estimated when the network was initially described, the poorly connected nodes may transmit an excessive degree of variance to these values and perhaps confound downstream applications.

This node and edge projection describes the grain metabolome observed in the methanolic extract from 210 inbred line varieties of maize. This network requires a minimum degree of connectivity between any two nodes (i.e. biochemical markers detected by mass spectrometry) that exceeds four standard deviations above the mean connectivity observed between detected markers. According to this threshold, 4,102 nodes are organized into 48 modules each represented by particular color. However, some modules have separated into multiple, distinct clusters as internal connectivity may fall beneath the 4 standard deviation cutoff, such that there are 101 objects in this projection.

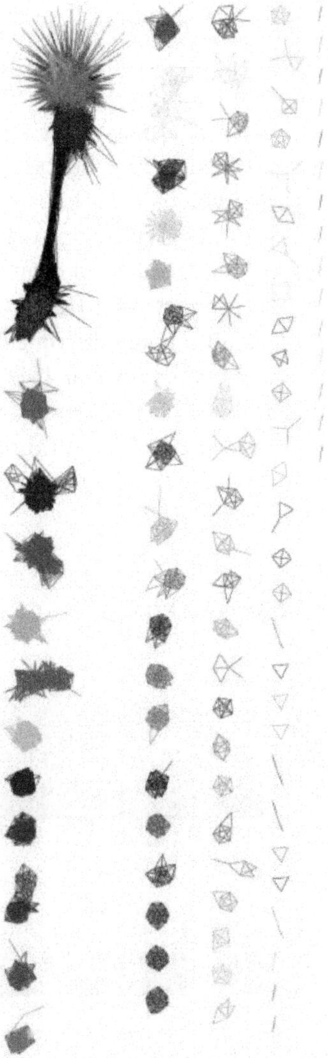

Figure 2: Visualization of maize grain metabolome.

Module Eigenvalues Drive Genome Wide Association Studies

We expected that using WGCNA for the analysis of mass spectrometry based non-targeted metabolite profiling data would accomplish two goals: (1) define the co-regulated networks of metabolites and peptides that contribute to maize kernel quality and composition and (2) reduce the number of variables for downstream analyses. One such analysis is a genome wide association study

(GWAS), to correlate particular genomic regions with phenotypes of interest. This approach has already been applied to maize but not on derived variables such as module eigenvalues, so far as we are aware. And while computational resources are improving, conducting GWAS with a SNP dataset as large as that available for the Buckler Diversity Panel using optimized procedures is still a time intensive procedure (0.5 hr/trait or >150 d for the original data) [12], [44]. Module eigenvalues for all 56 modules were analyzed, 19 of which found significant associations (FDR corrected p-value <0.05; Table S2). Modules that were detected under the most stringent membership conditions (>6SD) were more likely to produce significant GWAS outcomes than those present only under lesser requirements (14 of 27 versus 5 of 21; Table S2). However, modules with fewer connections at 4SD were more likely to identify significantly correlated SNPs with GWAS (χ^2=4.56, p=0.0328). While 4,830 SNPs were identified by GWAS, nearly two-thirds were associated with only two modules (plum2 and salmon). A variety of patterns were observed in the results, ranging from few to many SNPs and wide to narrow distribution across the genome (Figure 3).

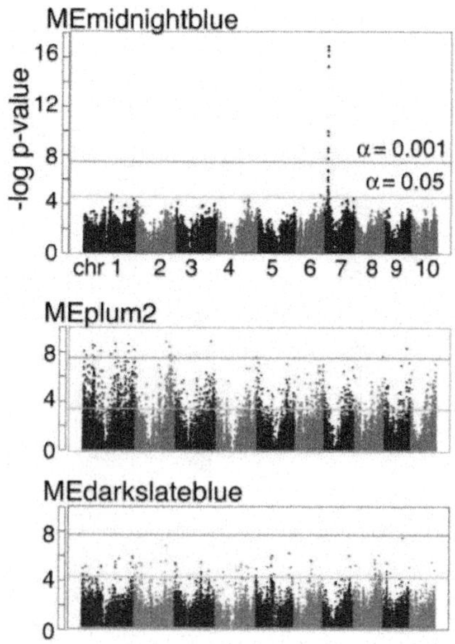

Figure 3: Genome-wide association studies on three module eigenvalues (ME).

Nineteen modules returned significantly correlated SNP markers according to GAPIT. Three are shown here. Significance thresholds were empirically calculated for each trait using GAPIT; FDR-corrected p-values at both a conservative (p<0.001; green line) and generous (p<0.05; aqua line) are displayed. MEmidnightblue identified one region of chromosome 7 with high confidence, with a second region of chromosome 1 with lower confidence. MEplum2 identified multiple genomic intervals with high confidence. MEdarkslateblue identified no significant regions at the conservative threshold, but several regions at the lower threshold.

The strongest associations between SNPs and module eigenvalues were found with the midnightblue module. Nearly all of the significant SNPs were identified at both conservative and relaxed FDR corrected p-value thresholds and were located in a single region of chromosome 7. Most SNPs were identified with variants of the α-zein 19C2 seed storage protein, a result that is supported by analysis of the mass spectrometry data, which are consistent with a C-terminal peptide derived from α-zein 19C2 protein (Figure S2; Table 1; File S1). The plum2 module gave results as one might optimistically hope for, with a variety of genomic regions identified under conservative significance thresholds. SNPs associated with plum2 included those within a putative mitogen-activated protein kinase, which suggested that signal transduction pathway components were detected by GWAS, and also a cytochrome P450 with significant similarity to flavonoid 3-monooxygenases (Table 1). The darkslateblue module returned far fewer SNPs above the $\alpha=0.001$ significance threshold, but many above the 0.05 level. Much like plum2, these SNPs identified a mixture of genes with potential functions while others had no obvious connection to the regulation of maize kernel composition.

Table 1: Sample results from genome-wide association studies.

SNP	−log (p-value)	Module/Marker	Gene Annotation
S7_18857356	17.45	Midnightblue*	α-zein precursor 19C2^{CDS}
S7_18857356	8.74	1056.528_226.57*	α-zein precursor 19C2^{CDS}
S2_160151277	8.05	darkslateblue	α-amylase/protease inhibitor^{CDS}
S9_24144378	7.85	darkslateblue	Protein phosphatase 2A regulatory subunit^{CDS}
S2_184267091	9.59	plum2	Flavonoid 3-monooxygenase$^{6.6\ kb\text{-}5'}$
S1_264986642	9.06	plum2	Mitogen-activated protein kinase^{CDS}
S5_168853373	6.75	orange#	bZIP transcription factor$^{4.4\ kb\text{-}5}$
S5_168853373	9.28	138.092_70.421#	bZIP transcription factor$^{4.4\ kb\text{-}5}$

SNP indicates the chromosome and position (bp) within the maize genome (version 5b.60). The significance of the SNP association is indicated by the negative of the log for the false discovery rate corrected p-value. Module/marker indicates which module eigenvalue the SNP was associated. Single constituents of the midnightblue (*) and orange (#) modules were also analyzed by GWAS. Gene annotation describes the closest gene model relative to the SNP evaluated.

doi:10.1371/journal.pone.0057667.t001

A potential pitfall for the WGCNA procedure as a data condensation tool was the potential to excessively smooth the data, creating a false picture of the genetic regulation of the metabolome. In this instance, collapsing multiple metabolite markers into a single signal might obscure the effect of a particular locus for importance of a module constituent. To address this concern, we examined the orange module in greater detail. At 4SD, orange has 9 nodes, one of which we identified as tyramine (Figure S3). The abundance of tyramine alone was used as a trait for GWAS; this result was compared with the GWAS on the orange module eigenvalue (Figure 4). The orange module returned 27 significant SNPs, 7 of which were also identified as significant for tyramine (Table 1). While GWAS on tyramine returned more significant SNPs than for the orange module, our result does indicate that SNPs associated with a single compound can be identified from GWAS on the module eigenvalue.

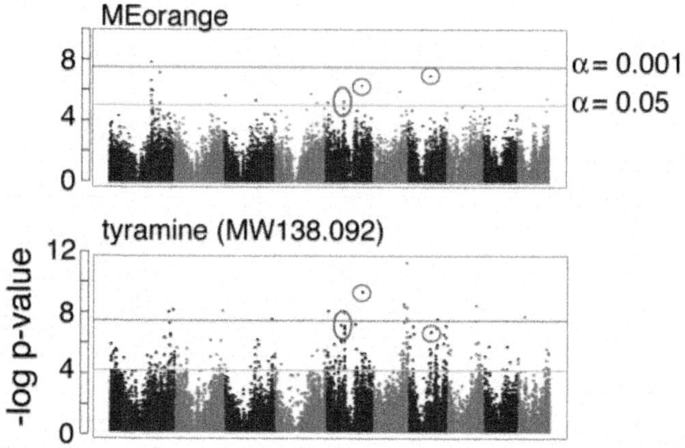

Figure 4: Module eigenvalues do not obscure the importance of single compounds.

MEorange was estimated from 81 molecular features, one of which was identified to be tyramine. GWAS on MEorange identified 27 significant SNPs at the FDR-corrected p<0.05 threshold. GWAS on tyramine alone identified 7 SNPs in common (red circles) with MEorange.

Leveraging Genomic Information to Assist Annotation of Mass Spectrometry Data

Compound identification in mass spectrometry based non-targeted metabolite profiling experiments represents a major challenge of this approach [45]. The utility of indiscriminant MS/MS (idMS/MS) was recently demonstrated to improve the rate and confidence of metabolite identification in non-targeted metabolite profiling experiments [37]. In the current study, the idMS/MS process was applied to selected features from modules with significant GWAS results. As described above, GWAS with the midnightblue module identified a region of maize chromosome 7 consistent with the α–zein 19C2 storage protein (UniProtKB P06677). Sixty-seven features were identified in this module with retention times between 226.35 and 227.06 sec, suggesting that they all represent the same compound. The reconstructed MS and idMS/MS spectra were highly suggestive of a peptide structure due the observation of multiple charge states. The molecular weight of the potential peptide was inferred from the two multiply charged isotope clusters in the MS spectra, and the corresponding idMS/MS spectrum was searched against the maize genome using the Mascot database search engine. This search returned a single peptide as the likeliest match (PAASYQQHIIGGALF), which represents the C-terminus for both the 19C1 and 19C2 variants of the α–zein storage protein. Taken together, these results suggest that *cis*-acting variation at the α-zein locus on chromosome 7 influences the quantitative expression of this protein and this variation is apparent in the cooked maize meal product. We achieved this peptide identification in spite of the fact that our MS data were collected with small molecules in mind and without the benefit of predictable proteolytic cleavage that most proteomic search engines rely upon. Further, we accomplished this from a single separation/MS experiment without the need for a second targeted MS/MS experiment, demonstrating the utility of the idMS/MS workflow. GWAS on this peptide alone returned significant SNPs common with the midnightblue module eigenvalue. However, midnightblue produced a far more significant p-value than the single feature, in contrast to the previous example (Table 1).

WGCNA alone, without the benefit of GWAS, can also assist in the annotation of mass spectra. The orange module contained 9 features with 4 different retention times at the 4SD network threshold, suggesting that there are only a few compounds contained in this module that behave similarly across this maize diversity panel. The first two eluting features demonstrated a strong similarity to tyramine, as described above. The third retention time group contained 3 isotopes of a molecular ion of 284.13, with a fragmentation patterns consistent with *p*-coumaric acid. A dehydration conjugation between tyramine and p-coumaric acid is consistent this molecular ion. The fourth retention time group contained 3 isotopes of the molecular ion 314.14; the idMS/MS patterns suggested tyramine, which lead us to hypothesize an additional tyramine conjugate. The molecular ion for 314.14 is consistent with a ferulic acid-tyramine dehydration conjugate. The final retention time group consisted of a single feature, where the idMS/MS spectra did not show strong matches in public databases. As a whole, the spectral annotations of the orange module members suggest that this cluster is focused on variation in tyramine and at least two of its phenylpropanoid conjugates. The module identification enabled annotation of the mass spectra, as we were able to restrict our decision space based on the data. Likewise, understanding the underlying chemistry should enable our analysis of the GWAS identified SNPs, to clarify how these genes would contribute to the synthesis and modification of tyramine.

Correlation of Module Eigenvalues with Potentially Related Phenotypes

One of the original applications of WGCNA estimated eigenvalues was correlation analysis with related phenotypes, such as using patterns of gene expression to predict disease risk [46]. WGCNA was used to condense large datasets into more computationally manageable ones, with the added advantage of creating new testable hypotheses regarding cause and effect between genetic networks and phenotypic outcomes. Here, we used the WGCNA estimated module eigenvalues for stepwise regression of kernel weight, a commonly studied quality trait. GWAS was applied to kernel weight, but failed to identify any significant SNPs at the α=0.05 significance threshold. The GWAS procedure, as implemented by GAPIT, estimated the heritability of kernel weight at 0.38, which suggested that genetic regulation of kernel weight was highly complex and resulted from the interaction of many genes of very small individual effect.

Table 2: Regression of kernel weight using module eigenvalues

Source	DF	SS	F-ratio	p-value	t Ratio
MEsalmon4^{gwas}	1	30242.822	22.7874	<0.0001	−4.77
MEroyalblue^{gwas}	1	21356.628	16.0918	<0.0001	−4.01
MEdarkslateblue^{gwas}	1	14387.453	10.8407	0.0012	3.29
MEsalmon^{gwas}	1	7384.471	5.564	0.0194	2.36
MEpink^{4SD}	1	36952.627	27.8431	<0.0001	5.28
MEblue^{4SD}	1	20934.891	15.774	0.0001	3.97
MEdarkgrey^{4SD}	1	11517.166	8.678	0.0037	2.95
MEgreen^{1SD}	1	53412.158	40.245	<0.0001	−6.34
MEred^{1SD}	1	31513.736	23.745	<0.0001	4.87
MEsaddlebrown^{1SD}	1	20056.221	15.112	0.0001	3.89
MEwhite^{1SD}	1	11504.719	8.6686	0.0037	−2.94
Error	173	229601.4			
Model	184	539795.71	21.2477	<0.0001	Adj r^2=0.548

Superscripts indicate whether the module eigenvalues gave significant correlations with GWAS and/or was included in the network using the 4SD threshold or the initial description of the network (1SD threshold).
doi:10.1371/journal.pone.0057667.t002

Stepwise regression for kernel weight that included all module eigenvalues reduced to a model using 11 modules and explained more than half of the observed variance (Table 2). These modules included those with significant GWAS associated SNPs, modules that were well defined according to the connectivity rules, and still others that were more diffuse. While the regression model may have overestimated the fraction of variance due to genetic factors (0.548 vs. 0.38), the two analyses were consistent in their overall findings. The eigenvalue regression model summarized the contributions of 11 modules, which conservatively contained at least 600 features (Table S2). Both statistical methods indicated that kernel weight was determined by the interplay of many genetic factors, however the module regression model did quantify the relative input from defined entities and provide a logical framework from which to build additional hypotheses.

DISCUSSION

One of the promises of systems biology is that through the integration of analytical technologies, a more comprehensive and complete view of biological processes can be achieved. Here, we utilized mass spectrometry based non-targeted metabolite profiling to characterize the maize kernel metabolome. We chose to profile cooked maize ground in a consumer-grade grain mill to better understand the variation present in food product that might reasonably be encountered by a consumer, rather than to estimate the maximal genetic potential found in these fractions. We used WGCNA to identify the patterns that help determine composition and quality, and to resolve the multiple testing problem and rebalance the number of observations to variables under analysis. This data condensation step allowed us leverage investments made in maize genetics and genomics to assist the annotation of our mass spectra, through the application of simple (i.e. correlation and regression) and complex (GWAS) statistical procedures. While we chose to profile maize kernels, the statistical and bioinformatic process outlined here is applicable to any biological system with sufficient genetic and genomic investment and should enhance the impact of systems biology approaches in plant, animal and microbial model organisms.

A second promise of systems biology is that of translational genomics, to apply our increasingly deep view of biological processes in more applied contexts and to produce positive outcomes for society. One such application is genomic selection or whole genome prediction, where all available genetic markers are used to predict phenotypes [15]. In a recent example, a SNP microarray was used to genotype a panel of diverse maize varieties that had also been evaluated using metabolite profiling and standard agronomic evaluation [15]. Both genetic and metabolomic markers gave high accuracy predictions of agronomically important traits such as biomass accumulation and flowering time. Additionally, the use of metabolomics allowed a light to shine into the "black box" of genomic selection, where the goal is merely to apply abundant and anonymous genetic markers to predict the phenotype of interest. Through simultaneous genetic and metabolomic profiling, it is possible to correlate micro-scale phenotypes (i.e. glucose) with macro-scale phenotypes (i.e. biomass) while also generating hypotheses to address causality. The methodology we describe here is consistent with that reported by Riedelsheimer and colleagues [14], [15], although the scale of genetic and phenotypic data available to us is considerably larger.

As our understanding of biology is an accretive process, there will always be more data to include in future analyses. One of the limitations of this study is the size of the diversity panel characterized by mass spectrometry. Statistical power increases as a function of the size of the study panel, such that if we had surveyed a larger fraction of the 282 varieties we should have been able to resolve genes with smaller individual function. One of the advantages of our workflow is that it encourages reexamination (Figure 1). As we gain additional phenotypes with the study panel, we can recalculate the network and start the correlation and GWAS over again. As we identify particular metabolite or protein markers, we can apply "guilt by association" to improve the annotation of other members of the same modules and the annotation of the genome itself. SNP detection technologies are also rapidly improving in scale and price, such that repeating the GWAS in a year's time will likely identify new regions of the maize genome that were not adequately covered in the present set of SNPs [12]. Finally, even with incomplete knowledge of the maize genome and inadequate statistical power, we were able to create a logical framework to explain an otherwise recalcitrant trait. We know that the basis of kernel weight is complicated [47], however, we can build testable hypotheses out of the module regression model that can be more fully explored in either larger diversity panels, to repeat GWAS with more power, or to choose biparental mapping populations, to test the effect of particular SNPs with the power advantages that simple mapping populations offer [48].

ACKNOWLEDGMENTS

The authors would like to thank Alex Lipka and Zhiwu Zhang for their input regarding GWAS. The authors would also like to thank Ted Thannhauser and Kevin Howe for technical assistance with liquid chromatography and mass spectrometry, Meghan den Bakker, Jon Hart, and the staff of the Musgrave Research Farm for assistance with the field related portion of the workflow, Paul Armstrong for building the weighing robot, and Matthew DiLeo for his original implementation of WGCNA to support metabolomics datasets. The U.S. Department of Agriculture (USDA) prohibits discrimination in all its programs and activities on the basis of race, color, national origin, age, disability, and where applicable, sex, marital status, familial status, parental status, religion, sexual orientation, genetic information, political beliefs, reprisal, or because all or part of an individual's income is derived from any public assistance program. (Not all prohibited bases apply to all programs.) Persons with disabilities who require alternative means for communication of program information (Braille, large print, audiotape, etc.) should contact USDA's TARGET Center at (202) 720–2600 (voice and TDD). To file a complaint of discrimination, write to USDA, Director, Office of Civil Rights, 1400 Independence Avenue, S.W., Washington, D.C. 20250-9410, or call (800) 795–3272 (voice) or (202) 720–6382 (TDD). USDA is an equal opportunity provider and employer.

AUTHOR CONTRIBUTIONS

Conceived and designed the experiments: OAH. Performed the experiments: MS GZ. Analyzed the data: MS CB EYC GZ IRB JEP OAH. Contributed reagents/materials/analysis tools: OAH CB IRB. Wrote the paper: MS CB JEP OAH

REFERENCES

1. Hunter DJ, Kraft P (2007) Drinking from the Fire Hose: Statistical Issues in Genomewide Association Studies. New England Journal of Medicine 357: 437–439. doi: 10.1056/nejmp078120
2. Langfelder P, Horvath S (2008) WGCNA: an R package for weighted correlation network analysis. BMC Bioinformatics 9: 559. doi: 10.1186/1471-2105-9-559
3. DiLeo MV, Strahan GD, den Bakker M, Hoekenga OA (2011) Weighted correlation network analysis (WGCNA) applied to the tomato fruit metabolome. PLoS ONE 6: e26683. doi: 10.1371/journal.pone.0026683
4. Hansey CN, Johnson JM, Sekhon RS, Kaeppler SM, de Leon N (2011) Genetic diversity of a maize association population with restricted phenology. Crop Science 51: 704–715. doi: 10.2135/cropsci2010.03.0178
5. Yan J, Shah T, Warburton ML, Buckler ES, McMullen MD, et al. (2009) Genetic characterization and linkage disequilibrium estimation of a global maize collection using SNP markers. PLoS ONE 4: e8451. doi: 10.1371/journal.pone.0008451

6. Buckler ES, Holland JB, Bradbury PJ, Acharya CB, Brown PJ, et al. (2009) The genetic architecture of maize flowering time. Science 325: 714–718. doi: 10.1126/science.1174276
7. Poland JA, Bradbury PJ, Buckler ES, Nelson RJ (2011) Genome-wide nested association mapping of quantitative resistance to northern leaf blight in maize. Proceedings of the National Academy of Sciences of the United States of America 108: 6893–6898. doi: 10.1073/pnas.1010894108
8. Kump KL, Bradbury PJ, Wisser RJ, Buckler ES, Belcher AR, et al. (2011) Genome-wide association study of quantitative resistance to southern leaf blight in the maize nested association mapping population. Nature Genetics 43: 163–168. doi: 10.1038/ng.747
9. R Development Team (2009) R: A Language and Environment for Statistical Computing. Vienna, Austria: R Foundation for Statistical Computing.
10. Buckler ES, Gaut BS, McMullen MD (2006) Molecular and functional diversity of maize. Current Opinion In Plant Biology 9: 172–176. doi: 10.1016/j.pbi.2006.01.013
11. Cook JP, McMullen MD, Holland JB, Tian F, Bradbury P, et al. (2012) Genetic architecture of maize kernel composition in the nested association mapping and inbred association panels. Plant Physiology 158: 824–834. doi: 10.1104/pp.111.185033
12. Chia JM, Song C, Bradbury PJ, Costich D, de Leon N, et al. (2012) Maize HapMap2 identifies extant variation from a genome in flux. Nature Genetics 44: 803–807. doi: 10.1038/ng.2313
13. Schnable PS, Ware D, Fulton RS, Stein JC, Wei F, et al. (2009) The B73 maize genome: complexity, diversity, and dynamics. Science 326: 1112–1115.
14. Riedelsheimer C, Lisec J, Czedik-Eysenberg A, Sulpice R, Flis A, et al. (2012) Genome-wide association mapping of leaf metabolic profiles for dissecting complex traits in maize. Proceedings of the National Academy of Sciences of the United States of America 109: 8872–8877. doi: 10.1073/pnas.1120813109
15. Riedelsheimer C, Czedik-Eysenberg A, Grieder C, Lisec J, Technow F, et al. (2012) Genomic and metabolic prediction of complex heterotic traits in hybrid maize. Nature Genetics 44: 217–220. doi: 10.1038/ng.1033
16. Cellini F, Chesson A, Colquhoun I, Constable A, Davies HV, et al. (2004) Unintended effects and their detection in genetically modified crops. Food and Chemical Toxicology 42: 1089–1125. doi: 10.1016/j.fct.2004.02.003

17. Hoekenga OA (2008) Using metabolomics to estimate unintended effects in transgenic crop plants: problems, promises, and opportunities. Journal of Biomolecular Techniques 19: 159–166.
18. Weckwerth W (2003) Metabolomics in systems biology. Annual Review Of Plant Biology 54: 669–689. doi: 10.1146/annurev.arplant.54.031902
19. Fernie AR, Schauer N (2009) Metabolomics-assisted breeding: a viable option for crop improvement? Trends in Genetics: TIG 25: 39–48. doi: 10.1016/j.tig.2008.10.010
20. Last RL, Jones AD, Shachar-Hill Y (2007) Towards the plant metabolome and beyond. Nature reviews Molecular Cell Biology 8: 167–174. doi: 10.1038/nrm2098
21. Dettmer K, Aronov PA, Hammock BD (2007) Mass spectrometry-based metabolomics. Mass spectrometry reviews 26: 51–78. doi: 10.1002/mas.20108
22. Yanes O, Tautenhahn R, Patti GJ, Siuzdak G (2011) Expanding coverage of the metabolome for global metabolite profiling. Analytical Chemistry 83: 2152–2161. doi: 10.1021/ac102981k
23. Dixon RA, Strack D (2003) Phytochemistry meets genome analysis, and beyond. Phytochemistry 62: 815–816. doi: 10.1016/s0031-9422(02)00712-4
24. Tolstikov VV, Fiehn O (2002) Analysis of highly polar compounds of plant origin: combination of hydrophilic interaction chromatography and electrospray ion trap mass spectrometry. Analytical Biochemistry 301: 298–307. doi: 10.1006/abio.2001.5513
25. Huhman DV, Sumner LW (2002) Metabolic profiling of saponins in Medicago sativa and Medicago truncatula using HPLC coupled to an electrospray ion-trap mass spectrometer. Phytochemistry 59: 347–360. doi: 10.1016/s0031-9422(01)00432-0
26. Lei Z, Huhman DV, Sumner LW (2011) Mass spectrometry strategies in metabolomics. The Journal of Biological Chemistry 286: 25435–25442. doi: 10.1074/jbc.r111.238691
27. Bolleddula J, Fitch W, Vareed SK, Nair MG (2012) Identification of metabolites in Withania sominfera fruits by liquid chromatography and high-resolution mass spectrometry. Rapid communications in mass spectrometry 26: 1277–1290. doi: 10.1002/rcm.6221
28. Fiehn O, Wohlgemuth G, Scholz M, Kind T, Lee do Y, et al. (2008) Quality control for plant metabolomics: reporting MSI-compliant studies. Plant Journal 53: 691–704. doi: 10.1111/j.1365-313x.2007.03387.x

29. Oliver MJ, Guo L, Alexander DC, Ryals JA, Wone BW, et al. (2011) A sister group contrast using untargeted global metabolomic analysis delineates the biochemical regulation underlying desiccation tolerance in Sporobolus stapfianus. Plant Cell 23: 1231–1248. doi: 10.1105/tpc.110.082800
30. Zhang N, Gibon Y, Gur A, Chen C, Lepak N, et al. (2010) Fine quantitative trait loci mapping of carbon and nitrogen metabolism enzyme activities and seedling biomass in the maize IBM mapping population. Plant Physiology 154: 1753–1765. doi: 10.1104/pp.110.165787
31. Fong C, Ko DC, Wasnick M, Radey M, Miller SI, et al. (2010) GWAS analyzer: integrating genotype, phenotype and public annotation data for genome-wide association study analysis. Bioinformatics 26: 560–564. doi: 10.1093/bioinformatics/btp714
32. Zeggini E, Scott LJ, Saxena R, Voight BF, Marchini JL, et al. (2008) Meta-analysis of genome-wide association data and large-scale replication identifies additional susceptibility loci for type 2 diabetes. Nature Genetics 40: 638–645. doi: 10.1038/ng.120
33. Gieger C, Geistlinger L, Altmaier E, Hrabe de Angelis M, Kronenberg F, et al. (2008) Genetics meets metabolomics: a genome-wide association study of metabolite profiles in human serum. PLoS Genetics 4: e1000282. doi: 10.1371/journal.pgen.1000282
34. McCarthy MI, Abecasis GR, Cardon LR, Goldstein DB, Little J, et al. (2008) Genome-wide association studies for complex traits: consensus, uncertainty and challenges. Nature Reviews Genetics 9: 356–369. doi: 10.1038/nrg2344
35. Inouye M, Ripatti S, Kettunen J, Lyytikainen LP, Oksala N, et al. (2012) Novel Loci for metabolic networks and multi-tissue expression studies reveal genes for atherosclerosis. PLoS Genetics 8: e1002907. doi: 10.1371/journal.pgen.1002907
36. Kettunen J, Tukiainen T, Sarin AP, Ortega-Alonso A, Tikkanen E, et al. (2012) Genome-wide association study identifies multiple loci influencing human serum metabolite levels. Nature Genetics 44: 269–276. doi: 10.1038/ng.1073
37. Broeckling CD, Heuberger AL, Prince JA, Ingelsson E, Prenni JE (2013) Assigning precursor-product ion relationships in indiscriminant MS/MS data from non-targeted metabolite profiling studies. Metabolomics 9: 33–43. doi: 10.1007/s11306-012-0426-4

38. Smith CA, Want EJ, O'Maille G, Abagyan R, Siuzdak G (2006) XCMS: processing mass spectrometry data for metabolite profiling using nonlinear peak alignment, matching, and identification. Analytical Chemistry 78: 779–787. doi: 10.1021/ac051437y

39. Horai H, Arita M, Kanaya S, Nihei Y, Ikeda T, et al. (2010) MassBank: a public repository for sharing mass spectral data for life sciences. Journal of mass spectrometry 45: 703–714. doi: 10.1002/jms.1777

40. Sana TR, Roark JC, Li X, Waddell K, Fischer SM (2008) Molecular formula and METLIN Personal Metabolite Database matching applied to the identification of compounds generated by LC/TOF-MS. Journal of biomolecular techniques 19: 258–266.

41. Kohl M, Wiese S, Warscheid B (2011) Cytoscape: software for visualization and analysis of biological networks. Methods in Molecular Biology 696: 291–303. doi: 10.1007/978-1-60761-987-1_18

42. Loiselle BA, Sork VL, Nason J, Graham C (1995) Spatial genetic-structure of a tropical understory shrub, *Psychotria officinalis (*Rubiaceae). American Journal of Botany 82: 1420–1425. doi: 10.2307/2445869

43. Pritchard JK, Stephens M, Rosenberg NA, Donnelly P (2000) Association mapping in structured populations. American Journal Of Human Genetics 67: 170–181. doi: 10.1086/302959

44. Lipka AE, Tian F, Wang Q, Peiffer J, Li M, et al.. (2012) GAPIT: Genome Association and Prediction Integrated Tool. Bioinformatics.

45. **45.**Patterson SD (2003) Data analysis–the Achilles heel of proteomics. Nature Biotechnology 21: 221–222. doi: 10.1038/nbt0303-221

46. Plaisier CL, Horvath S, Huertas-Vazquez A, Cruz-Bautista I, Herrera MF, et al. (2009) A systems genetics approach implicates USF1, FADS3, and other causal candidate genes for familial combined hyperlipidemia. PLoS Genetics 5: e1000642. doi: 10.1371/journal.pgen.1000642

47. Sabelli PA, Larkins BA (2009) The development of endosperm in grasses. Plant Physiology 149: 14–26. doi: 10.1104/pp.108.129437

48. Flint-Garcia SA, Thuillet AC, Yu J, Pressoir G, Romero SM, et al. (2005) Maize association population: a high-resolution platform for quantitative trait locus dissection. Plant Journal 44: 1054–1064. doi: 10.1111/j.1365-313x.2005.02591.x

Chapter 6

GENETIC GEOSTATISTICAL FRAMEWORK FOR SPATIAL ANALYSIS OF FINE-SCALE GENETIC HETEROGENEITY IN MODERN POPULATIONS: RESULTS FROM THE KORA STUDY

A. N. Diaz-Lacava[1,2,3] M. Walier[1] D. Holler[1] M. Steffens[1] C. Gieger[4,5] C. Furlanello[6] C. Lamina[7] H. E. Wichmann[8,9,10] and T. Becker[1,11]

[1]Institute for Medical Biometry, Informatics, and Epidemiology, University of Bonn, 53127 Bonn, Germany

[2]Cologne Center for Genomics, University of Cologne, 50931 Cologne, Germany

[3]DNA Analysis Unit, Official College of Pharmacists and Biochemists, C1184ABA Buenos Aires, Argentina

[4]Research Unit of Molecular Epidemiology, Helmholtz Zentrum München, German Research Center for Environmental Health, 85764 Neuherberg, Germany

[5]Institute of Epidemiology II, Helmholtz Zentrum München, German Research Center for Environmental Health, 85764 Neuherberg, Germany

[6]FBK, 38122 Trento, Italy

[7]Division of Genetic Epidemiology, Department of Medical Genetics, Molecular and Clinical Pharmacology, Medical University of Innsbruck, 6020 Innsbruck, Austria

[8]Institute of Medical Informatics, Biometry and Epidemiology, Chair of Epidemiology, Ludwig-Maximilians-University, 81377 Munich, Germany

[9]Institute of Epidemiology I, Helmholtz Zentrum München, German Research Center for Environmental Health, 85764 Neuherberg, Germany

[10]Institute of Medical Statistics and Epidemiology, Technical University Munich, 81675 Munich, Germany

[11]German Center for Neurodegenerative Diseases (DZNE), 53127 Bonn, Germany

ABSTRACT

Aiming to investigate fine-scale patterns of genetic heterogeneity in modern humans from a geographic perspective, a genetic geostatistical approach framed within a geographic information system is presented. A sample collected for prospective studies in a small area of southern Germany was analyzed. None indication of genetic heterogeneity was detected in previous analysis. Socio-demographic and genotypic data of German citizens were analyzed (212 SNPs; $n = 728$). Genetic heterogeneity was evaluated with observed heterozygosity (H_0). Best-fitting spatial autoregressive models were identified, using socio-demographic variables as covariates. Spatial analysis included surface interpolation and geostatistics of observed and predicted patterns. Prediction accuracy was quantified. Spatial autocorrelation was detected for both socio-demographic and genetic variables. Augsburg City and eastern suburban areas showed higher H_0 values. The selected model gave best predictions in suburban areas. Fine-scale patterns of genetic heterogeneity were observed. In accordance to literature, more urbanized areas showed higher levels of admixture. This approach showed efficacy for detecting and analyzing subtle patterns of genetic heterogeneity within small areas. It is scalable in number of loci, even up to whole-genome analysis. It may be suggested that this approach may be applicable to investigate the underlying genetic history that is, at least partially, embedded in geographic data.

INTRODUCTION

Accurate assessment of genetic heterogeneity is relevant to manifold fields, ranging from clinical research, pharmacogenetics, and statistical genetics, over forensic sciences up to evolution (for a review, cf. [1]). In planning genetic epidemiological studies or the collection of control cohorts for prospective studies it is crucial to prevent confounding effects due to undetected or disregarded population structure [2, 3]. Population-based association studies of unrelated individuals, involving case-control and cohort studies, are prone to population structure, which may lead to false positive results or to failure to reveal genuine associations [4, 5]. In family-based linkage analysis unknown population stratification may lower statistical power [6].

Uncovering the genetic basis of complex traits remains an immense and urgent challenge in genetic epidemiological research. Great efforts are set to establish well-designed cohorts and large control samples, intended to serve as basis for genetic epidemiological studies. Besides restricting recruitment to individuals of uniform ancestry, a common strategy applied to efficiently gain a representative sample of the inspected population and to control for potential unknown population substructure is to collect samples in smaller geographical areas, usually in medium to large urban centers (e.g., [7–9]).

Even well-characterized or supposedly homogeneous regions may still account for subtle genetic structure with potential geographical components [3, 10]. Sloan et al. [3] shortly reviewed studies related to geographic genetic structure of human populations and pointed out clear lack of research focusing on genetic heterogeneity of smaller geographic regions or those focused on more urban, highly admixed populations.

Most available well-standardized methods in geographical genetics [11] were developed for other research areas and may not be suited for assessing subtle genetic heterogeneity of modern populations inhabiting geographically restricted areas. Modern humans account per se for the lowest species genetic diversity among primates [12]. A typical western population inhabiting a geographically restricted area sets additional difficulties. Such populations are typically outbred and account for a large degree of admixture, product of older and recent regional, interregional, and even international migration. It is to expect that genetic evolutionary forces, such as selection, mutation, drift, or barriers to gene flow, would play a relatively insignificant role in modeling fine-scale variation of genetic heterogeneity. At this geographical scale, it is more likely that neighborhood preferences and modern mating behavior would have a central role in modeling recent admixture, consequently, having strong influence on the observed pattern of genetic variation of modern small areas (i.e., [13]). In other words, within modern western circumscribed areas, socio-demographic factors would probably explain a large proportion of the observed pattern of genetic heterogeneity.

With the aim of unveiling modest amounts of population substructure in a small, admixed area we (a) searched for subtle patters of genetic heterogeneity and (b) explored potential predictors of the observed patterns. To this end, we combined statistical genetics with spatial statistics (geostatistics) within the framework of a Geographic Information System (GIS). A GIS provides a computational environment designed for spatial analysis of geographic data, therefore the most suitable framework to detect, to model, and to analyze the geographic variation of genetic diversity.

We analyzed a well-characterized cohort collected for prospective studies in a small area of southern Germany. The sampling area included the middle-size city of Augsburg, the surrounding suburban area, and the neighboring countryside. As previously reported by Steffens et al. [14], the KORA S4 sample shows a minimal but measurable increase of the inbreeding factor ($8.4E - 5\%$ heterozygotes deficit) measured in terms of F_{IS} values [15], that is, withingroup deviation from expected heterozygosity but no indication of population substructure. Despite extensive search of potential population stratification with the software package STRUCTURE [16], in this cohort no signals could be detected [14]. The STRUCTURE program implements a model-based clustering method. It estimates the proportion of individuals' genome that may originate from differential populations, the probability that an individual belongs to a certain population as well as allele frequency differences in terms of Wright's F_{ST} statistics [17]. Regardless of intensive computations under several models, STRUCTURE results did not provide any indication of a potential pattern of genetic heterogeneity in the KORA S4 survey [14].

MATERIAL AND METHODS

Subjects and Genotypes

Our analysis is based on a subset of the KORA cohort (Kooperative Gesundheitsforschung in der Region Augsburg; in English: Cooperative Health Research in the Region of Augsburg; [7, 9]). The KORA survey is an ongoing study, which takes place in a circumscribed region of southern Germany: Augsburg City and the two neighboring districts. The KORA cohort was recruited for prospective studies. In 4 surveys (S1–S4), a total of 18,000 participants were randomly selected from the adult population of resident German citizens (25–74 years) [7, 9]. Phenotypic, socio-economic information, and residence locality were gathered. The KORA cohort is a sample of the extant German population in the region.

The analysis was conducted on a random set of the KORA S4 survey (n = 4261), recruited in the period between the years 1999 and 2001 [7, 9]. The data set consisted of 728 unrelated healthy German citizens, which included subjects born within and outside of Germany as well. The graphical method GRR (Graphical Representation of Relationships; [18]) was used to exclude the presence of biological relationship of individuals based on genetic data (see Supplement 7). In this paper we distinguished the portion of Germans citizens born outside of Germany as "immigrants" and those born in Germany as "natives." The immigrant group (n = 179) included subjects born in twenty worldwide distributed countries, half of these countries represented only once. Four countries, Czech Republic, Romania, Poland, and Ukraine, corresponded to the land of birth of 82 percent of all immigrants. The group of subjects born in one of these four countries was classified as "major immigrant group." Based on the information "land of birth" we differentiated between data sets: (a) ALL: the complete data set of 728 subjects (resident German citizens),

(b) GER: the total set of 549 natives, and (c) MAIN IMG: the subset of 146 immigrants, born in either Czech Republic, Romania, Poland, or Ukraine

The KORA S4 sample was genotyped for 212 single nucleotide polymorphisms (SNPs) (Supplementary Table S6 in supplementary material available online at http://dx.doi.org/10.1155/2015/693193) [14]. These SNPs can be differentiated in two sets. The first set includes 68 coding SNPs located in exons of functional genes. These SNPs either cause an amino acid exchange or an effective promotor alteration in respect to the resulting protein. Assuming evolutionary times these SNPs may be subject to selective forces. The second set comprises 144 neutral SNPs. These loci were chosen at random throughout the genome in putative "genomic deserts," pursuing to achieve uniform distribution across the genome (setting a minimum of 500 Kbp intermarker distance). For this selection only SNPs presenting a minor allele frequency between 10 and 50% in Caucasians were considered. The markers included in the final intergenic set were uniformly spaced and located >100 Kbp apart from any known genes and >1 Mbp apart from centromeres and telomeres. This procedure followed the set of rules proposed by Devlin and Roeder for genomic control markers [19]. Accordingly, these intergenic SNPs are assumed to be neutral to selection forces in the absence of any specific information. In this sense, these loci are expected to reflect the effects of demographic processes involving migration (gene flow) and even drift, if evolutionary times are considered. Steffens et al. [14] undertook an extensive quality assessment to this data set. The averaged call rate over all samples was 97.3%; intragenic SNPs achieved an average call rate of 96.2% and intergenic SNPs, an average call rate of 97.9% [14]. Details of the genetic properties of the full set of 212 SNPs are listed in Supplementary Table S6.

Study Area

The study area comprised three administrative regions: the municipality of Augsburg City and its two neighboring districts, Aichach-Friedberg District and Augsburg District (Figure 1). It covered an area of approximately 2,970 km^2. This is a surface comparable with the Grand Duchy of Luxembourg (Figure 1). The area is located in the Swabia administrative region of Bavaria, southern Germany, between the coordinates 10.491°E/48.091°N and 11.310°E/48.642°N. The population had approximately 630,000 members in 2004. The mean population density is 212 inhabitants/km^2, a figure that is comparable with the German average.

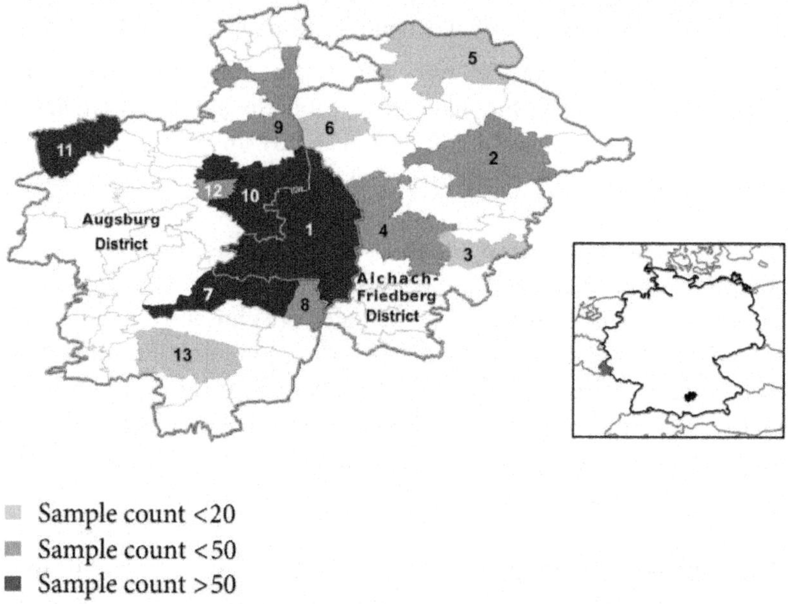

Figure 1: Study area. Land units: (1) Augsburg; (2) Aichach; (3) Eurasburg; (4) Friedberg; (5) Pottmes; (6) Rehling; (7) Bobin- ¨ gen; (8) Konigsbrunn; (9) Langweid (Meitingen); (10) Neus ¨ aß¨ (Gersthofen; Stadtbergen); (11) Altenmunster; (12) Aystetten; (13) ¨ Schwabmunchen. Location of the study area within Germany is ¨ indicated in the inset in black; German boundaries are displayed with a black line, neighboring-country boundaries are displayed with a gray line; Luxembourg, a country covering an extension similar to the study area, located on the western boundary of Germany, is displayed in gray.

Augsburg City is a typical middle-size German urban area. The Aichach-Friedberg and Augsburg Districts include a suburban area neighboring Augsburg City and a periurban area, a patchy pattern of smaller cities and villages widespread across a rural landscape.

Regionalization Methods

The spatial analysis required diverse types of regionalization of the study area. Three regionalization methods were applied: (a) a subdivision of the total study area into minimal representative spatial units of analysis; (b) a subdivision of the total study area into contiguous sampled units using a polygon-based method; and (c) modification of the first regionalization in order to achieve a set of contiguous spatial analytical units while retaining the original geometry defined in (a).

Land Units

Genetic landscapes, in this work referred to matricial representations of genetic variation in the geographic space, were created with geostatistic methods of surface interpolation (see Section 2.7.1). For this objective, it is convenient to define a minimal spatial unit of analysis which is representative of data coverage and it covers a spatial surface much smaller that the phenomenon of interest.

The basic spatial unit of analysis of the genetic landscapes was the postal area. The postal area corresponds to the smallest district or region defined by the German postal system (the German postal system divides Germany in ca 28.700 postal areas). We considered this an appropriate analytical area because the German postal system divides the territory into spatial units with a similar number of inhabitants, independent of the extension of the spatial unit. Similar population size among land units allows adequate comparisons from socio-demographic perspective. Postal areas include as well a population size large enough in order to guarantee subjects' anonymity. A finer geographical reference of subjects, that is, postal address, was not available and it would not be in agreement with local official restrictions in respect to personal anonymity. We considered the subdivision of the study area into postal areas adequate to identify and to analyze fine-scale patterns of genetic variation. The study area included a total of 64 postal areas. The spatial extension of the postal areas ranged from 1.8 km^2 to 93 km^2, with an average of 26 km^2.

The sampled area covered about 20% of the total study area, that is, approximately 600 km^2 (Figure 1). It included Augsburg City and 15 settlements located in AichachFriedberg District and Augsburg District. Each sampled settlement corresponded to one postal area, except for Augsburg City. Augsburg City itself contains 14 postal areas. In summary, out of a total of 64 postal areas, data was available in Augsburg City (including 14 postal areas) and in another 15 postal areas. Augsburg City samples were pooled together for frequency computations, since no information about postal area of residence was available for residents in this city. A subdivision of Augsburg City into postal areas was only considered in the step of spatial interpolation to improve interpolation results (see Section 2.7.1). Postal areas with a very low number of samples were aggregated to neighboring sampled areas in order to exclude bias due to low number of samples per land unit. Explicitly, the quarters Stadtbergen (n=7) and Gersthofen (n = 6) were aggregated to Neusaß¨ (n = 47); Meitingen (n = 12) was aggregated to Langweid (n = 22) (Figure 1). In the final geostatistical analysis the sampled area included 13 analytical land units. In this work analytical land units (areal representing sampled data) defined on the basis of the geographical coverage of postal areas are further referred to as land unit (LU). LUs were labeled with the sampling-location name; aggregated LUs were labeled with the name of the location accounting for the largest number of samples. Augsburg City included the maximum number of samples (n = 359). The remaining 15 sampled postal areas (aggregated into 12 LUs) included a total of 369 samples. Letting aside Augsburg City, sample size per LU ranged between n=9 samples (Eurasburg) and n = 60 samples (Neusaß). The mean sample size per LU was 30.8 and ¨ the standard deviation was 17.4 (Table 1).

Table 1: Description of LUs and sampling locations, total count of samples, natives, immigrants, MAIN_IMG (subset of German citizens born in either Czech Republic, Romania, Poland, or Ukraine), and values of MAIN_IMP (percentage of individuals out of the total count of immigrants per land unit corresponding to the major immigrant group). Sampling locations with a low number of samples, which were aggregated to a contiguous sampled unit, are indicated in square brackets.

LU-ID	LU/sampling location	All (n)	Natives (n)	Immigrants (n)	MAIN_IMG (n)	MAIN_IMP (%)
1	Augsburg	359	258	101	79	78
2	Aichach	23	18	5	3	60
3	Eurasburg	9	9	0	0	0
4	Friedberg	25	21	4	2	50
5	Pöttmes	12	12	0	0	0
6	Rehling	13	12	1	1	100
7	Bobingen	51	36	15	13	87
8	Königsbrunn	42	24	18	17	94
9	Langweid [Meitingen]	34	22	12	9	75
10	Neusäß [Gersthofen] [Stadtbergen]	60	52	8	8	100
11	Altenmünster	53	48	5	5	100
12	Aystetten	31	23	8	7	88
13	Schwabmünchen	16	14	2	2	100

Polygon-Based Regionalization

The implementation of the spatial autocorrelation tests performed in this study (see Section 2.7.2) required to count with a set of adjacent spatial analytical units. This means that only spatial units with at least one contiguous neighbor could be included in the analysis.

The total study area was divided into 13 Thiessen polygons (designation given to Voronoi diagrams used to analyze spatially distributed data) [20]. Each polygon corresponded to one LU defined in the first regionalization (see Section 2.3.1). We chose this simple type of regionalization since many natural patterns may be closely approximated to this type of areal structure. Thiessen polygons were delimited based on the centroids (polygon geometrical center) of the 13 sampled land units (Figure 2(a)). The Voronoi tessellation was created with the method v.voronoi of the open-source software package GRASS 6.4 (Geographic Resources Analysis Support System, http://grass.osgeo.org/).

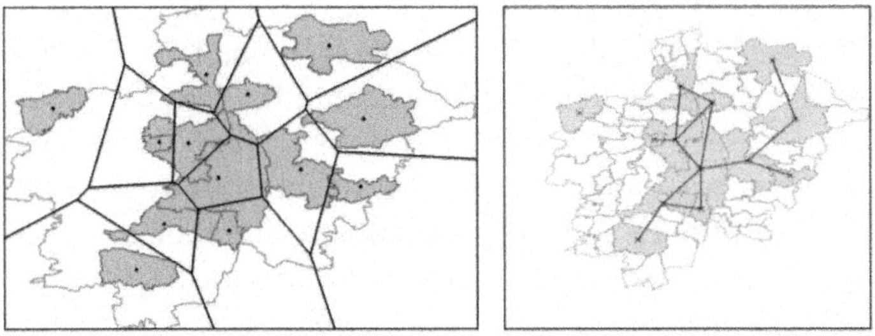

(a) (b)

Figure 2: (a) Regionalization of the study area in 13 Thiessen polygons; each polygon represents a LU. LU centroids were used to delimit the Thiessen polygons; (b) net of contiguous LUs; three LUs were spatially connected to the next sampled LU performing a geometrical correction of LUs' boundaries: Schwabmunchen was connected to Bobingen, Aichach to Friedberg, and P ¨ ottmes to Aichach; vectors show pairs of LU ¨ assigned a spatial weight equal to unity in the matrix of spatial weights; pairs of LUs not connected with vectors received a value equal to zero in this matrix.

Net of Contiguous Sampled Units

The implementation of the algorithms used in this study to search for best predictors of spatial variation fitting the data (see Section 2.7.3) required as well contiguous analytical units. For such more complex analysis, the coverage of each LU was retained. In this case the first step was to verify the presence of direct neighbors for all LUs. Four LUs did not account for contiguous neighbors: Aichach, Pöttmes, Schwabmünchen, and Altenmünster (Figure 1). Of these, the first four LUs were not further than 5 km away from the next closest LU border. This geographical distance was considered negligible in the context of connectivity and human interaction between modern settlements. In order to get maximal information of the available data we modified slightly the geometry of these four LUs and of their closest neighbors in order to meet the contiguity condition for at least 12 LUs. Schwabmünchen was connected to Bobingen, Aichach to Friedberg, and Pöttmes to Aichach (Figure 2(b)). With this step, 12 LUs conformed a continuous geographical space. The most peripheral sampled land unit, Altenmünster, without a close sampled contiguous neighbors (distance to the closest LU > 10 km), was not included in the computations (Figure 2(b)).

Matrix of Spatial Weights

Both implementations of spatial dependence analysis performed in this study required the definition of a matrix of spatial weights representing the interaction between LUs (see Sections 2.7.2 and 2.7.3).

On the basis of the previously defined regionalization, either the Thiessen polygons (Figure 2(a)) or the geometrically modified LUs, for each analytical unit the geographical central point (the centroid), were specified. These two tests were performed in this study based on a binary representation of the spatial weight matrix, which assigned a weight of unity for neighbors, and zero otherwise. A binary encoding was chosen since not enough information was available to set assumptions about the assumed spatial process. The function poly2nb was used to construct the neighbor list with default parameters and the function nb2listw to construct the weight matrix, setting the function parameter style = B for a binary system. Pairs of LUs assigned a spatial weight equal to unity are indicated in Figure 2 with a vector net. All other pairwise combinations of LUs received a spatial weight equal to zero.

Socio-Demographic Parameters

Socio-demographic information collected during recruitment included age, education years, degree of professional training and education, and place of birth. Age ranged between 25 and 74 years. Education years ranged between 8 and 17 years old. Detailed descriptions of demographic features are provided in Supplementary Table S1.

As described in the Introduction the KORA S4 survey mirrors the case of plenty of study designs in human genetic research, in which control cohorts are used for population studies. In the context of these studies it could be crucial to assume genetic homogeneity of controls. One strategy is to collect samples in small areas and to restrict recruitment to individuals of same ancestry (see Section 1). In the concrete case of the KORA S4 it could be verified that the presence of immigrants introduced a small but significant effect on the total amount of genetic variation (see Supplement 4). In this paper we focus on the case of a population that may be considered a priori to be genetically homogeneous and may account for subtle genetic substructure and if this is the case, which factors may be regarded as best predictors. To perform our study in accordance with these objectives and assumptions we worked with the two data sets. We first considered the total sample (ALL). We used this set with the purpose of inspecting the effect of immigrants on the total genetic variation among other factors. On the other side we excluded immigrants and analyzed the subset of natives (GER). This analysis is intended to specify best predictors of sublet genetic substructure and to estimate their effect in an admixed modern small population. The total set of immigrants did not include enough individuals of similar ancestry to perform further separated geostatistical analysis. Therefore no further group with homogeneous ancestry could be identified.

Measures related to age, education years, and education level were computed only on the native GER data set. Variables related to immigrant representation in the total sample were computed for the total data set (ALL). All measures were computed per land unit.

(I) Variables related to age, education years, and education level, computed only for the GER data set: AGE25_39: percentage of subjects in the age category of 25 to 39 years; AGE40_54: percentage of subjects in the age category of 40 to 54 years; AGE55_74: percentage of subjects in the age category of 55 to 74 years; AGE_MEAN: mean age; EY8_11: percentage of subjects achieving a maximum of 11 school years; this variable indicates the fraction of the sample which did not achieve the education level required to access to academic studies; EY_MEAN: mean years of school attendance; EDU_MEAN: mean education level, scored according to the degree of professional training and education, ranging from 0 = no school degree up to 9 = graduate degree (M.S. equivalent or higher).

(II) In order to inspect the effect of immigrant representation on total population from a model building perspective, variables related to birth land were included; these variables were computed for the ALL data set.

The representation of the total immigrant fraction in relation to the total sample was modeled with the variable: GER_P: percentage of natives over all subjects.

(III) on the same line and for purpose of ascertaining a potential effect related to the presence of major fraction of immigrants incoming from a reduced number of countries, which could be acting as a differentiated population within the migrant group, a further variable was included: MAIN_IMP: percentage of the major group of immigrants (subjects born in Czech Republic, Romania, Poland, or Ukraine) over all immigrants (German citizens born outside of Germany).

Measure of Genetic Diversity per Land Unit

We attempted to achieve a reduction in form of genetic landscapes of the complex georeferenced data available for this cohort (genotype per SNP and sample and geocoordinates of LUs). These genetic landscapes should allow visualization of the estimated distribution of genetic diversity across geographic space and further assessment of associations between spatial patterns of genetic diversity and other georeferenced data, for example, average socio-demographic characteristics per LU.

We chose to create maps based on indices calculated per LU. The reason for this is that we considered these types of genetic landscapes easier to interpret than those based on relative measures, for instance, the genetic differences between LUs. For this step it was necessary to select a genetic measure of diversity referred to each single LU, in opposition to relative measures such as Wright's F_{ST} [17] or alike, The average heterozygosity is a usual measure of the genetic variability of a group [21]. We chose this simple measure of genetic diversity, the observed heterozygosity (H_0) [22], to summarize sample genetic attributes of each LU. H_0, the observed frequency of heterozygotes averaged over loci, was estimated using

$$H_O = \frac{\sum_j^l h_j}{l}, \tag{1}$$

where l is the number of loci and h_j indicates the proportion of heterozygote individuals per locus j_{th} [22]. Observed heterozygosity was computed separately for each land unit with the total data set ((ALL)) and with the subset of natives (H_0(GER)).

Multivariate Analysis of Spatial Population Structure

In an attempt to frame the challenge embedded in this sample, further genetic measures were computed with well-standardized tools for detecting population structure. First exploratory analysis with geostatistical methods indicated a potential differentiation of the periurban areas from Augsburg city and its periphery (see Supplement 2). Potentially, the fine-scale patterns of genetic diversity observed in these first exploratory evaluations could be explained by various simple models of spatial variation. For instance, the observed pattern (Supplementary Figure S2a-b) may be the result of a simple process of isolation by distance. As a result, genetic diversity would follow a pattern of gradual variation (e.g., gradients of allele or genotype frequencies). It must be noted that in such case, the observed pattern would correspond to a small fraction of the geographical landscape where the process occurs. This is so because both geographic extension of the study area and evolutionary times of the study population (here it refers to the number of generations necessary to fix the effects of any gene-flow process [23, 24]) are jointly, most probably, not large enough to have generated a local process of fixation of gradual variation of genetic features. Spatial correlation methods (e.g., spatial autocorrelation) and Mantel tests would be the first methods of choice to detect spatial correlations of genetic distance with geographic distance.

The observed pattern (Supplementary Figure S2a-b) could as well be the product of undetected population clustering. In this case, individuals of similar genetic features tend to reside in distinct areas than individuals less similar. Clustering would also require that individuals of distinct groups present reduced interaction with individuals of other groups. At larger geographical scales this could be observed when cultural, linguistic, or political limits set a barrier to gene flow. It is important to note that this scenario is less probable. For instance, it is improbable that a modern western population inhabiting such a small area would be composed of several groups with reduced exchange (low migration rates among the subareas and low predisposition to mate with individuals of other groups). This situation is even less probable if considered that spatial patterns of genetic variation were even detected within the group of natives. The result is supported by previous analysis undertaken with this sample: an exhaustive evaluation of population clustering was conducted by Steffens et al. [14] with the well-known software STRUCTURE [16]. Despite the large number of runs with varying models and parameters there was no indication of any population substructure. Results indicated that the model assuming a number of populations equal to unity (k=1) showed the highest posterior probability for the KORA S4 data.

The following software packages were used in this step: (a) GENELAND [20, 25]; (b) EIGENSTRAT EIGENSOFT [26, 27]; (c) PLINK, version v0.99s (http://pngu.mgh.harvard.edu/purcell/plink; [28]), with additional multidimensional scaling using the R software package, version 2.12.1 (R Foundation for Statistical Computing, 2010); (d) SPAGeDI [29]. The wide-spread used software STRUCTURE [16] was not considered, since in a previous study [14] no evidence for genetic substructure was found with this tool. Methods (a) to (c), as well as STRUCTURE, share the possibility to search for groups of genetically similar individuals. SPAGeDI (d) is a tool for detecting dependency between genetic and geographic distances among individuals or populations. GENELAND (b) and SPAGeDI (d) are individual-based methods and require including in the computation the geographic reference of each individual. As mentioned above (see Section 2.3.1), available data and anonymity restrictions did not allow a more precise georeference of subjects than sampling location. For these reasons, all individuals sampled in one location were georeferenced to the same geographical coordinates. This data aggregation consequently involves loss of power when applying these methods. Therefore, in our case and as it most probably would occur in this type of human genetic studies, the full capabilities of software making use of individual geographical coordinates could not be exploited.

With each tool (a–d), several exploratory runs were performed. In each case runs were started with default parameters and recommended model assumptions. Following, multiple runs with varied parameter values and model assumptions were conducted. For computations demanding a priori definition of an assumed number of subpopulations, runs were repeated for incremental number of subpopulations not larger than ten.

Geostatistical Analysis

Geostatistical analysis was conducted using the open-source software package GRASS 6.4 and spatial packages contributed to R software package, version. 2.12.1 (R Foundation for Statistical Computing, 2010) within the GRASS environment.

Generation of Genetic Landscapes

In this framework, genetic landscapes were defined as matrix representations of genetic variation in the geographic space. Spatial matrices were created by the transformation of sampling-point data to an elevation surface by spatial interpolation. An elevation surface is a 3D layer of continuous data (grid or raster layer) with elevation information at each point of the area. GRASS defines this type of spatial object as 2.5 dimensions (2.5D). As a simplification, the usual denomination for this type of spatial object: "3 dimensions (3D)" is adopted. The elevation parameter characterizes the estimated statistic. We decided to perform interpolation based on spline function. Interpolation based on splines proved to be a better choice for phenomena which combine a random component as well as processes which minimize energy, as it could be considered socio-demographic processes [30]. We chose the function "regularized spline with tension" implemented in the GRASS-method v. surf.rst [31]. This method computes the continuous 3D layer (raster data) simulating a thin flexible plate passing through or close to the measured data points; it is the most general and accurate method available in GRASS [30].

In order to run v.surf.rst, point-data layers are required. For each LU, we first specified its geometrical center (centroid) with a GRASS basic module. Statistic values were linked to the centroids. We obtained one point-data layer for each measured statistic. In case of Augsburg City, which contains 14 postal areas and covers a disproportionately large area, centroids of all postal areas were used. Computed values for Augsburg City data were assigned to all its 14 centroids. With this step, we smoothed spatial interpolation results in the area of Augsburg City and surroundings, while avoiding interpolation artifacts. We modeled genetic landscapes based on tuned values of v.surf.rst parameters. In order to be able to adequately execute the v.surf.rst procedure the H_0 values computed using the raw data (H_{0raw}) were transformed into percentage as follows:

$$H_{O\,raw} \cdot 100 = H_O, \quad (2)$$

where $H_{O\,raw} \in [0, 1], H_O \in [0, 100]$.

Interpolation surfaces based on H_0 values were created for the following data sets: (ALL), and H_0(GER). Interestingly, since the KORA S4 genotypes conform a control population pool for genetic studies [7, 9], (ALL)landscape may be examined as a representative estimation of the spatial variation of genetic diversity of the extant population and H_0(GER) landscape of the native fraction in the region of Augsburg.

Spatial Autocorrelation

The presence of simple association between the variability of an attribute and the geographical space was tested by means of spatial autocorrelation. In this case, the null hypothesis is that the feature of interest is spatially distributed at random among other attributes within the study area. This analysis was based on the Moran's *I* tests. Spatial correlation measured with the test statistic Moran's *I* is inferential, which implies that results must be interpreted in dependence of the null hypothesis. For this analysis we used a Global Moran's *I* statistic, which means that we tested for spatial autocorrelation in the study area as a whole, assuming that the spatial process is the same everywhere.

Spatial autocorrelation of each of the genetic and sociodemographic variables defined in this study was tested with the R package spdep [32].

Moran's *I* tests were performed using the function implementations moran. test and moran.mc. Accounting for normality deviation of the data, moran. testwas run under the specification of randomization assumption in computing the variance of the statistic. This test specification allows relaxing the simpler normality assumption by introducing a correction term based on the kurtosis of the inspected variable.

The second implementation, moran.mc, is a permutation-based test. With this implementation spatial autocorrelation is evaluated independently of normality and randomization assumptions. The function moran.mc uses a Monte Carlo test, based on a permutation bootstrap. Observed values are randomly assigned to areal entities, and the value of the observed Moran's *I* is computed nsim times [33]. We set nsim = 10 000. These tests were run using a binary matrix of spatial weights (see Section 2.3.4). Both implementations, moran.test and moran.mc, were used to test for spatial autocorrelation in measures of genetic variation: (H_0(ALL), H_0(GER)), as well as on the socio-demographic variables: GER P, MAIN IMP, AGE25 39, AGE40 54, EA55 74, AGE MEAN, EY8 11, EY MEAN, and EDU MEAN.

Search of Best Predictors

Socio-demographic measures were inspected as predictors of the observed pattern of H_0(GER) under the assumption that socio-demography would provide useful indication of spatial arrangement of recent migration processes, specially regional and national migration, which we assumed that it must have had a strong influence on fine-scale genetic variation. The contribution of socio-demographic factors to explain the observed spatial pattern was analyzed under the assumption of spatial dependence. Best-fit spatial autoregressive models (SAR) predicting heterozygosity (H_m) were selected. A stepwise forward search was conducted using the function spautolm of the package R spdep [32].The function spautolm computes a regression on the values from the other areas to estimate the spatial dependence of the residuals of the specified linear predictor. The spatial dependence is estimated with a maximum likelihood test, computing a spatial autocorrelation parameter, λ. The p value of the likelihood ratio test compares the model with no spatial autocorrelation (λ=0) to the one which allows for it [33]. A binary matrix of spatial weights was used for this analysis (see Section 2.3.4).

Model selection was started with following parameters: GER P, AGE25 39, AGE40 54, EA55 74, AGE MEAN, EY8 11, EY MEAN, and EDU MEAN. In order to test if the spatial distribution of the major immigrant group (MAIN IMG) improves model prediction, the influence of the parameter MAIN IMP on the selected model was tested. A set of models best fitting the data were selected according to the p values of the covariates (p value < 0.05).

Evaluation of Model Accuracy

Finally, the goodness of fit of the selected SAR models was analyzed. In this step, the pixelwise divergence between predicted (H_m) and observed values (H_0) was quantified. Interpolation surfaces were created based on the predicted values (H_m) by each selected model. The pixelwise divergence in absolute values of these interpolation surfaces from the (GER) landscape was used to compare prediction accuracy among the selected SAR models. In order to facilitate comparison among models, a standardized difference was computed. The standardization was performed based on the maximal range of pixel values (max rg_{GER}) measured in the (GER) landscape. The parameter max rgGER was computed as follows:

$$\text{max\_rg}_{\text{GER}} = \text{zmax\_int}_{\text{GER}} - \text{zmin\_int}_{\text{GER}}, \tag{3}$$

where zmax int_{GER} is the maximal value measured in the H_0(GER) landscape and zmin intGER the minimal value. For each SAR model, a new elevation surface (raster layer) storing the respective pixelwise difference was created. Following pixelwise computation was performed with the GRASS basic module r.mapcalc:

$$\left(\frac{\text{abs}\left[H_O\,(\text{GER}) - H\,(m_n)\right]}{\text{max\_rg}_{\text{GER}}}\right)100, \tag{4}$$

where abs implies absolute value, H_0(GER) refers to the pixel values of the H_0(GER) landscape, $H(m_n)$ refers to the pixel values of the interpolated surface created on the bases of the predicted *H* values of the *mn* SAR model, and the parameter max rg_{GER} = 1.64 (zmin int = 43.36; zmax int = 45.00).

For each one of these elevation surfaces, spatial global statistics were computed. For this purpose, elevation surfaces were imported into the spatial R environment provided by the packages sp, rgdal, spdep, and spgrass6 (R Foundation for Statistical Computing, 2010, http://www.r-project .org/ foundation/). Mean, standard deviation (sd), median, minimum (min), and maximal values (max) of the elevation surfaces were computed with the R function summary(). These statistics were applied as global quantitative measures of prediction goodness of each selected SAR model and were used to select the model best fitting the data.

The model with the lowest global difference between observed and predicted *H* values was selected as the one best fitting the data. Based on this model, maps representing the spatial variation of predicted *H* values and the estimated divergence between observed and predicted *H* values were created. The former map represents the estimated variation in heterozygosity according to predictions obtained by the SAR model best fitting the data. The latter maps allows a visual estimation of the agreement between observed and predicted heterozygosity in each land unit as well as the estimated spatial variation of divergence in the total study area.

RESULTS

Descriptive statistics of all measures, including mean, standard deviation, median, minimum value, and maximum value, are presented in Table 2.

Table 2: Descriptive statistics of genetic diversity and socio-demographic measures per LU (mean, standard deviation, median, minimum value, and maximum value)

Variable	Mean	SD	Median	Min	Max
H_O (ALL)	43.76	0.54	43.70	42.96	45.02
H_O (GER)	43.84	0.49	43.68	43.36	45.02
GER_P	81.5	13.2	84.0	57.0	100.0
MAIN_IMP	71.7	35.5	87.0	0.0	100.0
AGE25_39[1]	26.9	11.3	22.7	12.5	44.4
AGE40_54[1]	21.3	8.5	21.4	9.5	33.3
AGE55_74[1]	51.8	13.1	52.4	25.0	75.0
AGE_MEAN[1]	52.0	3.6	53.2	44.8	55.5
EY8_11[1]	66.0	10.6	65.2	50.0	83.3
EY_MEAN[1]	11.3	0.7	11.7	10.1	12.1
EDU_MEAN[1]	4.3	0.6	4.6	3.2	4.8

[1] Age and education-related variables refer only to the native group (GER).

Spatial Variation of Socio-Demographic Factors

Agerelated parameters showed a heterogeneous spatial distribution (Figure 3). Younger individuals (25 to 39 years old) comprised more than 30 percent of the GER sample in the eastern sector and reached a proportion of 45 percent in Pottmes ¨ (Figure 3(a)). Subjects corresponding to the intermediate age category (40 to 54 years old) showed a lower proportion (less than 30 percent) in LUs contiguous to Augsburg City in the South and in the East (Figure 3(b)). The upper age category (55 to 74 years old) accounted for more than 50 percent in eight of the 13 LUs, presenting the higher proportions (>70%) southern from Augsburg City (Figure 3(c)). The lowest mean age values were recorded in Pottmes, Rehling, and Neus ¨ aß¨ (Figure 3(d)). The percentage of individuals of the intermediate and the upper age category (AGE40 54, AGE55 74) showed a deviation of the expected Moran's *I* value with both tests (moran.test, moran. mc) on a significance level of $\alpha = 0.05$. The percentage of individuals of the lower age category (AGE25 39) showed a significant deviation of the expected Moran's *I* value with moran.test, but with the MC permutation bootstrap test did not reach a significance at $\alpha = 0.05$ (Table 3).The mean age (AGE MEAN) did not show any indication of spatial dependence with either of both tests (Table 3).

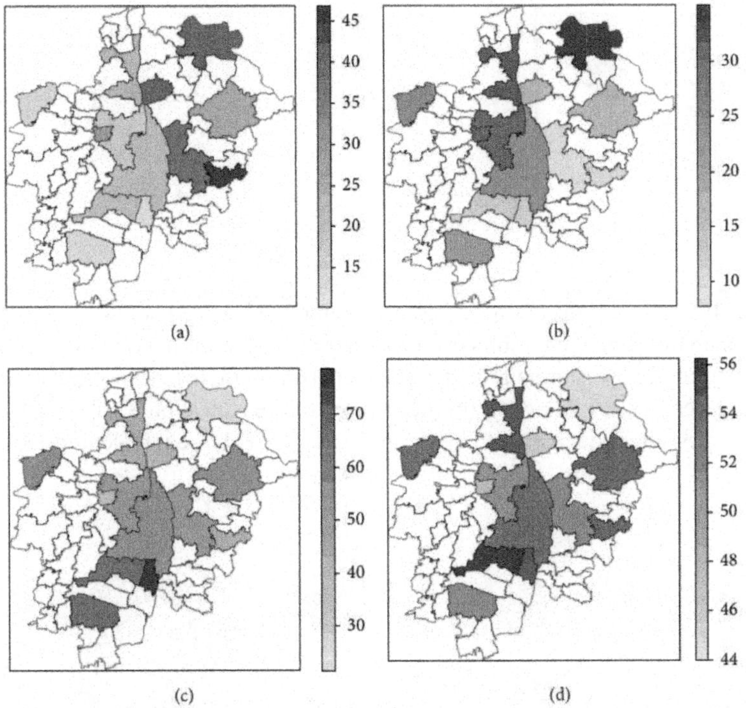

Figure 3: Values of age-related parameters; values refer only to German native subjects. (a) Percentage of subjects in the age category: 25–39 years (AGE25 39); (b) percentage of subjects in the age category: 40–54 years (AGE40 54); (c) percentage of subjects in the age category: 55–74 years (AGE55 74); (d) mean age per land unit (AGE MEAN).

With regard to the spatial distribution of education level in the study area, the largest values of education years, that is, lower values of EY8 11, were observed in Augsburg City and in neighboring LUs in the East (Friedberg) and in the West (Aystetten, Neusaß), as well as in the southern ¨ LU of Schwabmunchen (¨ Figure 4(a)). Both mean variables, means of education years and education level (EY MEAN, EDU MEAN), showed the largest values in the center and in the South of the study area, while the peripheral LUs Aichach and Altenmunster showed the lowest values (Figures ¨ 4(b) and 4(c)). Whereas education level (EDU MEAN) showed a significant deviation from the expected Moran's I value with both tests (moran.test, moran.mc) at $\alpha = 0.05$, the education years (EY MEAN) presented just an indication of potential spatial dependency at this significance level (Table 3).

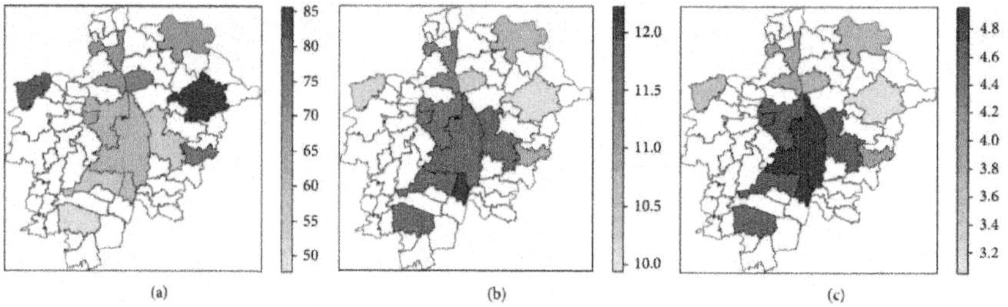

Figure 4: Values of education-related parameters; values refer only to German native subjects. (a) Percentage of subjects achieving a maximum of 11 school years, where the maximum in the sample is 17 school years (EY8 11); (b) mean years of school attendance per LU (EY MEAN); (c) mean score of the education level per LU, ranging from 0 = no school degree up to 9 = graduate degree (M.S. equivalent or higher).

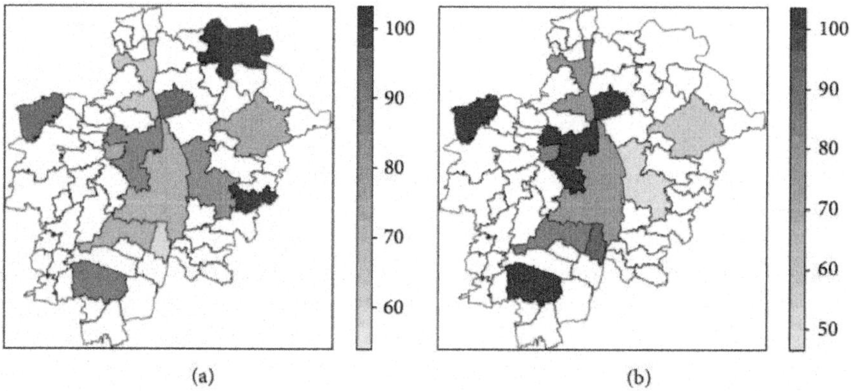

Figure 5: Spatial distribution of samples according to land of birth. (a) Spatial distribution of GER P; (b) spatial distribution of MAIN IMP.

The percentage of natives (GER P) showed a complex pattern (Figure 5(a)). Considering LUs with an intermediate number of samples, the percentage of natives decreased with an increase of the absolute number of samples per land unit with the exception of Altenmunster and Neus ¨ aß¨ (Table 1). In Augsburg City, with a quite larger number of samples in comparison with all other LUs (Table 1), natives composed ca. 70 percent of the total samples (Figure 5(a)). The lowest percentages were measured around Augsburg City, in Konigsbrunn, followed by Langweid. Bobingen, ¨ contiguous to Augsburg City on the South, and Aystetten, relatively peripheral to Augsburg City, presented the next lower frequencies of natives (Figure 5(a)). Samples included only natives in the eastern peripheral LUs, Pottmes and Eurasburg ¨ (Figure 5(a)), both accounting for the lowest sample counts as well (Table 1). The distribution of the proportion of natives (GER P) did not show any indication of a potential spatial dependency (Table 3).

Table 3: Estimated Moran's *I* values and *p* values of two Moran's *I* tests for spatial autocorrelation computed for all defined variables

Measure	Moran's *I*	*p* value[1]	*p* value (MC)[2]
H_O (ALL)	0.1132	*0.074*	*0.090*
H_O (GER)	0.2095	**0.015**	**0.024**
AGE25_39	0.1823	**0.039**	0.054
AGE40_54	0.3220	**0.004**	**0.010**
AGE55_74	0.2326	**0.013**	**0.020**
AGE_MEAN	−0.1092	0.571	0.550
EY8_11	0.0624	0.164	0.159
EY_MEAN	0.1425	*0.066*	*0.075*
EDU_MEAN	0.2509	**0.013**	**0.029**
GER_P	−0.1945	0.775	0.762
MAIN_IMP	0.1194	*0.072*	*0.091*

[1]Computed using the R package **spdep** `moran.test` based on a randomisation assumption.
[2]Computed using the R package **spdep** `moran.mc`, consisting of a Monte Carlo test, based on a permutation bootstrap test; *p* values were obtained on 10 000 runs.

The major group of immigrants (MAIN IMG) showed a higher ratio in the western LUs (Figure 5(b)). More than half of the units showed values of MAIN IMP larger than 80 percent. In Augsburg City, where a considerably larger total number of immigrants were sampled, the major immigrant group composed 78 percent of the total immigrant data set (Table 1). The results of both Moran's *I* tests of spatial autocorrelation (moran.test, moran.mc) pointed to a potential simple spatial dependency of the parameter MAIN IMP, which however did not reach a significance level of $\alpha = 0.05$ (Table 3).

Table 4: Best spatial autoregressive models fitting the data.The spatial autocorrelation left in the residuals (λ) and the p value of the likelihood ratio test, comparing the residuals of the fitted model with the one with no spatial autocorrelation (i.e., λ=0), are indicated for each model. In order to compare these four models, landscapes were created as the pixelwise difference between the observed and the predicted genetic landscape for each n model (H_0 (GER) $-H$ (m_n)). Differences were computed in percentage to the maximal range of values of the H_0 (GER) landscape. Mean, standard deviation (SD), and maximal values (Max) of the differences are indicated

	Model	λ	*p* value	Mean	SD	Max
m_1	40.768 + GER_P (0.014) + AGE40_54 (0.031) + EY8_11 (0.021)	0.203	0.449	17.0	14.9	70.6
m_2	48.903 + AGE25_39 (0.028) + EY8_11 (−0.036) + BILD_MN (−0.802)	−0.309	*0.018*	10.0	7.6	36.6
m_3	38.897 + GER_P (0.014) + EA55_74 (−0.013) + EY_MN (0.881) + BILD_MN (−1.293)	0.130	0.590	13.2	8.4	38.4
m_4	52.457 + EA55_74 (−0.031) + AGE_MN (0.090) + EY_MN (−0.811) + EY8_11 (−0.038)	−0.159	0.451	11.8	8.2	39.9

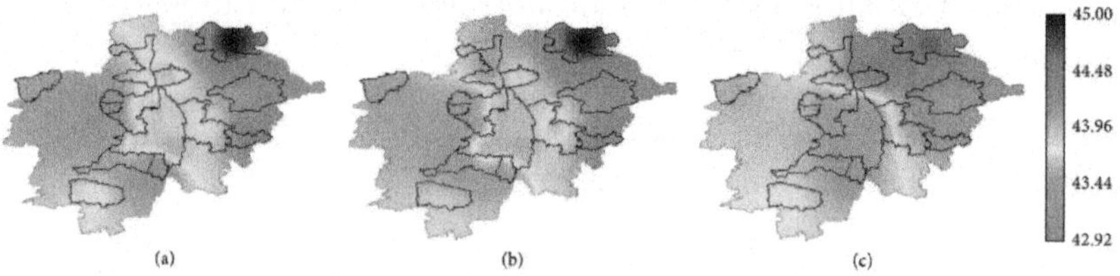

Figure 6: Landscapes estimating the genetic variation in the study area (a) H_0(ALL): observed heterozygosity of the total sample (officially registered German citizens); (b) H_0(GER) observed heterozygosity of the native data set (individuals born in Germany); (c) predicted heterozygosity according to the best-fit spatial autoregressive model (Table 4).

Geographic Variation of Genetic Diversity

The search for indications of spatial patterns of genetic heterogeneity with well-established procedures did not provide any positive results. Although a high number of different explorative runs with different parameters were performed, tests based either on EIGENSTRAT EIGENSOFT, PLINK, GENELAND, or SPAGeDI did not provide any indication of a potential geographic variation of genetic heterogeneity in the study area. A brief summary of a representative extract of these computations is presented in Supplement 5.

Geostatistical analysis based on the statistic observed heterozygosity (H_0) [22] provided indication of spatial patterning. Table 3 presents results of the spatial autocorrelation analysis performed with test statistic Moran's *I*. On the one side, the variable (ALL) showed an indication of association between genetic variation and geographic coordinates. On the other, in the native sample, tests of global spatial autocorrelation showed a significant deviation (on a significance level of $\alpha = 0.05$) of a random spatial distribution of (GER) values (Table 3). This result was obtained with both function implementations moran.test and moran.mc. Significant results obtained in the native sample may indicate that the additional genetic variability contributed by the immigrant fraction of the sample could introduce noise, which diluted a subtle patterning of the genetic attributes of the native sample.

(ALL) landscape, which estimates the variation of the genetic heterogeneity of the extant German population in the study area, presented a marked depression in the East (Figure 6(a)). The highest H_0 values were measured in the eastern area. Intermediate H_0 values covered the centralnorthern sectors, including some areas of Aichach-Friedberg District, Langweid, Augsburg City, and Schwabmunchen. ¨ The lowest H_0 values were observed in the western area. The minimum values were found in Neusaß and K ¨ onigsbrunn.

Figure 6: Landscapes estimating the genetic variation in the study area (a) : observed heterozygosity of the total sample (officially registered German citizens); (b) observed heterozygosity of the native data set (individuals born in Germany); (c) predicted heterozygosity according to the best-fit spatial autoregressive model (Table4).

The (GER) landscape, estimating the spatial variation of genetic heterogeneity of the native population, showed similar values to the H_0(ALL) landscape in the western and in the eastern periphery. In the central belt, running across the study area in north-south direction, this landscape showed higher values than the (ALL) landscape (Figure 6(b)).

Four spatial autoregressive models were selected according to the p value (p value < 0.05) of the covariates (Table 4). The four models included as covariates variables related to age and education; two models (m_1, m_3) included as well the proportion of natives per LU (GER P), which is at the same time an indication of the proportion of immigrants per LU. The inclusion of the variable MAIN IMP, which involves a differentiation of subgroups of immigrants, did not improve any of the selected models. Out of these four selected models m_2 showed a significant p value when the likelihood ratio of λ was tested, indicating left spatial correlation in the residuals (p value = 0.018). The $H(m_2)$ landscape (Figure 6(c)) showed as well the lowest mean and standard deviation of pixelwise difference to the H_0(GER) landscape (Table 4), which we used as indicators of model goodness. The pixelwise difference between (GER) and $H(m_2)$ surfaces showed a good agreement over a large area as well. Areas where both surfaces showed very similar values are indicated in Figure 7 with a white or light grey and correspond to a pixelwise H difference close to zero. The maximal differences were measured in Augsburg City and in Rehling, a small residential area in the country side. Best fitting was obtained in the peripheral ring surrounding Augsburg City (Figure 7).

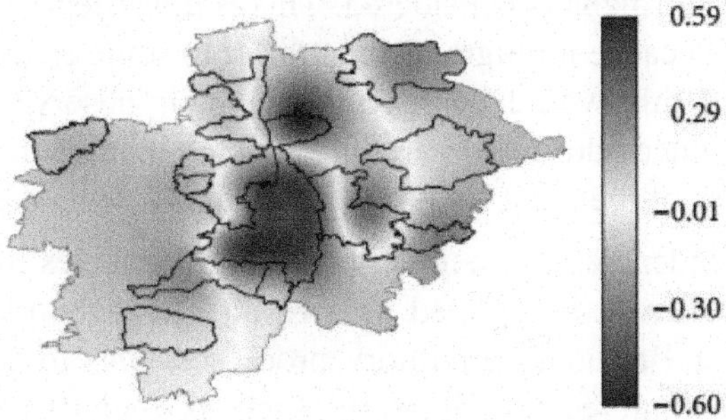

Figure 7: Difference between the interpolated surfaces of the observed and the predicted heterozygosity of the native subset ((GER) − $H(m_2)$)

DISCUSSION AND CONCLUSIONS

Fine-scale variation of genetic heterogeneity within a small region was detected and analyzed applying a geographic perspective. Population genetics and geostatistics were combined with the open-source geographic information system GRASS. The capabilities of this approach were tested on a subset of the KORA S4 survey [7,9], collected in southern Germany for prospective studies. Tests on this data set with the well-known software STRUCTURE previously reported by others [14] did not provide evidence for population substructure. We assumed that within small urbanized areas of modern western countries, as is the case in Germany, genetic composition may be strongly affected by migratory movements of the last half century, which may be still estimated by means of socio-demographic measures.

Genotypes (212 autosomal SNPs) and socio-demographic information (age, education, place of residence, and birth land) of 728 healthy German citizens were analyzed.

Socio-demographic and genetic measures showed heterogeneous distribution across the study area. The estimated values of the observed heterozygosity showed to some extent a cline of decreasing values from east to west. In a first step to analyze spatial processes controlling the observed patterns, Moran's tests for spatial autocorrelation were performed for all available parameters. Indications of global spatial dependencies were observed in socio-demographic variables related to age categories and education level. While significant deviation of the expected Moran's was obtained with the observed heterozygosity of the native subset (GER), the observed heterozygosity of the total data set showed just an indication of a potential deviation of the expected Moran's . In other words, these results could be interpreted as an indication that heterozygosity values of this small area may be regulated by a global spatial process, but such subtle process could only be observed when the subset of native Germans is considered. This is consistent with the elemental assumption that because immigrants may differ to some extent in their genetic background, and this would be reflected in their observed heterozygosity, they would introduce additional variability which hinders detection of spatial variability pattern of the most frequent group, the German natives.

Spatial dependencies of socio-demographic variables at this local level could be axiomatically interpreted as a result of neighborhood preferences of recent local and regional migratory processes. It is to expect that social preferences affecting recent migration, such as regional and international migratory movements of the last half century, may be reflected by socio-demographic parameters. Individuals would not choose new residence randomly and this may be reflected in similar socio-demographic attributes of contiguous neighborhoods. This expectation, which goes along with common sense, would also be in agreement with the fact that the level of admixture would associate to some extension with socio-demographic attributes of the location. This would be the result of new and recurrent resettlement of some areas or that inhabitants of areas presenting certain socio-demographic features would show higher predisposition to admixture. Indeed, the fact that patterns of socio-demographic structure and genetic admixture account for spatial autocorrelation could indicate the occurrence of further unobserved phenomena influencing both, as it could be "social preferences" or "official urban planning." Nevertheless, it went beyond the scope of this work to search for such unobserved processes.

Spatial autoregressive models (SAR) fitting the data were selected by a forward search. The four best-fit SAR models contained as covariates socio-demographic measures related to age, school-attendance years, and education level. Two of the four selected models included as well the percentage of natives per unit. This parameter directly accounts for degree of demographic admixture in sense of proportion of immigrants as measured by the parameter "land of birth." The prediction strength of the selected model was estimated with quantitative comparisons between the genetic landscapes created on basis of observed and predicted measures of heterozygosity. We found a good agreement between the predicted and the observed patterns, which supports the assumption of a certain relationship between genetic admixture degree and socio-demographic structure. Predicted and observed values showed the highest agreement in the surrounding belt of Augsburg City. The largest deviations were measured in Augsburg City itself and in two small residential areas in the eastern countryside. The differentiation of Augsburg City to the countryside (see Supplement 3) may be interpreted as the expected divergence of middle-size industrial settlements and countryside [34]. This effect could still be detected even after excluding immigrants, subjects born outside of Germany. This expected differentiation between urban and countryside areas in relation to genetic heterozygosity may be considered as a confirmation that the observed patterns are not artifacts.

The differentiated area in the countryside, Rehling, corresponded to a land unit with relatively low number of samples. The lower predictive capability of the best-fit model could be attributed to sampling bias. A replication would be necessary to verify this conjecture. A further possibility could be that this settlement offers an attractive residential area for individuals working in any of the larger urban centers located eastern from Augsburg, as it could be Ingolstadt or Munich. Both urban centers offer highly profitable working alternatives, in sense of carrier opportunities and higher income, and act as an attraction pole for domestic and foreign migration. As well, both cities are also among the ones with the most expensive living costs. Rehling showed also the largest proportion of younger adults. This may reinforce the idea that this location may be attractive for newly settled workers willing to commute between their working place and a less expensive residential area, relatively close to a middle-size urban center as Augsburg. If this would be the case, the inclusion of distance and accessibility to attractive urban centers could considerably improve model prediction. Further studies could test this possibility.

The exclusion of immigrants increased the global mean of observed heterozygosity. This effect was especially strong in areas accounting for the largest proportion of immigrants. About 15 percent of the samples were immigrants, born worldwide. Remarkably, almost 80 percent of them corresponded to individuals born either in Czech Republic, Romania, Poland, or Ukraine. Some areas situated in the periphery of Augsburg City showed a stronger component of immigrants, mostly or totally represented by this major immigrant group. Although this group involved four birth lands, it may be speculated that the concentration of these individuals is not casual. It could be expected that, within each provenance group, individuals could be originals of nearby regions or belonging to large related families. In total, a concentration of small groups, each one showing a higher degree of homozygosity, would stand out over a much more admixed group, as it is expected for the native German population.

Our additional analysis with vastly cited multivariate methods, GENELAND, EIGENSTRAT EIGENSOFT, PLINK, and SPAGeDI, did not also submit indication of population stratification in the study area. This outcome is consistent with our expectations. These tools proved to be successful in studies of groups with considerably larger genetic differences, significantly more polymorphic loci, or much larger number of loci than in the present work. The KORA study was carefully designed for prospective studies aiming to reduce any type of genetic structuring. Following this objective, only German citizens were included in the sample. The sampling area was kept very circumscribed as well. Consequently, the genetic differentiation of this subset of the KORA cohort is expected to be considerably lower than in humans studies conducted at broader geographical scales or in further studies acknowledging larger evolutionary histories and dimensions (i.e., nonhuman species or samples with large population differences). A further aspect to be considered is the population informativeness of the available SNPs. These loci [14] were not specifically selected by their informativeness for distinguishing among major regional groups (i.e., [35, 36]). For this reason, the number of available SNPs was probably too low for detecting fine-scale population differences with these standard tools.

Using full capabilities of tools such as SPAGeDI or GENELAND, both offering individual-based analysis tools, was not possible either since the search of patterns of variation of genetic heterogeneity of the KORA sample cannot be carried out on an individualized geographical basis.

Data of human genetic studies would most probably not include an individualized geographical reference. Official restrictions concerning personal anonymity forbid the use of data which could individualize a subject, such as postal address or any other precise geographical reference. Therefore, for human studies, precision of geographic references must be kept low with the consequence that data might be spatially aggregated. In opposite to studies analyzing other species with a continuous geographical distribution, the identification of spatial structures of humans inhabiting small areas may not be carried out on an individual basis. Such an approach would jeopardize personal anonymity and would go against most official restrictions of human studies. Accordingly, it is not surprising that well-established tools to detect human population stratification on a broader level or those which make use of geographic references on an individual basis (mostly developed for studies of other species) may not detect fine-scale patterns of genetic variation in small areas.

Most genetic studies on modern human groups address recruitment of control samples on extant populations or make use of available control cohorts. These samples may present some degree of heterogeneity even if recruitment was restricted to small areas, by citizenship, or in combination with homogeneous place of birth. The degree of spatial heterogeneity of small geographic areas, frequently assumed to be neglectful in the context of genetic studies of modern human groups, should be evaluated on a case-by-case basis. Based on our outcomes, it could be stated that genetic heterogeneity could not be automatically assumed to be negligible. These results support the elemental assumption that within multiethnic, urban, and suburban groups, as found in medium-sized German cities and surroundings, the socio-economic parameter "birth land" allows a first reduction of genetic heterogeneity.

As it was presented in this study (see Supplement 2), even after removing immigrants from the KORA S4 data set, the degree of genetic differentiation in natives still overlapped with the spatial frequency distribution of immigrants. If future studies verify our exploratory results obtained with visual examination of genetic landscapes based on analogous Reynolds' F_{ST} genetic distance,R potentially, the proportion of immigrants may be used as subrogate of degree of natives' admixture, which could actually reflect native's behavior in regard to choice of area of residence and tolerance or predisposition to admix.

Our results about a differentiation among urban, suburban, and periurban population are a suggestion of a true effect in the sense of subtle population differentiation (cf. [34]). As mentioned above in this section, it is to assume that such differentiation would not result from isolation between neighboring areas in the sense of evolutionary processes, but from differences in regard to the degree of migration, residential preferences and willingness to intermix.

However, it must be stressed that our analysis is preliminary and it is predominantly aimed at a methodological evaluation. In particular, the coverage of the area is patchy and far from complete. In order to make use of these results for further genetic studies, first, the postulated fine-scale variation of genetic heterogeneity should be confirmed with a larger data set. Second, the magnitude of the detected bias for the corresponding analysis should be evaluated. A comprehensive analysis on an augmented data set is in preparation.

Knowledge of fine-scale patterns of genetic variation could provide information about areas where expected genetic heterogeneity could introduce undesired bias. Areas with an observed higher genetic heterogeneity than tolerable could be avoided. In case that spatial heterogeneity would be assessed after recruitment, examining the spatial pattern of genetic heterogeneity could serve as a basis to decide about a stratified analysis (e.g., grouping samples according to residence or any other relevant spatial reference) or to correct for population stratification (cf. [26, 27]). Our vision is to further develop our approach in order to be capable of testing as well as detecting and correcting, if it is applicable, for spatial patterns of genetic heterogeneity within the study sample (cf. [26, 27]). In contrast to the method implemented in EIGENSOFT EIGENSTRAT [26,27], which infers strata based on genetic data alone, such approach would make use of information on subject area membership to define the strata. This usage of additional a priori information potentially leads to improve strata definition (cf. [37]).

Taking these findings of the KORA S4 sample altogether we can state that fine-scale spatial genetic variation may be assumed in the study area. Our results indicate that patterns of genetic heterogeneity can be present in small regions within Germany. In conclusion, it may be stated that the presented genetic geostatistical approach has the potential of being a powerful tool for detecting, modeling, and analyzing spatial patterns of genetic heterogeneity even within populations inhabiting small regions.

ACKNOWLEDGMENTS

The authors wish to thank Professor Dr. Thomas F. Wienker for helpful comments and criticisms. The study was supported by the German BMBF/GEM and national genotyping platform of the NGFN.

REFERENCES

1. M. J. Bamshad, S. Wooding, W. S. Watkins, C. T. Ostler, M. A. Batzer, and L. B. Jorde, "Human population genetic structure and inference of group membership," American Journal of Human Genetics, vol. 72, no. 3, pp. 578–589, 2003.
2. L. R. Cardon and L. J. Palmer, "Population stratification and spurious allelic association," The Lancet, vol. 361, no. 9357, pp. 598–604, 2003.
3. C. D. Sloan, E. J. Duell, X. Shi et al., "Ecogeographic genetic epidemiology," Genetic Epidemiology, vol. 33, no. 4, pp. 281–289, 2009.
4. M. L. Freedman, D. Reich, K. L. Penney et al., "Assessing the impact of population stratification on genetic association studies," Nature Genetics, vol. 36, no. 4, pp. 388–393, 2004.
5. J. Marchini, L. R. Cardon, M. S. Phillips, and P. Donnelly, "The effects of human population structure on large genetic association studies," Nature Genetics, vol. 36, no. 5, pp. 512–517, 2004.
6. M. D. Shriver, G. C. Kennedy, E. J. Parra et al., "The genomic distribution of population substructure in four populations using 8,525 autosomal SNPs," Human genomics, vol. 1, no. 4, pp. 274–286, 2004.
7. R. Holle, M. Happich, H. Löwel, and H. E. Wichmann, "KORA—a research platform for population based health research," Gesundheitswesen, vol. 67, supplement 1, pp. S19–S25, 2005.
8. Lindenberg, J. Brinkmeyer, N. Dahmen et al., "The German multi-centre study on smoking-related behavior-description of a population-based case-control study," Addiction Biology, vol. 16, no. 4, pp. 638–653, 2011.
9. H.-E. Wichmann, C. Gieger, and T. Illig, "KORA-gen—resource for population genetics, controls and a broad spectrum of disease phenotypes," Gesundheitswesen, vol. 67, supplement 1, pp. S26–S30, 2005.
10. Helgason, B. Yngvadòttir, B. Hrafnkelsson, J. Gulcher, and K. Stefánsson, "An Icelandic example of the impact of population structure on association studies," Nature Genetics, vol. 37, no. 1, pp. 90–95, 2005.
11. K. Epperson, Geographical Genetics, Princeton University Press, Princeton, NJ, USA, 2003.
12. H. Kaessmann, V. Wiebe, G. Weiss, and S. Pääbo, "Great ape DNA sequences reveal a reduced diversity and an expansion in humans," Nature Genetics, vol. 27, no. 2, pp. 155–156, 2001.

13. K. N. Laland, J. Odling-Smee, and S. Myles, "How culture shaped the human genome: bringing genetics and the human sciences together," Nature Reviews Genetics, vol. 11, no. 2, pp. 137–148, 2010.
14. M. Steffens, C. Lamina, T. Illig et al., "SNP-based analysis of genetic substructure in the German population," Human Heredity, vol. 62, no. 1, pp. 20–29, 2006.
15. S. Wright, Evolution and the Genetics of Populations, Vol. 2: Theory of Gene Frequencies, University of Chicago Press, Chicago, Ill, USA, 1969.
16. J. K. Pritchard, M. Stephens, and P. Donnelly, "Inference of population structure using multilocus genotype data," Genetics, vol. 155, no. 2, pp. 945–959, 2000.
17. S. Wright, "The genetical structure of populations," Annals of Eugenics, vol. 15, pp. 323–354, 1951.
18. G. R. Abecasis, S. S. Cherny, W. O. C. Cookson, and L. R. Cardon, "GRR: graphical representation of relationship errors," Bioinformatics, vol. 17, no. 8, pp. 742–743, 2001.
19. B. Devlin and K. Roeder, "Genomic control for association studies," Biometrics, vol. 55, no. 4, pp. 997–1004, 1999.
20. G. Guillot, A. Estoup, F. Mortier, and J. F. Cosson, "A spatial statistical model for landscape genetics,"Genetics, vol. 170, no. 3, pp. 1261–1280, 2005.
21. M. Nei and A. K. Roychoudhury, "Sampling variances of heterozygosity and genetic distance," Genetics, vol. 76, no. 2, pp. 379–390, 1974.
22. M. Nei, "Analysis of gene diversity in subdivided populations," Proceedings of the National Academy of Sciences, vol. 70, no. 12, pp. 3321–3323, 1973.
23. S. Wright, "Isolation by distance," Genetics, vol. 28, pp. 114–138, 1943.
24. S. Wright, "Isolation by distance under diverse systems of mating," Genetics, vol. 31, pp. 39–59, 1946.
25. G. Guillot, F. Santos, and A. Estoup, "Analysing georeferenced population genetics data with Geneland: a new algorithm to deal with null alleles and a friendly graphical user interface," Bioinformatics, vol. 24, no. 11, pp. 1406–1407, 2008.
26. N. Patterson, A. L. Price, and D. Reich, "Population structure and eigenanalysis," PLoS Genetics, vol. 2, p. e190, 2006.
27. L. Price, N. J. Patterson, R. M. Plenge, M. E. Weinblatt, N. A. Shadick, and D. Reich, "Principal components analysis corrects for stratification

in genome-wide association studies," Nature Genetics, vol. 38, no. 8, pp. 904–909, 2006.

28. S. Purcell, B. Neale, K. Todd-Brown et al., "PLINK: a tool set for whole-genome association and population-based linkage analyses," American Journal of Human Genetics, vol. 81, no. 3, pp. 559–575, 2007.
29. O. J. Hardy and X. Vekemans, "SPAGeDi: a versatile computer program to analyse spatial genetic structure at the individual or population levels," Molecular Ecology Notes, vol. 2, no. 4, pp. 618–620, 2002.
30. M. Neteler and H. Mitasova, Open Source GIS: A GRASS GIS Approach, Kluwer Academic, Boston, Mass, USA, 2004.
31. H. Mitášová and L. Mitáš, "Interpolation by regularized spline with tension: I. Theory and implementation," Mathematical Geology, vol. 25, no. 6, pp. 641–655, 1993.
32. R. Bivand, spdep: Spatial Dependence: Weighting Schemes, Statistics and Models, R package version 0.5-21, 2010, http://cran.r-project.org/web/packages/spdep/index.html.
33. R. S. Bivand, E. J. Pebesma, and V. Gomez-Rubio, Applied Spatial Data Analysis with R, Springer, New York, NY, USA, 2008.
34. V. Vitart, A. D. Carothers, C. Hayward et al., "Increased level of linkage disequilibrium in rural compared with urban communities: a factor to consider in association-study design," American Journal of Human Genetics, vol. 76, no. 5, pp. 763–772, 2005.
35. N. Liu, L. Chen, S. Wang, C. Oh, and H. Zhao, "Comparison of single-nucleotide polymorphisms and microsatellites in inference of population structure," BMC Genetics, vol. 6, supplement 1, article S26, 2005.
36. N. A. Rosenberg, L. M. Li, R. Ward, and J. K. Pritchard, "Informativeness of genetic markers for inference of ancestry," American Journal of Human Genetics, vol. 73, no. 6, pp. 1402–1422, 2003.
37. S. Manel, F. Berthoud, E. Bellemain et al., "A new individual-based spatial approach for identifying genetic discontinuities in natural populations," Molecular Ecology, vol. 16, no. 10, pp. 2031–2043, 2007.

Chapter 7

EXPLORING STATISTICAL TOOLS IN MEASURING GENETIC DIVERSITY FOR CROP IMPROVEMENT

C. O. Aremu

Department of Crop Production and Soil Science Ladoke Akintola University of Technology, Ogbomoso, Oyo state Landmark University, Omu-Aran Nigeria

INTRODUCTION

Increase in global numerical population especially in developing nations has gradually led to food shortage and hence increase in poverty. Addressing and tackling the issue and causes of poverty in the developing nations is one major challenge to breeders (Fu and Somers 2009). The different theories of econometircs have identified the human and material resources traceable to poverty, but fail to identify the crop improvement techniques in addressing world food shortage (Baudoin and Mergeai 2001). Crop improvement techniques therefore remains a major concern to plant breeders (Akbar and Kamran, 2006; Aremu et al, 2007a). Several factors affect crop improvement for specific or general environment performance. Such factors include climate, weather, soil, edaphic and biological and more importantly crop genotype (Aremu, et al, 2007b). Crop genotypes are composed of different crop forms including inbred or pure lines hybrids, landraces, wildraces germplasm accessions, cultivars or varieties. These crop genotypes have wide and diverse origin and genetic background known as genetic diversity. Genetic diversity study is a major breakthrough in understanding intraspecie crop performance leading to crop improvement (Aremu, 2005). Knowledge of crop performance in genetic diverse population reveals the differences in the nature of genetic materials used. Genetic diversity studies therefore, is a step wise process through which existing variations in the nature of individual or group of individual crop genotypes are identified using specific statistical method or combination of methods (Christini et al. 2009; Warburton and Crossa 2000; Aremu, 2005; Weir 1996). It is expected that the identified variations would form a pattern of genetic relationship useable in grouping genotypes. Several researchers including breeders have employed different data source and type from diverse crops in their methods to study genetic diversity. Such data source include morphological and agronomic, pedigree, proximate or biochemical and molecular data (Aremu, et al., 2007a in cowpea; Liu et al., 2000 in cotton; Mostafa et al., 2011 in wheat; Adewale et al., 2010 in African Yam bean; Christine et al., 2009 in bentgrass. The choice of statistical method to be used is dependent on the achievable objectives laid out in the studies. This chapter reveals the underlying importance of genetic diversity and reviews useable statistical techniques for identifying and grouping genotypes for intraspecies crop improvement.

NEED FOR GERMPLASM RESOURCE IN GENETIC DIVERSITY PRESERVATION

Crop genotypes sourced as germplasm accessions, landraces, breeding lines, wild species, have rich and variable genetic integrity explorable for breeding programmes. The first step of any meaningful breeding programme is to identify crop plants that exhibit exploitable variation for the trait(s) of interest. However, these genetic diverse crops are under threat. Continuous hybridization and crossing systems have reduced the genetic variations in cropping programmes and leave a dearth in harvesting and utilization of novel crop types with exploitable traits. Also, the continuous threat or loss of genetic diversity as a result of replacement of landraces, wild species and other primitive term of crop species by exotic high- yielding varieties remains an insurmountable problem to plant breeders. Another major source of loss of genetic diversity is by changes and or increase in population size, resulting in land use acts promoting deforestation, wars, industrialization, urbanization and other factors. According to Brown (1989), preservation of genetic diversity is possible when genetic or germplasm resource is realized as the most precious asset in conserving genetic diversity. Germplasm therefore is an essential resource for successful plant breeding. Certain areas of the world exhibit high level of genetic variability for crops (Vavilov, 1950). Falconer and Mackay (1996); Eivazi et al; (2007); reported that such areas are considered as regions or center of genetic diversity. Therefore genetic diversity in crop may be associated with the origin of the crop. This is supported by Christine et al. (2009), who reported genetic diversity to be associated with origin. Potter and Doyle, (1992) reported Tropical Africa to be the centre of diversity for African yam bean. Van Bueningen and Busch (1997), reported genetic diversity of wheat to be centered in North America. Ariyo and Odulaja (1991), found correlation between genetic diversity and eco-geographic background in okro. Some grouping methods in genetic diversity studies identified origin and geographical diversity not important in measuring genetic diversity. Nair et al. (1998) discovered diversity in sugarcane not to be associated with origin. Aremu et al. (2007a), discovered that center of origin is not a measure of genetic diversity in cowpea. If crop origin is somewhat not important in the measure of genetic diversity a resource centre is therefore needed to preserve and maintain the wide genetic sources exploitable in breeding programmes. Genetic relationship and diversity are useful for developing germplasm conservation strategies and utilization of crop genetic resources. The use of genetic diversity resource centre cannot be under estimated as earlier discussed.

IMPORTANCE OF GENETIC DIVERSITY STUDIES

Study on genetic diversity is critical to success in plant breeding. It provides information about the quantum of genetic divergence and serves a platform for specific breeding objectives (Thompson et al, 1998). It identifies parental combinations exploitable to create segregating progenies with maximum genetic potential for further selection, as proven by Akoroda (1987), Weir, (1996), Liu et al.(2000); Dje et al.(2000), (Aremu et al, 2007b). Genetic diversity exposes the genetic variability in diverse populations and provides justification for introgression and ideotype breeding programmes to enhance crop performance. Mostafa al et. (2011), postulated that genetic diversity studies provides the understanding of genetic relationships among populations and hence directs assigning lines to specific heterogeneous groups useable in identification of parents and hence choice selection for hybridization. Choice of parent has been identified to be the first basic step in meaningful breeding programme (Akoroda 1987); (Aremu et al. 2007a); (Islam 2004), (Rahim et al, 2010). Furthermore, the choice of parent selection in diversity studies is valuable because it is a means of creating useful variations in subsequent progenies.); Dje et al. (2000), discovered that the higher the genetic distance between parents, the higher the heterosis in the developed progenies. Hence the heterotic progenies can be further hybridized and selections based on transgressive segregation. Akbar and Kamran, (2006). exploited this parental selection technique in wheat breeding program through hybridization. Mostafa et al. (2011), investigated genetic distance among 36 winter wheat genotypes cultivated in different regions of Iran using principal component analysis and discovered five major groups in the genotypes to distantly related. Comprehensive and significant emphasis are made by researchers especially plant breeders on the analysis of genetic diversity in a number of field crops white and yellow yam, (Akoroda, 1987); cowpea, (Adewale and Aremu, 2010); African yam bean, (Baudoin and Mergeai 2001); Flax, (Mohammadi et al. 2010); wheat, (Mostafa et al. 2011) and several other crops. The diversity studies on these crops at their respective primitive levels (Landrace, wildtype, accessions, lines etc) led to the development of their widely distributed cultivars and varieties with proven characteristics based on stability and adaptability of performance with consistent tolerance to adverse weather conditions and resistant to diseases around the world. Fu and Somers (2009) supported that the use of identified wheat parents resistant to environmental stress under different growing conditions has led to increased world wheat production. The early report of Mohammadi and Prasna (2003) revealed that appropriate parent selection for hybridization in maize using a definite diversity study technique, Bohn et al (1999), identified six groups of wheat land races in the Western Iran that can be grown in different geographical locations for improved yield. Martin et al., (2008) discovered 42 cultivars of bentgrass in the mancet city and that only diversity studies would identify reliable and definite cultivar(s) with varietal purity and ensure protection of breeder and consumer rights. Understanding the inter and intra specie genetic relationships as provided by diversity studies has proven to increases hybrid vigor and reduce or avoid re-selection within existing germsplasm. It is worthy of note that existing cultivar populations have narrow genetic bases, hence need for creating variability within and among cultivars using genetic diversity methods.

GENETIC DIVERSITY MEASUREMENT TOOLS

Genetic diverse populations arising from pure lines, accessions, landraces, wild or weed races are analyzed using a number of methods. Such method can be single or in combination of two or more methods. Franco et al. (2001) stressed the need for careful considerations to be made when measuring genetic diversity within and between crop populations in research. Such considerations include:

1. Use of multivariate data collected from morphological or agronomic traits. Such data may effectively display discrete, continuous, binomial ordinal etc. variables
2. Use of multiple data sets arising from morphological, biochemical and DNA-based collections. The use of such multiple data sets in diversity study helps to reveal the adequacy in terms of strength and constraints in the choice of each of the data sets. The use of multiple data pose some puzzles including can analysis and result interpretation be based on individual or combined data sets? And more worrisome is the puzzle on how to effectively combine the different data sets and still achieve meaningful result. To provide answers to these puzzles, Wrigley et al. (1982), studied phylogenetic relationships among triticeae species using individual and combined analysis of data sets consisting of morphological and DNA-based traits and discovered divergent results in the analysed individual and combined data. The discrepancies in the

results may be attributed to the discrete nature of DNA-based data and the continuous variable nature of the morphological data. No wonder Hillis 1987; Chippindale and Wein (1994) suggested the assignment of specific numbers to both quantitative and qualitiative traits in morphological, biochemical and molecular data set. In view of this, Pedersen and Seberg (1998) advised that both individual and combined data sets can be analyzed in many possible and meaningful ways to draw conclusions on genetic divergence. In 1999 and 2001, Taba et al. and Franco et al., respectively utilized the modified Location Model (MLM) which combines all variables into one multinomial variable called "W" to classify maize accessions from the genetic resource centres of Latin America. Better still, this MLM can combine molecular and morphological data to classify data better than when individual data set is employed. Individual data from morphological, biochemical or molecular data set can be analyzed using one or a combination of techniques. These techniques shall be discussed.

3. Expected objective to be achieved. This dictates choice of statistical tool in measuring genetic distance and the level of clustering of the intragenic factors in use. Such objective(s) include to determine the quantum of variation and grouping such genotype based on genetic distance, identify action following parental selection. In essence, breeding focus determines applicable method in explaining the nature of genetic divergence.

Variations are recorded in the measurement of genetic diversity in genotype relationships based on genetic distances and grouping populations from individual genotypes such as accessions, lines, wild races etc. The recorded variations are primarily because of the differences in the nature of genetic materials. Therefore, the basis or genetic variance theories which identifies genotype relationships based on genetic distance estimating genetic diversity depends largely on statistical genetic variance theories which identifies genotype relationships based on genetic distance / variance.

THE USE OF MORPHOLOGICAL DATA TO MEASURE GENETIC DISTANCE

Nei, (1973), first defined Genetic distance as the difference between two entities that can be described by allelic variation. This definition was later in 1987, modified to "extent of gene differences among populations that are measured using numerical values. Betterstill, in 1998, Beaumont et al., provided a more comprehensive definition of genetic distance as any quantitative measure of genetic difference at either sequence or allele frequency level calculated between genotype individuals or populations. The first early work of Anderson (1957), proposed the use of metrogliph and index-score to study the pattern of morphological variations in individual data set. In the early seventies (Singh and Chaudhary 1985) used this method to study morphological variation in green gram. This method uses a range of variations arising from trait such that extent of trait variation is determined by the length of rays on the glyph. The performance of a genotype is adjudged by the value of the index score of that genotype. The score value determine the length of ray which may be small, medium or long Akoroda (1987); Ariyo and Odulaja (1991) and Van Bueningen and Busch (1997), extensively explored the use of metroglyph and index-score to morphological variations in yellow yam, Okro and wild rye accessions respectively. Similar to metroglyph and the score index is Euclidian Distance (ED) measurement. According to Nei (1987), Euclidian distance measures similarity between two genotypes, populations or individuals using using statistical measures where two individuals i and j, having observations on morphological traits (p) denoted by x_1, x_2, x_3,......x_n and y_1, y_2,......y_n for i and j individuals respectively. Metroglyph and index-score methods measures genetic distance by use of morphological traits. Euclidian distance measurements utilize both morphological and molecular based marker data sets. Smith et al. (1991), applied the following statistic to measure ED.

$$dij = \varepsilon[(T_{1(i)} - T_{2(i)}{}^2)/\sigma^2 T_{(i)}]^{1}/_{2}.$$

Where T_1 and T_2 are the values of the ith trait for 1 lines and 2 and σ^2 T(i) is the variance for the ith trait over all the lines used. Much later, Weir (1996) developed a formula for calculating genetic distance to be.

$$d(I,j) = [(x_1 - y_1)^2 + (x_2 - y_2)^2 + ...(x_p - y_p)^2]^{1/2}$$

where i and j is the ED between two individuals lines having morphological traits (p)

x_1, x_2......x_p is the traits for i individuals and y_1, y_2......x_p is the traits for j individuals

from here, the individual character distances are summed and then divided by the total number of characters scored in both individuals. ED measurement allows the use of both qualitative and quantitative data several workers identified genotype distances using ED. Van Bueningen and Busch (1997) in wheat, smith et al, 1987 in sorghum and Ajmone – Marsan (1998) in maize.

THE USE OF MOLECULAR DATA IN MEASURING GENETIC DISTANCES

The advent and explorations in molecular genetics led to a better definition of Euclidean distance by Beaumont et al; (1998) to mean a quantitative measure of genetic difference calculated between individuals, populations or species at DNA sequence level or allele frequency level. Various genetic distance measurements are proposed for analyzing DNA-based data for the purpose of genetic diversity studies. Powel et al. (1996), identified different DNA-based marker techniques to include Random Amplified Polymorphic DNA (RAPD), Amplified Fragment Length Polymorphism (AFLP), Restriction Fragment Length Polymorphic (RFLPs) and the most recent Simple Sequence Repeats (SSR) and Microsatellite (MT) of single nucleotide polymorphism (SNPs). The above nucleotide differences can be used effectively to run individual or combined data sets of morphological, biochemical or DNA based data. For DNA based data, where the amplification products are equated to alleles, the allele frequencies can be calculated and the genetic distance between i and j individuals estimated as follows.

$$d(ij) = 1 < \left[\sum^{n}(X_{ai} - X_{aj})\right]^{1/r}$$

Where X_{ai} is frequency of the allele a for individual I, and n is the number of alleles per loci; r is the constant based on the coefficient used. In its simple form, i.e r = 1, genetic distance can be calculated as:

$$d\int (ij) = 1/2\left[\sum^{n}(X_{ai} - X_{aj})\right]$$

Where r = 2, d(i,j) is referred to as Rogers (1972) measure of distance (RD), where

$$RD_{ij} = I/2[\,\Sigma(\,x_{ai}\text{-}x_{aj})\,^{2}]^{1/2}$$

Where allele frequencies are to be calculated for some of the molecular markers, the data must first generate a binary matrix for statistical analysis. Binary data has been long and widely used before the advent of molecular marker data to measure genetic distance by Rogers (1972); Nei and Chesser (1983) coefficient and known as GDMR and GDNL respectively. In the use of any given statistical formula to determine genetic diversity in molecular based data, one specific problem usually encountered is the failure of some genotypes to show amplification for some primer pairs. Robinson and Harris

(1999) noted that lack of amplification may be due to "null alleles". Most often, it is difficult to ascribe lack of amplification to "null allele". It is therefore the reposed confidence of the researcher, that a "null allele" status of a genotype will not be considered as missing data during computation of genetic similarity-distance matrix so as to avoid gross error during result interpretation. DNA based marker data have been successfully used to measure genetic distance in some crops (Pritchard et al. (2000) in pigeon pea; Beaumont et al. (1998) in wheat; Franco et al., (2001) in maize; Dje et al. (2000) in Sorghum.

GROUPING TECHNIQUES IN MEASURING GENETIC DIVERSITY

Genetic relationship among and with breeding materials can be identified and classified using multivariate grouping methods. The use of established multivariate statistical algorithms is important in classifying breeding materials from germplasm, accessions, lines, and other races into distinct and variable groups depending on genotype performance. Such groups can be resistant to diseases, earliness in maturity, reduced canopy drought resistant etc. The widely used techniques irrespective of the data source (morphological, biochemical and molecular marker data) are cluster analysis, Principal Component Analysis (PCA), Principal Coordinate Analysis (PCOA) Canonical Correlation and Multidimensional Scaling (MDS). Cluster analysis presents patterns of relationships between genotypes and hierarchical mutually exclusive grouping such that similar descriptions are mathematically gathered into same cluster (Hair et al. 1995); (Aremu 2005). Cluster analysis have five methods namely unweighted paired group method using centroids (UPGMA and UPGMC), Single Linkages (SLCA), Complete Linkage (CLCA) and Median Linkage (MLCA). UPGMA and UPAMC provide more accurate grouping information on breeding materials used in accordance with pedigrees and calculated results found most consistent with known heterotic groups than the other clusters (Aremu et al., (2007a). Principal components, canonical and multidimentional analyses are used to derive a 2-or 3- dimensonal scatter plot of individuals such that the geometrical distances among individual genotypes reflect the genetic distances among them. Wiley (1981), defined principal component as a reduced data form which clarify the relationship between breeding materials into interpretable fewer dimensions to form new variables. These new variables are visualized as different non correlating groups. Principal components analysis first determines Eigen values which explain the amount of total variation displayed on the component axes. It is expected that the first 3 axes will explain a large sum of the variations captured by the genotypes.

Cluster and principal component analysis can be jointly used to explain the variations in breeding materials in genetic diversity studies.

CONCLUSION

Genetic diversity studies is in no measure the first basic step in meaningful breeding programme and therefore require accurate and reliable means for estimation. Data sets sourced can morphological biochemical several workers successfully utilized various statistical tools in analysis diverse data sets and identified two major framework to really explain divergence in genotype performance. Genetic distance among and within individual data sets can be conveniently determined using specific tools while classificatory and cluster analysis require principal component and polymorphic sequence tools. Since each data set provide different molecular type of information, based marker data set is visualized to provide more reliable differentiate information on the genotypes. Analysis of data sets can be complex. Many software packages are available. There is still a need for a comprehensive and user-friendly software packages that would integrate different data set for analysis and generate reliable and useable information about genetic relationship. Equally important in genetic diversity studies is the need for a genetic resource centre. Studies should incorporate utilization of genetic diversity information in developing genetic resource centre accessible to breeders.

ACKNOWLEDGEMENT

Many thanks to Ibirinde Olalekan for the secretariat assistance. I also appreciate Olayinka Olabode the Head of Department of Agronomy LAUTECH for the technical contributions given to this chapter.

REFERENCES

1. Adewale, B.D.,Kehinde, O.B., Aremu, C. O. Popoola, and Dumet, J. 2010. Seed metrics for genetic and shape determinations in African yam bean [Fabaceae] (Sphenostylis stenocarpa) African Journal of Plant Science Vol. 4(4): 107-115
2. Ajmone-Marsan P., Carstiglioni, P., Fusari, F., Kuiper, M. and Motto, M. 1998. Genetic diversity and its relationship to hybrid performance in maize as revealed by RFLP and AFLP markers. Theor. Appl. Genet. 96:219-227.
3. Akbar, A.A. and Kamran, M. (2006). Relationship among yield components and selection criteria for yield improvement of Safflower – Carthamustinctorious L. J. Appl. Sci. 6:2853-2855.

4. Akoroda, M.O. 1987. Principal component analysised metroglyph of variation among Nigerian yellow yam. Euphytica 32:565-573.
5. Aremu C.O. 2005. Diversity selection and genotypes Environment interaction in cowpea unpublished Ph.D Thesis. University of Agriculture, Abeokuta, Nigeria. P. 210.
6. Aremu, C.O., Adebayo M.A., and Adeniji O.T. 2008: Seasonal performance of cowpea in humid tropics using GGE biplot analysis world jour. of Biol.. Research. 1: (1) 8-13.
7. Aremu, C.O., Adebayo, M.A., Ariyo O.J., and Adewale B.D. 2007b. Classification of genetic diversity and choice of parents for hydridization in cowpea vigna unguiculata (L) walip for humid savanna ecology. African journal of biotechnology 6: (20) 2333-2339.
8. Aremu, C.O., Adebayo, M.A., Oyegunle, M. and Ariyo, J.O.2007a. TRhe relative discriminatory abilities measuring Genotype by environment interaction in soybean (Glycine max). Agricultural journ. 2 (2).: 210-215
9. Ariyo, O.J. and Odulaja, A. 1991. Numerical analysis of variation among accessions of Okra. Ann. Bot. 67:527-531.
10. Barrett, B.A. and Kidwell, K.K. 1998. AFLP-based genetic diversity assessment among wheatcultivars from the pacific Northwest crop Sci. 38:1261-1271.
11. Baudoin, J. and Mergeai G. 2001. Yam bean sphenostylis stendcarpa. In: R.H. Raemaekers (ed) crop production in tropical Africa Directorate general for international (DGIC) Brussels, Belgium. Pp. 372 – 377.
12. Bearmont, M.A., Ibrahim, K.M., Boursot, P. and Bruqord, M.W. 1998. Measuring genetic distance. P. 315-325. In A. karp et al. (ed) molecular tools for screening biodiversity. Chapman and Hall, London.
13. Bohn, M.,H.F. Hutz and A.E.Melchinegr. 1999. Genetic similarities among winter wheat cultivars determined on the basis of RFLPs, AFLPs ans SSRs and their used ion predicting progeny variance. Crop Sci. 39; 228-237.
14. Brown, A.H.D. 1989. The case for core collection. P. 135-156. In A.H.D England Brown et al. (ed). The use of plant genetic resources. Cambridge Univ. press Cambridge.
15. Brown-Guedira, G.L; Thompson J. Ajnelson, R.L. and Warburton M.L. 2000. Evaluation of genetic diversity of soybean introduction and North American ancestors using RAPD and SSR markers crop Sci. 40:815-823.

16. Bueninger, L.T. and Busch, Ritt. 1997. Genetic diversity among North American spring wheat cultivars. Analysis of the coefficient of parentage matrix. Crop Sc. 37:570-579.

17. Chippindale, P.T. and Weins J.I. 1994. Weighting portioning and combining characters in phylogenetic analyses. Syst. Biol. 43:273-287.

18. Christine, Joshua, H., William, A., Stacy, A. 2009. Genetic diversity of creeping bentgrass cultivars using SSR markers. Intern. Turfgrass Soc. Research Jour. 11.

19. Dje, Y, Hevretz, M., Letebure, C. and Vekemans, X. 2000. Assessment of genetic diversity within and among germplasm accessions in cultivated sorghum using microsatellite marked theor. Appl. Gent. 100:918-925.

20. Eivazi, A.R, Naghavi, M.R, Hajheidari, M. Mohammadi, S.A, Majidi, S.A, Salakdeh, I. and Mardi, M. 2007. assessing wheat genetic diversity using quality traits, amplified fragment length polymorphism simple sequence repeats and proteome analysis Ann. Appl. Biol. 152:81-91.

21. Falconer, D.S. and Mackay T.F. 1996. Introduction to Quantitative Genetics,Longman, Harlow.

22. Franco, J., Crossa, J., Ribaot, M., Betran, J., Warburton, M. land Khairallah, M. 2001. A method for combinary molecular markers and phenotypic attributes for classifying plant genotype Theor-Appl. Gent. 103: 944-952.

23. Fu, Y. and Somers D. 2009. Genomo-wide reduction of genetic diversity in wheat breeding Crop Sci. 49:161-168.

24. Hair, J.R., Anderson, R.E., Tatham, R.L., and Black, W.C. 1995. Multivariatev data analysis with Readings. 4th edition, Prentice- Hall, Englewood Cliffs, NJ.

25. Hillis, D.M. 1987. Molecular versus morphological approaches to systematics. Annu. Rev. Ecol. Systems. 18:23-42.

26. Islam. M.R. 2004. Genetic diversity in irrigated rice pak. Jorn of Biol. Sci. 2:226-226.

27. Joshi, B.K, Mudwari, A., Bhatta, M.R, Ferrara, G.O. 2004. Genetic diversity in Nepalese wheat cultivars based on agro-morphological traits and coefficient of parentage. Nep Agric Res. J. 5:7-17.

28. Liu, S., Cantrell, R.G., Mccarty, J.C. and Stewart, M.D. 2000. simple sequence repeat based assessment of genetic diversity in cotton race accessions. Crop Sci. 40:1459-1469.

29. Martin, E., Cravero, V., Esposito A, Lozez F, Milanebi, L. and Cointry, E. 2008. Identification of markers linked to agronomic traits in globe artichoke. Aust. J. Crop Sci. 1:43-46.

30. Mohammadi A.A, Saeidi, G. and Arzuni, G. 2010. Genetic analysis of some agronomic traits in flax. Aust. J. Crop sci. 4:343-352.

31. Mostafa K., Mohammad, H. and Mohammad, M. 2011 genetic diversity of wheat genotype baspdon cluster and principal component analyses for breeding strategies Australian Jour. of Crop Sc. 5 (1): 17-24.

32. Nei, M. 1973. Analysis of gene diversity in subdivided populations. Proc. Natil. Acad. Sci. (USA) 70. 3321-3323.

33. Nei, M. and Chesser, R.K. (1983). Estimation of fixation indices and gene diversities. Ann. Hum. Genet. 47: 253.-259

34. Pedersen, G. and Seberg, O. 1998. Moleculesvs Morphology. P. 359-365. In A. karp et al (ed) Molecular tools for screening Biodiversity.. Chapman and Hall, London.

35. Potter, D. and Doyle, J. J. 1992. Origin of African yam bean (Sphenostylisstenocarpa, Leguminosae): evidence from morphology, isozymes,chloroplast DNA and Linguistics. Eco. Bot. 46: 276-292.

36. Powell, W., Morgante, M., Andre, C., Hanafey, M., Vogel, J.Tinjey, S. and Rafalsky, A. 1996. The comparison of RFLP, RAPD, AFLP and SSR markers for germplasm analysis. Mol.breed. 2: 225-238.

37. Rahim, M.A., Mia, A.A, Mahmud, F., Zeba, N. and Afrin, K. 2010. Genetic variability, character association and genetic divergence in murgbean platn Omic 3:1-6.

38. Rogers, J.S. 1972. Measures of genetic similarity and genetic distance studies in genetics. VII. Univ. Tex. Publ. 2713: 145-153.

39. Singh R.K. and Chaudhary B.D. 1985. Biometrical methods in quantitative genetic analysis.Kalyani publishers, New Delhi. India. P 38.

40. Smith, J.S., Paszkiewics, S. Smith,O.S.and Schaoffer, J. 1987. Electorphoretic, chromatographic and genetic techniques diversity among corn hybrids. P. 187-20. In Proc. Am.Seed Trade Assoc.Washington DC.

41. Smith, J.S., Smith, O.S., Boven, S.L., Tenburg, R.A.and Wall, S.J. 1991. The description and ssessment of distances between inbred lines of maize III: A revised scheme for the testing of distinctiveness between inbred lines utilizing DNA RFLPs. Maydica, 36. 213-226.

42. Swoftord D.L, Olsen, G.J. Wadell, P.J. and Hillis D.M. 1996. Phylogenetic inference P. 407- 514. In D.M. Hillis et al (ed) Molecualr systematics. 2nd edition Sinaver Associates, Sunderland, M.A.

43. Taba, S., Diaz, J., Franco, J., Crossa, J.and Eberhart, S.A. 1999. A core subject of LAMP, from the Latin American maize project. CD-ROM, CIMMYT, Mexico, D.F., Mexico.

44. Thompson, J.A., Nelson, R.L. and Vodkin, L.O.. 1998. Identification of diverse soybean germplasm using RAPD markers. Crop Sci. 38: 1348-1355.

45. Van Bueningen, L.T. and Busch, R.H. 1997. Genetic diversity among North American spring wheat cultivars: I. Analysis of the coefficient of parentage matrix. Crop Sci. 37: 570-579.

46. Vavilov, N.I. 1950. The origin, variation, immunity and breeding of cultivated plants. Chronica botanica, B. chronica Botanica company, Waltham, Massachusetts. P 364.

47. Warburton, M. and Crossa, J. 2000. Data analysis in the CIMMYT. Applied Biotechnology Center for fingerprinting and Genetic Diversity Studies. CIMMYT, Mexico.

48. Weir, B.S. 1996. Intraspecific differentiation P. 385-403. in D.M. Hillis et al. (ed). Molecular systematics 2nd edition sunderlands M.A.

49. Wiley, E.O., 1981. Phylogenetics: The theory and practice of phylogenetics and systemic.John Wiley, New York.

50. Wrigley, C.W, Autran, J.C. and Bushuk, W. 1982. Identification of cereal varieties by gel electorphoresis of the grain proteins. Adv. Cereal. Sci. Techolog. 5:211-259.

Chapter 8

GENETICS OF RHEUMATIC DISEASE

Alex Clarke and Timothy J Vyse
Imperial College London, Faculty of Medicine, Section of Molecular Genetics and Rheumatology, Hammersmith Hospital

ABSTRACT

Many of the chronic inflammatory and degenerative disorders that present to clinical rheumatologists have a complex genetic aetiology. Over the past decade a dramatic improvement in technology and methodology has accelerated the pace of gene discovery in complex disorders in an exponential fashion. In this review, we focus on rheumatoid arthritis, systemic lupus erythematosus and ankylosing spondylitis and describe some of the recently described genes that underlie these conditions and the extent to which they overlap. The next decade will witness a full account of the main disease susceptibility genes in these diseases and progress in establishing the molecular basis by which genetic variation contributes to pathogenesis.

GENETICS OF RHEUMATIC DISEASE

The spectrum of rheumatic disease is wide and includes conditions with diverse pathology, although most have in common a heritable risk with a complex genetic basis. There has therefore been intense effort to understand the contribution of genotype to the expression of disease in terms of both basic pathogenesis and clinical characteristics. Recent technical advances in genotyping and statistical analysis and international collaborations assembling large cohorts of patients have led to a wealth of new data. In this review we describe insights gained into the pathogenesis of autoimmune rheumatic disease by the techniques of modern genetics, in particular evidence from genome-wide association (GWA) studies, which provide support for the existence of a common genetic risk basis to several diseases. To reflect the new data from GWA studies, our discussion will be confined to rheumatoid arthritis (RA), systemic lupus erythematosus (SLE), and ankylosing spondylitis (AS), which in some cases share a common autoimmune pathogenesis. Osteoarthritis and osteoporosis are also complex genetic traits but limitations of space are such that these two conditions will not be considered in this review.

The concept of a systematic, GWA study became practical with the cataloguing of libraries of common polymorphisms. Currently, over 20 million single nucleotide polymorphisms (SNPs) have been identified [1] and platforms are available to type up to 1 million of these in a single reaction. Although not all SNPs are currently genotyped, as the human genome is arranged into haplotype blocks in linkage disequilibrium, it is only necessary to type so-called tag SNPs, which identify these areas of limited variability [2], to achieve good representation of the total amount of genetic variation. Most typed SNPs are relatively common (minor allele frequency of > 5%) and if associated with disease are likely, therefore, to have only modest pathogenic effects (odds ratios (ORs) usually between 1.2 and 2), as otherwise they would become depleted in a population due to natural selection. It is necessary, therefore, to invoke the 'common-disease common-variant' (CD-CV) model [3], which assumes an accumulation of risk caused by the carriage of multiple deleterious alleles, to explain current experimental findings.

One of the revolutionary advantages of the GWA study is the freedom from a required gene-centric hypothesis, which provides an unprecedentedly effective technique for risk gene discovery. Many disease-associated genes identified by GWA studies were completely unsuspected to be relevant - for example, the autophagy system in Crohn's disease [4]. However, because in essence up to 1 million independent hypotheses are being tested in each genotyping reaction, sample sizes powered to detect even the stronger associations must be large, and criteria for significance stringent. The general consensus is that significance can be defined as a *P*-value smaller than 5×10^{-7}, which in a cohort such as the Wellcome Trust Case Control Consortium (WTCCC) of 2,000 cases, for example, approximates to a power of 43% rising to 80% to detect alleles with ORs of 1.3 and 1.5, respectively [5]. However, the genome is subject to variation at more than the SNP level, and individuals also differ in the copy number of sections of DNA of greater than several kilobases in size, so called copy number variation (CNV), which in fact accounts for more total nucleotide difference between individuals than SNPs [6,7]. CNV can affect gene expression levels [8] and has been linked to autoimmune disease [9, 10], including SLE [11]. Whilst the latest genotyping platforms include assessment of CNV, earlier products actively excluded SNPs within regions of the most variation as they were more likely to fail quality control steps. Association studies based on CNV are, therefore, in their relative infancy. Finally, the genome is subject to modification without a change in DNA sequence; epigenetic mechanisms can have profound effects on gene expression. These include DNA methylation and changes in chromatin structure [12].

It has become apparent that SLE, RA, and AS, which have divergent clinical features, may share a common genetic risk framework, and we aim in our review to illustrate this.

THE MHC REGION AND ANTIGEN PROCESSING

The major histocompatibility complex (MHC) region on chromosome 6 contributes to the risk of almost all autoimmune diseases, and its role in immunity in mice was recognized over 60 years ago. In humans, the MHC locus is also known as the*HLA* (human leukocyte antigen) region, reflecting the initial identification of MHC gene products on the surface of white blood cells. The classical MHC extends over around 4 megabases, and comprises three clusters: class I, II, and III. Class I and II regions include genes that encode the α- and β-chains of the MHC I and II complexes, and flank the class III region, which contains an assortment of immunologically relevant genes. Despite extensive study, the mechanisms that link the MHC to disease are largely unknown, although it is supposed that variation in the MHC peptide binding cleft facilitates presentation of self-antigen to autoreactive lymphocytes.

These difficulties in understanding the MHC are not without reason; it contains some of the most polymorphic loci described in the genome, and has a highly complicated genetic architecture, with some regions exhibiting extended linkage disequilibrium [13].

In RA, the MHC accounts for around a third of the genetic liability [14]. Alleles at *HLA-DRB1* contribute much of this risk - for example, *DRB1*0401* carries an OR of 3. GWA studies confirm the strong association with MHC variants; risk alleles confer an OR of around 2 to 3 in homozygotes [15], with very high statistical significance ($P < 10^{-100}$). Additional loci contributing to the risk of RA identified by high-density genotyping include HLA-DP in patients with anti-cyclic citrullinated peptide antibodies [16]. SLE not only has strongly associated alleles in the class II region, *HLA-DR2 (DRB1*1501)* and *DR3 (DRB1*0301)* [14], with ORs of 2 [17], but also risk variants in the class III cluster, which encodes genes such as *TNF* and the complement components *C2*, *C4A* and *C4B*. C4 is crucial in the classical and mannose-binding lectin pathways of complement activation, and complete deficiency of C4 or indeed other components of the classical pathway are rare, but strong, risk factors for SLE [18]. The *C4* gene is subject to CNV and is of two isotypes, *C4A* and *C4B*. It is an attractive hypothesis that CNV at *C4* affects expression and contributes to SLE risk. However, it remains to be established whether haplotypes carrying partial C4 deficiency exert their risk via an influence on complement or through other genetic variants that are in linkage disequilibrium. Other loci in the class III region have been implicated in

SLE, including the *SKIV2L* gene, SNPs in which carry an OR of 2 in a family-based analysis [19]. *SKIV2L* encodes superkiller viralicidic activity 2-like, the human homologue of which is a DEAD box protein that may have nucleic acid processing activity. The second MHC III signal for SLE we will consider was identified in the International Consortium on the Genetics of Systemic Lupus Erythematosus (SLEGEN) GWA study [17, 20]. The SNP rs3131379 in mutS homologue 5 (*MSH5*) has an OR of 1.82. There is evidence that MSH5 has a role in immunoglobulin class switch variation [21]. Again, further work is required to definitively implicate this gene rather than variants in linkage disequilibrium, which include HLA-DRB1*0301 and *C4A* deletions.

Clearly, *HLA-B27* is the overwhelming association in AS, with an OR of 200 to 300. In the MHC, other genetic risk variants have been identified, including *HLA-B60* (OR 3.6) [22] and various *HLA-DR* genes with relatively minor contributions [23]. The pathogenic mechanism for these risk alleles is unknown. Outside of the MHC, two significant genes have so far been identified in AS: *ARTS1* and *IL-23R* [24], the latter of which will be discussed below and has been associated with several different autoimmune diseases. ARTS1 has two identified functions. Its first is in the processing of peptide for presentation via MHC I. It is localised in the endoplasmic reticulum, and is upregulated by IFNγ. It acts as an amino-terminal aminopeptidase and in mice is essential for the display of the normal peptide repertoire. In its absence, many unstable and highly immunogenic MHC-peptide complexes are presented [25]. A hypothetical connection with HLA-B27 can thus be drawn. Its other function is to downregulate signalling by IL-1, IL-6, and TNFα through surface receptor cleavage [26–28]. The most associated SNP rs30187 risk allele has an OR of 1.4, and is of unknown functional significance.

INNATE-ADAPTIVE INTERFACE

Interferon Signalling: *IRF5*

It is clear that type 1 interferons (IFNα and IFNβ) are of great importance in the pathogenesis of SLE. Patients with active disease have high levels of IFNα, which has multiple immunomodulatory actions [29], including the induction of dendritic cell differentiation, the upregulation of innate immune receptors such as toll-like receptors (TLRs), the polarization of T cells towards a T_H1 phenotype, and the activation of B cells. Type I interferons are produced by all cells in response to viral infection, but particularly by plasmacytoid dendritic cells in response to unmethylated CpG oligonucleotides binding to TLR-9, or RNA to TLR-7. Using a candidate gene approach targeting the IFN signalling pathway, the SNP rs2004640 in *IRF5*(interferon regulatory factor 5) was found to be significantly associated with SLE (OR 1.6) [30], a risk gene confirmed in several other studies [17, 31–35]. The functional consequences for IRF5 of the identified mutations are variable, but include the creation of a 5' donor splice site in an alternative exon 1, allowing the expression of several isoforms [35], a 30 base-pair in-frame insertion/deletion variant of exon 6, a change in the 3' untranslated region, and a CGGGG insertion-deletion (indel) polymorphism, the latter two affecting mRNA stability [32, 36]. Interestingly, these mutations may occur together in a haplotype, with varying degrees of associated risk. The exact role of IRF5 in IFN signalling has not been fully elucidated, but it is also critical for the gene induction programme activated by TLRs [37], providing further biological plausibility for its importance in the pathogenesis of SLE. Haplotypes of IRF5 are also implicated in RA, and may confer either protection (OR 0.76) or predisposition (OR 1.8) [38]. The same CGGGG indel allele described above also carries risk for multiple sclerosis and inflammatory bowel disease [36].

TNF-Associated Signalling Pathway: *TNFAIP3* and *TRAF1-C5*

TNF-associated signalling pathway genes play a prominent role in the risk for both SLE and RA, and associations with variants in *TNFAIP3*, and the *TRAF1-C5* locus have been identified [39, 40]. TNFα-induced protein-3 (TNFAIP3; also known as A20) is a ubiquitin editing enzyme that acts as a negative regulator of NFκB. A20 can disassemble Lys63-linked polyubiquitin chains from targets such as TRAF6 and RIP1. A second region of A20 catalyses Lys48-linked ubiquitination that targets the molecule for degradation by the proteasome [41]. A20 modifies key mediators in the downstream signalling of TLRs that use MyD88, TNF receptors, the IL-1 receptor family, and nucleotide-oligomerization domain protein 2 (NOD2) [42]. *Tnfaip3* knockout mice develop severe multi-organ inflammatory disease, and the phenotype is lethal [43]. The SNP rs10499194 in *TNFAIP3* carries an OR of 1.33 for RA, and rs5029939 an OR of 2.29 for SLE [44], the latter also conferring an increased risk of haematologic or renal complications [45].

On chromosome 9, the region containing *TRAF1* (TNF receptor associated factor 1) and *C5* (complement component 5) genes is associated with significant risk for RA (risk SNP OR of approximately 1.3) in most [15, 40, 46–48], but not all [5], studies. Due to linkage disequilibrium, the functional variant remains elusive. TRAF1 is principally expressed in lymphocytes, and inhibits NFκB signalling by TNF. This pathway is blocked in TRAF1 overexpression [49] whilst, conversely, *Traf1-/-* mice are sensitized to TNF and have exaggerated TNF-induced skin necrosis [50].

The complement system has long been known to be involved in the pathogenesis of RA. In the collagen-induced arthritis model of RA, C5 deficiency prevents disease *de novo* and ameliorates existing symptoms and signs [51, 52]. Interestingly, GG homozygotes at the *TRAF1-C5* SNP rs3761847 with RA have a significantly increased risk of death (hazard ratio 3.96, 95% confidence interval 1.24 to 12.6, $P = 0.02$) from malignancy or sepsis, potentially allowing identification of patients for appropriate screening [53].

Immunomodulatory Adhesion Molecule: *ITGAM*

Integrin-α-M (ITGAM), variants of which are strongly associated with SLE, forms a heterodimer with integrin-β-2 to produce $\alpha_M\beta_2$-integrin (also known as CD11b, Mac-1, or complement receptor-3), which mediates the adhesion of myeloid cells to the endothelium via ICAM-1 (Intercellular adhesion molecule-1) and recognizes the complement component iC3b. It not only has a role in cell trafficking and phagocytosis [54], but also has other immunomodulatory functions. Antigen-presenting cells produce tolerogenic IL-10 and transforming growth factor-β on iC3b binding to CD11b [55], and mice deficient in this receptor upregulate expression of IL-6, favouring a pro-inflammatory T_H17 response [56]. Despite its implication in defective immune complex clearance in SLE, experimental evidence for a role was lacking. GWA studies, however, demonstrate a strong and significant association [17, 33, 44], with an OR of 1.83 ($P = 7 \times 10^{-50}$) in meta-analysis [57]. The implicated SNP rs1143679 is non-synonymous, causing the substitution of histidine for arginine at amino acid 77, although this change does not affect the iC3b binding site [58]. Furthermore, although this SNP is disease associated in European and Hispanic patients, it is monomorphic in Japanese and Korean populations [59]; an explanation of its effect is therefore outstanding. It has been mentioned that CNV is important in C4 expression; the same is true for the Fcγ receptor IIIb (FCGR3B) [60], which relies on CD11b for function. Fcγ receptor IIIb is principally present on neutrophils and is important in the binding and clearance of immune complexes, therefore marking itself as a potential SLE risk gene. There is a significant association between low *FCGR3B* copy number and SLE. Patients with two or fewer copies of *FCGR3B* have an OR of 2.43 for SLE with nephritis, and 2.21 for SLE without nephritis [61].

LYMPHOCYTE DIFFERENTIATION

T cell Receptor Signalling: *PTPN22*

Outside the HLA region, the first reproducible genetic association for RA came with the implication of *PTPN22* from a candidate gene approach [62] based on linkage analysis identification of a susceptibility locus at 1p13 [63]. It has remained the strongest and most consistent association mapped by GWA studies in RA. A role in SLE has also been identified [17]. The OR for the risk allele is around 1.75 in RA, and 1.5 in SLE. However, it should be noted that this allele (encoding the R620W mutation) is monomorphic or not disease associated in Korean or Japanese patients [64, 65]. *PTPN22* encodes lymphoid tyrosine phosphatase (LYP), a protein tyrosine phosphatase that inhibits T cell receptor signalling, decreasing IL-2 production. The disease associated SNP is responsible for a change from arginine to tryptophan at position 620, which inhibits binding to the SH3 domain of carboxy-terminal Src kinase. This in turn appears to enhance dephosphorylation of tyrosine residues in the Src family kinases Lck, FynT, and ZAP-70 [66, 67]. The overall effect of the mutation is a reduction in T cell receptor signalling. The pathogenic effect of this is unclear, but may relate to impaired negative selection in the thymus, or lead to a reduction in regulatory T cells [68]. Conversely, the R623Q variant of *PTPN22*, which is a loss-of-function mutation affecting the phosphatase activity of LYP, is protective against SLE [69]. *PTPN22* does not appear to be a risk gene for AS [70].

Polarization Towards T_H1 and T_H17 Phenotypes: *STAT4* and *IL23R*

STAT4 encodes signal transducer and activation of transcription factor-4, responsible for signalling by IL-12, IL-23, and type 1 IFNs [71]. STAT4 polarizes T cells towards T_H1 and T_H17 phenotypes, which has the potential to promote autoimmunity [72]. In RA the OR for the risk allele of SNP rs7574865 is 1.32 in one case-control study [73], with a less strong disease association at rs11893432 in a meta-analysis of GWA studies (OR 1.14) [15]. There is convincing evidence that *STAT4* is a risk locus for SLE in multiple racial groups [33, 74], and it may be theorized that interference in type I IFN signalling may be the underlying pathogenic mechanism in this case. Distinctive disease pathways could, therefore, emerge from mutations in a single gene. The WTCCC AS study identified *IL23R* as a risk gene in AS [24]. IL-23 is instrumental in the development of T cells with the pro-inflammatory T_H17 phenotype [75], and *IL23R* has been linked to psoriasis, ulcerative colitis, and Crohn's disease in GWA studies [5, 76, 77]. An interesting connection between these conditions, all of which may share common clinical features, is thus made. In AS the risk SNP rs11209032 confers an OR of 1.3.

B cell Activation

B cells are a population long suspected to be important in autoimmune rheumatic disease, and the benefits of their depletion in RA and SLE has resurrected interest in their pathogenic role. The risk genes identified so far are involved in signalling from the B cell receptor (BCR). *BLK* encodes a Src family tyrosine kinase restricted to the B cell lineage and is poorly understood. Risk alleles in the region upstream of the transcription initiation site are associated with SLE (OR 1.39, $P = 1 \times 10^{-10}$) and reduce levels of *BLK* mRNA [33]. BANK1 (B cell scaffold protein with ankyrin repeats-1) undergoes tyrosine phosphorylation upon B cell activation by the BCR, leading to an increase in intracellular calcium through the inositol trisphosphate mechanism [78]. The non-synonymous SNP rs10516487 in *BANK1*, which substitutes histidine for arginine at amino acid 61, also has disease association (OR 1.38) [79]. The functional consequence of this may be higher affinity for the inositol trisphosphate receptor, as the substitution is located in the binding site.

Lyn, another Src tyrosine kinase, is important in determining signalling thresholds for myeloid and B cells. On BCR ligation, it phosphorylates tyrosine residues of Syk, an activating tyrosine kinase, CD19, and the immunoreceptor tyrosine-based activation motifs (ITAMs) of the Igα/Igβ subunits of the BCR. However, it also has a critical regulatory role, mediated by phosphorylation of the inhibitory motifs of CD22 and Fcγ RIIB, which in turn activate SH2-domain containing phosphatases, leading to dephosphorylation and deactivation of a number of signalling intermediaries [80]. *Lyn-/-* mice develop severe autoimmunity associated with glomerulonephritis [81]. An association between SNPs in *LYN* and SLE, identified initially in the SLEGEN GWA study [17], has been recently confirmed in a case-control study [82]. The most associated SNP, rs6983130, is near the primary transcription initiation site.

OX40L, a member of the TNF super-family encoded by *TNFSF4* (TNF superfamily 4), is associated with SLE. The cross-talk between B lymphocytes and dendritic cells expressing OX40L, and T cells that express its receptor, OX40, serves to enhance the adaptive immune response [83]. An upstream *TNFSF4* haplotype, associated with SLE, enhances gene expression *in vitro* [84, 85], although the mechanism responsible for the deleterious effects observed remains to be established.

Despite the importance of B cells in the pathogenesis of RA, none of the gene effects described above have been identified in the current generation of

GWA studies. However, variants at *CD40* in European patients do carry risk [15]. CD40 expressed on B cells, via interaction with its ligand CD154 on $CD4^+$ T cells, promotes immunoglobulin class switching, and germinal centre formation. B cells, however, also have a regulatory role, likely to be mediated by IL-10, and disruption of this function may be another route to autoimmune disease [86].

Post-Translational Modification: *PADI4*

Peptidyl arginine deiminase-4 (PADI4) is a member of the enzyme family responsible for the post-translational citrullination of arginine residues in RA synovium, subsequently recognized by anti-cyclic citrullinated protein antibodies. In Japanese [87] and Korean patients [88], case-control association studies have identified functional haplotypes of *PADI4* conferring risk of RA. However, in Caucasian populations this association is inconsistent [89–91].

CONCLUSION

Even with the proliferation of new genetic associations discovered in the past few years by GWA studies, only around 10 to 15% of the inherited risk for SLE and RA can be currently explained. This may be accounted for, in part, by a number of factors, some related to limitations of recent study design. As mentioned above, even the largest current GWA cohorts have limited power to detect associations with ORs < 1.3, potentially losing multiple risk genes. By definition, most genotyped SNPs are common, and so rare but causal variants have a tendency to be missed. These rarer SNPs may be either those with a low minor allele frequency (< 5%), or occur *de novo*, of which 200 to 500 non-synonymous SNPs are expected per individual [92]. In many cases, it is far from certain if the associated SNP is functional, or in linkage disequilibrium with the true cause. Finally, the great majority of GWA studies have been conducted on European populations, thereby excluding carriers of many potential risk variants from analysis. However, it is unfortunately the case that current genotyping platforms often have poor coverage of tagging SNPs within populations that exhibit low levels of genomic linkage disequilibrium, such as those of African ancestry [93]. For example, the latest high-density genotyping chips from Affymetrix (6.0) and Illumina (1 M) may capture fewer than half the SNPs identified through re-sequencing in Yoruban Nigerians [94]. Given that clear differences exist in the risk of autoimmune disease according to ethnicity, and that not all disease risk alleles are in common, it is imperative that full account of this variation is made. Structural genetic differences have only recently begun to be assessed by modern genotyping platforms, and the contribution of, for example, CNV to inherited disease risk is largely unquantified. Even more difficult to appreciate is the influence of heritable epigenetic factors, and the exact relationship between genotype and phenotype. Nevertheless, although it will probably not be possible to explain all the observed genetic risk in the near future, we are rapidly moving towards the ability to quickly and cheaply fully sequence individual genomes [95], with all the advantages that brings [96]. In the meantime, understanding the functional basis of the disease risk variants so far identified presents an outstanding challenge. Integration of genotypic with RNA and protein expression data in a systems biologic approach represents one potentially valuable methodology [97]. Exploring and therapeutically utilizing the genetic differences between individuals is axiomatic to personalized medicine, and will undoubtedly lead to better outcomes in the management of autoimmune disease.

REFERENCES

1. International HapMap Consortium, Frazer KA, Ballinger DG, Cox DR, Hinds DA, Stuve LL, Gibbs RA, Belmont JW, Boudreau A, Hardenbol P, Leal SM, Pasternak S, Wheeler DA, Willis TD, Yu F, Yang H, Zeng C, Gao Y, Hu H, Hu W, Li C, Lin W, Liu S, Pan H, Tang X, Wang J, Wang W, Yu J, Zhang B, Zhang Q, et al.: A second generation human haplotype map of over 3.1 million SNPs. Nature 2007, 449:851–861.
2. Risch N, Merikangas K: The future of genetic studies of complex human diseases. Science 1996, 273:1516–1517.
3. Rioux JD, Xavier RJ, Taylor KD, Silverberg MS, Goyette P, Huett A, Green T, Kuballa P, Barmada MM, Datta LW, Shugart YY, Griffiths AM, Targan SR, Ippoliti AF, Bernard EJ, Mei L, Nicolae DL, Regueiro M, Schumm LP, Steinhart AH, Rotter JI, Duerr RH, Cho JH, Daly MJ, Brant SR: Genome-wide association study identifies new susceptibility loci for Crohn disease and implicates autophagy in disease pathogenesis. Nat Genet 2007, 39:596–604.
4. Wellcome Trust Case Control Consortium: Genome-wide association study of 14,000 cases of seven common diseases and 3,000 shared controls. Nature 2007, 447:661–678.
5. Sebat J, Lakshmi B, Troge J, Alexander J, Young J, Lundin P, Månér S, Massa H, Walker M, Chi M, Navin N, Lucito R, Healy J, Hicks J, Ye K, Reiner A, Gilliam TC, Trask B, Patterson N, Zetterberg A, Wigler M: Large-scale copy number polymorphism in the human genome. Science 2004, 305:525–528.
6. Redon R, Ishikawa S, Fitch KR, Feuk L, Perry GH, Andrews TD, Fiegler H, Shapero MH, Carson AR, Chen W, Cho EK, Dallaire S, Freeman

JL, González JR, Gratacòs M, Huang J, Kalaitzopoulos D, Komura D, MacDonald JR, Marshall CR, Mei R, Montgomery L, Nishimura K, Okamura K, Shen F, Somerville MJ, Tchinda J, Valsesia A, Woodwark C, Yang F, et al.: Global variation in copy number in the human genome. Nature 2006, 444:444–454.

7. Stranger BE, Forrest MS, Dunning M, Ingle CE, Beazley C, Thorne N, Redon R, Bird CP, de Grassi A, Lee C, Tyler-Smith C, Carter N, Scherer SW, Tavaré S, Deloukas P, Hurles ME, Dermitzakis ET: Relative impact of nucleotide and copy number variation on gene expression phenotypes. Science 2007, 315:848–853.

8. Aitman TJ, Dong R, Vyse TJ, Norsworthy PJ, Johnson MD, Smith J, Mangion J, Roberton-Lowe C, Marshall AJ, Petretto E, Hodges MD, Bhangal G, Patel SG, Sheehan-Rooney K, Duda M, Cook PR, Evans DJ, Domin J, Flint J, Boyle JJ, Pusey CD, Cook HT: Copy number polymorphism in Fcgr3 predisposes to glomerulonephritis in rats and humans. Nature 2006, 439:851–855.

9. Hollox EJ, Huffmeier U, Zeeuwen PL, Palla R, Lascorz J, Rodijk-Olthuis D, Kerkhof PC, Traupe H, de Jongh G, den Heijer M, Reis A, Armour JA, Schalkwijk J: Psoriasis is associated with increased beta-defensin genomic copy number. Nat Genet 2008, 40:23–25.

10. Yang Y, Chung EK, Wu YL, Savelli SL, Nagaraja HN, Zhou B, Hebert M, Jones KN, Shu Y, Kitzmiller K, Blanchong CA, McBride KL, Higgins GC, Rennebohm RM, Rice RR, Hackshaw KV, Roubey RA, Grossman JM, Tsao BP, Birmingham DJ, Rovin BH, Hebert LA, Yu CY: Gene copy-number variation and associated polymorphisms of complement component C4 in human systemic lupus erythematosus (SLE): low copy number is a risk factor for and high copy number is a protective factor against SLE susceptibility in European Americans. Am J Hum Genet 2007, 80:1037–1054.

11. Felsenfeld G, Groudine M: Controlling the double helix. Nature 2003, 421:448–453.

12. Horton R, Wilming L, Rand V, Lovering RC, Bruford EA, Khodiyar VK, Lush MJ, Povey S, Talbot CC Jr, Wright MW, Wain HM, Trowsdale J, Ziegler A, Beck S: Gene map of the extended human MHC. Nat Rev Genet 2004, 5:889–899.

13. Fernando MM, Stevens CR, Walsh EC, De Jager PL, Goyette P, Plenge RM, Vyse TJ, Rioux JD: Defining the role of the MHC in autoimmunity: a review and pooled analysis. PLoS Genet 2008, 4:e1000024.

14. Raychaudhuri S, Remmers EF, Lee AT, Hackett R, Guiducci C, Burtt NP, Gianniny L, Korman BD, Padyukov L, Kurreeman FA, Chang M, Catanese JJ, Ding B, Wong S, Helmvan Mil AH, Neale BM, Coblyn J, Cui J, Tak PP, Wolbink GJ, Crusius JB, Horst-Bruinsma IE, Criswell LA, Amos CI, Seldin MF, Kastner DL, Ardlie KG, Alfredsson L, Costenbader KH, Altshuler D, et al.: Common variants at CD40 and other loci confer risk of rheumatoid arthritis. Nat Genet 2008, 40:1216–1223.

15. Ding B, Padyukov L, Lundström E, Seielstad M, Plenge RM, Oksenberg JR, Gregersen PK, Alfredsson L, Klareskog L: Different patterns of associations with anti-citrullinated protein antibody-positive and anti-citrullinated protein antibody-negative rheumatoid arthritis in the extended major histocompatibility complex region. Arthritis Rheum 2009, 60:30–38.

16. International Consortium for Systemic Lupus Erythematosus Genetics (SLEGEN), Harley JB, Alarcón-Riquelme ME, Criswell LA, Jacob CO, Kimberly RP, Moser KL, Tsao BP, Vyse TJ, Lange-feld CD, Nath SK, Guthridge JM, Cobb BL, Mirel DB, Marion MC, Williams AH, Divers J, Wang W, Frank SG, Namjou B, Gabriel SB, Lee AT, Gregersen PK, Behrens TW, Taylor KE, Fernando M, Zidovetzki R, Gaffney PM, Edberg JC, Rioux JD, et al.: Genome-wide association scan in women with systemic lupus erythematosus identifies susceptibility variants in ITGAM, PXK, KIAA1542 and other loci. Nat Genet 2008, 40:204–210.

17. Lewis MJ, Botto M: Complement deficiencies in humans and animals: links to autoimmunity. Autoimmunity 2006, 39:367–378.

18. Fernando MM, Stevens CR, Sabeti PC, Walsh EC, McWhinnie AJ, Shah A, Green T, Rioux JD, Vyse TJ: Identification of two independent risk factors for lupus within the MHC in United Kingdom families. PLoS Genet 2007, 3:e192.

19. Harley IT, Kaufman KM, Langefeld CD, Harley JB, Kelly JA: Genetic susceptibility to SLE: new insights from fine mapping and genome-wide association studies. Nat Rev Genet 2009, 10:285–290.

20. Sekine H, Ferreira RC, Pan-Hammarström Q, Graham RR, Ziemba B, de Vries SS, Liu J, Hippen K, Koeuth T, Ortmann W, Iwahori A, Elliott MK, Offer S, Skon C, Du L, Novitzke J, Lee AT, Zhao N, Tompkins JD, Altshuler D, Gregersen PK, Cunningham-Rundles C, Harris RS, Her C, Nelson DL, Hammarström L, Gilkeson GS, Behrens TW: Role for Msh5 in the regulation of Ig class switch recombination. Proc Natl Acad Sci USA 2007, 104:7193–7198.

21. Brown MA, Pile KD, Kennedy LG, Calin A, Darke C, Bell J, Wordsworth BP, Cornélis F: HLA class I associations of ankylosing spondylitis in the white population in the United Kingdom. Ann Rheum Dis 1996, 55:268–270.
22. Brown MA, Kennedy LG, Darke C, Gibson K, Pile KD, Shatford JL, Taylor A, Calin A, Wordsworth BP: The effect of HLA-DR genes on susceptibility to and severity of ankylosing spondylitis. Arthritis Rheum 1998, 41:460–465.
23. Wellcome Trust Case Control Consortium; Australo-Anglo-American Spondylitis Consortium (TASC), Burton PR, Clayton DG, Cardon LR, Craddock N, Deloukas P, Duncanson A, Kwiatkowski DP, McCarthy MI, Ouwehand WH, Samani NJ, Todd JA, Donnelly P, Barrett JC, Davison D, Easton D, Evans DM, Leung HT, Marchini JL, Morris AP, Spencer CC, Tobin MD, Attwood AP, Boorman JP, Cant B, Everson U, Hussey JM, Jolley JD, Knight AS, et al.: Association scan of 14,500 nonsynonymous SNPs in four diseases identifies autoimmunity variants. Nat Genet 2007, 39:1329–1337.
24. Hammer GE, Gonzalez F, James E, Nolla H, Shastri N: In the absence of aminopeptidase ERAAP, MHC class I molecules present many unstable and highly immunogenic peptides. Nat Immunol 2007, 8:101–108.
25. Cui X, Rouhani FN, Hawari F, Levine SJ: Shedding of the type II IL-1 decoy receptor requires a multifunctional aminopeptidase, aminopeptidase regulator of TNF receptor type 1 shedding. J Immunol 2003, 171:6814–6819.
26. Cui X, Rouhani FN, Hawari F, Levine SJ: An aminopeptidase, ARTS-1, is required for interleukin-6 receptor shedding. J Biol Chem 2003,278:28677–28685.
27. Cui X, Hawari F, Alsaaty S, Lawrence M, Combs CA, Geng W, Rouhani FN, Miskinis D, Levine SJ: Identification of ARTS-1 as a novel TNFR1-binding protein that promotes TNFR1 ectodomain shedding. J Clin Invest 2002, 110:515–526.
28. Ronnblom L, Eloranta ML, Alm GV: The type I interferon system in systemic lupus erythematosus. Arthritis Rheum 2006, 54:408–420.
29. Sigurdsson S, Nordmark G, Göring HH, Lindroos K, Wiman AC, Sturfelt G, Jönsen A, Rantapää-Dahlqvist S, Möller B, Kere J, Koskenmies S, Widén E, Eloranta ML, Julkunen H, Kristjansdottir H, Steinsson K, Alm G, Rönnblom L, Syvänen AC: Polymorphisms in the tyrosine kinase 2 and interferon regulatory factor 5 genes are associated with systemic lupus erythematosus. Am J Hum Genet 2005,76:528–537.

30. Sigurdsson S, Padyukov L, Kurreeman FA, Liljedahl U, Wiman AC, Alfredsson L, Toes R, Rönnelid J, Klareskog L, Huizinga TW, Alm G, Syvänen AC, Rönnblom L: Association of a haplotype in the promoter region of the interferon regulatory factor 5 gene with rheumatoid arthritis. Arthritis Rheum 2007, 56:2202–2210.

31. Graham RR, Kyogoku C, Sigurdsson S, Vlasova IA, Davies LR, Baechler EC, Plenge RM, Koeuth T, Ortmann WA, Hom G, Bauer JW, Gillett C, Burtt N, Cunninghame Graham DS, Onofrio R, Petri M, Gunnarsson I, Svenungsson E, Rönnblom L, Nordmark G, Gregersen PK, Moser K, Gaffney PM, Criswell LA, Vyse TJ, Syvänen AC, Bohjanen PR, Daly MJ, Behrens TW, Altshuler D: Three functional variants of IFN regulatory factor 5 (IRF5) define risk and protective haplotypes for human lupus. Proc Natl Acad Sci USA 2007, 104:6758–6763.

32. Hom G, Graham RR, Modrek B, Taylor KE, Ortmann W, Garnier S, Lee AT, Chung SA, Ferreira RC, Pant PV, Ballinger DG, Kosoy R, Demirci FY, Kamboh MI, Kao AH, Tian C, Gunnarsson I, Bengtsson AA, Rantapää-Dahlqvist S, Petri M, Manzi S, Seldin MF, Rönnblom L, Syvänen AC, Criswell LA, Gregersen PK, Behrens TW: Association of systemic lupus erythematosus with C8orf13-BLK and ITGAM-ITGAX. N Engl J Med 2008, 358:900–909.

33. Cunninghame Graham DS, Manku H, Wagner S, Reid J, Timms K, Gutin A, Lanchbury JS, Vyse TJ: Association of IRF5 in UK SLE families identifies a variant involved in polyadenylation. Hum Mol Genet 2007, 16:579–591.

34. Graham RR, Kozyrev SV, Baechler EC, Reddy MV, Plenge RM, Bauer JW, Ortmann WA, Koeuth T, González Escribano MF, Argentine and Spanish Collaborative Groups, Pons-Estel B, Petri M, Daly M, Gregersen PK, Martín J, Altshuler D, Behrens TW, Alarcón-Riquelme ME: A common haplotype of interferon regulatory factor 5 (IRF5) regulates splicing and expression and is associated with increased risk of systemic lupus erythematosus. Nat Genet 2006, 38:550–555.

35. Sigurdsson S, Göring HH, Kristjansdottir G, Milani L, Nordmark G, Sandling JK, Eloranta ML, Feng D, Sangster-Guity N, Gunnarsson I, Svenungsson E, Sturfelt G, Jönsen A, Truedsson L, Barnes BJ, Alm G, Rönnblom L, Syvänen AC: Comprehensive evaluation of the genetic variants of interferon regulatory factor 5 (IRF5) reveals a novel 5 bp length polymorphism as strong risk factor for systemic lupus erythematosus. Hum Mol Genet 2008, 17:872–881.

36. Takaoka A, Yanai H, Kondo S, Duncan G, Negishi H, Mizutani T, Kano S, Honda K, Ohba Y, Mak TW, Taniguchi T: Integral role

of IRF-5 in the gene induction programme activated by Toll-like receptors. Nature 2005, 434:243–249.

37. Dieguez-Gonzalez R, Calaza M, Perez-Pampin E, de la Serna AR, Fernandez-Gutierrez B, Castañeda S, Largo R, Joven B, Narvaez J, Navarro F, Marenco JL, Vicario JL, Blanco FJ, Fernandez-Lopez JC, Caliz R, Collado-Escobar MD, Carreño L, Lopez-Longo J, Cañete JD, Gomez-Reino JJ, Gonzalez A: Association of interferon regulatory factor 5 haplotypes, similar to that found in systemic lupus erythematosus, in a large subgroup of patients with rheumatoid arthritis. Arthritis Rheum 2008, 58:1264–1274.

38. Plenge RM, Cotsapas C, Davies L, Price AL, de Bakker PI, Maller J, Pe›er I, Burtt NP, Blumenstiel B, DeFelice M, Parkin M, Barry R, Winslow W, Healy C, Graham RR, Neale BM, Izmailova E, Roubenoff R, Parker AN, Glass R, Karlson EW, Maher N, Hafler DA, Lee DM, Seldin MF, Remmers EF, Lee AT, Padyukov L, Alfredsson L, Coblyn J, et al.: Two independent alleles at 6q23 associated with risk of rheumatoid arthritis. Nat Genet 2007, 39:1477–1482.

39. Plenge RM, Seielstad M, Padyukov L, Lee AT, Remmers EF, Ding B, Liew A, Khalili H, Chandrasekaran A, Davies LR, Li W, Tan AK, Bonnard C, Ong RT, Thalamuthu A, Pettersson S, Liu C, Tian C, Chen WV, Carulli JP, Beckman EM, Altshuler D, Alfredsson L, Criswell LA, Amos CI, Seldin MF, Kastner DL, Klareskog L, Gregersen PK: TRAF1-C5 as a risk locus for rheumatoid arthritis - a genomewide study. N Engl J Med 2007, 357:1199–1209.

40. Komander D, Barford D: Structure of the A20 OTU domain and mechanistic insights into deubiquitination. Biochem J 2008, 409:77–85.

41. Sun SC: Deubiquitylation and regulation of the immune response. Nat Rev Immunol 2008, 8:501–511.

42. Lee EG, Boone DL, Chai S, Libby SL, Chien M, Lodolce JP, Ma A: Failure to regulate TNF-induced NF-kappaB and cell death responses in A20-deficient mice. Science 2000, 289:2350–2354.

43. Graham RR, Cotsapas C, Davies L, Hackett R, Lessard CJ, Leon JM, Burtt NP, Guiducci C, Parkin M, Gates C, Plenge RM, Behrens TW, Wither JE, Rioux JD, Fortin PR, Graham DC, Wong AK, Vyse TJ, Daly MJ, Altshuler D, Moser KL, Gaffney PM: Genetic variants near TNFAIP3 on 6q23 are associated with systemic lupus erythematosus. Nat Genet 2008, 40:1059–1061.

44. Bates JS, Lessard CJ, Leon JM, Nguyen T, Battiest LJ, Rodgers J, Kaufman KM, James JA, Gilkeson GS, Kelly JA, Humphrey MB,

Harley JB, Gray-McGuire C, Moser KL, Gaffney PM: Meta-analysis and imputation identifies a 109 kb risk haplotype spanning TNFAIP3 associated with lupus nephritis and hematologic manifestations. Genes Immun 2009, 10:470–477.

45. Kurreeman FA, Padyukov L, Marques RB, Schrodi SJ, Seddighzadeh M, Stoeken-Rijsbergen G, Helmvan Mil AH, Allaart CF, Verduyn W, Houwing-Duistermaat J, Alfredsson L, Begovich AB, Klareskog L, Huizinga TW, Toes RE: A candidate gene approach identifies the TRAF1/C5 region as a risk factor for rheumatoid arthritis. PLoS Med 2007, 4:e278.

46. Kurreeman FA, Rocha D, Houwing-Duistermaat J, Vrijmoet S, Teixeira VH, Migliorini P, Balsa A, Westhovens R, Barrera P, Alves H, Vaz C, Fernandes M, Pascual-Salcedo D, Michou L, Bombardieri S, Radstake T, van Riel P, Putte L, Lopes-Vaz A, Prum B, Bardin T, Gut I, Cornelis F, Huizinga TW, Petit-Teixeira E, Toes RE, European Consortium on Rheumatoid Arthritis Families: Replication of the tumor necrosis factor receptor-associated factor 1/complement component 5 region as a susceptibility locus for rheumatoid arthritis in a European family-based study. Arthritis Rheum 2008, 58:2670–2674.

47. Barton A, Thomson W, Ke X, Eyre S, Hinks A, Bowes J, Gibbons L, Plant D, Wellcome Trust Case Control Consortium, Wilson AG, Marinou I, Morgan A, Emery P, YEAR consortium, Steer S, Hocking L, Reid DM, Wordsworth P, Harrison P, Worthington J: Re-evaluation of putative rheumatoid arthritis susceptibility genes in the post-genome wide association study era and hypothesis of a key pathway underlying susceptibility. Hum Mol Genet 2008, 17:2274–2279.

48. Carpentier I, Beyaert R: TRAF1 is a TNF inducible regulator of NF-kappaB activation. FEBS Lett 1999, 460:246–250.

49. Tsitsikov EN, Laouini D, Dunn IF, Sannikova TY, Davidson L, Alt FW, Geha RS: TRAF1 is a negative regulator of TNF signaling. enhanced TNF signaling in TRAF1-deficient mice. Immunity 2001, 15:647–657.

50. Wang Y, Rollins SA, Madri JA, Matis LA: Anti-C5 monoclonal antibody therapy prevents collagen-induced arthritis and ameliorates established disease. Proc Natl Acad Sci USA 1995, 92:8955–8959.

51. Wang Y, Kristan J, Hao L, Lenkoski CS, Shen Y, Matis LA: A role for complement in antibody-mediated inflammation: C5-deficient DBA/1 mice are resistant to collagen-induced arthritis. J Immunol 2000, 164:4340–4347.

52. Panoulas VF, Smith JP, Nightingale P, Kitas GD: Association of the TRAF1/C5 locus with increased mortality, particularly from malignancy or sepsis, in patients with rheumatoid arthritis. Arthritis Rheum 2009, 60:39–46.

53. Larson RS, Springer TA: Structure and function of leukocyte integrins. Immunol Rev 1990, 114:181–217.

54. Sohn JH, Bora PS, Suk HJ, Molina H, Kaplan HJ, Bora NS: Tolerance is dependent on complement C3 fragment iC3b binding to antigen-presenting cells. Nat Med 2003, 9:206–212.

55. Ehirchiou D, Xiong Y, Xu G, Chen W, Shi Y, Zhang L: CD11b facilitates the development of peripheral tolerance by suppressing Th17 differentiation. J Exp Med 2007, 204:1519–1524.

56. Han S, Kim-Howard X, Deshmukh H, Kamatani Y, Viswanathan P, Guthridge JM, Thomas K, Kaufman KM, Ojwang J, Rojas-Villarraga A, Baca V, Orozco L, Rhodes B, Choi CB, Gregersen PK, Merrill JT, James JA, Gaffney PM, Moser KL, Jacob CO, Kimberly RP, Harley JB, Bae SC, Anaya JM, Alarcón-Riquelme ME, Matsuda K, Vyse TJ, Nath SK: Evaluation of imputation-based association in and around the integrin-alpha-M (ITGAM) gene and replication of robust association between a non-synonymous functional variant within ITGAM and systemic lupus erythematosus (SLE). Hum Mol Genet 2009, 18:1171–1180.

57. Nath SK, Han S, Kim-Howard X, Kelly JA, Viswanathan P, Gilkeson GS, Chen W, Zhu C, McEver RP, Kimberly RP, Alarcón-Riquelme ME, Vyse TJ, Li QZ, Wakeland EK, Merrill JT, James JA, Kaufman KM, Guthridge JM, Harley JB: A nonsynonymous functional variant in integrin-alpha(M) (encoded by ITGAM) is associated with systemic lupus erythematosus. Nat Genet 2008, 40:152–154.

58. Han S, Kim-Howard X, Deshmukh H, Kamatani Y, Viswanathan P, Guthridge JM, Thomas K, Kaufman KM, Ojwang J, Rojas-Villarraga A, Baca V, Orozco L, Rhodes B, Choi CB, Gregersen PK, Merrill JT, James JA, Gaffney PM, Moser KL, Jacob CO, Kimberly RP, Harley JB, Bae SC, Anaya JM, Alarcón-Riquelme ME, Matsuda K, Vyse TJ, Nath SK: Evaluation of imputation-based association in and around the integrin-alpha-M (ITGAM) gene and replication of robust association between a non-synonymous functional variant within ITGAM and systemic lupus erythematosus (SLE). Hum Mol Genet 2009, 18:1171–180.

59. Willcocks LC, Lyons PA, Clatworthy MR, Robinson JI, Yang W, Newland SA, Plagnol V, McGovern NN, Condliffe AM, Chilvers ER, Adu D, Jolly EC, Watts R, Lau YL, Morgan AW, Nash G, Smith KG: Copy number

of FCGR3B, which is associated with systemic lupus erythematosus, correlates with protein expression and immune complex uptake. J Exp Med 2008, 205:1573–1582.

60. Fanciulli M, Norsworthy PJ, Petretto E, Dong R, Harper L, Kamesh L, Heward JM, Gough SC, de Smith A, Blakemore AI, Froguel P, Owen CJ, Pearce SH, Teixeira L, Guillevin L, Graham DS, Pusey CD, Cook HT, Vyse TJ, Aitman TJ: FCGR3B copy number variation is associated with susceptibility to systemic, but not organ-specific, autoimmunity. Nat Genet 2007, 39:721–723.

61. Begovich AB, Carlton VE, Honigberg LA, Schrodi SJ, Chokkalingam AP, Alexander HC, Ardlie KG, Huang Q, Smith AM, Spoerke JM, Conn MT, Chang M, Chang SY, Saiki RK, Catanese JJ, Leong DU, Garcia VE, McAllister LB, Jeffery DA, Lee AT, Batliwalla F, Remmers E, Criswell LA, Seldin MF, Kastner DL, Amos CI, Sninsky JJ, Gregersen PK: A missense single-nucleotide polymorphism in a gene encoding a protein tyrosine phosphatase (PTPN22) is associated with rheumatoid arthritis. Am J Hum Genet 2004, 75:330–337.

62. Jawaheer D, Seldin MF, Amos CI, Chen WV, Shigeta R, Etzel C, Damle A, Xiao X, Chen D, Lum RF, Monteiro J, Kern M, Criswell LA, Albani S, Nelson JL, Clegg DO, Pope R, Schroeder HW Jr, Bridges SL Jr, Pisetsky DS, Ward R, Kastner DL, Wilder RL, Pincus T, Callahan LF, Flemming D, Wener MH, Gregersen PK, North American Rheumatoid Arthritis Consortium: Screening the genome for rheumatoid arthritis susceptibility genes: a replication study and combined analysis of 512 multicase families. Arthritis Rheum 2003, 48:906–916.

63. Lee HS, Korman BD, Le JM, Kastner DL, Remmers EF, Gregersen PK, Bae SC: Genetic risk factors for rheumatoid arthritis differ in Caucasian and Korean populations. Arthritis Rheum 2009, 60:364–371.

64. Ikari K, Momohara S, Inoue E, Tomatsu T, Hara M, Yamanaka H, Kamatani N: Haplotype analysis revealed no association between the PTPN22 gene and RA in a Japanese population. Rheumatology (Oxford) 2006, 45:1345–1348.

65. Cloutier JF, Veillette A: Cooperative inhibition of T-cell antigen receptor signaling by a complex between a kinase and a phosphatase. J Exp Med 1999, 189:111–121.

66. Gjörloff-Wingren A, Saxena M, Williams S, Hammi D, Mustelin T: Characterization of TCR-induced receptor-proximal signaling events negatively regulated by the protein tyrosine phosphatase PEP. Eur J Immunol 1999, 29:3845–3854.

67. Vang T, Miletic AV, Arimura Y, Tautz L, Rickert RC, Mustelin T: Protein tyrosine phosphatases in autoimmunity. Annu Rev Immunol2008, 26:29–55.

68. Orrú V, Tsai SJ, Rueda B, Fiorillo E, Stanford SM, Dasgupta J, Hartiala J, Zhao L, Ortego-Centeno N, D›Alfonso S, Italian Collaborative Group, Arnett FC, Wu H, Gonzalez-Gay MA, Tsao BP, Pons-Estel B, Alarcon-Riquelme ME, He Y, Zhang ZY, Allayee H, Chen XS, Martin J, Bottini N: A loss-of-function variant of PTPN22 is associated with reduced risk of systemic lupus erythematosus. Hum Mol Genet2009, 18:569–579.

69. Orozco G, García-Porrúa C, López-Nevot MA, Raya E, González-Gay MA, Martín J: Lack of association between ankylosing spondylitis and a functional polymorphism of PTPN22 proposed as a general susceptibility marker for autoimmunity. Ann Rheum Dis 2006,65:687–688.

70. Watford WT, Hissong BD, Bream JH, Kanno Y, Muul L, O›Shea JJ: Signaling by IL-12 and IL-23 and the immunoregulatory roles of STAT4. Immunol Rev 2004, 202:139–156.

71. Mathur AN, Chang HC, Zisoulis DG, Stritesky GL, Yu Q, O›Malley JT, Kapur R, Levy DE, Kansas GS, Kaplan MH: Stat3 and Stat4 direct development of IL-17-secreting Th cells. J Immunol 2007, 178:4901–4907.

72. Remmers EF, Plenge RM, Lee AT, Graham RR, Hom G, Behrens TW, de Bakker PI, Le JM, Lee HS, Batliwalla F, Li W, Masters SL, Booty MG, Carulli JP, Padyukov L, Alfredsson L, Klareskog L, Chen WV, Amos CI, Criswell LA, Seldin MF, Kastner DL, Gregersen PK: STAT4 and the risk of rheumatoid arthritis and systemic lupus erythematosus. N Engl J Med 2007, 357:977–986.

73. Namjou B, Sestak AL, Armstrong DL, Zidovetzki R, Kelly JA, Jacob N, Ciobanu V, Kaufman KM, Ojwang JO, Ziegler J, Quismorio FP Jr, Reiff A, Myones BL, Guthridge JM, Nath SK, Bruner GR, Mehrian-Shai R, Silverman E, Klein-Gitelman M, McCurdy D, Wagner-Weiner L, Nocton JJ, Putterman C, Bae SC, Kim YJ, Petri M, Reveille JD, Vyse TJ, Gilkeson GS, Kamen DL, et al.: High-density genotyping of STAT4 reveals multiple haplotypic associations with systemic lupus erythematosus in different racial groups. Arthritis Rheum 2009, 60:1085–1095.

74. Kastelein RA, Hunter CA, Cua DJ: Discovery and biology of IL-23 and IL-27: related but functionally distinct regulators of inflammation. Annu Rev Immunol 2007, 25:221–242.

75. Nair RP, Duffin KC, Helms C, Ding J, Stuart PE, Goldgar D, Gudjonsson JE, Li Y, Tejasvi T, Feng BJ, Ruether A, Schreiber S, Weichenthal M,

Gladman D, Rahman P, Schrodi SJ, Prahalad S, Guthery SL, Fischer J, Liao W, Kwok PY, Menter A, Lathrop GM, Wise CA, Begovich AB, Voorhees JJ, Elder JT, Krueger GG, Bowcock AM, Abecasis GR, Collaborative Association Study of Psoriasis: Genome-wide scan reveals association of psoriasis with IL-23 and NF-kappaB pathways. Nat Genet 2009, 41:199–204.

76. Silverberg MS, Cho JH, Rioux JD, McGovern DP, Wu J, Annese V, Achkar JP, Goyette P, Scott R, Xu W, Barmada MM, Klei L, Daly MJ, Abraham C, Bayless TM, Bossa F, Griffiths AM, Ippoliti AF, Lahaie RG, Latiano A, Paré P, Proctor DD, Regueiro MD, Steinhart AH, Targan SR, Schumm LP, Kistner EO, Lee AT, Gregersen PK, Rotter JI, et al.: Ulcerative colitis-risk loci on chromosomes 1p36 and 12q15 found by genome-wide association study. Nat Genet 2009, 41:216–220.

77. Yokoyama K, Su Ih IH, Tezuka T, Yasuda T, Mikoshiba K, Tarakhovsky A, Yamamoto T: BANK regulates BCR-induced calcium mobilization by promoting tyrosine phosphorylation of IP(3) receptor. EMBO J 2002, 21:83–92.

78. Kozyrev SV, Abelson AK, Wojcik J, Zaghlool A, Linga Reddy MV, Sanchez E, Gunnarsson I, Svenungsson E, Sturfelt G, Jönsen A, Truedsson L, Pons-Estel BA, Witte T, D›Alfonso S, Barizzone N, Danieli MG, Gutierrez C, Suarez A, Junker P, Laustrup H, González-Escribano MF, Martin J, Abderrahim H, Alarcón-Riquelme ME: Functional variants in the B-cell gene BANK1 are associated with systemic lupus erythematosus. Nat Genet 2008, 40:211–216.

79. Kurosaki T, Hikida M: Tyrosine kinases and their substrates in B lymphocytes. Immunol Rev 2009, 228:132–148.

80. Hibbs ML, Tarlinton DM, Armes J, Grail D, Hodgson G, Maglitto R, Stacker SA, Dunn AR: Multiple defects in the immune system of Lyn-deficient mice, culminating in autoimmune disease. Cell 1995, 83:301–311.

81. Lu R, Vidal GS, Kelly JA, Delgado-Vega AM, Howard XK, Macwana SR, Dominguez N, Klein W, Burrell C, Harley IT, Kaufman KM, Bruner GR, Moser KL, Gaffney PM, Gilkeson GS, Wakeland EK, Li QZ, Langefeld CD, Marion MC, Divers J, Alarcón GS, Brown EE, Kimberly RP, Edberg JC, Ramsey-Goldman R, Reveille JD, McGwin G Jr, Vilá LM, Petri MA, Bae SC, et al.: Genetic associations of LYN with systemic lupus erythematosus. Genes Immun 2009, 10:397–403.

82. Croft M: The role of TNF superfamily members in T-cell function and diseases. Nat Rev Immunol 2009, 9:271–285.

83. Cunninghame Graham DS, Graham RR, Manku H, Wong AK, Whittaker JC, Gaffney PM, Moser KL, Rioux JD, Altshuler D, Behrens TW, Vyse TJ: Polymorphism at the TNF superfamily gene TNFSF4 confers susceptibility to systemic lupus erythematosus. Nat Genet2008, 40:83–89.

84. Manku H, Graham DS, Vyse TJ: Association of the co-stimulator OX40L with systemic lupus erythematosus. J Mol Med 2009, 87:229–234.

85. Fillatreau S, Gray D, Anderton SM: Not always the bad guys: B cells as regulators of autoimmune pathology. Nat Rev Immunol 2008,8:391–397.

86. Suzuki A, Yamada R, Chang X, Tokuhiro S, Sawada T, Suzuki M, Nagasaki M, Nakayama-Hamada M, Kawaida R, Ono M, Ohtsuki M, Furukawa H, Yoshino S, Yukioka M, Tohma S, Matsubara T, Wakitani S, Teshima R, Nishioka Y, Sekine A, Iida A, Takahashi A, Tsunoda T, Nakamura Y, Yamamoto K: Functional haplotypes of PADI4, encoding citrullinating enzyme peptidylarginine deiminase 4, are associated with rheumatoid arthritis. Nat Genet 2003, 34:395–402.

87. Kang N, Clarke AJ, Nicholson IA, Chard RB: Circulatory arrest for repair of postcoarctation site aneurysm. Ann Thorac Surg 2004,77:2029–2033.

88. Gandjbakhch F, Fajardy I, Ferré B, Dubucquoi S, Flipo RM, Roger N, Solau-Gervais E: A functional haplotype of PADI4 gene in rheumatoid arthritis: positive correlation in a French population. J Rheumatol 2009, 36:881–886.

89. Martinez A, Valdivia A, Pascual-Salcedo D, Lamas JR, Fernández-Arquero M, Balsa A, Fernández-Gutiérrez B, de la Concha EG, Urcelay E:PADI4 polymorphisms are not associated with rheumatoid arthritis in the Spanish population. Rheumatology (Oxford) 2005,44:1263–1266.

90. Caponi L, Petit-Teixeira E, Sebbag M, Bongiorni F, Moscato S, Pratesi F, Pierlot C, Osorio J, Chapuy-Regaud S, Guerrin M, Cornelis F, Serre G, Migliorini P, ECRAF: A family based study shows no association between rheumatoid arthritis and the PADI4 gene in a white French population. Ann Rheum Dis 2005, 64:587–593.

91. Ng PC, Levy S, Huang J, Stockwell TB, Walenz BP, Li K, Axelrod N, Busam DA, Strausberg RL, Venter JC: Genetic variation in an individual human exome. PLoS Genet 2008, 4:e1000160.

92. Jakobsson M, Scholz SW, Scheet P, Gibbs JR, VanLiere JM, Fung HC, Szpiech ZA, Degnan JH, Wang K, Guerreiro R, Bras JM, Schymick JC, Hernandez DG, Traynor BJ, Simon-Sanchez J, Matarin M, Britton A, Leemput J, Rafferty I, Bucan M, Cann HM, Hardy JA, Rosenberg

NA, Singleton AB: Genotype, haplotype and copy-number variation in worldwide human populations. Nature 2008, 451:998–1003.

93. Bhangale TR, Rieder MJ, Nickerson DA: Estimating coverage and power for genetic association studies using near-complete variation data. Nat Genet 2008, 40:841–843.

94. Bentley DR, Balasubramanian S, Swerdlow HP, Smith GP, Milton J, Brown CG, Hall KP, Evers DJ, Barnes CL, Bignell HR, Boutell JM, Bryant J, Carter RJ, Keira Cheetham R, Cox AJ, Ellis DJ, Flat-bush MR, Gormley NA, Humphray SJ, Irving LJ, Karbelashvili MS, Kirk SM, Li H, Liu X, Maisinger KS, Murray LJ, Obradovic B, Ost T, Parkinson ML, Pratt MR, et al.: Accurate whole human genome sequencing using reversible terminator chemistry. Nature 2008, 456:53–59.

95. Kryukov GV, Shpunt A, Stamatoyannopoulos JA, Sunyaev SR: Power of deep, all-exon resequencing for discovery of human trait genes. Proc Natl Acad Sci USA 2009, 106:3871–3876.

96. Chen Y, Zhu J, Lum PY, Yang X, Pinto S, MacNeil DJ, Zhang C, Lamb J, Edwards S, Sieberts SK, Leonardson A, Castellini LW, Wang S, Champy MF, Zhang B, Emilsson V, Doss S, Ghazalpour A, Horvath S, Drake TA, Lusis AJ, Schadt EE: Variations in DNA elucidate molecular networks that cause disease. Nature 2008, 452:429–435.

Chapter 9

STATISTICAL DESIGN OF PERSONALIZED MEDICINE INTERVENTIONS: THE CLARIFICATION OF OPTIMAL ANTICOAGULATION THROUGH GENETICS (COAG) TRIAL

Benjamin French[1] , Jungnam Joo[2], Nancy L Geller[2], Stephen E Kimmel[1], Yves Rosenberg[3], Jeffrey L Anderson[4], Brian F Gage[5], Julie A Johnson[6], Jonas H Ellenberg[1]

[1]Department of Biostatistics and Epidemiology, University of Pennsylvania School of Medicine, 423 Guardian Drive, Philadelphia, Pennsylvania 19104 USA

[2] Office of Biostatistics Research, National Heart, Lung and Blood Institute, 6701 Rockledge Drive MSC 7913, Bethesda, Maryland 20892 USA

[3] Atherothrombosis and Coronary Artery Disease Branch, National Heart, Lung and Blood Institute, 6701 Rockledge Drive MSC 7956, Bethesda, Maryland 20892 USA

[4] JL Sorenson Heart and Lung Center, Intermountain Medical Center, 5121 S Cottonwood St, Murray, Utah 84107 USA

[5] Division of General Medical Sciences, Washington University School of Medicine, 660 S Euclid Ave, St. Louis, Missouri 63110 USA

[6] Department of Pharmacotherapy and Translational Research, University of Florida College of Pharmacy, Box 100486, Gainesville, Florida 32610 USA.

ABSTRACT

Background

There is currently much interest in pharmacogenetics: determining variation in genes that regulate drug effects, with a particular emphasis on improving drug safety and efficacy. The ability to determine such variation motivates the application of personalized drug therapies that utilize a patient's genetic makeup to determine a safe and effective drug at the correct dose. To ascertain whether a genotype-guided drug therapy improves patient care, a personalized medicine intervention may be evaluated within the framework of a randomized controlled trial. The statistical design of this type of personalized medicine intervention requires special considerations: the distribution of relevant allelic variants in the study population; and whether the pharmacogenetic intervention is equally effective across subpopulations defined by allelic variants.

Methods

The statistical design of the Clarification of Optimal Anticoagulation through Genetics (COAG) trial serves as an illustrative example of a personalized medicine intervention that uses each subject's genotype information. The COAG trial is a multicenter, double blind, randomized clinical trial that will compare two approaches to initiation of warfarin therapy: genotype-guided dosing, the initiation of warfarin therapy based on algorithms using clinical information and genotypes for polymorphisms in *CYP2C9* and *VKORC1*; and clinical-guided dosing, the initiation of warfarin therapy based on algorithms using only clinical information.

Results

We determine an absolute minimum detectable difference of 5.49% based on an assumed 60% population prevalence of zero or multiple genetic variants in either *CYP2C9* or *VKORC1* and an assumed 15% relative effectiveness of genotype-guided warfarin initiation for those with zero or multiple genetic variants. Thus we calculate a sample size of 1238 to achieve a power level of 80% for the primary outcome. We show that reasonable departures from these assumptions may decrease statistical power to 65%.

Conclusions

In a personalized medicine intervention, the minimum detectable difference used in sample size calculations is not a known quantity, but rather an unknown quantity that depends on the genetic makeup of the subjects enrolled. Given the possible sensitivity of sample size and power calculations to these key assumptions, we recommend that they be monitored during the conduct of a personalized medicine intervention.

BACKGROUND

Personalized Medicine Interventions

The recent availability of lower-cost genetic testing has motivated medical researchers to determine whether patient care and safety is improved by using a patient's genetic information to initiate and manage drug therapy [1]. To evaluate scientific hypotheses regarding a personalized medicine intervention, a randomized clinical trial can be used to contrast outcomes between subjects randomized to receive genotype-guided drug therapy and those randomized to receive an identical therapy without reference to their genetic characteristics [2]. However, because not all subjects may benefit from the pharmacologic intervention due to their genetic makeup, genotype-guided therapy may not benefit the entire study population. Hence, any putative difference between treatment groups will be attenuated, which may adversely impact key components of the statistical design, such as sample size and statistical power. Therefore, the primary statistical challenge of designing a personalized therapy intervention is to accommodate the potential differential effectiveness of genotype-guided therapy across subpopulations defined by allelic variation.

Although interventions that use a subject's clinical factors, gene expression profile, or perhaps other factors can also be considered as personalized medicine, we restrict our attention to interventions that use genotype. In addition, personalized medicine interventions may be evaluated using several different study designs. For example, in a targeted design, frequently used to evaluate genetic-based therapies for cancer, study eligibility may be restricted to a marker-positive subset of the population anticipated to benefit from therapy based on their genetic characteristics [3]. We focus on untargeted designs, such as those that have been used to evaluate genotype-guided dosing of warfarin, in which all subjects are enrolled regardless of their genetic characteristics.

Genotype-Guided Dosing of Warfarin

Warfarin sodium is the most common oral anticoagulant used for the prevention and treatment of thromboembolism, the formation of a clot in a blood vessel or cardiac chamber that may be carried by the blood stream and obstruct another vessel. Initiation of warfarin therapy is usually based on empiric dosing, which may put patients at an increased risk for either major bleeding complications due to over-anticoagulation or thromboembolic events due to under-anticoagulation. Therefore, initiation of warfarin therapy at an improper dose may be associated with increased costs and higher morbidity [4].

Many patient-specific clinical factors impact warfarin dose-response. In addition, two genes influence warfarin dose: the cytochrome P-450 family 2 subfamily C polypeptide 9 enzyme (*CYP2C9*) gene effects pharmacokinetics, i.e., the effects of the body on the drug; and the vitamin K epoxide reductase complex 1 (*VKORC1*) gene effects pharmacodynamics, i.e., the effects of the drug on the body. Thus, *CYP2C9* variants alter S-warfarin metabolism [5]; *VKORC1* variants alter warfarin response [6]. Both *CYP2C9* and *VKORC1* have proven useful in algorithms to predict the ultimate maintenance dose for optimal warfarin therapy [7]. However, they have not yet been proven to be beneficial in choosing the initial warfarin dose or to impact clinical outcomes.

The goal of this manuscript is to provide practical guidance on the statistical design of a personalized medicine intervention that uses each subject›s genotype information in an untargeted design. The statistical design of the COAG trial serves as an illustrative example. We briefly summarize the clinical rationale and the general study design for the COAG trial. We use power and sample size calculations to illustrate the primary statistical challenge of designing a personalized therapy intervention: to accommodate the potential differential effectiveness of genotype-guided therapy across subpopulations defined by allelic variation. We provide a sensitivity analysis to quantify the extent to which power and sample size calculations may be sensitive to key assumptions required in the statistical design of a personalized medicine intervention. We conclude with general recommendations for the statistical design of personalized medicine interventions.

METHODS

The objective of the COAG trial (clinicaltrials.gov identifier: NCT00839657) is to conduct a multicenter, double blind, randomized clinical trial that compares two approaches to initiation of warfarin therapy:

Genotype-guided dosing, the initiation of warfarin therapy based on algorithms using clinical information and genotypes for polymorphisms in two genes known to influence warfarin response (*CYP2C9* and *VKORC1*); and

Clinical-guided dosing, the initiation of warfarin therapy based on algorithms using only clinical information.

Both approaches will include a baseline dose-initiation algorithm [8] and a dose-revision algorithm [9] applied after four or five days of warfarin therapy. Subsequent doses will be determined using a standard dose-titration algorithm, which is identical for both groups. By comparing the efficacy of genotype-guided dosing to that of clinical-guided dosing, the COAG trial will determine whether the incremental use of genetic information improves stability of anticoagulation during the early treatment period. Future studies could then determine whether such an improvement leads to significantly reduced costs and lower morbidity.

Eligible subjects will be recruited from at least 12 clinical sites in the United States. Clinical and genotype data will be collected on all subjects. Subjects will be randomized to initiate warfarin therapy either using genotype-guided or clinical-guided dosing. All subjects will receive their warfarin on a standard-of-care schedule. Study investigators, clinicians, and subjects will be blinded to treatment assignment and warfarin dose for the first four weeks of the trial. After four weeks of therapy, subjects will be unblinded to dose and followed for up to an additional five months. The Institutional Review Board of all participating institutions approved the COAG trial. Written informed consent will be obtained from all patients who participate in the trial.

The primary outcome of the COAG trial is the percentage of time that participants spend within a therapeutic range for anticoagulation (PTTR) during the first four weeks of therapy. The therapeutic range is defined using the International Normalized Ratio (INR), which reflects the ratio of a patient's prothrombin time to that for a control sample. An INR between 2.0 and 3.0, inclusive, is typically considered to be within the therapeutic range. To calculate the PTTR for each subject, we will use a standard interpolation method that assumes a linear change in INR from one measurement to the next [10]. Figure 1 illustrates the linear interpolation method for a hypothetical subject whose therapeutic INR range is between 2.0 and 3.0, with a corresponding PTTR of 60%.

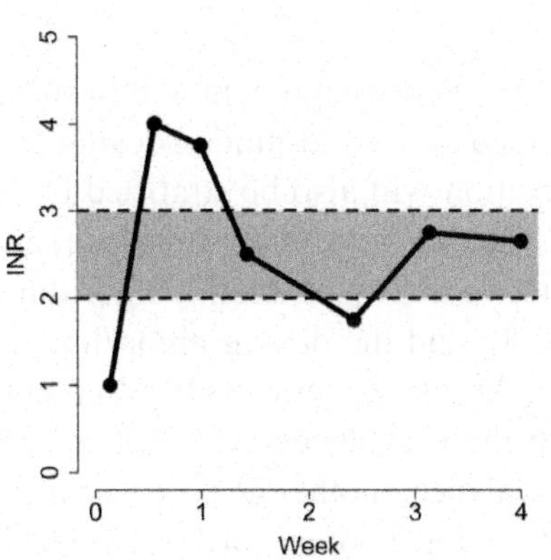

Figure 1: INR measurements (solid circles) and linear interpolations (solid lines) for a hypothetical subject with 60% of time within the therapeutic INR range (shaded region).

Analysis of the primary outcome will be by intention-to-treat [11]. It will not be possible for subjects to switch from their assigned treatment group, but there might be crossovers due to the unavailability of genetic information at the time that the initial dose is dispensed. Every attempt will be made to determine a subject's genotype prior to administration of the initial dose. Given recent technologies, same-day genotyping for warfarin is now possible in practice. In the COAG trial, clinical sites are using one of two genotyping platforms; each has a rapid turnaround time. Both platforms have been FDA approved, have high call and concordance rates, very low failure rates, and the ability to genotype the SNPs needed for the selected dosing algorithms.

For those subjects assigned to the genotype-guided dosing group whose genetic information is not available prior to the initial dose, the initial dose will be determined using the clinical dose-initiation algorithm. Once genetic information becomes available, the dose for these subjects will be determined using the genetic dose-initiation and dose-revision algorithms. The genotype-guided dose-initiation algorithm on day one only uses information on *VKORC1* (not *CYP2C9*) [8]. Therefore, we expect the dose differences on day one to be small relative to the dose differences after the first day. The genotype-guided dose-initiation algorithm on day two, as well as the genotype-guided dose-revision algorithm on days four and five [9], uses information from both *VKORC1* and *CYP2C9*, so that the availability of genetic information by day two will allow the full use of the subject's genetic information to determine their dose for days two through five. We fully expect genotype information to be available on almost all subjects within 24 hours, and certainly by the time of the dose-revision calculations on days four and five.

Randomization

To provide balance in treatment assignment within sites, random assignment to either the genotype-guided or clinical-guided dosing group will be stratified by clinical site. Randomization will also be stratified by race (African American versus not, including Caucasian and Asian American) because race is associated with differential predictive ability of dosing algorithms, with lesser accuracy in African Americans [8], and the dosing algorithms used in the trial predict dose differently among African Americans [9]. In addition, African-American race is associated with the prevalence *CYP2C9* and*VKORC1* variants and is associated with the prevalence of other genetic variants that influence warfarin dose-response. Finally, some clinical sites may recruit a small number of African Americans due to the demographic makeup of their surrounding community.

We will use a block-randomized procedure to assign the treatment groups. Blocking ensures that there will be a balance in the number of patients in each treatment group within each clinical site. Thus, we will use permuted blocks with block sizes of four and six, randomly chosen, which will minimize any imbalances in treatment group assignment. The RANUNI function in SAS 9.2 will be used to generate the randomization numbers within each site for two strata [12].

Sample Size and Statistical Power

A critical element of the statistical design of a randomized clinical trial is to determine a sample size so that a statistical test has adequate power to detect a clinically relevant difference in the primary outcome between treatment groups. The parameters considered in the estimation of sample size include: a minimum detectable difference in the primary outcome between groups; an assumed level of significance for the statistical test of the primary outcome; a measure of variability for the primary outcome in the study population; and the percentage of subjects, if any, expected to drop out of the trial. The sample size parameters for a personalized medicine intervention require additional considerations: the distribution of relevant allelic variants in the study population; and whether the intervention is equally effective across subpopulations defined by allelic variants (e.g., if patients with particular genotypes are not expected to benefit from genotype-guided drug therapy, as we illustrate for warfarin). Due to uncertainly in the distribution of allelic variants and uncertainty in the effectiveness of the intervention across subpopulations, careful attention is required in the design of a personalized medicine to ensure that the study will have adequate power to detect a clinically relevant minimum detectable difference.

In the statistical design of the COAG trial, we focused on the difference in the relative effectiveness of genotype-guided across two genetic subpopulations: those with a single genetic variant versus zero or multiple genetic variants in either*CYP2C9* or *VKORC1*. We viewed the primary outcome of PTTR in each treatment group as a weighted average of PTTR and the corresponding treatment effect (Δ) across subpopulations defined by 1 versus 0, > 1 variants, in which the weights (w) are determined by the populations prevalences that sum to 1:

$$\mathrm{PTTR_C} = w_1 \times \mathrm{PTTR}_1 + w_{0,>1} \times \mathrm{PTTR}_{0,>1}; \tag{1}$$

$$\mathrm{PTTR_G} = w_1 \times \mathrm{PTTR}_1 \times \Delta_1 + w_{0,>1} \times \mathrm{PTTR}_{0,>1} \times \Delta_{0,>1}; \tag{2}$$

where $PTTR_C$ and $PTTR_G$ denote the PTTR in the clinical-guided and genotype-guided dosing groups, respectively. It is straightforward to generalize this approach to more than two subpopulations of interest. Adding additional terms into the weighted average, given the population prevalence and the anticipated treatment effect in each subpopulation, could accommodate more than two subpopulations. Indeed, this approach is generalizeable to any setting in which treatment effects are expected to differ across any number of subpopulations. Specific assumptions are discussed in the following section.

Minimum Detectable Difference

We considered the distribution of *CYP2C9* and *VKORC1* variants and whether genotype-guided dosing of warfarin is equally effective across groups defined by *CYP2C9* and *VKORC1* variants. Current evidence suggests that there will be a subgroup with certain genotypes that will not benefit from genotype-guided dosing [13], most likely because their predicted dose from genotype-guided dosing algorithms will not meaningfully differ from predicted dosing with clinical dosing. We based sample size estimates on the comparison of PTTR between the genotype-guided and clinical-guided dosing groups:

$$PTTR_C = 0.4 \times 73\% + 0.6 \times 61\% = 65.80\%; \tag{3}$$

$$PTTR_G = 0.4 \times 73\% \times 1 + 0.6 \times 61\% \times 1.15 = 71.29\%; \tag{4}$$

where:

The proportion in the population who possess a single genetic variant (in either *CYP2C9* or *VKORC1*) and who possess zero or multiple variants is assumed to be 0.4 and 0.6, respectively;

A PTTR of 73% and 61% is assumed for those who possess a single genetic variant and for those who possess zero or multiple variants, respectively;

A 0% relative difference in PTTR is assumed for those with a single genetic variant; and

A 15% relative difference in PTTR for those with zero or multiple variants is assumed to be a clinically relevant difference between the genotype-guided and clinical-guided dosing groups [14].

We assumed that subjects who possess a single genetic variant (in either *CYP2C9* or *VKORC1*) would not benefit from clinical-guided dosing because previous data suggest that the genotype-guided algorithm will predict essentially the same dose as the clinical-guided algorithm. These subjects are expected to attain the same PTTR regardless of their treatment assignment and thus attenuate the mean difference in PTTR between the two groups. To wit, the assumed 15% relative difference in PTTR between the genotype-guided and clinical-guided dosing groups is attenuated to an absolute difference of 5.49% ($PTTR_G$ - $PTTR_C$). Therefore, we assumed an overall minimum detectable difference of 5.49% between groups in the full cohort for sample size calculations. If we had ignored the fact that the intervention is not equally effective across subpopulations defined by genetic variants and assumed a minimum detectable difference of 15%, then the trial would have chosen an inadequately small sample size to achieve adequate power.

The assumed proportion of 0.4 who possess a single genetic variant is based on the Couma-Gen trial [13] and the International Warfarin Pharmacogenetics Consortium (IWPC) [15]. We considered the sensitivity of sample size calculations to a range a population proportions. The assumption that those who possess a single genetic variant will not benefit from genotype-guided dosing, while suggested by the Couma-Gen trial, was not supported in another clinical trial in which all patients benefited from dosing based on *CYP2C9* [16]. In the latter study, the effect of a pharmacogenetic dosing algorithm was similar regardless of the number of *CYP2C9* variants present. Therefore, we believe that our assumptions are conservative.

Level of Significance

To determine the level of significance (α) for the statistical test of PTTR between the genotype-guided and clinical-guided dosing groups, we considered an alpha-allocation approach [17–19]. In this approach, a portion (α_A) of the overall level of significance is used to test the comparison in the full cohort; the remaining portion (α_S) is used to test the comparison in a pre-defined primary subgroup. The alpha-allocation approach facilitates a traditional primary analysis to assess a statistically significant difference between the treatment groups, as well as a predefined primary subgroup analysis that is not relegated to a secondary analysis, as in a traditional analysis.

We defined the primary subgroup based on subjects whose predicted initial dose employing the genetic and clinical dose-initiation algorithms differs by $\geq$ 1.0 mg, a factor known at the time of randomization and therefore not a post-randomization selection. We posited that the subgroup of participants with a larger difference between the predicted initial doses should have a larger separation in PTTR between the two groups. If the improvement in PTTR is related to the magnitude of difference in dosing between the genotype-guided and clinical-guided dosing groups, then the primary subgroup comparison should reflect a larger absolute difference than the 5.49% assumed for the full cohort analysis. We assumed that a clinically relevant absolute difference to detect in the primary subgroup is 9.15%, from a PTTR of 61% to 70.15% in Equation (4), reflecting a 15% relative difference.

We selected $\alpha_A = 0.04$ for the full cohort analysis and $\alpha_S = 0.01$ for the primary subgroup analysis, for an overall type-I error rate of $\alpha = 0.05$. However, allocating alpha so that sum of α_A and α_S is equal to α is a conservative Bonferroni-type correction, which may be unnecessarily conservative if there is a positive correlation between the tests in the full cohort and in the primary subgroup [20, 21]. The correlation between the two tests will be obtained under the null hypothesis when the size of the primary subgroup is known. The correlation will then be incorporated to obtain $\alpha_S > \alpha - \alpha_A$ given that α_A is fixed, so that the overall type-I error rate is controlled at α.

Other assumptions in the computation of sample size were the standard deviation of the PTTR in the study population and the percentage of subjects expected to drop out before reaching the primary endpoint. The within-study variability of PTTR in the literature varied across study designs and populations under study. However, there was a reasonable consistency of variability for the genetic-guided and clinical-guided dosing groups in the studies reviewed. We assumed a standard deviation of 25% based on a study of dose-refinement algorithms in which the standard deviation averaged 23% [22]. We also assumed that 10% of subjects would drop out before reaching the primary endpoint and increased the sample size by dividing the calculated sample size by the square of one minus the drop-out rate [23].

Primary Analysis

The null hypothesis for the primary outcome is that the percent of time that subjects spend within the therapeutic INR range (PTTR) during the first four weeks of therapy is equal between the genotype-guided and clinical-guided dosing groups. We will estimate the difference in mean PTTR between the genotype-guided and clinical-guided dosing groups using a linear regression model, both for the full cohort and for the primary subgroup whose predicted initial dose employing the genetic and clinical dose-initiation algorithms differs by ≥ 1 mg. Inference will be based on a Wald test with a level of significance of 0.05 allocated between the full cohort analysis and the primary subgroup analysis. Because randomization will be stratified by site and race, these variables will be included in the linear regression model. We will perform additional analyses in subgroups defined *a priori* by allelic variation (zero versus a single versus multiple *CYP2C9*or *VKORC1* variants) and by race (African American versus not).

Additional genetic factors may be considered in secondary analyses. Specifically, because *CYP2C9* and *VKORC1* genotypes may not be the only genetic variants that determine optimal warfarin dosing, it is possible that more variants will be identified during the trial. To adjust for additional genetic factors in secondary analyses, we will include them as covariates in a linear regression model if their prevalence differs between the clinical and genetic groups, and also consider possible interactions with *CYP2C9* and/or *VKORC1*.

RESULTS

Table 1 provides the sample size required for the full cohort analysis using a two-sample *t*-test with $\alpha_A = 0.04$ (two-sided), assuming various proportions with a single genetic variant (0.4, 0.5, and 0.6), estimates for the standard deviation of PTTR (20%, 25%, and 30%), and power levels (80% and 90%), and drop-out rate (10%). A sample size of 1140 would provide 90% power to detect an absolute difference of 5.49% in the full cohort, given that the proportion with a single genetic variant is 0.4 and the standard deviation is 25%. We selected a sample size of 1238 to protect against departures from the assumed proportion with a single genetic variant, study drop-out rate, and standard deviation of PTTR. For example, if the proportion with a single genetic variant is 0.5 and the standard deviation is 25%, then there is 80% power to detect an absolute difference in PTTR of 4.58%. If the proportion with a single genetic variant is 0.4 and standard deviation is 30%, then there is 80% power to detect the assumed 5.49% absolute difference.

Table 1: Sample size estimates for the full cohort analysis; *p* denotes the proportion with a single genetic variant and Δ denotes the corresponding minimum detectable difference

		Standard Deviation of PTTR					
		20%		**25%**		**30%**	
		Power		**Power**		**Power**	
p	**Δ**	**80%**	**90%**	**80%**	**90%**	**80%**	**90%**
0.4	5.49%	550	730	860	1140	1238	1642
0.5	4.58%	792	1050	1238	1642	1782	2364
0.6	3.67%	1238	1642	1932	2564	2782	3692

A sample size of 1238 provides sufficient power for the primary subgroup analysis using a two-sample *t*-test with $\alpha_S = 0.01$ (two-sided). Recall that the size of the primary subgroup is determined by the percentage of subjects whose predicted initial dose employing the genetic and clinical dose-initiation algorithms differs by ≥ 1 mg. If the relative size of the primary subgroup is 50% and the standard deviation of PTTR is 25%, then there is 93.6% power to detect a 9.15% absolute difference. In addition, if the relative size of the primary subgroup is 60% and the standard deviation is 30%, then there is 87.8% power. In fact, the power will be higher because α_S will be increased according to the correlation between the tests in the full cohort and in the primary subgroup.

Sensitivity Analysis

In the statistical design of the COAG trial, there was a concern that the genotype-guided and clinical-guided dosing algorithms may not produce sufficiently differentiable doses between the treatment groups, which may lead to an underestimation of the minimum detectable difference in PTTR between groups. We assumed that any difference between the two groups would arise from the subgroup of patients with either zero or multiple genetic variants. (Recall that the assumed relative difference in the genotype-guided dosing group was 15% for those with zero or multiple variants.) For subjects in this allelic subgroup, if the difference between the two algorithm predictions is negligible or clinically irrelevant, then it is reasonable to expect no difference in PTTR. In this case the PTTR for the genotype-guided dosing group can be expressed as:

$$\mathrm{PTTR}_G = 0.4 \times 73\% + 0.6 \times 61\% \times [1 + (0.15 \times d)], \tag{5}$$

where d is the proportion of subjects with zero or multiple genetic variants in whom there is a clinically meaningful difference between the predicted dose determined by the genotype-guided and clinical-guided dosing algorithms. Hence the expected 15% difference would be diluted by a factor d and it would be more difficult to detect a clinically relevant difference between groups.

To explore the impact of dilution of the treatment effect, we examined the distribution of the differences between the predicted doses among groups defined by allelic variation in the IWPC cohort [15] and calculated the difference between the rounded predicted doses. An absolute dose difference < 1.0 mg per day was defined as the ‹same› predicted dose; an absolute dose difference of ≥ 1.0 mg per day was defined as a ‹different› predicted dose. The rationale for the 1.0 mg cut-point is that the average initial dose is 5.0 milligrams; therefore, a 1.0 mg absolute difference represents a clinically relevant 20% difference, on average. Approximately 9% of IWPC participants in the (0, >1) allelic variant group would have received the ‹same› initial dose, i.e., $d = 0.91$. With this dilution of the treatment effect, in order to detect an overall effect size of 5.49% in PTTR, the relative effect size in the (0, > 1) group would need to be 16.5%.

Table 2 provides power estimates for the test of the full cohort analysis for a range of diluted treatment effects corresponding to the parameter d, the proportion of subjects with zero or multiple genetic variants in whom there is a difference between the predicted doses. There is sufficient power when $d > 0.9$.

We are not highly confident in our estimate of how frequently the predicted dose will differ between the two algorithms and therefore have not taken this potential dilution effect into account in our calculations for sample size and power. However, given the potential impact of the dilution effect on the sample size requirements of the study seen in Table 2, we have planned to monitor this factor during the operation of the trial.

Table 2: Power estimates for the full cohort analysis in which the treatment effect is diluted; *d* denotes the proportion of subjects with zero or multiple genetic variants in whom there is a difference between the predicted initial doses

d	**Treatment Effect**	**Power**
0.7	3.84	65%
0.8	4.39	77%
0.9	4.94	86%
1.0 (Undiluted)	5.49	93%

DISCUSSION

In this manuscript we provided practical guidance on the statistical design of a personalized medicine intervention that uses each subject's genotype information in an untargeted design. We used power and sample size calculations to illustrate the primary statistical challenge of designing this type of personalized therapy intervention: to accommodate the potential differential effectiveness of genotype-guided therapy across subpopulations defined by allelic variation. To determine a minimum detectable difference in PTTR between groups, we assumed that 40% of enrolled subjects would have a single genetic variant and would therefore not benefit from genotype-guided warfarin therapy. Hence, the minimum detectable difference used in sample size calculations is not a known quantity, but rather an unknown quantity that depends on the genetic makeup of the subjects enrolled. In addition, the sample size for the primary subgroup analysis depends on the proportion of subjects whose predicted initial dose employing the genetic and the clinical dose-initiation algorithms differs by $\geq$ 1.0 mg. Due to the importance of these parameters for adequate sample size and statistical power to detect a clinically meaningful difference, they will be monitored during the course of the trial.

As shown in Table 1, the sample size is sensitive to the standard deviation of PTTR. The Data Safety and Monitoring Board (DSMB) may suggest an 'internal pilot study' in which an estimate of the standard deviation will be obtained using the first half of the observed data and the sample size calculations will be updated based on the new estimate [24, 25]. The pre-planned sample size will be assumed to represent a minimum sample size (i.e., the final sample size based on the 'internal pilot study' will not be less than the pre-planned sample size). In this case, the 'internal pilot study' is known as restricted. For restricted designs, the disparity in the type-I error rate in testing the primary hypothesis is negligible [26]. Therefore, it will not be necessary to adjust the type-I error rate of any hypothesis tests regarding the primary outcome. In assessing the need for a sample size increase, data will neither be unblinded nor assessed for the primary outcome. In addition, a sample size adjustment will not impact the overall design of the study. Because the DSMB will not monitor efficacy during the conduct on the COAG trial, there is no conflict between any interim sample size adjustment and interim measures of efficacy.

In our sensitivity analysis, we examined the dilution of the treatment effect due to a clinically irrelevant difference between the predicted doses (employing the genetic and clinical dose-initiation algorithms) for subjects with zero or multiple genetic variants. However, we did not consider the impact of a clinically relevant difference between the predicted doses for subjects with a single variant. In this situation the PTTR for the genotype-guided dosing group can be expressed as:

$$\text{PTTR}_G = 0.4 \times 73\% \times [1 + (0.15 \times d')] + 0.6 \times 61\% \times [1 + (0.15 \times d)], \quad (6)$$

where d' is the proportion of subjects with a single genetic variant in whom there is a meaningful difference between the predicted doses and d is defined in Equation (5). For example, in the IWPC cohort, approximately 26% of subjects with a single genetic variant would have received a ‹different› initial dose, i.e., $d' = 0.26$. For these subjects, we expect that there would be a difference in PTTR, which would increase the power of the full cohort analysis. Because we were not highly confident in this estimate, we did not examine the increase in power associated with this allelic subgroup. Therefore, our sensitivity analysis is conservative.

An individual's genetic information could be used prior to randomization to identify subjects who are potentially unresponsive to either drug therapy or the pharmacologic intervention, motivating researchers to decide whether to include or exclude those subjects from the trial [27]. For example, in a targeted design, study eligibility may be restricted to subjects who, based on their genetic characteristics, are predicted to be responsive [28]. By excluding potentially unresponsive individuals, a targeted design will require a smaller sample size to detect a statistically significant effect. Conversely, in a traditional (untargeted) design, particularly of an intervention designed to select dose, subjects for whom genetic-based drug therapy is not effective are eligible, because they would still receive drug treatment regardless of their genetic makeup. For example, subjects in the COAG trial would receive warfarin therapy regardless of their *CYP2C9* and*VKORC1* variants. As we have shown with the statistical design of the COAG trial, by including potentially unresponsive subjects, a larger sample size may be required. Cost-benefit considerations regarding the cost of genetic screening for eligibility versus the cost of enrolling potentially unresponsive subjects may be useful to determine which design is more practical in specific applications.

For the COAG trial, we favored including all participants, regardless of their genetic variants. First, the assumption that those who possess a single genetic variant will not benefit from genotype-guided dosing, while suggested by the Couma-Gen trial [13], was not supported in another clinical trial in which all patients benefited from dosing based on *CYP2C9* [16]. Therefore, if we excluded subjects who may not benefit from genotype-guided dosing, we would be unable to evaluate our assumptions. Second, all subjects are genotyped prior to randomization, so that much of the cost is already incurred in screening. Third, including subjects potentially unresponsive to genotype-guided dosing allows the results of the trial to be more generalizable. That is, if the COAG trial indicates that genotype-guided dosing provides increased efficacy compared to clinical-guided dosing, then it motivates consideration of the policy question of whether all patients prescribed warfarin should be genotyped to predict the drug›s efficacy.

We recommend that the statistical design of a personalized medicine intervention that uses each subject›s genotype information, within the framework of a randomized clinical trial, consider the distribution of relevant allelic variants in the study population and whether the intervention is equally effective across subpopulations defined by allelic variants. In the statistical design of the COAG trial, we considered the distribution of *CYP2C9* and *VKORC1* variants and whether genotype-guided dosing of warfarin therapy would provide an equal improvement in efficacy across populations defined by genetic variants. We assumed that subjects with a single genetic variant would not benefit from genotype-guided dosing, thus attenuating the postulated 15% relative difference between the two treatment groups to a 5.49% absolute difference. In our sample size calculations, if we ignored the fact that the genotype-guided dosing is not equally effective across subpopulations defined by genetic variants and assumed a minimum detectable difference of 15%, then the COAG trial would likely have chosen an inadequately small sample size to achieve adequate power. We also recommend that key assumptions regarding sample size and statistical power be monitored during the conduct of the trial, to inform any requisite increase in the sample size needed to detect a clinically relevant difference in the primary outcome between treatment groups. Further research is required to determine whether an interim sample size adjustment based on the observed proportion of allelic variants increases the type-I error rate.

CONCLUSIONS

In summary, we found that sample size and power calculations may be sensitive to key assumptions required in the design of a personalized medicine intervention: the distribution of relevant allelic variants in the study population; and whether the pharmacogenetic intervention is equally effective across subpopulations defined by allelic variants. Given the novelty of pharmacogenetic research, we recommend that these assumptions be monitored during the conduct of a personalized medicine intervention.

ACKNOWLEDGEMENTS

We gratefully acknowledge the National Heart, Lung and Blood Institute (N01 HV88210) and the University of Pennsylvania for supporting this research, and two reviewers for comments that greatly improved the manuscript.

COAG Investigators: Sherif Abdel-Rahman, University of Texas; Robert J Desnick and Jonathan L Halperin, Mount Sinai School of Medicine; Margaret C Fang, University of California, San Francisco; Brian F Gage, Washington University School of Medicine; Richard B Horenstein, University of Maryland School of Medicine; Julie A Johnson, University of Florida; Scott Kaatz, Henry Ford Hospital; Robert D McBane, Mayo Clinic College of Medicine; Emile R Mohler III, Hospital of the University of Pennsylvania; James A S Muldowney III, Vanderbilt University; Scott M. Stevens, Intermountain Medical Center; Steven Yale, Marshfield Clinical Research Foundation.

AUTHORS' CONTRIBUTIONS

All authors made substantial contributions to conception and design. BF and JJ drafted the manuscript. All authors revised the manuscript critically for important intellectual content and approved the final manuscript.

REFERENCES

1. Terra SG, Johnson JA: Pharmacogenetics, pharmacogenomics, and cardiovascular therapies: The way forward. Am J Cardiovasc Drugs 2002, 2:287–296.
2. Wang SJ, O›Neill RT, Hung HM: Approaches to evaluation of treatment effect in randomized clinical trials with genomic subset.Pharm Stat 2007, 6:227–244.

3. Sargent DJ, Conley BA, Allegra C, Collette L: Clinical trial designs for predictive marker validation in cancer treatment trials. J Clin Oncol 2005, 23:2020–2027.

4. Garcia D, Regan S, Crowther M, Hughes RA, Hylek EM: Warfarin maintenance dosing patterns in clinical practice: Implications for safer anticoagulation in the elderly population. Chest 2005, 127:2049–2056.

5. Sanderson S, Emery J, Higgins J: CYP2C9 gene variants, drug dose, and bleeding risk in warfarin-treated patients: A HuGEnet systematic review and meta-analysis. Genet Med 2005, 7:97–104.

6. Rieder MJ, Reiner AP, Gage BF, Nickerson DA, Eby CS, McLeod HL, Blough DK, Thummel KE, Veenstra DL, Rettie AE: Effect of VKORC1haplotypes on transcriptional regulation and warfarin dose. N Engl J Med 2005, 352:2285–2293.

7. Schelleman H, Chen J, Chen Z, Christie J, Newcomb CW, Brensinger CM, Price M, Whitehead AS, Kealey C, Thorn CF, Samaha FF, Kimmel SE: Dosing algorithms to predict warfarin maintenance dose in Caucasians and African Americans. Clin Pharmacol Ther 2008,84:332–339.

8. Gage BF, Eby D, Johnson JA, Deych E, Rieder MJ, Ridker PM, Milligan PE, Grice G, Lenzini P, Rettie AE, Aquilante CL, Grosso L, Marsh S, Langaee T, Farnett LE, Voora D, Veenstra DL, Glynn RJ, Barrett A, McLeod HL: Use of pharmacogenetic and clinical factors to predict the therapeutic dose of warfarin. Clin Pharmacol Ther 2008, 84:326–331.

9. Lenzini P, Wadelius M, Kimmel S, Anderson JL, Jorgensen A, Pirmohamed M, Caldwell MD, Limdi N, Burmester JK, Dowd MB, Angchaisuksiri P, Bass AR, Chen J, Eriksson N, Rane A, Lindh JD, Carlquist JF, Horne BD, Grice G, Milligan PE, Eby C, Shin J, Kim H, Kurnik D, Stein CM, McMillin G, Pendleton RC, Berg RL, Deloukas P, Gage BF: Integration of genetic, clinical, and laboratory data to refine warfarin dosing. Clin Pharmacol Ther 2010, 87:572–578.

10. Rosendaal FR, Cannegieter SC, van der Meer FJ, Briët E: A method to determine the optimal intensity of oral anticoagulant therapy.Thromb Haemost 1993, 69:236–239.PubMed

11. Ellenberg JH: Intention-to-treat analysis. In Encyclopedia of Biostatistics. Edited by: Armitage P, Colton T. New York: John Wiley & Sons; 1998:2056–2060.

12. Fishman GS, Moore LR: A statistical evaluation of multiplicative congruential random number generators with modulus $2^{31} - 1$. J Amer Statist Assoc 1982, 77:129–136.View Article

13. Anderson JL, Horne BD, Stevens SM, Grove AS, Barton S, Nicholas ZP, Kahn SF, May HT, Samuelson KM, Muhlestein JB, Carlquist JF, Couma-Gen Investigators: Randomized trial of genotype-guided versus standard warfarin dosing in patients initiating oral anticoagulation. Circulation 2007, 116:2563–2570.

14. Dolan G, Smith LA, Collins S, Plumb JM: Effect of setting, monitoring intensity and patient experience on anticoagulation control: A systematic review and meta-analysis of the literature. Curr Med Res Opin 2008, 24:1479–1472.View Article

15. The International Warfarin Pharmacogenetics Consortium: Estimation of the warfarin dose with clinical and pharmacogenetic data.N Engl J Med 2009, 360:753–764.View Article

16. Caraco Y, Blotnick S, Muszkat M: CYP2C9 genotype-guided warfarin prescribing enhances the efficacy and safety of anticoagulation: A prospective randomized controlled study. Clin Pharmacol Ther 2008, 83:460–470.

17. Moyé LA: P-value interpretation and alpha allocation in clinical trials. Ann Epidemiol 1998, 8:351–357.

18. Moyé LA, Deswal A: Trials within trials: Confirmatory subgroup analyses in controlled clinical experiments. Control Clin Trials 2001,22:605–619.

19. Coats AJ: CAPRICORN: A story of alpha allocation and beta-blockers in left ventricular dysfunction post-MI. Int J Cardiol 2001,78:109–113.

20. Alosh M, Hugue MF: A flexible strategy for testing subgroups and overall population. Stat Med 2009, 15:3–23.View Article

21. Joo J, Geller NL, French B, Kimmel SE, Rosenberg Y, Ellenberg JE: Prospective alpha allocation in the Clarification of Optimal Anticoagulation through Genetics (COAG) trial. Clin Trials 2010, 7:597–604.

22. Lenzini PA, Grice GR, Milligan PE, Dowd MB, Subherwal S, Deych E, Eby CS, King CR, Porche-Sorbet RM, Murphy CV, Marchand R, Millican EA, Barrack RL, Clohisy JC, Kronquist K, Gatchel SK, Gage BF: Laboratory and clinical outcomes of pharmacogenetic vs. clinical protocols for warfarin initiation in orthopedic patients. J Thromb Haemost 2008, 6:1655–1662.

23. Lachin JM: Introduction to sample size determination and power analysis for clinical trials. Control Clin Trials 1981, 2:93–113.

24. Wittes J, Brittain E: The role of internal pilot studies in increasing the efficiency of clinical trials. Stat Med 1990, 9:65–71.

25. Betensky RA, Tierney C: An examination of methods for sample size recalculation during an experiment. Stat Med 1997, 16:2587–2598.
26. Wittes J, Schabenberger O, Zucker D, Brittain E, Proschan M: Internal pilot studies I: Type I error rate of the naive t-test. Stat Med1999, 18:3481–91.
27. Simon R: The use of genomics in clinical trial design. Clin Cancer Res 2008, 14:5984–5993.
28. Simon R, Maitouram A: Evaluating the efficiency of targeted designs for randomized clinical trials. Clin Cancer Res 2004, 10:6759–6763.

Chapter 10

FROM BIOPHYSICS TO EVOLUTIONARY GENETICS: STATISTICAL ASPECTS OF GENE REGULATION

Michael Lässig
Institut für Theoretische Physik, Universität zu Köln

ABSTRACT

This is an introductory review on how genes interact to produce biological functions. Transcriptional interactions involve the binding of proteins to regulatory DNA. Specific binding sites can be identified by genomic analysis, and these undergo a stochastic evolution process governed by selection, mutations, and genetic drift. We focus on the links between the biophysical function and the evolution of regulatory elements. In particular, we infer fitness landscapes of binding sites from genomic data, leading to a quantitative evolutionary picture of regulation.

INTRODUCTION

Genomic functions often cannot be understood at the level of single genes but require the study of gene networks. This systems biology credo is nearly commonplace by now. Evidence comes from the comparative analysis of entire genomes: Current estimates put, for example, the number of human genes at around 22000, hardly more than the 14000 of the fruit fly, and not even an order of magnitude higher than the 6000 of baker's yeast. The complexity and diversity of higher animals therefore cannot be explained in terms of their gene numbers. If, however, a biological function requires the concerted action of several genes, and conversely, a gene takes part in several functional contexts, an organism may be defined less by its individual genes but by their interactions. The emerging picture of the genome as a strongly interacting system with many degrees of freedom brings new challenges for experiment and theory, many of which are of a statistical nature. And indeed, this picture continues to make the subject attractive to a growing number of statistical physicists.

Genes encode proteins, and proteins perform functions in the cell. Hence, a gene takes part in a biological function only if it is expressed, i.e., if the protein produced from it is present in the cell. Genes interact by regulation: the protein of one gene can influence the production of protein from another gene. Gene regulation can take place during transcription, the process by which the cell reads the information contained in a gene and copies it to messenger RNA (which is subsequently used to make a functional protein). This is the most fundamental level of interactions between genes: the transcription of one gene may be enhanced or reduced by the expression of other genes. Transcriptional regulation is thus a good starting point for theory. We should keep in mind, however, that it is not the only mode of gene interactions. Especially in eukaryotes, additional regulation mechanims involving histones, chromatin, micro-RNAs etc. become relevant, which are just entering the stage of model building. An excellent introduction to the biology of regulation can be found in [1].

This article is a primer on theoretical aspects of gene interactions, and we limit ourselves to transcriptional regulation. Clearly, the subject has rather diverse aspects:

(1) Transcription is a biophysical process, which involves the interaction of DNA and proteins. Its regulation takes place through the binding of proteins to DNA at specific loci in the vicinity of the gene to be regulated. Already at this level, this process is rather complex and not yet fully understood. What enables the protein to find one or a few specific functional sites in a genome of up to billions of base pairs, bind there with sufficient strength to influence transcription, and leave again once its task is performed?

(2) Given that the protein can find its functional sites, can we as well? If that is possible, we can predict the specific gene interactions building regulatory networks from sequence data. The analysis of regulatory DNA is a major topic of research in bioinformatics, with the aim of identifying statistical characteristics of functional loci and of building search algorithms.

(3) Regulation is also becoming an important part of evolutionary biology [2, 3]. If regulatory networks are to explain the differentiation of higher animals, there must be efficient modes of evolution for the interactions between genes. At the level of regulary DNA, these modes remain largely to be explored. It is clear, however, that the underlying evolutionary dynamics is the basis of a quantitative understanding of regulatory networks.

All three aspects of regulation contribute to a unified theoretical picture. Key concepts such as the biophysical binding energy, the bioinformatic scoring function, and the evolutionary fitness turn out to be rather deeply related. We will focus on these crosslinks between different fields, which are likely to become important for future research. A challenge for an introductory presentation is the diversity of relevant background material, only a rather ecclectic account of which can be presented here. Yet, I hope it transpires even from this short introduction that present quantitative genomics is an area of science shaped by a remarkable confluence of ideas from different disciplines.

BIOPHYSICS OF TRANSCRIPTIONAL REGULATION

The fundamental step in the regulatory interaction between two genes is a binding process: the protein produced by the first gene acts as a transcription factor for the second gene, i.e., it binds to a functional site on the DNA close to the second gene and thereby enhances or suppresses its transcription. Binding sites are short, typically segments of 10 to 15 base pairs in prokaryotes and even shorter segments in eukaryotes. They are primarily located in the cis-regulatory region of a gene, which lies just upstream of its protein-coding sequence and extends over hundreds of base pairs in prokaryotes and over thousands of base pairs in eukaryotes.

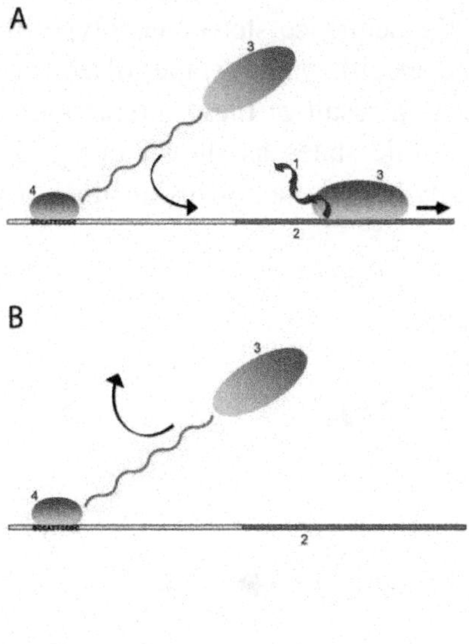

Figure 1: Transcriptional regulation. Transcription is the synthesis of messenger RNA (1) whose genetic code is a copy of the coding DNA (2) of a gene, by means of RNA

polymerase (3). A transcription factor (4) bound to a DNA target site interacts with RNA polymerase molecules, (a) enhancing or (b) reducing the transcription rate of a nearby gene.

The scenario of transcriptional regulation is sketched in Fig. 1. A transcription factor bound to a functional binding site regulates the downstream gene by recruiting or repelling RNA polymerase. This protein-protein interaction catalyzes or suppresses the process of transcription of the gene. All these binding processes should not be understood as on or off; they happen with certain probabilities, which are determined by the binding energies and the numbers of the molecules involved.

Factor-DNA Binding Energies

The interaction of a transcription factor protein with DNA is two-fold: There is a position-unspecific attraction with energy E_u and a specific interaction, whose energy depends on the particular locus where the factor binds. The unspecific part is the electrostatic interaction between the positively charged protein and the negatively charged DNA backbone, while the specific part involves hydrogen bonds between the binding domain of the protein and the nucleotides of the binding locus. A locus is specified by its starting position r and its length ℓ (with relevant values ℓ of order 10). The specific binding energyE(r) depends on ℓ consecutive nucleotides $\mathbf{a} = (a_1,..., a_\ell)$ counted downstream from the starting position, the sequence stateor genotype of that locus. Switching between unspecific and specific binding takes place via a conformation change of the factor protein. As a result of these interactions, the factor protein can be in three thermodynamic states as shown in fig. 2: unbound (i.e., freely diffusing), unspecifically bound (i.e., diffusing along the DNA backbone), and specifically bound.

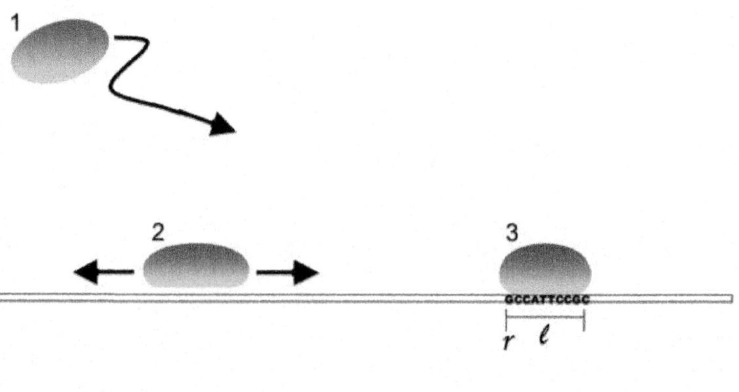

Figure 2: Thermodynamic states of a transcription factor. (1) Unbound state, with three-dimensional diffusion. (2) Unspecific bound state, with one-dimensional diffu-

sion along the DNA backbone. (3) Specific bound state. The binding energy depends on the genotype at the binding locus, which has length ℓ and whose position is specified by the coordinate r.

The biophysics of factor-DNA binding has been established in a series of seminal papers [4–7]. More recently, the characteristics of specific binding have been measured for some bacterial transcription factors [8–12]. These can be summarized as follows:

(a) The single nucleotides of a binding locus $\mathbf{a} = (a_1, ..., a_\ell)$ give approximately independent contributions to the binding energy,

$$E(\mathbf{a}) = \sum_{i=1}^{\ell} \varepsilon_i(a_i). \tag{1}$$

(b) At each position i, there is typically one preferred nucleotide a_i^* with $\varepsilon_i(a_i^*) = \min_a \varepsilon_i(a)$. Hence, there is a unique "ground state" sequence $\mathbf{a}^*$ a_1^* ,...a_ℓ^*) with minimal binding energy($\mathbf{a}^*$), i.e., with strongest binding.

(c) ismatches with respect to the minimum-energy sequence involve energy costs $\varepsilon_i(a) - \varepsilon_i(a_i^*) \approx 1 - 3\ k_B T$ per nucleotide.

(d) There is an energy difference $E_u - E^* \sim 15\ k_B T$ between unspecific and strongest specific binding. Experimental data for the binding energies $\varepsilon_i(a)$ are known only for a few transcription factors.

Approximate values for these energies can also be inferred from nucleotide frequencies in functional binding sites [10]. A promising recent approach is to infer binding energies from large-throughput expression data [13]. For order-of-magnitude estimates, one often uses the so-called two-state approximation [7], which is homogeneous in the nucleotide positions and distinguishes only between match and mismatch:

$$\varepsilon_i(a) - \varepsilon_i(a_i^*) = \begin{cases} \varepsilon & \text{if } a_i \neq a_i^* \\ 0 & \text{if } a_i = a_i^* \end{cases} \tag{2}$$

with $\varepsilon \approx 2k_B T$. In this approximation, the binding energy of a sequence $\mathbf{a}$ is simply related to the Hamming distance d($\mathbf{a}$,$\mathbf{a}^*$), i.e., the number of nucleotide mismatches between $\mathbf{a}$ and $\mathbf{a}^*$,

$$E(\mathbf{a}) = E^* + \varepsilon \cdot d(\mathbf{a}, \mathbf{a}^*). \tag{3}$$

Energy Distribution in the genome

Fig. 3(a) shows the sequence of energy values E(r) found in a segment of the E. coli genome for a specific transcription factor, the cAMP response protein

(CRP) This «energy landscape» looks quite random, i.e., consecutive energy values are approximately uncorrelated. The distribution $W_{dat}(E)$ of energies over the entire noncoding part of the E. coli genome is shown in fig. 3(b). We can compare this with the distribution $W_0(E)$ obtained from a random sequence with the same nucleotide frequencies (i.e., from a scrambled genome). According to eq. (1), the binding energy E is then a sum of independent random variables ε_i, and its distribution becomes approximately Gaussian by the law of large numbers. Fig.3(b) shows that the actual distribution $W_{dat}(E)$ is indeed of the same form as $W_0(E)$ for most energies. However, a closer look at the low-energy tail of the distribution shows that there are significantly more strong binding sites than expected from a random sequence [14–16]. So at least some of them are there not by chance but for a reason.

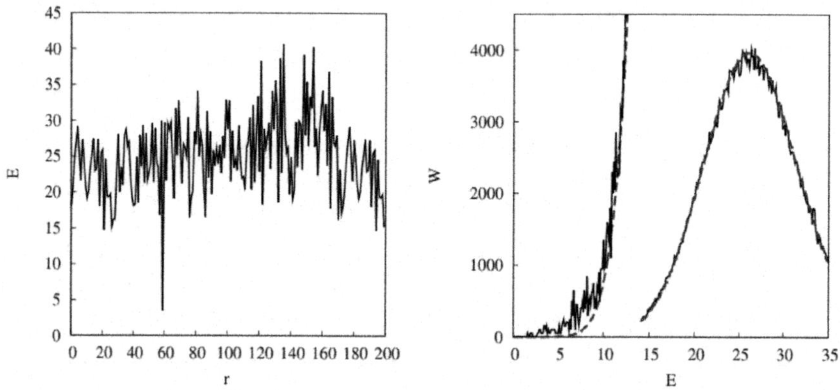

Figure 3: Transcription factor binding energies of the E. coli genome. (a) Energy "landscape" E(r) for specific binding of the CRP factor at 200 consecutive positions r in an intergenic region, with a binding site at position 59. (b) Count histogram $W_{dat}(E)$ with energy bins of width 0.1 obtained from all intergenic regions, together with the distribution $W_0(E)$ for a random sequence (dashed line, shown with a 30 fold zoom into the region $E < 14$). From [16].

Search Kinetics

All three thermodynamic modes of a factor molecule – free diffusion, unspecific binding, and specific binding – are important for the search kinetics towards a functional site [4–6]. The unspecific attraction causes the transcription factor to be bound to DNA with a finite probability, i.e., a given molecule spends about equal amounts of time on and off the DNA backbone. Hence, the search process is a mixture of effectively one-dimensional diffusion along the DNA backbone and three-dimensional diffusion in the surrounding medium. This proves more efficient than purely one- or three-dimensional diffusion.

In the 1D mode, the factor diffuses in a flat energy landscape if it is in the conformation of unspecific binding, or in the landscape E(r) if it is in the conformation of specific binding. In this way, it can sample the low-energy part of the landscape E(r) while avoiding its barriers. The main obstacles on its way to a functional site are spurious binding sites, which have a low energy E(r) by chance and act as traps. We lack a completely satisfactory picture of the search kinetics, which is an area of current research [14, 17]. However, this process proves to be remarkably fast. Typical search times are less than a minute, i.e., substantially shorter than typical functional intervals in a cell cycle of at least minutes. Therefore, the regulatory effect of a site is related to its probability of binding a factor molecule at equilibrium, which can be evaluated by standard thermodynamics.

Thermodynamics of Factor Binding

We start with the idealized but instructive problem of a single factor protein interacting with a genome of length $L \gg 1$, which contains a single functional site, while the rest of the sequence is random. Since the protein is bound to the DNA with a probability of about 1/2, we neglect the unbound state for the subsequent probability estimates and study only the bound protein, which is at equilibrium between specific and unspecific binding. At each position r, the likelihood of these two states is given by the Boltzmann factors $\exp[-E(r)/k_B T]$ and $\exp[-E_u/k_B T]$, respectively. Hence, the partition function for a single protein has the form

$$Z = \sum_{r=1}^{L} e^{-E(r)/k_B T} + L e^{-E_u/k_B T}. \tag{4}$$

The functional site, which is assumed to be positioned at $r = r_f$, must have a low specific binding energy $E = E(r_f)$. We now single out this position and write

$$\begin{aligned} Z &= e^{-E/k_B T} + \sum_{r \neq r_f} e^{-E(r)/k_B T} + L e^{-E_u/k_B T} \\ &\approx e^{-E/k_B T} + Z_0, \end{aligned} \tag{5}$$

where Z_0 is the partition function of a completely random sequence. The probability of the factor being bound specifically at the functional site is then

$$p(E) = \frac{e^{-E/k_B T}}{Z} = \frac{1}{1 + e^{(E-F_0)/k_B T}}, \tag{6}$$

where $F_0 = -k_B T \log Z_0$ is the free energy for a random genome. Thus, the binding probability depends on the binding energy in a sigmoid way, with a threshold energy $E = F_0$ between strong and weak binding.

This strongly nonlinear dependence is known to physicists as a Fermi function.

It is easy to generalize the thermodynamic formalism to more than one factor molecule. Ignoring the overlap between close sites, each position r can be empty or be occupied either by an unspecifically or by a specifically bound factor. Using a chemical potential σ, the many-factor partition function can hence be written as

$$Z(\sigma) = \prod_{r=1}^{L} Z(\sigma, r), \tag{7}$$

where

$$Z(\sigma, r) = 1 + e^{\sigma - E(r)/k_B T} + e^{\sigma - E_u/k_B T} \tag{8}$$

is a sum over the three thermodynamic states at position r: no factor bound, one factor bound specifically or unspecifically. The chemical potential σ is determined by the number of factor molecules, n, via the relation $n = (d/d\sigma) \log Z(\sigma)$. For actual transcription factor numbers, which are of order $1 - 10^4$, this relation is well approximated by [14]

$$\sigma = \frac{F_0}{k_B T} + \log n. \tag{9}$$

The functional site is now occupied by a specifically bound factor with probability

$$p(E) = \frac{e^{\sigma - E/k_B T}}{Z(\sigma, r_f)} = \frac{1}{1 + e^{(E - F_0)/k_B T - \log n}}. \tag{10}$$

The binding probability – and hence the effects of the functional site on the regulated gene – are thus determined by the binding energy, the number of factor molecules, and on the genomic background (via the free energy F_0). The dependencep(E) is a Fermi function with threshold energy $E = F_0 + k_B T \log n$, which is shifted with respect to the single-molecule case. Clearly, p is also a Fermi function of log n at fixed binding energy, with a threshold at $\log n = (E - F_0)/k_B T$. If there is more than one functional site in the genome, the calculation remains unaffected as long as their number is much smaller than n.

Sensitivity and Genomic Design of Regulation

The regulatory machinery can be very efficient: in bacteria, it has been shown that single factor molecules can have regulatory effects. We can use eq. (6) to enquire how the cell can reach this high level of sensitivity, following mostly ref. [14]. We assume a minimal genome which has a single functional site of maximum binding strength E* and is otherwise random. If a single factor molecule is to affect regulation, its binding to the functional site must not be overwhelmed by the remainder of the genome. This leads to a criterion on the signal-to-noise ratio of regulatory interactions,

$$F_0 \gtrsim E^*, \qquad (11)$$

which in turn imposes a number of constraints on the design of regulatory DNA:

(a) In a random genome, there must be at most a number of order one minimum-energy binding sites. Estimating the probability to find such a site at a given position as $(1/4)^\ell$, we obtain the condition

$$L(1/4)^\ell \lesssim 1. \qquad (12)$$

This gives a lower bound on the site length, $\ell \gtrsim \log L/\log 4$. For a bacterial genome ($L \sim 10^6$), we obtain $\ell \gtrsim 10$, which gives the right length of functional binding sites. However, this bound is not fulfilled in eukaryotes. Indeed, eukaryotic genomes use a different design with groups of adjacent binding sites.

(b) For each minimum-energy site, there are ℓ suboptimal sites of Hamming distance 1 from the minimum-energy sequence. These must not suppress the binding to the minimum-energy site, i.e.,

$$\exp(-E^*/k_B T) \gtrsim \ell \exp\left[-(E^* + \varepsilon)/k_B T\right] \qquad (13)$$

in the two-state approximation. This gives a lower bound on the binding energy per nucleotide, $\varepsilon/k_B T \gtrsim \log \ell \approx 2 - 3$.

(c) Finally, the unspecific binding in the entire genome must not suppress the specific binding to a minimum-energy site, i.e.,

$$\exp(-E^*/k_B T) \gtrsim L \exp(-E_u/k_B T). \qquad (14)$$

This produces a lower bound on the energy gap between unspecific and optimal specific binding, $(E_u - E^*)/k_B T \gtrsim \log L \approx 15$.

Quite remarkably, these bounds are fulfilled as approximate equalities in bacteria. Hence, the machinery of transcriptional regulation operates just at the treshold of single-molecule sensitivity, i.e, $F_0 \approx E^*$.

Programmability and Evolvability of Regulatory Networks

Of course, not every regulatory interaction is equally sensitive. To switch genes on or off, the cell uses the dependencies of the binding probability both on factor numbers and on binding energies. During the cell cycle, the level of n can vary over several orders of magnitude, say, between a few and tens of thousands of molecules. At a given value of n, the effects on the regulated genes differ since their functional sites have different values of E. The binding energies can change on evolutionary time scales by mutations of the site sequence, which leads to regulatory differences between individuals and, ultimately, between species. Both parameters are thus necessary to encode pathways in regulatory networks. This is most flexible if minimum-energy sites are indeed sensitive to a single factor molecule as discussed above. Differentialprogrammability as a network design principle [14] thus favors complicated molecular structures with longer binding sites and larger binding energies. However, this competes with the evolvability of the system by a stochastic evolution process [18]. We have seen that the single-molecule sensitivity is just marginally reached in bacteria. This indicates that the actual machinery may result from a compromise between programmability and evolvability: binding sites are just complicated enough to work. It also indicates that genomic structures can only be understood from their evolution. This aspect will be developed further below, after we have introduced sequence analysis aspects in the next section.

Bioinformatics of Regulatory DNA

Predicting regulatory interactions between genes is clearly a key problem in bioinformatics, which is as important as the analysis of individual genes and proteins. It is not surprising that this problem is very difficult since, as we have discussed in the previous section, targeting regulatory input in a large genome is a tremendous signal-to-noise problem even for the cell itself. Its solution via the analysis of regulatory DNA requires finding statistical criteria to distinguish between functional binding sites and background sequence. A general introduction to the relevant sequence statistics can be found in ref. [19].

Markov Model for Background Sequence

We begin by specifying a stochastic model for the nonfunctional segments of intergenic DNA. These are assumed to be Markov sequences with uniform single-nucleotide frequencies $p_0(a)$ (a = A, C, G, T). Hence, the probability of finding a given sequence has the factorized form

$$P_0(a_1,\ldots,a_k) = \prod_{i=1}^{k} p_0(a_i). \tag{15}$$

This assumption should not be taken too literally. The term "nonfunctional" refers to binding of a particular transcription factor. Intergenic DNA contains plenty of non-random elements with other functions (e.g., binding sites for other factors) or without known function (such as repeat elements). The salient point is, however, that most of intergenic DNA is well approximated by a Markov sequence with respect to binding of a given transcription factor. To make this more precise, we project the distribution $P_0(\mathbf{a})$ for segments of length ℓ onto the binding energy E as independent variable. Denoting the projected distribution for simplicity with the same letter P_0, we have

$$P_0(E) \equiv \sum_{\mathbf{a}} P_0(\mathbf{a})\delta(E - E(\mathbf{a})). \tag{16}$$

This distribution is close to the actual genomic distribution $W_{dat}(E)$ for most values of E, as we have seen in fig. 3. It is possible to improve the background model by introducing small frequency couplings between neigboring letters [15, 16].

Probabilistic Model for Functional Sites

The sequences $\mathbf{a} = (a_1,\ldots, a_\ell)$ at functional sites of a given transcription factor are assumed to be drawn from a different distribution $Q(\mathbf{a})$. We write this distribution in the form

$$Q(\mathbf{a}) = P_0(\mathbf{a}) \exp [S(\mathbf{a})]. \tag{17}$$

The quantity $S(\mathbf{a})$, which is called the relative log likelihood score of the distributions P_0 and Q, will turn out to have an important evolutionary meaning as well.

The single-nucleotide distribution $q_i(a)$ at a given position i within functional loci is obtained by summing the full distribution Q over all other positions

$$q_i(a) = \sum_{a_1,\ldots,a_{i-1},a_{i+1},\ldots,a_\ell} Q(\mathbf{a}). \tag{18}$$

The set of these marginal distributions, $q_i(a)$ ($i = 1,\ldots, \ell$; a = A, C, G, T) is called the position weight matrix for binding sites of a given factor [20]. If the score function is additive in the nucleotide positions, $S(\mathbf{a}) = \sum_{i=1}^{\ell} s_i(a_i)$, the Q distribution has a factorized form, Q(a)=∏ℓi=1qi(ai)$Q(\mathbf{a}) = \prod_{i=1}^{\ell} q_i(a_i)$ with

$$q_i(a) = p_0(a) \exp [s_i(a)]. \tag{19}$$

This additivity assumption is made in most of the existing literature since the position weight matrix (18) can be inferred from a sample of known functional site sequences, which in turn determines directly the single nucleotide scores (19). This scoring is the basis for a number of site prediction methods in single species and by cross-species analysis; see, e.g., refs. [20–24].

Here we treat functional sites as coherent statistical units and do not make the assumption of additivity of the score function [16]. As will be discussed in the next section, functionality imposes correlations between the nucleotide frequencies within a functional site, preventing factorization of the Q distribution. Of course, it is not possible to reconstruct the full distribution Q(**a**), which lives on a 4^{ℓ}-dimensional sequence space, from a limited sample of experimentally known functional sites. However, we can again project this distribution onto the binding energy as independent variable, $Q(E) = \sum_{a} Q(\mathbf{a})\delta(E - E(\mathbf{a}))$. Since all regulatory effects of a functional site depend on its sequence a only via the binding energy, we can also write the score as a function of the energy, $S(\mathbf{a}) = S(E(\mathbf{a}))$ (this will become obvious in the next section). Hence, the relationship (17) has the same form for the projected distributions,

$$Q(E) = P_0(E) \exp [S(E)]. \tag{20}$$

Bayesian Model for Genomic loci

Assuming that functional loci are distributed randomly with a small probability λ, we now combine the models for background sequence and for functional sites into a model for the full distribution of sequences **a** in intergenic DNA,

$$W(\mathbf{a}) = (1 - \lambda) P_0(\mathbf{a}) + \lambda Q(\mathbf{a}). \tag{21}$$

(At the moment, we are ignoring the possible overlap between functional sites). In the language of statistics, this is a probabilistic model with hidden variables. The output of this model consists of pairs (m, **a**): First, the model variable m is randomly drawn, labelling a locus as nonfunctional (m = 0) with probability 1 - λ or as functional (m = 1) with probability λ. Then the sequence is drawn from the corresponding distribution $P_0(\mathbf{a})$ or $Q(\mathbf{a})$. However, only the sequence counts **a** are available data. The "hidden" variable m can be inferred from the data in a probabilistic way using Bayes' formula, which expresses the joint probability distribution of data and model in terms of its conditional and its marginal distributions

$$\text{prob}(\mathbf{a}, m) = \text{prob}(\mathbf{a}|m)\, \text{prob}(m) = \text{prob}(m|\mathbf{a})\, \text{prob}(\mathbf{a}) \tag{23}$$

with $\text{prob}(\mathbf{a}) = \sum_{m} \text{prob}(\mathbf{a}|m)\text{prob}(m)$. We can solve for the conditional probability of the model for given data **a**,

$$\text{prob}(m \mid \mathbf{a}) = \frac{\text{prob}(\mathbf{a} \mid m)\text{prob}(m)}{\sum_m \text{prob}(\mathbf{a} \mid m)\text{prob}(m)}. \quad (23)$$

For the probability of functionality, $\rho_f(\mathbf{a}) = \text{prob}(m = 1|\mathbf{a})$, this formula reads

$$\rho_f(\mathbf{a}) = \frac{\lambda Q(\mathbf{a})}{W(\mathbf{a})} = \frac{1}{1 + \exp\left[-S(\mathbf{a}) + \log\frac{1-\lambda}{\lambda}\right]}. \quad (24)$$

The dependence on S has again the form of a Fermi function. Its threshold value $S = \log[(1 - \lambda)/\lambda]$ separates sequences that are more likely to be functional or more likely to be background.

The full Bayesian model (21) can again be projected onto the energy variable,

$W(E) = (1 - \lambda)P_0(E) + \lambda\, Q(E)$.

In this form, it can be tested against genomic data [16]. To plot the distributions P_0, Q, and W as functions of E, we use eq. (1) with an energy matrix $\varepsilon_i(a) = \varepsilon_0 \log[q_i(a)/p_0(a)]$ estimated from the position weight matrix up to an overall constant ε_0[10]. For our example of the CRP transcription factor, the distribution Q(E) can be estimated from the about 50 known binding sites in the E. coli genome. Using this Q distribution and a probability of functionality $\lambda \approx 6 \times 10^{-4}$, the full distribution W(E) produces an excellent fit of the count histogram $W_{dat}(E)$ over the entire range of energies; see fig. 4(a). The log likelihood score function $S(E) = \log[Q(E)/P_0(E)]$ is shown in fig. 4(b), shifted such that the curve has its zero at a point $E_s \approx 13$ beyond which binding becomes negligible.

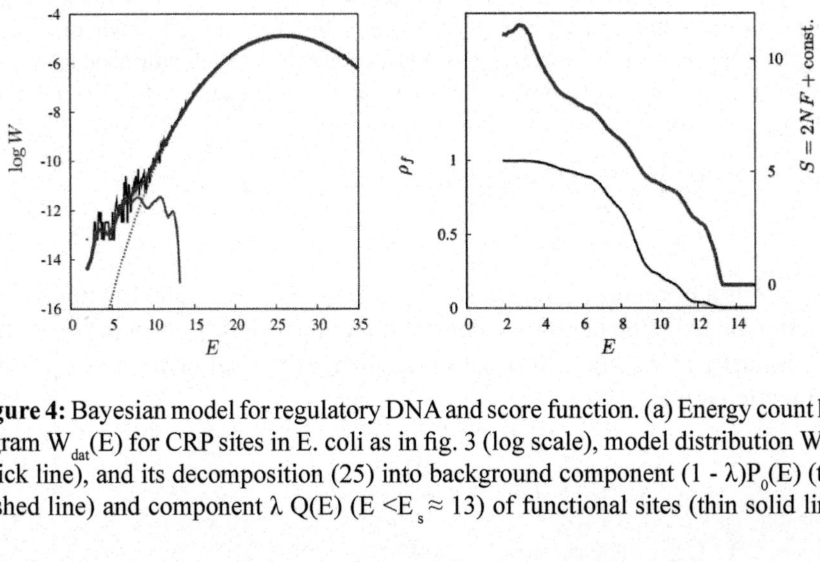

Figure 4: Bayesian model for regulatory DNA and score function. (a) Energy count histogram $W_{dat}(E)$ for CRP sites in E. coli as in fig. 3 (log scale), model distribution W(E) (thick line), and its decomposition (25) into background component $(1 - \lambda)P_0(E)$ (thin dashed line) and component $\lambda\, Q(E)$ ($E < E_s \approx 13$) of functional sites (thin solid line).

(b) Log-likelihood score $S(E) = \log [Q(E)/P_0(E)]$ (shifted by a constant, thick line) and probability of functionality $\rho_f(E)$ (thin line). From [16].

The resulting probability of functionality ρf(E) as given by eq. (24) is also shown in fig. 4(b). This indicates the dilemma for the prediction of individual binding sites based on sequence data from a single species. Many functional sites have energies in the "twilight" region between the ensembles λQ and $(1 - \lambda)P_0$, where ρf takes values around 1/2. Hence, depending on the energy cutoff chosen, any prediction is torn between many false negatives or many false positives.

Dynamic Programming and Sequence Analysis

It is straightforward to generalize the Bayesian approach to longer segments of intergenic DNA, which are covered by an unknown number s of non-overlapping functional sites as shown in fig. 5[22]. The hidden variables are now the sequence of left initial positions $\mathbf{r}_f \equiv (r_1,..., r_s)$ of the functional sites (with the no-overlap constraint $r_{v+1} \geq r_v + \ell$ for $v = 1,..., s - 1$). The full sequence distribution in a segment of length L has the form

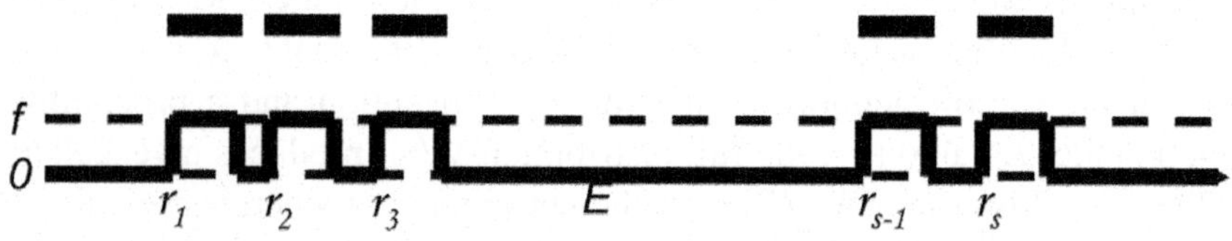

Figure 5: Analysis of regulatory sequences. A configuration of s nonoverlapping binding sites is given by the sequence of left initial positions $r_f = (r_1,..., r_s)$ (with $r_{v+1} - r_v \geq \ell$ for $v = 1, 2,..., s - 1$). It can be associated with a path m(r) which takes the values m = 1 at the nucleotide positions of binding sites and m = 0 elsewhere. Dynamic programming algorithms based on a Bayesian model (27) of genomic sequences assign to each site configuration a probability of occurence $\rho(r|a_1,..., a_L)$ for given sequence data $a_1,..., a_L$; see eq. (29).

$$W_L(a_1,\ldots,a_L) = Z^{-1}\sum_{\mathbf{r}_f} \tilde{\lambda} W_L(a_1,\ldots,a_L \mid \mathbf{r}_f), \tag{26}$$

where Z is a normalization factor, $\lambda \sim \tilde{\lambda} = \lambda + O(\lambda^2)$ is a weight factor for each functional locus (the negligible correction terms originate from the no-overlap constraint), and $W_L(a_1,..., a_L|\mathbf{r}_f)$ is the sequence distribution for given positions of functional loci,

$$W_L(a_1,\ldots,a_L \mid \mathbf{r}_f) =$$
$$p_0(a_1)\ldots p_0(a_{r_1-1})\prod_{\nu=1}^{s} Q(a_{r_\nu},\ldots,a_{r_\nu+\ell-1})p_0(a_{r_\nu+\ell})\ldots p_0(a_{r_{\nu+1}-1}) =$$
$$p_0(a_1)\ldots p_0(a_L)\exp\left[\sum_{\nu=1}^{s} S(a_{r_\nu},\ldots,a_{r_\nu+\ell-1})\right] \tag{27}$$

with $r_{n+1} = L + 1$. The sum over sequences $\mathbf{r}_f$ of arbitrary length s seems formidable at first, but W_L is easy to compute from the recursion

$$W_r(a_1,\ldots, a_r) = (1 - \lambda\hat{\ })p_0(a_r)W_{r-1}(a_1,\ldots, a_{r-1}) + \lambda\tilde{\ }Q(a_{r-\ell+1},\ldots, a_r)\, W_{r-\ell}(a_1,\ldots, a_{r-\ell}) \tag{28}$$

with the initial condition $W_0 = 1$ and $\lambda\hat{\ }\hat{\lambda} = \lambda\tilde{\ }\tilde{\lambda} + O(\lambda\tilde{\ }\tilde{\lambda}^2)$. This type of recursion relation is usually called a dynamic programming algorithm in computer science. In physics, it is known as a transfer matrix, and the sum (27) is recognized as the corresponding discrete path integral in imaginary time r, if we interpret $\mathbf{r}_f$ as encoding a path m(r) that takes the value m = 1 at the nucleotide positions $r_\nu,\ldots, r_\nu + \ell - 1$ ($\nu = 1,\ldots, s$) within functional loci and m = 0 otherwise (see fig. 5). Both concepts prove very useful also in more general problems of sequence alignment.

In analogy to (24), the probability of a set $\mathbf{r}_f$ of functional loci for given sequence data is

$$p(\mathbf{r}_f \mid a_1,\ldots,a_L) = \frac{W_L(a_1,\ldots,a_L \mid \mathbf{r}_f)}{W_L(a_1,\ldots,a_L)}. \tag{29}$$

The most likely set r∗f$\mathbf{r}_f^*$ can be obtained by the following "backward" algorithm: Given the sequence ($W_1,\ldots, W_L$) obtained from the «forward» recursion (28), we can decide for every point r whether it is more likely to be a background position or the endpoint of a functional locus, ignoring all sequence information from positions > r. This depends on whether the leading contribution to W_r comes from the first or second term on the r.h.s. of (28) and defines the local optimum modelm*(r). The global optimum set of functional loci respecting the no-overlap constraint is then r∗f$\mathbf{r}_f^*$ = {r|b(r) = 1}, where b(r) is given by the recursion b(r) = ℓ if b(r + 1) ≤ 1 &m*(r) = 1 and b(r) = max(b(r + 1) - 1, 0) otherwise, with the initial condition b(L + 1) = 0.

The Bayesian model can easily be extended to sequences containing several types of binding sites, which bind different transcription factors and are distinguished by their Q distributions. Dynamic programming algorithms can thus predict the likely coverage of a sequence with binding sites of known type [22]. This is the first step in extending the statistical analysis from single binding sites to entire regions of regulatory DNA. Indeed, models of this kind have been applied successfully to predict regulatory elements in eukaryotes, which typically consist of functional groups of adjacent binding sites. In the algorithms currently used, however, the scoring in (27) is strictly additive for groups of non-overlapping binding sites: it does not take into account dependencies between the sites within one functional group or overlapping sites within one sequence.

Evolution of regulatory DNA

In the statistical picture developed so far, background sequences and functional sites are reduced to ensembles P_0 and Q. This picture is incomplete in two ways. On one hand, it is quite disconnected from the biophysical aspects discussed before: the specific function of binding sites hardly enters the standard formalism of position weight matrices. On the other hand, there is not yet any notion of time and dynamics. Sequences change by various mutation processes, and the observed sequence ensembles derive from this evolutionary dynamics. The evolution of functional loci is fundamentally different from that of background sequence: it is subject to natural selection, that is, the fitness of an organism depends on its genotype **a** at a functional locus via the effects on the regulated gene. At this point, the biophysics of binding enters the evolution of functional sequences [25–27]. Moreover, it becomes clear that the statistical framework has to be extended from individual sequences to distributions of genotypes in a population. In this section, we develop an evolutionary picture of regulatory DNA, from which we obtain expressions for the sequence ensembles P_0, Q, and the score function S. The next four paragraphs are a self-contained introduction to the underlying concepts of population genetics.

Deterministic Population Dynamics and Fitness

We start by describing the evolution of a large population, which contains individuals of different genotypes **a**. Each genotype is assumed to produce a specific phenotype, which may influence the reproductive success of the individuals carrying it. With respect to factor binding, the phenotype can be associated with the binding energy E(**a**), since presumably all organismic effects of a locus depend on its genotype only via the binding energy. However, the discussion in the following paragraphs is more general.

We first assume that the subpopulations of a given genotype reproduce separately, i.e., there neither transitions between genotypes through mutations nor (in a sexually reproducing population) mixing through genomic recombination. Writing the dynamics of the subpopulations in the form of simple growth laws,

$$\frac{d}{dt}N_{\mathbf{a}}(t) = F_{\mathbf{a}}(t)N_{\mathbf{a}}(t), \tag{30}$$

defines the (Malthusian) fitness $F_a(t)$ of each genotype. For notational simplicity, we now limit ourselves to the case of just two genotypes **a** and **b**, where (30) can be written as growth laws for the total population size $N(t) = N_a(t) + N_b(t)$ and for the population fraction $x(t) = N_b(t)/N(t)$ of genotype **b**,

$$\frac{d}{dt}N(t) = \bar{F}(t)N(t), \tag{31}$$

$$\frac{d}{dt}x(t) = \Delta F_{\mathbf{ab}}(t)x(t)[1 - x(t)], \tag{32}$$

with $\bar{F}(t) \equiv [1 - x(t)]F_a(t) + x(t)F_b(t)$ and $\Delta F_{ab}(t) \equiv F_b(t) - F_a(t)$. This decomposition is useful since the overall growth rate $\bar{F}(t)$ is often strongly time-dependent due to external conditions (e.g., seasonality), while fitness differences, which reflect intrinsic properties of the phenotypes, are more stable. Different genotypes coexisting in a population frequently produce the same or very similar phenotypes and thus have equal fitness ($\Delta F_{ab} = 0$).

Assuming ΔF_{ab} to be constant over the time of observation, the solution of eq. (32) is the evolutionary trajectory

$$x(t) = \frac{x_0 \exp[\Delta F_{\mathbf{ab}}(t - t_0)]}{1 - x_0(\exp[\Delta F_{\mathbf{ab}}(t - t_0)] - 1)} \tag{33}$$

with the initial condition $x(t_0) = x_0$, shown in fig. 6(a). For $\Delta F_{ab} \neq 0$, the fixed points of this dynamics are the monomorphic population states $x = 0$, and $x = 1$, of which $x = 1$ is stable for $\Delta F_{ab} > 1$ and $x = 0$ for $\Delta F_{ab} < 1$. The approach to the stationary state takes place on a characteristic time scale $\tau_d = 1/\Delta F_{ab}$.

In the important case of neutral evolution ($\Delta F_{ab} = 0$), the evolutionary outcome remains indefinite. These results, which can readily be generalized to more than two phenotypes, are a simple version of Fisher›s fundamental theorem of natural selection: any population with initially coexisting phenotypes of different fitness will evolve towards a state where only the fittest phenotype is present.

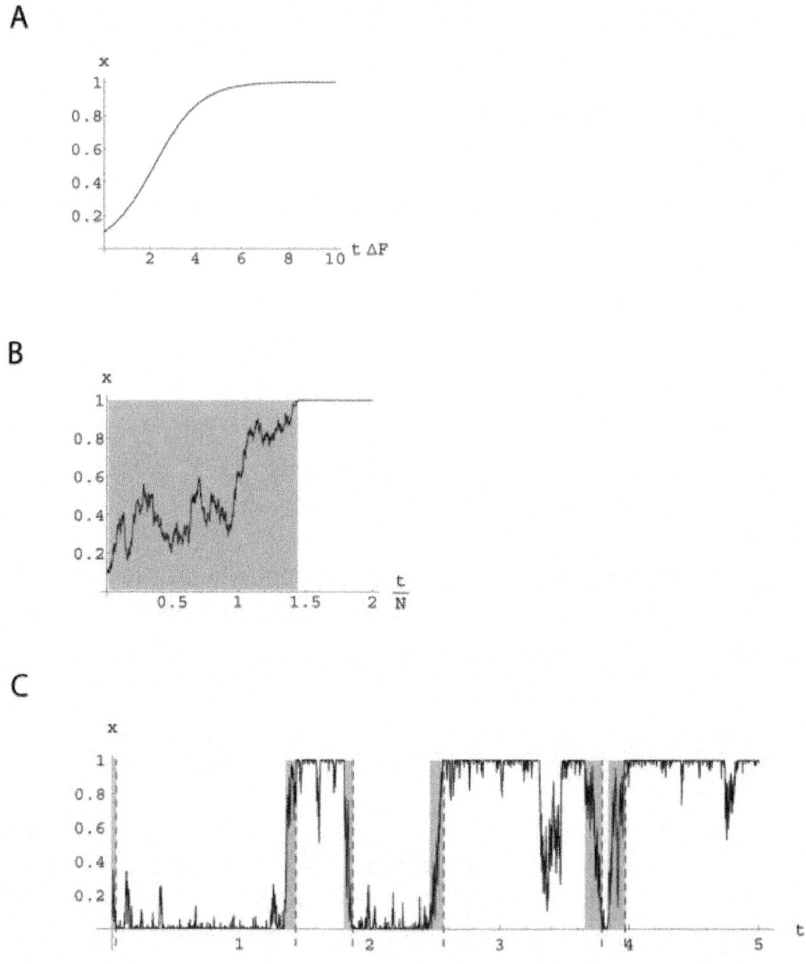

Figure 6: Evolution of genotype composition x (t). (a) Deterministic evolution with fitness difference $\Delta F_{ab} > 0$, leading to certain fixation of genotype b (time is shown in units of $\tau_d = 1/\Delta F_{ab}$). (b) Stochastic evolution with selection and genetic drift, leading to fixation of one of the genotypes. The time to fixation (grey shading) is of order τ_s ($N \Delta F_{ab} = 0.5$, time is shown in units of N). (c) Stochastic evolution with selection, genetic drift, and mutations in the regime $N\mu \ll 1$, leading to a substitution dynamics with rates $u_{a \to b}$ and $u_{b \to a}$ given by (49). Substitution events are marked by dashed lines. The typical time between initial mutation and fixation (grey shading) for a given substitution, τ_s, is much shorter than the time between subsequent substitutions, $1/u_{a \to b}$ resp. $1/u_{b \to a}$ ($N \Delta F_{ab} = 0.5$, $N\mu = 0.05$, time is shown in units of $1/\mu$).

Fisher's theorem seems to prove the popularized Darwinian notion of the "survival of the fittest". However, it rests on very restrictive assumptions that are never fulfilled in a natural population. The deterministic growth law (32) neglects mutations and recombinations, as well as the reproductive fluctuations present in any population due to its finite number of individuals. These other evolutionary forces have to be incorporated in our theoretical picture before we can even define fitness as a measurable quantity and before the theory can address the important case of neutral evolution.

Stochastic Dynamics and Genetic Drift

Stochastic fluctuations of the reproduction process in a large but finite population have been studied extensively in population genetics, see [28, 29]. They are called genetic drift, an unfortunate name which may falsely suggest a deterministic effect. To take these fluctuations into account, we replace eq. (30) by a stochastic growth law,

$$\frac{d}{dt} N_{\mathbf{a}}(t) = F_{\mathbf{a}}(t) N_{\mathbf{a}}(t) + \chi_{\mathbf{a}}(t), \tag{34}$$

where $\chi_a(t)$ are Gaussian random variables with $\overline{\chi_{\mathbf{a}}(t)} = 0$ and

$$\overline{\chi_{\mathbf{a}}(t)\chi_{\mathbf{b}}(t')} = N_{\mathbf{a}}(t)\delta(t - t')\delta_{\mathbf{a},\mathbf{b}}. \tag{35}$$

This form of noise is simply due to the law of large numbers, and the continuum dynamics (34) emerges as an effective large-N description for a plethora of discrete evolution models, which are defined at the level of individuals and have finite generation times. In the application to real populations, N has to be interpreted as the so-called effective population size, which can be inferred from genome data and is in general smaller than the actual population size.

In the case of two genotypes, eq. (34) can again be projected onto the population fraction x,

$$\frac{d}{dt} x(t) = \Delta F_{\mathbf{ab}}(t) x(t)[1 - x(t)] + \chi_x(t), \tag{36}$$

where $\chi_x(t) = (\partial x/\partial N_a)\chi_a(t) + (\partial x/\partial N_b)\chi_b(t)$ are Gaussian random variables with zero mean and

$$\overline{\chi_x(t)\chi_x(t')} = \frac{x(1-x)}{N}\delta(t - t'). \tag{37}$$

This dynamics produces stochastic evolutionary trajectories x(t) as shown in fig. 6(b). To capture their statistics, we convert the Langevin equation (36) into a Fokker-Planck equation for the probability distribution of the genotype composition [28,30],

$$\frac{\partial}{\partial t}\mathcal{P}(x,t) = \frac{1}{2N}\frac{\partial^2}{\partial x^2}x(1-x)\mathcal{P}(x,t) - \Delta F_{\mathbf{ab}}(t)\frac{\partial}{\partial x}x(1-x)\mathcal{P}(x,t). \tag{38}$$

The mathematical subtlety of this equation lies in the x-dependent diffusion "constant" x(1 - x)/2N, which reflects the multiplicative nature of the reproduction process. As a consequence, the two monomorphic population states x = 0 and x = 1 are also fixed points also of the stochastic dynamics. Any evolutionary trajectory x(t) will eventually lead to one of these states with probability 1; this is called the fixation of the corresponding genotype in the population. In other words, the Fokker-Planck equation (38) describes diffusion in the interval (0, 1) with absorbing boundaries. There is a family of stationary states

$$\mathcal{P}(x) = (1 - \varphi)\delta(x) + \varphi\delta(1 - x), \tag{39}$$

parametrized by the fixation probability φ of genotype **b**. The value of φ depends on the initial condition x_0 and can be computed by solving the backward diffusion equation

$$\frac{\partial}{\partial t}\mathcal{P}(x,t \mid x_0,t_0) = x_0(1-x_0)\left(\frac{1}{2N}\frac{\partial^2}{\partial x_0^2} + \Delta F_{\mathbf{ab}}(t)\frac{\partial}{\partial x_0}\right)\mathcal{P}(x,t \mid x_0,t_0). \tag{40}$$

For time-independent ΔF_{ab}, the stationary solution $\varphi(x_0) = \lim_{t\to\infty} P\mathcal{P}(x = 1, t|x_0, t_0)$ has the form [28, 30]

$$\phi(x_0,\Delta F_{\mathbf{ab}},N) = \frac{1-\exp(-2N\Delta F_{\mathbf{ab}}x_0)}{1-\exp(-2N\Delta F_{\mathbf{ab}})}, \tag{41}$$

which for near-neutral evolution ($N\Delta F_{ab} \ll 1$) reduces to

$$\varphi(x_0, 0, N) = x_0 + N\Delta F_{ab}\, x_0(1 - x_0) + .. \tag{42}$$

The characteristic time τ_s of the stochastic dynamics interpolates between the diffusive scale N and the deterministic scale:$\tau_s \approx \min(N, \tau_d)$. It determines the typical time of the evolution process up to fixation, shown shaded in fig. 6(b).

Hence, the stochastic population dynamics depends no longer only on the fitness difference of the genotypes as in the deterministic case, but also on the initial state of the population and the the population size. Yet, our evolutionary picture is still incomplete. Population states with coexisting genotypes enter the dynamics as initial conditions, but since mutations are neglected, the model does not explain how this coexistence is generated and maintained.

Mutation Processes and Evolutionary Equilibria

At the level of an individual, mutations are rare stochastic genotype changes **a** → **b**, which take place with rates $\mu_{a\to b}$, often coupled to the reproduction process. (These rates are all of the same order of magnitude, in estimates we therefore omit the indices.) We include mutations into the population dynamics (34) by their systematic effect on the genotype subpopulations,

$$\frac{d}{dt}N_{\mathbf{a}}(t) = F_{\mathbf{a}}(t)N_{\mathbf{a}}(t) + \sum_{\mathbf{b}}[\mu_{\mathbf{b}\to\mathbf{a}}N_{\mathbf{b}}(t) - \mu_{\mathbf{a}\to\mathbf{b}}N_{\mathbf{a}}(t)] + \chi_{\mathbf{a}}(t), \tag{43}$$

while their stochastic effect (whose variance is of order $N\mu$) is neglected since it is small against the reproductive sampling noise $\chi_a(t)$. In the case of two different genotypes, this dynamics can again be projected onto the variable x,

$$\frac{d}{dt}x(t) = \Delta F_{\mathbf{ab}}(t)x(t)[1 - x(t)] + \mu_{\mathbf{a}\to\mathbf{b}}[1 - x(t)] - \mu_{\mathbf{b}\to\mathbf{a}}x(t) + \chi_x(t), \tag{44}$$

which leads to the Fokker-Planck equation [31]

$$\begin{aligned}\frac{\partial}{\partial t}\mathcal{P}(x,t) &= \frac{1}{N}\frac{\partial^2}{\partial x^2}x(1-x)\mathcal{P}(x,t) - \Delta F_{\mathbf{ab}}(t)\frac{\partial}{\partial x}x(1-x)\mathcal{P}(x,t) \\ &\quad -\mu_{\mathbf{a}\to\mathbf{b}}\frac{\partial}{\partial x}(1-x)\mathcal{P}(x,t) + \mu_{\mathbf{b}\to\mathbf{a}}\frac{\partial}{\partial x}x\mathcal{P}(x,t).\end{aligned} \tag{45}$$

For time-independent ΔF_{ab}, this equation has a single stable stationary state,

$$\mathcal{P}(x) = \frac{1}{Z}x^{-1+N\mu_{\mathbf{a}\to\mathbf{b}}}(1-x)^{-1+N\mu_{\mathbf{b}\to\mathbf{a}}}\exp(2N\Delta F_{\mathbf{ab}}x) \tag{46}$$

with a normalization constant Z that can be expressed in terms of Bessel and Gamma functions [32].

Substitution Dynamics

Here we are interested in the stochastic evolution (45) and its equilibrium state (46) for $N\mu \ll 1$, which is the relevant dynamical regime for nuclear DNA in eukaryotes and in most prokaryotes (but not in viral systems). In this regime, the mutation term in (45) is small against the diffusion term except for values of x close to the boundaries 0 or 1. In this region, the continuum approximation of eq. (45) is no longer valid, and (46) has to be replaced by a stationary solution $P^{\mathcal{P}}_d(N_a)$ of the underlying discrete evolution model, which gives the probability that the population contains N_a individuals of genotype **a** (with $N_a = N - N_b = 0, 1, ..., N$). The discrete solution is easily shown to have the singularity $\mathcal{P}_d(0) \simeq (N\mu_{\mathbf{a}\to\mathbf{b}})^{-1}P_d(1)$. This singularity is correctly captured if we use the approximation $P_d(N_{\mathbf{a}}) \simeq \int_{N_{\mathbf{a}}/N}^{(N_{\mathbf{a}}+1)/N} dx\,\mathcal{P}(x)$ for all N_a (except at the other boundary, where there is a similar singularity $\mathcal{P}_d(N) \simeq (N\mu_{\mathbf{b}\to\mathbf{a}})^{-1} P_d(N-1)$) [33].

From this solution, we read off the following characteristics of the evolutionary dynamics at equilibrium, which are illustrated by the trajectory of fig. 6(c)[32]:

(a) For sufficiently small values of μ, the population remains monomorphic for most of the time. Using the shorthands Q(**a**) = $\mathcal{P}_d(N_a = 0)$ and Q(**b**) = $\mathcal{P}_d(N_a = N)$, we have

$$Q(\mathbf{a}) + Q(\mathbf{b}) = 1 - O(\mu N \log N). \tag{47}$$

(b) The ratio of probabilities for the two monomorphic population states is given by the ratio of "forward" and "backward" mutation rate, the fitness difference, and the effective population size:

$$\frac{Q(\mathbf{b})}{Q(\mathbf{a})} = \frac{\mu_{\mathbf{a}\to\mathbf{b}}}{\mu_{\mathbf{b}\to\mathbf{a}}} \exp(2N\Delta F_{\mathbf{ab}}) + O(N_\mu). \tag{48}$$

(c) The monomorphic population states x = 0 and x = 1 are unstable due to mutations even at arbitrarily small values of μ, which cause occasional transitions of the entire population from genotype **a** to **b**, and vice versa. These so-called substitutions are marked by dashed lines in fig. 6(c). The substitution rate $u_{a\to b}$ can be evaluated as the product of creating a single mutant of genotype **b** in an initially monomorphic **a** population, Nμa→b $N\mu_{\mathbf{a}\to\mathbf{b}}$, and its probability of fixation, φ (x_0 = 1/N, ΔF_{ab}, N). The time between initial mutation and fixation (shown by grey shading in fig. 6(c)) is still of order τ_s and thus much shorter than the time scale 1/μ, on which mutation effects become important. Hence, the fixation probability φ is given to leading order by (41), which has been derived forμ = 0. Together we have [28, 30]

$$u_{\mathbf{a}\to\mathbf{b}} = N\mu_{\mathbf{a}\to\mathbf{b}} \frac{1-\exp(-2\Delta F_{\mathbf{ab}})}{1-\exp(-2N\Delta F_{\mathbf{ab}})}. \tag{49}$$

Hence, the substitution rate $u_{a\to b}$ is enhanced over $\mu_{a\to b}$ for $\Delta F_{ab} > 0$ and suppressed for $\Delta F_{ab} < 0$, as shown in Fig. 7. For weak selection ($N|\Delta F_{ab}| \ll 1$), eq. (49) becomes

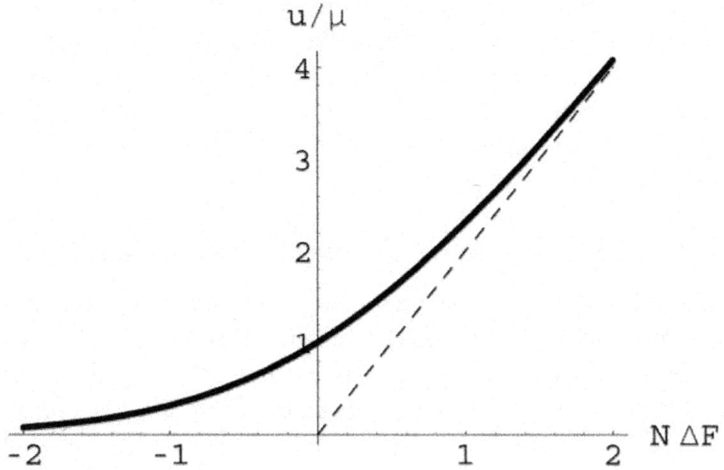

Figure 7: Substitution rate in a population versus mutation rate in an individual. The ratio of these rates, $u_{a\to b}/\mu_{a\to b}$, depends on the product N ΔF_{ab} of effective population size and fitness difference between the genotypes (in the relevant regime $N \gg 1$, $\Delta F_{ab} \ll 1$, $N\,\Delta F_{ab}$ finite). The substitution rate $u_{a\to b}$ is equal to μ_{ab} for neutral mutations ($\Delta F_{ab} = 0$), reduced for deleterious mutations ($\Delta F_{ab} < 0$), and enhanced for advantageous mutations ($\Delta F_{ab} > 0$).

$u_{a\to b} = \mu_{a\to b}(1 + N\Delta F_{ab} + ...)$.

This reproduces Kimura›s famous original result: for neutral evolution, the substitution rate equals the mutation rate in an individual, independently of the population size. For this reason, the rates $\mu_{a\to b}$ are referred to as neutral mutation rates. For strong selection ($N|\Delta F_{ab}| \gg 1 \gg |\Delta F_{ab}|$), eq. (49) takes the asymptotic forms

$$u_{a\to b} = \mu_{a\to b}\begin{cases} 2N\,|\,\Delta F_{ab}\,|\,\exp(2N\Delta F_{ab}) & (2N\Delta F_{ab} \ll 1), \\ 2N\Delta F_{ab} & (2N\Delta F_{ab} \gg 1). \end{cases} \quad (51)$$

The backward substitution rate $u_{b\to a}$ is given by a formula similar to (49) with $\Delta F_{ba} = -\Delta F_{ab}$. Forward and backward substitution rate have the simple ratio

$$\frac{u_{a\to b}}{u_{b\to a}} = \frac{\mu_{a\to b}}{\mu_{b\to a}}\exp(2N\Delta F_{ab}) \quad (52)$$

for $N \gg 1$. Comparing with (48), we obtain the consistency condition

$$\frac{u_{a\to b}}{u_{b\to a}} = \frac{Q(b)}{Q(a)}. \quad (53)$$

Hence, for sufficiently small mutation rates ($\mu N \log N \ll 1$), a simple picture emerges: The evolution of a population can be described as a sequence of transitions between monomorphic genotype states (substitutions). The substitution rate u is determined by the corresponding mutation rate in an individual, the fitness difference between the genotypes, and the effective population size.

Neutral Dynamics in Sequence Space, Sequence Entropy

This evolutionary picture can be generalized to multiple genotypes, for example, the 4^{ℓ} dimensional sequence space of genomic loci $\mathbf{a} = (a_1,..., a_{\ell})$. Transitions between different sequence states are point mutations $\mathbf{a} \rightarrow \mathbf{b}$, which change exactly one nucleotide. (We neglect here insertion and deletion processes, which change the length of the sequence). We first discuss neutral evolution, where the substitution rate $u_{\mathbf{a}\rightarrow\mathbf{b}}$ equals the mutation rate in an individual, $\mu_{\mathbf{a}\rightarrow\mathbf{b}}$, for all elementary transitions $\mathbf{a} \rightarrow \mathbf{b}$. Bona fide neutral mutation rates can be inferred from DNA sequence alignments of sufficiently close species, recent insights have also come from studying repeat elements.

We assume the neutral dynamics has an equilibrium distribution $P_0(\mathbf{a})$ which obeys detailed balance, i.e., the relation

$$\frac{\mu_{\mathbf{a}\rightarrow\mathbf{b}}}{\mu_{\mathbf{b}\rightarrow\mathbf{a}}} = \frac{P_0(\mathbf{b})}{P_0(\mathbf{a})} \tag{54}$$

holds for each pair of sequence states linked by an elementary transition process $\mathbf{a} \rightarrow \mathbf{b}$. This says that the probability current at equilibrium, $\mu_{\mathbf{a}\rightarrow\mathbf{b}}P_0(\mathbf{a}) - \mu_{\mathbf{b}\rightarrow\mathbf{a}}P_0(\mathbf{b})$, vanishes for each elementary transition. Clearly, any distribution $P_0(\mathbf{a})$ satisfying the conditions (54) is stationary under the dynamics with rates $\mu_{\mathbf{a}\rightarrow\mathbf{b}}$, but not every such dynamics has a stationary distribution which satisfies (54) (the simplest counterexample involving three states and a circular probability current $\mathbf{a} \rightarrow \mathbf{b} \rightarrow \mathbf{c}$ at stationarity).

However, as will be verified below, detailed balance is a good approximation for the genomic substitution dynamics at least in prokaryotes. (There are known violations at CpG islands in eukaryotes [34]). In the simplest type of models, every nucleotide a mutates independently of all other positions with uniform rates $\mu_{a\rightarrow b}$(i.e., $\mu_{\mathbf{a}\rightarrow\mathbf{b}} = \mu_{a\rightarrow b}$ for any two sequences $\mathbf{a}$= (..., a ,...) and $\mathbf{b}$ = (..., b, ...) differing by exactly one nucleotide). This produces a factorized equilibrium distribution $P_0(\mathbf{a})$ of the form (15).

We can project the equilibrium distribution onto a measurable quantity as independent variable. For binding site sequences, a convenient choice is the binding energy E, and the projected distribution $P_0(E)$ has the form (16). Hence we can define the sequence entropy [35]

$$S_0(E) = \log P_0(E), \qquad (55)$$

which counts the log density of sequence states **a** at energy E, weighed by the distribution $P_0(\mathbf{a})$.

Dynamics Under selection, the Score-Fitness Relation

The dynamics of substitutions can be studied in the same way for evolution under selection, which is specified at the level of genotypes by an arbitrary fitness function F(**a**) [18, 36]. This generalizes the results of [37] for a model with selection acting independently at different nucleotide positions, $F(\mathbf{a}) = \sum_{i=1}^{\ell} f_i(a_i)$. For each elementary transition $\mathbf{a} \to \mathbf{b}$, the substitution rate $u_{\mathbf{a}\to\mathbf{b}}$ is determined by the neutral rate $\mu_{\mathbf{a}\to\mathbf{b}}$, the fitness difference $\Delta F_{\mathbf{ab}}$, and the effective population size N according to (49). Given the detailed balance (54) of neutral evolution and the relation (52) between forward and backward rates, it then follows immediately that the evolutionary dynamics under selection also obeys detailed balance, as given by (53) with an equilibrium distribution Q(**a**) of the form (48). Thus we have [18, 36]:

The equilibrium distribution Q(**a**) of fixed genotypes generated by a substitution dynamics (49) with fitness function F(**a**) is related to its neutral counterpart $P_0(\mathbf{a})$ by

$$Q(\mathbf{a}) = P_0(\mathbf{a}) \exp\,[2NF(\mathbf{a}) + \text{const.}], \qquad (56)$$

with the constant given by normalization.

We can project eq. (56) onto the fitness as independent variable. Defining the distribution $Q(F) \equiv \sum_{\mathbf{a}} Q(\mathbf{a})\delta(F(\mathbf{a}) - F)$, similarly $P_0(F)$, and the sequence entropy $S_0(F) = \log P_0(F)$, the projected identity takes the form

$$Q(F) = \exp\,[2NF + S_0(F) + \text{const.}] \qquad (57)$$

For binding site sequences, we have a similar projection on the binding energy, $Q(E) = \exp\,[2NF(E) + S_0(E) + \text{const.}]$, since all genotypes with the same "phenotype" E have the same fitness, i.e., the same score S. The projected identities express the equilibrium distribution under selection in terms of fitness and sequence entropy, reflecting the balance between stochasticity (genetic drift) and selection [18]. For strong selection, the exponent $2NF - S_0$ is dominated by the fitness term, and Q(F) takes appreciable values only at points of near-maximal fitness, i.e., where $F_{max} - F \lesssim 1/2N$. For moderate selection, there is a nontrivial balance between both terms, and for weak selection, the Q distribution can be approximated by its neutral counterpart $P_0 = \exp(S_0)$. Clearly, the roles of fitness and sequence entropy are formally analogous to those of energy and entropy in statistical physics of thermodynamic systems, if 2N is identified with the inverse temperature $1/k_B T$. Some consequences of this analogy are discussed in ref. [38].

The dynamics of substitutions establishes a rather general evolutionary grounding of genome statistics, if we identify the equilibrium distributions $P_0(\mathbf{a})$ and $Q(\mathbf{a})$ with the genomic distributions discussed in the previous section, as already anticipated by our notation. Comparing eqs. (56) and (17) gives a relation between fitness and score [16, 18]:

The log-likelihood score $S(\mathbf{a}) = \log [Q(\mathbf{a})/P_0(\mathbf{a})]$ equals the fitness function multiplied by twice the effective population size up to a constant,

$$S(\mathbf{a}) = 2NF(\mathbf{a}) + \text{const.}. \tag{58}$$

This relation allows us to use sequence data of a given genome to infer quantitative patterns of its evolution. We now discuss specific consequences for the evolution of regulatory DNA; an application to protein evolution can be found in ref. [37].

Measuring Selection for Binding Sites

We first give a precise definition of functionality for regulatory (and other) elements: A binding locus is functional if the genotype at that locus is under selection (for binding of the corresponding factor). Nonfunctional loci have evolutionarily neutral genotypes. This definition asks whether binding at a given locus makes a difference to the organism or not. It is weaker than that of a functional binding site, which is a functional locus with a sequence **a** that is likely to actually bind the factor. A functional locus can lose its binding sequence due to deleterious mutations, leading to suboptimal fitness of the organism. Conversely, a nonfunctional locus can have by chance a sequence which does bind the factor: this is a spurious binding site without consequences for the organism. To measure the selection on functional sites in silico, we apply the identity (58) to the genomic distributions $P_0(\mathbf{a})$ and $Q(\mathbf{a})$. (Assuming equilibrium for most loci seems to be justified for our example of CRP binding sites in E. coli since we find very similar distributions in the distant bacterial species Salmonella typhimurium, and the factor protein itself is highly conserved between these species.) After projection onto the energy, the fitness landscape 2NF(E) for CRP binding sites is thus given by fig. 4(b)[16]. The fitness is constant in the no-binding region ($E \gtrsim E_s \approx 13$) since the evolution is always neutral in that region. This constant is set to 0 in our normalization, i.e., F(E) measures the fitness gain of functional sites due to factor binding. Loci with strong binding are also under strong selection, with effective fitness values 2NF of order 10. Genetic drift counteracts selection, producing also loci with weaker binding and reduced effective fitness. This fitness "landscape" is thus qualitatively of the form predicted from the underlying biophysics [18, 25]. Of course, it should be kept in mind that this landscape results from averaging over a family of binding sites, which may have a spectrum of individual selection coefficients and selected binding strengths.

Nucleotide Frequency Correlations

A further consequence of (57) is the generic occurence of nucleotide frequency correlations within functional loci [18]. If the fitness function F(**a**) is not additive in the nucleotide positions, nucleotide frequencies are correlated in selected genotypes even if they are independent under neutral evolution. This happens quite generically since selection acts on the entire genotype **a** as a functional unit and not on its single nucleotides. For binding sites, fitness effects follow from the expression level of the regulated gene, which depends on the sequence **a** via the binding probability of the corresponding transcription factor. While the binding energy is often approximately additive in the nucleotide positions as given by (1), the binding probability (10) is a strongly nonlinear function of the energy. This introduces correlations between nucleotide frequencies at any two positions within functional loci, preventing factorization of the distribution Q(**a**).

Stationary Evolution of Binding Sites

Functional loci with a substantial level of selection (as found for the CRP binding sites in E. coli) evolve in a way quite different from background sequence. This is quantified in fig. 8(a), which shows pairs of binding energies (E_1, E_2) for experimentally verified CRP binding sites in E. coli and the corresponding sites regulating orthologous genes in S. typhimurium [16, 27]. The evolutionary distance t between the two species and characteristics of the neutral mutation process can be inferred from alignments of background sequence. The "phenotypic" evolution of CRP binding is quantified by the energy transition probabilities $G_0(E_2|E_1)$ under neutral evolution and $G_f(E_2|E_1)$ under stationary selection [16]. These are readily obtained by simulating the substitution dynamics over a time interval t for given initial value E_1, both with neutral rates $\mu_{a\to b}$ and with rates $u_{a\to b}$ given by (49) and the fitness function 2NF(E) measured in E. coli. The resulting conditional expectation values $\langle G_0(E_2|E_1)\rangle$ and $\langle G_f(E_2|E_1)\rangle$ for the binding energy in S. typhimurium are also shown in fig.8(a). The data conform to the selection model, showing a substantially stronger conservation of binding energy than expected for neutral evolution [16, 27, 39].

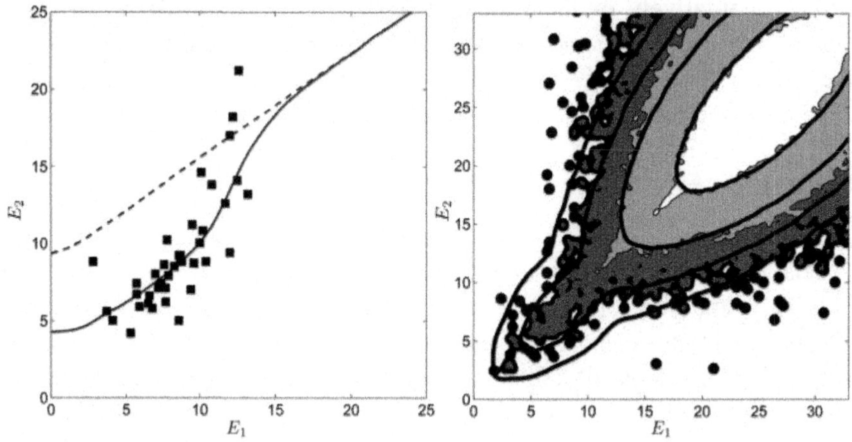

Figure 8: Evolution of binding sites. (a) Binding energy pairs (E_1, E_2) for 32 experimentally verified CRP binding sites in E. coli from the DPInteract database [57] and their aligned orthologs in S. typhimurium (dots). Conditional expectation value for the binding energy in S. typhimurium under neutral evolution, $\langle G_0(E_2|E_1)\rangle$ (dashed line), and under selection, $\langle G_f(E_2|E_1)\rangle$ (solid line). (b) Distribution of energy pair counts $W_{dat}(E_1, E_2)$ (filled contours), compared to the distribution $W(E_1, E_2)$ given by the Bayesian model (62). The symmetry of these distributions under exchange of E_1 and E_2 reflects detailed balance of the substitution dynamics. From [16,39].

We can now build a probabilistic model for cross-species comparisons [16]. It is based on the joint distributions of energy pairs

$$P_0(E_1, E_2) = G_0(E_2|E_1)\, P_0(E_1) \qquad (59)$$

under neutral evolution and

$$Q(E_1, E_2) = G_f(E_2|E_1)\, Q(E_1) \qquad (60)$$

under stationary selection, which are determined by the corresponding distributions in one species and the energy transition probabilities. Detailed balance of the substitution dynamics implies

$$\frac{P_0(E_2)}{P_0(E_1)} = \frac{G_0(E_2 \mid E_1)}{G_0(E_1 \mid E_2)} \text{ and } \frac{Q(E_2)}{Q(E_1)} = \frac{G_f(E_2 \mid E_1)}{G_f(E_1 \mid E_2)}, \qquad (61)$$

i.e., the joint distributions $P_0(E_1, E_2)$ and $Q(E_1, E_2)$ must be symmetric functions of their arguments. These distributions combine into a model for pairs of aligned loci, which generalizes the single-species model (25) and takes the form

$$W(E_1, E_2) = (1 - \lambda)P_0(E_1, E_2) + \lambda Q(E_1, E_2). \qquad (62)$$

(This model can be extended further to include non-stationary selection.) The distribution $W(E_1, E_2)$ with a fraction of functionality $\lambda = 0.0018$ is in excellent agreement with the count distribution $W_{dat}(E_1, E_2)$ obtained from E. coli and S. typhimurium, as shown in fig. 8(b). The symmetry of W_{dat} is thus consistent with the underlying assumption of detailed balance. Analogous Bayesian models can be defined for more than two species related by a phylogeny. This approach has been applied to binding site prediction in bacteria [16]; a related study of several species of funghi has been reported in ref. [40].

Adaptive Evolution of Binding Sites

What does this picture say about the adaptive evolution of transcriptional regulation in response to a newly arising selection pressure? The evolution from a genotype with marginal binding ($E(\mathbf{a}) \approx E_s$) to strong binding requires only about three uphill point mutations in the fitness landscape of fig. 4(b), i.e., there is an effective fitness gain $2N\Delta F \approx 3$ per mutation. Hence, according to (51), the rate of uphill substitutions per locus is enhanced by a factor $2N\Delta F \cdot d(\mathbf{a}, \mathbf{a}^*)$ at least of order 10 over the neutral point mutation rate per nucleotide. At the same time, the downhill rate is strongly suppressed. This shows that the adaptive formation of a binding site from background sequence can indeed be a rapid mode of regulatory evolution, due to the substantial level of selection [18].

However, this mode is only efficient if adaptation can set in immediately after the selection pressure is established. In larger regulatory regions, the exact position of a binding site is often not important. We assume the initial genome contains a set of $\tilde{L}$ shadow sites, i.e., positions $r_1, \ldots, r_{\tilde{L}}$ where a given sequence **a** would have the same regulatory effect. If one of these shadow sites has already a genotype with marginal binding, it acts as a "seed" for the onset of adaptation [41]. On the other hand, if all shadow sites of the initial genome have energy $E > E_s$, there is typically a substantial waiting time of neutral evolution before one of them reaches the threshold energy E_s. Assuming the initial genome to be entirely background sequence, it will contain at least one such seed if $\int_{E<E_s} p_0(E)\,dE \gtrsim 1/\tilde{L}$, which is a joint condition on $\tilde{L}$ and the site length ℓ: the shadow regulatory region must be long enough and binding sites must be short enough. The example shows that the evolvability of regulation imposes constraints on genome architecture [18]. Adaptive point substitution may thus be a feasible mode for the formation of a single binding site, but will hardly explain the groups of adjacent sites characteristic of eukaryotic promoters. These may originate from repeat duplication by slippage, which

has recently been shown to be an efficient source of sequence innovation in intergenic regions of Drosophila.

Towards a dynamical picture of the genome

The relationship S = 2NF + const. between score and fitness is a cornerstone of the theoretical picture developed so far, which links its population genetic, bioinformatic and biophysical arches. It relates a key evolutionary variable with the statistics of genomic frequency counts. The physical binding energy is an appropriate phenotypic variable on which fitness and score depend, because molecular function is determined by binding interactions.

We have discussed this picture for transcription factor binding sites, but it can be applied more generally to functional elements in genomes. It relates the statistics of these elements in one genome with their evolutionary dynamics, which is observed in cross-species comparisons. This dynamics is shaped by selection: The components of functional elements are coupled by a common fitness function; such fitness interactions are called epistasis. Hence, functional correlations lead to evolutionary correlations. These can be traced in the Q distribution over fixed genomes of a functional element. A more detailed statistical analysis using the statistics of polymorphisms within a population is briefly sketched below.

Thus, the picture of the genome as a system with multiple interactions has a fundamental dynamical significance. This is important since it allows us to trace functional modules from evolutionary patterns. We conclude the article with a brief outlook on functional integration of regulatory sequences at various and its dynamical implications.

Evolutionary Interactions between Sites

Regulatory function is often determined not by single binding sites, but jointly by a group of sites in the same regulatory region [42]. An important mechanism is binding cooperativity, i.e., the formation of a protein complex between two (or more) factors bound to their corresponding DNA sites. The binding energy of this complex has the form $E = E_1 + E_2 + \Delta E_{12}$, where E_1 and E_2 are the energies of the factors bound individually and $\Delta E_{12} < 0$ is the energy gain due to the protein-protein interaction, which is of the order of a few $k_B T$. Cooperative binding has a number of functional effects [1]:

(a) It increases the signal-to-noise ratio for the targeting of regulatory input to a specific gene, which is important in larger eukaryotic genomes, where single spurious binding sites are abundant in background sequence.

(b) It sharpens the response of the binding probability to variations in the factor concentrations around their threshold value. This follows from the thermodynamics of two factors, which is a straightforward generalization of the case of a single factor discussed above.

(c) It implements logical connections between regulatory input signals to a given gene. The simplest example is an AND connection between two factors, where the regulated gene is affected only if both factors are simultaneously present. This happens if the binding energies and factor concentrations are such that individual binding is weak but joint binding is strong. Larger groups of binding sites can encode a whole repertoire of more complicated logical functions [43].

Regulatory modules with several jointly acting binding sites are frequently found in eukaryotes. The functional coupling of sites in a module translates into interactions between these sites in their sequence evolution. The genomic functional element, i.e., the subset of the regulatory region on which selection acts, is the module as a whole. Its fitness $F(E_1, E_2, \Delta E_{12},...)$ is a joint function of the binding energies as the relevant phenotypic variables [18, 25]. The evolutionary dynamics under this selection allows for a large number of compensatory changes, i.e., pairs of correlated substitutions changing two binding energies such that the fitness remains constant. These lead to nucleotide frequency correlations between different sites. Such compensatory changes have indeed been observed in experiments on Drosophila promoters [44].

Site-Shadow Interactions

In larger regulatory regions, there is a number of shadow sites where a binding sequence **a** would have a similar regulatory effect as at the functional sites present. In that case, the genomic functional element contains not only the functional binding sites but also the shadow sites. Once a functional site has disappeared due to deleterious mutations, a shadow site can turn functional by adaptive evolution as described in the last section. The resulting evolutionary dynamics leads to sequence turnover with the actual binding sites present at different but functionally equivalent positions [36]. Substantial sequence turnover has been observed in a number of case studies [44–49]. Also the number of actual sites is subject to evolutionary variation since the same regulatory effect, i.e., the same fitness, can be distributed over fewer stronger or more weaker sites. With increasing number $L^{\sim}\tilde{L}$ of shadow positions, one expects that the number of actual sites grows while individual sites get weaker [36].

Gene Interactions

Evolutionary interactions are not limited to regulatory elements for the same gene. An example are gene duplications and the subsequent evolution of the daughter genes. Selection acts jointly on this pair of genes [50], which have initially identical functions, eventually leading to either loss of one of them or to subfunctionalization, which has been argued to be an important mode of genome evolution in eukaryotes [51, 52]. This process can take place by regulation, i.e., via a correlated distribution of the regulatory elements on the daughter genes. More generally, the evolution of genes in a regulatory network is correlated if their functions are coupled either in series (i.e., one gene acts on the other) or in parallel (i.e., they are part of alternative pathways for the same function). Although some regulatory networks in model organisms – e.g. the embryonic development in the sea urchin [53] – have been studied in detail, we lack a coherent view of their functional evolution to date.

Interactions and Time-Dependent Selection

The functional integration of regulation at multiple levels and the resulting fitness interactions (epistasis) imply that the selection at one genomic site is influenced by changes at other sites. A recent analysis of single-nucleotide polymorphisms and substitutions in Drosophila provides indeed evidence on a genome-wide scale that selection is time-dependent: at individual loci, changes in the direction of selection occur at nearly the rate of neutral evolution [54, 55]. At the same time, selection is sufficiently strong so that the adaptive response can keep up with the rate of selection changes. This rate is faster in non-coding DNA, which points towards the role of regulation in the adaptive differentiation between species. Genomic evolution emerges as a complex stochastic process, shaped jointly be the driving force of time-dependent selection, fitness interactions between sites, and the ongoing background of near-neutral changes. Much more remains to be learned about the interplay of these evolutionary forces: in a large and strongly coupled system, one external signal can trigger an avalanche of subsequent compensatory responses. This dynamics seems now within reach of genomic sequence analysis.

Evolutionary Innovations

Under stationary selection, functional elements are more conserved than background sequence, and the score-fitness relation quantifies the amount of conservation. But evolution is, of course, not limited to conservation. On one hand, there is typically a multitude of different genotypes yielding the same molecular function, and the evolutionary dynamics continuously plays with these alternatives. On the other hand, organisms face long-term changes of their environment, which lead to new selection pressures and a response by adaptive evolution of new functions. If regulation is to account for a large part of the diversification in higher eukaryotes, loss or gain of regulatory function should be an important mode of molecular evolution. Changes in regulatory DNA leading to new functions of gene networks have been observed [56], and it is possible to extend the statistical models described in the previous section to include evolutionary gain or loss of function of individual binding sites [16]. On a broader scale, time-dependent selection and fitness couplings appear act as a major driving forces of genomic change, triggering avalanches of evolutionary innovation. Understanding this molecular basis of innovations is a major challenge for theory and experiment in the coming years. It will profoundly change our dynamical view of the genome.

ACKNOWLEDGEMENTS

This article has been published as part of BMC Bioinformatics Volume 8 Supplement 6, 2007: Otto Warburg International Summer School and Workshop on Networks and Regulation. The full contents of the supplement are available online athttp://www.biomedcentral.com/1471-2105/8?issue=S6

REFERENCES

1. Ptashne M, Gann A: Genes and Signals. 2002, Cold Spring Harbor Laboratory Press
2. Tautz D: Evolution of transcriptional regulation. Curr Opin Genet Dev. 2000, 10 (5): 575-579. 10.1016/S0959-437X(00)00130-1.
3. Wray GA, Hahn MW, Abouheif E, Balhoff JP, Pizer M, Rockman MV, Romano LA: The evolution of transcriptional regulation in eukaryotes. Mol Biol Evol. 2003, 20 (9): 1377-1419. 10.1093/molbev/msg140. [http://dx.doi.org/10.1093molbev/msg140]
4. Berg OG, Winter RB, von Hippel PH: Diffusion-driven mechanisms of protein translocation on nucleic acids. 1. Models and theory. Biochemistry. 1981, 20 (24): 6929-6948. 10.1021/bi00527a028.
5. Winter RB, von Hippel PH: Diffusion-driven mechanisms of protein translocation on nucleic acids. 2. The Escherichia coli repressor-operator interaction: equilibrium measurements. Biochemistry. 1981, 20 (24): 6948-6960. 10.1021/bi00527a029.
6. Winter RB, Berg OG, von Hippel PH: Diffusion-driven mechanisms of protein translocation on nucleic acids. 3. The Escherichia coli lac

repressor-operator interaction: kinetic measurements and conclusions. Biochemistry. 1981, 20 (24): 6961-6977. 10.1021/bi00527a030.

7. von Hippel PH, Berg OG: On the specificity of DNA-protein interactions. Proc Natl Acad Sci USA. 1986, 83 (6): 1608-1612. 10.1073/pnas.83.6.1608.
8. Sarai A, Takeda Y: Lambda repressor recognizes the approximately 2-fold symmetric half-operator sequences asymmetrically. Proc Natl Acad Sci USA. 1989, 86 (17): 6513-6517. 10.1073/pnas.86.17.6513.
9. Fields DS, He Y, Al-Uzri AY, Stormo GD: Quantitative specificity of the Mnt repressor. J Mol Biol. 1997, 271 (2): 178-194. 10.1006/jmbi.1997.1171.
10. Stormo GD, Fields DS: Specificity, free energy and information content in protein-DNA interactions. Trends Biochem Sci. 1998, 23 (3): 109-113. 10.1016/S0968-0004(98)01187-6.
11. Oda M, Furukawa K, Ogata K, Sarai A, Nakamura H: Thermodynamics of specific and non-specific DNA binding by the c-Myb DNA-binding domain. J Mol Biol. 1998, 276 (3): 571-590. 10.1006/jmbi.1997.1564. [http://dx.doi.org/10.1006/jmbi.1997.1564]
12. Omagari K, Yoshimura H, Takano M, Hao D, Ohmori M, Sarai A, Suyama A: Systematic single base-pair substitution analysis of DNA binding by the cAMP receptor protein in cyanobacterium Synechocystis sp. PCC 6803. FEBS Lett. 2004, 563 (1–3): 55-58. 10.1016/S0014-5793(04)00248-0. [http://dx.doi.org/10.1016/S0014-5793(04)00248-0]
13. Foat BC, Houshmandi SS, Olivas WM, Bussemaker HJ: Profiling condition-specific, genome-wide regulation of mRNA stability in yeast. Proc Natl Acad Sci USA. 2005, 102 (49): 17675-17680. 10.1073/pnas.0503803102. [http://dx.doi.org/10.1073/pnas.0503803102]
14. Gerland U, Moroz JD, Hwa T: Physical constraints and functional characteristics of transcription factor-DNA interaction. Proc Natl Acad Sci USA. 2002, 99 (19): 12015-12020. 10.1073/pnas.192693599. [http://dx.doi.org/10.1073/pnas.192693599]
15. Djordjevic M, Sengupta AM, Shraiman BI: A biophysical approach to transcription factor binding site discovery. Genome Res. 2003, 13 (11): 2381-2390. 10.1101/gr.1271603. [http://dx.doi.org/10.1101/gr.1271603]
16. Mustonen V, Lässig M: Evolutionary population genetics of promoters: predicting binding sites and functional phylogenies. Proc Natl Acad Sci USA. 2005, 102 (44): 15936-41. 10.1073/pnas.0505537102. [http://dx.doi.org/10.1073/pnas.0505537102]

17. Slutzky M, Mirny L: Kinetics of protein-DNA interaction: facilitated target location in a sequence-dependent potential. Biophys J. 2004, 87 (6): 4021-4035. 10.1529/biophysj.104.050765.

18. Berg J, Willmann S, Lässig M: Adaptive evolution of transcription factor binding sites. BMC Evol Biol. 2004, 4: 42-10.1186/1471-2148-4-42. [http://dx.doi.org/10.1186/1471-2148-4-42]

19. Durbin R, Eddy SR, Krogh A, Mitchison G: Biological Sequence Analysis. 1998, Cambridge University Press

20. Stormo GD, Hartzell GW: Identifying protein-binding sites from unaligned DNA fragments. Proc Natl Acad Sci USA. 1989, 86 (4): 1183-1187. 10.1073/pnas.86.4.1183.

21. Hertz GZ, Stormo GD: Identifying DNA and protein patterns with statistically significant alignments of multiple sequences. Bioinformatics. 1999, 15 (7–8): 563-577. 10.1093/bioinformatics/15.7.563.

22. Rajewsky N, Socci ND, Zapotocky M, Siggia ED: The evolution of DNA regulatory regions for proteo-gamma bacteria by interspecies comparisons. Genome Res. 2002, 12 (2): 298-308. 10.1101/gr.207502. Article published online before print in January 2002. [http://www.genome.org/cgi/content/full/12/2/298]

23. van Nimwegen E, Zavolan M, Rajewsky N, Siggia ED: Probabilistic clustering of sequences: inferring new bacterial regulons by comparative genomics. Proc Natl Acad Sci USA. 2002, 99 (11): 7323-7328. 10.1073/pnas.112690399. [http://dx.doi. org/10.1073/pnas.112690399]

24. Lenhard B, Sandelin A, Mendoza L, Engström P, Jareborg N, Wasserman WW: Identification of conserved regulatory elements by comparative genome analysis. J Biol. 2003, 2 (2): 13-10.1186/1475-4924-2-13. [http://dx.doi.org/10.1186/1475-4924-2-13]

25. Gerland U, Hwa T: On the selection and evolution of regulatory DNA motifs. J Mol Evol. 2002, 55 (4): 386-400. 10.1007/s00239-002-2335-z. [http://dx.doi.org/10.1007/s00239-002-2335-z]

26. Moses AM, Chiang DY, Kellis M, Lander ES, Eisen MB: Position specific variation in the rate of evolution in transcription factor binding sites. BMC Evol Biol. 2003, 3: 19-10.1186/1471-2148-3-19. [http://dx.doi.org/10.1186/1471-2148-3-19]

27. Brown CT, Callan CG: Evolutionary comparisons suggest many novel cAMP response protein binding sites in Escherichia coli. Proc Natl Acad Sci USA. 2004, 101 (8): 2404-2409. 10.1073/pnas.0308628100.

28. Kimura M, Crow J: An Introduction to Population Genetics Theory. 1970, Harper & Row, New York

29. Kimura M: The Neutral Theory of Molecular Evolution. 1983, Cambridge University Press

30. Kimura M: On the probability of fixation of mutant genes in a population. Genetics. 1962, 47: 713-719.

31. Kimura M, Ohta T: The Average Number of Generations until Fixation of a Mutant Gene in a Finite Population. Genetics. 1969, 61 (3): 763-771.

32. Rouzine IM, Rodrigo A, Coffin JM: Transition between stochastic evolution and deterministic evolution in the presence of selection: general theory and application to virology. Microbiol Mol Biol Rev. 2001, 65: 151-185. 10.1128/MMBR.65.1.151-185.2001. [http://dx.doi.org/10.1128/MMBR.65.1.151-185.2001]

33. Grün D, Lässig M: to be published.

34. Arndt PF, Hwa T: Identification and measurement of neighbor-dependent nucleotide substitution processes. Bioinformatics. 2005, 21 (10): 2322-2328. 10.1093/bioinformatics/bti376. [http://dx.doi.org/10. 1093/bioinformatics/bti376]

35. Peliti L: Quasispecies evolution in general mean-field landscapes. Europhys Lett. 2002, 57: 745-51. 10.1209/epl/i2002-00526-5.

36. Berg J, Lässig M: Stochastic evolution of transcription factor binding sites. Biophysics (Moscow). 2003, 48 (Suppl 1): S36-S44.

37. Halpern AL, Bruno WJ: Evolutionary distances for protein-coding sequences: modeling site-specific residue frequencies. Mol Biol Evol. 1998, 15 (7): 910-917.

38. Sella G, Hirsh AE: The application of statistical physics to evolutionary biology. Proc Natl Acad Sci USA. 2005, 102 (27): 9541-9546. 10.1073/pnas.0501865102. [http://dx.doi.org/10.1073/pnas.0501865102]

39. Mustonen V, Lässig M: to be published.

40. Moses AM, Chiang DY, Pollard DA, Iyer VN, Eisen MB: MONKEY: identifying conserved transcription-factor binding sites in multiple alignments using a binding site-specific evolutionary model. Genome Biol. 2004, 5 (12): R98-10.1186/gb-2004-5-12-r98. [http://dx.doi.org/10.1186/gb-2004-5-12-r98]

41. MacArthur S, Brookfield JFY: Expected rates and modes of evolution of enhancer sequences. Mol Biol Evol. 2004, 21 (6): 1064-1073. 10.1093/molbev/msh105. [http://dx.doi.org/10. 1093/molbev/msh105]

42. Arnosti DN: Analysis and function of transcriptional regulatory elements: insights from Drosophila. Annu Rev Entomol. 2003, 48: 579-602.

10.1146/annurev.ento.48.091801.112749. [http://dx.doi.org/10.1146/annurev.ento.48.091801.112749]

43. Buchler NE, Gerland U, Hwa T: On schemes of combinatorial transcription logic. Proc Natl Acad Sci USA. 2003, 100 (9): 5136-5141. 10.1073/pnas.0930314100. [http://dx.doi.org/10.1073/pnas.0930314100]

44. Ludwig MZ, Bergman C, Patel NH, Kreitman M: Evidence for stabilizing selection in a eukaryotic enhancer element. Nature. 2000, 403 (6769): 564-567. 10.1038/35000615. [http://dx.doi.org/10.1038/35000615]

45. Ludwig MZ, Patel NH, Kreitman M: Functional analysis of eve stripe 2 enhancer evolution in Drosophila: rules governing conservation and change. Development. 1998, 125 (5): 949-958.

46. McGregor AP, Shaw PJ, Hancock JM, Bopp D, Hediger M, Wratten NS, Dover GA: Rapid restructuring of bicoid-dependent hunchback promoters within and between Dipteran species: implications for molecular coevolution. Evol Dev. 2001, 3 (6): 397-407. 10.1046/j.1525-142X.2001.01043.x.

47. Dermitzakis ET, Bergman CM, Clark AG: Tracing the evolutionary history of Drosophila regulatory regions with models that identify transcription factor binding sites. Mol Biol Evol. 2003, 20 (5): 703-14. 10.1093/molbev/msg077. [http://dx.doi.org/10.1093/molbev/msg077]

48. Scemama JL, Hunter M, McCallum J, Prince V, Stellwag E: Evolutionary divergence of vertebrate Hoxb2 expression patterns and transcriptional regulatory loci. J Exp Zool. 2002, 294 (3): 285-99. 10.1002/jez.90009. [http://dx.doi.org/10.1002/jez.90009]

49. Costas J, Casares F, Vieira J: Turnover of binding sites for transcription factors involved in early Drosophila development. Gene. 2003, 310: 215-20. 10.1016/S0378-1119(03)00556-0.

50. Wagner A: Selection and gene duplication: a view from the genome. Genome Biol. 2002, 3 (5): reviews1012-10.1186/gb-2002-3-5-reviews1012.

51. Lynch M, Conery JS: The evolutionary demography of duplicate genes. J Struct Funct Genomics. 2003, 3 (1–4): 35-44. 10.1023/A:1022696612931.

52. Lynch M, Conery JS: The origins of genome complexity. Science. 2003, 302 (5649): 1401-4. 10.1126/science.1089370. [http://dx.doi.org/10.1126/science.1089370]

53. Davidson E: A view from the genome: spatial control of transcription in sea urchin development. Curr Opin Genet Dev. 1999, 9 (5): 530-41. 10.1016/S0959-437X(99)00013-1.

54. Mustonen V, Lässig M: Adaptations to fluctuating selection in Drosophila. Proc Natl Acad Sci USA. 2007, 104 (7): 2277-82. 10.1073/pnas.0607105104. [http://dx.doi.org/10.1073/pnas.0607105104]

55. Mustonen V, Lässig M: Sequence evolution under quenched selection fluctuations. preprint. 2006

56. Gasch AP, Moses AM, Chiang DY, Fraser HB, Berardini M, Eisen MB: Conservation and evolution of cis-regulatory systems in ascomycete fungi. PLoS Biol. 2004, 2 (12): e398-10.1371/journal.pbio.0020398. [http://dx.doi.org/10.1371/journal.pbio.0020398]

57. Robison K, McGuire AM, Church GM: A comprehensive library of DNA-binding site matrices for 55 proteins applied to the complete Escherichia coli K-12 genome. J Mol Biol. 1998, 284 (2): 241-254. 10.1006/jmbi.1998.2160. [http://dx.doi.org/10.1006/jmbi.1998.2160]

Chapter 11

STATISTICAL ANALYSIS IN GENETIC STUDIES OF MENTAL ILLNESSES

Heping Zhang

Yale School of Public Health, Yale University, 60 College Street, New Heaven, Connecticut 06520-8034, USA

ABSTRACT

Identifying the risk factors for mental illnesses is of significant public health importance. Diagnosis, stigma associated with mental illnesses, comorbidity, and complex etiologies, among others, make it very challenging to study mental disorders. Genetic studies of mental illnesses date back at least a century ago, beginning with descriptive studies based on Mendelian laws of inheritance. A variety of study designs including twin studies, family studies, linkage analysis, and more recently, genomewide association studies have been employed to study the genetics of mental illnesses, or complex diseases in general. In this paper, I will present the challenges and methods from a statistical perspective and focus on genetic association studies.

INTRODUCTION

Mental illnesses affect the health and well-being of all populations and all ages. Schizophrenia—a chronic, severe, and disabling brain disorder—is one of these mental illnesses, affecting about 1.1 percent of the U.S. population age 18 and older in a given year. People with schizophrenia sometimes hear voices others do not hear, believe that others are broadcasting their thoughts to the world, or become convinced that others are plotting to harm them. These experiences can make them fearful and withdrawn and cause difficulties when these people try to have relationships with others (http://www.nimh.nih.gov). Emil Kraepelin (1856– 1926) described "Dementia Praecox" as an inherited disorder in his influential "Textbook of Psychiatry" (1899). Dementia Praecox, coined "schizophrenia," was first used by Arnold Pick (1851–1924)—a professor of psychiatry at the German branch of Charles University in Prague—to describe a patient with a psychotic disorder resembling hebephrenia in

1891. Nearly a century ago, Cannon and Rosanoff (1911) made an attempt to understand whether there are any forms of nervous and mental diseases that are transmitted from generation to generation in concordance with Mendelian laws.

They examined the families of 11 neuropathetic patients, which are now referred to as probands in pedigrees. Using Mendelian laws as their theoretical expectation, they concluded that the neuropathetic make-up is recessive to normal. Although the report was indeed "preliminary," a few things are noteworthy. First, they noted that "any form of insanity or even all the forms of hereditary insanity do not constitute an independent hereditary character." This raised an early sign of the complexity associated with studying mental disorders compared to the characterization of the disorders and their comorbidity. Here, comorbidity refers to more than one disease condition in the same patient. Second, they remarked "should larger accumulations of such data in the future give similar results, we shall be able" to confirm their result.

The requirement for more samples and replication is another challenge in studies of complex diseases. Last, but not the least, while they said "let us test,..., the hypothesis ..." they did not mean a statistical test. However, the idea of the χ^2-test is evident. Despite this early work, it was not until the 1960s that the researchers began to use scientifically rigorous designs and methods to study the inheritance of mental illnesses. For example, the key idea in adoption studies lies in the belief that any links between an adopted child and the biological parents are attributable to genetics, and any links between that child and adoptive parents can be attributed to environment (Plomin et al., 1997).

This enables us to separate the confounding environment (i.e., a family) from genetic contribution. Consequently, there are two strategies in adoption studies. One approach compares the risk of developing schizophrenia in the adopted children of schizophrenic parents to the risk of adopted children whose parents do not have schizophrenia. Several studies including Heston (1966), Rosenthal (1972) and Tienari (1991) used this approach to study schizophrenia. Each study found an elevated risk in adopted-away children of schizophrenic parents, supporting the role of genetics in the transmission of schizophrenia. The origin of this approach is the schizophrenic parents. Another approach backtracks from adopted children who have developed schizophrenia and compares the risks of schizophrenia in their adoptive and biological families. Kety, Rosenthal and Wender (1978) and others found that the risk was significantly higher in the biological relatives than in the adoptive families, again underscoring the role of genetics as a risk factor. While these schizophrenia adoption studies are in- fluential in understanding the role of genetics in mental disorders, the majority of the genetic factors associated with mental disorders are based on family and twin studies. By comparing the concordance in the risk between identical (monozygotic) and fraternal (dizygotic) twins, twin studies arguably provide the most compelling results about genetic and environmental effects.

For example, the concordance in monozygotic twins for Tourette's syndrome, a complex disorder characterized by repetitive, sudden and involuntary movements or noises called tics, was reported to be about 50% whereas it is less than 10% in dizygotic twins. Twin studies are most helpful in demonstrating the magnitude of genetic effect, but they do not provide insight into the inheritance pattern of a condition. Thus, family studies can offer information that twin studies cannot. Thus, Cannon and Rosanoff (1911) employed a small-scale, simple family study. Using the Mendelian laws, not only might we find evidence of genetics, but also infer the mode of transmission, as Cannon and Rosanoff (1911) concluded for the heredity of insanity. Although twin and family studies continue to be useful for understanding the genetics of complex diseases, different studies are needed to locate a specific gene on a chromosome that may underlie the disease. Gene mapping in humans through linkage analysis emerged in the 1930s, but it was Morton (1955) who laid the foundation for the methodology. It was only during the 1970s and 1980s, when the Elston-Steward (1971) algorithm was developed and implemented (Ott, 1974), that the method thrived as a common tool of genetic studies.

These initial and subsequent developments allowed for linkage analyses of multiple markers simultaneously. In light of the sheer number of genes and that we do not know which specific gene we are looking for, we typically genotype 300 to 400 "landmarks" that cover the 22 pairs of autosomes and the X chromosome. By inferring the transmission patterns of these markers, then linking them to the disease status, we can obtain information about the most probable region where the gene of interest resides. While linkage studies have had some successes (e.g., BRCA1), they have generated many more premature excitements. In the late 1980s, two particular studies attracted significant public attention after they reported that bipolar affective disorders were linked to DNA markers on chromosome 11, and that a susceptibility locus for schizophrenia was located on chromosome 5. Unfortunately, these findings were not replicated. Replications in genetic studies of mental disorders do not come easily. For example, Abelson et al. (2005) identified mutations involving the SLITRK1 gene (13q31.1) in a small number of people with Tourette's syndrome. However, most people with Tourette's syndrome do not have a mutation in the SLITRK1 gene. Because the mutations were reported in so few people with this condition, the association of the SLITRK1 gene with this disorder could not be confirmed.

In fact, Scharf et al. (2008) reported a lack of the association between SLITRK1var321 and Tourette's syndrome in a large family-based sample. Various reasons have been suggested to explain the difficulties detouring progress in genetic studies using linkage analysis. A key concept underlying linkage analysis is the recombination fraction, which re- flects the distance between any two markers, such as a DNA marker and the disease locus. There may be limited information in the data, however, diminishing the power of the linkage study. Furthermore, complex diseases are polygenic, involving multiple genes (Carter and Chung, 1980). Linkage analyses, however, are generally under the assumption of one major gene. Additionally, heterogeneity in the diagnosis and comorbidity of mental illnesses make linkage analysis considerably more difficult, if even possible at all. Many investigators have adopted association analyses to take advantage of the advent of high-throughput genotyping technologies. Recent efforts have identified genes that contribute to a number of complex human traits using the ultra-dense genetic markers (Arking et al., 2006; Klein et al., 2005; Duerr et al., 2006; Chen et al., 2007). Trios (one affected offspring and two parents) have been an effective design for association studies, particularly with the development of the elegant transmission/disequilibrium test (TDT) (Spielman, McGinnis and Ewens, 1993).

The central idea of this test is that each affected child serves as his or her own matched case and control. This acts to control for all potential confounding issues and examines alleles that both are and are not transmitted from the parents. In the absence of association between the affective status and the gene, the distributions of the transmitted and non-transmitted alleles are expected to be the same. Deviations in distribution as evaluated by a $\chi2$-test indicate the existence of association. Trios are the simplest example of nuclear family, but when other siblings are available, the trio design is not cost-effective. As a result, family-based association tests (FBAT) including sibships (Spielman and Ewens, 1998; Horvath and Laird, 1998; Knapp, 1999), nuclear families (Weinberg, 1999; Lunetta et al., 2000; Rabinowitz and Laird, 2000) and general pedigrees (Martin, Monks, Warren and Kaplan, 2000) have been developed. Another restriction in the use of trios is the requirement of defining the affective status of a disease. Consequently, association tests have been proposed for quantitative traits (Allison, 1997; Rabinowitz, 1997), traits with distribution belonging to an exponential family (Liu, Tritchler and Bull, 2002), ordinal traits (Zhang, Wang and Ye, 2006; Wang, Ye and Zhang, 2006) and multiple traits (Lange et al., 2003; Zhang, Liu and Wang, 2010).

Since the early success in identifying the complement factor H polymorphism in age-related macular degeneration (Klein et al., 2005), case-control association studies have intensified, and many genetic variants have been identified and catalogued (Hindorff et al., 2009). Despite the enormous investment, the intense attention to the genetics of diseases, the rapid improvement in technology, and the increasingly large sample sizes in many studies, it remains challenging to identify disease genes, especially those underlying mental illnesses. Some of the common genetic variants that have been identified for complex diseases only account for a small portion of the genetic risk, which may vary across populations (Goldstein, 2009).

For example, Kopp et al. (2008) and Kao et al. (2008) identified several variations in the MYH9 gene as major contributors to excess risk of kidney disease among AfricanAmericans. They found that 60 percent of AfricanAmericans carry the risk variants as opposed to 4 percent of white Americans. Technology will continue to improve and the amount of genetic data will increase. The purpose of this article is to review some of the progress from a statistical perspective and discuss some of the potential challenges. Obviously, it would take volumes or series to do justice to all of the work in statistical genetics. Instead of taking on that impossible task, this article is oriented toward the publications directly related to my own recent work.

METHODS

Since 1952, the American Psychiatric Association has published four editions of the Diagnostic and Statistical Manual of mental disorders (DSM) and plans to release its fifth edition in 2013. While widely used, the use and development of the DSM has not gone without controversy and criticism. Unlike diseases for which the diagnoses are well accepted by physicians and patients, such as cancer, the diagnosis of mental disorders must reflect biological factors (e.g., gender and racial disparities), non-biological factors such as culture that are not specific to one person, and it also must reflect the natural variation within the same person.

Ordinal Traits

It is clear from the above discussion that a simple dichotomous diagnosis (e.g., yes or no), or a well distributed continuous trait, is unlikely to characterize the state of mental disorders. In fact, the questions used in the diagnosis of mental disorders, such as DSM-IV, are usually posed in terms of severity or frequency and hence in an ordinal scale. Statistical methods for genetic analysis are well established for both quantitative (continuous) and binary traits (see, e.g., Blackwelder and Elston, 1985; Goldgar, 1990; Schork, 1993; Amos, 1994; Risch and Zhang, 1995; Kruglyak et al., 1996; Blangero and Almasy, 1997; Ott, 1999). While there has been some progress in the analysis of ordinal traits (e.g., Heath et al., 2002; Steinke, Borish and Rosenwasser, 2003; Vergne et al., 2003; Zhang, Feng and Zhu, 2003; Feng, Leckman and Zhang, 2004; Zhang, Liu and Wang, 2010), especially in plant science (Rao and Xu, 1998; Xu and Xu, 2006), insufficient attention has been paid to addressing the unique challenges of analyzing ordinal traits. Some researchers have recognized that it is difficult to conduct genetic analyses of ordinal traits because such traits cannot be directly characterized by a linear function of genetic and environmental effects (Rao and Xu, 1998). To fill in this methodological gap, we have made a systematic effort to develop statistical methods for segregation analysis (Zhang, Feng and Zhu, 2003), linkage analysis (Feng, Leckman and Zhang, 2004) and association analysis (Zhang, Wang and Ye, 2006) of ordinal traits (for family studies and case-control studies).

Analysis of Family Data

Long before the era of genomics, researchers collected data in families, also called pedigrees as illustrated in Figure 1. Although the ascertainment process for families varies, Figure 1 depicts a representative three-generation pedigree. The proband is the first person who enters into the study according to defined inclusion and exclusion criteria: such criteria are related to the disease of interest. Other members of the proband's family are included and directly or indirectly assessed, depending on the circumstance. The key idea in analyzing family data is that if a gene is a major driving force behind a disease, a trace in the concordance of diseases in family members would reflect the transmission pattern of a gene under the Mendelian laws. This is the fundamental concept that Cannon and Romanoff (1911) employed. This type of analysis is referred to as . The Elston–Stewart (1971) algorithm set up the quintessential framework to analyze data from general pedigrees through a technique called peeling. The main complication in analyzing pedigree data is the complex relationship among family members, making it difficult to express the likelihood function in an easily computed form. The peeling algorithm makes use of the conditional independence embedded in the pedigree resulting from the Mendelian laws, and so peels off the complete likelihood function into smaller pieces before putting them back together.

Other methods have been relatively recently developed using the concept of latent random variables (Hopper, 1989; Babiker and Cuzick, 1994; Li and Thompson, 1997; Siegmund and McKnight, 1998; Zhang and Merikangas, 2000), which are closely related to the classic ousiotype models of Cannings, Thompson and Skolnick (1978) in pedigree analysis. The basic idea is to use latent variables to represent the contribution of unobserved factors including a major gene, residual genetic factors and common environmental factors. As discussed by Zhang and Merikangas (2000), the computation involving pedigrees is similar to the peeling algorithm. Advantages of using latent variable based models are that the interactions between underlying genetic effects and the observed covariates (e.g., demographic variables) can be considered. Additionally, more relevant to this article, we can accommodate ordinal traits in the latent variable framework.

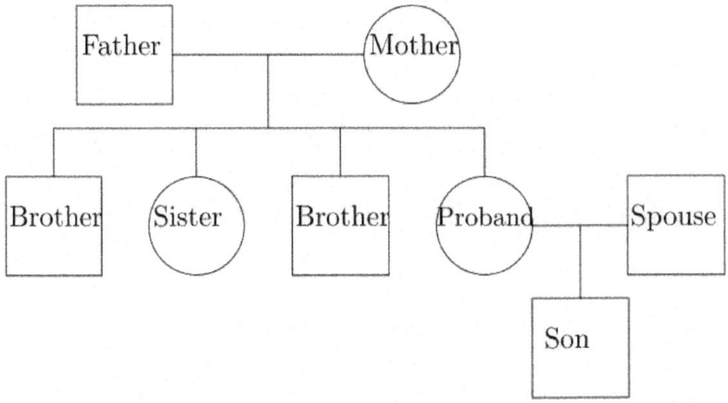

Figure 1: A three-generation pedigree.

A Latent Variable Model

We follow the notation of Zhang and Merikangas (2000) and Zhang, Feng and Zhu (2003). Consider a trait, Y, that takes an ordinal value of 0, 1,...,K. Let x be a p-vector of covariates that is also available for each study subject. Three types of latent random variables Ui 1,Ui 2 and Ui 3 are introduced within family i to represent, respectively, (a) common, unmeasured environmental factors; (b) genetic susceptibility of the family founders (a founder refers to a subject whose parents are not a part of the observed pedigree, e.g., father, mother and spouse in Figure 1); and (c) the transmission of susceptibility genes from a parent to an offspring. The concept of latent variables is straightforward, but the interesting and difficult part lies in the specification of their distributions. They need to be interpretable and convenient. The following are the assumptions that we found useful:

- U_1^i follows Bernoulli distributions $P\{U_1^i = 1\} = 1 - P\{U_1^i = 0\} = \theta_1$, where θ_1 is an unknown parameter.
- $U_2^i = (U_{2,1}^i, U_{2,2}^i, \ldots, U_{2,2n_i-1}^i, U_{2,2n_i}^i)'$, where n_i is the size of pedigree i. Here, $P\{U_{2,2j-1}^i = 1\} = 1 - P\{U_{2,2j}^i = 0\} = \theta_2$ when $U_{2,2j-1}^i$ and $U_{2,2j}^i$ are the U_2^i-variables of a founder.
- $U_3^i = (U_{3,1}^i, \ldots, U_{3,s_i}^i)'$. According to the Mendelian laws, $P\{U_{3,j}^i = 1\} = P\{U_{3,j}^i = 0\} = \frac{1}{2}, j = 1, \ldots, s_i$, and s_i is the number of parent-offspring pairs in family i. U_3^i facilitates the transmission of U_2^i-variables from the founders to the offspring. For example, if a parent of subject j has U_2^i-variables, $U_{2,2k-1}^i$ and $U_{2,2k}^i$, and the U_3^i-variable for this parent-offspring pair is $U_{3,l}^i$, then one of subject j's U_2^i-variables is $U_{2,2j-1}^i = U_{2,2k-1}^i U_{3,l}^i + U_{2,2k}^i(1 - U_{3,l}^i)$.
- All latent variables are independent.

U^i_1 is a simple "switch" indicating the presence or absence of a shared environment factor within family i. U^i_2 is assigned independently to each of founders who are the source for any gene to enter into a family, and thus mimics the transmission of a single major susceptibility locus with alleles A and a of frequencies θ_2 and $1 - \theta_2$, respectively. Conditional on all of the latent variables, denoted by U^i, within family i, the probability distribution for member j is assumed to be

$$P\{Y_j^i \le k | U^i\} = \frac{\exp(\mathbf{x}^i_j \beta + \alpha_k + \mathbf{a}^i_j \gamma)}{1 + \exp(\mathbf{x}^i_j \beta + \alpha_k + \mathbf{a}^i_j \gamma)}, \quad k = 0, \ldots, K - 1, \tag{2.1}$$

where $\mathbf{a}^i_j = (U_1^i, U_{2,2j-1}^i + U_{2,2j}^i, U_{2,2j-1}^i U_{2,2j}^i)^T$, and β and γ are p- and 3-vectors of parameters. The αk is the trait level dependent intercept, k = 0,...,K − 1. As Zhang, Feng and Zhu (2003) pointed out, the β parameters measure the strength of association between the trait and the covariates, conditional on the latent variables. The γ parameters indicate the familial and genetic contributions to the trait. The mode of inheritance can be inferred from γ. For example, $\gamma_2 = 0$ and $\gamma_3 \neq 0$ suggests a recessive effect.

The likelihood function can be derived from (2.1). Due to the presence of latent variables, the EM algorithm (Dempster, Laird and Rubin, 1977) is the most convenient choice for parameter estimation (Guo and Thompson, 1992; Zhang and Merikangas, 2000; Zhang, Feng and Zhu, 2003). Although Zhang and Merikangas (2000) and Zhang, Feng and Zhu (2003) presented an effective solution (e.g., a modified likelihood), we should note that the lack of concavity in the likelihood function makes it a challenging task to find the maximum likelihood estimates of the model parameters. In addition, the θ's and γ's are not fully identifiable. The identifiability issue not only causes computational problems, but also presents theoretical challenges in statistical inference. Another important, yet understudied, issue is the validation of the assumptions on the distributions of the latent variables. But, how useful is the latent variable model (2.1)? First, it provides a regression framework to assess familial aggregation and genetic contribution, and possibly interactions between measured covariates and latent factors. Using data from a family study of substance use (Merikangas et al., 1998), Zhang and Merikangas (2000) were able to present extremely significant evidence of familial aggregation p-value < 10–9 for alcohol dependence. This study additionally demonstrated that transmission does not follow a major locus pattern. In retrospect, their findings predicted the difficulty of identifying major genes associated with alcoholism. In addition, Zhang and Merikangas (2000) presented simulation examples to delineate when the absence of latent variables in (2.1) affects the estimates of the effects by the measured covariates. For example, hypothetically, if the greater presence of females in a family has an impact on the well-being of the family, ignoring the familial latent variables is likely to result in a biased estimate of the sex difference.

Not only is it important to include the latent factors, but also it is important to adjust for covariates. To further illustrate this point, Zhang, Feng and Zhu (2003) reported the following simulation. Ten thousand data sets were generated from model (2.1) with θ1 = 0.3, β chosen from 0, 1, 5 or 10, γ_1 from 0, 1 or 2, $\alpha_0 = -1$ and α1 = 1. To focus on the difference of having or not having covariates, they set $\gamma_2 = \gamma_3 = 0$. Each data set consists of 200 families with 7 family members (similar to Figure 1). One covariate x was generated as follows. For family i, U^i_1 were generated according to whether a random number ri1 from the uniform (0, 1) was greater than 0.3 or not. For member j in family i, an independent random number r_{ij2} from the uniform (0, 1) was generated. Then, $x_{ij} = 0.9r_{ij2} + 0.2r_{i1}$.

To evaluate the performance of the test statistic, the covariate was deliberately ignored in the test. When β = 0, the covariate played no role in the data generating process. The row corresponding to β = 0 in Table 1 displays the p-value (the column corresponding to $\gamma_1 = 0$) and the power for two values of γ_1 (1 or 2).

Table 1: The probability estimates of rejecting $\gamma_1 = 0$ at the significance level of 0.05. The covariate is omitted from the testing despite the fact that its coefficient β may not be zero

	$\gamma_1 = 0$	$\gamma_1 = 1$	$\gamma_1 = 2$
$\beta = 0$	0.0494	0.9503	1.0
$\beta = 1$	0.0534	0.9843	1.0
$\beta = 5$	0.1667	0.9971	1.0
$\beta = 10$	0.3828	0.9890	1.0

When β 0, the covariate plays a role in the data generating model. The data in Table 1 reveal the consequence of ignoring the covariate, which is more severe when the effect of the covariate is greater.

Linkage Analysis

While linkage analysis has a long history, it only became a common practice after the availability of several convenient computing programs (Ott, 1974; Kruglyak et al., 1996; Almasy and Blangero, 1998). For statisticians, some of the common terminologies in linkage analysis are puzzling, including the so-called LOD-score method and nonparametric method. Morton (1955) first introduced the term "LODscore." LOD stands for "the logarithm (base 10) of odds." The "odds" is a probability ratio, or likelihood ratio, of the probability under an alternative hypothesis to the probability under the null hypothesis. The LODscore method is essentially a log-likelihood ratio test with two fundamental differences: (a) the use of the base 10 logarithm versus the natural logarithm; (b) the log-likelihood ratio statistic has a multiplier of 2 conforming to a χ^2 distribution under certain regularity conditions. Specifically, the LOD-score is the log(base 10)-ratio of the likelihood when the recombination fraction is less than 1/2 (i.e., two loci are not on the same chromosome, or called unlinked), to the likelihood when the recombination fraction is 1/2 (no linkage).

The recombination fraction is the frequency that a chromosomal crossover occurs between two loci (or genes) during meiosis; 1% of combination frequency is termed the distance of one centimorgan (cM) in a genetic linkage map. Because the LOD-score is in base 10, a score of 3 indicates 1000 to 1 odds in favor of the linkage, which is the conventional threshold for declaring the evidence for linkage. If we convert a LOD-score of 3 into the standard log-likelihood ratio statistic, it yields a p-value of 2×10^{-4} under χ^2_1. By Bonnferoni correction, it corresponds to a genomewide p-value of 0.05 for 250 markers. This number is in the range for the number of microsatellites used in typical linkage studies. In order to compute the LOD-score, we first need a number of parameters that determine the likelihood for a given recombination fraction. Then use the maximum likelihood over the recombination fraction for the likelihood under the alternative hypothesis. The parameters that are required include the mode of inheritance, penetrance, and disease allele frequency.

These parameters are generally unknown and difficult to estimate for complex diseases including mental illnesses. For example, using segregation analysis (see Section 2.1.1) Pauls and Leckman (1986) examined specific genetic hypotheses about the mode of transmission of Gilles de la Tourette's syndrome, by performing segregation analyses in 30 nuclear families (two-generation pedigrees). They concluded that Tourette's syndrome is inherited as an autosomal dominant trait (one copy of the abnormal allele is sufficient to cause the disease). The penetrance (the probability of having the disease for a given genotype) was reported at 0.71 in males and 1.0 in females with at least one abnormal allele. After several decades of research, no major genetic variant has been identified for Tourette's syndrome, and most likely this syndrome involves multiple genes, interacting with environmental factors. This reality makes it difficult to infer the mode of inheritance, penetrance, and disease allele frequency, and conceptually, this may not make sense for complex diseases (non-Mendelian inheritance). This difficulty is somewhat alleviated since the LOD-score method has been found to work reasonably well (e.g., Abreu, Greenberg and Hodge, 1999) under various parameter settings. There have been some efforts to improve the robustness of the method (Gastwirth, 1966, 1985; Whittemore, 1996). See Zheng et al. (2009) for a thorough review. Existing methods do not extend to the case of ordinal traits.

The effectiveness of the robust methods remains to be studied. Naturally, nonparametric linkage methods have been developed to avoid specification of the genetic model parameters. In statistics, "nonparametric" methods typically refers to distribution-free methods such as rank-based tests and methods based on the empirical distribution. In linkage analysis, however, "nonparametric" does not mean "distribution-free," but instead refers to the replacement of true genetic model parameters with the parameters of inheritance of markers, hypothesized to be close to the disease locus. Thus, with nonparametric linkage methods, we still need to compute the likelihood. Two core algorithms are used to compute the likelihood: the Elston–Steward algorithm (1971) and the Lander–Green (1987) algorithm.

As previously discussed, the Elston–Steward algorithm (1971) is a peeling algorithm that makes the computation in a large pedigree feasible by splitting it into small pieces. This algorithm was implemented in early versions of linkage analysis programs (e.g., LIPED and LINKAGE); computational time increased linearly in family size, but exponentially with the number of loci. More recent programs (e.g., GENEHUNTER) use the Lander– Green (1987) algorithm that has first-order complexity in the number of loci, but unfortunately exponential in the family size. Although Markov chain Monte Carlo methods have been used to accommodate linkage analysis of large families and a large number of markers (Guo and Thompson, 1992), in practice, one may have to break large pedigrees apart in order to run programs such as GENEHUNTER. We should note that there had not been a linkage analysis program to handle ordinal traits until the release of LOT (Zhang et al., 2008). Typically, the methods for linkage analysis can be divided into two main steps; only the second step involves the trait (Kruglyak et al., 1996). The first step infers how genetic information travels in a family as represented by the so-called "inheritance vector." We will use the pedigree in Figure 1 to illustrate this concept.

The two parents and spouse are the founders of the family, meaning that their parents are not in the current pedigree. The four siblings and the child are nonfounders. The inheritance pattern at marker locus t is completely described by an inheritance vector $v(t) = (v_1, v_2, v_3, v_4, \ldots, v_9, v_{10})$. In other words, we devote two elements for every nonfounder. The founders are not included because they are the sources of the genes in the family and the inheritance vector is conditional on their genes. The paired elements describe the outcomes of the paternal and maternal meioses transmitted to the nonfounders. Specifically, $v_{2j-1} = 1$ or 2 according to whether the grand paternal or grand maternal allele is transmitted in the paternal meiosis to the j th nonfounder. v_{2j} carries the similar information for the corresponding maternal meiosis, namely, $v_{2j} = 3$ or 4 according to whether the grand paternal or grand maternal allele was transmitted in the maternal meiosis to the j th nonfounder.

In practice, the genetic markers do not always allow us to determine the true inheritance vector. In this case, the inheritance distribution is the conditional probability distribution over the possible inheritance vectors that conform with the alleles observed at t, which we denote by $p\{v(t) = w\}$ for all inheritance vectors $w \in V$; here V is the set of all possible inheritance vectors. In the absence of any genotypic information, all inheritance vectors are equally likely according to Mendel's first law; the probability distribution is uniform. For segregation analysis, we employed latent variables to reflect the "imaginative" genetic effects in (2.1). In linkage analysis, we have genetic markers that flow through the inheritance vector.

Thus, we can still use (2.1) for linkage analysis except that a_{ij} should be $(U_1^i, U_{2,v_{2j-1}}^i + U_{2,v_{2j}}^i)$. On one hand, we have a reduced number of latent variables. On the other hand, many of the latent variables depend on each other through the inheritance vectors. The computation of the likelihood would be summed over all inheritance vectors w in V, in addition to the probability space of the remaining independent latent variables. Because of this connection and distinction, the challenges in the linkage analysis of ordinal traits are, to a great extent, similar to those in segregation analysis of ordinal traits, for example, the asymptotic mixture of χ^2- distributions and the need to introduce the penalized likelihood (Liang and Rathouz, 1999; Zhang, Feng and Zhu, 2003).

Association test

As discussed above, linkage analysis focuses on testing the position of a marker, although it has been difficult to replicate findings in linkage studies of mental disorders. An association analysis, however, tests whether a genetic variant, including particular allele or genotype of a marker and a haplotype in several markers, is associated with a trait. Some study cohorts recruited for linkage studies have been re-genotyped for genomewide association analyses. For binary or quantitative traits, many methods have been developed and implemented. Two commonly used programs are PLINK (Purcell et al., 2007) and FBAT (Rabinowitz and Laird, 2000). To analyze an ordinal trait, Zhang, Wang and Ye (2006) introduced the following proportional odds model:

$$\text{logit}\{P(y_{ij} \le k|G_{ij})\} = \alpha_k + \beta c_{ij}, \tag{2.2}$$

where $\alpha_0,\ldots,\alpha_{K-1}$ are non-descending level parameters, β is the genetic effect. The genetic factor c_{ij} can be chosen to reflect the underlying mode of inheritance such as the number of the risk allele. Under model (2.2), the null hypothesis is H_0: β = 0. The score statistic is

$$S = \sum_{i,j}[R^+(y_{ij}) - R^-(y_{ij})]A_{ij}, \tag{2.3}$$

where $R^+(y_{ij})$ and $R^-(y_{ij})$ are the counts of offspring in the entire sample whose trait values are greater or less than y_{ij} , respectively, and Aij is the number of copies of transmitted alleles at the marker locus. Thus, Zhang, Wang and Ye (2006) proposed the following OTDT test based on the score statistic:

$$\frac{[S - E(S|Y)]^2}{\text{Var}(S|Y)},$$

which follows a χ^2_1 -distribution asymptotically. For a case-control study, $R^+(y_{ij})$ and $R-(y_{ij})$ are the numbers of subjects whose trait values are greater or less than y_{ij} , respectively. If we rewrite the statistic in (2.3) in a general form as $\Sigma_{i,j} w_{ij} A_{ij}$, this yields the classic TDT when $w_{ij} = 1$ and the QTDT (Rabinowitz, 1997) when $w_{ij} = y_{ij} - \bar{y}$, where $\bar{y}$ is the average of all y_{ij} 's. In other words, all of these tests are a weighted function of the number of transmitted alleles at the marker locus, and the choice of the weights depends on the property of the trait. With this observation, after the proper weights are computed, the existing FBAT software can be used to test the association between any trait and alleles at a marker locus.

In the following, we describe a unified method to choose weights for any kind of trait. It is straightforward to categorize a quantitative trait into any reasonable number of categories (such as deciles) and induce an ordinal scaled trait. This would allow the use of the O-TDT for a quantitative trait. In their simulation studies, Zhang, Wang and Ye (2006) demonstrated that this strategy has comparable power to the QTDT for quantitative traits. This is due to the fact that the number of categories is enough to capture most of the information in the data (e.g., following Cochran's rule; Cochran, 1977). The advantage is that the ordinal scaled test is not affected by the nonnormal distribution of a quantitative trait, and so, the unified approach is robust. One limitation of the test proposed by Zhang, Wang and Ye (2006) is that it does not adjust for covariates. Environmental factors or covariates, such as gender and age, may confound the association of interest. In a subsequent work, Wang, Ye and Zhang (2006) generalized model (2.2) to include covariates as follows:

$$\text{logit}\{P(y_{ij} \le k|G_{ij}), z_{ij}\} = \alpha_k + \beta c_{ij} + \delta' z_{ij}, \tag{2.4}$$

where z_{ij} denotes the covariates and δ is the vector of the corresponding coefficients. Consequently, the score statistic becomes

$$S = \sum_{i,j} [\hat{\gamma}(y_{ij}, z_{ij}) - \hat{\gamma}(y_{ij} - 1, z_{ij})] A_{ij}, \tag{2.5}$$

where

$$\hat{\gamma}(k, z) = \frac{\exp(\hat{\alpha}_k + \hat{\delta}' z_{ij})}{1 + \exp(\hat{\alpha}_k + \hat{\delta}' z_{ij})},$$

which is the estimated probability of having a trait value no greater than k. Thus, the weight function in (2.5) is the difference between the probability of having a trait value greater than y_{ij} and the probability of having a trait value less than yij . Not surprisingly, this is in essence the same as the weight function in (2.3) where we used counts instead of frequency (or probability). It is important to note that association analysis does not directly equate to a causal relationship. In well-designed genetic association studies, an observed association is expected to result from either a causal functional variant of a gene, or the linkage disequilibrium between the marker and a susceptibility gene. In population-based case-control studies, there are typically attempts to match cases and controls by important demographic and/or baseline information.

It is not wise to over-match subjects. Alternatively, we can collect potentially important environmental variables and consider them in the association analysis. We can also use principal component analysis on the genotypes to explore whether there are "clusters" in the study cohorts that are not appropriately reflected in the environmental variables. In family-based studies, the association tends to be conditional on parental genotypes and all phenotypes.

Unique Challenges in Analyzing Ordinal Traits

Understanding the genetic mechanisms for complex diseases is challenging regardless of whether we analyze binary, ordinal or continuous traits. Any challenges that exist for analyzing binary and continuous traits remain for ordinal traits. What are the unique challenges in analyzing ordinal traits? The key difference is that there is not a simple distribution function for ordinal traits. For continuous traits, the assumption is that the traits can somehow be treated under normality, by transformation if needed. For binary traits, through a link function (e.g., logit) we only need to deal with a Bernoulli distribution. However, for ordinal traits, the two typical approaches are (a) to assume a reliability variable or a continuous latent variable or (b) to assume a proportional odds model as we presented above. The first challenge is in the estimation. The likelihood function is complicated, and based on the numerical results, it has multiple local maxima. In addition, due to identifiability (or near-identifiability), the likelihood function may be relatively flat. Combinations of the EM and other algorithms can provide practical solutions, but finding a more efficient algorithm is an open problem.

The second challenge is in the inference. When latent variables or mixture distributions are used, some of the commonly assumed regularity conditions do not hold. One solution is to use a penalized likelihood function (Zhang, Feng and Zhu, 2003) that prevents the parameters from being near the singularity points. Finally, model diagnostics are difficult. For example, how do we know the latent variable-based model or the proportional odds model provides an adequate fit to the data? Although the models and methods presented above do not address this and other questions, they provide a foundation for further research and improvement.

Comorbidity

The methods described above only deal with a single trait. However, comorbidity is the rule rather than the exception in studies of mental and behavioral disorders. For example, a patient may suffer from both anxiety and depression (Li and Burmeister, 2009), and the same patient may also be addicted to nicotine, alcohol, or other substances (Merikangas et al., 1998; True et al., 1999). From a data analysis perspective, we need to consider how important it is to accommodate multiple diseases/traits. In a real-data example, Chen et al. (2011) analyzed a data set from the Study of Addiction: Genetics and Environment (SAGE). By simply considering addiction to at least two of the six substances (addiction to nicotine, alcohol, marijuana, cocaine, opiates or other drugs), we were able to identify the PKNOX2 gene that reached genomewide signifi- cance level among European-origin females. Interestingly, the PKNOX2 gene has been previously identi- fied as one of the cis-regulated genes for alcohol addiction in mice (Mulligan et al., 2006). To further delineate the benefit of considering multivariate traits, Zhu and Zhang (2009) conducted comprehensive simulation studies, considered the correlations of 0.2 and −0.2 among three quantitative traits, and demonstrated that testing correlated traits jointly is more powerful than testing a single trait at a time. Using generalized estimation equation, Lange et al. (2003) developed a family-based association test for multivariate quantitative traits (FBAT-GEE). Recently, Zhang, Liu and Wang (2010) constructed a nonparametric test based on the generalized Kendall's tau to accommodate any combination of dichotomous, ordinal, and quantitative traits.

Kendall's tau

Kendall's τ is a rank-based correlation between two variables. It contracts the probability of observing the two variables in the same order in two observations with the probability of observing the two variables in the opposite order. Specifically, for a sample of n observations $(X_1, Y_1),...,(X_n,Y_n)$, two observations (X_i,Y_i) and (X_j,Y_j) are called concordant if $(X_i - X_j)(Y_i - Y_j) > 0$ and discordant if $(X_i - X_j)(Y_i - Y_j) < 0$. Then Kendall's τ is based on the difference between the numbers of concordant pairs and discordant pairs. We introduce a kernel function,

$$\begin{aligned}
&\phi((X_i, Y_i), (X_j, Y_j)) \\
&= \text{sign}\{(X_i - X_j)(Y_i - Y_j)\} \\
&= \begin{cases} 1, & \text{if } (X_i - X_j)(Y_i - Y_j) > 0, \\ -1, & \text{if } (X_i - X_j)(Y_i - Y_j) < 0, \\ 0, & \text{if } (X_i - X_j)(Y_i - Y_j) = 0, \end{cases}
\end{aligned}$$

and define a U-statistic

$$U = \binom{n}{2}^{-1} \sum_{i<j} \phi((X_i, Y_i), (X_j, Y_j)). \tag{2.6}$$

Then, Kendall's τ is

$$\tau = \frac{U}{\sqrt{\mathrm{Var}_0(U)}}, \tag{2.7}$$

where $\mathrm{Var}_0(U)$ is the variance of U under the null hypothesis of no correlation between X and Y, and equal to n(n−1)(2n+5)/18 if X and Y are continuous variables (Hollander and Wolfe, 1999).

Generalized Kendall's tau

To test the association between genetic markers and comorbidity, Zhang, Liu and Wang (2010) generalized Kendall's tau as follows. For individuals i and j, let Ti and T_j be their vectors of traits, respectively. Then, a trait kernel is defined as

$$F_{ij} = (f_1(T_i^{(1)} - T_j^{(1)}), \ldots, f_p(T_i^{(p)} - T_j^{(p)}))',$$

where function fk(·) is the identity function for a quantitative or binary trait (Rabinowitz, 1997), or the sign function for an ordinal trait (Zhang, Wang and Ye, 2006).

Also, recall that, as in Section 2.1.4, c is the number of any chosen allele for marker genotype and let C_i refer to the C for the ith subject. Then, Zhang, Liu and Wang (2010) defined a marker kernel as

$$D_{ij} = c_i - c_j.$$

Their U-statistic is defined as

$$U = \binom{n}{2}^{-1} \sum_{i<j} D_{ij} F_{ij}. \tag{2.8}$$

The association test statistic, or generalized Kendall's tau, is U Var^{-1}_0 (U) U, where Var0(U) is the variance of U under the null hypothesis that there is no association between marker alleles and any linked locus that influences the trait T . The test statistic follows an asymptotic χ^2-distribution under the null hypothesis. Obviously, the statistic in (2.8) does not incorporate covariate effects. This is relatively straightforward for a single trait as was done in (2.4). Here, the traits can be a hybrid of different traits. An alternative is to impose different weights for each pair of samples in the statistic (2.8) according to the information of their covariates. The weight, denoted by $w(z_i, z_j)$ for the pair (i, j), reflects the relative importance attributed by the covariates when we derive the statistic. Zhu, Jiang and Zhang (2010) examined the following weight function.

Write $z = (z^{co}, z^{ca})'$ with $z^{co} = (z^{(1)}, \ldots, z^{(l_1)})'$ for the continuous covariates and $z^{ca} = (z^{(l_1+1)}, \ldots, z^{(l)})'$ for the categorical covariates. They defined the weight function $w(z_i, z_j)$ as

$$w(z_i, z_j) = W(\|z_i^{co} - z_j^{co}\|) I(z_i^{ca} = z_j^{ca}), \tag{2.9}$$

where W(·) is a positive and decreasing function, for example, W(u) = exp(−u2/2h2), and I(·) is the indicator function. Then a weighted test statistic is given by

$$S = \binom{n}{2}^{-1} \sum_{i<j} D_{ij} F_{ij} w(z_i, z_j). \tag{2.10}$$

Zhang, Liu and Wang (2010) and Zhu, Jiang and Zhang (2010) showed that under the null hypothesis, the test statistic S (weighted or not) has the following asymptotic distribution conditional on all phenotypes and parental genotypes:

$$\text{Var}_0^{-1/2}(S)[S - E_0(S)] \xrightarrow{d} N(0, I_p),$$

where

$$E_0(S) = \frac{2}{n-1} \sum_{i=1}^{n} \bar{u}_i E_0(C_i | M_i^{pa}),$$

$$\text{Var}_0(S) = \frac{4}{(n-1)^2} \cdot \sum_{i=1}^{n} \sum_{j=1}^{n} \bar{u}_i \bar{u}'_j \text{Cov}_0(C_i, C_j | M_i^{pa}, M_j^{pa}).$$

Consequently, the following test statistic

$$\chi^2_{tau} = [S - E_0(S)]' \text{Var}_0^{-1}(S)[S - E_0(S)]$$

converges to χ^2_p in distribution under the null hypothesis provided that Var0(S) is full rank. In a case-control study, we do not have the markers from parents and hence the conditional expectations are replaced with the unconditional ones. Thus, the key difference in the test statistics between family studies and population studies lies in the conditioning on the parental markers. The conditioning on the parental markers gives the family studies a major advantage in removing the effect of population admixture, but family studies tend to be more difficult and expensive to carry out. Under the alternative hypothesis, the test statistic χ^2_{tau} can be written as a weighted sum of non-central

$\chi_1^2 = \sum_{i=1}^{p} e_i \chi_1^2(\phi_i)$, where $e_1 \geq \cdots \geq e_p$ are the nonnegative eigenvalues of $\boldsymbol{\Sigma}_1^{1/2}\boldsymbol{\Sigma}_0^{-1}\boldsymbol{\Sigma}_1^{1/2}$.

$\varphi_i = \mu^2_{Ri}$ and μR_i is the ith component of $\mu_p = Q\boldsymbol{\Sigma}_1^{-1/2}\boldsymbol{\mu}$, where Q is an orthonormal matrix such that $Q\boldsymbol{\Sigma}_1^{1/2}\boldsymbol{\Sigma}_0^{-1}\boldsymbol{\Sigma}_1^{1/2}Q' = \text{diag}(e_1, \ldots, e_p)$. $\boldsymbol{\mu}$ is the difference in the means of S under the alternative and null hypotheses. Using the approximation theory of Pearson (1959), Solomon and Stephens (1977) and Liu, Tang and Zhang (2009), we can find a certain degree of freedom l and non-centrally parametric υ such that the distribution of χ^2_{tau} can be closely approximated by $\chi^2_l(\upsilon)$. Through simulation studies, Zhu et al. con- firmed that this approximation is accurate enough for power calculation.

It is noteworthy that the weight function in (2.9) is restrictive with respect to categorical covariates, especially so for ordinal covariates. The use of genomic propensity score can give rise to an alternative weight function. Specifically, for a di-allelic marker G (e.g., SNP), the genomic propensity score is the conditional probability $p_g(z) = P(G = g|Z = z)$. This probability can be fitted by a logistic regression model or proportional odds model depending on whether G is chosen as an allele type or genotype. In the latter choice, the model also depends on the mode of inheritance. In the current genomewide association studies, we usually only have genotypes and cannot distinguish the phases of individual alleles. Thus, we have to construct genomic propensity scores by considering various modes of inheritance. Once the genomic propensity score is estimated, it can be treated as a numerical covariate and then we can use (2.9) again.

Examples

Zhang, Liu and Wang (2010) reanalyzed a data set from the Collaborative Study on the Genetics of Alcoholism (COGA) (Begleiter, 1995; Edenberg et al., 2005). The data came from a multicenter (9 sites) consortium that recruited study participants by requiring every proband to meet two alcohol dependence diagnostic criteria based on DSM-IVR (American Psychiatric Association, 1994). The firstdegree relatives of the probands were invited into the study. Zhang, Liu and Wang (2010) included a total of 1614 individuals from 143 families. They considered three phenotypes: (1) alcohol DX-DSM3R+Feighner; (2) maximum number of drinks in a 24-hour period; and (3) the response to "spent so much time drinking, had little time for anything else." Using the first phenotype alone, the p-value of the association between a peak marker D7S679 on chromosome 7 and the trait was 0.0019. However, when the three traits are analyzed together, D7S679 remains the peak marker, and the p-value is reduced to 0.00055, demonstrating the possibility that the other two phenotypes enhanced the association signal. If the other two phenotypes are analyzed alone, the analysis did not lead to anything worthy of further attention.

In the analysis cited above, the association was assessed without considering covariates. In a follow-up analysis, Zhu, Jiang and Zhang (2010) considered two important covariates: age at interview and sex. When these two covariates were controlled for, the p-value of the association between the peak marker D7S679 and the three phenotypes went down further to 0.000313.

DISCUSSION

Studying comorbidity is a significant issue in mental and behavioral research, dating back to a century ago (Cannon and Rosanoff, 1911). This is challenging due to a lack of statistical methods that accommodate the complexity of comorbidity. While dealing with comorbidity in genetic studies is the focus of this review, it is achieved through gradual development, and accumulation of methods.

Various challenges are dealt with along the way. Although I focused on the analysis of ordinal traits and applications in mental health, the presented methods are closely related to robust and rank-based methods for binary and quantitative traits. Furthermore, ordinal traits arise in studies of diseases besides mental illnesses, such as cancer (specifically, different stages). From the statistical perspective, the methods that are presented here have broad applications beyond genetic association studies.

From college admissions, to job searches, to scientific investigations, we make inferences based on multidimensional data. It is important and imperative to consider and develop inferential tools for multivariate outcomes, particularly when the outcomes are discrete. There is extensive literature on the statistical analysis of multivariate normal variables as well as on nonparametric tests for a single variable of nonnormal distribution. However, few options are available for the inference when we have multiple nonnormally distributed variables and potential hybrids of continuous and discrete variables. To overcome this challenge, I presented several useful statistical techniques such as the rank-based U-statistics and the kernel-based weighted statistics to accommodate the mix of continuous and discrete outcomes and the presence of important covariates.

ACKNOWLEDGMENTS

This work is supported in part by National Institute on Drug Abuse Grant R01DA016750. The author wishes to thank Professor David Madigan for his encouragement on this review. He also thanks Dr. Gang Zheng and Professor Joseph Gastwirth for helpful discussions and comments, and Jennnifer Brennan for careful reading and comments.

REFERENCES

1. Abelson, J. F., Kwan, K. Y., O'Roak, B. J., Baek, D. Y., Stillman, A. A., Morgan, T. M., Mathews, C. A., Pauls, D. L., Rasin, M.-R., Gunel, M., Davis, N. R., Ercan-Sencicek, A. G., Guez, D. H., Spertus, J. A., Leckman, J. F., Leon S. Dure, t., Kurlan, R., Singer, H. S., Gilbert, D. L., Farhi, A., Louvi, A., Lifton, R. P., Sestan, N. and State, M. W. (2005). Sequence variants in SLITRK1 are associated with Tourette's syndrome. *Science* 310 317–320.
2. Abreu, P. C., Greenberg, D. A. and Hodge, S. E. (1999). Direct power comparisons between simple lod scores and NPL scores for linkage analysis in complex diseases. *Am. J. Hum. Genet.* 65 847–857.
3. Allison, D. B. (1997). Transmission-disequilibrium tests for quantitative traits. *Am. J. Hum. Genet.* 60 676–690.
4. Almasy, L. and Blangero, J. (1998). Multipoint quantitative-trait linkage analysis in general pedigrees. *Am. J. Hum. Genet.* 62 1198–1121.
5. American Psychiatric Association (1994). *Diagnostic and Statistical Manual of Mental Disorders*, 4th ed. American Psychiatric Association Press, Washington, DC.
6. Amos, C. I. (1994). Robust variance-components approach for assess-ing genetic linkage in pedigrees. *Am. J. Hum. Genet.* 54 535–543.
7. Arking, D. E., Pfeufer, A., Post, W., Kao, W. H. L., Newton-Cheh, C., Ikeda, M., West, K., Kashuk, C., Akyol, M., Perz, S., Jalilzadeh, S., Illig, T., Gieger, C., Guo, C.-Y., Larson, M. G., Wichmann, H. E., Marbán, E., O'Donnell, C. J., Hirschhorn, J. N., Kääb, S., Spooner, P. M., Meitinger, T. and Chakravarti, A. (2006). A common genetic variant in the NOS1 regulator NOS1AP modulates cardiac repolarization. *Nat. Genet.* 38 644–651.
8. Babiker, A. and Cuzick, J. (1994). A simple frailty model for family studies with covariates. *Stat. Med.* 13 1679–1692.
9. Begleiter, e. a. H. (1995). The collaborative study on the genetics of alcoholism. *Alcohol Health Res. World* 19 228–236.
10. Blackwelder, W. C. and Elston, R. C. (1985). A comparison of sib-pair linkage tests for disease susceptibility loci. *Genet. Epidemiol.* 285–97.

11. Blangero, J. and Almasy, L. (1997). Multipoint oligogenic linkage analysis of quantitative traits. *Genet. Epidemiol.* 14 959–964.
12. Cannings, C., Thompson, E. A. and Skolnick, M. H. (1978). Probability functions on complex pedigrees. *Adv. Appl. Probab.* 10 26–61.
13. Cannon, G. L. and Rosanoff, A. J. (1911). Preliminary report of a study of heredity in insanity in the light of the Mendelian laws. Reprinted from *J. Nervous and Mental Disorders* 38 272–279.
14. Carter, C. L. and Chung, C. S. (1980). Segregation analysis of schizophrenia under a mixed genetic model. *Hum. Hered.* 30 350–356.
15. Chen, X., Liu, C. T., Zhang, M. Z. and Zhang, H. P. (2007). A forest-based approach to identifying gene and gene–gene interactions.*Proc. Natl. Acad. Sci. USA* 104 19199–19203.
16. Chen, X., Cho, K., Singer, B. H. and Zhang, H. P. (2011). The nuclear transcription factor PKNOX2 is a candidate gene for substance dependence in European-origin women. *PLoS ONE* 6 e16002.
17. Cochran, W. G. (1977). *Sampling Techniques*, 3rd ed. Wiley, New York.
18. Dempster, A. P., Laird, N. M. and Rubin, D. B. (1977). Maximum likelihood from incomplete data via the EM algorithm (with discussion).*J. Roy. Statist. Soc. Ser. B* 39 1–38.
19. Duerr, R. H., Taylor, K. D., Brant, S. R., Rioux, J. D., Silverberg, M. S., Daly, M. J., Steinhart, A. H., Abraham, C., Regueiro, M. and Griffiths, A. et al. (2006). A genomewide association study identifies IL23R as an inflammatory bowel disease gene. *Science* 314 1461–1463.
20. Edenberg, H. J., Bierut, L. J., Boyce, P., Cao, M., Cawley, S., Chiles, R. and Doheny, K. F. (2005). Description of the data from the Collaborative Study on the Genetics of Alcoholism (COGA) and single-nucleotide polymorphism genotyping for Genetic Analysis Workshop 14. *BMC Genetics* 6 S2.
21. Elston, R. C. and Steward, J. (1971). A general model for the analysis of pedigree data. *Hum. Hered.* 21 523–542.
22. Feng, R., Leckman, J. and Zhang, H. P. (2004). Linkage analysis of ordinal traits for pedigree data. *Proc. Natl. Acad. Sci. USA* 10116739–16744.
23. Gastwirth, J. L. (1966). On robust procedures. *J. Amer. Statist. Assoc.* 61 929–948.
24. Gastwirth, J. L. (1985). The use of maximin efficiency robust tests in combining contingency tables and survival analysis. *J. Amer. Statist. Assoc.* 80 381–384.

25. Goldgar, D. E. (1990). Multipoint analysis of human quantitative genetic variation. *Am. J. Hum. Genet.* 47 957–967.

26. Goldstein, D. B. (2009). Common genetic variation and human traits. *N. Eng. J. Med.* 360 1696–1698.

27. Guo, S. W. and Thompson, E. A. (1992). A Monte Carlo method for combined segregation and linkage analysis. *Am. J. Hum. Genet.* 511111–1126.

28. Heath, A. C., Todorov, A. A., Nelson, E. C., Madden, P. A. F., Bucholz, K. K. and Martin, N. G. (2002). Gene–environment interaction effects on behavioral variation and risk of complex disorders: The example of alcoholism and other psychiatric disorders. *Twin Research*5 30–37.

29. Heston, L. L. (1966). Psychiatric disorders in foster home reared children of schizophrenic mothers. *Bristish J. Psychiatry* 112 819–825.

30. Hindorff, L. A., Sethupathy, P., Junkinsa, H. A., Ramosa, E. M., Mehtac, J. P., Collinsb, F. S. and Manolioa, T. A. (2009). Potential etiologic and functional implications of genome-wide association loci for human diseases and traits. *Proc. Natl. Acad. Sci. USA* 1069362–9367.

31. Hollander, M. and Wolfe, D. A. (1999). *Nonparametric Statistical Methods*, 2nd ed. Wiley, New York.

32. Hopper, J. L. (1989). Modelling sibship environment in the regressive logistic model for familial disease. *Genet. Epidemiol.* 6 235–240.

33. Horvath, S. and Laird, N. M. (1998). A discordant-sibship test for disequilibrium and linkage: No need for parental data. *Am. J. Hum. Genet.* 63 1886–1897.

34. Kao, W. H. et al. (2009). MYH9 is associated with nondiabetic end-stage renal disease in African-Americans. *Nat. Genet.* 40 1185–1192.

35. Kety, S. S., Rosenthal, D. and Wender, P. (1978). Genetic relationships within the schizophrenia spectrum: Evidence from adoption studies. In *Critical Issues in Psychiatric Diagnosis* (R. L. Spitzer and D. F. Klein, eds.) 213–223. Raven Press, New York.

36. Klein, R. J., Zeiss, C., Chew, E. Y., Tsai, J.-Y., Sackler, R. S., Haynes, C., Henning, A. K., SanGiovanni, J. P., Mane, S. M., Mayne, S. T., Bracken, M. B., Ferris, F. L., Ott, J., Barnstable, C. and Hoh, J. (2005). Complement factor H polymorphism in age-related macular degeneration. *Science* 308 385–389.

37. Knapp, M. (1999). Using exact *P* values to compare the power between the reconstruction-combined transmission/disequilibrium test and the sib transmission/disequilibrium test. *Am. J. Hum. Genet.* 65 1208–1210.

38. Kopp, J. B., Smith, M. W., Nelson, G. W., Johnson, R. C., Freedman, B. I., Bowden, D. W., Oleksyk, T., McKenzie, L. M., Kajiyama, H., Ahuja, T. S., Berns, J. S., Briggs, W., Cho, M. E., Dart, R. A., Kimmel, P. L., Korbet, S. M., Michel, D. M., Mokrzycki, M. H., Schelling, J. R., Simon, E., Trachtman, H., Vlahov, D. and Winkler, C. A. (2008). MYH9 is a major-effect risk gene for focal segmental glomerulosclerosis. *Nat. Genet.* 40 1175–1184.
39. Kraepelin, E. (1899). *Psychiatrie: Ein Lehrbuch fur Studirende und Aerzte.* Barth, Leipzig.
40. Kruglyak, L., Daly, M. J., Reeve-Daly, M. P. and Lander, E. S. (1996). Parametric and nonparametric linkage analysis: A unified multipoint approach. *Am. J. Hum. Genet.* 58 1347–1363.
41. Lander, E. S. and Green, P. (1987). Construction of multilocus genetic linkage maps in humans. *Proc. Natl. Acad. Sci. USA* 84 2363–2367.
42. Lange, C., Silverman, E. K., Xu, X., Weiss, S. T. and Laird, N. M. (2003). A multivariate family-based association test using generalized estimating equations: FBAT-GEE. *Biostatistics* 4 195–306.
43. Li, H. Z. and Thompson, E. (1997). Semiparametric estimation of major gene and family-specific random effects for age of onset. *Biometrics* 53 282–293.
44. Li, M. D. and Burmeister, M. (2009). New insights into the genetics of addiction. *Nat. Rev. Genet.* 10 225–231.
45. Liang, K.-Y. and Rathouz, P. J. (1999). Hypothesis testing under mixture models: Application to genetic linkage analysis. *Biometrics* 5565–74.
46. Liu, H., Tang, Y. and Zhang, H. H. (2009). A new chi-square approximation to the distribution of non-negative definite quadratic forms in non-central normal variables. *Comput. Statist. Data Anal.* 53 853–856.
47. Liu, Y., Tritchler, D. and Bull, S. B. (2002). A unified framework for transmission-disequilibrium test analysis of discrete and continuous traits. *Genet. Epidemiol.* 22 26–40.
48. Lunetta, K. L., Farone, S. V., Biederman, J. and Laird, N. M. (2000). Family based tests of association and linkage that used unaffected sibs, covariates, and interactions. *Am. J. Hum. Genet.* 66 605–614.
49. Martin, E. R., Monks, S. A., Warren, L. L. and Kaplan, N. L. (2000). A test for linkage and association in general pedigrees: The pedigree disequilibrium test. *Am. J. Hum. Genet.* 67 146–154.
50. Merikangas, K. R., Stolar, M., Stevens, D. E., Goulet, J., Preisig, M. A., Fenton, B., Zhang, H., O'Malley, S. S. and Rounsaville, B. J.

(1998). Familial transmission of substance use disorders. *Arch. Gen. Psychiatry* 55 973–979.

51. Morton, N. E. (1955). Sequential tests for the detection of linkage. *Am. J. Hum. Genet.* 7 277–318.
52. Mulligan, M. K., Ponomarev, I., Hitzemann, R. J., Belknap, J. K., Tabakoff, B., Harris, R. A., Crabbe, J. C., Blednov, Y. A., Grahame, N. J., Phillips, T. J., Finn, D. A., Hoffman, P. L., Iyer, V. R., Koob, G. F. and Bergeson, S. E. (2006). Toward understanding the genetics of alcohol drinking through transcriptome meta-analysis. *Proc. Natl. Acad. Sci. USA* 103 6368–6373.
53. Ott, J. (1974). Estimation of the recombination fraction in human pedigrees: Efficient computation of the likelihood for human linkage studies. *Am. J. Hum. Genet.* 26 588–597.
54. Ott, J. (1999). *Analysis of Human Genetic Linkage*, 3rd ed. Johns Hopkins Univ. Press, Baltimore, MD.
55. Pauls, D. L. and Leckman, J. F. (1986). The inheritance of Gilles de la Tourette's syndrome and associated behaviors. Evidence for autosomal dominant transmission. *New Eng. J. Med.* 315 993–997.
56. Pearson, E. S. (1959). Note on an approximation to the distribution of non-central ^{2}. *Biometrika* 46 364.
57. Plomin, R., DeFries, J. C., McClearn, G. E. and Rutter, M. (1997). *Behavioral Genetics*, 3rd ed. Freeman, New York.
58. Purcell, S., Neale, B., Todd-Brown, K., Thomas, L., Ferreira, M. A. R., Bender, D., Maller, J., Sklar, P., de Bakker, P. I. W., Daly, M. J. and Sham, P. C. (2007). PLINK: A toolset for whole-genome association and population-based linkage analysis. *Am. J. Hum. Genet.* 81559–575.
59. Rabinowitz, D. (1997). A transmission disequilibrium test for quantitative trait loci. *Hum. Hered.* 47 342–350.
60. Rabinowitz, D. and Laird, N. M. (2000). A unified approach to adjusting association tests for population admixture with arbitrary pedigree structure and arbitrary missing marker information. *Hum. Hered.* 50 211–223.
61. Rao, S. and Xu, S. (1998). Mapping quantitative trait loci for ordered categorical traits in four-way crosses. *Heredity* 81 214–224.
62. Risch, N. R. and Zhang, H. P. (1995). Extreme discordant sib pairs for mapping quantitative trait loci in humans. *Science* 268 1584–1589.
63. Rosenthal, D. (1972). Three adoption studies of heredity in the schizophrenic disorders. *Internat. J. Mental Health* 1 63–75.

64. Scharf, J. M., Moorjani, P., Fagerness, J., Platko, J. V., Illmann, C., Galloway, B., Jenike, E., Stewart, S. E., Pauls, D. L. and The Tourette Syndrome International Consortium for Genetics (2008). Lack of association between SLITRK1var321 and Tourette syndrome in a large family-based sample. *Neurology* 70 1495–1496.

65. Schork, N. J. (1993). Extended multipoint identity-by-descent analysis of human quantitative traits: Efficiency, power, and modeling considerations. *Am. J. Hum. Genet.* 53 1306–1319.

66. Siegmund, K. and McKnight, B. (1998). Modeling hazard functions in families. *Genet. Epidemiol.* 15 147–171.

67. Solomon, H. and Stephens, M. A. (1977). Distribution of a sum of weighted chi-square variables. *J. Amer. Statist. Assoc.* 72 881–885.

68. Spielman, R. S. and Ewens, W. J. (1998). A sibship rest for linkage in the presence of association: The sib transmission/disequilibrium test. *Am. J. Hum. Genet.* 62 450–458.

69. Spielman, R. S., McGinnis, R. E. and Ewens, W. J. (1993). Transmission test for linkage disequilibrium: The insulin gene region and insulin-dependent diabetes mellitus (IDDM). *Am. J. Hum. Genet.* 52 506–516.

70. Steinke, J. W., Borish, L. and Rosenwasser, L. J. (2003). Genetics of hypersensitivity. *J. Allergy Clin. Immunol.* 111 S495–S501.

71. Tienari, P. (1991). Interaction between genetic vulnerability and family environment: The Finnish adoptive family study of schizophrenia.*Acta Psychiatr. Scand.* 84 460–465.

72. True, W. R., Heath, A. C., Scherrer, J. F., Xian, H., Lin, N., Eisen, S. A., Lyons, M. J., Goldberg, J. and Tsuang, M. T. (1999). Interrelationship of genetic and environmental influences on conduct disorder and alcohol and marijuana dependence symptoms. *Am. J. Med. Genet.* 88 391–397.

73. Vergne, L., Bourgeois, A., Mpoudi-Ngole, E., Mougnutou, R., Mbuagbaw, J., Liegeois, F., Laurent, C., Butel, C., Zekeng, L., Delaporte, E. and Peeters, M. (2003). Biological and genetic characteristics of HIV infections in Cameroon reveals dual group M and O infections and a correlation between SI-inducing phenotype of the predominant CRF02_AG variant and disease stage. *Virology* 310 254–266.

74. Wang, X. Q., Ye, Y. Q. and Zhang, H. P. (2006). Family-based association tests for ordinal traits adjusting for covariates. *Genet. Epidemiol.* 30 728–736.

75. Weinberg, C. R. (1999). Allowing for missing parents in genetic studies of case-parental triads. *Am. J. Hum. Genet.* 64 1186–1193.

76. Whittemore, A. S. (1996). Genome scanning for linkage: An overview. *Am. J. Hum. Genet.* 59 704–716.
77. Xu, S. and Xu, C. (2006). A multivariate model for ordinal trait analysis. *Heredity* 97 409–417.
78. Zhang, H. P., Feng, R. and Zhu, H. (2003). A latent variable model of segregation analysis for ordinal traits. *J. Amer. Statist. Assoc.* 981023–1034.
79. Zhang, H. P., Liu, C.-T. and Wang, X. (2010). An association test for multiple traits based on the generalized Kendall's tau. *J. Amer. Statist. Assoc.* 105 473–481.
80. Zhang, H. P. and Merikangas, K. (2000). A frailty model of segregation analysis: Understanding the familial transmission of alcoholism. *Biometrics* 56 815–823.
81. Zhang, H. P., Wang, X. Q. and Ye, Y. Q. (2006). Detection of genes for ordinal traits in nuclear families and a unified approach for association studies. *Genetics* 172 693–699.
82. Zhang, M., Feng, R., Chen, X., Hu, B. and Zhang, H. (2008). LOT: A tool for linkage analysis of ordinal traits for pedigree data. *Bioinformatics* 24 1737–1739.
83. Zheng, G., Joo, J., Zaykin, D., Wu, C. and Geller, N. (2009). Robust tests in genome-wide scans under incomplete linkage disequilibrium. *Statist. Sci.* 24 503–516.
84. Zhu, W. S., Jiang, Y. and Zhang, H. P. (2010). Covariate-adjusted association tests and power calculations based on the generalized Kendall's tau. Technical report.
85. Zhu, W. and Zhang, H. (2009). Why do we test multiple traits in genetic association studies? *J. Korean Statist. Soc.* 38 1–10.
86. Zhu, X., Cooper, R., Kan, D., Cao, G. and Wu, X. (2005). A genome-wide linkage and association study using COGA data. *BMC Genet.* 6 S128.

Chapter 12

USING EQTL WEIGHTS TO IMPROVE POWER FOR GENOME-WIDE ASSOCIATION STUDIES: A GENETIC STUDY OF CHILDHOOD ASTHMA

Lin Li[1], Michael Kabesch[2], Emmanuelle Bouzigon[3,4], Florence Demenais[3,4], Martin Farrall[5], Miriam F. Moffatt[6], Xihong Lin[1] and Liming Liang[1,7]

[1]Department of Biostatistics, Harvard School of Public Health, Boston, MA, USA

[2]Department of Pediatric Pneumology and Allergy, KUNO University Children's Hospital Regensburg, Regensburg, Germany

[3]INSERM, Genetic Variation and Human Diseases Unit, U946, Paris, France

[4]Sorbonne Paris Cité, Institut Universitaire d'Hématologie, Université Paris Diderot, Paris, France

[5]Wellcome Trust Centre for Human Genetics, Oxford, UK

[6]Molecular Genetics and Genomics Section, National Heart and Lung Institute, Imperial College London, London, UK

[7]Department of Epidemiology, Harvard School of Public Health, Boston, MA, USA

ABSTRACT

Increasing evidence suggests that single nucleotide polymorphisms (SNPs) associated with complex traits are more likely to be expression quantitative trait loci (eQTLs). Incorporating eQTL information hence has potential to increase power of genome-wide association studies (GWAS). In this paper, we propose using eQTL weights as prior information in SNP based association tests to improve test power while maintaining control of the family-wise error rate (FWER) or the false discovery rate (FDR). We apply the proposed methods to the analysis of a GWAS for childhood asthma consisting of 1296 unrelated individuals with German ancestry. The results confirm that eQTLs are enriched for previously reported asthma SNPs. We also find that some SNPs are insignificant using procedures without eQTL weighting, but become significant using eQTL-weighted Bonferroni or Benjamini–Hochberg procedures, while controlling the same FWER or FDR level. Some of these SNPs have been reported by independent studies in recent literature. The results suggest that the eQTL-weighted procedures provide a promising approach for improving power of GWAS. We also report the results of our methods applied to the large-scale European GABRIEL consortium data.

INTRODUCTION

Asthma is a disorder characterized by inflamed mucosa of small airways of lung, causing wheezing and shortness of breath (Moffatt et al., 2010). Among the most common chronic diseases of childhood, asthma has been reported to affect more than 10% of children in many westernized societies (Cookson, 2004). It is caused by a combination of genetic and environmental factors (Cookson, 2004; Moffatt et al., 2007), and several genome-wide association studies (GWAS) have been conducted to study the genetic basis underlying the complex disorder. More than 50 single nucleotide polymorphisms (SNPs) have been reported to be associated with asthma, according to the GWAS catalog (www.genome.gov/gwastudies, accessed on January 15, 2013). Remarkably, the recent report (Moffatt et al., 2010) from the GABRIEL (A Multidisciplinary Study to Identify the Genetic and Environmental Causes of Asthma in the European Community) consortium identified several SNPs reaching genome-wide significance through a large-scale meta-analysis.

Prior biological information, often available in practice, has potential to increase power of GWAS. The common practice of GWAS, "agnostic" in some sense, assumes no prior information about any of the SNPs under investigation, meaning that all the SNPs have an equal likelihood of being causal. Some recent studies have taken advantage of information from linkage analysis (Roeder et al., 2006) and gene expression (Xiong et al., 2012) in genome-wide association scans. In genetic studies of etiology of asthma, it is of our particular interest to employ similar approaches and explore potentials of power gain in identifying asthma-associated SNPs by incorporating expression quantitative trait loci (eQTL) information.

Catalogs of eQTLs in multiple tissues have been made publicly available, resulting from recent efforts of GWAS of gene expressions (Stranger et al., 2005, 2007, 2012; Dixon et al., 2007; Dimas et al., 2009; Yang et al., 2010). eQTLs provide insight into biology of transcription regulation. It has been shown that eQTLs are enriched for SNPs associated with complex diseases and traits using GWAS (Cookson et al., 2009; Nicolae et al., 2010). eQTL results can be used to provide functional interpretation for findings from GWAS (Moffatt et al., 2007; Heid et al., 2010; Hsu et al., 2010;Lango Allen et al., 2010; Speliotes et al., 2010; Chu et al., 2011; Wu et al., 2012) and prioritize genes in an association region for carrying out functional experiments using animal models (Teslovich et al., 2010). Focusing on eQTLs may also be useful to identify genetic pathways associated with the risk of complex diseases and traits, such as basal cell carcinoma in a skin cancer GWAS (Zhang et al., 2012) and type 2 diabetes (Zhong et al., 2010). Other results show that many cis eQTLs are shared across tissues (Ding et al., 2010) and that a comprehensive eQTL catalog in one tissue might be used to increase the power of capturing relevant transcripts for other diseases (including those that are only weakly or incidentally expressed in tissues where eQTL information was collected).

As single-SNP analysis still remains the most popular in GWAS, we focus on those methods designed for this type of analysis. Single-SNP analysis tests one SNP at a time for association by scanning across the whole genome, and hence involves a large number of hypotheses. To correct for multiple comparisons, statistical methods have been proposed and applied to control for the family-wise error rate (FWER) (Bonferroni, 1936; Holm, 1979) or the false discovery rate (FDR) (Benjamini and Hochberg, 1995; Storey and Tibshirani, 2003). Recent advances in statistical methodology make it possible to incorporate prior information through weighted hypothesis testing. In several of such methods (Genovese et al., 2006; Roeder et al., 2006, 2007), hypotheses are up-weighted or down-weighted based on prior likelihood of association with the trait of interest. While keeping the FWER or FDR under control, the procedures can improve power with informative weights and suffer small loss in power with uninformative weights (Genovese et al., 2006; Roeder and Wasserman, 2009). This feature is appealing as compared to prescreening SNPs based on prior information (e.g., to consider only eQTLs for association testing). In this paper, we propose to use eQTLs as prior information, and apply these weighted hypothesis testing methods to reanalyze the MAGICS (Multicentre Asthma Genetics in Childhood Study) data of asthma GWAS (Moffatt et al., 2007) as well as the GABRIEL meta-study of asthma (Moffatt et al., 2010).

RESULTS

Published Asthma Associations are Enriched with eQTLs

We extracted published asthma associations from the GWAS catalog maintained by the National Human Genome Research Institute. As of January 15, 2013, 52 distinct reference SNPs in or near more than 40 genes have been reported to be associated with asthma (Table A1). According to the eQTL database (described in Materials and Methods), 20 of these 52 SNPs (38.5%) are eQTLs. Using the proxy SNP search tool SNAP (Johnson et al., 2008), we then obtained an extended list of 506 SNPs that were either in the GWAS catalog or in strong linkage disequilibrium (LD) with the 52 SNPs ($r^2 \geq 0.8$). We called all these 506 SNPs the extended set of asthma-associated SNPs.

We calculated an eQTL enrichment p-value (Hosack et al., 2003) using the MAGICS data. There are 300,821 SNPs that passed quality control in the MAGICS data. Among these SNPs, 29 SNPs are in the GWAS catalog, and 64 SNPs are among the 506 extended asthma-associated SNPs defined previously. To account for the LD between SNPs in the calculation of enrichment p-value, we conducted LD pruning with the r^2 threshold of 0.8 on the 300,821 SNPs. This resulted in 251,826 SNPs and 38 of them are extended asthma-associated SNPs according to the GWAS catalog. According to the eQTL database, 22,922 SNPs (9.1% of 251,826 SNPs) are eQTLs, and 13 asthma associated SNPs (34.2% of 38 SNPs) are eQTLs. The corresponding enrichment p-value is 6.78×10^{-5}, suggesting the asthma associations are enriched with eQTLs in the MAGICS data. Note that other analyses considered all the SNPs rather than the pruned set of SNPs.

These results are in line with the previous findings (Nicolae et al., 2010), which studied the eQTLs in lymphoblastoid cell lines (LCL) from the HapMap samples and the GWAS catalog. Their results suggest that SNPs associated with complex traits are more likely to be eQTLs.

Weights Using eQTL Information

We calculated two kinds of weights using eQTL information for the MAGICS data. All the 300,821 SNPs passing quality control were considered. First, we defined a SNP as an eQTL SNP if it was labeled as an eQTL in the eQTL database (see details in Materials and Methods). There are 31,781cis eQTL SNPs (10.6% of 300,821 SNPs) according to the definition, and for each of them, we retrieved an eQTL p-value p_{eQTL}. Next, we considered two choices of weights, the general weight and the binary weight. The general weight is $w_g = \sqrt{-\log_{10} p_{eQTL}}$ for an eQTL SNP, and $w_g = 1$ otherwise. The binary weight takes only two possible values, $w_b = 3.70$ for any eQTL SNP and $w_b = 0.68$ otherwise. The two values of the binary weight were chosen to maximize the minimum power while keeping at least 10.6% (also the percentage of eQTL SNPs) of all the hypotheses with a power of 60%. The parameters for calculating the binary weight are $\epsilon = 0.106$, $\alpha = 0.05$, and $\beta = 0.4$ (see details in Materials and Methods). Last, both weights were normalized to have the mean equal to 1 which is necessary for the weighted hypothesis testing methods to maintain the correct FWER or FDR (Genovese et al., 2006). After normalization, the general weight w_g has a mean of 2.44 and a median of 2.21 among eQTL SNPs, while the binary weight w_b is 3.70 for all eQTL SNPs (Figure 1).

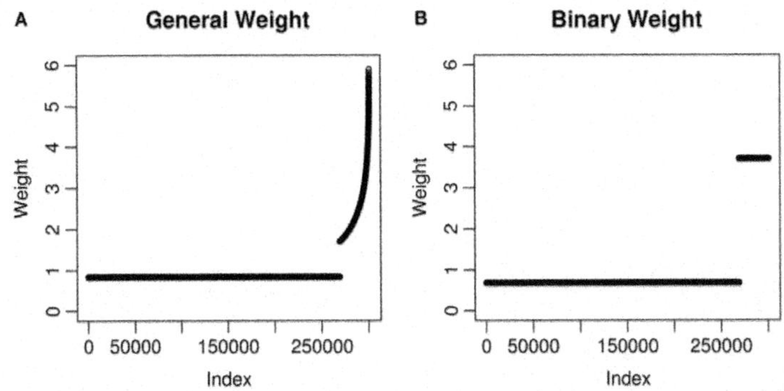

Figure 1: Weights used in the MAGICS analysis. Each weight corresponds to a SNP and a hypothesis. The weights have been normalized to have mean 1 and shown in the ascending order. **(A)** The weights are based on the square root of $-\log_{10} p_{eQTL}$ where p_{eQTL} is the eQTL p-value; **(B)** the weights take only two possible values, which are decided using the method described in Materials and Methods.

Weighted Hypothesis Testing

We applied the weighted hypothesis testing methods (Genovese et al., 2006; Roeder and Wasserman, 2009) using the general weight w_g and the binary weight w_b to the MAGICS data. For each of the 300,821 SNPs, we calculated the trait association p-value, p, from the single-SNP association test on the phenotypes of asthma status, as well as the weighted p-values $Q_g = p/w_g$ and$Q_b = p/w_b$. Multiple testing adjustments were done for both the original p-values (p) and the weighted p-values (Q_g and Q_b). Bonferroni (1936) and Holm's (1979) methods were considered to control for FWER, and Benjamini and Hochberg's (1995) method was used to control for FDR.

We first ranked the SNPs using their p-values in the ascending order, and compared the ranks based on the weighted p-values with those based on the original p-values (Figure 2). Since only 10.6% SNPs are eQTL SNPs according to the eQTL database, and hypotheses for eQTL SNPs are up-weighted, eQTLs generally have higher ranks after weighting, and non-eQTLs' ranks are lower but the magnitude of changes is small. This is true for both the general and binary weights. When restricting to the 29 asthma-associated SNPs reported in the GWAS catalog, we also observed similar behaviors, suggesting that weighting hypotheses may improve power using informative weights, and sacrifice a little power using uninformative weights (Roeder and Wasserman, 2009).

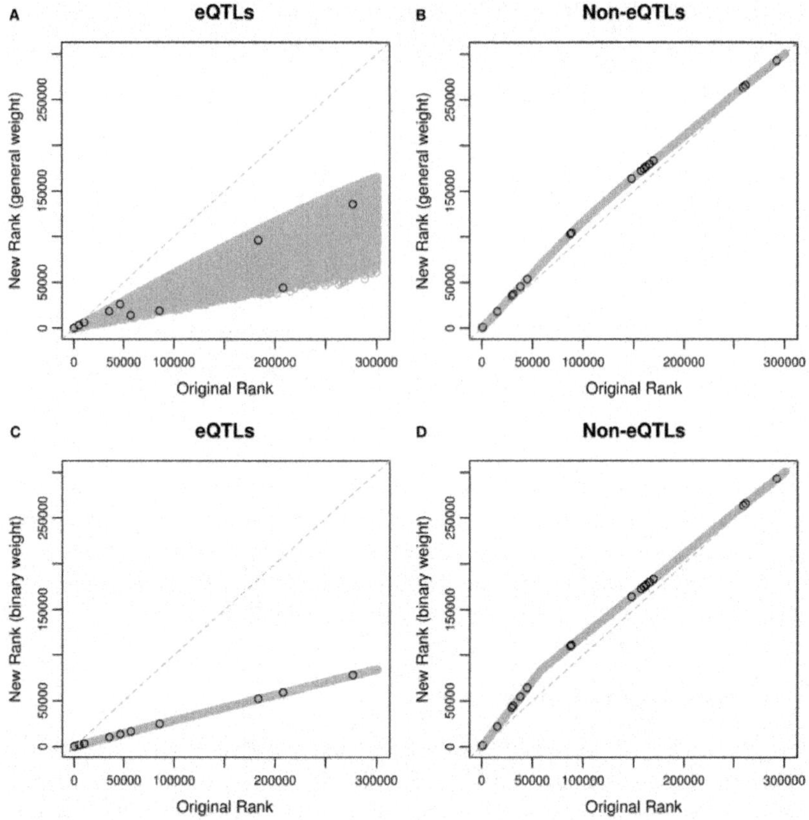

Figure 2: Rankings of the SNPs based on original p-values and weighted p-values in the MAGICS analysis. (A) Original ranks of eQTLs compared to their new ranks based on the general weight; (B) original ranks of non-eQTLs compared to their new ranks based on the general weight; (C) original ranks of eQTLs compared to their new ranks based on the binary weight; (D) original ranks of non-eQTLs compared to their new ranks based on the binary weight. The black circles represent the reported asthma-associated SNPs in the GWAS catalog, and the gray circles represent the rest of the SNPs in the data.

We then looked at the Q–Q plots of the original and weighted p-values. For p-values greater than 0.0001, the Q–Q curves (Figure 3) are similar between the original and weighted p-values, regardless of the weights used. For those p-values less than 0.0001, some weighted p-values are smaller than original ones, and the difference is larger using the binary weight. We also observed that 3 asthma-associated SNPs in the GWAS catalog are among the top SNPs with original p-values less than 10^{-6}. The weighted p-values for all the 3 asthma-associated SNPs are smaller than original ones.

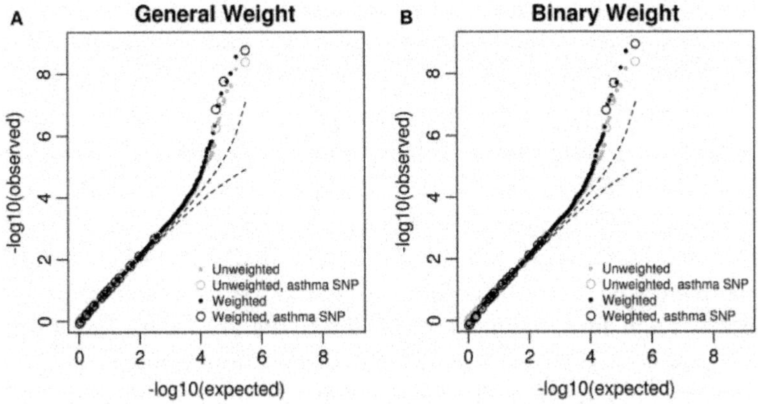

Figure 3: Q–Q plots of original p-values and weighted p-values in the MAGICS analysis. The weighted p-values are based on (A) the general weight, or (B) the binary weight. The reported asthma-associated SNPs in the GWAS catalog are shown in circles.

Next, we applied the methods to control for the FWER. An effective ratio of 0.791 (Li et al., 2012) was used to calculate the effective number of SNPs ($300{,}821 \times 0.791$). Controlling for an FWER level of 0.05, we obtained significant SNPs using both the original and weighted p-values. Both Bonferroni and Holm's methods gave the same results, and both weights (binary and general) also gave the same results (Tables A2, A3). The unweighted hypothesis testing claimed 6 SNPs to be significant, all on chromosome 17, including 2 asthma-associated SNPs (rs3894194 with GSDMA, and rs7216389 with ORDML3) that have been reported previously (Moffatt et al., 2007, 2010). After applying the weighted hypothesis testing, we obtained 9 significant SNPs including all the 6 SNPs identified by the unweighted method, although the ranks are not exactly the same. The 3 SNPs additionally identified by eQTL weighting were rs3902025, rs4795405, and rs2305480. The SNP rs2305480, a missense SNP in the gene GSDMB, was not reported in the previous GWAS study (Moffatt et al., 2007) but has been reported as an asthma-associated SNP in a later larger scale study by the GABRIEL consortium (including the MAGIC data, Moffatt et al., 2010) and was found to be strongly interacting with exposure to tobacco smoke in early life (Bouzigon et al., 2008). We found that rs2305480 is actually in LD ($r^2 = 0.702$, $D' = 0.926$) with rs7216389 that was identified by the unweighted methods, suggesting that rs2305480 may not represent a new association. Using a stringent r^2 threshold of 0.4, we found that the other two SNPs, rs3902025 and rs4795405, are also in LD with at least one SNP identified by the unweighted methods. So there is no new association identified by the weighted methods in this particular analysis.

Besides controlling for FWER, we also used Benjamini and Hochberg's (BH) procedure (Benjamini and Hochberg, 1995) to control for a FDR level of 0.05 (Table A4). Based on the original p-values without weighting, the BH procedure gave 11 positive results (SNPs). The weighted BH procedures based on the general weight and the binary weight resulted in 7 and 8 additional positive results (SNPs), respectively. Using a stringent r^2 threshold of 0.4, we found that 5 SNPs (Table 1) are not in LD with any of the SNPs identified without weighting. Although none of the 5 SNPs are, or in LD with, any asthma-associated SNPs according to the GWAS catalog, there are some SNPs that seem interesting. Some of the SNPs are in or close to the genes PGAP3 and STARD3 on chromosome 17, and interestingly, rs2941504 has been reported in a recent independent study (Anantharaman et al., 2011) to be associated with asthma, although it does not meet the criteria for inclusion in the GWAS catalog. This suggests that the reanalysis using the eQTL weighting approaches is promising and potentially useful.

Table 1: Additional significant SNPs or positive results identified by eQTL weighting methods after accounting for linkage disequilibrium in the MAGICS analysis

Chr	SNP	Gene	*p*-value	Q_g	Q_b	Method
17	rs1877031	*STARD3*	4.32×10^{-6}	1.54×10^{-6}	1.17×10^{-6}	BH
17	rs931992	*TCAP, STARD3*	5.61×10^{-6}	1.96×10^{-6}	1.52×10^{-6}	BH
17	rs1565922	*PGAP3*	7.28×10^{-6}	2.55×10^{-6}	1.97×10^{-6}	BH
17	rs2941504	*PGAP3*	7.64×10^{-6}	2.68×10^{-6}	2.06×10^{-6}	BH
10	rs11191325	*SUFU*	8.49×10^{-6}	3.37×10^{-6}	2.29×10^{-6}	BH

Q_g and Q_b are the eQTL weighted p-values based on the general weight and the binary weight, respectively. Using both weights result in the same set of SNPs.

Reanalysis of the GABRIEL Data

As another application, we reanalyzed the GABRIEL data using the eQTL weighted approaches. Since only the p-values are necessary for the use of eQTL weighting, we took the p-values of the meta analysis of 37 studies that were calculated based on imputed data. In total, there are 2,473,850 SNPs and their p-values available in the GABRIEL study, which include 267,350 out of 268,204 eQTL SNPs in the eQTL database. The weights based on eQTL information were calculated in the similar way to the MAGICS data analysis.

We applied Bonferroni and Holm's methods with an FWER level of 0.05, as well as the BH procedure with an FDR level of 0.05. An effective ratio of 0.30 (Li et al., 2012) was used to calculate the effective number of SNPs ($2,473,850 \times 0.30$). After obtaining the lists of significant SNPs using different methods, we report any SNPs identified by eQTL weighting that are not in LD with any SNPs identified by unweighted methods using an r^2 threshold of 0.4. Such SNPs may be informative and suggest new associations. Tables 2 and 3 show the SNPs that were identified based on the general weight and the binary weight, respectively.

Table 2: Additional significant SNPs or positive results identified by eQTL weighting methods after accounting for linkage disequilibrium in the GABRIEL analysis

Chr	SNP	Gene	p-value	Q_g	Q_b	Method
5	rs244749	*MEF2C*	6.25×10^{-5}	2.51×10^{-5}	1.54×10^{-5}	BH
5	rs10075941		6.13×10^{-5}	3.03×10^{-5}	1.51×10^{-5}	BH
17	rs7503195		7.66×10^{-5}	3.07×10^{-5}	1.88×10^{-5}	BH
17	rs17637472		6.13×10^{-5}	3.56×10^{-5}	1.51×10^{-5}	BH
9	rs7047575		6.87×10^{-5}	3.92×10^{-5}	1.69×10^{-5}	BH

The results are based on the general weight.

Table 3: Additional significant SNPs or positive results identified by eQTL weighting methods after accounting for linkage disequilibrium in the GABRIEL analysis

Chr	SNP	Gene	p-value	Q_g	Q_b	Method
5	rs736801		2.15×10^{-7}	8.81×10^{-8}	5.29×10^{-8}	Bonferroni, Holm
6	rs2596450	*HCG26*	2.72×10^{-7}	1.16×10^{-7}	6.69×10^{-8}	Bonferroni, Holm
5	rs10075941		6.13×10^{-5}	3.03×10^{-5}	1.51×10^{-5}	BH
17	rs17637472		6.13×10^{-5}	3.56×10^{-5}	1.51×10^{-5}	BH
5	rs244749	*MEF2C*	6.25×10^{-5}	2.51×10^{-5}	1.54×10^{-5}	BH
9	rs7047575		6.87×10^{-5}	3.92×10^{-5}	1.69×10^{-5}	BH
17	rs7503195		7.66×10^{-5}	3.07×10^{-5}	1.88×10^{-5}	BH
6	rs9273363	*HLA_DQB1*	8.38×10^{-5}	4.41×10^{-5}	2.06×10^{-5}	BH
5	rs4351182		9.98×10^{-5}	4.91×10^{-5}	2.45×10^{-5}	BH
2	rs13391794		1.01×10^{-4}	5.62×10^{-5}	2.49×10^{-5}	BH
5	rs10044342	*MEF2C*	1.10×10^{-4}	4.59×10^{-5}	2.71×10^{-5}	BH
2	rs6751196		1.14×10^{-4}	6.22×10^{-5}	2.80×10^{-5}	BH
6	rs176095	*PBX2, GPSM3*	1.25×10^{-4}	5.12×10^{-5}	3.08×10^{-5}	BH
2	rs2675073		1.34×10^{-4}	5.40×10^{-5}	3.29×10^{-5}	BH
2	rs1913621		1.46×10^{-4}	7.92×10^{-5}	3.60×10^{-5}	BH
2	rs10497621	*NUP35*	1.59×10^{-4}	7.55×10^{-5}	3.91×10^{-5}	BH
5	rs244750	*MEF2C*	1.70×10^{-4}	7.08×10^{-5}	4.17×10^{-5}	BH
6	rs9366689	*POM121L2*	1.72×10^{-4}	8.73×10^{-5}	4.23×10^{-5}	BH
6	rs7775759		1.81×10^{-4}	6.43×10^{-5}	4.46×10^{-5}	BH
6	rs7741091		1.81×10^{-4}	6.43×10^{-5}	4.46×10^{-5}	BH
8	rs6601649		1.90×10^{-4}	9.68×10^{-5}	4.68×10^{-5}	BH
6	rs204994		1.94×10^{-4}	8.02×10^{-5}	4.76×10^{-5}	BH

The results are based on the binary weight.

Size Simulations on FWER

We conducted simulations using 5000 permutations based on the MAGICS data, and calculated the percentage of having at least one false positive claimed by Bonferroni and Holm's methods ($\alpha = 0.05$, with an effective ratio of 0.791). In fact, any SNPs claimed significant using the two methods would be a false positive. The calculated percentages (Table 4) provide estimates of the FWER. Bonferroni and Holm's methods give the same results. The results suggest that, under the null hypothesis, the FWER level is controlled for the methods based on both the original (unweighted) and the weighted p-values. The simulations confirm the validity of the weighted hypothesis method (Genovese et al., 2006).

Table 4: Family-wise error rate estimates in 5000 permutations

P-value	FWER
Original	0.0454
Weighted by general weight	0.0458
Weighted by binary weight	0.0460

Both Bonferroni and Holm's methods gave the same results in the same scenarios.

DISCUSSION

It is of substantial interest to enhance the power for identifying associations in the era of post-GWAS. Besides meta-analysis that has been proved successful in power gain (Moffatt et al., 2010), incorporating prior information has also received increasing attention. Such information can be obtained from various sources and levels, such as linkage analysis (Roeder et al., 2006), gene expression (Yang et al., 2010), and annotation information of variants (Adzhubei et al., 2010), genes (Saccone et al., 2007), and pathways (Wang et al., 2007). The so-called "agnostic" GWAS may benefit from incorporating useful prior information. In our study of asthma, gene expression information is of particular interest, as a recent study (Moffatt et al., 2007) identified several eQTLs associated with asthma.

In the reanalysis of the MAGICS data (Moffatt et al., 2007), we applied recently developed statistical methods that can improve power by weighting hypothesis (Genovese et al., 2006; Roeder and Wasserman, 2009). Using eQTL information obtained from an independent dataset, we employed weighted procedures that up-weighted eQTL SNPs and down-weighted non-eQTL SNPs while controlling for the FWER or the FDR. It has been proved (Genovese et al., 2006) that any set of nonnegative weights can guarantee substantial power gain given informative weights and little power loss for uninformative weights. The property implies that the weighted procedures are robust to informativeness of weights and to the uneven coverage of genes and expression targets on the genome. We took advantage of this robustness and applied the procedures to an asthma study. We found additional SNPs that were significantly associated with asthma according to the weighting hypothesis methods. Some of them were interesting after we accounted for LD and compared them to literature. Our analysis was the first application of this approach to asthma GWAS studies, and the results successfully illustrated the use of eQTL weighting in the context of asthma studies. As another application, we also reanalyzed the GABRIEL meta-analysis p-values and reported corresponding results.

It is noted that the weighted procedures can utilize eQTL information from a reference database. Multiple choices of eQTL databases have already been made available (e.g., Yang et al., 2010; Liang et al., 2013), and future efforts may provide even better reference of eQTL information. For example, the eQTL information considered in our reanalysis was obtained through a single platform (Affymetrix HG-U133 Plus 2.0), and better coverage of gene expression profiling may be achieved through RNA-Seq technologies or by combining information from various platforms.

Besides the weighted procedures, an alternative method of using eQTL information is to simply test association between eQTL SNPs and the trait of interest. Such a method is not recommended in GWAS as it excludes non-eQTL completely and relies on the prior information too heavily. By contrast, weighted procedures make it possible to consider eQTLs and non-eQTLs simultaneously. More importantly, they can possibly increase power if the prior information is useful and are able to maintain the type I error under the null.

Applying the weighted procedures in our reanalysis only requires p-value of eQTL SNPs. This flexibility means that such analyses can be applied to any existing GWAS data, even if they do not have accompanying gene expression data. Although gene expression may have tissue-specific patterns, a substantial fraction of eQTLs may be shared across tissues (Ding et al., 2010). Hence eQTLs developed from tissues that are not directly relevant to the outcome of interest, such as those from publicly available eQTL databases based on LCL, can be used to improve power on GWAS. It is possible that using eQTL information from relevant tissues may result in even more power gain, if such information is available.

Besides the particular weighting hypothesis method (Genovese et al., 2006) we adopted, Bayesian methods are potentially alternative strategies to incorporate eQTL information. The use of Bayes factors has been applied to genetic association studies (Wellcome Trust Case Control Consortium, 2007; Stephens and Balding, 2009). In single SNP analysis, a prior is assumed for each SNP effect [e.g., $N(0, 0.2^2)$ under a model of association in Wellcome Trust Case Control Consortium (2007)]. eQTL information can be naturally incorporated into the prior, although it may be challenging to choose a realistic yet tractable alternative model and to assess error rates (Hoggart et al., 2008), especially with the eQTL weight. One possible choice is through modifying the variance of the prior, for example assuming a prior $N(0.2^2w)$, where w is a weight of eQTL signal. Another possible choice is to keep the variance the same and increase the probability of association for eQTLs a priori. It is of interest in future research to explore these possibilities and consider the extension of Bayesian methods to incorporate eQTL information.

In our analysis, we took into account the LD between SNPs by considering the effective number of SNPs (Li et al., 2012). As an alternative, testing SNP sets for association has potential of improving power and reducing the correlation between tests. Since the focus of this paper is to demonstrate the use of eQTL information in association testing, we will consider the weighted correlated hypothesis in future research.

Two choices of weights were applied in our analysis including a binary weight and a weight using strength of eQTLs, and the results using the two weights were similar in our analysis. Theoretical results exist (Roeder and Wasserman, 2009) for the optimal binary weight, which provide guidance in choosing the values of the weight. The weight taking advantage of the eQTL strength may possibly provide more useful information, and what is the best choice of weights is still under research.

Through an application to an asthma GWAS, we demonstrated the usefulness of eQTL weights in GWAS. Although results may vary depending on the traits of interest and the underlying biological mechanism, the potentials of increasing power and little investment required for reanalysis make the eQTL-weighted procedures desirable for reanalysis of existing GWAS data and useful for design and analysis of future studies.

MATERIALS AND METHODS

The MAGICS Asthma GWAS Samples and Data

The MAGICS (Multicentre Asthma Genetics in Childhood Study) study data (Moffatt et al., 2007), part of the GABRIEL consortium, were reanalyzed by incorporating eQTL information. Quality control procedures were conducted similarly to a published protocol (Anderson et al., 2010). Individuals with missing phenotypes, elevated missing rates ($\geq 5\%$), or outlying heterozygosity rate were removed. Markers with an excessive missing rate ($\geq 5\%$), low MAF ($<5\%$), or failing in the HWE test (p-value $< 10^{-5}$) were all excluded as well. The remaining dataset contains 1296 individuals (647 affected and 649 unaffected) genotyped across 300,821 SNPs.

To account for possible divergent ancestry and population stratification, principal component analysis (PCA) was conducted using EIGENSOFT 4.2 (Patterson et al., 2006; Price et al., 2006). The genotype data were pruned for LD prior to the PCA. The PCA result (Figure A1) suggests that no obvious stratification exists, and the signal of the first principal component is very weak.

In the subsequent analysis, we still included the first principal component as a covariate.

LD pruning was considered only in the calculation of enrichment p-value. It was conducted using PLINK (v1.07, downloaded from http://pngu.mgh.harvard.edu/purcell/plink/) (Purcell et al., 2007). A moving window with a width of 50 SNPs and a step size of 5 SNPs was considered, and pairwise LDs were calculated and pruned if $r^2 > 0.8$ (corresponding PLINK arguments: "–indep-pairwise 50 5 0.8").

The GABRIEL Meta-Analysis p-Values

Association testing results, including SNP ID and p-values, were obtain from a reanalysis of the GABRIEL consortium data using imputed SNPs (Bouzigon et al., personal communication). The meta-analysis considered imputation of SNP genotypes using the HapMap 2 reference data for 37 studies, and calculated a meta-analysis p-value for each SNP using available data. Imputed SNPs were kept for analysis if their imputation scores (Rsq) were ≥ 0.5 and if their minor allele frequencies were ≥ 1%. In total there were 2,473,850 SNPs that passed the quality control. Only the SNP ID and the p-values of these SNPs were obtained and used for the reanalysis described in this paper.

Expression Quantitative Trait Loci Data

An eQTL database (http://www.hsph.harvard.edu/liming-liang/software/eqtl/) resulting from an independent dataset was used as prior information to be incorporated in the GWAS. The sample contains 405 siblings from a panel of families of British descent (MRC-A) (Dixon et al., 2007). Global gene expression in LCLs was measured using Affymetrix HG-U133 Plus 2.0 chips. All siblings were genotyped using the Illumina Sentrix HumanHap300 BeadChip (ILMN300K) and/or the Illumina Sentrix Human-1 Genotyping BeadChip (ILMN100K). The SNP genotype data were further imputed using the MaCH program, and each SNP was tested for association with probes in the gene expression data. Restricting to cis eQTLs (1 Mb region) and controlling for the FDR of 1%, there are 515,947 tests with logarithm of odds (LOD) scores greater than 3.172, corresponding to 268,204 unique SNPs. In case a SNP has multiple p-values reported for associations with different probes, the minimal p-value was used for that SNP. These 268,204 SNPs are considered as eQTLs, and the database contains information of their physical positions, LOD scores, p-values, and residing or nearby genes. Details of the database are described by Liang et al. (2013).

Genetic Association Analysis

Genetic association analysis of the MAGICS data was conducted in PLINK. Logistic regression was used to test for disease-trait SNP association while adjusted for gender and the first principal component. Meta-analysis on GABRIEL data was carried out by combining association results from 37 studies using a random effect model, and all computations were done using Stata software.

p-Value Weighting Methods

Consider m hypotheses $H_1, \ldots, H_m$ and their test statistic p-values, $P_1, \ldots, P_m$. Suppose there are weights $W_1, \ldots, W_m$ available for the m tests, respectively, satisfying $W_i > 0$ and $\sum_{i=1}^{m} W_i = m$. Define $Q_i = P_i/W_i$ and let $Q_{(1)} \leq \ldots \leq Q_{(m)}$ be the sorted values. Let $P_{(1)}, \ldots, P_{(m)}$ and $W_{(1)}, \ldots, W_{(m)}$ be the values in the corresponding order. Q_i is sometimes referred to a "weighted p-value" (e.g.,Roeder and Wasserman, 2009), although it is not a p-value.

The weighted Bonferroni procedure is to reject any hypothesis H_j $(1 \leq j \leq m)$ that satisfies $Q_j \leq \alpha/m$, where α is the desired level of FWER. Genovese et al. (2006) showed that this procedure controls FWER at level no greater than α.

Holm's weighted procedure (1979) is carried out as follows: given the desired α level of FWER, if$Q_{(1)} \geq \alpha/m$, no hypothesis is rejected; otherwise, find the largest j that satisfies $Q_{(i)} \leq \alpha/\sum_{k=i}^{m} W_{(k)}$for all $i \leq j$, and reject the hypotheses corresponding to the j smallest Q_j's. Genovese et al. (2006)also prove that this procedure can work for a general setting of weights.

We also consider Benjamini and Hochberg's procedure (1995) for controlling FDR. Given the desired level α, find the largest j such that $Q_{(j)} \leq \alpha \cdot j/m$, and reject the hypotheses corresponding to the j smallest Q_j's. Genovese et al. (2006) prove that this procedure controls FDR at level α.

eQTL Information as Weights

The eQTL p-values were used to construct weights for the SNPs in the asthma GWAS reanalysis. We considered two kinds of weights, the binary weight w_b and the general weight w_g. The binary weight takes only two possible values that are predefined, denoted by w_{eQTL} and $w_{non-eQTL}$. For mhypotheses, a binary weight is defined as $w_b = (w_{b,1}, \ldots, w_{b,m})$ where $w_{b,j} = w_{eQTL}$ if the jth SNP is an eQTL SNP, and $w_{b,j} = w_{non-eQTL}$ if it is not an eQTL SNP. Given the values of α, β, and ϵ, the optimal values of w_{eQTL} and $w_{non-eQTL}$ were chosen (Roeder and Wasserman, 2009) to maximize the minimum power among all the hypotheses while having at least a fraction ϵ with high power $1-\beta$. Here α is either the level of FWER or FDR. We also considered a general weight, where the weight $w_g = (w_{g,1}, \ldots, w_{g,m})$ has $w_{g,j} = \sqrt{-\log_{10} p_{eQTL}}$ if the jth SNP is an eQTL SNP with the eQTL p-value p_{eQTL}, and $w_{g,j} = 1$ otherwise. The particular form was intuitively chosen prior to the reanalysis of the GWAS data in consideration of avoiding up-weighting top eQTL SNPs too much. Both w_b and w_g were then normalized such that the means equal to 1, i.e., $\overline{w}_B = 1$ and $\overline{w}_G = 1$.

Reported Associations in the GWAS Catalog

Asthma-associated SNPs and genes reported in publications were retrieved from the online catalog of published GWAS on January 15, 2013. The catalog limits the associations to those with p-values less than 1.0×10^{-5} and records only one SNP with a gene or region of high LD unless there was evidence of independent association. The reported associations were compared against the findings in the asthma GWAS data we reanalyzed.

Linkage Disequilibrium Information

To account for LD between SNPs, LD information based on HapMap 2 was obtained. The SNAP proxy search tool (http://www.broadinstitute.org/mpg/snap/ldsearch.php) was used to obtain the information, based on the HapMap 2 (rel22) reference and a distance limit of 500kb.

Size Simulation Using the Asthma GWAS Data

Besides analyzing the MAGICS asthma GWAS data, we also conducted size simulations by permuting the disease status in the data. Logistic regression was considered where the dependent variable was the disease status (affected or unaffected) and the independent variables included a single SNP effect, gender, and the first principal component. The regression was applied to all the ~300,000 SNPs across the whole genome. Five thousand permutations were done by permuting the disease status among all the individuals, and then the model was refitted for each SNP. In the end of simulations, 5000 permutation p-values were obtained for each of the ~300,000 SNPs.

Conflict of Interest Statement

The authors declare that the research was conducted in the absence of any commercial or financial relationships that could be construed as a potential conflict of interest.

ACKNOWLEDGMENTS

Lin Li and Xihong Lin's research was supported by grants from National Cancer Institute (R37 CA076404 and P01 CA134294). Martin Farrall is supported by the British Heart Foundation Centre for Research Excellence in Oxford and the Wellcome Trust core award [090532/Z/09/Z]. We thank the members of the GABRIEL consortium groups for providing data and summary results (a full list of the GABRIEL groups can be found in the Supplement of Moffatt et al., 2010).

RESOURCES

R functions have been made available for users' convenience to compute the eQTL weighted p-values in order to conduct weighted procedures such as Bonferroni, Holm's and Benjamini–Hochberg procedures. The functions can be found at http://www.hsph.harvard.edu/liming-liang/eqtl-weighted-gwas.

REFERENCES

1. Adzhubei, I. A., Schmidt, S., Peshkin, L., Ramensky, V. E., Gerasimova, A., Bork, P., et al. (2010). A method and server for predicting damaging missense mutations. Nat. Methods 7, 248–249. doi: 10.1038/nmeth0410-248
2. Anantharaman, R., Andiappan, A. K., Nilkanth, P. P., Suri, B. K., Wang, D. Y., and Chew, F. T. (2011). Genome-wide association study identifies PERLD1 as asthma candidate gene. BMC Med. Genet. 12:170. doi: 10.1186/1471-2350-12-170
3. Anderson, C. A., Pettersson, F. H., Clarke, G. M., Cardon, L. R., Morris, A. P., and Zondervan, K. T. (2010). Data quality control in genetic case-control association studies. Nat. Protoc. 5, 1564–1573. doi: 10.1038/nprot.2010.116
4. Benjamini, Y., and Hochberg, Y. (1995). Controlling the false discovery rate: a practical and powerful approach to multiple testing. J. R. Stat. Soc. Ser. B 57, 289–300. doi: 10.2307/2346101
5. Bonferroni, C. E. (1936). Teoria statistica delle classi e calcolo delle probabilità. Pubblicazioni del R Istituto Superiore di Scienze Economiche e Commerciali di Firenze 8, 3–62.
6. Bouzigon, E., Corda, E., Aschard, H., Dizier, M. H., Boland, A., Bousquet, J., et al. (2008). Effect of 17q21 variants and smoking exposure in early-onset asthma. N. Engl. J. Med. 359, 1985–1994. doi: 10.1056/NEJMoa0806604

7. Chu, X., Pan, C.-M., Zhao, S.-X., Liang, J., Gao, G.-Q., Zhang, X.-M., et al. (2011). A genome-wide association study identifies two new risk loci for Graves' disease. Nat. Genet. 43, 897–901. doi: 10.1038/ng.898

8. Cookson, W., Liang, L., Abecasis, G., Moffatt, M., and Lathrop, M. (2009). Mapping complex disease traits with global gene expression. Nat. Rev. Genet. 10, 184–194. doi: 10.1038/nrg2537

9. Cookson, W. (2004). The immunogenetics of asthma and eczema: a new focus on the epithelium. Nat. Rev. Immunol. 4, 978–988. doi: 10.1038/nri1500

10. Dimas, A. S., Deutsch, S., Stranger, B. E., Montgomery, S. B., Borel, C., Attar-Cohen, H., et al. (2009). Common regulatory variation impacts gene expression in a cell type-dependent manner. Science 325, 1246–1250. doi: 10.1126/science.1174148

11. Ding, J., Gudjonsson, J. E., Liang, L., Stuart, P. E., Li, Y., Chen, W., et al. (2010). Gene expression in skin and lymphoblastoid cells: refined statistical method reveals extensive overlap in cis-eQTL signals. Am. J. Hum. Genet. 87, 779–789. doi: 10.1016/j.ajhg.2010.10.024

12. Dixon, A. L., Liang, L., Moffatt, M. F., Chen, W., Heath, S., Wong, K. C. C., et al. (2007). A genome-wide association study of global gene expression. Nat. Genet. 39, 1202–1207. doi: 10.1038/ng2109

13. Genovese, C. R., Roeder, K., and Wasserman, L. (2006). False discovery control with p-value weighting. Biometrika 93, 509–524. doi: 10.1093/biomet/93.3.509

14. Heid, I. M., Jackson, A. U., Randall, J. C., Winkler, T. W., Qi, L., Steinthorsdottir, V., et al. (2010). Meta-analysis identifies 13 new loci associated with waist-hip ratio and reveals sexual dimorphism in the genetic basis of fat distribution. Nat. Genet. 42, 949–960. doi: 10.1038/ng.685

15. Hoggart, C. J., Clark, T. G., De Iorio, M., Whittaker, J. C., and Balding, D. J. (2008). Genome-wide significance for dense SNP and resequencing data. Genet. Epidemiol. 32, 179–185. doi: 10.1002/gepi.20292

16. Holm, S. (1979). A simple sequentially rejective multiple test procedure. Scand. J. Stat. 6, 65–70. doi: 10.2307/4615733

17. Hosack, D. A., Dennis, G. Jr. Sherman, B. T., Lane, H. C., and Lempicki, R. A. (2003). Identifying biological themes within lists of genes with EASE. Genome Biol. 4:R70. doi: 10.1186/gb-2003-4-10-r70

18. Hsu, Y.-H., Zillikens, M. C., Wilson, S. G., Farber, C. R., Demissie, S., Soranzo, N., et al. (2010). An integration of genome-wide association study and gene expression profiling to prioritize the discovery of

novel susceptibility Loci for osteoporosis-related traits. PLoS Genet. 6:e1000977. doi: 10.1371/journal.pgen.1000977

19. Johnson, A. D., Handsaker, R. E., Puilt, S., Nizzari, M. M., O'Donnell, C. J., and de Bakker, P. I. W. (2008). SNAP: a web-based tool for identification and annotation of proxy SNPs using HapMap. Bioinformatics 24, 2938–2939. doi: 10.1093/bioinformatics/btn564
20. Lango Allen, H., Estrada, K., Lettre, G., Berndt, S. I., Weedon, M. N., Rivadeneira, F., et al. (2010). Hundreds of variants clustered in genomic loci and biological pathways affect human height. Nature 467, 832–838. doi: 10.1038/nature09410
21. Li, M. X., Yeung, J. M. Y., Cherny, S. S., and Sham, P. C. (2012). Evaluating the effective numbers of independent tests and significant p-value thresholds in commercial genotyping arrays and public imputation reference datasets. Hum. Genet. 131, 747–756. doi: 10.1007/s00439-011-1118-2
22. Liang, L., Morar, N., Dixon, A. L., Lathrop, G. M., Abecasis, G. R., Moffatt, M. F., et al. (2013). A cross-platform catalogue of 14, 177 expression quantitative trait loci derived from lymphoblastoid cell lines. Genome Res. 23, 716–726. doi: 10.1101/gr.142521.112
23. Moffatt, M. F., Kabesch, M., Liang, L., Dixon, A. L., Strachan, D., Heath, S., et al. (2007). Genetic variants regulating ORMDL3 expression contribute to the risk of childhood asthma. Nature 448, 470–473. doi: 10.1038/nature06014
24. Moffatt, M. F., Gut, I. G., Demenais, F., Strachan, D. P., Bouzigon, E., Heath, S., et al. (2010). A large-scale, consortium-based genomewide association study of asthma. N. Engl. J. Med. 363, 1211–1221. doi: 10.1056/NEJMoa0906312
25. Nica, A. C., Parts, L., Glass, D., Nisbet, J., Barrett, A., Sekowska, M., et al. (2011). The architecture of gene regulatory variation across multiple human tissues: the MuTHER study. PLoS Genet. 7:e1002003. doi: 10.1371/journal.pgen.1002003
26. Nicolae, D. L., Gamazon, E., Zhang, W., Duan, S., Dolan, M. E., and Cox, N. J. (2010). Trait-associated SNPs are more likely to be eQTLs: annotation to enhance discovery from GWAS. PLoS Genet. 6:e1000888. doi: 10.1371/journal.pgen.1000888
27. Patterson, N., Price, A. L., and Reich, D. (2006). Population structure and eigenanalysis. PLoS Genet. 2:e190. doi: 10.1371/journal.pgen.0020190
28. Price, A. L., Patterson, N. J., Plenge, R. M., Weinblatt, M. E., Shadick, N. A., and Reich, D. (2006). Principal components analysis corrects for

stratification in genome-wide association studies. Nat. Genet. 38, 904–909. doi: 10.1038/ng1847

29. Purcell, S., Neale, B., Todd-Brown, K., Thomas, L., Ferreira, M. A. R., Bender, D., et al. (2007). PLINK: a tool set for whole-genome association and population-based linkage analyses. Am. J. Hum. Genet. 81, 559–575. doi: 10.1086/519795

30. Roeder, K., and Wasserman, L. (2009). Genome-wide significance levels and weighted hypothesis testing. Stat. Sci. 24, 398–413. doi: 10.1214/09-STS289

31. Roeder, K., Bacanu, S.-A., Wasserman, L., and Devlin, B. (2006). Using linkage genome scans to improve power of association in genome scans. Am. J. Hum. Genet. 78, 243–252. doi: 10.1086/500026

32. Roeder, K., Devlin, B., and Wasserman, L. (2007). Improving power in genome-wide association studies: weights tip the scale.Genet. Epidemiol. 31, 741–747. doi: 10.1002/gepi.20237

33. Saccone, S. F., Hinrichs, A. L., Saccone, N. L., Chase, G. A., Konvicka, K., Madden, P., et al. (2007). Cholinergic nicotinic receptor genes implicated in a nicotine dependence association study targeting 348 candidate genes with 3713 SNPs. Hum. Mol. Genet. 16, 36–49. doi: 10.1093/hmg/ddl438

34. Speliotes, E. K., Willer, C. J., Berndt, S. I., Monda, K. L., Thorleifsson, G., Jackson, A. U., et al. (2010). Association analyses of 249, 796 individuals reveal 18 new loci associated with body mass index. Nat. Genet. 42, 937–948. doi: 10.1038/ng.686

35. Stephens, M., and Balding, D. (2009). Bayesian statistical methods for genetic association studies. Nat. Rev. Genet. 10, 681–690. doi: 10.1038/nrg2615

36. Storey, J. D., and Tibshirani, R. (2003). Statistical significance for genomewide studies. Proc. Natl. Acad. Sci. U.S.A. 100, 9440–9445. doi: 10.1073/pnas.1530509100

37. Stranger, B. E., Forrest, M. S., Clark, A. G., Minichiello, M. J., Deutsch, S., Lyle, R., et al. (2005). Genome-wide associations of gene expression variation in humans. PLoS Genet. 1:e78. doi: 10.1371/journal.pgen.0010078

38. Stranger, B. E., Nica, A. C., Forrest, M. S., Dimas, A., Bird, C. P., Beazley, C., et al. (2007). Population genomics of human gene expression. Nat. Genet. 39, 1217–1224. doi: 10.1038/ng2142

39. Stranger, B. E., Montgomery, S. B., Dimas, A. S., Parts, L., Stegle, O., Ingle, C. E., et al. (2012). Patterns of cis regulatory variation in diverse

human populations. PLoS Genet. 8:e1002639. doi: 10.1371/journal.pgen.1002639

40. Teslovich, T. M., Musunuru, K., Smith, A. V., Edmondson, A. C., Stylianou, I. M., Koseki, M., et al. (2010). Biological, clinical and population relevance of 95 loci for blood lipids. Nature 466, 707–713. doi: 10.1038/nature09270
41. Wang, K., Li, M., and Bucan, M. (2007). Pathway-based approaches for analysis of genomewide association studies. Am. J. Hum. Genet. 81, 1278–1283. doi: 10.1086/522374
42. Wellcome Trust Case Control Consortium. (2007). Genome-wide association study of 14, 000 cases of seven common diseases and 3, 000 shared controls. Nature 447, 661–678. doi: 10.1038/nature05911
43. Wu, C., Miao, X., Huang, L., Che, X., Jiang, G., Yu, D., et al. (2012). Genome-wide association study identifies five loci associated with susceptibility to pancreatic cancer in Chinese populations. Nat. Genet. 44, 62–66. doi: 10.1038/ng.1020
44. Xiong, Q., Ancona, N., Hauser, E. R., Mukherjee, S., and Furey, T. S. (2012). Integrating genetic and gene expression evidence into genome-wide association analysis of gene sets. Genome Res. 22, 386–397. doi: 10.1101/gr.124370.111
45. Yang, T.-P., Beazley, C., Montgomery, S. B., Dimas, A. S., Gutierrez-Arcelus, M., Stranger, B. E., et al. (2010). Genevar: a database and Java application for the analysis and visualization of SNP-gene associations in eQTL studies. Bioinformatics 26, 2474–2476. doi: 10.1093/bioinformatics/btq452
46. Zhang, M., Liang, L., Morar, N., Dixon, A. L., Lathrop, G. M., Ding, J., et al. (2012). Integrating pathway analysis and genetics of gene expression for genome-wide association study of basal cell carcinoma. Hum. Genet. 131, 615–623. doi: 10.1007/S00439-011-11047-8
47. Zhong, H., Yang, X., Kaplan, L. M., Molony, C., and Schadt, E. E. (2010). Integrating pathway analysis and genetics of gene expression for genome-wide association studies. Am. J. Hum. Genet. 86, 581–591. doi: 10.1016/j.ajhg.2010.02.020

CITATION

CHAPTER 1

He T, Sa J, Zhong P-S, Cui Y (2014) Statistical Dissection of Cyto-Nuclear Epistasis Subject to Genomic Imprinting in Line Crosses. PLoS ONE 9(3): e91702. doi:10.1371/journal.pone.0091702.

CHAPTER 2

Bernd Genser, Philip J Cooper, Maria Yazdanbakhsh, Mauricio L Barreto and Laura C Rodrigues, A guide to modern statistical analysis of immunological data, DOI: 10.1186/1471-2172-8-27.

CHAPTER 3

Fenger M (2014) Next generation genetics. *Front. Genet.* **5**:322. doi: 10.3389/fgene.2014.00322.

CHAPTER 4

Li Y, Guo Y, Wang J, Hou W, Chang MN, Liao D, et al. (2011) A Statistical Design for Testing Transgenerational Genomic Imprinting in Natural Human Populations. PLoS ONE 6(2): e16858. doi:10.1371/journal.pone.0016858.

CHAPTER 5

Shen M, Broeckling CD, Chu EY, Ziegler G, Baxter IR, Prenni JE, et al. (2013) Leveraging Non-Targeted Metabolite Profiling via Statistical Genomics. PLoS ONE 8(2): e57667. doi:10.1371/journal.pone.0057667.

CHAPTER 6

A. N. Diaz-Lacava, M. Walier, D. Holler, et al., "Genetic Geostatistical Framework for Spatial Analysis of Fine-Scale Genetic Heterogeneity in Modern Populations: Results from the KORA Study," International Journal of Genomics, vol. 2015, Article ID 693193, 15 pages, 2015. doi:10.1155/2015/693193.

CHAPTER 7

C. O. Aremu, Exploring Statistical Tools in Measuring Genetic Diversity for Crop Improvement, http://cdn.intechopen.com/pdfs/31485.pdf.

CHAPTER 8

Alex Clarke and Timothy J Vyse, Genetics of rheumatic disease, DOI: 10.1186/ar2781.

CHAPTER 9

Benjamin FrenchEmail author, Jungnam JooEmail author, Nancy L Geller, Stephen E Kimmel, Yves Rosenberg, Jeffrey L Anderson, Brian F Gage, Julie A Johnson, Jonas H Ellenberg and the COAG (Clarification of Optimal Anticoagulation through Genetics) Investigators, Statistical design of personalized medicine interventions: The Clarification of Optimal Anticoagulation through Genetics (COAG) trial, DOI: 10.1186/1745-6215-11-108.

CHAPTER 10

Michael Lässig, From biophysics to evolutionary genetics: statistical aspects of gene regulation, DOI: 10.1186/1471-2105-8-S6-S7.

CHAPTER 11

Zhang, Heping. Statistical Analysis in Genetic Studies of Mental Illnesses. Statist. Sci. 26 (2011), no. 1, 116--129. doi:10.1214/11-STS353. http://projecteuclid.org/euclid.ss/1307626569.

CHAPTER 12

Li L, Kabesch M, Bouzigon E, Demenais F, Farrall M, Moffatt MF, Lin X and Liang L (2013) Using eQTL weights to improve power for genome-wide association studies: a genetic study of childhood asthma. Front. Genet. 4:103. doi: 10.3389/fgene.2013.00103.

INDEX

Encyclopaedia of MODERN GENETICS ANALYSIS INTEGRATING GENES AND GENOMES

Encyclopaedia of MODERN GENETICS ANALYSIS INTEGRATING GENES AND GENOMES

Volume I: Genetics: Analysis of Genes and Genomes

Contributors

Jianliang Ni, Weidong Mao., and Tao He et al.

AURIS
Reference

www.aurisreference.com

Encyclopaedia of Modern Genetics Analysis Integrating Genes and Genomes

Volume I: Genetics: Analysis of Genes and Genomes

Contributors: Jianliang Ni, Weidong Mao., and Tao He al.

Published by Auris Reference Limited

www.aurisreference.com

United Kingdom

Printed in 2017 for Sale in the Indian Subcontinent

Genetics: Analysis of Genes and Genomes

ISBN: 978-1-78154-772-4 (Set of 3 Volumes)

British Library Cataloguing in Publication Data
A CIP record for this book is available from the British Library

Printed in the United Kingdom

Exclusively distributed by CBS Publishers & Distributors Pvt. Ltd.

Sales & Distribution Rights only for India, Pakistan, Bangladesh, Sri Lanka, Nepal and Bhutan.This book is not to be sold outside these territories.

Contents

List of Abbreviations

AIM	Ancestry informative markers
AP	Alternative pathway
CHA	Chinese Americans
CRY2	Cryptochrome2
EASTASAIMS	East Asian substructure ancestry informative markers
FDR	False discovery rate
FGF	Fibroblast growth factor
GBM	Glioblastoma
GG	Gene-gene
GMDR	Generalized Multifactor Dimensionality Reduction
GO	Gene Ontology
GSEA	Gene Set Enrichment Analysis
GSEA	Gene Set Enrichment Analysis
GWAS	Genome-wide association studies
HMM	Hidden Markov model
KEGG	Kyoto Encyclopedia of Genes and Genomes
LD	Linkage disequilibrium
LGT	Lateral gene transfer
MDR	Multifactor Dimensionality Reduction), 1
NIAID	National Institute of Allergy and Immunologic Diseases
OG	Orthologue groups
OV	Ovarian cancer
PCA	Principal Components Analyses
PCA	Principal-component analysis
PCR	Polymerase chain reaction
SBP	Simulated quantitative trait
SBP	Systolic blood pressure
SLE	Ystem lupus erythamatosus
SNP	Single nucleotide polymorphism
SNPQ	Single nucleotide polymorphisms
TR	Transcription rate
TSG	Tumor suppressor genes
TSGD	Teleost-specific genome duplication
Tsnp	Tagging single nucleotide polymorphism
WGD	Whole-genome duplication

List of Contributors

Jianliang Ni
Tongde Hospital of Zhejiang Province, Zhejiang, China

Shuangfei Hu
Zhejiang Provincial People's Hospital, Zhejiang, China

Jiangtao Zhang
Tongde Hospital of Zhejiang Province, Zhejiang, China

Wenxin Tang
Hangzhou Seventh People's Hospital, Zhejiang, China

Weihong Lu
Schizophrenia Program, Shanghai Mental Health Center, Shanghai Jiao Tong University School of Medicine, Shanghai, China

Chen Zhang
Schizophrenia Program, Shanghai Mental Health Center, Shanghai Jiao Tong University School of Medicine, Shanghai, China

David Montaner
Department of Bioinformatics and Genomics, Centro de Investigacio´n Prı´ncipe Felipe (CIPF), Valencia, Spain
Functional Genomics Node (INB), Centro de Investigacio´n Prı´ncipe Felipe (CIPF), Valencia, Spain

Joaquı´n Dopazo
Department of Bioinformatics and Genomics, Centro de Investigacio´n Prı´ncipe Felipe (CIPF), Valencia, Spain
Functional Genomics Node (INB), Centro de Investigacio´n Prı´ncipe Felipe (CIPF), Valencia, Spain
CIBER de Enfermedades Raras (CIBERER), Valencia, Spain

Wei Liu
Division of Animal Infectious Diseases, State Key Laboratory of Agricultural Microbiology, College of Veterinary Medicine, Huazhong Agricultural University, Wuhan, People's Republic of China

Liurong Fang
Division of Animal Infectious Diseases, State Key Laboratory of Agricultural Microbiology, College of Veterinary Medicine, Huazhong Agricultural University, Wuhan, People's Republic of China

Mao Li
Division of Animal Infectious Diseases, State Key Laboratory of Agricultural Microbiology, College of Veterinary Medicine, Huazhong Agricultural University, Wuhan, People's Republic of China

Sha Li
Division of Animal Infectious Diseases, State Key Laboratory of Agricultural Microbiology, College of Veterinary Medicine, Huazhong Agricultural University, Wuhan, People's Republic of China

Shaohua Guo
Division of Animal Infectious Diseases, State Key Laboratory of Agricultural Microbiology, College of Veterinary Medicine, Huazhong Agricultural University, Wuhan, People's Republic of China

Rui Luo
Division of Animal Infectious Diseases, State Key Laboratory of Agricultural Microbiology, College of Veterinary Medicine, Huazhong Agricultural University, Wuhan, People's Republic of China

Zhixin Feng
Institute of Veterinary Medicine, Jiangsu Academy of Agricultural Sciences, Nanjing, People's Republic of China

Bin Li
Institute of Veterinary Medicine, Jiangsu Academy of Agricultural Sciences, Nanjing, People's Republic of China

Zhemin Zhou
Environmental Research Institute, University College Cork, Cork, Ireland

Guoqing Shao
Institute of Veterinary Medicine, Jiangsu Academy of Agricultural Sciences, Nanjing, People's Republic of China

Huanchun Chen
Division of Animal Infectious Diseases, State Key Laboratory of Agricultural Microbiology, College of Veterinary Medicine, Huazhong Agricultural University, Wuhan, People's Republic of China

Shaobo Xiao
Division of Animal Infectious Diseases, State Key Laboratory of Agricultural Microbiology, College of Veterinary Medicine, Huazhong Agricultural University, Wuhan, People's Republic of China

Yongkang Kim
Department of Statistics, Seoul National University, Seoul, South Korea

Taesung Park
Department of Statistics, Seoul National University, Seoul, South Korea
Interdisciplinary Program in Bioinformatics, Seoul National University, Seoul, 151–741, South Korea

Pingzhao Hu
The Centre for Applied Genomics, The Hospital for Sick Children, 686 Bay Street, Toronto, ON, M5G 0A4, Canada
Department of Biochemistry and Medical Genetics and George and Fay Yee Centre for Healthcare Innovation, University of Manitoba,745 Bannatyne Avenue, Winnipeg, MB, R3E 0W3, Canada

Andrew D Paterson
The Centre for Applied Genomics, The Hospital for Sick Children, 686 Bay Street, Toronto, ON, M5G 0A4, Canada
rogram in Genetics and Genome Biology, The Hospital for Sick Children, 686 Bay Street, Toronto, ON, M5G 0A4, Canada
Dalla Lana School of Public Health, University of Toronto, Health Sciences Building, 155 College St, Toronto, ON, M5T 3M7, Canada

Peng Jiang
Department of Biostatistics and Computational Biology, Dana-Farber Cancer Institute, Harvard T.H. Chan School of Public Health, Boston, MA 02215, USA

Hongfang Wang
Department of Pathology, Brigham and Women's Hospital, Boston, MA 02115, USA

Wei Li
Department of Biostatistics and Computational Biology, Dana-Farber Cancer Institute, Harvard T.H. Chan School of Public Health, Boston, MA 02215, USA

Chongzhi Zang
Department of Biostatistics and Computational Biology, Dana-Farber Cancer Institute, Harvard T.H. Chan School of Public Health, Boston, MA 02215, USA

Bo Li
Department of Biostatistics and Computational Biology, Dana-Farber Cancer Institute, Harvard T.H. Chan School of Public Health, Boston, MA 02215, USA

Yinling J. Wong
Department of Pathology, Brigham and Women's Hospital, Boston, MA 02115, USA

Cliff Meyer
Department of Biostatistics and Computational Biology, Dana-Farber Cancer Institute, Harvard T.H. Chan School of Public Health, Boston, MA 02215, USA

Jun S. Liu
Department of Statistics, Harvard University, Cambridge 200092, USA

Jon C. Aster
Department of Pathology, Brigham and Women's Hospital, Boston, MA 02115, USA

X. Shirley Liu
Department of Biostatistics and Computational Biology, Dana-Farber Cancer Institute, Harvard T.H. Chan School of Public Health, Boston, MA 02215, USA
School of Life Science and Technology, Tongji University, Shanghai, MA 02138, China

Maximo Rivarola
Institute for Genome Sciences, University of Maryland School of Medicine, Baltimore, Maryland, United States of America

Jeffrey T. Foster
Center for Microbial Genetics and Genomics, Northern Arizona University, Flagstaff, Arizona, United States of America

Agnes P. Chan
J. Craig Venter Institute, Rockville, Maryland, United States of America

Amber L. Williams
Department of Biological Sciences, Environmental Genetics and Genomics Laboratory, Northern Arizona University, Flagstaff, Arizona, United States of America

Danny W. Rice
Department of Biology, Indiana University, Bloomington, Indiana, United States of America

Xinyue Liu
Institute for Genome Sciences, University of Maryland School of Medicine, Baltimore, Maryland, United States of America

Admasu Melake-Berhan
J. Craig Venter Institute, Rockville, Maryland, United States of America

Heather Huot Creasy
Institute for Genome Sciences, University of Maryland School of Medicine, Baltimore, Maryland, United States of America

Daniela Puiu
J. Craig Venter Institute, Rockville, Maryland, United States of America

M. J. Rosovitz
J. Craig Venter Institute, Rockville, Maryland, United States of America

Hoda M. Khouri
Institute for Genome Sciences, University of Maryland School of Medicine, Baltimore, Maryland, United States of America

Stephen M. Beckstrom-Sternberg
Center for Microbial Genetics and Genomics, Northern Arizona University, Flagstaff, Arizona, United States of America
Pathogen Genomics Division, Translational Genomics Research Institute, Phoenix, Arizona, United States of America

Gerard J. Allan
Department of Biological Sciences, Environmental Genetics and Genomics Laboratory, Northern Arizona University, Flagstaff, Arizona, United States of America

Paul Keim
Center for Microbial Genetics and Genomics, Northern Arizona University, Flagstaff, Arizona, United States of America

Jacques Ravel
Institute for Genome Sciences, University of Maryland School of Medicine, Baltimore, Maryland, United States of America
Department of Microbiology and Immunology, University of Maryland School of Medicine, Baltimore, Maryland, United States of America

Pablo D. Rabinowicz
Institute for Genome Sciences, University of Maryland School of Medicine, Baltimore, Maryland, United States of America
J. Craig Venter Institute, Rockville, Maryland, United States of America
Department of Biochemistry and Molecular Biology, University of Maryland School of Medicine, Baltimore, Maryland, United States of America

Chao Tian
Rowe Program in Human Genetics, Departments of Biochemistry and Medicine, University of California Davis, Davis, California, United States of America

Roman Kosoy
Rowe Program in Human Genetics, Departments of Biochemistry and Medicine, University of California Davis, Davis, California, United States of America

Annette Lee
The Robert S. Boas Center for Genomics and Human Genetics, Feinstein Institute for Medical Research, North Shore LIJ Health System, Manhasset, New York, United States of America

Michael Ransom
Rowe Program in Human Genetics, Departments of Biochemistry and Medicine, University of California Davis, Davis, California, United States of America

John W. Belmont
Department of Molecular and Human Genetics, Baylor College of Medicine, Houston, Texas, United States of America

Peter K. Gregersen
The Robert S. Boas Center for Genomics and Human Genetics, Feinstein Institute for Medical Research, North Shore LIJ Health System, Manhasset, New York, United States of America

Michael F. Seldin
Rowe Program in Human Genetics, Departments of Biochemistry and Medicine, University of California Davis, Davis, California, United States of America

Matthew J. Loza
Center for Human Genomics, Department of Internal Medicine, Wake Forest University School of Medicine, Winston-Salem, North Carolina, United States of America

Charles E. McCall
Department of Internal Medicine, Wake Forest University School of Medicine, Winston-Salem, North Carolina, United States of America

Liwu Li
Department of Biology, Virginia Polytechnic Institute and State University, Blacksburg, Virginia, United States of America

William B. Isaacs
Department of Urology, Johns Hopkins University Medical Institutions, Baltimore, Maryland, United States of America
Department of Oncology, Johns Hopkins University Medical Institutions, Baltimore, Maryland, United States of America

Jianfeng Xu
Center for Human Genomics, Department of Epidemiology and Prevention, Wake Forest University School of Medicine, Winston-Salem, North Carolina, United States of America

Bao-Li Chang
Center for Human Genomics, Department of Pediatric Medicine, Wake Forest University School of Medicine, Winston-Salem, North Carolina, United States of America

Fadhl M. Al-Akwaa
Biomedical Eng. Dept., Univ. of Science & Technology, Sana'a, Yemen

Peng Xu
Centre for Applied Aquatic Genomics, Chinese Academy of Fishery Sciences, Beijing, China
These authors contributed equally to this work. Correspondence should be addressed to X.S

Xiaofeng Zhang
Heilongjiang River Fisheries Research Institute, Chinese Academy of Fishery Sciences, Harbin, China
These authors contributed equally to this work. Correspondence should be addressed to X.S

Xumin Wang
Chinese Academy of Sciences Key Laboratory of Genome Sciences and Information, Beijing Institute of Genomics, Chinese Academy of Sciences, Beijing, China
These authors contributed equally to this work. Correspondence should be addressed to X.S

Jiongtang Li
Centre for Applied Aquatic Genomics, Chinese Academy of Fishery Sciences, Beijing, China
These authors contributed equally to this work. Correspondence should be addressed to X.S

Guiming Liu
Chinese Academy of Sciences Key Laboratory of Genome Sciences and Information, Beijing Institute of Genomics, Chinese Academy of Sciences, Beijing, China

These authors contributed equally to this work. Correspondence should be addressed to X.S

Youyi Kuang
Heilongjiang River Fisheries Research Institute, Chinese Academy of Fishery Sciences, Harbin, China
These authors contributed equally to this work. Correspondence should be addressed to X.S

Jian Xu
Centre for Applied Aquatic Genomics, Chinese Academy of Fishery Sciences, Beijing, China
These authors contributed equally to this work. Correspondence should be addressed to X.S

Xianhu Zheng
Heilongjiang River Fisheries Research Institute, Chinese Academy of Fishery Sciences, Harbin, China
These authors contributed equally to this work. Correspondence should be addressed to X.S

Lufeng Ren
Chinese Academy of Sciences Key Laboratory of Genome Sciences and Information, Beijing Institute of Genomics, Chinese Academy of Sciences, Beijing, China

Guoliang Wang
Chinese Academy of Sciences Key Laboratory of Genome Sciences and Information, Beijing Institute of Genomics, Chinese Academy of Sciences, Beijing, China

Yan Zhang
Centre for Applied Aquatic Genomics, Chinese Academy of Fishery Sciences, Beijing, China

Linhe Huo
Chinese Academy of Sciences Key Laboratory of Genome Sciences and Information, Beijing Institute of Genomics, Chinese Academy of Sciences, Beijing, China

Zixia Zhao
Centre for Applied Aquatic Genomics, Chinese Academy of Fishery Sciences, Beijing, China

Dingchen Cao
Heilongjiang River Fisheries Research Institute, Chinese Academy of Fishery Sciences, Harbin, China

Cuiyun Lu
Heilongjiang River Fisheries Research Institute, Chinese Academy of Fishery Sciences, Harbin, China

Chao Li
Heilongjiang River Fisheries Research Institute, Chinese Academy of Fishery Sciences, Harbin, China

Yi Zhou
Stem Cell Program, Division of Hematology and Oncology, Boston Children's Hospital and Dana-Farber Cancer Institute, Harvard Medical School, Boston, Massachusetts, USA

Zhanjiang Liu
Centre for Applied Aquatic Genomics, Chinese Academy of Fishery Sciences, Beijing, China
Fish Molecular Genetics and Biotechnology Laboratory, Department of Fisheries and Allied Aquacultures, Auburn University, Auburn, Alabama, USA

Zhonghua Fan
Chinese Academy of Sciences Key Laboratory of Genome Sciences and Information, Beijing Institute of Genomics, Chinese Academy of Sciences, Beijing, China

Guangle Shan
Chinese Academy of Sciences Key Laboratory of Genome Sciences and Information, Beijing Institute of Genomics, Chinese Academy of Sciences, Beijing, China

Xingang Li
Chinese Academy of Sciences Key Laboratory of Genome Sciences and Information, Beijing Institute of Genomics, Chinese Academy of Sciences, Beijing, China

Shuangxiu Wu
Chinese Academy of Sciences Key Laboratory of Genome Sciences and Information, Beijing Institute of Genomics, Chinese Academy of Sciences, Beijing, China

Lipu Song
Chinese Academy of Sciences Key Laboratory of Genome Sciences and Information, Beijing Institute of Genomics, Chinese Academy of Sciences, Beijing, China

Guangyuan Hou
Centre for Applied Aquatic Genomics, Chinese Academy of Fishery Sciences, Beijing, China

Yanliang Jiang
Centre for Applied Aquatic Genomics, Chinese Academy of Fishery Sciences, Beijing, China

Zsigmond Jeney
Research Institute for Fisheries, Aquaculture and Irrigation, Szarvas, Hungary

Dan Yu
Chinese Academy of Sciences Key Laboratory of Genome Sciences and Information, Beijing Institute of Genomics, Chinese Academy of Sciences, Beijing, China

Li Wang
Chinese Academy of Sciences Key Laboratory of Genome Sciences and Information, Beijing Institute of Genomics, Chinese Academy of Sciences, Beijing, China

Changjun Shao
Chinese Academy of Sciences Key Laboratory of Genome Sciences and Information, Beijing Institute of Genomics, Chinese Academy of Sciences, Beijing, China

Lai Song
Chinese Academy of Sciences Key Laboratory of Genome Sciences and Information, Beijing Institute of Genomics, Chinese Academy of Sciences, Beijing, China

Jing Sun
Chinese Academy of Sciences Key Laboratory of Genome Sciences and Information, Beijing Institute of Genomics, Chinese Academy of Sciences, Beijing, China

Peifeng Ji
Centre for Applied Aquatic Genomics, Chinese Academy of Fishery Sciences, Beijing, China

Jian Wang
Centre for Applied Aquatic Genomics, Chinese Academy of Fishery Sciences, Beijing, China

Qiang Li
Centre for Applied Aquatic Genomics, Chinese Academy of Fishery Sciences, Beijing, China

Liming Xu
Centre for Applied Aquatic Genomics, Chinese Academy of Fishery Sciences, Beijing, China

Fanyue Sun
Fish Molecular Genetics and Biotechnology Laboratory, Department of Fisheries and Allied Aquacultures, Auburn University, Auburn, Alabama, USA

Jianxin Feng
Henan Academy of Fishery Science, Zhengzhou, China

Chenghui Wang
College of Fisheries and Life Science, Shanghai Ocean University, Shanghai, China

Shaolin Wang
Department of Psychiatry and Neurobiology Science, University of Virginia, Charlottesville, Virginia, USA

Baosen Wang
Centre for Applied Aquatic Genomics, Chinese Academy of Fishery Sciences, Beijing, China

Yan Li
Centre for Applied Aquatic Genomics, Chinese Academy of Fishery Sciences, Beijing, China

Yaping Zhu
Centre for Applied Aquatic Genomics, Chinese Academy of Fishery Sciences, Beijing, China

Wei Xue
Centre for Applied Aquatic Genomics, Chinese Academy of Fishery Sciences, Beijing, China

Lan Zhao
Centre for Applied Aquatic Genomics, Chinese Academy of Fishery Sciences, Beijing, China

Jintu Wang
Centre for Applied Aquatic Genomics, Chinese Academy of Fishery Sciences, Beijing, China

Ying Gu
Heilongjiang River Fisheries Research Institute, Chinese Academy of Fishery Sciences, Harbin, China

Weihua Lv
Heilongjiang River Fisheries Research Institute, Chinese Academy of Fishery Sciences, Harbin, China

Kejing Wu
Chinese Academy of Sciences Key Laboratory of Genome Sciences and Information, Beijing Institute of Genomics, Chinese Academy of Sciences, Beijing, China

Jingfa Xiao
Chinese Academy of Sciences Key Laboratory of Genome Sciences and Information, Beijing Institute of Genomics, Chinese Academy of Sciences, Beijing, China

Jiayan Wu
Chinese Academy of Sciences Key Laboratory of Genome Sciences and Information, Beijing Institute of Genomics, Chinese Academy of Sciences, Beijing, China

Zhang Zhang
Chinese Academy of Sciences Key Laboratory of Genome Sciences and Information, Beijing Institute of Genomics, Chinese Academy of Sciences, Beijing, China

Jun Yu
Chinese Academy of Sciences Key Laboratory of Genome Sciences and Information, Beijing Institute of Genomics, Chinese Academy of Sciences, Beijing, China

Xiaowen Sun
Centre for Applied Aquatic Genomics, Chinese Academy of Fishery Sciences, Beijing, China
Heilongjiang River Fisheries Research Institute, Chinese Academy of Fishery Sciences, Harbin, China

Chol-Hee Jung
Plant Molecular Biology and Biotechnology Laboratory, ARC Centre of Excellence for Integrative Legume Research, Melbourne School of Land and Environment, The University of Melbourne, Parkville, Victoria, Australia

Chui E. Wong Mohan B. Singh
Plant Molecular Biology and Biotechnology Laboratory, ARC Centre of Excellence for Integrative Legume Research, Melbourne School of Land and Environment, The University of Melbourne, Parkville, Victoria, Australia

Prem L. Bhalla
Plant Molecular Biology and Biotechnology Laboratory, ARC Centre of Excellence for Integrative Legume Research, Melbourne School of Land and Environment, The University of Melbourne, Parkville, Victoria, Australia

Wen-Yan Li
State Key Laboratory of Crop Stress Biology in Arid Areas, College of Life Sciences, Northwest A&F University, Yangling, Shaanxi, China

Xiang Wang
State Key Laboratory of Crop Stress Biology in Arid Areas, College of Life Sciences, Northwest A&F University, Yangling, Shaanxi, China

Ri Li
State Key Laboratory of Crop Stress Biology in Arid Areas, College of Life Sciences, Northwest A&F University, Yangling, Shaanxi, China

Wen-Qiang Li
State Key Laboratory of Crop Stress Biology in Arid Areas, College of Life Sciences, Northwest A&F University, Yangling, Shaanxi, China

Kun-Ming Chen
State Key Laboratory of Crop Stress Biology in Arid Areas, College of Life Sciences, Northwest A&F University, Yangling, Shaanxi, China

David Montaner
Department of Bioinformatics and Genomics, Centro de Investigacio´n Prı´ncipe Felipe (CIPF), Valencia, Spain,
Functional Genomics Node (INB), Centro de Investigacio´n Prı´ncipe Felipe (CIPF), Valencia, Spain,

Joaquı´n Dopazo
Department of Bioinformatics and Genomics, Centro de Investigacio´n Prı´ncipe Felipe (CIPF), Valencia, Spain,
Functional Genomics Node (INB), Centro de Investigacio´n Prı´ncipe Felipe (CIPF), Valencia, Spain,
CIBER de Enfermedades Raras (CIBERER), Valencia, Spain

Preface

Encyclopaedia of Modern Genetics Analysis Integrating Genes and Genomes deals with genetics analysis of genes and genomes. Genetic analysis is the overall process of studying and researching in fields of science that involve genetics and molecular biology. The text *Genetics Analysis of Genes and Genomes* provides a clear, balanced, and comprehensive introduction to genetics and genomics. Genomics is a discipline in genetics that applies recombinant DNA, DNA sequencing methods, and bioinformatics to sequence, assemble, and analyze the function and structure of genomes. A preliminary genetic analysis of complement three gene and schizophrenia has been presented in first chapter. In second chapter, we present a multidimensional logistic model that allows studying the relationship of gene modules with different genome-scale. Third chapter employs comparative evolutionary analysis of twenty mycoplasma genomes to gain an improved understanding of essential genes. Fourth chapter proposes a robust GMDR estimation method (based on the L-estimator and M-estimator estimation methods) in an attempt to reduce the effects caused by outlying traits. In fifth chapter, we evaluate the associations between rare and common genetic variants and the simulated quantitative trait (SBP) measured at three time points at the gene and pathway levels. Sixth chapter provides a general method for gene essentiality analysis in functional genomic experiments. In seventh chapter, we report the compete castor bean chloroplast and mitochondrion genome sequences generated from a whole genome shotgun (WGS) sequencing project of the cultivar. The analysis of East Asia genetic substructure using genome-wide SNP arrays has been presented in eighth chapter. In ninth chapter, we review various phases of inflammation responses, including the development of immune cells, sensing of danger, influx of cells to sites of insult, activation and functional responses of immune and non-immune cells, and resolution of the immune response. Tenth chapter focuses on analysis of gene expression data using biclustering algorithms. Genome sequence and genetic diversity of the common carp have been discussed in eleventh chapter. Twelfth chapter provides a framework for the soybean flowering pathway and insights into the relationship and evolution of flowering genes between a short-day soybean and the long-day plant, Arabidopsis. Last chapter provides a comparative genomic analysis that identified 74 NADK gene homologs from 24 species representing the eight major plant lineages within the supergroup Plantae: glaucophytes, rhodophytes, chlorophytes, bryophytes, lycophytes, gymnosperms, monocots and eudicots. Phylogenetic and structural analysis classified these NADK genes into four well-conserved subfamilies with considerable variety in the domain organization and gene structure among subfamily members.

Chapter 1

A PRELIMINARY GENETIC ANALYSIS OF COMPLEMENT 3 GENE AND SCHIZOPHRENIA

Jianliang Ni[1], Shuangfei Hu[2], Jiangtao Zhang[1], Wenxin Tang[3], Weihong Lu[4], Chen Zhang[4]

[1] Tongde Hospital of Zhejiang Province, Zhejiang, China

[2] Zhejiang Provincial People's Hospital, Zhejiang, China

[3] Hangzhou Seventh People's Hospital, Zhejiang, China

[4] Schizophrenia Program, Shanghai Mental Health Center, Shanghai Jiao Tong University School of Medicine, Shanghai, China

ABSTRACT

Complement pathway activation was found to occur frequently in schizophrenia, and complement 3 (C3) plays a major role in this process. Previous studies have provided evidence for the possible role of C3 in the development of schizophrenia. In this study, we hypothesized that the gene encoding C3 (C3) may confer susceptibility to schizophrenia in Han Chinese. We analyzed 7 common single nucleotide polymorphisms (SNPs) of C3 in 647 schizophrenia patients and 687 healthy controls. Peripheral *C3* mRNA expression level was measured in 23 drug-naïve patients with schizophrenia and 24 controls. Two SNPs (rs1047286 and rs2250656) that deviated from Hardy-Weinberg equilibrium were excluded for further analysis. Among the remaining 5 SNPs, there was no significant difference in allele and genotype frequencies between the patient and control groups. Logistic regression analysis showed no significant SNP-gender interaction in either dominant model or recessive model. There was no significant difference in the level of peripheral *C3* expression between the drug-naïve schizophrenia patients and healthy controls. In conclusion, the results of this study do not support *C3* as a major genetic susceptibility factor in schizophrenia. Other factors in AP may have critical roles in schizophrenia and be worthy of further investigation.

INTRODUCTION

Schizophrenia is a chronic, severe and disabling brain disorder that affects approximately 1% of worldwide population. In the past decades, schizophrenia has been regarded as a neurodevelopment disorder. Early literature reported that adverse conditions may result in abnormal brain development during the perinatal period, whilst schizophrenic symptoms appear in later life after the synaptic pruning process [1,2]. However, the pathophysiology of schizophrenia remains unknown [3].

Although heritability estimates for schizophrenia reach 80%, twin concordance is around 50% [4]. Hence, non-genetic factors also play an important role in this disorder [5]. It has been well-documented that maternal virus infection is one of the most consistently identified environmental risk factors for schizophrenia [6]. On the other hand, clinical observations indicated that schizophrenia and certain autoimmune diseases share some key clinical, epidemiological and genetic features [7]. Such findings suggested that immune abnormalities may be implicated with the pathophysiology of schizophrenia [8].

Complement acts as a rapid and efficient immune surveillance system that serves to protect the body against the invasion and proliferation of various microorganisms [9,10]. Complement pathway activation was reported to occur frequently in schizophrenia, in which complement 3 (C3) regulates the process [11]. C3 is a protein of the immune system that plays a central role in the complement cascade and contributes to innate immunity. Recent observations have demonstrated that C3 is a critical mediator for synaptic refinement and plasticity in neurodevelopment [12,13]. In comparison with healthy controls, Hakobyan et al. [14] observed a significant higher level of C3 protein in schizophrenia patients. As such, the above findings provide interesting clues for the potential role of C3 in schizophrenia.

At the molecular level, the gene encoding C3 (*C3*) is located at chromosome 19, which has been reported to be a genetic schizophrenia susceptibility region [15]. However, few genetic studies have been carried out to investigate the association of *C3* with schizophrenia and yielded inconsistent results [16,17,18]. It is known that *C3* contains 41 exons and spreads over 41kb. One weakness for the early genetic studies is too few polymorphisms tested. In the present study, we aimed to examine whether the region of *C3* is associated with schizophrenia. A total of 7 polymorphisms were selected for a better coverage of this region. As a secondary aim, prior study reported that *C3* has a gender-specific effect [19], which may underlie differential susceptibility to schizophrenia [20]. So we attempted to examine whether there was any gender difference in the association of *C3* with schizophrenia. Data has shown that *C3*polymorphisms result in alternations in its protein function. To validate previous findings, we opted to measure the serum *C3* expression level among drug-naïve schizophrenia patients and healthy controls.

METHODS

Subjects

All subjects provided written informed consent prior to performing any of the procedures related to this study. All procedures were reviewed and approved by the ethical committees at Tongde Hospital of Zhejiang Province and Hangzhou Seventh People's Hospital, and performed in strict accordance with the Declaration of Helsinki, and other relevant national and international regulations.

For the genetic analysis, a total of 647 schizophrenia patients recruited from Tongde Hospital of Zhejiang Province and The Seventh People's Hospital of Hangzhou. The inclusion criteria for this study were according to our previous ones [20,21,22]. All patients (1) met the Diagnostic and Statistical Manual of Mental Disorders, Fourth Edition (DSM-IV) criteria for schizophrenia; (2) were not first-episode; (3) had no chronic physical disease or other psychiatric disorder aside from schizophrenia. Prior to analysis, all diagnosis and review of psychiatric case records were independently checked and verified by two senior psychiatrists. The control group comprised of 687 Han Chinese enrolled from the local community in Hangzhou. Before sampling, the volunteers self-reported that they were in good physical health and have no family history of psychiatric disorders. Those who have medical illnesses or drug and alcohol abuse/dependence were excluded. Demographic and clinical characteristics were presented inS1 Table.

For the expression analysis, twenty-three drug-naïve patients with first-episode schizophrenia were recruited from Tongde Hospital of Zhejiang Province. The patients were diagnosed according to the DSM-IV criteria for schizophrenia and had no physical disease. Twenty-four healthy subjects from Hangzhou city were also recruited for control group. Basic blood and urine tests were performed prior to recruitment in order to exclude any current physical illness. Patients and controls did not significantly differ for age, gender, BMI and smoking status. Detailed information was presented in S2 Table.

SNP Selection

We retrieved CHB data from the HapMap database (http://www.hapmap.org) and defined linkage disequilibrium (LD) blocks using Haploview

4.2 (Broad Institute, Cambridge, MA, USA) to set inclusion criteria for tagging SNPs. Haplotype-tagging single nucleotide polymorphisms (htSNPs) with R^2 cutoff>0.8 and minor allele frequency (MAF)>0.1 were selected. In total, three tag SNPs of *C3* were selected for genotyping. Four potential *C3* functional SNPs (rs7951, rs2230199, rs2250656 and rs11672613) [23,24,25,26] were also examined in this study (S3 Table).

Genotyping

Genomic DNA of all participants was extracted from peripheral blood using a Tiangen DNA Isolation Kit (Tiangen Biotech, Beijing, China). All 7 SNPs were amplified independently via polymerase chain reaction (PCR) and then genotyped via direct sequencing on an ABI PRISM 3730 Genetic Analyzer (Perkin-Elmer Applied Biosystems). S4 Table detailed the primers information. Genotyping was carried out according to the methods described in our previous studies [27,28]. PCR amplification was performed in a volume of 25 μL containing primer pair for each SNP. PCR primers were also used for sequencing. Sequencing results were handled using DNAStar package (DNA Star Inc., USA), and the original sequencing chromatograms of each sample were then manually checked.

Quantitative Real-Time Polymerase Chain Reaction (qRT-PCR)

We carried out the *C3* mRNA expression analysis using qRT-PCR as previously described [29,30]. Peripheral blood was collected and mononuclear cells were separated by Ficoll-Paque PLUS density gradient centrifugation (GE Healthcare, Amersham, NJ, USA) within 2 hour, placed in TRIzol (Invitrogen, Carlsbad, CA, USA) and stored at -80°C. The total RNA was isolated from peripheral blood mononuclear cells according to the manufacturer's protocol, and 2 μg total RNA was re-transcribed into complementary DNA with reverse transcription (ReverTra Ace, Toyobo, Osaka, Japan) according to the manufacturer's instruction. Relative*C3* mRNA expression levels were assessed by real-time PCR with commercially available TaqMan gene expression assays for target gene *C3* and glyceraldehydes-3-phosphate dehydrogenase (*GAPDH*) as reference gene (Applied Biosystems, CA, USA). All experiments were conducted in optical 384-well reaction microtiter plates on an ABI Prism 7900HT Sequence Detection System (Applied Biosystems, CA, USA). PCR was performed in a total volume of 10μL containing 1×TaqMan Universal Master Mix with AmpErase UNG, 1×Assay Mix (Applied Biosystems, CA, USA) and complementary DNA template at cycle conditions: 95°C for 15 min, followed by 40 cycles at 95°C for 15s and 60°C for 60s. All reactions were run in triplicate. In each sample, the expression of *C3* was normalized to the expression of the reference gene. Results were reported in fold change using $2^{-\Delta\Delta Ct}$.

Statistical Analysis

The Hardy-Weinberg equilibrium testing and individual SNP association analyses were conducted using SHEsis (http://analysis.bio-x.cn). The odds ratio (OR) and corresponding 95% confidence interval (CI) were calculated with the major allele as reference. Pairwise linkage disequilibrium of all pairs of htSNPs was performed using HaploView 4.2 (Broad Institute, Cambridge, MA, USA), and the extent of linkage disequilibrium (LD) was measured by the standardized D' and R^2. Referring to the previous report [19], logistic regression was performed with SNP-gender interaction to adjust the effect of gender on SNPs. For the expression analysis, ANCOVA was carried out with age, gender, smoking status and BMI as covariates controlled in the model, to minimize the potential effect of these factors on the expression levels of *C3* mRNA. The ANCOVA analysis was performed using SPSS 17.0 (SPSS Inc., Chicago, IL, USA). To adjust for multiple testing, the level of significance was corrected via Bonferroni correction. Power calculations were carried out using Quanto 1.2.3 (http://hydra.usc.edu/GxE).

Results

For the genetic analysis, there was no significant difference between the schizophrenia and control groups in term of age and gender. Seven SNPs were genotyped to investigate the association of *C3* with schizophrenia. Two SNPs (rs1047286 and rs2250656) that deviated from Hardy-Weinberg equilibrium were excluded for the further analysis. Among the remaining 5 SNPs, no deviation from the Hardy-Weinberg was observed in genotype distribution. Table 1showed that there was no significant difference in allele and genotype frequencies between the patient and control groups. After calculating LD for all pairs of SNPs, we found a low R^2 in *C3*(S1 Fig), indicating that no specific haplotype block could be identified. We further investigate whether there was any gender difference in the association of *C3* with schizophrenia. Logistic regression analysis showed no significant SNP-gender interaction in either dominant model or recessive model (Tables 2 and 3). A total of 23 drug-naïve patients with schizophrenia and 24 well-matched healthy controls were recruited for the *C3* expression study. As shown in Fig 1, there was no significant difference in the level of peripheral *C3* mRNA expression between the drug-naïve schizophrenia patients and healthy controls. On the basis of the genotype data, the statistical power of the 5 SNPs within *C3* was more than 85% (= 0.05) for schizophrenia samples under the assumption of a moderate effect size (OR = 1.5), a log additive model, and the prevalence of schizophrenia (≈1%).

Table 1: Comparison of genotype and allele frequencies of 5 *C3* SNPs in 647 schizophrenia patients and 687 healthy controls

		Genotype, n (%)						Allele, n (%)				
rs2277984	N	G/G	G/A	A/A	P^a	P^b	N	G	A	OR (95%CI)	P^a	P^c
Case	647	132 (20.4)	336 (51.9)	179 (27.7)	0.95	0.26	1294	600 (46.4)	694 (53.6)	1.02 (0.88–1.19)	0.76	
Control	687	137 (19.9)	355 (51.7)	195 (28.4)		0.28	1374	629 (45.8)	745 (54.2)			
rs7951		T/T	T/C	C/C				T	C			
Case	647	7 (1.1)	122 (18.9)	518 (80.1)	0.42	0.95	1294	136 (10.5)	1158 (89.5)	0.91 (0.71–1.16)	0.45	
Control	687	5 (0.7)	147 (21.4)	535 (77.9)		0.13	1374	157 (11.4)	1217 (88.6)			
rs11672613		C/C	C/T	T/T				C	T			
Case	647	98 (15.1)	307 (47.4)	242 (37.4)	0.22	0.97	1294	503 (38.9)	791 (61.1)	0.90 (0.77–1.05)	0.17	
Control	687	109 (15.9)	352 (51.2)	226 (32.9)		0.15	1374	570 (41.5)	804 (58.5)			
rs2230205		A/A	A/G	G/G				A	G			
Case	647	141 (21.8)	346 (53.5)	160 (24.7)	0.09	0.07	1294	628 (48.5)	666 (51.5)	1.18 (1.01–1.37)	0.035	0.175
Control	687	127 (18.5)	357 (52.0)	203 (29.5)		0.17	1374	611 (44.5)	763 (55.5)			
rs2230199		G/G	G/C	C/C				G	C			
Case	647	0 (0.0)	10 (1.5)	637 (98.5)	0.50	0.84	1294	10 (0.8)	1284 (99.2)	0.76 (0.33–1.71)	0.50	
Control	687	0 (0.0)	14 (2.0)	673 (98.0)		0.79	1374	14 (1.0)	1360 (99.0)			

[a] *P* values were not adjusted by Bonferroni correction
[b] *P* values were calculated for Hardy-Weinberg equilibrium
[c] *P* values were adjusted by Bonferroni correction.

Table 2: Logistic regression analysis of *C3* SNPs×gender interaction in dominant model

Variables	B	S.E	Wals	OR (95%CI)	P^a	P^b
rs2277984×gender	0.54	0.22	5.87	1.71 (1.11–2.65)	0.015	0.075
rs7951×gender	-0.40	0.24	2.78	0.67 (0.42–1.07)	0.10	
rs11672613×gender	-0.23	0.21	1.26	0.79 (0.53–1.19)	0.26	
rs2230205×gender	0.21	0.21	0.95	1.23 (0.81–1.87)	0.33	
rs2230199×gender	-0.21	0.27	0.62	0.81 (0.47–1.38)	0.43	

[a] *P* values were not adjusted by Bonferroni correction
[b] *P* values were adjusted by Bonferroni correction.

Table 3: Logistic regression analysis of *C3* SNPs×gender interaction in recessive model

Variables	B	S.E	Wals	OR (95%CI)	P^a
rs2277984×gender	0.04	0.24	0.02	1.04 (0.65–1.64)	0.88
rs7951×gender	0.35	1.25	0.08	1.42 (0.12–16.26)	0.78
rs11672613×gender	-0.17	0.26	0.43	0.84 (0.60–1.41)	0.79
rs2230205×gender	-0.14	0.23	0.37	0.87 (0.55–1.37)	0.55
rs2230199×gender	NA	NA	NA	NA	NA

[a] *P* values were not adjusted by Bonferroni correction
NA, Not applicable.

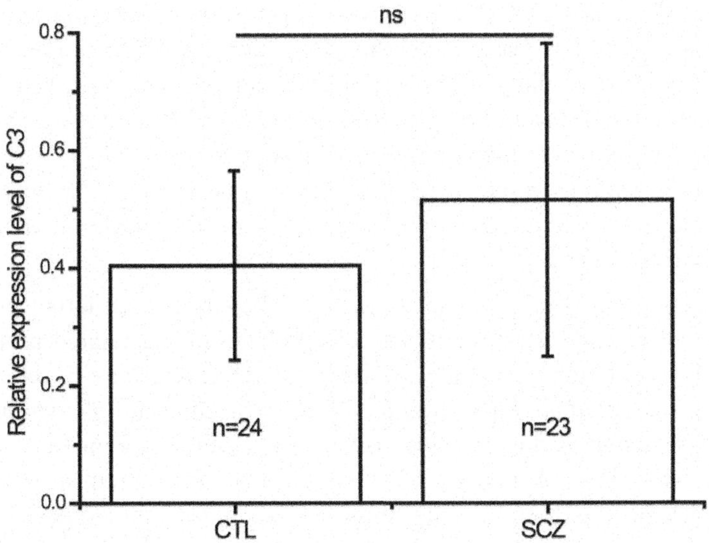

Figure 1: Expression levels of *C3* mRNA in peripheral blood in drug-naïve schizophrenia patients and healthy subjects.

C3 mRNA was normalized to that of *GAPDH*. CTL, control subjects (n = 24); SCZ, schizophrenia patients (n = 23); ns, no significance.

DISCUSSION

As a key component of innate immunity, accumulating evidence has indicated that abnormalities in the complement system are implicated in the etiology of schizophrenia [23]. We have examined the association of schizophrenia with the gene encoding C4-binding protein (*C4BPB/C4BPA*), a potent circulating soluble inhibitor of the classical and lectin pathways of complement. However, our results did not support the involvement of *C4BPB/C4BPA* in schizophrenia [22]. Here, we aimed to investigate the association of *C3*, another critical factor in complement system, with schizophrenia in Chinese Han population.

In this study, we did not observe any significant difference of allele and genotype frequencies between the schizophrenia patients and healthy controls. The statistical power of our study was also enough to detect an association between the variants and schizophrenia. Although Liu et al. [19] reported a SNP-gender interaction in *C3*, we did not have such findings in our sample. These results demonstrated that there is no genetic association between *C3* and schizophrenia, at least in Han Chinese. However, a recent study showed that increased levels of C3, acting as activation of complement system, can be found in schizophrenia patients when compared with healthy controls [31]. In contrast, Wong et al. [32] found a lower level of C3 in schizophrenia patients than that in controls. We noticed that the patients in both studies were those with chronic schizophrenia [31,32]. The aforementioned inconsistent results prompted us to determine the expression of *C3* in drug-naïve patients with first-episode schizophrenia. Our results showed no significant difference in the level of *C3* expression between schizophrenia patients and healthy controls. Therefore, our findings suggested that *C3* may not confer susceptibility to schizophrenia in Han Chinese.

Recently, Li et al. [33] performed a label-free quantitative proteomics analysis to identify 27 proteins as being schizophrenia related proteins, and found dysregulation of the alternative complement pathway in schizophrenia patients. The alternative pathway (AP) is one of three complement pathways, which is initiated by the spontaneous hydrolysis of C3. A number of molecules are involved in the occurrence of AP. Even though no association of *C3* with schizophrenia was found in this study, we could not exclude possible role of AP in the development of schizophrenia.

On the other side, cytokines are believed to play a vital role in coordinating immunologic and inflammatory responses in physiological and pathological conditions [34]. Therefore, cytokines may be critical mediators of the cross-talk between immune system and neuropsychiatric disorders [35]. Miller et al. [36] meta-analyzed 40 studies on cytokines and schizophrenia, and observed significant alternations of cytokine network in schizophrenia. Therefore, imbalance of cytokine network may be involved in the pathophysiology of schizophrenia. Prior literature indicated that complement activation products, such as C3a and C3a desArg, may enhance cytokine synthesis and inhibit the systemic synthesis of proinflammatory cytokines [37]. It is known that schizophrenia results from the cumulative impact of multiple common small-effect genetic variants and interactions between genes with small effect may contribute a larger heritable proportion to the overall risk of this disorder [38]. Therefore, we assumed that interaction of *C3* with genes encoding cytokines may be more sensitive to account for its susceptibility to schizophrenia. There is a need for further investigations to validate this hypothesis.

This study has some limitations that should be noted. First, the lack of a significant association may be caused by the modest sample size, possibly resulting in a type II error. Second, we did not psychiatrically screen the control subjects. Third, the principal hypothesis underlying this study is that common SNPs within *C3* may confer susceptibility to schizophrenia. Therefore, we did not sequence the *C3* to assess the influence of rare variants on schizophrenia, and this prevented us to detect their active role in the development of this disorder. Fourth, the case-control association analyses have the potential for population stratification, although all participants were ethnically matched in our sample. Finally, Wong et al. [32] reported lower level of C3 protein in schizophrenia patients in comparison to healthy controls only in male subjects, suggesting that there might be interesting to test *C3* expression separately in male and female subjects. However, we recruited only 47 individuals with or without schizophrenia in the *C3* expression study. The small sample size limited us to further analyze the gender-specific effect of *C3* expression on schizophrenia. Meanwhile, this also limited us to detect the association of studied *C3* SNPs with *C3* mRNA.

In conclusion, the results of this study do not support *C3* as a major genetic susceptibility factor in schizophrenia. Other factors in AP may have critical roles in schizophrenia and be worthy of further investigation.

ACKNOWLEDGMENTS

We are deeply grateful to all participants. We thank the anonymous reviewers for their insightful comments.

AUTHOR CONTRIBUTIONS

Conceived and designed the experiments: JN CZ. Performed the experiments: JN JZ. Analyzed the data: JN SH CZ. Contributed reagents/materials/analysis tools: JN JZ WT WL CZ. Wrote the paper: JN CZ.

REFERENCES

1. Weinberger DR (1996) On the plausibility of "the neurodevelopmental hypothesis" of schizophrenia. Neuropsychopharmacology 14: 1S–11S. pmid:8866738 doi: 10.1016/0893-133x(95)00199-n
2. Schmitt A, Malchow B, Hasan A, Falkai P (2014) The impact of environmental factors in severe psychiatric disorders. Front Neurosci 8: 19. doi: 10.3389/fnins.2014.00019. pmid:24574956
3. Zhang C, Fang Y, Xie B, Cheng W, Du Y, Wang D, et al. (2009) DNA methyltransferase 3B gene increases risk of early onset schizophrenia. Neurosci Lett 462: 308–311. doi: 10.1016/j.neulet.2009.06.085. pmid:19576953
4. Cannon TD, Kaprio J, Lonnqvist J, Huttunen M, Koskenvuo M (1998) The genetic epidemiology of schizophrenia in a Finnish twin cohort. A population-based modeling study. Arch Gen Psychiatry 55: 67–74. pmid:9435762 doi: 10.1001/archpsyc.55.1.67

5. Avramopoulos D, Pearce BD, McGrath J, Wolyniec P, Wang R, Eckart N, et al. (2015) Infection and inflammation in schizophrenia and bipolar disorder: a genome wide study for interactions with genetic variation. PLoS One 10: e0116696. doi: 10.1371/journal.pone.0116696. pmid:25781172

6. Schmitt A, Malchow B, Hasan A, Falkai P (2014) The impact of environmental factors in severe psychiatric disorders. Frontiers in Neuroscience 8. doi: 10.3389/fnins.2014.00019

7. Benros ME, Nielsen PR, Nordentoft M, Eaton WW, Dalton SO, Mortensen PB. (2011) Autoimmune Diseases and Severe Infections as Risk Factors for Schizophrenia: A 30-Year Population-Based Register Study. Am J Psychiatry 168: 1303–1310. doi: 10.1176/appi.ajp.2011.11030516. pmid:22193673

8. Leza JC, Bueno B, Bioque M, Arango C, Parellada M, Do K, et al. (2015) Inflammation in schizophrenia: A question of balance. Neurosci Biobehav Rev 55: 612–626. doi: 10.1016/j.neubiorev.2015.05.014. pmid:26092265

9. Ricklin D, Hajishengallis G, Yang K, Lambris JD (2010) Complement: a key system for immune surveillance and homeostasis. Nat Immunol 11: 785–797. doi: 10.1038/ni.1923. pmid:20720586

10. Tichaczek-Goska D (2012) Deficiencies and excessive human complement system activation in disorders of multifarious etiology. Adv Clin Exp Med 21: 105–114. pmid:23214307

11. Severance EG, Gressitt KL, Halling M, Stallings CR, Origoni AE, Vaughan C, et al. (2012) Complement C1q formation of immune complexes with milk caseins and wheat glutens in schizophrenia. Neurobiol Dis 48: 447–453. doi: 10.1016/j.nbd.2012.07.005. pmid:22801085

12. Fourgeaud L, Boulanger LM (2007) Synapse remodeling, compliments of the complement system. Cell 131: 1034–1036. pmid:18083091 doi: 10.1016/j.cell.2007.11.031

13. Michailidou I, Willems JG, Kooi EJ, van Eden C, Gold SM, Geurts JJ, et al. (2015) Complement C1q-C3 associated synaptic changes in multiple sclerosis hippocampus. Ann Neurol 77: 1007–1026. doi: 10.1002/ana.24398. pmid:25727254

14. Hakobyan S, Boyajyan A, Sim RB (2005) Classical pathway complement activity in schizophrenia. Neurosci Lett 374: 35–37. pmid:15631892 doi: 10.1016/j.neulet.2004.10.024

15. Francks C, Tozzi F, Farmer A, Vincent JB, Rujescu D, St Clair D, et al. (2010) Population-based linkage analysis of schizophrenia and bipolar case-control cohorts identifies a potential susceptibility locus on 19q13.

Mol Psychiatry 15: 319–325. doi: 10.1038/mp.2008.100. pmid:18794890

16. Rudduck C, Beckman L, Franzen G, Lindstrom L (1985) C3 and C6 complement types in schizophrenia. Hum Hered 35: 255–258. pmid:4029965 doi: 10.1159/000153555
17. Fananas L, Moral P, Panadero MA, Bertranpetit J (1992) Complement genetic markers in schizophrenia: C3, BF and C6 polymorphisms. Hum Hered 42: 162–167. pmid:1511994 doi: 10.1159/000154060
18. Blackwood DHR, Muir WJ, Stephenson A, Wentzel J, Adhiah A, Walker MJ, et al. (1996) Reduced expression of HLA-B35 in schizophrenia. Psychiatr Genet 6: 51–59. pmid:8840390 doi: 10.1097/00041444-199622000-00004
19. Liu K, Lai TYY, Chiang SWY, Chan VCK, Young AL, Tam PO, et al. (2014) Gender specific association of a complement component 3 polymorphism with polypoidal choroidal vasculopathy. Sci Rep 4: 7018. doi: 10.1038/srep07018. pmid:25388911
20. Tang WX, Cai J, Yi ZH, Zhang Y, Lu WH, Zhang C. (2014) Association study of common variants within the G protein-coupled receptor kinase 6 gene and schizophrenia susceptibility in Han Chinese. Hum Psychopharmacol 29: 100–103. doi: 10.1002/hup.2375. pmid:24302161
21. Zhu YL, Wang ZL, Ni JL, Zhang Y, Chen MJ, Cai J, et al. (2015) Genetic variant in NDUFS1 gene is associated with schizophrenia and negative symptoms in Han Chinese. J Hum Genet 60: 11–16. doi: 10.1038/jhg.2014.94. pmid:25354934
22. Wang SH, Lu HQ, Ni JL, Zhang JT, Tang WX, Lu WH, et al. (2015) An evaluation of association between common variants in C4BPB/C4BPA genes and schizophrenia. Neurosci Lett 590: 189–192. doi: 10.1016/j.neulet.2015.02.005. pmid:25660618
23. Miyagawa H, Yamai M, Sakaguchi D, Kiyohara C, Tsukamoto H, Kimoto Y, et al. (2008) Association of polymorphisms in complement component C3 gene with susceptibility to systemic lupus erythematosus. Rheumatology 47: 158–164. doi: 10.1093/rheumatology/kem321. pmid:18174230
24. Yu QQ, Yao Y, Zhu J, Bao X, Xie TH, Sun C, et al. (2015) Nonsynonymous single nucleotide polymorphisms in the complement component 3 gene are associated with risk of age-related macular degeneration: A meta-analysis. Gene 561: 249–255. doi: 10.1016/j.gene.2015.02.039. pmid:25688879
25. Phillips CM, Goumidi L, Bertrais S, Ferguson JF, Field MR, Kelly ED, et al. (2009) Complement component 3 polymorphisms interact with

polyunsaturated fatty acids to modulate risk of metabolic syndrome. Am J Clin Nutr 90: 1665–1673. doi: 10.3945/ajcn.2009.28101. pmid:19828715

26. Rhodes B, Hunnangkul S, Morris DL, Hsaio LC, Graham DSC, Nitsch D, et al. (2009) The heritability and genetics of complement C3 expression in UK SLE families. Genes Immun 10: 525–530. doi: 10.1038/gene.2009.23. pmid:19387462
27. Zhang Y, Chen MJ, Wu ZG, Chen J, Yu SY, Fang Y, et al. (2013) Association Study of Val66Met Polymorphism in Brain-Derived Neurotrophic Factor Gene with Clozapine-Induced Metabolic Syndrome: Preliminary Results. PLoS One 8: e72652. doi: 10.1371/journal.pone.0072652. pmid:23967328
28. Zhang Y, Chen M, Chen J, Wu Z, Yu S, Fang Y, et al. (2014) Metabolic syndrome in patients taking clozapine: prevalence and influence of catechol-O-methyltransferase genotype. Psychopharmacology (Berl) 231: 2211–2218. doi: 10.1007/s00213-013-3410-4
29. Zhang C, Wang Z, Hong W, Wu Z, Peng D, Fang Y. (2015) ZNF804A Genetic Variation Confers Risk to Bipolar Disorder. Mol Neurobiol. doi: 10.1007/s12035-015-9193-3.
30. Zhang C, Wu Z, Hong W, Wang Z, Peng D, Chen J, et al. (2014) Influence of BCL2 gene in major depression susceptibility and antidepressant treatment outcome. J Affect Disord 155: 288–294. doi: 10.1016/j.jad.2013.11.010. pmid:24321200
31. Soria LD, Gubert CD, Ceresser KM, Gama CS, Kapczinski F (2012) Increased serum levels of C3 and C4 in patients with schizophrenia compared to eutymic patients with bipolar disorder and healthy. Revista Brasileira De Psiquiatria 34: 119–120. pmid:22392401 doi: 10.1590/s1516-44462012000100022
32. Wong CT, Tsoi WF, Saha N (1996) Acute phase proteins in male Chinese schizophrenic patients in Singapore. Schizophrenia Research 22: 165–171. pmid:8958601 doi: 10.1016/s0920-9964(96)00037-0
33. Li Y, Zhou KJ, Zhang Z, Sun LY, Yang JL, Zhang M, et al. (2012) Label-free quantitative proteomic analysis reveals dysfunction of complement pathway in peripheral blood of schizophrenia patients: evidence for the immune hypothesis of schizophrenia. Mol Biosyst 8: 2664–2671. doi: 10.1039/c2mb25158b. pmid:22797129
34. Deleidi M, Jaggle M, Rubino G (2015) Immune aging, dysmetabolism, and inflammation in neurological diseases. Front Neurosci 9: 172. doi: 10.3389/fnins.2015.00172. pmid:26089771

35. Potvin S, Stip E, Sepehry AA, Gendron A, Bah R, Kouassi E. (2008) Inflammatory cytokine alterations in schizophrenia: a systematic quantitative review. Biol Psychiatry 63: 801–808. pmid:18005941 doi: 10.1016/j.biopsych.2007.09.024

36. Miller BJ, Buckley P, Seabolt W, Mellor A, Kirkpatrick B (2011) Meta-analysis of cytokine alterations in schizophrenia: clinical status and antipsychotic effects. Biol Psychiatry 70: 663–671. doi: 10.1016/j.biopsych.2011.04.013. pmid:21641581

37. Takabayashi T, Vannier E, Clark BD, Margolis NH, Dinarello CA, Burke JF, et al. (1996) A new biologic role for C3a and C3a desArg: regulation of TNF-alpha and IL-1 beta synthesis. J Immunol 156: 3455–3460. pmid:8617973

38. McClellan JM, Susser E, King MC (2007) Schizophrenia: a common disease caused by multiple rare alleles. Br J Psychiatry 190: 194–199. pmid:17329737 doi: 10.1192/bjp.bp.106.025585

Chapter 2

MULTIDIMENSIONAL GENE SET ANALYSIS OF GENOMIC DATA

David Montaner[1,2], Joaquı´n Dopazo[1,2,3]

[1] Department of Bioinformatics and Genomics, Centro de Investigacio´n Prı´ncipe Felipe (CIPF), Valencia, Spain

[2] Functional Genomics Node (INB), Centro de Investigacio´n Prı´ncipe Felipe (CIPF), Valencia, Spain

[3] CIBER de Enfermedades Raras (CIBERER), Valencia, Spain

ABSTRACT

Understanding the functional implications of changes in gene expression, mutations, etc., is the aim of most genomic experiments. To achieve this, several functional profiling methods have been proposed. Such methods study the behaviour of different gene modules (e.g. gene ontology terms) in response to one particular variable (e.g. differential gene expression). In spite to the wealth of information provided by functional profiling methods, a common limitation to all of them is their inherent unidimensional nature. In order to overcome this restriction we present a multidimensional logistic model that allows studying the relationship of gene modules with different genome-scale measurements (e.g. differential expression, genotyping association, methylation, copy number alterations, heterozygosity, etc.) simultaneously. Moreover, the relationship of such functional modules with the interactions among the variables can also be studied, which produces novel results impossible to be derived from the conventional unidimensional functional profiling methods. We report sound results of gene sets associations that remained undetected by the conventional one-dimensional gene set analysis in several examples. Our findings demonstrate the potential of the proposed approach for the discovery of new cell functionalities with complex dependences on more than one variable.

INTRODUCTION

The development of new genomic technologies, such as microarrays of gene expression, genotyping or array-CGH, along with the new next-generation sequencing techniques is increasing the volume of data throughput amazingly. As a direct consequence of this, the bottleneck in functional genomics has shifted from the data production phase to the data analysis steps. In particular, the necessity for providing a functional interpretation at molecular level that accounts for the genome-scale experimental designs has promoted the development of different methods for the functional analysis of this type of data in the last years [1], [2].

It is widely accepted that most of the biological functionality of the cell arises from complex interactions among their molecular components that define operational interacting entities or modules [3]. Functions collectively performed by such modules have conceptually been represented in different ways. Gene ontology (GO) [4] and KEGG pathways [5] are the most popular and widely used module definitions although many other are available in different repositories (e.g., Reactome [6], Biocarta, etc.) For practical purposes, functional modules are henceforth defined as sets of genes sharing functional annotations extracted from any of these repositories. Functional profiling methods exploit different definitions of modules in an attempt of understanding the functional basis of high-throughput experimental results [7]. Initially, functional enrichment methods, in different implementations [7], [8], have been used for this purpose. More sensitive approaches, generically known as gene-set analysis (GSA) methods, pioneered by the Gene Set Enrichment Analysis (GSEA) [9], were later proposed [1], [10]. In the original formulation, GSA methods aimed to directly detect sets of functionally related genes (modules) with a coordinate and significant over- or under-expression across the complete list of genes ranked according to their differential expression [9], [11], [12], [13], [14],[15]. GSA methods can detect such modules even if their gene components are not significantly differentially expressed when tested individually. GSA has been successfully applied to the analysis of microarray experiments and has contributed to the adoption of a systems-biology perspective in distinct fields such as cancer [16]. Recent findings, brought about by the application of GSA methods on microarray experiments [17] are consistent with the idea that pathways, rather than individual genes, appear to govern the course of tumorigenesis [18]. The use of GSA has been extended to other areas beyond transcriptomics, such as evolution [19], QTL analysis [20] or genotyping [21].

Nevertheless, the different versions of GSA published to date [1], [2], [10] are inherently one-dimensional. Its application to the analysis of genomic datasets is at present limited to the study of a unique variable measured for the genes. The experimental conditions studied, even if corrected by other variables (e.g. age, gender, treatments, etc.), are typically summarized into a unique value for each gene (e.g., differential expression in a case-control, risk in the case of survival analysis, etc.) which is used to rank them accordingly.

Nowadays, the extensive use of different high-throughput methodologies allows the obtention of different measurements for the genes such as methylation status, splicing variants, linkage to diseases, etc., in a straightforward manner. As an illustration of this, a pilot study by The Cancer Genome Atlas (http://cancergenome.nih.gov/) consortium on glioblastomas has recently been published [22]. In it, different types of transcriptomic and genomic profiling were obtained and analyzed in an example of application of different genomic methodologies that would become routine soon. In addition, different measurements of the same type in different experimental contexts can easily be done. For instance, gene expression measurements in case-controls of different, but mechanistically related experimental conditions, phenotypes, diseases, treatments, etc. can be easily obtained. In such scenario, more than one measurement could be obtained to rank the genes involved in the study. Under the conventional GSA paradigm the different ranked lists of genes could be analyzed one at a time and still a good deal of information might be obtained. Nevertheless, by taking this approach any list of ranked genes is considered independent from each other and, consequently, behaviour of functional modules which are dependent on the combination of the studied ranking variables will, most likely, remain undetected.

Here we focus on a conceptually different strategy for GSA by extending the gene set based functional analysis to a multidimensional scenario in which more than one variable or genomic measurement is available for all genes in the study. Logistic regression allows for fitting models that include more than one variable. We show here, by means of several examples, how the application of the multidimensional GSA (MD-GSA) uncovers biological processes activated by different combinations of parameters (measured for all the genes and derived from microarray of other experiments) that would have remained undetected if the parameters would have been analysed one at a time, independently.

RESULTS

Gene-Set Activation Dependent on the Transcription Rates and MRNA Activities In Yeast

Gene expression is a process that involves two steps of synthesis which end when the appropriate level of protein required for performing a given function is reached. Some processes in the cell require of a quick activation and/or deactivation, while others remain in activity for longer periods and their activation processes do not involve any urgency. Thus, it is expectable different cell functionalities will use different strategies of gene and protein expression and degradation. Measurements of these parameters can be found in a recent genome-wide analysis on common gene expression strategies in yeast [23]. Using these data, we have studied two relevant and opposite biological processes that account for the steady-state mRNA level in the cell: transcription and stability [24]. The authors used a functional enrichment strategy [25] to test the GO terms associated to the parameters measured and to their correlations. Essentially, they used quintiles as cut-off values and tested for enrichments in the genes showing a high or low correlation in rates (transcription and translation) or abundances (mRNA and protein copy number), finding a total of 22 GO terms significantly over-represented at different combinations of rates and abundances. Nevertheless, other interesting situations in which the measurements are not correlated (e.g. transcription rate and mRNA stability) could not be analysed with this approach that, in addition, has the disadvantage of requiring an arbitrary threshold.

Here we analysed the dependences of GO terms on two measurements, transcription rate (TR) and mRNA stability (RS), as well as on the interaction between them. When the logistic model was applied to the mRNA stability and to the transcription rate independently, we obtained 170 and 80 GO terms significantly associated to extreme values of these variables (see Table S1). This increase in the number of GO terms found was due to the well known fact that GSA strategies are much more sensitive than threshold-based functional enrichment strategies [1],[10]. Actually, similar results were obtained when other equivalent GSA strategies were used (data not shown) [11], [19].

Nevertheless, the most interesting aspect of this study is the analysis of the interaction between both variables. Table 1 shows 18 GO terms which were significantly associated to the interaction between transcription rate and mRNA stability. Figure S1 depicts the GO terms within the GO hierarchy. Nine of these GO terms could only be detected when the model takes into account simultaneously both parameters. In most of the cases, the GO was associated to both low transcription rate and mRNA stability (pattern *q3i*, see methods for an explanation of the patterns) such as *sister chromatid segregation* (Figure 1 top) in a subtle way that can only be detected when both parameters are included in the model. On the other hand, other processes, such as *DNA packaging*, *Chromatin assembly* (Figure 1 bottom), *Chromatin assembly or disassembly* and *Establishment and/or maintenance of chromatin architecture*(which are related terms, see File S1), or *protein-DNA complex assembly* are associated to high transcription rates but low mRNA stability (pattern *q4i,* see methods). This last strategy, opposite to the first one, suggest a transient necessity of these processes, whose genes are produced at a fast rate but quickly discarded after their functions have been carried out.

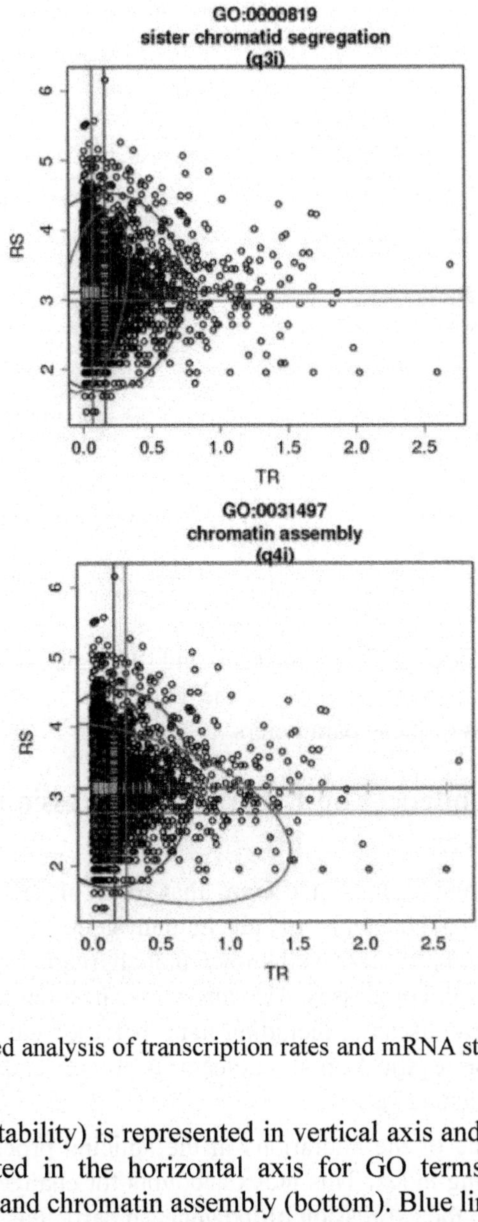

Figure 1: Combined analysis of transcription rates and mRNA stability in yeast with the logistic model.

RS (mRNA stability) is represented in vertical axis and TR (transcription rate) is represented in the horizontal axis for GO terms sister chromatid segregation (top) and chromatin assembly (bottom). Blue lines intersect in the mean of the distribution of all the values and red lines intersect in the mean of the distribution of values of the genes corresponding to the GO term analysed. Blue ellipse delimits the confidence interval for all the values and red ellipse delimits the confidence interval for the GO term analysed. The red ellipse marks the trend of the relationship between both variables. MD-GSA assigns patterns q3i and q4i respectively to these functional modules.

Table 1: Significant GO terms when transcription rate and mRNA stability are taken into account in the model

	Log odds ratio (model coefficients)			Adjusted p-value					
GO Id	TR	RS	inter	TR	RS	inter	pattern	new	GO name
GO:0019953	−11.87	−0.82	3.29	0.04	0.01	0.02	q3i	yes	sexual reproduction
GO:0051704	−11.98	−0.69	3.23	0.04	0.02	0.02	q3i	yes	multi-organism process
GO:0000819	−30.49	−0.87	7.1	0.02	0.03	0.02	q3i	yes	sister chromatid segregation
GO:0006260	−20.35	−0.97	4.99	0	0	0.01	q3i	no	DNA replication
GO:0006261	−25.15	−1.31	6.28	0	0	0.01	q3i	no	DNA-dependent DNA replication
GO:0022613	−4.69	−1.78	1.61	0.08	0	0.03	q3i	no	ribonucleoprotein complex biogenesis and assembly
GO:0042254	−5.05	−1.91	1.75	0.09	0	0.03	q3i	no	ribosome biogenesis
GO:0000746	−11.48	−0.73	3.17	0.06	0.02	0.03	q3i	yes	conjugation
GO:0000747	−11.39	−0.74	3.16	0.06	0.02	0.03	q3i	yes	conjugation with cellular fusion
GO:0042221	−6.65	−0.12	2.05	0.02	0.6	0.01	q3i	yes	response to chemical stimulus
GO:0000070	−30.23	−0.78	7.01	0.03	0.07	0.03	q3i	yes	mitotic sister chromatid segregation
GO:0019725	−9.13	−0.38	2.71	0.02	0.15	0.01	q3i	yes	cellular homeostasis
GO:0042592	−8.75	−0.3	2.59	0.02	0.27	0.01	q3i	yes	homeostatic process
GO:0006325	8.01	−0.47	−3.09	0	0.03	0.01	q4i	no	establishment and/or maintenance of chromatin architecture
GO:0065004	12.12	−0.49	−4.6	0	0.21	0.02	q4i	no	protein-DNA complex assembly
GO:0006323	12.63	−0.48	−4.96	0	0.15	0.01	q4i	no	DNA packaging
GO:0006333	12.44	−0.4	−4.84	0	0.23	0.01	q4i	no	chromatin assembly or disassembly
GO:0031497	12.51	−0.44	−4.84	0	0.2	0.01	q4i	no	chromatin assembly

A total of 18 GO terms were found as significant at FDR-adjusted $p<0.05$, nine of them were also found by the multivariate analysis. Column new indicates if the term as been found only because of the interaction factor (yes) or if it was found also in the univariate analysis in one or both dimensions independently.
doi:10.1371/journal.pone.0010348.t001

Different strategies of production and degradation, corresponding to different biological requirements of the cell, can be thus detected by the combined analysis of these parameters.

Gene-Set Dependences on Differential Expression and Splicing Index

Recent studies have shown that more that 70% of the multi-exon genes, corresponding to about 50% of all human genes, are predicted to be alternatively spliced [26]. It is well known that alternative splicing participates in many pathways and processes. Also alterations in splicing function has been implicated in many diseases, including neuropathological conditions such as Alzheimer disease, cystic fibrosis, defects in growth and development, and many human cancers [27].

The magnitude of the alterations in the splicing process can be studied through the splicing index. This index accounts for changes at the exon level that are relative to the expression of the gene. In particular, the intensity value of an exon's probeset is divided by an estimate of the expression level of the transcript cluster to which the exon belongs to. In this way, a gene-level-normalized intensity that can be compared across samples or conditions is created. Changes in this value between case and control samples provide a quantitative measure of alternative splicing between the two conditions [28]. Thus each gene in the data set can be studied both in terms of its differential expression and its alternative splicing. Our multidimensional logistic model can be used to explore this two dimensional gene space.

Here we reanalyze data obtained using Affymetrix exon arrays [29] in which human breast cancer cell lines are compared to non tumorigenic human breast epithelial cell lines. The parameters studied by means of the multidimensional logistic model are: differential gene expression estimates obtained upon the application of a t-test for the above mentioned comparison and a splicing index, that accounts for changes at the exon level that are relative to the expression of the gene [30].

A total of 141 GO terms were found to be significantly associated to high values of the differential gene expression dimension (pattern yh, yl; see methods section). These terms are equivalent to those that would be found by conventional one-dimensional GSA methods and, as expected, GO definitions related to cell proliferation, cell signalling, apoptosis, cellular adhesion, etc., were found among them. One significant GO term, *regulation of viral reproduction*, was significant in the splicing index dimension alone. The trend of the enrichment was towards the positive values of the splicing index (pattern *xh*; see methods section) meaning that genes in the GO term are "subordinately" more spliced in the tumour than in the normal tissue (see File S2A).

Another 12 terms were found by the MD-GSA (see Table 2), whose relationships within the GO hierarchy is depicted in Figure S2. The processes discovered here were related (but yet undetected) to other processes already detected by the conventional analysis of differential expression (see File S2A). For example, *positive regulation of cell adhesion* and its parent*regulation of cell adhesion* are descendants of *cell adhesion*, and two sister processes (*cell-matrix adhesion* and *cell-substrate adhesion*) were found by the model when the two variables were taken into account, and would have remained undetected if a conventional, unidimensional GSA approach would have been used. The patterns for these terms are bimodal in the two dimensional space (pattern *b24*, see methods section) indicating that the genes annotated to them behave as if they were in two sub-modules. For example, *positive regulation of cell adhesion* and its parent processes *regulation of cell adhesion*, which are known to be related to cancer, show a bimodal pattern towards the quadrants 2 and 4 (pattern*b24*). This means that part of the annotated genes are more spliced but underexpressed in the tumour samples while the other part is more spliced but underexpressed in the control samples (see Figure 2).

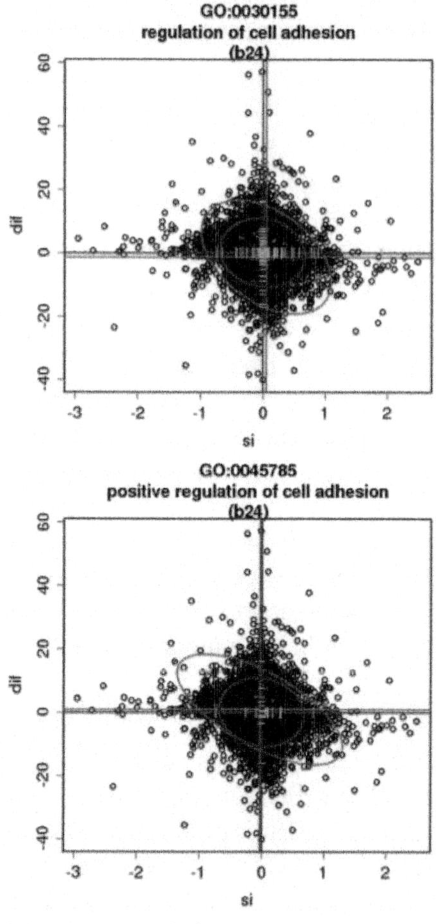

Figure 2: Combined analysis of differential gene expression and splicing index with the logistic model.

Differential expression is represented in vertical axis and splicing index is represented in the horizontal axis for GO terms positive regulation of cell adhesion (bottom) and its parent processes regulation of cell adhesion (top). Blue lines intersect in the mean of the distribution of all the values and red lines intersect in the mean of the distribution of values of the genes corresponding to the GO term. Blue ellipse delimits the confidence interval for all the values and red ellipse delimits the confidence interval for the GO term analysed. The red ellipse marks the trend of the relationship between both variables. MD-GSA assigns a bimodal pattern b24 to these functional modules.

doi:10.1371/journal.pone.0010348.g002

Table 2: Significant GO terms when differential expression and splicing index are taken into account in the model

	Log odds ratio (model coefficients)			Adjusted p-value				
GO id	**splicing**	**diff.exp**	**inter**	**splicing**	**diff.exp**	**inter**	**pattern**	**GO name**
GO:0006767	0.15	−0.15	0.14	1	0.61	0.04	b13	water-soluble vitamin metabolic process
GO:0045216	0.29	−0.04	0.17	1	0.95	0.02	b13	cell-cell junction assembly and maintenance
GO:0007043	0.38	−0.03	0.18	1	0.97	0.02	b13	cell-cell junction assembly
GO:0048706	0.2	0.08	0.17	1	0.89	0.03	b13	embryonic skeletal development
GO:0007034	0.32	−0.18	0.17	1	0.65	0.02	b13	vacuolar transport
GO:0007041	0.32	−0.1	0.18	1	0.86	0.01	b13	lysosomal transport
GO:0048704	0.23	0.12	0.19	1	0.84	0.02	b13	embryonic skeletal morphogenesis
GO:0048705	0.17	0.1	0.17	1	0.85	0.02	b13	skeletal morphogenesis
GO:0016197	0.08	0.1	0.15	1	0.79	0.02	b13	endosome transport
GO:0030155	0.01	−0.16	−0.15	1	0.43	0.01	b24	regulation of cell adhesion
GO:0045785	−0.04	0.06	−0.18	1	0.94	0.02	b24	positive regulation of cell adhesion
GO:0030032	−0.16	−0.17	−0.18	1	0.72	0.03	b24	lamellipodium biogenesis

A total of 12 GO terms were found as significant in the interaction at FDR-adjusted $p<0.05$.
doi:10.1371/journal.pone.0010348.t002

An equivalent analysis for KEGG can be found in File S2B.

Gene-Sets Differentially Activated In Related Diseases: A Case Study with Psoriasis and Dermatitis.

The study of gene expression at genomic level in both psoriasis [31] and dermatitis [32] and further functional analysis reveals a considerable number of deregulated pathways when both diseases are compared to their corresponding healthy samples. Thus, when the multivariate logistic model was applied to gene lists arranged by differential expression 172 GO terms were found to be significant only for dermatitis (patterns *xh, xl*; see methods section) and 202 only for psoriasis (patterns *yh, yl*). Another 77 GO terms were found to be significant in both, dermatitis and psoriasis but did not show an interaction effect (patterns *q1f, q2f, q3f, q4f*) Most of this terms will also be found by the independent unidimensional analysis of the dermatitis dataset and the psoriasis dataset. In the case of dermatitis, terms related to signalling, cell proliferation, immune system and development of epidermis were found, among others (see Files S3A and S3B). Similar terms can be found in psoriasis with some variations (see Files S3A and S3B). A detailed comparative functional analysis of these diseases is beyond the scope of this manuscript and we will only focus on the results obtained when both diseases are simultaneously analysed.

Table 3 shows the GO terms that are significant when both diseases are taken into account in the logistic model (column labelled with "inter"). Figure S3 shows the GO terms within the GO hierarchy. The GO terms *M phase of mitotic cell cycle* (and their parent terms *M phase* and *cell cycle phase*) and *cell division* where associated to both diseases in their main effects and also in their interaction effect (pattern *q1i,* seemethods) reinforcing their relevance in the biological mechanisms underlying both skin syndromes. Some other GO terms are only significant in the interaction effect. Their genes show a bimodal behaviour as if the functional module was composed of two sub-units (pattern *b13, b24*; see methods). For instance, GO terms*phosphoinositide-mediated signaling* and *response to reactive oxygen species* have a positive interaction coefficient, which means that some of the genes of the module are being coordinately over-expressed in both diseases while the remaining genes in the GO term are under-expressed also in both diseases. In a symmetric way, *negative regulation of lymphocyte proliferation* (and the parent process *negative regulation of mononuclear cell proliferation*) shows a negative interaction. Part of the genes in these modules increase their expression in dermatitis but decrease it in psoriasis while the rest of them present the opposite behaviour. The reduced cutaneous IFNalpha2 transcription which has been described as a differential characteristic of dermatitis with respect to psoriasis [32] could be causing this effect detectable in the analysis when the two variables are included in the model. All this bimodal terms highlight antagonistic effect, detectable only trough the combined analysis of both diseases.

Table 3: Significant GO terms when differential expression of dermatitis and psoriasis are taken into account in the model

	Log odds ratio (model coefficients)			Adjusted p-value				
GO id	dermatitis	psoriasis	inter	dermatitis	psoriasis	inter	pattern	GO name
GO:0022403	−0.13	0.36	0.11	0.11	0	0.01	q1i	cell cycle phase
GO:0000279	−0.06	0.37	0.12	0.55	0	0.03	q1i	M phase
GO:0051301	−0.1	0.25	0.15	0.36	0	0	q1i	cell division
GO:0000087	−0.11	0.4	0.12	0.32	0	0.05	q1i	M phase of mitotic cell cycle
GO:0048015	0.08	0.07	0.16	0.72	0.68	0.05	b13	phosphoinositide-mediated signaling
GO:0000302	0.24	−0.06	0.29	0.59	0.85	0	b13	response to reactive oxygen species
GO:0032945	0.43	0.33	−0.79	0.26	0.39	0	b24	negative regulation of mononuclear cell proliferation
GO:0050672	0.43	0.33	−0.79	0.26	0.39	0	b24	negative regulation of lymphocyte proliferation
GO:0048589	−0.19	−0.06	−0.59	0.53	0.91	0.04	b24	developmental growth
GO:0007028	0.21	−0.11	−0.75	0.47	0.83	0	b24	cytoplasm organization and biogenesis
GO:0007043	0.07	−0.5	−0.91	0.86	0.22	0	b24	cell-cell junction assembly
GO:0045216	0.12	−0.26	−0.86	0.75	0.59	0	b24	cell-cell junction assembly and maintenance

A total of 12 GO terms were found as significant in the interaction at FDR-adjusted $p<0.05$.
doi:10.1371/journal.pone.0010348.t003

Combined analysis of several genomic measurements: a case study with genotyping, gene expression and copy number alterations in breast cancer

It is known that mutations or alteration in copy number are related to cancer and tumour development [33], [34]. Current microarray technologies allow for the measurement of SNP variation and copy number estimation at the same time [35], [36] and have been used to gain insights into breast cancer [37], [38], [39], among other diseases.

Using the multidimensional logistic model proposed we have re-analyzed here data from several separated studies previously collected by us in an integrative analysis of breast cancer disease [38]. In particular we provide a combined description of GO and KEGG relationship to different parameters such as SNP association, copy number alteration and differential gene expression in connection to disease outcome (all the data were taken from the additional information of the above mentioned study, see methods).

When analyzing SNP association data and copy number in luminal B tumours by the proposed MD-GSA, *basal cell carcinoma* KEGG pathway raised up (File S4B) showing a bimodal pattern towards quadrants 1 and 3 (*b13*, see methods). This indicates that the genes in the pathway highly associated to disease are also increased in their copy number, and that genes not associated to disease do not have an increased copy number (they may even have a reduced copy number what would fit with the no association or even protection of the SNPs to disease). Most probably, the SNPs are markers associated either to regions undergoing copy number alterations or to other mutations that affect the *basal cell carcinoma* pathway, which obviously underlies breast cancer disease. The same analysis using the GO reported some negative bimodal terms (Table 4 and File S4B) like *L-amino acid transport* which is known to be involved proliferation processes [40]. A similar analysis with GO terms can be found in File S4A. Figure S4 displays the GO terms in Table 4 within the GO hierarchy.

Table 4: Significant GO terms when copy number and gene association to the disease (see text) are taken into account in the model

	Log odds ratio (model coefficients)			Adjusted p-value				
GO id	association	copy number	inter	association	copy number	inter	pattern	GO name
GO:0015807	−0.09	−0.85	−0.59	0.98	0.46	0.04	b24	L-amino acid transport
GO:0032228	−0.63	−1.21	−0.68	0.65	0.24	0.01	b24	regulation of synaptic transmission, GABAergic
GO:0050805	−0.94	−1.24	−0.63	0.22	0.24	0.04	b24	negative regulation of synaptic transmission
GO:0051932	−0.82	−1.35	−0.67	0.49	0.17	0.02	b24	synaptic transmission, GABAergic
GO:0042398	−0.77	−0.02	0.12	0.04	0.99	1	xl	amino acid derivative biosynthetic process
GO:0042401	−0.93	0.12	0.2	0.01	0.98	1	xl	biogenic amine biosynthetic process
GO:0030216	0.2	0.41	−0.03	0.8	0.03	1	yh	keratinocyte differentiation
GO:0031424	0.29	0.59	−0.01	0.81	0	1	yh	keratinization

A total of 8 GO terms were found as significant at FDR-adjusted $p<0.05$.
doi:10.1371/journal.pone.0010348.t004

We also applied the MD-GSA to the variables prognosis and differential expression in tumours. In the representation (File S5A), high values in the differential expression dimension indicate under-expression in tumour while low values indicate over-expression. Conversely, high values in the prognosis dimension indicate bad prognosis (if the gene is expressed) while low values in the prognosis dimension indicate good prognosis (if the gene is expressed).

Table 5 (more details in File S5A) show results obtained from the application of the MD-GSA using modules defined with GO terms. The relationships among them within the GO hierarchy are depicted in Figure S5. Most of the GO terms related to *cell division* and *cell cycle* show a*q2i* pattern (see methods) indicating a significant convergence of their genes in the prognosis and differential expression dimensions. From the relatively high prognosis value associated to the genes annotated to this GO terms we know that, if over expressed they indicate bad prognosis. From the low values in the t-statistic we know these GO terms are enriched in the tumours samples. Hence the multivariate logistic model is pointing out those modules which are dangerous to the patient if they are activated, and, that are certainly know to be activated in luminal B tumours. This extended functional analysis provides the researcher not only with a quick an easy interpretation of the combined data but also with the additional information of the interaction term in the model. It is worth pointing out here that better and more detailed results are obtained by combining both datasets under the proposed methodology than by applying independently the univariant methodology to any of the datasets and summing up the results obtained. The equivalent MD-GSA for KEGG pathways can be found in File S5B.

Table 5: Significant GO terms when differential expression and prognosis are taken into account in the model

	Log odds ratio (model coefficients)			Adjusted p-value				
GO id	diff.exp	prognosis	inter	diff.exp	prognosis	inter	pattern	GO name
GO:0000087	−0.45	−0.08	−0.42	0.01	0.81	0	q2i	M phase of mitotic cell cycle
GO:0000279	−0.53	−0.07	−0.38	0.04	0.85	0	q2i	M phase
GO:0000910	−0.27	−0.09	−0.57	0.01	0.95	0	q2i	cytokinesis
GO:0007067	−0.47	−0.07	−0.4	0.04	0.9	0	q2i	mitosis
GO:0022618	−0.22	−0.33	−0.42	0.03	0.21	0	q2i	ribonucleoprotein complex assembly
GO:0051301	−0.38	0	−0.38	0.01	0.99	0	q2i	cell division
GO:0051726	−0.01	0.05	−0.22	0.03	0.91	0.01	q2i	regulation of cell cycle
GO:0045638	0.09	−0.35	−0.6	0.01	0.65	0.04	q4i	negative regulation of myeloid cell differentiation
GO:0000226	−0.08	0.16	−0.31	0.11	0.47	0.02	b24	microtubule cytoskeleton organization and biogenesis
GO:0000278	−0.34	0.04	−0.28	0.11	0.94	0	b24	mitotic cell cycle
GO:0007346	−0.3	−0.08	−0.39	0.07	0.9	0	b24	regulation of mitotic cell cycle
GO:0022403	−0.42	0	−0.31	0.09	0.99	0	b24	cell cycle phase
GO:0042254	−0.4	−0.45	−0.42	0.19	0.1	0.01	b24	ribosome biogenesis
GO:0006412	0.06	−0.28	−0.2	0.02	0.01	0.07	q4f	translation
GO:0006414	0.45	−1.12	−0.43	0	0	0.28	q4f	translational elongation
GO:0042312	0.45	0.08	−0.51	0.03	0.97	0.22	xh	regulation of vasodilation
GO:0000209	−0.25	0.55	0.13	0.94	0.01	1	yh	protein polyubiquitination
GO:0006066	0.08	0.2	−0.02	0.97	0.02	1	yh	alcohol metabolic process
GO:0010033	0.05	0.29	0	0.99	0.02	1	yh	response to organic substance
GO:0032944	−0.17	−0.7	0.06	0.97	0.02	1	yl	regulation of mononuclear cell proliferation
GO:0042098	−0.18	−0.61	0.08	0.95	0.04	1	yl	T cell proliferation
GO:0042110	0.03	−0.38	0.14	0.75	0.03	0.86	yl	T cell activation
GO:0042129	−0.33	−0.74	−0.02	0.99	0.05	1	yl	regulation of T cell proliferation
GO:0045321	−0.04	−0.28	0.06	0.92	0.03	1	yl	leukocyte activation
GO:0046649	−0.06	−0.33	0.07	0.89	0.02	1	yl	lymphocyte activation
GO:0046651	−0.19	−0.49	−0.05	0.99	0.05	1	yl	lymphocyte proliferation
GO:0050670	−0.17	−0.7	0.06	0.97	0.02	1	yl	regulation of lymphocyte proliferation
GO:0051249	−0.06	−0.44	0.24	0.52	0.04	0.71	yl	regulation of lymphocyte activation

Terms were significant at FDR-adjusted $p<0.05$.
doi:10.1371/journal.pone.0010348.t005

Advantages and Limitations of the Logistic Regression Methodology

The major advantage of the logistic regression methodology is it flexibility. It can be used in any genomic context in which certain biological characteristic of a gene is measured using a numerical scale. This numerical scale may be a continuous "ranking statistic" as described previously [41] or in this paper, but it may also be a categorical variable [42].

Moreover, many modifications of the logistic model with potential applications in biology are already statistically developed and can be used straight forward. Here, for instance we showed how to extend the methodology to study not one but two gene characteristics at a time. It is also straightforward to include the interaction in the model as we showed here. A unidimensional binary logistic model can be used instead the conventional 2×2 contingency table alternative because the logistic model easily allows for weighting genes [42]. This simplicity of extension is not at all intrinsic to most other GSA approaches, what makes the logistic model worth to be explored.

Another advantage of the method is that it does not start from the original observed data set (gene expression matrix for instance) but from a ranking statistic that already summarizes the relevant characteristic under study. This makes the methodology useful in many genomic contexts beyond the microarray paradigm. One example of ranking statistic we have discussed is the classical t-test which, perhaps with some modification, is underneath most GSA methodologies. For each gene, this statistic measures the biological characteristic of "how much" the gene is differentially expressed in a particular biological experiment. But we also exemplified how the ranking statistic can be a hazard ratio form a Cox model or other gene-wise variable[19]. In the case of the hazard ratio, the biological characteristic measured for each gene by the statistic is the association of expression and risk disease. The GSA for this second example can be directly carried out using the logistic methodology and software. On the contrary, most GSA approaches will require major modifications of their methods and software to be applied in a case other than differential gene expression in a class comparison experiment.

Virtually any gene-wise variable can be explored from a GSA perspective using the logistic regression model. In this paper we presented examples for the analysis of transcription rates, mRNA stabilities, splicing, SNP association to disease and copy number estimation. The analysis of other measurements is possible, including the evolutionary selective pressure in the human genome or a study of gene connectivity in the interactome [19]. Other publications also discuss on the advantage of a methodology that starts form a single ranking statistic and not from the whole expression data matrix [42], [43].

Having said that, some remarks and warnings should be given related mainly with the null hypothesis that underpin the method and p-value computation.

In Sator's logistic regression approach [41] and in the extension we are proposing here, the distribution of the ranking statistic within each module is compared to that of its complement. Thus, following Goeman's nomenclature they are "competitive" tests [10]. Also, the way p-values are computed in the logistic model make of this approach a "gene sampling model" methodology [10].

It has been shown that, in general contexts of gene expression, where gene measurements are correlated within modules, GSA approaches that test "competitive" hypothesis based on "gene sampling models" are anticonservative [10]. This undesirable property also applies to the main effects of the bivariate logistic model as we could confirm in simulation studies (only in the case of internal correlation in the gene sets, which is the case of gene expression but not of the rest of the measurements used in this study). Interestingly, the consequence of gene correlation over the interaction effect, which is the main contribution of the proposed methodology, was the opposite and makes the method more conservative (see File S6). One way to avoid the bias of the particular context of gene expression would be to compute p-values based on a subject sampling permutation.

Care should be taken also when interpreting p-values from the method proposed here due to its "competitive" nature and the fact that it starts from a ranking statistic instead of the original data. Consequently, p-values test whether the distribution of the ranking statistic within each module is different to that of the whole genome. Therefore p-values do not extrapolate directly to the individual level class comparison which was done in order to compute the ranking statistic.

DISCUSSION

Functional annotations, such as GO or KEGG pathways, have been used for the definition of modules of genes, carrying out common functional roles, in functional profiling methods [1], [2]. All these methods, including the most recent versions, such as the GSA, can only deal with data that have been preselected or arranged by a unique variable (e.g. differential gene expression between cases and controls, etc.) The approach we are presenting here constitutes a novel and conceptually different proposal for the functional analysis of genomic experiments. It allows the simultaneous analysis of several variables, which can account for different properties of the genes.

This approach can detect interactions between these variables that account for functional roles dependent on several genomic properties or measurements.

We have used for this purpose a logistic model. It has recently been shown that the application of the logistic model to one single variable (differential gene expression in this case) produces results conceptually similar to the outcome of any conventional GSA method [41]. The aim here is not to improve the one dimensional detection of gene modules related to the measurement, but to look for gene modules that have complex dependences on several genomic variables or measurements. Thus, in the first example we show how some functional GO categories depend on particular combinations of their transcription rates and mRNA stabilities. Different strategies can be used by the cellular machinery to ensure, for example, a rapid activation or a long lasting of a particular team of genes that cannot be explained with only one variable. Thus, combinations of several variables (e.g. a rapid transcription rate and a low mRNA stability can be useful for a rapid release and a rapid deactivation of a transient function) are on the root of many biological processes. The variables used can be properties of the genes or can be also measurements of behaviours such as their expression in a given condition. In the second case example we have analyzed a combination of gene property (splicing index) and gene behaviour (differential gene expression). The MD-GSA was able of detecting biological processes that depend on combinations of both variables and would remain undetected if the variables were independently analyzed. Finally, we applied the same concept to the same type of measurement (differential gene expression) in two different but related scenarios: a case control of dermatitis and another case-control of psoriasis. In this example we were able of finding common and distinctive altered functionalities of both related diseases that remained otherwise undetected with the conventional one-dimensional GSA. The combination of measurements that can be studied under this framework and their biological relevance is unimaginable. Thus the relation of biological roles to combinations of different parameters of different types, such as gene intrinsic properties (e.g. mRNA stability), gene behaviours (e.g. level of expression) or gene states (e.g. methylation, SNPs, copy number), etc., can be easily be studied using this approach.

Summarizing, MD-GSA constitutes a novel approach to the functional profiling of genome scale experiments that paves the way for a higher level understanding of the behaviour of functional modules in the cell.

MATERIALS AND METHODS

Datasets and Data Preprocessing

Transcription Rates and mRNA Stabilities in Yeast

Genome-wide values for the transcription rates (TR) and mRNA stabilities (RS) of the genes of yeast used in the first sub-section of results can be found in the supplementary material of the manuscript by Garcia-Martinez et al. [23].

Gene Expression and Splicing Index.

Okoniewski & Miller [44] used exon arrays to compare breast cancer cell line MCF7 (fetal calf serum) to non tumorgenic breast epithelial cell line MCF10A (horse serum). They estimated differential gene expression using standard t-statistics and alternative splicing using the splicing index described in [30]. Since the splicing index is defined for each exon, we have used here median values to provide splicing measurements at a gene level. Thus, we have two numerical variables recorded for each gene in the study. The first one assesses the variation in the general expression level. The second one quantifies the change in splicing pattern of the gene, independently of its expression levels.

Differential Expression in Psoriasis and Dermatitis

Expression data from two separated case control experiments where combined in this analysis. The first experiment consisted of the comparison of lessional and non lessional skin samples in atopic dermatitis patients [32] (data were obtained from the GEO database, accession: GSE5667). The second experiment compared affected and unaffected skin in psoriatic patients[31] (GEO database, accession: GSE6710). Separated gene expression analyses of these two datasets were performed using standard methods: RMA algorithm [45] was used to normalize data within each of the experiments. The limma package [46] from Bioconductor [47] was used to estimate, separately for each of the studies, differential gene expression between diseased and non-diseased skin. Hence, two experimental measurements (limma t-statistics) where generated for each gene and used in the analysis: a first measurement of differential gene expression in dermatitis and a second measurement of differential gene expression in psoriasis.

Combined Analysis of Several Breast Cancer Genomic Measurements

Data used in the combined analysis of genomic measurements, in the results section, were taken from the supplementary material of [38]. SNP association to disease was measured using Odds Ratio (OR) of their corresponding minor allele frequencies. Then, the magnitude of the association of each gene to the disease was obtained as the value of association of the SNP more associated to the disease among all the SNPs mapping in the gene (or near the gene and being in linkage disequilibrium) [21], [38]. Differences in gene expression between tumour and normal breast tissues where estimated using t-statistics. Cox regression models where used to correlate survival time and gene expression, yielding a "prognosis" value for each gene (genes with "high" hazard ratios in the Cox model are associated to poor prognosis; genes with "low" hazard ratios associated to good prognosis). Another genomic measurements used was the average copy number for each gene in luminal B tumours, obtained from the hybridization intensity of the probesets corresponding to each gene (taken from the additional material of our study [38]).

Annotation Data

Functional modules are defined according the annotations of the GO [4] and the KEGG Pathway [48] repositories. Functional modules of more than 500 genes where considered to be too general to be informative so they where filtered out. Functional modules having less than 10 genes annotated to them where considered to be too small to be properly fitted by the multivariate logistic model and where also discarded.

Multi dimensional GSA (MD-GSA) using a logistic model that considers more than one variable

Logistic regression is a well established statistical methodology used to model the probability of occurrence of a binary event as a function of some other independent variables [49]. In the context of genomic studies, univariate logistic models have been shown to be suitable to perform gene set enrichment analysis [41].

Modelling functional class membership in terms of some measurement, X, of differential gene expression between two conditions as follows:

$$\ln\left(\frac{P(g \in F)}{P(g \notin F)}\right) = K + \alpha X \quad (1)$$

we can call the gene set F enriched in one of the conditions a significant estimate of the α coefficient is obtained [41].

In this paper we extend the use logistic models to perform a multidimensional gene set enrichment analysis. Our model describes the probability of a gene belonging to a functional class as a function of not one, but several experimental measurements. For two of those measurements the model will be as follows:

$$\ln\left(\frac{P(g\in F)}{P(g\notin F)}\right) = K + \alpha X + \beta Y + \gamma XY \qquad (2)$$

where α and β are the main effects and γ is the interaction effect.

In a case-control study measuring, for instance, gene expression and genotype, we could model the probability of genes being annotated to a GO term as a function of both, differential gene expression (X) and allelic association to disease (Y).

Modelling not only the additive effects but also the interaction term, we accurately describe how the genes in a gene set are related to both measurements X and Y together, allowing for the detection of enrichment patterns which will remain unnoticed in two independent univariate analyses.

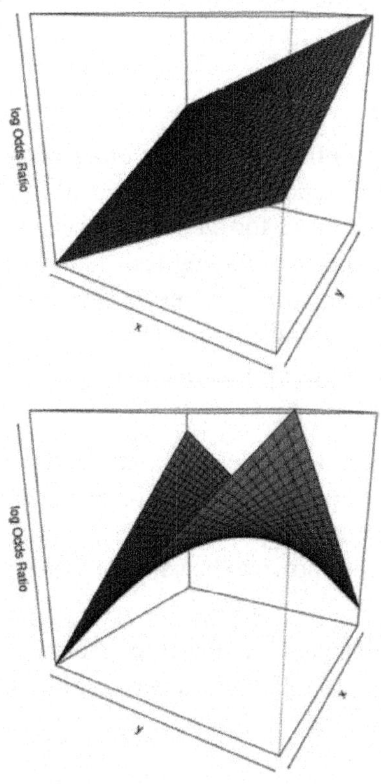

Figure 3: Surfaces described by the logistic model.

The model in equation (2) describes the log odds ratio of a gene being annotated to functional module F as a function of two variables, X and Y. The shape of this surface when embedded in a 3D space is that of a plane if the interaction coefficient γ is zero (Figure 3, top), or a hyperbolic paraboloid, also called saddle surface, when the estimate of γ is different from zero (Figure 3, bottom). Hence, from the sign and significance of the fitted coefficients, we can find the direction in the two dimensional space XY in which the genes annotated to the function F are more likely to be found.

The surface described by the logistic model is a plane when the interaction term (γ) is 0 (top) and a hyperbolic paraboloid when the interaction term (γ) is not zero (bottom).

When γ is zero the sign of the coefficients α and β describe the slopes of the plane and therefore, the direction towards which the probability of genes being annotated is greater.Figure 4 describes the areas where genes belonging to a functional module are more likely to be found, depending on the estimated α and β coefficients of the logistic model (2) and provided that the estimate of γ is not significantly different from zero.

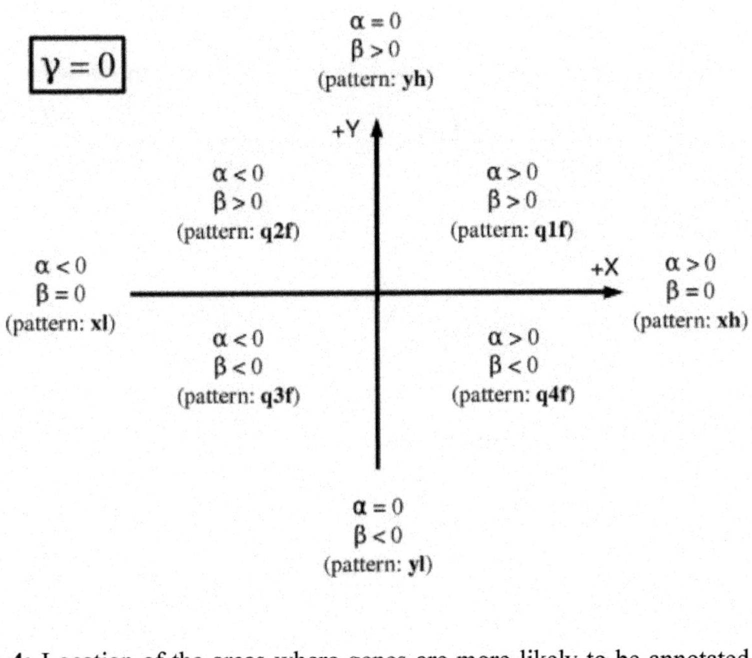

Figure 4: Location of the areas where genes are more likely to be annotated to the function F depending on the coefficients of the fitted model.

When γ=0 the fitted surface is a plane which slope grows towards the area.

When γ is different from zero the interaction dominates the growth of the log odds ratio while the saddle point in the surface has the coordinates $(-\beta/\gamma, -\alpha/\gamma)$. If for instance, for a particular functional module F, all estimated coefficients are positive, then, the saddle point of the hyperbolic paraboloid will be in the third quadrant and the surface will grow to the infinite in the first quadrant. As the surface represents how likely we are to find genes annotated to module F in the plane XY, we will conclude that the module F is located towards the firs quadrant. Moreover, as the interaction effect is positive we know that the evidence of this localization is greater than the one we will get from separated analysis of each one of the dimensions X and Y on their own (following equation 1). Then, biological interpretation can be done recalling the meaning of the X and Y quantities. Figure 5 (top) describes the areas where genes belonging to a functional module are more likely to be found, depending on the estimates of α, β and γ and when γ is estimated to be different from zero.

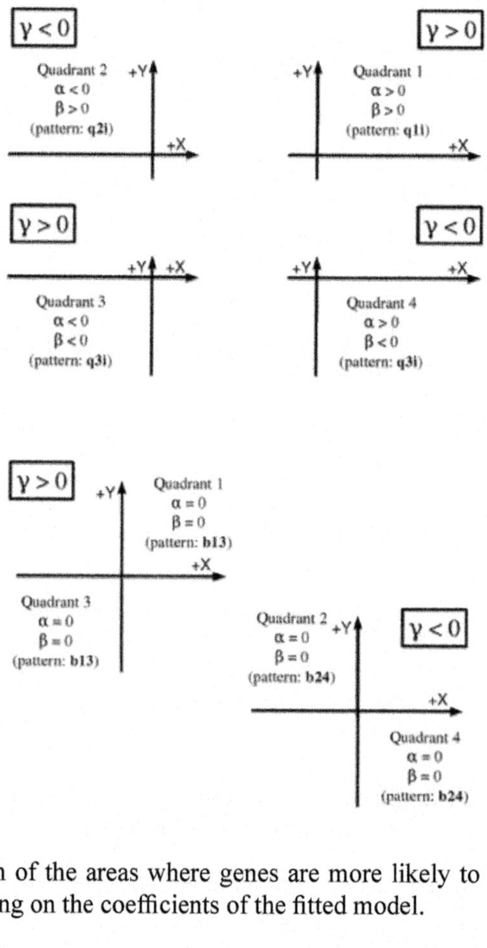

Figure 5: Location of the areas where genes are more likely to be annotated to the function F depending on the coefficients of the fitted model.

If γ≠0 the fitted surface is a hyperbolic paraboloid, when α≠0 and β≠0 (top part) the most likely area to find genes annotated to F is the quadrant opposite to the saddle point of the surface. When α=0 and β=0 (bottom part) the saddle point of the surface is in the (0,0) and the genes annotated to the function F are more likely to be found in two opposite quadrants, reflecting the bimodality of the function F.

If it was the case that just the interaction coefficient γ would be different from zero, then the saddle point will be the (0, 0) and the genes annotated to functional module F will be allocated to opposite quadrants of the XY space; the first and the third quadrant if γ>0; the second and the fourth quadrants if γ<0. In this latest case we will call the functional module F bimodal and the biological interpretation will be that, genes in F are effectively spited up in two groups of opposite patterns. Figure 5 (bottom) describes the areas where genes belonging to a functional module are more likely to be found, if the estimates of α and β are zero.

Table 6 shows how to interpret all possible combinations of α, β and γ estimates.

Table 6: Interpretation of all relevant combinations of α, β and γ estimates

α	β	γ	pattern identifier	pattern	description
+	+	+	q1i	Quadrant 1 with interaction	F is allocated towards one of the quadrants and the evidence is greater than just the additive evidences from the univariate analysis.
+	0	+			
0	+	+			
−	−	+	q3i	Quadrant 3 with interaction	
−	0	+			
0	−	+			
−	+	−	q2i	Quadrant 2 with interaction	
−	0	−			
0	+	−			
+	−	−	q4i	Quadrant 4 with interaction	
+	0	−			
0	−	−			
0	0	+	b13	Bimodal + (quadrants 1 and 3)	F is split in two opposite quadrants.
0	0	+	b24	Bimodal − (quadrants 2 and 4)	
+	+	0	q1f	Quadrant 1 flat	F is allocated towards one of the quadrants and the evidence is similar to the additive evidences from the univariate analysis.
−	−	0	q3f	Quadrant 3 flat	
−	+	0	q2f	Quadrant 2 flat	
+	−	0	q4f	Quadrant 4 flat	
+	0	0	xh	X high (+) values	F is enriched just in the first condition.
−	0	0	xl	X low (−) values	
0	+	0	yh	Y high (+) values	F is enriched just in the second condition.
0	−	0	yl	Y low (−) values	

doi:10.1371/journal.pone.0010348.t006

Wald statistics to test the main effect coefficients and the interaction effects [41]. Other approaches like likelihood ratio tests could also have been used.

As one logistic regression model needs to be fit for each functional module in the analysis, multiple testing occurs and p-value correction must be performed. In this paper we use Benjamini and Hochberg [50] approach to correct all p-values of the same parameter of the model α, β or γ.

Implementation

The proposed algorithm has been implemented as an R library available athttp://bioinfo.cipf.es/supplementary/multidimensional_GSA, released under the GPL license.

AUTHOR CONTRIBUTIONS

Conceived and designed the experiments: DM JD. Performed the experiments: DM. Analyzed the data: DM. Wrote the paper: JD.

REFERENCES

1. Dopazo J (2009) Formulating and testing hypotheses in functional genomics. Artif Intell Med 45: 97–107.
2. Huang DW, Sherman BT, Lempicki RA (2008) Bioinformatics enrichment tools: paths toward the comprehensive functional analysis of large gene lists. Nucleic Acids Res 37: 1–13.
3. Hartwell LH, Hopfield JJ, Leibler S, Murray AW (1999) From molecular to modular cell biology. Nature 402: C47–52.
4. Ashburner M, Ball CA, Blake JA, Botstein D, Butler H, et al. (2000) Gene ontology: tool for the unification of biology. The Gene Ontology Consortium. Nat Genet 25: 25–29.
5. Kanehisa M, Goto S, Kawashima S, Okuno Y, Hattori M (2004) The KEGG resource for deciphering the genome. Nucleic Acids Res 32: D277–280.
6. Vastrik I, D'Eustachio P, Schmidt E, Joshi-Tope G, Gopinath G, et al. (2007) Reactome: a knowledge base of biologic pathways and processes. Genome Biol 8: R39.
7. Dopazo J (2006) Functional interpretation of microarray experiments. Omics 10: 398–410.
8. Khatri P, Draghici S (2005) Ontological analysis of gene expression data: current tools, limitations, and open problems. Bioinformatics 21: 3587–3595.
9. Mootha VK, Lindgren CM, Eriksson KF, Subramanian A, Sihag S, et al. (2003) PGC-1alpha-responsive genes involved in oxidative

phosphorylation are coordinately downregulated in human diabetes. Nat Genet 34: 267–273.

10. Goeman JJ, Buhlmann P (2007) Analyzing gene expression data in terms of gene sets: methodological issues. Bioinformatics 23: 980–987.
11. Al-Shahrour F, Diaz-Uriarte R, Dopazo J (2005) Discovering molecular functions significantly related to phenotypes by combining gene expression data and biological information. Bioinformatics 21: 2988–2993.
12. Goeman JJ, van de Geer SA, de Kort F, van Houwelingen HC (2004) A global test for groups of genes: testing association with a clinical outcome. Bioinformatics 20: 93–99.
13. Subramanian A, Tamayo P, Mootha VK, Mukherjee S, Ebert BL, et al. (2005) Gene set enrichment analysis: a knowledge-based approach for interpreting genome-wide expression profiles. Proc Natl Acad Sci U S A 102: 15545–15550.
14. Clark AG, Eisen MB, Smith DR, Bergman CM, Oliver B, et al. (2007) Evolution of genes and genomes on the Drosophila phylogeny. Nature 450: 203–218.
15. Kim SY, Volsky DJ (2005) PAGE: parametric analysis of gene set enrichment. BMC Bioinformatics 6: 144.
16. Kitano H (2004) Cancer as a robust system: implications for anticancer therapy. Nat Rev Cancer 4: 227–235.
17. Bentink S, Wessendorf S, Schwaenen C, Rosolowski M, Klapper W, et al. (2008) Pathway activation patterns in diffuse large B-cell lymphomas. Leukemia 22: 1746–1754.
18. Bardelli A, Velculescu VE (2005) Mutational analysis of gene families in human cancer. Curr Opin Genet Dev 15: 5–12.
19. Al-Shahrour F, Arbiza L, Dopazo H, Huerta-Cepas J, Minguez P, et al. (2007) From genes to functional classes in the study of biological systems. BMC Bioinformatics 8: 114.
20. Wu C, Delano DL, Mitro N, Su SV, Janes J, et al. (2008) Gene set enrichment in eQTL data identifies novel annotations and pathway regulators. PLoS Genet 4: e1000070.
21. Medina I, Montaner D, Bonifaci N, Pujana MA, Carbonell J, et al. (2009) Gene set-based analysis of polymorphisms: finding pathways or biological processes associated to traits in genome-wide association studies. Nucleic Acids Res 37: W340–344.

22. McLendon R, Friedman A, Bigner D, Van Meir EG, Brat DJ, et al. (2008) Comprehensive genomic characterization defines human glioblastoma genes and core pathways. Nature.

23. Garcia-Martinez J, Gonzalez-Candelas F, Perez-Ortin JE (2007) Common gene expression strategies revealed by genome-wide analysis in yeast. Genome Biol 8: R222.

24. Perez-Ortin JE (2007) Genomics of mRNA turnover. Brief Funct Genomic Proteomic 6: 282–291.

25. Al-Shahrour F, Diaz-Uriarte R, Dopazo J (2004) FatiGO: a web tool for finding significant associations of Gene Ontology terms with groups of genes. Bioinformatics 20: 578–580.

26. Johnson JM, Castle J, Garrett-Engele P, Kan Z, Loerch PM, et al. (2003) Genome-wide survey of human alternative pre-mRNA splicing with exon junction microarrays. Science 302: 2141–2144.

27. Faustino NA, Cooper TA (2003) Pre-mRNA splicing and human disease. Genes Dev 17: 419–437.

28. Srinivasan K, Shiue L, Hayes JD, Centers R, Fitzwater S, et al. (2005) Detection and measurement of alternative splicing using splicing-sensitive microarrays. Methods 37: 345–359.

29. Bitton DA, Okoniewski MJ, Connolly Y, Miller CJ (2008) Exon level integration of proteomics and microarray data. BMC Bioinformatics 9: 118.

30. Clark TA, Schweitzer AC, Chen TX, Staples MK, Lu G, et al. (2007) Discovery of tissue-specific exons using comprehensive human exon microarrays. Genome Biol 8: R64.

31. Reischl J, Schwenke S, Beekman JM, Mrowietz U, Sturzebecher S, et al. (2007) Increased expression of Wnt5a in psoriatic plaques. J Invest Dermatol 127: 163–169.

32. Plager DA, Leontovich AA, Henke SA, Davis MD, McEvoy MT, et al. (2007) Early cutaneous gene transcription changes in adult atopic dermatitis and potential clinical implications. Exp Dermatol 16: 28–36.

33. Wood LD, Parsons DW, Jones S, Lin J, Sjoblom T, et al. (2007) The genomic landscapes of human breast and colorectal cancers. Science 318: 1108–1113.

34. Pinkel D, Albertson DG (2005) Array comparative genomic hybridization and its applications in cancer. Nat Genet 37: SupplS11–17.

35. Bignell GR, Huang J, Greshock J, Watt S, Butler A, et al. (2004) High-resolution analysis of DNA copy number using oligonucleotide microarrays. Genome Res 14: 287–295.

36. Peiffer DA, Le JM, Steemers FJ, Chang W, Jenniges T, et al. (2006) High-resolution genomic profiling of chromosomal aberrations using Infinium whole-genome genotyping. Genome Res 16: 1136–1148.

37. Easton DF, Pooley KA, Dunning AM, Pharoah PD, Thompson D, et al. (2007) Genome-wide association study identifies novel breast cancer susceptibility loci. Nature 447: 1087–1093.

38. Bonifaci N, Berenguer A, Diez J, Reina O, Medina I, et al. (2008) Biological processes, properties and molecular wiring diagrams of candidate low-penetrance breast cancer susceptibility genes. BMC Med Genomics 1: 62.

39. Hunter DJ, Kraft P, Jacobs KB, Cox DG, Yeager M, et al. (2007) A genome-wide association study identifies alleles in FGFR2 associated with risk of sporadic postmenopausal breast cancer. Nat Genet 39: 870–874.

40. Singh RK, Rinehart CA, Kim JP, Tolleson-Rinehart S, Lawing LF, et al. (1996) Tumor cell invasion of basement membrane in vitro is regulated by amino acids. Cancer Invest 14: 6–18.

41. Sartor MA, Leikauf GD, Medvedovic M (2008) LRpath: A logistic regression approach for identifying enriched biological groups in gene expression data. Bioinformatics 25: 211–217.

42. Montaner D, Minguez P, Al-Shahrour F, Dopazo J (2009) Gene set internal coherence in the context of functional profiling. BMC Genomics 10: 197.

43. Pavlidis P, Qin J, Arango V, Mann JJ, Sibille E (2004) Using the gene ontology for microarray data mining: a comparison of methods and application to age effects in human prefrontal cortex. Neurochem Res 29: 1213–1222.

44. Okoniewski MJ, Miller CJ (2008) Comprehensive analysis of affymetrix exon arrays using BioConductor. PLoS Comput Biol 4: e6.

45. Irizarry RA, Hobbs B, Collin F, Beazer-Barclay YD, Antonellis KJ, et al. (2003) Exploration, normalization, and summaries of high density oligonucleotide array probe level data. Biostatistics 4: 249–264.

46. Smyth GK (2004) Linear models and empirical bayes methods for assessing differential expression in microarray experiments. Stat Appl Genet Mol Biol 3: Article3.

47. Gentleman RC, Carey VJ, Bates DM, Bolstad B, Dettling M, et al. (2004) Bioconductor: open software development for computational biology and bioinformatics. Genome Biol 5: R80.

48. Kanehisa M, Araki M, Goto S, Hattori M, Hirakawa M, et al. (2008) KEGG for linking genomes to life and the environment. Nucleic Acids Res 36: D480–484.

49. Agresti A (2002) Categorical data analysis. Hoboken, New Jersey: John Wiley and Sons.

50. Benjamini Y, Hochberg Y (1995) Controlling the false discovery rate: a practical and powerful approach to multiple testing. Journal of the Royal Statistical Society Series B 57: 289–300.

Chapter 3

COMPARATIVE GENOMICS OF MYCOPLASMA: ANALYSIS OF CONSERVED ESSENTIAL GENES AND DIVERSITY OF THE PANGENOME

Wei Liu[1] , Liurong Fang[1] , Mao Li[1] , Sha Li[1] , Shaohua Guo[1] , Rui Luo[1] , Zhixin Feng[2] , Bin Li[2] , Zhemin Zhou[3] , Guoqing Shao[2] , Huanchun Chen[1] , Shaobo Xiao[1]

[1] Division of Animal Infectious Diseases, State Key Laboratory of Agricultural Microbiology, College of Veterinary Medicine, Huazhong Agricultural University, Wuhan, People's Republic of China

[2] Institute of Veterinary Medicine, Jiangsu Academy of Agricultural Sciences, Nanjing, People's Republic of China

[3] Environmental Research Institute, University College Cork, Cork, Ireland

ABSTRACT

Mycoplasma, the smallest self-replicating organism with a minimal metabolism and little genomic redundancy, is expected to be a close approximation to the minimal set of genes needed to sustain bacterial life. This study employs comparative evolutionary analysis of twenty*Mycoplasma* genomes to gain an improved understanding of essential genes. By analyzing the core genome of mycoplasmas, we finally revealed the conserved essential genes set for mycoplasma survival. Further analysis showed that the core genome set has many characteristics in common with experimentally identified essential genes. Several key genes, which are related to DNA replication and repair and can be disrupted in transposon mutagenesis studies, may be critical for bacteria survival especially over long period natural selection. Phylogenomic reconstructions based on 3,355 homologous groups allowed robust estimation of phylogenetic relatedness among mycoplasma strains. To obtain deeper insight into the relative roles of molecular evolution in pathogen adaptation to their hosts, we also analyzed the positive selection pressures on particular sites and lineages. There appears to be an approximate correlation between the divergence of species and the level of positive selection detected in corresponding lineages.

INTRODUCTION

Mycoplasmas are widespread in nature as parasites of humans, mammals, reptiles, fish, arthropods, and plants [1]. As a conditional pathogenic organism, it associates with various diseases, including pneumonia, arthritis, meningitis and chronic urogenital tract disease [2]. Although they are the smallest self-replicating organisms, both commensal forms and pathogenic forms are diverse. With a minimal metabolism and little genomic redundancy, the genome of *Mycoplasma* is expected to be a close approximation to the minimal set of genes needed to sustain bacterial life [3]. An early projection proposed a minimal gene set composed of 206 genes based on the analysis of eight free-living and endosymbiotic bacterial genomes[4]. More recently, Glass *et al.* [5] performed a global transposon mutagenesis study and identified 100 putatively nonessential genes in *M. genitalium.* Logically, the remaining 387 genes presumably constitute the set of essential genes. However, these data greatly exceed theoretical projections of how many genes comprise a minimal genome, as proposed by Gil *et al.* [4].

Natural selection leads to the fixation of essential genes and can delete nonessential genes in a wide range of species [6], [7]. This process is similar, but more robust than manual mutagenesis studies. Through long term evolution form a more conventional progenitor in the*Firmicutes* taxon [8], *Mycoplasmas* have undergone a process of massive genome reduction[1]. These wall-less bacteria are obligate parasites that live in relatively unchanging niches requiring little adaptive capability. *M. genitalium*, a human urogenital pathogen, is the extreme manifestation of this genomic parsimony, having only 482 protein-coding genes and the smallest genome of any known free-living organism capable of being grown in axenic culture[9]. Thus, with little genomic redundancy and contingencies for different environmental conditions, *Mycoplasmas* are regarded as optimal microbes to perform genes essentiality studies [5]. Along with the burgeoning increase in *Mycoplasma* genome sequence data, this would appear to be the right time to explore gene essentiality from a comparative genomics perspective.

An enormous genetic diversity exists in the mycoplasmas, yet how much diversity is functional, and what are the important adaptations that serve to partition species into different niches? *M. hyopneumoniae* and *M. hyorhinis* are the causal agents of swine mycoplasmosis. The former causes a mild, chronic pneumonia of swine and results in deactivation of mucociliary functions[10]. This agent is infective for a single host species, but the mechanisms of host specificity are unknown. The latter is responsible for respiratory tract and arthritis disease in swine [1]. *M. hyorhinis* is generally considered a swine pathogen, yet is most commonly infect laboratory cell lines, implying that it can thrive among different species of cell lines [11]. A strong link between*M. hyorhinis* and human cancer was reported recently by Huang *et al.* [12], who used a monoclonal antibody against the unique *M. hyorhinis*–specific protein p37 to detect mycoplasma in over 600 carcinoma tissues from a variety of organs. The study indicated that up to 56% of gastric carcinoma and 55% of colon carcinoma biopsies were positive for *M. hyorhinis* [12]. With a similar genome size, *M. hyorhinis* and *M. hyopneumoniae* exhibit high levels of functional diversity. Interest has therefore shifted to questions of why *M. hyorhinis* can thrive among different species of cell lines.

This paper communicates the results of three major analyses. In the first analysis, we present the details of a comparative analysis of twenty *Mycoplasma* strains and investigate the conserved essential genes set for mycoplasma survival. For the second analysis, phylogenomic reconstructions based on 3,355 homologous groups allows robust estimation of phylogenetic relatedness among mycoplasma strains. The third analysis employs the branch-site method to assess positive selection pressures on particular sites and lineages. There appears to be an approximate correlation between the divergence of species and the level of positive selection detected in corresponding lineages.

RESULTS

Diversity of *Mycoplasmataceae* Family: Core Genome *vs.* Flexible Gene Pool

The number of protein coding genes per genome within the various strains and species of mycoplasmas is relatively similar (ranging from 475 to 1,037; Table 1), but the gene composition of these genomes is much more variable. Based on the gene content table (obtained as described in Materials and Methods; Table S1), three *M. hyopneumoniae* strains share about 95% of their genes, and three different species of mycoplasmas share only around 71% of their genes (Figure 1). This latter result appears to be independent of the particular strains or species involved in the comparison. Even with the inclusion of 20 genomes, the pan-genome size of *Mycoplasmas* appears not to be determined, and we estimate that the size probably surpasses 8,000 genes. This huge pan-genome size may be a reflection of their different lifestyles in distinct ecological niches. Within species, the pan-genome size also remains uncertain, although our estimates suggest that the pan-genome size of *M. hyopneumoniae* is smaller.

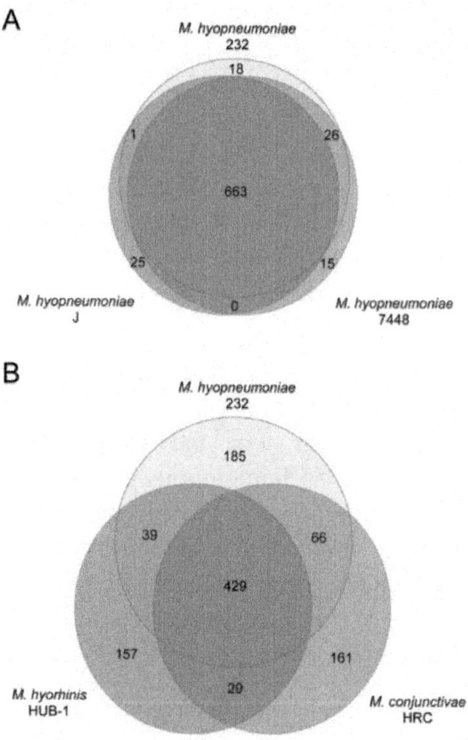

Figure 1: Venn diagram for two sets of three taxa.

Above are taxa of the same species and below are taxa of different species. The surfaces are approximately proportional to the number of genes.

Table 1: Bacterial Strains Used in This Study

Mollicute strains	Host	CDS	Genome size (bp)	Accession	Citation
U. urealyticum serovar 10 str. ATCC 33699	Human	646	874478	CP001184	-
U. parvum serovar 3 str. ATCC 700970	Human	611	751719	AF222894	[45]
U. parvum serovar 3 str. ATCC 27815	Human	609	751679	CP000942	-
M. synoviae 53	Bird	659	799476	NC_007294	[46]
M. pulmonis UAB CTIP	Rodent	-	963879	AL445566	[47]
M. pneumoniae M129	Human	688	816394	U00089	[48]
M. penetrans HF-2	Human	1037	1358633	BA000026	[49]
M. mycoides subsp. mycoides SC str. PG1	Ruminant	1016	1211703	BX293980	[50]
M. mobile 163K	Fish	635	777079	AE017308	[51]
M. hyorhinis HUB-1	Swine	658	839615	NC_014448	[52]
M. hyopneumoniae J	Swine	657	897405	NC_007295	[46]
M. hyopneumoniae 7448	Swine	657	920079	NC_007332	[46]
M. hyopneumoniae 232	Swine	691	892758	NC_006360	[53]
M. genitalium G37	Human	475	580076	NC_000908	[5]
M. gallisepticum str. R(low)	Bird	763	1012800	AE015450	[54]
M. crocodyli MP145	Crocodile	689	934379	CP001991	-
M. conjunctivae HRC	Sheep and goats	696	846214	FM864216	[55]
M. capricolum subsp. ATCC 27343	Ruminant	812	1010023	CP000123	-
M. arthritidis 158L3-1	Rats and mice	631	820453	NC_011025	[56]
M. agalactiae PG2	Sheep and goats	759	877438	CU179680	-

doi:10.1371/journal.pone.0035698.t001

The extent of the pan-genome is opposed to the core. Genes that are in common between the different species within the family *Mycoplasmataceae* comprised our core genome - the set of orthologous genes determined the common properties of this family. In this work, the tribeMCL program was used to cluster orthologous genes, and a total of 13,654 predicted proteins were grouped into 3,355 clusters, each cluster representing a group of putative orthologs. The 1,481 genes that are present in single genomes (Figure 2) represent lineage specific genes. In addition, the 196 genes shared by all the 20 strains comprised our core genome (Figure 3).

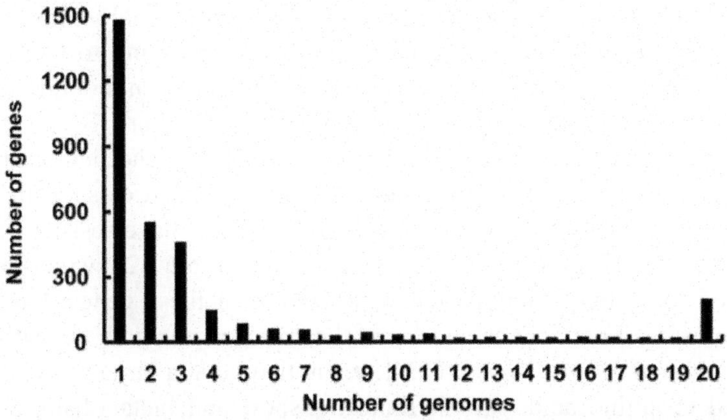

Figure 2: Frequency of genes within the 20 genomes included in this analysis.

Genes present in a single genome represent lineage specific genes, while at the opposite end of the scale, genes found in all 20 genomes represent the *Mycoplasma*core genome.

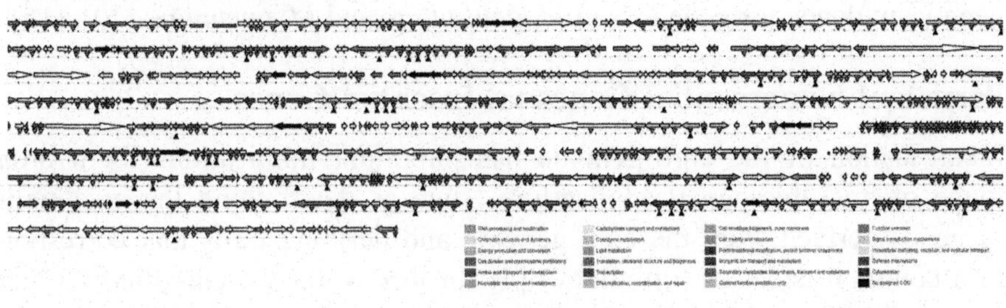

Figure 3: Viable core and pan genes distribution in the *M. hyorhinis* HUB-1 map.

The *M. hyorhinis* HUB-1 genome is shown at a scale of 100 kb per line. Colored arrows above the line indicate annotated genes. Genes are colored according to their functional category, as indicated in the key. rRNAs and tRNAs are denoted by red and purple rectangles, respectively. Red triangles above the line represent the core genes of*Mycoplasmataceae*. Blue triangles above the line represent the genes shared by all the species in clade III, but which are absent from other lineages. Black triangles below the line represent the disrupted genes documented in *M. genitalium* G37, mapped onto the*M. hyorhinis* HUB-1 genome [5].

Functional Characterization of the Core Genome

The use of the core genome concept has led to important insights into the evolution of bacterial species and identification of potentially important novel genes [13]. In terms of functional assignments according to COGs, almost half (42.3%) of the proteins observed from the core genome are devoted to translation, ribosomal structure, and biogenesis (Figure S1). Our results support the analysis of Ouzounis and Kyrpides [14], who demonstrated that genetic processes such as translation are conserved and close to the original form. Strikingly, 10.6% of the observed core genes have resisted functional assignments according to COGs classification (Table S2), which highlights the need for better functional characterization of these genes. Furthermore, by comparing functional categories of the core genome with the categories of the genome of *M. hyorhinis* HUB-1, we noticed that a large array of proteins devoted to amino acid, carbohydrate transport and metabolism, as well as defense mechanisms, were sharply reduced. Our results support the analysis of Fraser *et al.* [9] and Himmelreich *et al.* [15], who demonstrated that both *M. genitalium* and *M. pneumoniae* lost all the genes involved in amino acid synthesis, and their survival is totally dependent on an exogenous supply of the complete spectrum of amino acids. Beyond this, the pronounced reduction of those functional categories observed in the core genome might be the further genetic evidence for gene loss in *M. genitalium* and *M. pneumonia* [9], [15].

Persistent Nonessential Genes *vs.* Essential Genes

Identification of the core genome has important implications for a broad range of microbiological applications, such as determining the essentiality of genes derived from the core genome and deriving traits that correspond to a common ancestor (orthology) [4]. In this work, we classified the core genome into two classes according to persistence and essentiality: persistent nonessential genes and conserved essential genes (Figure 3). Glass *et al.* [5]performed a global transposon mutagenesis study and identified 100 putatively nonessential genes in *M. genitalium*. We mapped those nonessential genes onto the *M. hyorhinis* HUB-1 genome and 24 of them were persistent among *Mycoplasma* genomes.

Focusing on gene persistence, the essentiality of a gene is relative to a set of experimental conditions. It is quite different for a cell to survive in a laboratory setting, with plenty of supplied metabolites, compared to thriving in the wild, where it competes with other organisms for limited resources. Starvation or stresses are omnipresent, and the fitness effect of persistent genes may be essential for survival under transition from one environmental condition to another [16]. After transposon mutagenesis, disrupted genes may not be essential for growth in rich media, but their loss may lead to such a low fitness that its deletion will never be fixed in natural populations [16]. For example, Glass *et al.* [5] isolated six mutants involved in recombination and DNA repair: *recA*, *recU*, Holliday junction DNA helicases *uvrA* and *uvrB*, formamidopyrimidine-DNA glycosylase *mutM*, which excises oxidized urines from DNA, and a likely DNA damage inducible protein gene. Interestingly, we noticed that these six disrupted genes occur in the core genome set, which suggests that these disrupted persistent genes might be critical for growth in variable environments over long periods. The survival of these mutants is probably due to their tolerances of IS element within a relatively short period. According to Glass *et al.*, these six mutants grew more poorly after repeated passage, which indicates that those genes are also critical for bacterial survival [5]. Our results in this regard generally agree with the analysis by Glass *et al.* This analysis prompted us to explore gene essentiality by combining both the experimental approaches and comparative genomics analysis.

Genome-Based Reconstruction of *Mycoplasmataceae* Phylogeny

We constructed robust phylogenies for the family *Mycoplasmataceae* based on whole genome analysis. The supertree contains five major, distinct clades (Figure 4). In clade I, three*Ureaplasma* strains with a common host are clustered on a single branch. In clade II, six mycoplasma species (*M. pulmonis* etc.) with various lifestyles formed the cluster. In clade III, the strain HRC appears adjacent to strain HUB-1 and three isolates of *M. hyopneumoniae*strains are clustered closely, indicative of a recent common ancestry. In clade IV, *M. capricolum*and *M. mycoides* are clustered closely on a single branch and both of them are the agents of ruminant mycoplasmosis. In clade V, *M. genitalium* and *M. pneumoniae* are closely related, together with *M. penetrans* and *M. gallisepticum*, forming the cluster.

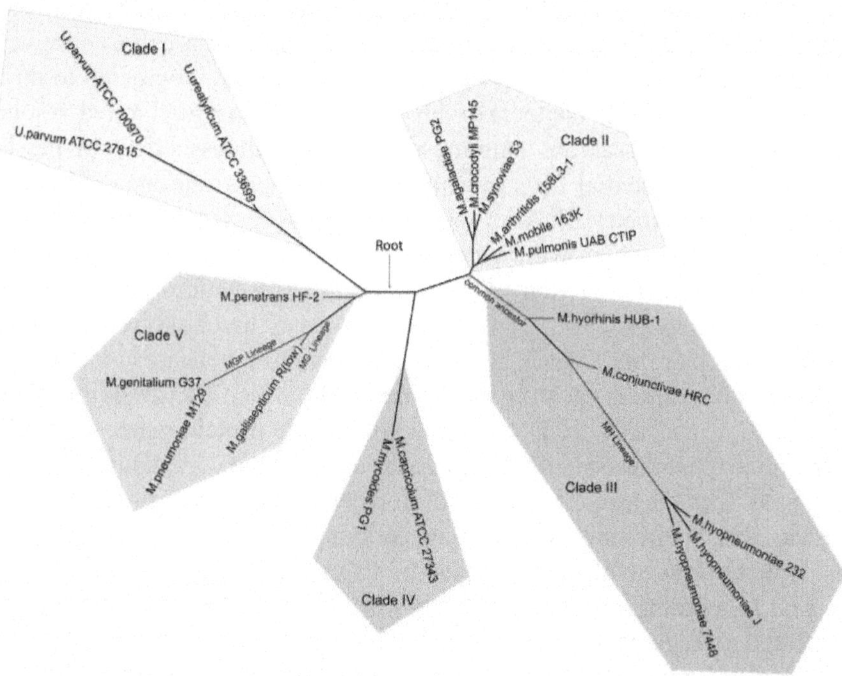

Figure 4: Phylogenetic tree of *Mycoplasmataceae.*

The phylogenetic relationship was estimated and tested in one thousand bootstrap samples using TREE-PUZZLE version 5.2 with a BIONJ model (see Materials and Methods). This supertree shows five major distinct clades. The four lineages that were used as foreground in the branch-site model positive selection test are highlighted in red.

Although single gene trees have been used extensively to estimate the species tree, evidence has shown that single gene trees may have particular difficulty in representing prokaryotic species phylogeny, because lateral gene transfer (LGT) occurs among prokaryotic genomes, and LGT may obscure the phylogenetic signal [17]. In this work, the 16S rRNA-based phylogeny has also been reconstructed for the family *Mycoplasmataceae* (Figure 5), which shows almost the same topology as the supertree. However, these two trees differ in the placement of *M. penetrans* HF-2 and *M. mobile* 163.

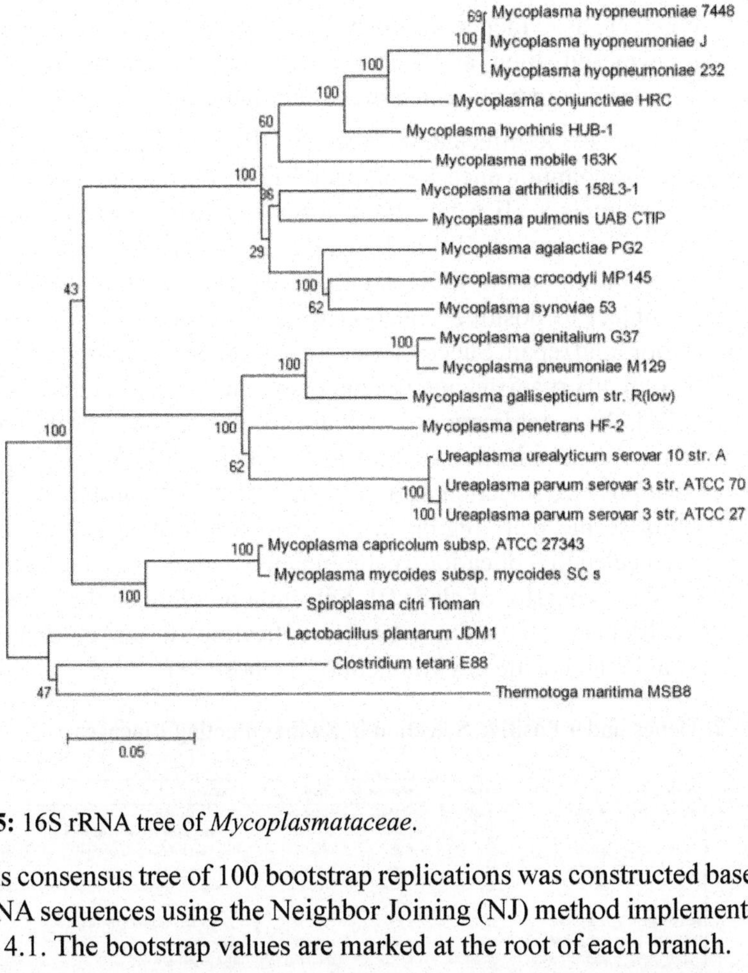

Figure 5: 16S rRNA tree of *Mycoplasmataceae*.

This consensus tree of 100 bootstrap replications was constructed based on 16S rRNA sequences using the Neighbor Joining (NJ) method implemented in MEGA 4.1. The bootstrap values are marked at the root of each branch.

A recent study proposed that the "Tree of Life" may be resolved by concatenation of 31 orthologs occurring in 191 species [18], and an analogous approach has been applied to infer*Mycoplasmas* phylogeny by random concatenation of 91 protein sequences shared by 16*mollicutes* [19]. The placement of *M. penetrans* in the concatenation-based phylogeny is consistent with the supertree (Figure 4).

Positive Selection Analysis

We employed the branch-site test of Yang and Nielsen [20], [21] implemented in the program HY-PHY (http://www.datam0nk3y.org/hyphy/doku.php) to assess positive selection [22], [23] at particular sites in particular lineages. This method compares synonymous and nonsynonymous substitution rates in protein coding genes and regards a nonsynonymous rate elevated above the synonymous rate as evidence for positive selection. For the *Mycoplasma* data set, both swine-infecting and human-infecting lineages were tested. The branches we tested are highlighted in red on Figure 4.

In the four swine-infected lineages (HUB-1 lineage, MC lineage, MH lineage, and the common ancestor of clade III), 661 orthologous groups shared by all species in clade III were tested. A total of 23 genes were identified to be under positive selection (Table 2). These genes were assigned to functional categories according to the COG database. We found that a large fraction of the genes subject to positive selection were connected to DNA replication, recombination, and repair. Successful genome replication is essential for growth and survival of an organism, and polymerase complexes often fail to complete this task [24]. Also, replication is thought to contribute to proliferation and efficiency of the colonization of hostile environments[25]. Interestingly, we detected that positive selection occurs in the both of the *dnaA* and *dnaN*genes, which comprise the *oriC* region in *M. hyorhinis*, and in several copies of proteins connected to replication, recombination and repair in both MC lineage and the common ancestor of clade III. Selection pressure on these genes could reflect constraints on efficient genome replication during colonization and proliferation in the hostile environment of the host[25].

Table 2: Genes under Positive Selection in Swine-infecting Lineages

Lineage	Gene	dN/dS[a]	Sequence %	COG(s)	Product
MH Lineage	mhp623	543.871	16.60%	COG1744R	ABC transporter
	mhp388	306.029	5.00%	COG3037S	ascorbate-specific PTS system enzyme IIC
	mhp603	4.42	0.00%	COG0195K	transcription elongation factor NusA
	mhp368	4.773	9.70%	COG0531E	putative membrane lipoprotein
	mhp595	515.577	13.40%	COG0266L	formamidopyrimidine-DNA glycosylase
	mhp480	1913.82	15.20%	-	hypothetical protein
MC Lineage	MCJ_005740	525.955	9.30%	COG1196D	putative ABC transporter ATP-binding protein P
	MCJ_003040	547.097	7.90%	COG0013J	alanyl-tRNA synthetase
	MCJ_007160	227.357	8.10%	COG2176L	DNA polymerase III PolC
	MCJ_002410	488.408	7.20%	COG4608E	oligopeptide ABC transporter ATP-binding protein
	MCJ_000340	316.719	5.10%	COG0556L	excinuclease ABC subunit B
HUB-1 Lineage	MHR_0001	2.122	7.10%	COG0593L	Chromosomal replication initiator protein dnaA
	MHR_0002	786.654	25.60%	COG0592L	DNA polymerase III beta subunit
	MHR_0009	8.155	5.50%	COG0525J	Valyl tRNA synthetase
	MHR_0132	306.624	7.30%	COG0060J	Isoleucyl tRNA synthetase
	MHR_0148	732.204	14.30%	COG0006E	Xaa-pro aminopeptidase
	MHR_0248	2.637	3.60%	COG0187L	DNA gyrase subunit B
	MHR_0318	180.115	16.40%	-	ABC transporter permease protein
	MHR_0377	6.345	9.70%	COG0202K	DNA-directed RNA polymerase subunit alpha
	MHR_0443	33.397	7.70%	COG0178L	Excinuclease ATPase subunit-like protein
	MHR_0609	14.416	16.70%	COG0544O	Trigger factor
Common Ancester[b]	MHR_0128	133.038	23.40%	COG0322L	Excinuclease ABC subunit C
	MHR_0131	33.679	13.50%	COG0188L	Topoisomerase IV subunit A
	MHR_0310	754.512	15.80%	COG2274V	ABC transporter ATP-binding and permease protein
	MHR_0486	58.164	12.90%	COG0587L	DNA polymerase III alpha subunit
	MHR_0489	3.669	29.10%	COG0532J	Translation initiation factor IF-2
	MHR_0639	8298.46	31.60%	-	Lipoprotein
	MHR_0363	6343.51	14.00%	COG1164E	Oligoendopeptidase F
	MHR_0356	6343.51	14.00%	COG1164E	Oligoendopeptidase F

[a]Ratio of the nonsynonymous to the synonomous mutation rate (dN/dS) measures the strength of selection, where values >1 indicate positive selection, and larger values indicate stronger selection.
[b]In the common ancestor lineage: a single gene of *M. hyorhinis* HUB-1 was used to represent each ortholog group (Table S1). Genes of *M. conjunctivae* HRC and three *M. hyopneumoniae* strains in the same ortholog group are also under positive selection.
doi:10.1371/journal.pone.0035698.t002

In the case of human-infecting lineages, the lineage that stood out from the rest with regard to host specificity was *M. gallisepticum*, which is significantly associated with chronic respiratory disease in chickens [26]. Not surprisingly, this lineage was identified to be under the strongest selection pressure in clade V (Table S3). However, we failed to notice any selection pressure on the *dnaA* and *dnaN* genes in the MG lineage. A large number of genes related to DNA repair, RNA processing, Amino acid transport and metabolism were found under positive selection. Selection pressure on these genes may facilitate evolutionary flexibility in the MG lineage, hence its ability to adapt to new environments.

Discussion

Numerous global transposon mutagenesis studies of minimal genomes have been performed to identify essential genes [5], [27]. Long-term natural selection can also delete nonessential genes in a wide range of species [6], [7], which is similar to, but more robust than, manual mutagenesis studied. With a minimal metabolism and little genomic redundancy, mycoplasmas are regarded as optimal microbes for the identification of essential genes [1]. It is believed that the *Mycoplasmas* evolved from a more conventional progenitor in the *Firmicutes* taxon by a process of massive genome reduction [16]. We found that each of the species in*Mycoplasmataceae* has undergone a similar process. These species have undergone various natural selection pressures in different environments. Most of the core genes remaining for such a long time should be considered as the essential genes needed by all the species within this family. Generally speaking, our comparative analysis was highly consistent with the studies by Glass *et al.* (Figure 6); however, we identified more genes that may have been deleted due to natural selection. The six genes differing between the two studies are all key genes of DNA replication and DNA repair: *recA*, *recU*, Holliday junction DNA helicases *uvrA* and *uvrB*, *mutM*and a likely DNA damage inducible protein gene. Those genes are core genes, but were disrupted in transposon mutagenesis studies [5]. Interestingly, Glass *et al.* stated that these six mutants grew more poorly after repeated passage, probably due to an accumulation of cell damage over time. This indicates that these six disrupted core genes may be critical for bacterial survival, especially over long periods. Therefore, we classified these genes as truly essential genes.

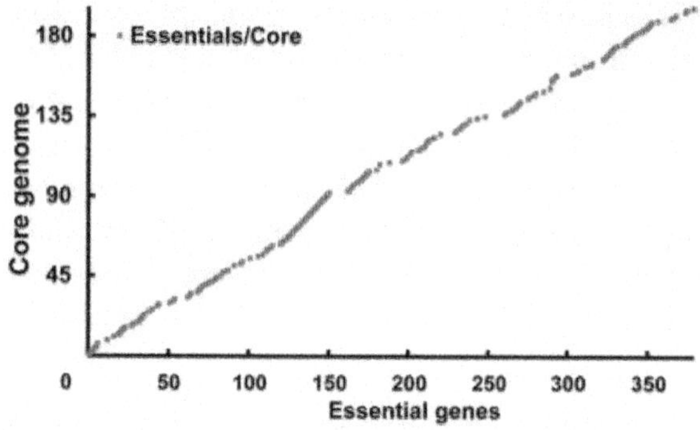

Figure 6: Similarity relationship between core and essential genes.

Genes on x-axis represent the essential genes documented in *M. genitalium* G37, while genes on y-axis represent the core genome identified in this study. Both the genes on x-axis and y-axis are distributed in terms of genomic coordinates. On-diagonal, genes that are both essentials and core are indicated as a series of colored dots.

Although transposon mutagenesis has proven to be a useful method and has been used extensively to determine the essentiality of genes [27], this method is highly dependent on environmental conditions. For the most part, mutagenesis studies performed under nutrient-rich conditions provide a substantial underestimate of the number of genes that are essential under host environmental conditions [28]. In reality, it is quite different for a cell to survive in a laboratory setting and to thrive in the wild. Transposon mutagenesis might misclassify nonessential genes that slow growth without arresting it but can also miss essential genes that tolerate transposon insertions [4]. However, comparative genomics analysis has also limitations, since it is likely to underestimate the core genome because it takes into account only the genes that have remained similar enough during the course of evolution to be recognized as true orthologues. Therefore, it will not include genes with a high rate of evolution, which may not show their relationship in comparisons of distant taxons [4]. Taken together, both the experimental approaches and comparative genomics analysis should be taken into account when addressing questions of essentiality. Besides, the core genes set proposed in the current study are only essential for most species within this family. We failed to estimate the conserved genes needed for single species. Each member of a particular species was maintained in a distinct ecological niche, in which some of the genes that were not present in core genome may also be important for the mycoplasma survival. These genes may be termed “lineage specific” essential genes. As more genome sequence becomes available in the future, there will be an opportunity to explore more properties of species special core genes set using comparative genomic tools.

To date, most mycoplasma phylogenies have been derived from single gene comparisons, or from concatenated alignments of a small number of genes. The increasing availability of mycoplasma genomes presents an opportunity to reconstruct evolutionary events using entire genomes [29], [30]. As a tool for future comparative phylogenetic studies, we used both supertrees and single gene alignments to infer relationships between 20 strains of the family*Mycoplasmataceae*. Our supertree and 16S rRNA phylogenies are consistent in most of their branches. However, there are conflicts regarding whether *M. penetrans* is clustered with the*Ureaplasma* lineage or with Clade V, as well as the placement of *M. mobile* 163. We also compared our trees with a recent study, in which the phylogeny of the *Mycoplasmas* was reconstructed by random concatenation of 91 protein sequences shared by 16 *mollicutes* [19]. The placement of *M. penetrans* in the concatenation-based phylogeny is consistent with the supertree, while the bootstrap value (=62) of the branch of *M. penetrans* and *Ureaplasma* in the 16S rRNA tree is low. Therefore, we placed *M. penetrans* into clade V. The location of *M. mobile* 163 is different among all three trees, which indicates a complicated phylogenic history for this strain, which may have involved recombination or other LGT events.

Phylogenetic reconstruction based upon concatenation of multiple orthologous genes can generate a more accurate tree than that done with a single gene [31], [32]. The supertree is even better than the concatenation-based tree, because it is immune to long-branch attraction artifacts [33], [34]. Thus, a robust supertree was constructed in this study to present the phylogeny of the family *Mycoplasmataceae*. The supertree was then used as a foreground for further analysis. Based on the supertree of *Mycoplasmataceae*, we classified these twenty strains into five different clades, between which the host specificity varies. All three strains in Clade I and three of the four strains in Clade V (except *M. gallisepticum*) were identified to be agents of human infection; therefore, these two clades form the human-infecting lineage. Both of the two sequenced strains in clade IV are the agent of ruminant infection, and thus represent the ruminant-infecting lineage. Four of the five strains in Clade III (except *M. conjunctivae*) are associated with swine mycoplasmosis and form the swine-infecting lineage. Briefly, most species with the same host specificity clustered together, forming a separate clade. There appears to be an approximate correlation between the divergence of species and the level of positive selection detected in different lineages. We suspect that host specificity was determined after the emergence of *Mycoplasma* species. Subsequent host jumping events may have been caused by a series of natural selection events during evolution.

To gain deeper insights into the molecular evolution events underlying natural selection, we employed the branch-site method to assess positive selection in swine-infecting and human-infect lineages. According to Petersen *et al.* [35], two categories of genes, immune-related and environmental adaptation related genes, are expected to show strong evidence for positive selection. Our analysis revealed that a number of genes related to DNA replication and repair (*dnaA*, *dnaN*, *gyrB*, *uvrA*, *polC*, *uvrB*, *uvrC*, *parC*, *dnaE*), show remarkably strong evidence for positive selection. These genes were unevenly distributed across HUB-1, the MC lineage and the common ancestor of clade III. Notably, both the *dnaA* and *dnaN* genes, which compose the*oriC* region, were identified to be under positive selection in the HUB-1 lineage. Previous studies have already demonstrated that replication may contribute to proliferation and efficiency of the colonization of hostile environments [25]. Therefore, we suspected that selection pressure on *oriC* may be one of the reasons why *M. hyorhinis* can thrive among different species of cell lines. This research provides a better insight into, and understanding of, persistent nonessential genes, and encourages exploration of essential genes by combining both the experimental approaches and comparative genomics analysis. This study also provides a comparative genomics method for addressing questions of essentiality. With the increasing number of genome sequences available for the same species in the future, this method will be useful for exploring species-specific essential genes.

MATERIALS AND METHODS

Bacterial Strains and Genome Sequences

M. hyorhinis strain HUB-1 was isolated from the respiratory tract of swine in China and confirmed to be an *M. hyorhinis* strain by verifying the 16S rRNA region. The main characteristics of 20 *Mycoplasmas* strains with freely available genomes at the time of the study are presented in Table 1. These genomes were used for comparative analysis.

Assignment of Orthologs and Phylogenetic Analysis

We analyzed *M. hyorhinis* HUB-1 and 19 other *Mycoplasmataceae* genomes from the NCBI databases. To ensure consistency, the annotations of all genomes were verified based on the similarity with *M. hyorhinis* HUB-1, using the tBLASTn algorithm [36]. The sets of orthologous protein-coding genes were defined as mutual fully transitive reciprocal BLASTP [37] hits (with E-value<10^{-4}) [38]. Co-ortholog groups were identified by the method similar to Inparanoid [39]and ortholog gene clusters were obtained using the tribeMCL program [40]. The nucleic acid sequence of each ortholog group was aligned using the CLUSTALW program version 1.83 [41]. For each data set, the phylogenetic relationship was estimated and tested in one thousand bootstrap samples using TREE-PUZZLE version 5.2 (general time reversible (GTR) +Γ4+I model of evolution with a BIONJ starting tree) [42]. The bi-partitions with at least 70% support from the bootstrap test in each data set were recorded as "0/1" status and used to reconstruct the consensus sequence. The phylogenetic relationship of the consensus sequence was built using the SplitsTREE 4 with the BioNJ model.

Positive Selection Analysis

We employed the branch-site test of Yang and Nielsen [20], implemented in the program HY-PHY, to assess positive selection at particular sites and lineages. Briefly, the likelihood of a model that does not allow positive selection is compared to one allowing positive selection on some specified lineages. The model allowing positive selection is tested using a likelihood ratio test (LRT) [43] that is compared to a $\chi 2$ statistic with two degrees of freedom. Likelihoods were estimated on the genes or species trees. For the *Mycoplasma* data set, both swine-infecting and human-infecting lineages were tested (Figure 4). To avoid the interference of recombination, only genes that support all four lineages in their gene trees (with >70% bootstrap support) were used. In total, 661 genes were tested. Finally, p values were corrected for multiple hypotheses testing using the Bonferroni method [44].

ACKNOWLEDGMENTS

We thank Dr. Maojun Liu (Jiangsu Academy of Agricultural Sciences, Nanjing, China), Dr. Lei Wang (Nankai University, Tianjin, China), Mr. Feng Li (Huazhong agricultural university, Wuhan, China) assistance with sequencing and analysis.

AUTHOR CONTRIBUTIONS

Conceived and designed the experiments: WL SX LF HC. Performed the experiments: WL ML SL SG ZF. Analyzed the data: WL ZZ GS RL BL. Contributed reagents/materials/analysis tools: WL GS RL BL. Wrote the paper: WL LF SX.

REFERENCES

1. Razin S, Yogev D, Naot Y (1998) Molecular biology and pathogenicity of mycoplasmas. Microbiol Mol Biol Rev 62: 1094–1156.
2. Waites KB, Katz B, Schelonka RL (2005) *Mycoplasmas* and *ureaplasmas* as neonatal pathogens. Clin Microbiol Rev 18: 757–89.
3. Morowitz HJ, Tourtellotte ME (1962) The smallest living cells. Sci Am 206: 117–126.
4. Gil R, Silva FJ, Pereto J, Moya A (2004) Determination of the core of a minimal bacterial gene set. Microbiol Mol Biol Rev 68: 518–537.
5. Glass JI, Assad-Garcia N, Alperovich N, Yooseph S, Lewis MR, et al. (2006) Essential genes of a minimal bacterium. Proc Natl Acad Sci U S A 103: 425–430.
6. Parkhill J, Sebaihia M, Preston A, Murphy LD, Thomson N, et al. (2003) Comparative analysis of the genome sequences of *Bordetella pertussis*, *Bordetella parapertussis*and *Bordetella bronchiseptica*. Nat Genet 35: 32–40.
7. Jin Q, Yuan ZH, Xu JG, Wang Y, Shen Y, et al. (2002) Genome sequence of *Shigella flexneri* 2a: insights into pathogenicity through comparison with genomes of *Escherichia coli* K12 and O157. Nucleic Acids Res 30: 4432–4441.
8. Wolf M, Muller T, Dandekar T, Pollack JD (2004) Phylogeny of Firmicutes with special reference to Mycoplasma (Mollicutes) as inferred from phosphoglycerate kinase amino acid sequence data. Int J Syst Evol Micr 54: 871–875.
9. Fraser CM, Gocayne JD, White O, Adams MD, Clayton RA, et al. (1995) The minimal gene complement of *Mycoplasma genitalium*. Science 270: 397–403.
10. DeBey MC, Ross RF (1994) Ciliostasis and loss of cilia induced by *Mycoplasma hyopneumoniae* in porcine tracheal organ cultures. Infect Immun 62: 5312–5318.
11. Kotani H, Butler GH, Tallarida D, Cody C, McGarrity GJ (1990) Microbiological cultivation of *Mycoplasma hyorhinis* from cell cultures. In Vitro Cell Dev Biol 26: 91–96.
12. Huang S, Li JY, Wu J, Meng L, Shou CC (2001) Mycoplasma infections and different human carcinomas. World J Gastroenterol 7: 266–269.
13. Lefebure T, Stanhope MJ (2007) Evolution of the core and pan-genome of*Streptococcus*: positive selection, recombination, and genome

composition. Genome Biol 8: R71.

14. Ouzounis C, Kyrpides N (1996) The emergence of major cellular processes in evolution. FEBS Lett 390: 119–123.
15. Himmelreich R, Hilbert H, Plagens H, Pirkl E, Li BC, et al. (1996) Complete sequence analysis of the genome of the bacterium *Mycoplasma pneumoniae*. Nucleic Acids Res 24: 4420–4449.
16. Fang G, Rocha E, Danchin A (2005) How essential are nonessential genes? Mol Biol Evol 22: 2147–2156.
17. Bapteste E, Boucher Y, Leigh J, Doolittle WF (2004) Phylogenetic reconstruction and lateral gene transfer. Trends Microbiol 12: 406–411.
18. Ciccarelli FD, Doerks T, von Mering C, Creevey CJ, Snel B, et al. (2006) Toward automatic reconstruction of a highly resolved tree of life. Science 311: 1283–1287.
19. Sirand-Pugnet P, Citti C, Barre A, Blanchard A (2007) Evolution of mollicutes: down a bumpy road with twists and turns. Res Microbiol 158: 754–766.
20. Zhang J, Nielsen R, Yang Z (2005) Evaluation of an improved branch-site likelihood method for detecting positive selection at the molecular level. Mol Biol Evol 22: 2472–2479.
21. Yang Z, Nielsen R (2002) Codon-substitution models for detecting molecular adaptation at individual sites along specific lineages. Mol Biol Evol 19: 908–917.
22. Kosiol C, Vinar T, da Fonseca RR, Hubisz MJ, Bustamante CD, et al. (2008) Patterns of positive selection in six Mammalian genomes. PLoS Genet 4: e1000144.
23. Shapiro BJ, Alm EJ (2008) Comparing patterns of natural selection across species using selective signatures. PLoS Genet 4: e23.
24. McGlynn P, Lloyd RG (2001) Rescue of stalled replication forks by RecG: simultaneous translocation on the leading and lagging strand templates supports an active DNA unwinding model of fork reversal and Holliday junction formation. Proc Natl Acad Sci U S A 98: 8227–8234.
25. Anisimova M, Bielawski J, Dunn K, Yang Z (2007) Phylogenomic analysis of natural selection pressure in *Streptococcus* genomes. BMC Evol Biol 7: 154.
26. Szczepanek SM, Tulman ER, Gorton TS, Liao X, Lu Z, et al. (2010) Comparative genomic analyses of attenuated strains of *Mycoplasma gallisepticum*. Infect Immun 78: 1760–1771.

27. Hutchison CA, Peterson SN, Gill SR, Cline RT, White O, et al. (1999) Global transposon mutagenesis and a minimal Mycoplasma genome. Science 286: 2165–2169.

28. Papp B, Pal C, Hurst LD (2004) Metabolic network analysis of the causes and evolution of enzyme dispensability in yeast. Nature 429: 661–664.

29. Snel B, Bork P, Huynen MA (1999) Genome phylogeny based on gene content. Nat Genet 21: 108–110.

30. Kunin V, Ahren D, Goldovsky L, Janssen P, Ouzounis CA (2005) Measuring genome conservation across taxa: divided strains and united kingdoms. Nucleic Acids Res 33: 616–621.

31. Daubin V, Gouy M, Perriere G (2002) A phylogenomic approach to bacterial phylogeny: evidence of a core of genes sharing a common history. Genome Res 12: 1080–1090.

32. Rokas A, Williams BL, King N, Carroll SB (2003) Genome-scale approaches to resolving incongruence in molecular phylogenies. Nature 425: 798–804.

33. Fitzpatrick DA, Logue ME, Stajich JE, Butler G (2006) A fungal phylogeny based on 42 complete genomes derived from supertree and combined gene analysis. BMC Evol Biol 6: 99.

34. Gadagkar SR, Rosenberg MS, Kumar S (2005) Inferring species phylogenies from multiple genes: concatenated sequence tree versus consensus gene tree. J Exp Zool B Mol Dev Evol 304: 64–74.

35. Petersen L, Bollback JP, Dimmic M, Hubisz M, Nielsen R (2007) Genes under positive selection in *Escherichia coli*. Genome Res 17: 1336–1343.

36. Iguchi A, Thomson NR, Ogura Y, Saunders D, Ooka T, et al. (2009) Complete genome sequence and comparative genome analysis of enteropathogenic *Escherichia coli*O127:H6 strain E2348/69. J Bacteriol 191: 347–354.

37. Altschul SF, Madden TL, Schaffer AA, Zhang J, Zhang Z, et al. (1997) Gapped BLAST and PSI-BLAST: a new generation of protein database search programs. Nucleic Acids Res 25: 3389–3402.

38. Zhaxybayeva O, Gogarten JP (2002) Bootstrap, Bayesian probability and maximum likelihood mapping: exploring new tools for comparative genome analyses. BMC Genomics 3: 4.

39. O'Brien KP, Remm M, Sonnhammer EL (2005) Inparanoid: a comprehensive database of eukaryotic orthologs. Nucleic Acids Res 33: D476–480.

40. Enright AJ, Van Dongen S, Ouzounis CA (2002) An efficient algorithm

for large-scale detection of protein families. Nucleic Acids Res 30: 1575–1584.

41. Thompson JD, Higgins DG, Gibson TJ (1994) CLUSTAL W: improving the sensitivity of progressive multiple sequence alignment through sequence weighting, position-specific gap penalties and weight matrix choice. Nucleic Acids Res 22: 4673–4680.

42. Schmidt HA, Strimmer K, Vingron M, von Haeseler A (2002) TREE-PUZZLE: maximum likelihood phylogenetic analysis using quartets and parallel computing. Bioinformatics 18: 502–504.

43. Yang Z (1998) Likelihood ratio tests for detecting positive selection and application to primate lysozyme evolution. Mol Biol Evol 15: 568–573.

44. Guilbaud O (2007) Bonferroni parallel Gatekeeping - Transparent generalizations, adjusted P-values, and short direct proofs. Biometrical J 49: 917–927.

45. Glass JI, Lefkowitz EJ, Glass JS, Heiner CR, Chen EY, et al. (2000) The complete sequence of the mucosal pathogen *Ureaplasma urealyticum*. Nature 407: 757–762.

46. Vasconcelos ATR (2005) Swine and poultry pathogens: the complete genome sequences of two strains of *Mycoplasma hyopneumoniae* and a strain of *Mycoplasma synoviae* (vol 187, pg 5568, 2005). J Bacteriol 187: 7548–7548.

47. Chambaud I, Heilig R, Ferris S, Barbe V, Samson D, et al. (2001) The complete genome sequence of the murine respiratory pathogen *Mycoplasma pulmonis*. Nucleic Acids Res 29: 2145–2153.

48. Dandekar T, Huynen M, Regula JT, Ueberle B, Zimmermann CU, et al. (2000) Re-annotating the *Mycoplasma pneumoniae* genome sequence: adding value, function and reading frames. Nucleic Acids Res 28: 3278–3288.

49. Sasaki Y, Ishikawa J, Yamashita A, Oshima K, Kenri T, et al. (2002) The complete genomic sequence of *Mycoplasma penetrans*, an intracellular bacterial pathogen in humans. Nucleic Acids Res 30: 5293–5300.

50. Westberg J, Persson A, Holmberg A, Goesmann A, Lundeberg J, et al. (2004) The genome sequence of *Mycoplasma mycoides* subsp. mycoides SC type strain PG1T, the causative agent of contagious bovine pleuropneumonia (CBPP). Genome Res 14: 221–227.

51. Jaffe JD, Stange-Thomann N, Smith C, DeCaprio D, Fisher S, et al. (2004) The complete genome and proteome of *Mycoplasma mobile*.

Genome Res 14: 1447–1461.

52. Liu W, Fang L, Li S, Li Q, Zhou Z, et al. (2010) Complete genome sequence of*Mycoplasma hyorhinis* strain HUB-1. J Bacteriol 192: 5844–5845.

53. Minion FC, Lefkowitz EJ, Madsen ML, Cleary BJ, Swartzell SM, et al. (2004) The genome sequence of *Mycoplasma hyopneumoniae* strain 232, the agent of swine mycoplasmosis. J Bacteriol 186: 7123–7133.

54. Papazisi L, Gorton TS, Kutish G, Markham PF, Browning GF, et al. (2003) The complete genome sequence of the avian pathogen *Mycoplasma gallisepticum* strain R(low). Microbiology 149: 2307–2316.

55. Calderon-Copete SP, Wigger G, Wunderlin C, Schmidheini T, Frey J, et al. (2009) The*Mycoplasma conjunctivae* genome sequencing, annotation and analysis. BMC Bioinformatics 10: Suppl 6S7.

56. Voelker LL, Dybvig K (1999) Sequence analysis of the *Mycoplasma arthritidis*bacteriophage MAV1 genome identifies the putative virulence factor. Gene 233: 101–107.

Chapter 4

ROBUST GENE-GENE INTERACTION ANALYSIS IN GENOME WIDE ASSOCIATION STUDIES

Yongkang Kim[1], Taesung Park[1,2]

[1] Department of Statistics, Seoul National University, Seoul, South Korea

[2] Interdisciplinary Program in Bioinformatics, Seoul National University, Seoul, 151–741, South Korea

ABSTRACT

Genome-wide association studies (GWAS) have successfully discovered hundreds of associations between genetic variants and complex traits. Most GWAS have focused on the identification of single variants. It has been shown that most of the variants that were discovered by GWAS could only partially explain disease heritability. The explanation for this missing heritability is generally believed to be gene-gene (GG) or gene-environment (GE) interactions and other structural variants. Generalized multifactor dimensionality reduction (GMDR) has been proven to be reasonably powerful in detecting GG and GE interactions; however, its performance has been found to decline when outlying quantitative traits are present. This paper proposes a robust GMDR estimation method (based on the L-estimator and M-estimator estimation methods) in an attempt to reduce the effects caused by outlying traits. A comparison of robust GMDR with the original MDR based on simulation studies showed the former method to outperform the latter. The performance of robust GMDR is illustrated through a real GWA example consisting of 8,577 samples from the Korean population using the Homeostasis Model Assessment of Insulin Resistance (HOMA-IR) level as a phenotype. Robust GMDR identified the KCNH1 gene to have strong interaction effects with other genes on the function of insulin secretion.

INTRODUCTION

A genome-wide association (GWA) study has become a common approach for testing the association between a single nucleotide polymorphism (SNP)

and a complex trait of interest [1]. There have been many successful results from genome-wide association studies (GWAS). However, SNPs that were identified by GWAS have been shown to explain only a small fraction of disease etiology, because the relatedness between complex diseases and multiple genes and/or their interactions are ignored. For this reason, the analysis of gene-gene (GG) and gene-environment (GE) interactions have been emphasized as a new alternative for understanding the etiology of common complex traits. However, GG and GE interactions are hard to detect and characterize by using traditional parametric statistical methods, for the following reasons [2]. First, high dimensional genetic data may be of a sparse nature. The issue of data sparseness can be addressed by using exponentially large sample sizes when parametric statistical methods are used for determining GG and GE interactions. Second, detecting GG and GE interactions using traditional procedures may lead to an increase in type II errors and a decrease in power. As a result, detecting GG and GE interactions has been a well-known challenge in statistics and data mining areas [3].

Ritchie et al. proposed a Multifactor Dimensionality Reduction (MDR) method for detecting GG interaction [4]. This method is not affected by data sparsity because it allows multilocus genotype combinations with very few or no data points. However, the original MDR method can only be used in association studies for qualitative traits. The MDR method was extended to more general types of traits, including quantitative traits, by Lou et al., who proposed the Generalized Multifactor Dimensionality Reduction (GMDR) method [5]. This method is quite flexible in handling various types of traits and is even able to account for individual covariates.

Since the initial introduction of the original MDR and GMDR methods, numerous extensions of MDR have been developed. Calle et al. proposed model-based MDR (MB-MDR) for the parametric extension of the MDR method [6]. Recently, Quantitative MDR (QMDR) was proposed for handling quantitative traits [7].

These MDR methods have brought many remarkable successes in GG interaction analysis. However, the effects of outlying trait observations on the results of MDR methods have not been studied. Outlying trait observations are those traits that diverge from the global distribution of traits. Bennett et al., Most et al., and Yang et al. discussed the treatment of outliers in GWAS and proposed simple approaches, such as removing outlying observations and transforming traits, including log-transformation or inverse-normalization [8, 9, 10]. The R-package QCGWAS includes a process for detecting outliers by skewness or kurtosis. However, these QC methods are very sensitive to the underlying distributional assumption and may lead to the wrong conclusions when the distributional assumption is not met. In this paper, we first demonstrate the effects of outlying trait observations on GMDR analysis. Next, we propose robust GMDR for reducing the effects caused by outlying traits. Robust GMDR uses robust estimation based on L-estimator and M-estimator. Simulation studies were used to compare the robust GMDR developed in this work to GMDR, and was shown to outperform GMDR. The performance of robust GMDR is illustrated through a real GWA example consisting of 8,577 samples from the Korean population.

MATERIALS AND METHODS

KARE Data

The Korea Association Resource (KARE) is a project for gathering large-scale GWA analyses in the Gyeonggi Province of South Korea [11]. Two community cohorts participated in the KARE project: the Ansung and Ansan cohorts. The number of participants in the Ansung cohort is 5,018 and that in Ansan cohort is 5,020. The age of the participants were between 40 and 69, and more than 260 traits have been extensively examined through epidemiological surveys, physical examinations, and laboratory tests. In this paper, the Homeostasis Model Assessment of Insulin Resistance (HOMA-IR) levels are used, which are widely employed to estimate insulin resistance. As HOMA-IR has a nonnegative skewness distribution, a gamma distribution is commonly assumed [12]. DNA samples were genotyped on the Affymetrix Genome-Wide Human SNP array 5.0, which is able to genotype 500,568 SNPs. The quality control processes are well described in a study by Cho et al. [13]. For the purposes of our work, samples without Body Mass Index (BMI) or HOMA-IR scores were disregarded. After removing the samples with the missing phenotypes, a total of 8,577 individuals and 327,872 SNPs remained in the study.

Review of MDR Method

The evaluation of genetic effects using the MDR method requires several steps. The first step involves selecting genetic factors according to their effects on the phenotype. If GG interactions of the nth order are required, n SNPs are selected and a 3^n contingency table is created. Each cell of the contingency table contains the number of cases and controls with the same genetic factor combinations. The second step calculates the ratio of the number of cases to the number of controls in each cell, and uses this ratio to determine whether a cell is "high risk" or "low risk". If the ratio of case to control in a cell exceeds the threshold (which is commonly specified as 1), then those cells are labeled "high risk"; otherwise they are labeled "low risk". This process enables each cell to be classified as either "high risk" or "low risk". In the third step, the ability of selected SNP sets to perform these classifications based on the case / control ratio are evaluated through accuracy and balanced accuracy (BA) measures, where BA is a measure for the prediction ability of the model and is defined as follows:

$$BA = \frac{1}{2}(sensitivity + specificity) \quad (1)$$

The larger the BA produced by the model, the more accurate the model's prediction ability [14]. These measures are computed from the 2×2 confusion table in which the first variable represents the case / control ratio and the second variable high risk and low risk. Then, the sensitivity is defined as the ratio of the number of cases to the number of high risk samples and the specificity is defined as the ratio of the number of controls to low risk samples. All these steps would be exhaustively performed for all SNP combinations.

The problem of over-fitting is usually avoided by performing a 10-fold cross-validation. It is also possible to use the cross-validation consistency (CVC) as a measure for selecting SNP combinations. This selection is based on the number of times the same SNP combination is selected as best out of 10 cross-validated samples. Alternatively, the SNP combination with the largest CVC measure can also be selected as the best SNP combination. BA and CVC measures have been widely used in MDR analysis

Review of GMDR Method

Although MDR has been used extensively in GG interaction analysis, it has several limitations. First, the MDR method is only capable of handling qualitative traits such as case and control data, because the method classifies each cell by using the case / control ratio. Second, it is not possible for MDR to adjust for other individual covariates such as environmental factors. Therefore, using MDR to analyze traits that are affected by environmental factors may produce biased results. In an attempt to overcome these limitations, Lou et al. proposed the Generalized Multifactor Dimensionality Reduction (GMDR) method [5], each step of which is very similar to those of the original MDR method. However, instead of using the counts of individuals, the GMDR method uses a residual-based score.

Let y_i denote the phenotype of individual i, which can be either quantitative or qualitative. Let$E(y_i) = \mu_i$. Consider the following generalized linear model (GLM):

$$l(\mu_i) = \alpha + x_i^T \beta + z_i^T \gamma \quad (2)$$

where $l(\mu_i)$ is the link function, α is the intercept, and x_i is a vector that expresses possible genotype combinations of interest. z_i is a vector representing environmental factors, and β and γ are coefficient vectors.

The first step when using GMDR is to fit the null model only with environmental factors under the null hypothesis, i.e., β = 0. The second step is to classify each SNP combination using the residuals from the null model. For each cell of the contingency table, the sum of the residuals is computed. Cells with a positive residual sum are classified as "high risk"; otherwise, they are classified as "low risk". Subsequent GMDR cross-validation steps are performed similar to those of MDR.

It is noteworthy that, in this GMDR procedure, the distinction between high and low classification is based on the sum of the residuals. Thus, a few outlying observations may have a large impact on the performance of GMDR, which provided the motivation for considering robust GMDR. Although GLM can handle both quantitative and qualitative traits, our efforts were focused on a quantitative trait with outlying observations. Thus, an identity link function was used with the supposition that y_i has an independent normal distribution with equal variance.

Proposed Robust GMDR Method

The effect of outlying observations can be reduced by considering robust scores such as an L-estimator and an M-estimator type score. The L-estimator is a linear combination of order statistics that have to be measured. A well-known example of the L-estimator is least trimmed square (LTS) [15]. LTS regression is formulated as follows.

$$\arg\min_\beta \sum_{i=1}^{k} |\gamma_{(i)}(\beta)|^2, \; where\ \gamma_i(\beta) = (y_i - f(x_i, \beta)),\ i = 1, \ldots, \text{n}. \quad (3)$$

Here, k is the number of samples that is actually used in the regression and n is the total sample size. $\gamma_i(\beta)$ is the ith residual for a fixed β that is ordered according to the absolute value of the residuals. Bickel and Lehmann considered trimmed expectations to be the only ones which are both robust and whose estimators have guaranteed high efficiency [16]. LTS is known to outperform other least square estimators. Using LTS regression, our L-estimator GMDR uses a studentized trimmed residual as a score for GMDR, where the trimmed studentized residual is defined as follows:

$$\dot{\gamma}_i(\beta) = \begin{cases} \dfrac{y_i - f(x_i,\beta)}{\sqrt{\mathrm{var}(y_i - f(\widehat{x_i,\beta}))}} & \text{if } \left|\dfrac{y_i - f(x_i,\beta)}{\sqrt{\mathrm{var}(y_i - f(\widehat{x_i,\beta}))}}\right| \le T \\ 0 & \text{if } \left|\dfrac{y_i - f(x_i,\beta)}{\sqrt{\mathrm{var}(y_i - f(\widehat{x_i,\beta}))}}\right| > T \end{cases} \qquad (4)$$

where T is the threshold value to determine whether the *i*th sample should be used, and it is possible to adjust T to determine the extent to which outliers would have to be trimmed.

The second robust GMDR uses an M-estimator type score, which is derived as the minima of the sums of the functions of the data. Least-squares estimators and many maximum-likelihood estimators are examples of M-estimators. The M-estimator was first derived for the purpose of introducing it into robust regression by Huber [17]. Tukey subsequently proposed a biweight function, which is a type of M-estimator for robust regression. The M-estimator using Tukey's biweight function is defined as follows.

$$\rho(\gamma_i(\beta)) = \begin{cases} \dfrac{1}{6}[1-(1-\gamma_i(\beta))^2)^3] & \text{if } |\gamma_i(\beta)| \le 1 \\ \dfrac{1}{6} & \text{if } |\gamma_i(\beta)| > 1 \end{cases}, \qquad (5)$$

$$\arg\min_\beta \sum_{i=1}^{k} \rho(\gamma_i(\beta), \text{ where } \gamma_i(\beta) = (y_i - f(x_i,\beta)),\ i = 1,\ldots,\mathrm{n}. \qquad (6)$$

Solving the M-estimator requires differentiation of the ρ function, which usually requires selection of an appropriate shape for the biweight function.

Using the biweight function, our M-estimator GMDR defines the threshold residual score as:

$$\dot{\gamma}_i(\beta) = \begin{cases} \dfrac{y_i - f(x_i,\beta)}{\sqrt{\mathrm{var}(y_i - f(\widehat{x_i,\beta}))}} & \text{if } \left|\dfrac{y_i - f(x_i,\beta)}{\sqrt{\mathrm{var}(y_i - f(\widehat{x_i,\beta}))}}\right| \le T \\ T \cdot sign(y_i - f(x_i,\beta)) & \text{if } \left|\dfrac{y_i - f(x_i,\beta)}{\sqrt{\mathrm{var}(y_i - f(\widehat{x_i,\beta}))}}\right| > T \end{cases} \qquad (7)$$

In this formula, the threshold T is used to shrink all residuals exceeding T. As this function does not require differentiation; the biweight function is modified by simply reducing the weight caused by extreme loss.

RESULTS

Simulation Study

First, we simulated the phenotypes from the Normal distribution. In this case, samples were generated by using the following model.

$$y = envir_1 + \beta_1 SNP_1 + \beta_2 SNP_2 + \beta_{1\times 2} SNP_1 SNP_2 + \varepsilon,$$
$$envir_1 \sim N(0,1), SNP_1, \ldots, SNP_p \sim Bin(2, MAF), \varepsilon \sim N(0,1). \quad (8)$$

The effects of MAFs were checked by considering two MAF values: 0.1 and 0.3. Next, outlying samples were generated by using the following model

$$y \sim N(0,400) \quad (9)$$
$$SNP_1, \ldots, SNP_p \sim Bin(2, MAF).$$

A series of designs were created by using different settings for the purpose of data simulation, and the settings are summarized in Table 1. The first three columns of Table 1 denote the effect of the size of SNP1, SNP2, and the interaction in pure samples, respectively. The fourth column lists the mixed proportion, i.e., the proportion of outlying samples as generated by Model (9). Larger mixed proportions indicate that the particular simulation design generates more outlying observations. The fifth and sixth columns of Table 1 denote the heritability of each simulation setting when MAF = 0.1 and 0.3, respectively. The detailed equations for calculating the heritability are described in the Appendix.

Table 1: Models of simulation with Normal distribution

	SNP1	SNP2	interaction	mixed proportion	Heritability (MAF = 0.1)	Heritability (MAF = 0.3)
Design1	0.2	0.2	0.18	0.067	0.010	0.041
Design2	0.2	0.2	0.18	0.033	0.010	0.041
Design3	0.4	0.4	-0.3	0.067	0.022	0.027
Design4	0.4	0.4	-0.3	0.033	0.022	0.027
Design5	0.25	0.25	0.15	0.067	0.014	0.048
Design6	0.25	0.25	0.15	0.033	0.014	0.048
Design7	0.2	0.2	0.18	0	0.010	0.041

The first to third columns denote the effect sizes of SNP1, SNP2 and interaction in pure samples respectively. The fourth column of Table 1 indicates mixed proportion. The fifth and sixth columns of Table 1 indicates heritability and skewness respectively.

doi:10.1371/journal.pone.0135016.t001

Second, we performed simulation studies for the skewed distribution. The samples were generated from the gamma distribution given in (10). The MAF of SNPs was set to 0.3. The outlying observations were generated by the shifted Gamma distribution in (11). As the Gamma distribution has a long tail to the right, outlying observations were only generated for the right-hand side of the distribution. A series of designs are summarized in Table 2

$$y \sim Gamma(\alpha, \beta) \quad (10)$$

$$y \sim 0.1 + Gamma(\alpha, \beta) \quad (11)$$

$\alpha = (envir_1 + \beta_1 SNP_1 + \beta_2 SNP_2 + \beta_3 SNPS_1 NP_2)^2, \beta = 1/(envir_1 + \beta_1 SNP_1 + \beta_2 SNP_2 + \beta_3 SNPS_1 NP_2),$
$envir_1 \sim N(0,1), SNP_1, \ldots, SNP_p \sim Bin(2, 0.3)$'.

Table 2: Models of simulation with Gamma distribution

	SNP1	SNP2	Interaction	mixed proportion	Heritability	Skewness
Design 1	0.6	0.6	0.3	0.067	0.117	0.770
Design 2	0.6	0.6	0.3	0.033	0.117	0.8
Design 3	1.2	0.4	0.3	0.067	0.197	0.617
Design 4	1.2	0.4	0.3	0.033	0.197	0.587
Design 5	0.8	-0.4	-0.2	0.067	0.074	0.965
Design 6	0.8	-0.4	-0.2	0.033	0.074	0.801
Design 7	0.6	0.6	0.3	0	0.117	0.602

The first column shows MAF of each design. Second to fourth columns denote the effect sizes of SNP1, SNP2 and interaction in pure samples respectively. The fifth and sixth columns of Table indicates heritability and skewness respectively.

doi:10.1371/journal.pone.0135016.t002

For these Normal and Gamma distributions, we considered two sample sizes of 1,000 and 3,000 to evaluate the effect of sample size, and assumed the number of SNPs to be 100 and 1000. Ordinary GMDR, L-estimator-based GMDR, and M-estimator-based GMDR were compared in terms of their power of detecting a true SNP pair. We used Eq (12) for calculating the power, where N refers to the total number of iterations.

$$\frac{1}{N}\sum_{i=1}^{N} I(\text{set } \{SNP_1, SNP_2\} \text{ is selected in } i\text{th iteration}). \quad (12)$$

The number of iterations was 1,000 when the number of SNPs is 100, while it was 100 when the number of SNPs is 1000. The best SNP pair was selected by using the average CVC through a 10-fold cross validation.

We also computed the false detection rate (FDER) [18], which was done by assuming that a targeted non-causal SNP pair, say SNP_1 and SNP_2, exists. To detect this pair, we simulated 500,000 null datasets by permuting the response variables. Because there were at most 100 SNPs in each dataset, the randomly selected rate of each SNP pair is at least 1/499,500. We counted how many times the pair of SNP_1 and SNP_2 was selected as the best model. We found all three methods selected the pair of SNP_1 and SNP_2 at most twice when the number of SNPs is 1000, which shows that the three methods have a very low FDER.

Fig 1 shows the results of the power comparison when SNPs are generated by Bin(2,0.3) and the phenotypes are obtained by Normal distributions.

Designs 1 to 6 show the results when outlying observations exist and Design 7 when outlying observations do not exist. For Design 7, the ordinary GMDR and M-estimator GMDR outperform L-estimator GMDR. As L-estimator GMDR uses a smaller number of samples than other methods, the power of L-estimator GMDR is less than that of the others. For all the other designs, the performance of L-estimator GMDR and M-estimator GMDR exceeds that of ordinary GMDR.

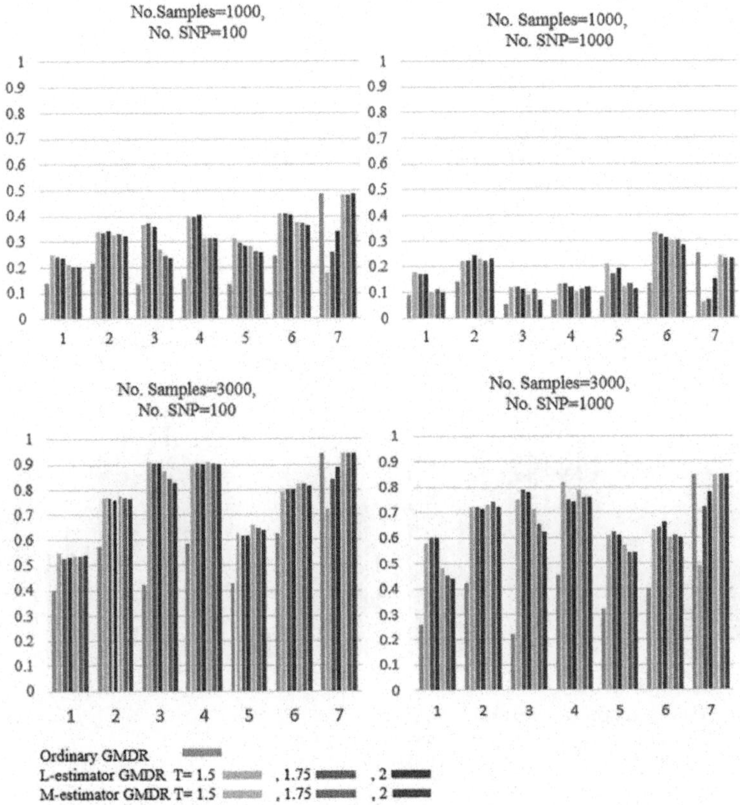

Figure 1: Results of Simulation: Normal distribution with MAF 0.3.

The upper and lower figures show the results when the number of samples is equal to 1,000 and 3,000, respectively. These four figures almost show the same patterns, but the simulation power is different.

doi:10.1371/journal.pone.0135016.g001

Fig 2 shows the results of the power comparison when SNPs are generated by Bin(2,0.1) and the phenotypes are obtained by Normal distributions. The patterns are quite similar to those inFig 1. For Designs 1 to 6, when outlying observations exist, the two robust GMDRs were found to outperform ordinary GMDR. For Design 7, when outlying observations do not exist, the performance of ordinary GMDR and M-estimator GMDR exceeds that of L-estimator GMDR.

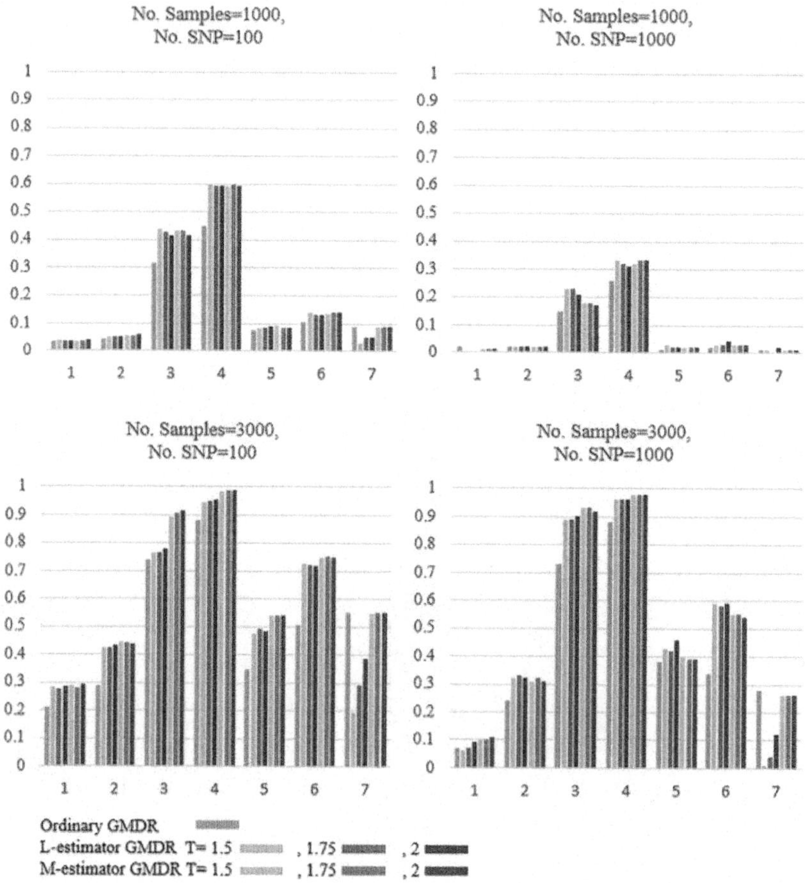

Figure 2: Results of Simulation: Normal distribution with MAF 0.1.

The upper figures show the results when the number of samples is equal to 1000 and the lower figures show the results when the number of samples equals 3,000. These four figures almost show the same patterns, but the simulation power is different.

Fig 3 shows the results of the power comparison when SNPs are generated by Bin(2,0.3) and the phenotypes are obtained by Gamma distributions. The residuals were calculated by generalized linear models under the Gamma distributional assumption. For Designs 1 to 6, when outlying observations exist, the two robust GMDRs were found to outperform ordinary GMDR to a large extent. In particular, L-estimator GMDR performed better than M-estimator GMDR. Note that the performance of L-estimator is highly dependent on the threshold values, whereas this is not the case for M-estimator. For Design 7, when outlying observations do not exist, all GMDRs perform similarly, probably due to the large SNP size effect.

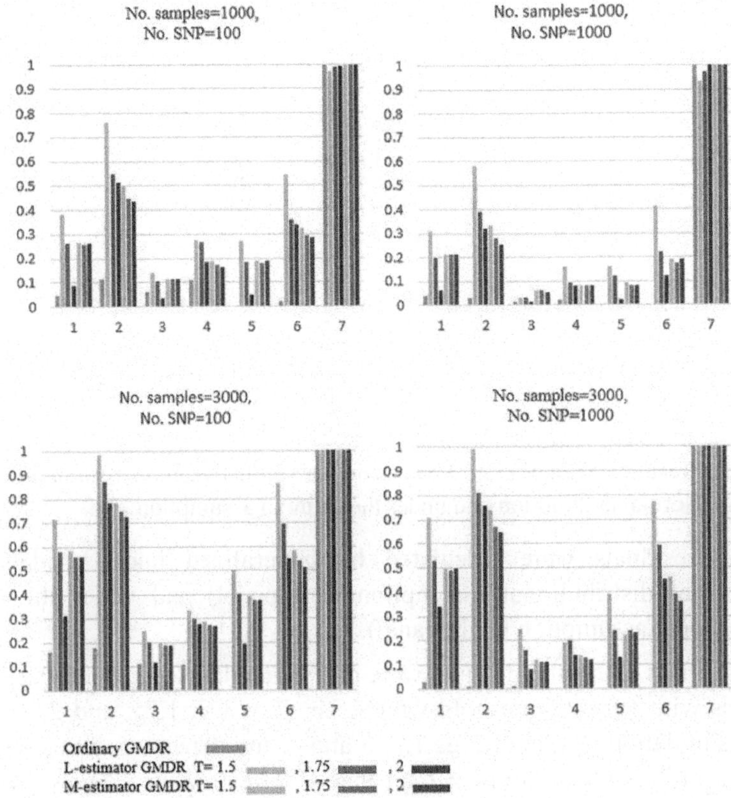

Figure 3: Results of Simulation: Gamma distribution with MAF 0.3.

The upper figures show the results when the number of samples is equal to 1000 and the lower figures show the results when the number of samples equals 3,000. These four figures almost show the same patterns, but the simulation power is different.

doi:10.1371/journal.pone.0135016.g003

Fig 4 shows the results of the power comparison when SNPs are generated by Bin(2,0.3) and the phenotypes obtained by Gamma distributions. The residuals were calculated by generalized linear models under the Normal distributional assumption (left panel) and under the Gamma distributional assumption (right panel). In general, the use of GMDR in combination with the correct distributional assumption is much more powerful than with an incorrect distributional assumption. For Designs 1 to 6, when outlying observations exist, two robust GMDRs perform much better than ordinary GMDR. In particular, L-estimator GMDR outperforms M-estimator GMDR.

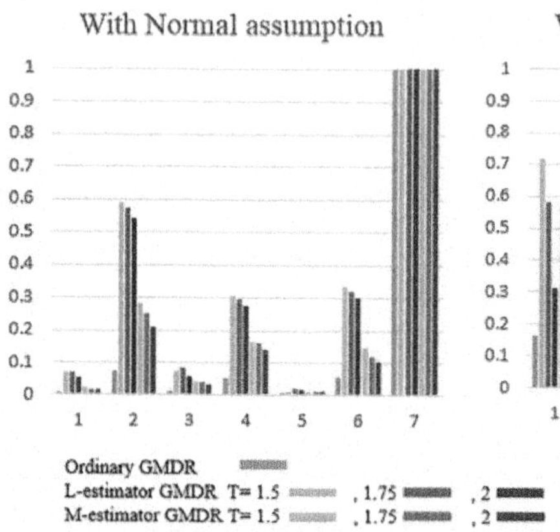

Figure 4: Results of Simulation: Power comparison between the residuals calculated under the Normal assumption and under the Gamma assumption.

The residuals were calculated by generalized linear models under the Normal distributional assumption (left panel) and under the Gamma distributional assumption (right panel).

The effect of threshold values were investigated by using the robust GMDR methods with various threshold values, i.e., T = 1.5, 1.75, and 2. As shown in our simulation results in Figs 1,2, 3 and 4, the dependence of L-estimator GMDR on T is much stronger than that of M-estimator GMDR. When the distribution of phenotypes is non-skewed (e.g., a Normal distribution), the effect of T is not large. On the other hand, when the distribution of phenotypes is skewed (e.g., a Gamma distribution), the effect of T is quite large. Although our simulation setting is limited, T = 1.5 consistently performed better than or equal to other threshold values in the presence of outlying observations. Thus, for practical applications, we recommend using T = 1.5.

Real Data Analysis

Our proposed robust GMDR was applied to KARE data with a HOMA-IR phenotype. Fig 5displays the boxplots of HOMA-IR without and with log-transformation. The skewed distribution of Homeostasis Model Assessment of

Insulin Resistance (HOMA-IR) levels prompted many researchers to perform a log-transform before conducting the regression analysis (shown in the upper panel of Fig 5) [12]. Thus, a log-transform was also performed in our work and its boxplot is given in the lower panel of Fig 5. In spite of performing a log-transformation, the distribution of HOMA-IR levels still remained skewed, with many samples apparently qualifying as outliers.Fig 6 shows the normal quantile-quantile (QQ) plots of the HOMA-IR levels: the left panel represents the case without log-transformation and the right panel the case with log-transformation. The distribution that was obtained after the log-transformation appears to be more symmetrical than without log-transformations. Thus, our GMDR analysis was performed using log-transformed HOMA-IR levels.

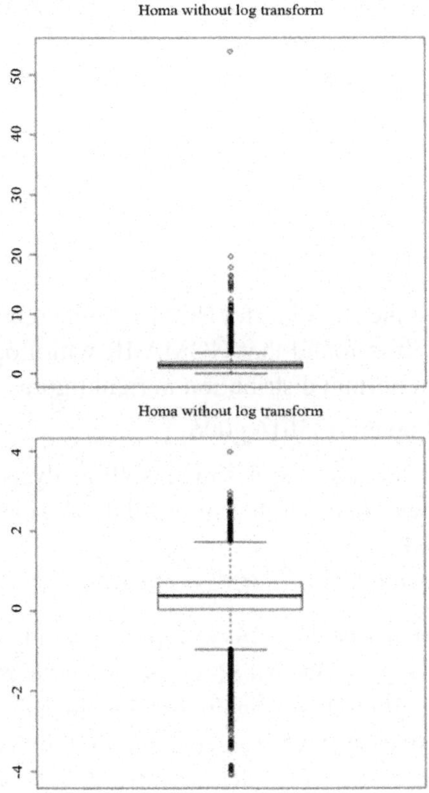

Figure 5: Boxplots of HOMA-IR.

Before log transformation to HOMA-IR, HOMA-IR has much skewed distribution. In contrasts, after log transformation, HOMA-IR has almost symmetric distribution.

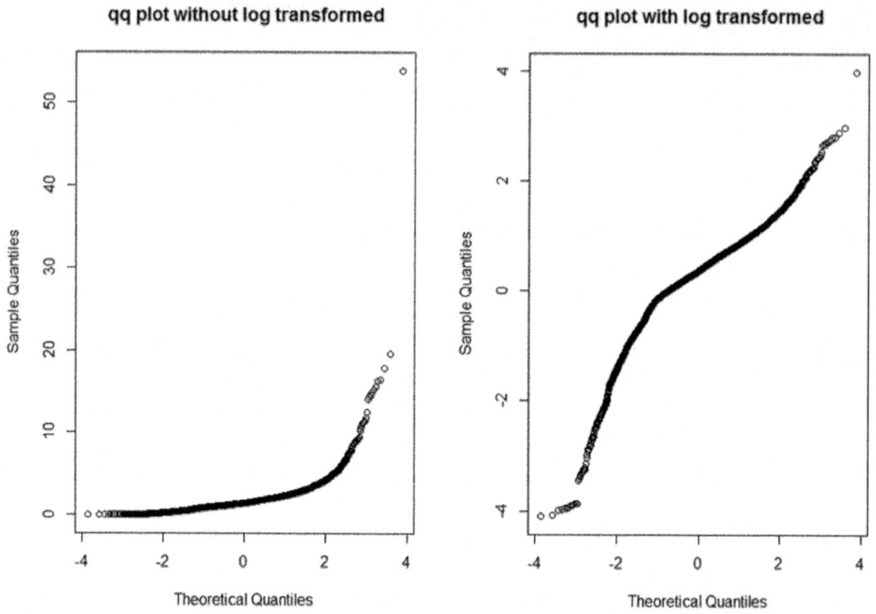

Figure 6: QQ-plot for HOMA-IR data.

This figure shows the QQ-plots for HOMA-IR data comparing with standard normal distribution. It is found that HOMA-IR with Log-transformation acts more like samples of normal distribution in right figure.

doi:10.1371/journal.pone.0135016.g006

Before the GMDR analysis, a single SNP analysis was first performed to reduce the computational burden on GMDR. A single SNP analysis was performed for all SNPs.

$$\log(HOMA - IR_i) = \alpha_0 + \alpha_1 SEX_i + \alpha_2 AGE_i + \gamma_3 AREA_i + \alpha_4 BMI_i + \alpha_{5j} SNP_{ij} + \varepsilon_i, \quad (13)$$

Noteworthy is that i (= 1,…,8577) represents individuals; whereas, j (= 1,…,327,872) represents SNPs. All SNPs are ranked in the order of ascending p-values. SNPs are filtered out again by the linkage disequilibrium (LD) coefficient r^2. Only one SNP was selected among the SNP pairs with $r^2 \geq 0.8$. Thus, the SNPs can be treated independently. After filtering SNPs via LD, we set the total number of selected SNPs as n / log(n) ≈ 585 by the sure independence screening (SIS) criterion for the purpose of performing the GMDR analysis [19].

The same model that was used for selecting SNPs was used for the GMDR analysis. Sex, age, area, and BMI were used as environmental covariates. The regression models for GMDR are given as follows:

$$\log(HOMA - IR_i) = \beta_0 + \beta_1 SEX_i + \beta_2 AGE_i + \beta_3 AREA_i + \beta_4 BMI_i + \varepsilon_i, \tag{14}$$

The residual scores were then calculated based on this model. The threshold value of T was assumed as 1.5, 1.75, or 2 in the L-estimator and M-estimator. Table 3 shows the results obtained with the three GMDR methods. Each method provided a different SNP combination. These three pairs of SNPs have not previously been reported in the literature in relation to HOMA-IR. When we performed the analysis using the real data, the computational times required by the M-estimator GMDR were similar to those required for ordinary GMDR analysis. However, the L-estimator required slightly more computational time, because it contains an additional step to compute residuals.

Table 3: GMDR Analyses for log-transformed HOMA-IR in KARE data

Method	Best Combination	CVC	Average Test BA
Ordinary GMDR	rs576563 rs2920792	4	0.512535
L-estimator GMDR	rs4915657 rs7500315	9	0.518679
M-estimator GMDR	rs11125090 rs9353581	9	0.522633

Table 3 shows the results of three GMDR methods. Each GMDR method provided different SNP combination.

doi:10.1371/journal.pone.0135016.t003

We performed the network analysis with the top 100 SNP pairs with the highest BAs. Each node represents an SNP and each edge connecting two nodes represents an SNP pair. We defined the hub genes as genes with at least four edges. Table 4 summarizes the list of hub genes for log-transformed HOMA-IR identified by robust GMDRs and ordinary GMDR. The KCNH1 gene, which has the largest number of edges, was reported to have the function of insulin secretion [20]. Both M-estimator GMDR and L-estimator GMDR successfully identified the KCNH1 gene as a hub gene regardless of the threshold values. In addition, M-estimator GMDR with a threshold of 1.5 and L-estimator GMDR with all three threshold values identified the RYR2 gene as a hub gene. This gene is reported to be related with diabetes [21]. The PBX1 was found by M-estimator GMDR with a threshold of 2 and an ordinary GMDR. It was also reported that the PBX1 gene is related to type 2 diabetes [22]. In summary, the network analysis that was performed with the top 100 SNP pairs showed supporting evidence that the robust GMDR methods identified a larger number of hub genes that were reported in the literature than the ordinary GMDR method.

Table 4: Hub-genes for log-transformed HOMA-IR in KARE data with several methods

Method	threshold = 1.5	No. nodes	threshold = 1.75	No. nodes	threshold = 2	No. nodes
L-estimator	KCNH1*	4	KCNH1*	9	KCNH1*	13
	RYR2*	7	RYR2*	5	RYR2*	9
	LAMB3	5	LAMB3	4	HPCAL4	4
	C1orf168	3	SMYD3	5	SMYD3	4
	LMX1A*	3				
M-estimator	KCNH1*	43	KCNH1*	52	KCNH1*	52
	RYR2*	6			PBX1*	4
	KIF26B	4				
	TXNRD2	3				
Ordinary GMDR	PBX1*	31				
	TMEM51	5				
	VSIG2	4				
	KIF26B	3				

Table 4 shows the hub-genes for log-transformed HOMA-IR with different threshold. KCNH1 and RYR2 are commonly detected in L-estimator and M-estimator GMDR. However, Ordinary GMDR could not detect those genes.

doi:10.1371/journal.pone.0135016.t004

For a selected SNP pair, we investigated the high /low risk prediction results of individual cells.Fig 7 shows the boxplots of residuals that were calculated by ordinary GMDR, M-estimator GMDR, and L-estimator GMDR, respectively. Each cell displays three boxplots of GMDR methods. The red boxplot represents the high-risk group with the sum of the residuals larger than 0, whereas the blue boxplot represents the low-risk group with the sum of the residuals smaller than 0. The high/low classification pattern of L-estimator GMDR is the same as that of M-estimator GMDR. However, the pattern of ordinary GMDR differs from that of robust GMDRs. For example, in the first cell, when the numbers of minor alleles of rs4915657 and rs7500316 are (0,0), respectively, ordinary GMDR classified this cell as a low-risk group, whereas the two robust GMDRs classified it as a high-risk group. Note that the sum of residuals for ordinary GMDR becomes negative, which is due to the negative outlying observations in the tail part. Similarly, when the numbers of minor alleles in rs4915657 and rs7500316 are (0,1), (1,1), and (2,0), the cells showed the same tendency. The different results between the robust GMDRs and ordinary GMDR could be explained by the negative outlying residuals in the tail part. Because ordinary GMDR uses the original outlying residuals as they are, the sum of the residuals is highly affected by these outlying residuals. Thus, these outlying observations can cause errors in high/low classification. This investigation demonstrates the usefulness of robust GMDRs.

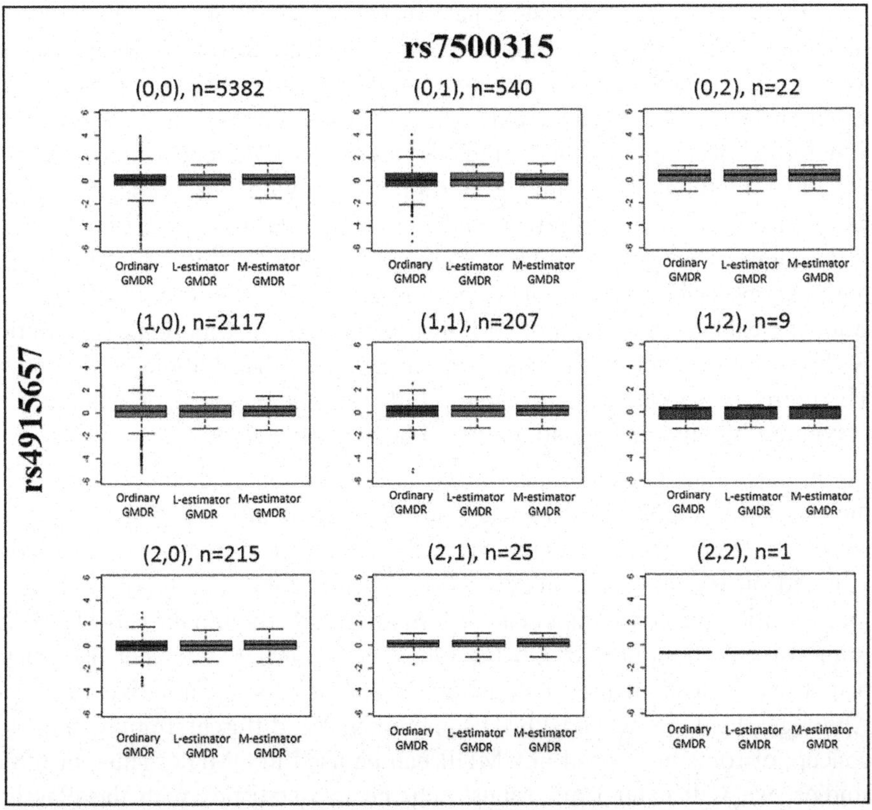

Figure 7: Boxplots of residuals calculated by ordinary GMDR, M-estimator GMDR and L-estimator GMDR.

Each cell displays three boxplots of the GMDR methods divided by the combination of rs4915657 and rs7500316. The red and blue boxplots represents the high-risk and low-risk groups with the sum of the residuals larger than and smaller than 0, respectively.

Discussion

When searching for hidden heritability, it is important to use an efficient method to establish the effects of GG interactions. To this effect, GMDR is a powerful method. However, using a simple example, it is demonstrated that the power of the GMDR method may decline in the presence of outlying observations. This problem was addressed by proposing the robust GMDR: L-estimator GMDR and M-estimator GMDR.

Simulation studies indicated that outlying observations caused the power of ordinary GMDR to decline critically. In contrast, the L-estimator and M-estimator GMDR methods were found to perform reasonably well even when outlying observations existed. When the samples are generated by assuming a Normal distribution, the L-estimator GMDR appears to outperform the other methods when the proportion of outlying observations is large. Otherwise, M-estimator GMDR and L-estimator GMDR perform similarly. If the samples are generated by assuming a Gamma distribution and there are outlying observations, L-estimator performs best. The results of the simulation study indicate that the power of GMDR with a correct distributional assumption is shown to be much larger than with an incorrect distributional assumption. Therefore, if we know the distribution of y_i, a parametric approach, such as maximum likelihood estimation, would be recommended.

In our robust GMDR analysis, we used BA as an evaluation measure for the prediction ability. BA uses less information than continuous evaluation measures when used for quantitative traits. However, as the GMDR approach is based on transforming quantitative traits into binary high/low responses, incorporating the continuous scale information directly into the robust GMDR framework is complicated. On the other hand, QMDR does not use binary responses; instead, it uses t-statistics to measure for quantitative traits [7]. Although the GMDR and QMDR methods use different frameworks, the concept of robustness used in GMDR can be used in QMDR. Thus, in future studies, we will extend our robust scheme to t-statistics with the classical statistical method.

APPENDIX

Without loss of generality, we could consider all kinds of 2nd order genetic models with a normal distribution by using the following formula (15).

$$y = \sum_{j=0}^{2}\sum_{i=0}^{2} \beta_{ij} I(SNP_1 = i, SNP_2 = j) + \varepsilon \tag{15}$$

Then, we could consider the variance of y to be represented by the following equations.

$$\begin{aligned} \mathrm{var}(y) &= \sigma^2 = \sigma^2_{envir} + \sigma^2_{genetic} = E(\mathrm{var}(y|SNP_1, SNP_2)) + \mathrm{var}(E(y|SNP_1, SNP_2)) \\ &= E(\sigma^2_{envir}) + \mathrm{var}\Big(\sum_{j=0}^{2}\sum_{i=0}^{2} \beta_{ij} I(SNP_1 = i, SNP_2 = j)\Big) \end{aligned} \tag{16}$$

With this equation, we could calculate the narrow sense heritability h^2 as:

$$h^2 = \frac{\sigma^2_{genetic}}{\sigma^2} = \frac{\text{var}(\sum_{j=0}^{2}\sum_{i=0}^{2}\beta_{ij}I(SNP_1 = i, SNP_2 = j))}{\sigma^2} \tag{17}$$

AUTHOR CONTRIBUTIONS

Conceived and designed the experiments: TP YK. Performed the experiments: YK. Analyzed the data: YK. Contributed reagents/materials/analysis tools: YK TP. Wrote the paper: YK TP.

REFERENCES

1. Oh S, Lee J, Kwon MS, Weir B, Ha K, Park T (2012) A novel method to identify high order gene-gene interactions in genome-wide association studies: Gene-based MDR,BMC Bioinformatics, 13(Suppl 9):S5 doi: 10.1186/1471-2105-13-S9-S5. pmid:22901090
2. Gilbert-Diamond D, Moore JH (2011) Analysis of Gene-Gene Interactions. Current Protocols in Human Genetics. 70:1.14.1–1.14.12. doi: 10.1002/0471142905.hg0114s70
3. Freitas AA (2001) Understanding the crucial role of attribute interaction in data mining. Artif. Intel.Rev. 16:177–99.
4. Ritchie MD, Hahn LW, Roodi N, Bailey LR, Dupont WD, Parl FF et al. (2001) Multifactor-dimensionality reduction reveals high-order, Am J Hum Genet. Jul;69(1):138–47. Epub 2001 Jun 11 interactions among estrogen-metabolism genes in sporadic breast cancer. pmid:11404819 doi: 10.1086/321276
5. Lou XY, Chen GB, Yan L, Ma JZ, Zhu J, Elston RC et al. (2007) A Generalized Combinatorial Approach for Detecting Gene-by-Gene and Gene-by-Environment Interactionswith Application to Nicotine Dependence,The American Journal of Human Genetics, Volume 80, Issue 6, 1125–37, 1 June 2007 pmid:17503330 doi: 10.1086/518312
6. Cattaert T, Calle ML, Dudek SM, John JM, Lishoutet FV, Urrea V et al. (2011) Model-based multifactor dimensionality reduction for detecting epistasis in case-control data in the presence of noise.Ann Hum Genet. 2011 Jan;75(1):78–89. doi: 10.1111/j.1469-1809.2010.00604.x. pmid:21158747

7. Gui J, Moore JH, Williams SM, Andrews P, Hillege HL, van der Harst P et al. (2013) A Simple and Computationally Efficient Approach to Multifactor Dimensionality Reduction Analysis of Gene-Gene Interactions for Quantitative Traits. PLoS ONE 8(6): e66545. doi: 10.1371/journal.pone.0066545. pmid:23805232
8. Bennett SN, Caporaso N, Fitzpatrick AL, Agrawal A, Barnes K, Boyd HA et al. (2011) Phenotype Harmonization and Cross-Study Collaboration in GWAS Consortia: The GENEVA Experience, Genetic Epidemiology 35: 159–73. doi: 10.1002/gepi.20564. pmid:21284036
9. Most PJ, Vaez A, Prins BP, Munoz ML, Snieder H, Alizadeh BZ et al. (2014) QCGWAS: A flexible R package for automated quality control of genome-wide association results, Bioinformatics, 30 (8): 1185–86. doi: 10.1093/bioinformatics/btt745
10. Yang J, Loos RJ, Powell JE, Medland SE, Speliotes EK, Chasman DI et al. (2012) FTO genotype is associated with phenotypic variability of body mass index, Nature, 267–73.
11. Cho YS, Go MJ, Kim YJ, Heo JY, Oh JH, Ban HJ et al. (2009) A large-scale genome-wide association study of Asian populations uncovers genetic factors influencing eight quantitative traits. Nature Genetics. May;41(5):527–34 doi: 10.1038/ng.357. pmid:19396169
12. Gayoso-Diz P, Otero-Gonzalez A, Rodriguez-Alvarez MX, Gude F, Cadarso-Suarezet C, García F et al. (2011) Insulin resistance index (HOMA-IR) levels in a general adult population: curves percentile by gender and age. The EPIRCE study. Diabetes Res Clin Pract. Oct;94(1):146–55 doi: 10.1016/j.diabres.2011.07.015. pmid:21824674
13. Cho S, Kim K, Kim YJ, Lee JK, Cho YS, Lee JY et al. (2010) Joint Identification of Multiple Genetic Variants via Elastic-Net Variable Selection in a Genome-Wide Association Analysis. Ann Hum Genet. Sep 1;74(5):416–28 doi: 10.1111/j.1469-1809.2010.00597.x. pmid:20642809
14. Velez DR, White BC, Motsinger AA, Bush WS, Ritchie MD, Williams SM et al. (2007) A balanced accuracy function for epistasis modeling in imbalanced datasets using multifactor dimensionality reduction. Genet. Epidemiol., 31: 306–15. pmid:17323372 doi: 10.1002/gepi.20211
15. Koenker R, Zhao Q (1994) L-estimatton for linear heteroscedastic models, Journal of Nonparametric Statistics, 3:3–4, 223–35 doi: 10.1080/10485259408832584
16. Bickel PJ, Lehmann EL (1975) Descriptive Statistics for Nonparametric Models 11: Location, Ann. Statist., 3, 1045–69 doi: 10.1214/aos/1176343240

17. Owen M (2010) Tukey's Biweight Correlation and the Breakdown. Master's thesis. Pomona College.
18. Fan J, Lv J (2008) Sure independence screening for ultrahigh dimensional feature space, JRSSB, 70(5), 849–911. doi: 10.1111/j.1467-9868.2008.00674.x
19. Namkung J, Kim K, Yi S, Chung W, Kwon MS, Park T (2009) New evaluation measures for multifactor dimensionality reduction classifiers in gene–gene interaction analysis, Bioinformatics, 25 (3), 338–45. doi: 10.1093/bioinformatics/btn629. pmid:19164302
20. Yan L, Figueroa DJ, Austin CP, Liu Y, Bugianesi RM, Slaughter RS et al. (2004) Expression of voltage-gated potassium channels in human and rhesus pancreatic islets, Diabetes 53, 597–607. pmid:14988243 doi: 10.2337/diabetes.53.3.597
21. Yaras N, Ugur M, Ozdemir S, Gurdal H, Purali N, Lacampagne A et al. (2005) Effects of Diabetes on Ryanodine Receptor Ca Release Channel (RyR2) and Ca2+ Homeostasis in Rat Heart, Diabetes, 54(11) 3082–88 pmid:16249429 doi: 10.2337/diabetes.54.11.3082
22. Wang H, Chu W, Wang X, Zhang Z, Elbein SC (2005) Evaluation of sequence variants in the pre-B cell leukemia transcription factor 1 gene: a positional and functional candidate for type 2 diabetes and impaired insulin secretion, Mol Genet Metab. 86(3):384–91. Epub 2005 Sep 2 pmid:16140554 doi: 10.1016/j.ymgme.2005.07.008

Chapter 5

DYNAMIC PATHWAY ANALYSIS OF GENES ASSOCIATED WITH BLOOD PRESSURE USING WHOLE GENOME SEQUENCE DATA

Pingzhao Hu[1,4] and Andrew D Paterson[1,2,3]

[1] The Centre for Applied Genomics, The Hospital for Sick Children, 686 Bay Street, Toronto, ON, M5G 0A4, Canada

[2] Program in Genetics and Genome Biology, The Hospital for Sick Children, 686 Bay Street, Toronto, ON, M5G 0A4, Canada

[3] Dalla Lana School of Public Health, University of Toronto, Health Sciences Building, 155 College St, Toronto, ON, M5T 3M7, Canada

[4] Department of Biochemistry and Medical Genetics and George and Fay Yee Centre for Healthcare Innovation, University of Manitoba,745 Bannatyne Avenue, Winnipeg, MB, R3E 0W3, Canada

ABSTRACT

Groups of genes assigned to a pathway, also called a module, have similar functions. Finding such modules, and the topology of the changes of the modules over time, is a fundamental problem in understanding the mechanisms of complex diseases. Here we investigated an approach that categorized variants into rare or common and used a hierarchical model to jointly estimate the group effects of the variants in a pathway for identifying enriched pathways over time using whole genome sequencing data and blood pressure data. Our results suggest that the method can identify potentially biologically meaningful genes in modules associated with blood pressure over time.

Background

It has long been recognized that genetic analysis of longitudinal phenotypic data is important for understanding the genetic architecture and biological variations of complex diseases. The analysis can help identify the stage of disease development at which specific genetic variants play a role. However, the statistical methods to analyze longitudinal genetic data are limited. A commonly used approach is to analyze the longitudinal genetic traits by averaging multiple response measurements obtained at different time points from the same individual. This approach may miss a lot of useful information related to the variability of repeated genetic traits, although it is simple and computationally less expensive. Linear mixed models have also been used for repeated measures data [1].

Recently, there has been a shift to testing rare variants, mostly using next-generation sequence technologies, for association with complex diseases. We explored dynamic pathway-based analysis of genes associated with blood pressure over time using whole genome sequencing data. We first performed gene-based association analysis at each of the 3 time points by stratifying the variants into rare and common. Then we performed pathway enrichment analysis separately at each time point. Finally, we built pathway crosstalk network maps using the enriched pathways to identify potential subnetworks associated with blood pressure over time.

METHODS

Data Description

For genotype data, we analyzed sequencing data of the 142 unrelated individuals on chromosome 3, which includes 1,215,120 variants. For phenotype data, we analyzed the simulated phenotypes of replicate 1. We analyzed 2 quantitative traits: systolic blood pressure (SBP) and Q1. SBP was measured at 3 time points (T1, T2, and T3), and was close to normally distributed (data not shown) after treatment effect adjustment (see below). There are 31 functional loci (genes) on chromosome 3 that influence the simulated SBP. Q1 was simulated as a normally distributed phenotype but not influenced by any of the genotyped single-nucleotide polymorphisms. It also has no correlation with SBP measured at T1, T2, and T3. The Pearson correlation of SBP at the 3 time points with Q1 based on the 142 unrelated individuals is −0.09 (p value = 0.27), −0.02 (p value = 0.78), and −0.006 (p value = 0.94), respectively. Q1 was generated primarily to facilitate assessment of type 1 error.

Adjust Treatment Effects

It has been shown that the association between measured blood pressure and underlying genotype is potentially confounded by antihypertensive treatment [2]. Following Cui et al [3], we adjusted SBP of subjects receiving antihypertensive medications by adding a constant value of 10 mm Hg at the 3 study exams (n = 22, 51, and 73; 15.5%, 35.9%, and 51.4%, respectively). Such a strategy had higher power than the alternatives [2].

Analyze Common and Rare Genetic Variants Using Hierarchical Models

We applied an extended hierarchical generalized linear model [4] to simultaneously analyze rare and common variants at the gene level. The model can be summarized as follows: Assume that the observed values of a given quantitative trait (SBP or Q1) are denoted as $y = (y_1, \cdots, y_n)$ and the predictor variables, that are variants, can be categorized into 2 groups: rare (minor allele frequency <1%) and common variants (minor allele frequency ≥1%). The number of variants in the rare and common groups are J_1 and J_2, respectively. The extended hierarchical generalized linear model to fit the rare and common variants in a given gene can be expressed as a multiplicative form for the linear predictor η of individual *i*:

$$\eta_i = \beta_0 + \sum_{k=1}^{2} g_k \sum_{j \in G_k}^{J_k} \beta_j Z_{ij}$$

where z_{ij} is the predictor of main-effect for individual *i* at genetic variant *j* in group G_k, equaling to the number of minor alleles for an additive coding and $Z_i = (z_{i1}, \ldots, z_{iJ})$, where $J = \sum_{k=1}^{2} J_k$ is the total number of variants. g_k represents the group effect for J_k variants in group G_k. β is a vector of all the coefficients and the intercept β.

The mean of y is related to the linear predictor η via a link function *h*:

$$E(y_i|Z_i) = h^{-1}(Z_i\beta)$$

The data distribution is expressed as $p(y|Z\beta, \theta) = \prod_{i=1}^{n} p(y_i|Z_i\beta, \theta)$, where θ is a dispersion parameter, and the distribution $p(y_i|Z_i\beta, \theta)$ takes normal distribution. Because there are many highly correlated variants in a given gene in next-generation sequencing studies, a hierarchical framework is constructed for priors of the distributions of coefficients (*g* and *β*) in the model. The method was implemented in R package BhGLM.

We assigned the genetic variants to a gene if they were in the gene or within 10 kilobases (kb) of either side of the gene. We performed 2 analyses to evaluate the association between genotype and SBP at each study exam separately. First, we divided the variants within a gene into rare ($k = 1$) and common variants ($k = 2$). Separately we analyzed all the genetic variants in a gene, irrespective of allele frequency. Our main objective was to estimate gene effects g_k and to test the hypothesis $g_k = 0$, $k = 1$ (rare variants) and $k = 2$ (common variants) for the first analysis and $k = 1$ (rare and common variants) for the second analysis. We corrected for multiple testing using the Benjamini and Hochberg method [5].

Dynamic Pathway Analysis

We mapped the approximately 1200 genes on chromosome 3 to the c2 curated pathways (version 3) from the Broad Institute, which includes 2934 gene sets collected from 186 Kyoto Encyclopedia of Genes and Genomes, 430 Reactome, 217 BioCarta pathways, 880 canonical pathways, 825 biological process, and 396 molecular function gene ontology terms. We kept only the pathways with at least 5 genes in our data set, which left 531 pathways for analysis.

There are different ways to test for genes associated with an excess of SBP in the same pathway. We used the "gene set enrichment test" implemented in the limma R package [6]. The approach uses the Wilcoxon signed rank test to compute a*p* value to test the hypothesis that a given gene set tends to be more highly ranked than would be expected by chance. The ranking is based on a *t*-like test statistic, and here we used the *z* statistics from the hierarchical model described in above Section (Analyze common and rare genetic variants using hierarchical models). The test is essentially a streamlined version of the gene set enrichment analysis approach introduced by Subramanian et al [7].

We performed dynamic pathway crosstalk analysis between each pair of time points using the enriched pathways with a nominal p value of <0.05. Two pathways were considered to crosstalk if they shared at least 1 functional locus (gene). This ensures that each of pathway and its crosstalk has biological meaningfulness. We built pathway crosstalk subnetworks using Cytoscape

RESULTS

Given a false discovery rate (FDR) of 0.05 at the gene-level analysis, we identified 116, 57, 2, and 0 significant genes for SBP measured at T1, T2, T3, and Q1, respectively, using rare variants. However, there were no significant genes for SBP measured at the 3 time points and Q1 using common variants. Of those significant genes from the rare variant analysis, 4, 1, 0, and 0 were true positives (Table 1) for SBP measured at T1, T2, T3, and Q1, respectively. We observed that these 4 genes had significant positive associations with SBP (Table 1). For the gene-level analysis irrespective of allele frequency we identified many significant genes (468, 415, 306, and 214 for SBP measured at T1, T2, T3, and Q1, respectively) (Table 2). However, the vast majority of them were false positives (see false-positive rate analysis below), implying that, irrespective of allele frequency, the analysis strategy had a grossly inflated type I error, possibly as a result of linkage disequilibrium between variants. We calculated the false-positive rate and false-negative rare for T1, T2, T3, and Q1 using the 2 analysis approaches. We defined the positive as those genes that had an adjusted p value smaller than 0.05, and the negative as those genes that had an adjusted *p* value larger than or equal to 0.05. As shown in Table 2, irrespective of allele frequency, the analysis approach had many false positives compared with the approach that stratified by allele frequency. The genes based on common variants used in the first analysis had no power to detect the genetic association between genotypes and SBP.

Table 1: Significant association of causal genes with rare variants with SBP at T1 and T2 based on chromosome 3 gene-based tests

Time period and gene		Rare variants				Common variants			
		Estimate	SE	*z* Value	Adjusted *p* value	Estimate	SE	*z* Value	Adjusted *p* value
T1	*ABTB1*	0.71	0.24	3.01	0.0321	−0.15	0.21	−0.70	0.99
	SCAP	0.87	0.22	4.00	0.0058	−0.02	0.08	−0.27	0.99
	PROK2	3.30	0.90	3.66	0.011	−0.19	0.21	−0.93	0.99
	MUC13	0.65	0.22	2.96	0.035	0.00	0.08	0.02	1.0
T2	*SCAP*	1.06	0.27	3.89	0.0076	−0.08	0.08	−1.00	0.95

Note: There were no significant associations for causal genes including only either rare or common variants at T3 (or for Q1).

Table 2: False-positive rate (FPR) and false-negative rate (FNR) of gene-based analyses

Time period or trait	Gene-based stratified by allele frequency				Gene based	
	Rare variants		Common variants		Rare and common variants	
	FPR (%)	FNR (%)	FPR (%)	FNR (%)	FPR (%)	FNR (%)
T1	9.6	86.7	0.0	100.0	39.2	53.3
T2	4.8	96.7	0.0	100.0	34.9	63.3
T3	0.2	100.0	0.0	100.0	25.6	66.7
Q1	0.0	100.0	0.0	100.0	17.8	76.7

To further evaluate the whole-spectrum of power of the approaches to identify causal genes, we drew the receiver operating characteristic curves for each type of strategy at each time point and estimated its area under the curve. We found that each analysis strategy has different power to identify causal genes at different time points. Overall, the analysis based on rare variants had the largest power at T3, which also had larger power to detect disease-causing genes than common variants at T2.

Using the *z* statistic obtained from modeling rare variants in a gene, we did not find significant pathways associated with SBP at FDR of 0.05 level based on gene set enrichment analysis. However, given a nominal *p* value cutoff of 0.05, we identified the same 3 enriched pathways for SBP measured at 3 time points but not for Q1 (Table 3). Each of these 3 pathways included 1 «functional» gene (*FLNB*), which had 286 rare variants with 1 functional variant (chr3: 58109162, explained 0.00273 of the variance for SBP).

Table 3: Enriched pathways found in T1, T2, and T3

Pathway names	No. of genes	No. of genes on chr. 3	No. of function-al loci	*p* Value of T1	*p* Value of T2	*p* Value of T3
Actin filament-based process	114	10	1	0.021	0.016	0.018
Actin cytoskel-eton organization and biogenesis	104	10	1	0.021	0.016	0.018
Cytoskeleton organization and biogenesis	205	12	1	0.030	0.013	0.022

To identify pathway crosstalk, we built 2 pathway subnetworks (Figure 1) for the pathways with nominal *p* value smaller than 0.05. Two pathways crosstalk if at 2 time points they included at least 1 common true gene. As shown in Figure 1, we found that there was a subnetwork formed by the «actin filament based process,» the «actin cytoskeleton organization and biogenesis,» and the «cytoskeleton organization and biogenesis and organelle organization and biogenesis» pathways that were consistently enriched across adjacent time periods (T1 → T2 and T2 → T3).

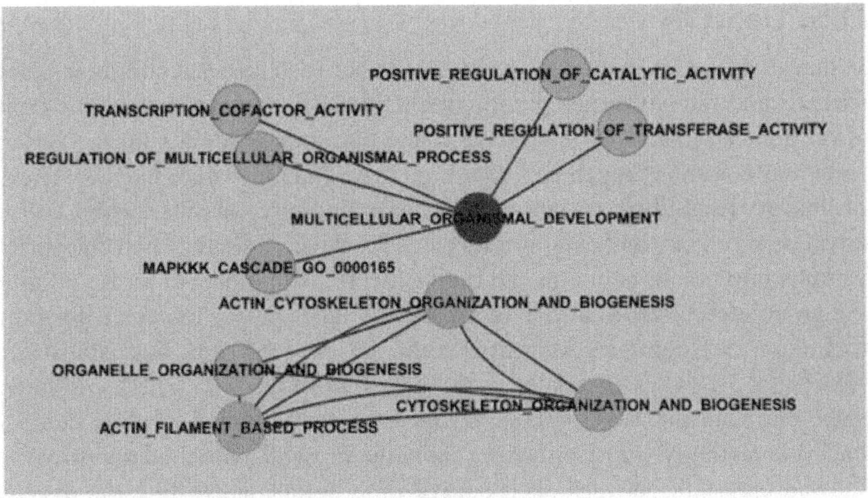

(A)

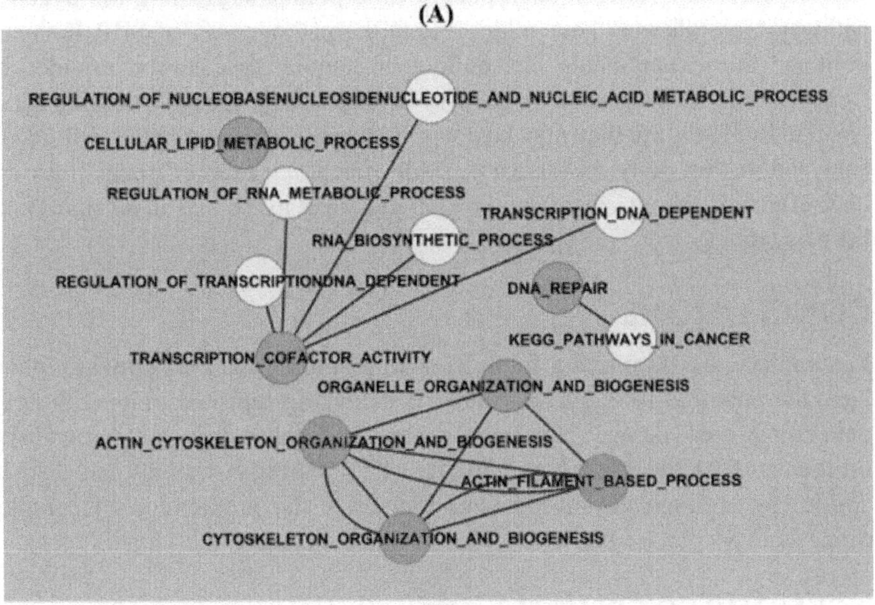

(B)

Figure 1: Dynamic pathway crosstalk: (A) T1 and T2; (B) T2 and T3. Pathways with blue, red and yellow are enriched at T1, T2, and T3, respectively. Pathways with green are enriched at both T1 and T2 (A) and at both T2 and T3 (B). A single line between 2 pathways indicates that each of the 2 pathways is enriched in only 1 of the 2 time points; a double-line between 2 pathways indicates that both pathways are enriched in both T1 and T2 (A) and in both T2 and T3 (B).

DISCUSSION

In this study, we evaluated the associations between rare and common genetic variants and the simulated quantitative trait (SBP) measured at 3 time points at the gene and pathway levels. We found that joint modeling all the variants (rare and common) together had a high type I error, which may be a result of linkage disequilibrium between common and rare variants, or the average effect between rare and common variants. However, a strategy that categorized variants into rare or common and used a hierarchical model to jointly estimate the group effects showed rare variants had higher power to detect functional loci than did common variants. Although we did not find statistically significant pathways associated with SBP (FDR of the 0.05 level), we showed some enriched pathways shared across time at a nominal p value cutoff of 0.05. Interestingly, we also found a subnetwork with 3 enriched pathways that showed crosstalk between each pair of time points, suggesting the dynamic pathway crosstalk may have a key role in the pathogenesis of SBP. It should be noted the «›functional» loci defined in simulation answers provided by Genetic Analysis Workshop 18 (GAW18) organizers were polymorphic based on all individuals, but they may be not polymorphic in the unrelated individuals analyzed in this study. In this case, some functional loci (or genes) may not have effects in the unrelated data, which may lead to the bias in calculation of false negatives.

CONCLUSIONS

In summary, we proposed a framework to identify dynamic pathways which have the potential in regulating SBP via analyzing repeated traits with next-generating sequencing. This can generate insights into the progressive mechanisms of the underlying disease. This analysis strategy can also be applied to examine the mechanisms that drive the progression of complex diseases.

ACKNOWLEDGEMENTS

This work was supported by a grant from Genome Canada through the Ontario Genomics Institute. ADP holds a Canada Research Chair in Genetics of Complex Diseases.

The GAW18 whole genome sequence data were provided by the T2D-GENES Consortium, which is supported by NIH grants U01 DK085524, U01 DK085584, U01 DK085501, U01 DK085526, and U01 DK085545. The other genetic and phenotypic data for GAW18 were provided by the San Antonio Family Heart Study and San Antonio Family Diabetes/Gallbladder Study, which are supported by NIH grants P01 HL045222, R01 DK047482, and R01 DK053889. The Genetic Analysis Workshop is supported by NIH grant R01 GM031575.

This article has been published as part of *BMC Proceedings* Volume 8 Supplement 1, 2014: Genetic Analysis Workshop 18. The full contents of the supplement are available online at Publication charges for this supplement were funded by the Texas Biomedical Research Institute.

AUTHORS' CONTRIBUTIONS

PH designed the study, performed the data analysis, and drafted the manuscript. ADP participated in designing the study. ADP supervised the study. All authors helped revise the manuscript. All authors read and approved the final manuscript.

REFERENCES

1. Paterson AD, Waggott D, Boright AP, Hosseini SM, Shen E, Sylvestre MP, Wong I, Bharaj B, Cleary PA, Lachin JM, et al.: A genome-wide association study identifies a novel major locus for glycemic control in type 1 diabetes, as measured by both A_{1C} and glucose.Diabetes 2010, 59:539–549.
2. Tobin MD, Sheehan NA, Scurrah KJ, Burton PR: Adjusting for treatment effects in studies of quantitative traits: antihypertensive therapy and systolic blood pressure. Stat Med 2005, 24:2911–2935.
3. Cui JS, Hopper JL, Harrap SB: Antihypertensive treatments obscure familial contributions to blood pressure variation. Hypertension2003, 41:207–210.
4. Yi N, Liu N, Zhi D, Li J: Hierarchical generalized linear models for multiple groups of rare and common variants: jointly estimating group and individual-variant effects. PLoS Genet 2011, 7:e1002382.
5. Benjamini Y, Hochberg Y: Controlling the false discovery rate: a practical and powerful approach to multiple testing. J R Stat Soc Series B Stat Methodol 1995, 85:289–300.
6. Smyth GK: Linear models and empirical Bayes methods for assessing differential expression in microarray experiments. Stat Appl Genet Mol Biol 2004., 3:: Article 3
7. Subramanian A, Tamayo P, Mootha VK, Mukherjee S, Ebert BL, Gillette MA, Paulovich A, Pomeroy SL, Golub TR, Lander ES, Mesirov JP:Gene set enrichment analysis: a knowledge-based approach for

interpreting genome-wide expression profiles. Proc Natl Acad Sci USA 2005, 102:15545–15550.

Chapter 6

NETWORK ANALYSIS OF GENE ESSENTIALITY IN FUNCTIONAL GENOMICS EXPERIMENTS

Peng Jiang[1] , Hongfang Wang[2] , Wei Li[1] , Chongzhi Zang[1] , Bo Li[1] , Yinling J. Wong[2] , Cliff Meyer[1] , Jun S. Liu[3] , Jon C. Aster[2] and X. Shirley Liu[1,4]

[1]Department of Biostatistics and Computational Biology, Dana-Farber Cancer Institute, Harvard T.H. Chan School of Public Health, Boston, MA 02215, USA

[2] Department of Pathology, Brigham and Women's Hospital, Boston, MA 02115, USA

[3] Department of Statistics, Harvard University, Cambridge 200092, USA

[4] School of Life Science and Technology, Tongji University, Shanghai, MA 02138, China

ABSTRACT

Many genomic techniques have been developed to study gene essentiality genome-wide, such as CRISPR and shRNA screens. Our analyses of public CRISPR screens suggest protein interaction networks, when integrated with gene expression or histone marks, are highly predictive of gene essentiality. Meanwhile, the quality of CRISPR and shRNA screen results can be significantly enhanced through network neighbor information. We also found network neighbor information to be very informative on prioritizing ChIP-seq target genes and survival indicator genes from tumor profiling. Thus, our study provides a general method for gene essentiality analysis in functional genomic experiments.

Background

Essential genes are those genes critical for cell viability under certain contexts. Recent years have seen the rapid development of functional genomics techniques for studying gene essentiality genome-wide. For example, large-scale shRNA screens have been used to search for essential genes in diverse cell lines [1]. If a specific transcription factor drives the cell viability under certain condition, ChIP-seq technique can be used to profile the regulatory targets to further find essential genes [2]. Many computational methods have also been developed to predict context specific gene essentiality through integration of gene expression, molecular alterations, and biological pathways [3].

Recently, the CRISPR (clustered regularly interspaced short palindromic repeats) screen emerged as an exciting new approach to profile gene essentiality at genome scale [4–11]. In the CRISPR system, single-guide RNAs (sgRNA) direct Cas9 nucleases to induce double-strand breaks (DSB) at targeted genomic regions [12, 13]. When the error-prone non-homologous end-joining mechanism repairs the DSBs, insertions and deletions occur with high frequency, which produce a non-functional protein. Catalytically inactive Cas9 fused with a transcriptional activator or repressor has also been used to modulate gene expression at targeted loci [8, 9, 14–17]. Combined with lentiviral delivery method, CRISPR systems enable genome-scale functional screening in a cost-effective manner [4–11]. In CRISPR screens, sgRNAs targeting candidate genes are synthesized, and viral integration enables readout through next-generation sequencing [18]. The relative abundances of each integrated sgRNA between different conditions are compared and the importance of sgRNA target gene is inferred according to its sgRNAs' effect on cell growth.

The progress of CRISPR screen technology enabled systematic and reliable determination of gene essentiality under diverse conditions. The high quality gene essentiality profiles from CRISPR could enable a better comparison among essentiality prediction methods and better identification of distinct features of the essential genes. Such features not only facilitate a better understanding of the CRISPR screen data, but also can help prioritize the leads from CRISPR screens. From the analysis of yeast protein interactions, it is well known that highly connected proteins in a network (degree hubs) are more likely to be essential for viability [19–21]. Thus, we hypothesize that the gene essentiality outcome in CRISPR screens might depend on the gene connectivity in biological networks. Protein interaction networks have been integrated to improve the quality of RNAi screen results, which are very noisy due to off-target effect and low knockdown efficiency [22–25]. These previous works on RNAi screen indicate that the CRISPR screen result quality may also be improved by integration with protein interaction networks.

In this study, we took a network perspective and developed a method called NEST (Network Essentiality Scoring Tool) to systematically analyze the recent genome wide CRISPR screen data. We found that gene essentiality determined by CRISPR screen largely depends on the expression level of interacting genes in the biological network. Moreover, the quality of CRISPR and shRNA screen data can be further improved by NEST after considering the gene neighborhood screen outcome. Besides applications on CRISPR and shRNA screens, NEST is also generally applicable on many other types of genomics data analysis, such as ChIP-seq target gene prioritization and survival gene identification from tumor profiling data.

RESULTS AND DISCUSSION

NEST Predicts Gene Essentiality in CRISPR Screen

We first collected recently published CRISPR loss-of-function screen data [4, 5, 8], and selected three cell lines (K562, HL60, and A375) with publicly available gene expression data [26–28]. The significant CRISPR screen gene hits are called with software MAGeCK [29]. In CRISPR screens for growth phenotype, most significant genes are negatively selected, which means these genes are essential in the corresponding experimental condition (Additional file 1: Figure S1). To identify distinct features of gene essentiality in CRISPR screens, we developed a network-based method called NEST (Network Essentiality Scoring Tool), and found the following metric to give reliable performance.

For each gene, NEST calculates neighbor expression measure as the sum of normalized expression of its neighbor genes connected in the protein interaction network, weighted by the interaction confidence (Fig. 1a). The calculation of NEST score can also be formulated as product between connectivity matrix, which is composed of interaction weights between protein pairs, and gene expression vector. Each gene's relative expression in one cell is normalized against its average expression across all cell lines, and the protein interaction network information is from STRING [30] (Additional file 1: Figure S2). For essential genes selected by CRISPR screen, we defined the gold standard set as the genes hits called by MAGeCK with FDR threshold 0.05 [29]. For each measure, such as NEST score or network degree, all genes were ranked by their values in descending order. Receiver operating characteristic (ROC) curve was used to test the performance of predicting the CRISPR screen gold standard set (Fig. 1b).

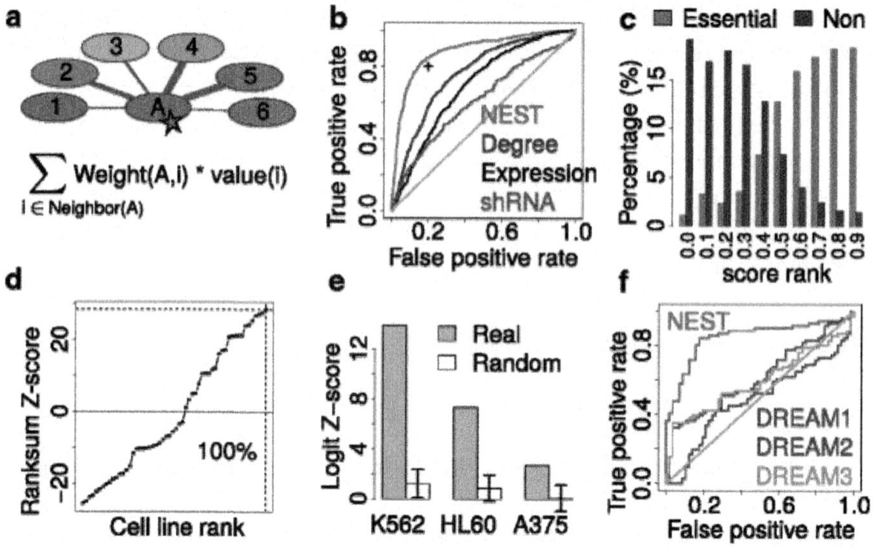

Figure. 1: Prediction of CRISPR screen outcome. **a** NEST calculates the neighbor expression of a gene as the sum of expression values of its neighbor genes connected in the network, weighted by the interaction weight. **b** Receiver operating characteristic (ROC) curve is used to test the performance of predicting gene essentiality determined by K562 CRISPRi screen. The performance of NEST score, network degree, gene expression, and shRNA screen are shown. The black point represents false positive rate 0.2 and true positive rate 0.8. **c** The NEST scores are converted to rank percentiles from 0 to 1, and shown for essential genes and non-essential genes determined in K562 CRISPRi screen. **d** For each Roadmap expression profile, we calculated the prediction power of NEST score on gene essentiality in K562 screen by Wilcoxon rank-sum test. The rank-sum Z-scores for all cell lines are ranked and the K562 profile has the largest value. **e** The STRING network was randomized 1,000 times, and the NEST scores were calculated for random networks. We used multivariate logistic regression to test the association of NEST score with gene essentiality after controlling the effects of network degree and gene expression (Table 1). The Logit Z-scores are shown for real and random networks. **f** In DREAM gene essentiality prediction challenge, A375 cell line also has CRISPR screen data available. Using essential genes selected in CRISPR screen as gold standard, the prediction performance is compared between NEST (red) and the top three winners in DREAM

For gene essentiality prediction in K562 CRISPRi screen, NEST achieved a false positive rate of 0.2 and a true positive rate of 0.8, with an area under the ROC curve (AUC) of 0.89. The AUC of NEST score is consistently better than network degree, gene expression, and shRNA screen data from the Achilles project [1] (Delong *P* value <1e-10 for all comparisons). Similar performance differences were also observed in CRISPR screen in HL60 and

A375 (Additional file 1: Figure S3a). To visualize the CRISPR prediction performance in an intuitive way, we plotted the rank percentile of NEST scores for essential genes and non-essential genes in CRISPR screen (Fig. 1c and Additional file 1: Figure S3B). The NEST ranks are significantly higher for essential genes than non-essential genes (Wilcoxon rank-sum *P* value <1e-10 for cell lines). Besides STRING network, we also used other large-scale networks for CRISPR outcome prediction. However, we did not find any performance improvement using either other network or merged network among several data sources (Additional file 1: Figure S4).

The results above suggest that if a gene's network neighbors are over-expressed in some conditions, the gene itself becomes more essential. We also found that genes with high NEST scores are tightly clustered in protein interaction network. The STRING network genes were grouped into 2,271 dense complexes using SPICi [31]. Gene with high NEST scores tend to stay in fewer number of STRING clusters than clusters with gene names shuffled (Additional file 1: Figure S5). Thus, a high NEST score may indicate the gene to be member of an active protein complex.

To test the prediction specificity, we applied NEST for gene expression profiles of 56 cell lines profiled by Roadmap project [26]. Measured by rank-sum test Z-scores, K562 CRISPRi screen data achieved the highest association with NEST score in K562 cell than all other cell lines (Fig. 1d). Similarly, HL60 and A375 CRISPR screen data also achieved higher associations with NEST scores in the same cell line (Additional file 1: Figure S3C). Housekeeping genes, such as ribosome members are often selected as essential genes in CRISPR screens [5, 8]. Thus, we further tested that the high prediction power of NEST scores was not purely derived from the same set of housekeeping genes. The prediction performance of NEST remains high after removal of housekeeping genes annotated previously [32] (Additional file 1: Figure S6). Notably, the majority of essential genes selected in CRISPR screen do not overlap between K562, HL60, and A375 (Additional file 1: Table S1). Thus, our NEST score is an orthogonal feature of CRISPR selected gene essentiality other than the universal housekeeping genes shared across conditions.

Since gene network degree and gene expression are also predicative of gene essentiality (Fig. 1b), we then tested whether the prediction performance of NEST is simply an additive effect of network degree and gene expression (Table 1). Using the gene essentiality in CRISPR screen as responsive variable, we did a multivariate logistic regression among all three covariates (NEST score, gene network degree, and expression). While all covariates are predictive of gene essentiality jointly, NEST has the largest statistical significance defined as Logit Z-score (Table 1). Moreover, the logistic regression fitted value, combining all covariates together, did not improve the CRISPR prediction performance comparing to NEST score alone (Additional file 1: Figure S7). As a further control, we randomized the STRING network but preserved the network degree for each gene [33]. The Logit Z-scores for the NEST score in random networks are significantly smaller than in real data (Fig. 1e, empirical *P* value <0.001 for K562 and HL60, and *P* value=0.003 for A375).

Table 1: Confounding factors for NEST prediction performance

Covariate	**Coefficient**	**Standard error**	***Z*-score**	***P* value**
NEST	0.02329	0.001748	13.32	1.72e-40
Degree	0.00415	0.000846	4.91	9.33e-07
Expression	0.12937	0.054223	2.39	1.70e-02
A. K562				
NEST	0.03494	0.00505	6.91	4.73e-12
Degree	0.00608	0.00146	4.16	3.13e-05
Expression	0.33873	0.16595	2.04	4.12e-02
B. HL60				
NEST	0.07296	0.02483	2.94	0.00329
Degree	0.00792	0.00357	2.22	0.02647
Expression	1.12266	0.48343	2.32	0.02022
C. A375				

The prediction power of NEST score on gene essentiality decided by CRISPR screen is tested through logistic regression with gene network degree and gene expression as covariates. The Logit Z-score is defined as Coef/Stderr. The *P* value is calculated by Ward test. The result is shown for (A) K562, (B) HL60, and (C) A375

There have been many previous methods developed for gene essentiality prediction. Since CRISPR screen measures the gene essentiality, any previous methods can be predictive for CRISPR outcome. In a recent DREAM challenge, contenders were asked to develop algorithms to predict cell specific gene essentiality [3]. Among cell lines included in the DREAM challenge, A375 has CRISPR screen data available. We compared the CRISPR outcome prediction performance between our method and the top three methods from the DREAM challenge, and found NEST to consistently outperform all DREAM winners (Fig. 1f). Besides the methods in DREAM, we also compared the performance of NEST with other methods using gene expression and network to predict gene essentiality [34, 35]. NEST significantly outperformed all other methods (Additional file 1: Figure S8 and Additional file 1: Methods).

Besides gene expression, we also used H3K27ac histone mark data to compute NEST scores and tested the gene essentiality prediction performance. Previously, we developed a method to calculate the regulatory potential (RP) scores of a histone modification on each gene promoter from the ChIP-seq profile [36, 37]. Based on our previous method, gene level RP scores in K562 cell were computed using the Roadmap H3K27ac ChIP-seq profile [26]. For each gene, NEST computed neighbor H3K27ac score as the sum of H3K27ac RP scores of its neighbor genes connected in the protein interaction network, weighted by the interaction confidence (Fig. 1a). H3K27ac NEST scores could also reliably predict the gene essentiality in K562 CRISPRi screen (Additional file 1: Figure S3), suggesting the applicability of NEST analysis on histone modification data.

NEST Enhances the Quality of CRISPR Screen Results

Since early CRISPR screens might have inefficient sgRNA selection and few sgRNA per gene, these screens may not give very strong hits. Encouraged by the prediction performance, we checked whether the network neighbor information could enhance the quality of CRISPR screen results. To measure the quality of a screen data, we need to know the expected outcome. In a K562 CRISPRi screen, the authors performed a genome-scale selection for genes that modulate sensitivity to Cholera/Diphtheria toxin [8]. For genes that work with the toxin, their knock out will protect the cell against the toxin and induce a positive gene selection in screen For genes that are targeted by toxin, their knock out will sensitize the cell for toxin effect and induce a negative gene selection The positively selected genes, which played a protective role against toxin, were enriched in KEGG pathways 'Vibrio Cholerae Infection' and 'Glycosphingolipid Biosynthesis' [8]. The negatively selected genes, which played a sensitizing role for toxin, were enriched in 'Ribosome' and 'Proteasome' pathways [8]. We used these enriched pathway genes as gold standard and tested how well network interaction could improve the CRISPR screen result (Fig. 2).

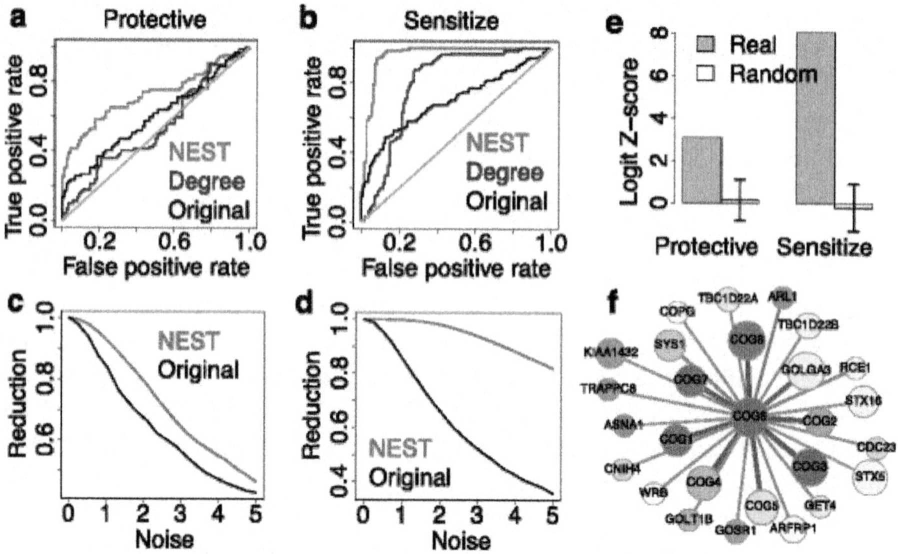

Figure. 2: Enhancement of CRISPR screen result. **a** For K562 CRISPRi screen under Cholera/Diptheria toxin selection, the gold standard of toxin protective genes comes from KEGG pathways 'Vibrio Cholera Infection' and 'Glycosphingolipid biosynthesis'. For each gene, NEST computes the neighbor CRISPR score as the sum of CRISPR screen fold change scores of neighbor genes connected in the network. The prediction performance is compared between NEST and original CRISPR scores. **b** The gold standard of toxin sensitizing genes comes from KEGG pathways 'Ribosome' and 'Proteasome'. The prediction performances of NEST and original CRISPR values are compared. **c** The original CRISPR values were randomized by Gaussian white noise. The standard deviation of all original CRISPR values was used as base level. At each noise level relative to the base level, the area under ROC curve (AUC) of prediction is compared with the initial AUC for toxin protective genes in K562 Cholera toxin screen. The reduction ratios were plotted for NEST and original CRISPR scores. **d** The reduction ratios were plotted for toxin sensitizing genes. **e** The STRING network was randomized and the NEST scores were calculated for 1,000 random networks. We used multivariate logistic regression to test the association of NEST scores with gold standards, after controlling the effects from network degree and original CRISPR score. The Logit Z-scores are shown for real and random networks. **f** As an example of high NEST score gene, *COG6* is a component of Golgi complex, and connected with several other components in Golgi complex. The thickness of each edge represents the interaction weight. The color of each gene represents the CRISPR screen fold change score; red color indicates toxin protective and blue color indicates toxin sensitizing

For each gene, NEST calculated a neighbor CRISPR score by adding up the CRISPR fold change scores among neighbor genes connected in the STRING network, weighted by the interaction confidence. This NEST score is significantly more predictive on the gold standard outcome than the original CRISPR scores for both protective and sensitizing genes (Fig. 2a, b, Delong test *P* value=0.010 for protective genes, *P* value=9.92e-14 for sensitizing genes). Moreover, when we put different levels of Gaussian white noise into CRISPR screen scores, the prediction performance of NEST score diminishes slower than original CRISPR scores (Fig. 2c, d). As a control, if we calculated the NEST scores from randomized network, the prediction power became significantly worse (Fig. 2e, *P* value <0.001 for both protective and sensitizing genes). Thus, through the connectivity of protein interaction network, NEST can enhance the quality of CRISPR screen result.

As an example of gene with high NEST score, *COG6* is a member of Golgi complex and its NEST score is significantly larger than expected (permutation test *P* value <0.001). *COG6* is connected with many other members of Golgi complex (Fig. 2f), and most of them have positive CRISPR screen fold change scores. Since they are connected with each other in network, they mutually boosted each other's NEST scores. Our result is consistent with the knowledge that cholera toxin needs to enter host cells and travel through the trans-Golgi network to take effect [38].

The above results suggest that if a gene's network neighbors are under CRISPR screen selection, the gene itself is more likely to be under CRISPR screen selection in the same direction. Besides CRISPR screen, we applied NEST on the Achilles shRNA screen data [1]. NEST can also significantly enhance the quality of shRNA screen result (Additional file 1: Figure S9). Thus, in general, the quality of functional genomic screen result can be improved by considering the gene network neighbor information.

Previously, there were methods developed to improve the quality of RNAi screen results from integration with protein interaction networks [25]. For CRISPR enhancement, we compared our method NEST against NePhe, which was a leading method on RNAi screen network analysis [24]. Using K562 toxin screen as the gold standard, we found that NePhe and NEST show similar performance as measured by ROC curves (Additional file 1: Figure S10AB). However, while NePhe used 14 GB memory and 6.2 h running time, NEST only used 8.3 MB memory and 10.8 s (Additional file 1: Figure S10C). Thus, NEST maintains reliable screen enhancement performance of previous method with much better computational efficiency.

NEST Prioritizes ChIP-Seq Essential Targets

Besides functional genomic CRISPR/shRNA screen, many other genomic experimental techniques can be used to search for essential genes. For example, if a specific transcription factor (TF) drives the cell viability under certain condition, ChIP-seq technique can be used to profile its regulatory targets to further find essential genes [2]. The previous analyses demonstrate that NEST can identify the essential genes in a CRISPR screen. We further explored whether NEST can help prioritize key target genes in a ChIP-seq experiment. ChIP-seq often finds tens of thousands in vivo binding sites for a TF. Since target genes can be regulated by TF binding through long range DNA looping, often thousands of genes near the TF binding sites can be putative targets, and it is hard to prioritize the functional target genes directly from a ChIP-seq experiment. We therefore investigated using network neighbor information to prioritize the functional TF target genes.

Our previous studies of NOTCH1 ChIP-seq and gene expression profiles in the T-lymphoblastic leukemia (TLL) cell line CUTLL1 identified 1,012 differential NOTCH1 binding sites between the *NOTCH* on and off conditions [2]. Based on the ChIP-seq peaks, we calculated a regulatory potential (RP) score for each gene [36, 37, 39], a distance-weighted sum of binding sites measuring the overall regulatory impact of differential NOTCH1 binding on target genes. We set the KEGG NOTCH signaling pathway members as the gold stand, and tested the prediction performance of expression, ChIP-seq RP and NEST scores (Fig. 3). In addition to NEST scores computed from gene expression (NEST E), we also computed NEST scores from ChIP-seq (NEST C), which measures the sum of ChIP-seq RP scores of neighbor genes connected in network. While expression and ChIP-seq measures are barely better than random, NEST scores can predict the annotated KEGG NOTCH signaling pathway members at AUC 0.90 and 0.95 (Fig. 3). It suggests that if a gene's network neighbors are enriched in the binding target of a TF, the gene itself is more likely to be regulated by the same TF.

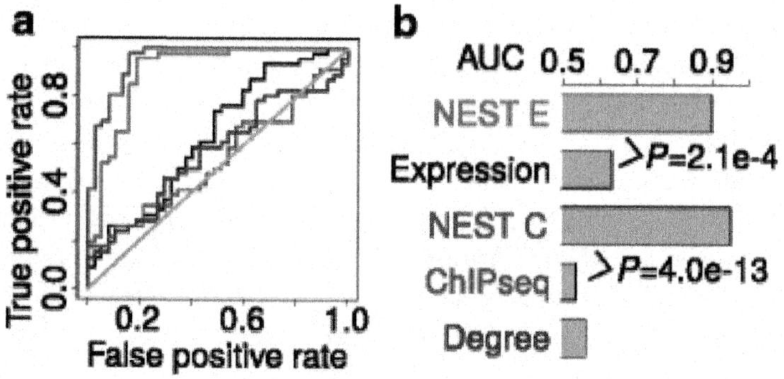

Figure. 3: Prediction of NOTCH signaling pathway members. **a** The differential gene expression between *NOTCH* on and off conditions is used to calculate the NEST E score. The NOTCH1 ChIP-seq regulatory potential score for each gene is used to calculate the NEST C score. The KEGG Notch Signaling pathway members are

used as gold standard and the prediction performances of all measures are shown by ROC curves. **b** The area under ROC curve (AUC) is shown for each measure. The comparison between AUCs is done by Delong test

NEST Predicts Cancer Patient Survival

Encouraged by the above analyses in cell lines, we checked whether NEST could facilitate the analysis of tumor profiling data. There have been previous studies integrating biological networks with cancer (or disease) biology data to understand the mechanisms of pathogenesis [35, 40–43]. Inspired by these studies, we examined the TCGA tumor profiling data to see whether NEST score computed from gene expression can better predict patient survival than gene expression. For example, over-activation of oncogene *EGFR* is a key feature of Glioblastoma (GBM) [44]. In TCGA GBM profiles [45], while*EGFR* over-expression does not correlate with worse survival (Fig. 4a, Cox-PH *P* value = 0.109), higher *EGFR* NEST score is significantly associated with worse survival (Fig. 4b, Cox-PH *P* value = 0.001).

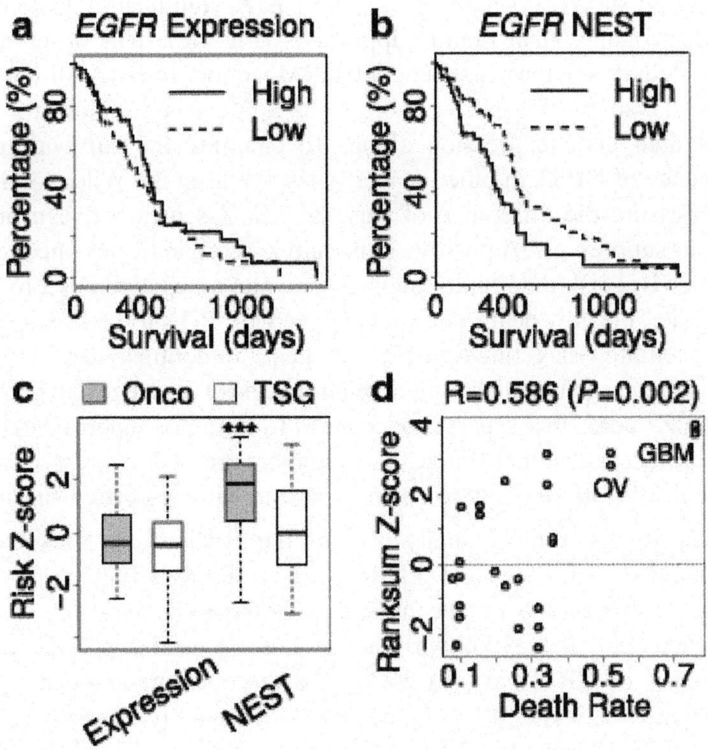

Figure. 4: Prediction of patient survival. **a** All TCGA Glioblastoma (GBM) patients are ranked by *EGFR* expression; the top half patients are assigned as high group and

the lower half are assigned as low. Kaplan Meier (KM) survival plot is shown for two groups. **b** The survival analysis is done in the same way as A for *EGFR* NEST scores. **c** For each gene, we calculated a death risk Z-score by Cox-PH model from either gene expression or NEST score. We compared the Z-scores for oncogenes (Onco) and tumor suppressor genes (TSG) based on the annotation from Vogelstein et al. The bottom and top of the boxes are the 25th and 75th percentiles (interquartile range). Whiskers on the top and bottom represent the maximum and minimum data points within the range represented by 1.5 times the interquartile range. The distribution of Z-scores is compared by Wilcoxon rank-sum test and three stars represent P value <0.001. **d** For each TCGA cohort, the difference of risk Z-scores computed from NEST was tested by Wilcoxon rank-sum test. The rank-sum Z-scores are plotted against the death rate of each cancer type, with Spearman's rank correlation as title

To systematically evaluate the survival prediction performance, we hypothesized that a good gene-wise survival predictor should show significant higher death risk for oncogenes than for tumor suppressors. We tested this hypothesis on all the annotated oncogenes and tumor suppressors [46] using the TCGA GBM data (Fig. 4c). While gene expression showed no significant difference on survival risk Z-scores, NEST gave significantly higher survival risk for oncogenes than tumor suppressors (Fig. 4c). This observation was corroborated in another independent GBM cohort [47] (Additional file 1: Figure S11), suggesting NEST score to be a much better indicator of GBM survival than gene expression alone. To examine the survival prediction performance of NEST in other cancer types, we used the Wilcoxon rank-sum test to measure the difference of survival risk Z-scores between oncogenes and tumor suppressors. A positive rank-sum Z-score indicates oncogenes with higher survival risk than tumor suppressors, and a negative Z-score indicates the opposite. For low death rate cancers, the Cox-PH survival regression may not get accurate risk estimation for each gene. In contrast, cancer types with high death rate, such as GBM and ovarian cancer (OV), seemed to give positive rank-sum Z-scores that separate oncogenes from tumor suppressors (Fig. 4d). These results suggest that if a gene's neighbors are over expressed in tumors, the gene itself is more likely to be an oncogene with associated survival risk.

We conducted pathway analysis on all the genes whose NEST scores are associated with GBM survival (FDR<= 0.05), and found 'cytokine cytokine-receptor interaction' to be the most enriched KEGG pathway (Additional file 1: Table S2). It was known that cytokines played a pivotal role in the pathogenesis of GBM [48], so we plotted the outcome-associated cytokine genes using CytoScape [49] (Fig. 5). Many of them are known oncogenes in GBM, such as *EGFR* and *CSF1R* [46], and several also have known targeted drugs from Drug Bank [50]. For example, the inhibitors of *EGFR*, *CSF1R*, and *CXCR4* were shown to reduce the invasiveness of Glioma cells or block

GBM progression [51–53]. Besides the known druggable genes, many other genes in our prediction could serve as promising targets. For example, NEST predicted *KITLG* as indicator of poor GBM survival which is consistent with the finding that downregulation of *KITLG* inhibits angiogenesis and Glioma growth [54]. Thus, our predictions could sketch a general landscape to investigate therapeutic possibilities for GBM and other cancers.

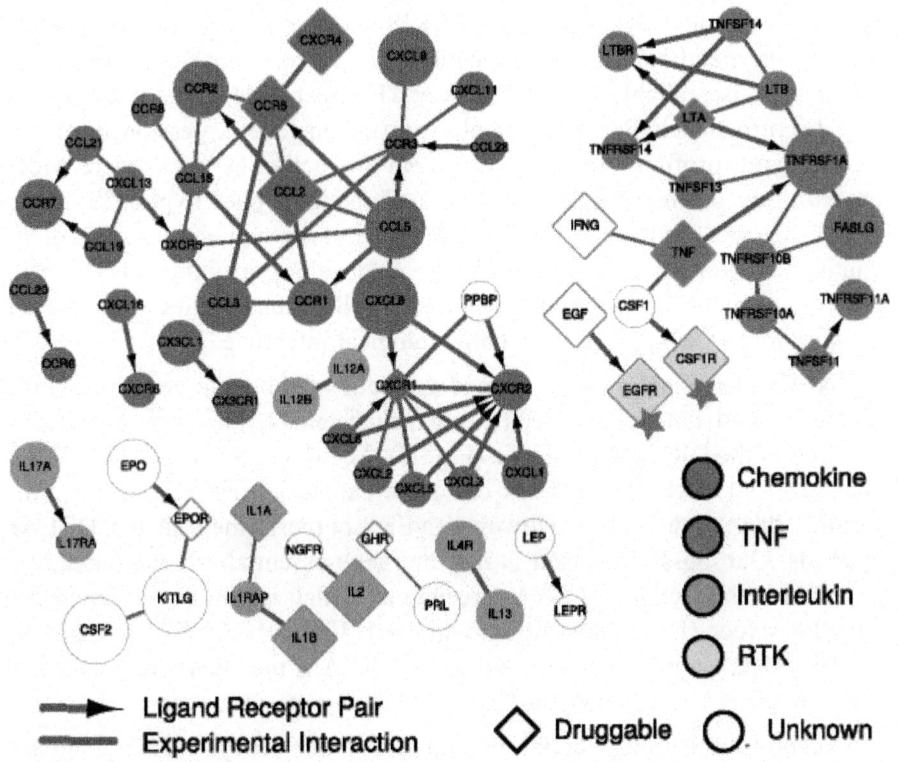

Figure. 5: Cytokine receptor interaction network for GBM survival. For all members in KEGG pathway Cytokine Cytokine-Receptor Interaction, we selected the genes indicating death risk with FDR threshold 0.05. The pathway members are colored by their gene family categories, including Chemokine, Tumor Necrosis Factor (TNF), Interleukin, and Receptor Tyrosine Kinase (RTK). A diamond shape indicates the gene to be drug targetable in Drug Bank. The node size is proportional to the NEST score averaged among all GBM patients. Stars are used to label known oncogenes annotated by Vogelstein et al. The directed edge indicates a cytokine-to-receptor relation in KEGG, and undirected edge indicates an experimental protein interaction curated by STRING. The thickness of each edge indicates the interaction confidence

Conclusion

To identify distinct features of gene essentiality in CRISPR screens, we developed a network-based method called NEST (Network Essentiality Scoring Tool). We found that essential genes selected in CRISPR screens showed characteristic higher expression level of neighbor genes connected in protein interaction network. Our analysis of Cholera toxin screen in K562 cell also suggests that the quality of CRISPR screen result can be enhanced through the neighbor CRISPR selection score. For a ChIP-seq experiment, NEST can also reliably identify the key TF target genes. Last but not least, NEST score can better predict patient survival than gene expression alone from TCGA tumor profiles. Historically, protein interaction networks were widely used to infer discrete labels such as gene functions, phenotypes [55–57], or gene categories [58]. Our study is different from these previous works in that continuous expression or screen change fold values are integrated with the protein networks. Despite these differences, all of these studies indicate that network information can greatly help biological inference.

NEST significantly outperformed previous methods on gene essentiality prediction and functional screen result enhancement, including all winning methods in the DREAM challenge (Fig. 1f). According to the rule of DREAM challenge, all DREAM methods can gene expression as well as any other features they could utilize. However, NEST outperformed all top DREAM methods. One possible reason is that the gene essentiality gold standard of DREAM is the Achilles shRNA screen data, which is poorly correlated with CRISPR screen (Fig. 1b and Additional file 1: Figure S3A). Because we used CRISPR data as gold standard, those top DREAM methods, optimized to fit Achilles shRNA screen, may not have satisfactory performance.

Several limitations should be noted for our study. NEST computed gene activity is based on network interaction partners, which could have either an activating or a repressive effect. Meanwhile, for compensating interaction such as synthetic lethality, the activation of interaction partners indicates gene loss of function. For example, *PLK1*, an interaction partner of*TP53* in STRING network, was consistently upregulated in cancer cells with inactivated *TP53* compared with those with wild type [59]. We currently summed all neighbor values without distinguishing between activating, repressive, or synthetic lethal relations. Thus, further categorization of network interaction types will be critical for better gene prioritization. Another limitation of our study is that current data on protein interaction network only covered a subset of well-studied genes [60]. Because of the dependence on interaction knowledge, our method may not reliably infer the activity for under-studied genes. As a third limitation, we only tested NEST on gene loss-of-function CRISPR screens. However, for CRISPRa gain of function screen [8, 9], it remains to see whether network-based analysis can bring any predictive power and result enhancement.

In summary, we derived a network-based method, NEST, to interpret and enhance the outcome of genome-wide CRISPR screens, and NEST showed significantly better performance than previous related methods. We recommend researchers using NEST to calculate neighbor CRISPR values from their CRISPR screen result. Moreover, the candidate essential genes in a cell condition might be prioritized before running a large-scale screen to reduce the total number of genes under the screen, which might improve the results and practicality of in vivo CRISPR screens. Besides CRISPR analysis, our method can also identify key targets from ChIP-seq experiments, and find clinical outcome associated genes from tumor profiling data. Thus, we foresee NEST as generally applicable to many applications related with gene essentiality prioritization.

MATERIALS AND METHODS

Availability

The web application and source code of NEST are freely available under the GNU Public License v3 at . The source code and testing data of NEST are additionally deposited at .

Data Collection

For CRISPR screen data, we searched published studies with data publicly available and sgRNA coverage on genome scale for human cell lines until 1 June 2015. There are three studies fulfilling our criterion. In K562 cell, growth phenotype and toxin selection phenotype are screened with CRISPRi technology [8]. In HL60 and A375 cell lines, growth phenotype is screened on genome scale with CRISPR technology [4, 5]. Significant gene hits are called from these datasets by MAGeCK 0.5 with default parameters and FDR threshold 0.05 [29]. For gene essentiality prediction in each cell line, only negatively selected gene hits were considered as gold standard, because most significant gene hits are negatively selected in collected CRISPR experiments (Additional file 1: Figure S1). For gold standard control set, we extracted the same number of genes ranked by MAGeCK on bottom.

For K562, the gene expression profile was downloaded from the Roadmap project [26]. For HL60, the gene expression profile by exon-array was downloaded from the ENCODE project [27] and converted to gene level values by JETTA [61]. For A375, the gene expression profile was downloaded from the CCLE project [28]. For each gene, we normalized the expression value by subtracting the mean across all samples in each cohort. Compared to absolute expression level, the normalized expression value can achieve a better CRISPR prediction performance of NEST (Additional file 1: Figure S12). The TCGA tumor gene expression data was downloaded from TCGA Data Portal on 27 July 2014. Only cohorts that are not embargoed are used. For each cancer cohort, the expression values of all normal control samples were averaged as background, and the difference of gene expression between tumor sample and normal background was analyzed. For NOTCH signaling pathway analysis, the NOTCH off condition is defined as gamma secretase inhibitors (GSI) treatment 3 days, and NOTCH on condition is defined as GSI wash 4 h [62]. The differential expression value between on/off conditions was analyzed [62]. The NOTCH1 ChIP-seq data are generated in our previous work, and the dynamic binding peaks between NOTCH on/off conditions were used [2].

For H3K27ac ChIP-seq profiles, we downloaded data from the Roadmap project [26]. Among all cell lines with CRISPR data collected, K562 is the only one having H3K27ac profile available. Previously, we developed a BETA method to calculate the regulatory potential (RP) on gene promoters from the ChIP-seq profile of a transcription factor or histone modification [36,37]. We used the implementation in RABIT package with default parameters to calculate the H3K27ac RP scores [39]. For each gene, the RP scores were normalized, by subtracting the mean across all cell lines profiled.

Network Randomization and Permutation Test

We used stub rewiring method to randomize unweighted STRING network, which preserves gene degree [33]. The edges from each gene are first detached from its partners, and then randomly connected with each other. Since we do not allow self-interaction and duplicated edges, the connection process may fail to finish. In this case, we restart the rewiring process until 98 % edges are reconnected.

Based on random networks, we derived a permutation test to access whether the NEST score of each gene is significantly larger (or smaller) than expected. For each random network, we calculated the NEST values as random NEST. For each gene, we computed the Z-score as (real NEST – average random NEST)/(Stderr of random NEST). If the Z-score is positive, we computed the *P* value as the fraction of random NEST scores that are larger than or equal to the real NEST score. If the Z-score is negative, we computed the *P* values as the fraction of random NEST scores that are smaller than or equal to the real NEST score.

Survival Analysis

We used Cox-PH model to analyze the effect of gene expression or NEST scores on survival. For GBM, there are several factors that affect the survival and we included them as covariates in survival regression, including age, gender, G-CIMP status, and treatment status [45]. So the final survival effect was corrected with the effects of these confounding factors. For TCGA pan-cancer analysis, we only included cancer types with more than 50 patients and 5 % death rate. In the Cox-PH regression, we only included age, gender, and stage (if available) to enable uniform comparison among different cancer types.

ACKNOWLEDGEMENTS

The authors would like to thank Michael Love, Eric Severson, Han Xu, Yiwei Chen, and Jing Mi for helpful discussions. The project was supported by the U01 CA180980 grant from NIH and the Claudia Adams Barr Award in Innovative Basic Cancer Research from the Dana-Farber Cancer Institute.

AUTHORS' CONTRIBUTIONS

PJ, HW, WL, CZ, BL, CM, YW, JSL, JCA, and XSL designed the study and interpreted the results. PJ developed the algorithm and performed the analyses. PJ and XSL wrote the paper. All authors read and approved the final manuscript.

REFERENCES

1. Cowley GS, Weir BA, Vazquez F, Tamayo P, Scott JA, Rusin S, et al. Parallel genome-scale loss of function screens in 216 cancer cell lines for the identification of context-specific genetic dependencies. Sci Data. 2014;1:140035.
2. Wang H, Zang C, Taing L, Arnett KL, Wong YJ, Pear WS, et al. NOTCH1-RBPJ complexes drive target gene expression through dynamic interactions with superenhancers. Proc Natl Acad Sci U S A. 2014;111:705–10.
3. Broad-DREAM Gene Essentiality Prediction Challenge. Available at:

4. Shalem O, Sanjana NE, Hartenian E, Shi X, Scott DA, Mikkelsen TS, et al. Genome-scale CRISPR-Cas9 knockout screening in human cells. Science. 2014;343:84–7.
5. Wang T, Wei JJ, Sabatini DM, Lander ES. Genetic screens in human cells using the CRISPR-Cas9 system. Science. 2014;343:80–4.
6. Koike-Yusa H, Li Y, Tan EP, Velasco-Herrera Mdel C, Yusa K. Genome-wide recessive genetic screening in mammalian cells with a lentiviral CRISPR-guide RNA library. Nat Biotechnol. 2014;32:267–73.
7. Zhou Y, Zhu S, Cai C, Yuan P, Li C, Huang Y, et al. High-throughput screening of a CRISPR/Cas9 library for functional genomics in human cells. Nature. 2014;509:487–91.
8. Gilbert LA, Horlbeck MA, Adamson B, Villalta JE, Chen Y, Whitehead EH, et al. Genome-scale CRISPR-mediated control of gene repression and activation. Cell. 2014;159:647–61.
9. Konermann S, Brigham MD, Trevino AE, Joung J, Abudayyeh OO, Barcena C, et al. Genome-scale transcriptional activation by an engineered CRISPR-Cas9 complex. Nature. 2015;517:583–8.
10. Chen S, Sanjana NE, Zheng K, Shalem O, Lee K, Shi X, et al. Genome-wide CRISPR screen in a mouse model of tumor growth and metastasis. Cell. 2015;160:1246–60.
11. Shi J, Wang E, Milazzo JP, Wang Z, Kinney JB, Vakoc CR. Discovery of cancer drug targets by CRISPR-Cas9 screening of protein domains. Nat Biotechnol. 2015;33:661–7.
12. Jinek M, Chylinski K, Fonfara I, Hauer M, Doudna JA, Charpentier E. A programmable dual-RNA-guided DNA endonuclease in adaptive bacterial immunity. Science. 2012;337:816–21.
13. Gasiunas G, Barrangou R, Horvath P, Siksnys V. Cas9-crRNA ribonucleoprotein complex mediates specific DNA cleavage for adaptive immunity in bacteria. Proc Natl Acad Sci U S A. 2012;109:E2579–86.
14. Maeder ML, Linder SJ, Cascio VM, Fu Y, Ho QH, Joung JK. CRISPR RNA-guided activation of endogenous human genes. Nat Methods. 2013;10:977–9.
15. Qi LS, Larson MH, Gilbert LA, Doudna JA, Weissman JS, Arkin AP, et al. Repurposing CRISPR as an RNA-guided platform for sequence-specific control of gene expression. Cell. 2013;152:1173–83.
16. Gilbert LA, Larson MH, Morsut L, Liu Z, Brar GA, Torres SE, et al. CRISPR-mediated modular RNA-guided regulation of transcription in eukaryotes. Cell. 2013;154:442–51.

17. Zalatan JG, Lee ME, Almeida R, Gilbert LA, Whitehead EH, La Russa M, et al. Engineering complex synthetic transcriptional programs with CRISPR RNA scaffolds. Cell. 2015;160:339–50.

18. Shalem O, Sanjana NE, Zhang F. High-throughput functional genomics using CRISPR-Cas9. Nat Rev Genet. 2015;16:299–311.

19. Jeong H, Mason SP, Barabasi AL, Oltvai ZN. Lethality and centrality in protein networks. Nature. 2001;411:41–2.

20. Han JD, Bertin N, Hao T, Goldberg DS, Berriz GF, Zhang LV, et al. Evidence for dynamically organized modularity in the yeast protein-protein interaction network. Nature. 2004;430:88–93.

21. Hahn MW, Kern AD. Comparative genomics of centrality and essentiality in three eukaryotic protein-interaction networks. Mol Biol Evol. 2005;22:803–6.

22. Kaplow IM, Singh R, Friedman A, Bakal C, Perrimon N, Berger B. RNAiCut: automated detection of significant genes from functional genomic screens. Nat Methods. 2009;6:476–7.

23. Tu Z, Argmann C, Wong KK, Mitnaul LJ, Edwards S, Sach IC, et al. Integrating siRNA and protein-protein interaction data to identify an expanded insulin signaling network. Genome Res. 2009;19:1057–67.

24. Wang L, Tu Z, Sun F. A network-based integrative approach to prioritize reliable hits from multiple genome-wide RNAi screens in Drosophila. BMC Genomics. 2009;10:220.

25. Ma X, Chen T, Sun F. Integrative approaches for predicting protein function and prioritizing genes for complex phenotypes using protein interaction networks. Brief Bioinform. 2014;15:685–98.

26. Roadmap Epigenomics C, Kundaje A, Meuleman W, Ernst J, Bilenky M, Yen A, et al. Integrative analysis of 111 reference human epigenomes. Nature. 2015;518:317–30.

27. Consortium EP, Bernstein BE, Birney E, Dunham I, Green ED, Gunter C, et al. An integrated encyclopedia of DNA elements in the human genome. Nature. 2012;489:57–74.

28. Barretina J, Caponigro G, Stransky N, Venkatesan K, Margolin AA, Kim S, et al. The Cancer Cell Line Encyclopedia enables predictive modelling of anticancer drug sensitivity. Nature. 2012;483:603–7.

29. Li W, Xu H, Xiao T, Cong L, Love MI, Zhang F, et al. MAGeCK enables robust identification of essential genes from genome-scale CRISPR/Cas9 knockout screens. Genome Biol. 2014;15:554.

30. Franceschini A, Szklarczyk D, Frankild S, Kuhn M, Simonovic M, Roth A, et al. STRING v9.1: protein-protein interaction networks, with increased coverage and integration. Nucleic Acids Res. 2013;41:D808–15.
31. Jiang P, Singh M. SPICi: a fast clustering algorithm for large biological networks. Bioinformatics. 2010;26:1105–11.
32. Eisenberg E, Levanon EY. Human housekeeping genes are compact. Trends Genet. 2003;19:362–5.
33. Milo R, Shen-Orr S, Itzkovitz S, Kashtan N, Chklovskii D, Alon U. Network motifs: simple building blocks of complex networks. Science. 2002;298:824–7.
34. Chuang HY, Lee E, Liu YT, Lee D, Ideker T. Network-based classification of breast cancer metastasis. Mol Syst Biol. 2007;3:140.
35. Hofree M, Shen JP, Carter H, Gross A, Ideker T. Network-based stratification of tumor mutations. Nat Methods. 2013;10:1108–15.
36. Tang Q, Chen Y, Meyer C, Geistlinger T, Lupien M, Wang Q, et al. A comprehensive view of nuclear receptor cancer cistromes. Cancer Res. 2011;71:6940–7.
37. Wang S, Sun H, Ma J, Zang C, Wang C, Wang J, et al. Target analysis by integration of transcriptome and ChIP-seq data with BETA. Nat Protoc. 2013;8:2502–15.
38. Wernick NL, Chinnapen DJ, Cho JA, Lencer WI. Cholera toxin: an intracellular journey into the cytosol by way of the endoplasmic reticulum. Toxins (Basel). 2010;2:310–25.
39. Jiang P, Freedman ML, Liu JS, Liu XS. Inference of transcriptional regulation in cancers. Proc Natl Acad Sci U S A. 2015;112:7731–6.
40. Goh KI, Cusick ME, Valle D, Childs B, Vidal M, Barabasi AL. The human disease network. Proc Natl Acad Sci U S A. 2007;104:8685–90.
41. Menche J, Sharma A, Kitsak M, Ghiassian SD, Vidal M, Loscalzo J, et al. Disease networks. Uncovering disease-disease relationships through the incomplete interactome. Science. 2015;347:1257601.
42. Greene CS, Krishnan A, Wong AK, Ricciotti E, Zelaya RA, Himmelstein DS, et al. Understanding multicellular function and disease with human tissue-specific networks. Nat Genet. 2015;47:569–76.
43. Wang X, Wei X, Thijssen B, Das J, Lipkin SM, Yu H. Three-dimensional reconstruction of protein networks provides insight into human genetic disease. Nat Biotechnol. 2012;30:159–64.

44. Hatanpaa KJ, Burma S, Zhao D, Habib AA. Epidermal growth factor receptor in glioma: signal transduction, neuropathology, imaging, and radioresistance. Neoplasia. 2010;12:675–84.

45. Brennan CW, Verhaak RG, McKenna A, Campos B, Noushmehr H, Salama SR, et al. The somatic genomic landscape of glioblastoma. Cell. 2013;155:462–77.

46. Vogelstein B, Papadopoulos N, Velculescu VE, Zhou S, Diaz Jr LA, Kinzler KW. Cancer genome landscapes. Science. 2013;339:1546–58.

47. Gravendeel LA, Kouwenhoven MC, Gevaert O, de Rooi JJ, Stubbs AP, Duijm JE, et al. Intrinsic gene expression profiles of gliomas are a better predictor of survival than histology. Cancer Res. 2009;69:9065–72.

48. Zhu VF, Yang J, Lebrun DG, Li M. Understanding the role of cytokines in Glioblastoma Multiforme pathogenesis. Cancer Lett. 2012;316:139–50.

49. Shannon P, Markiel A, Ozier O, Baliga NS, Wang JT, Ramage D, et al. Cytoscape: a software environment for integrated models of biomolecular interaction networks. Genome Res. 2003;13:2498–504.

50. Law V, Knox C, Djoumbou Y, Jewison T, Guo AC, Liu Y, et al. DrugBank 4.0: shedding new light on drug metabolism. Nucleic Acids Res. 2014;42:D1091–7.

51. Taylor TE, Furnari FB, Cavenee WK. Targeting EGFR for treatment of glioblastoma: molecular basis to overcome resistance. Curr Cancer Drug Targets. 2012;12:197–209.

52. Pyonteck SM, Akkari L, Schuhmacher AJ, Bowman RL, Sevenich L, Quail DF, et al. CSF-1R inhibition alters macrophage polarization and blocks glioma progression. Nat Med. 2013;19:1264–72.

53. Ehtesham M, Winston JA, Kabos P, Thompson RC. CXCR4 expression mediates glioma cell invasiveness. Oncogene. 2006;25:2801–6.

54. Sun L, Hui AM, Su Q, Vortmeyer A, Kotliarov Y, Pastorino S, et al. Neuronal and glioma-derived stem cell factor induces angiogenesis within the brain. Cancer Cell. 2006;9:287–300.

55. Schwikowski B, Uetz P, Fields S. A network of protein-protein interactions in yeast. Nat Biotechnol. 2000;18:1257–61.

56. Hishigaki H, Nakai K, Ono T, Tanigami A, Takagi T. Assessment of prediction accuracy of protein function from protein--protein interaction data. Yeast. 2001;18:523–31.

57. Sharan R, Ulitsky I, Shamir R. Network-based prediction of protein function. Mol Syst Biol. 2007;3:88.

58. Warde-Farley D, Donaldson SL, Comes O, Zuberi K, Badrawi R, Chao P, et al. The GeneMANIA prediction server: biological network integration for gene prioritization and predicting gene function. Nucleic Acids Res. 2010;38:W214–20.

59. Sur S, Pagliarini R, Bunz F, Rago C, Diaz Jr LA, Kinzler KW, et al. A panel of isogenic human cancer cells suggests a therapeutic approach for cancers with inactivated p53. Proc Natl Acad Sci U S A. 2009;106:3964–9.

60. Rolland T, Tasan M, Charloteaux B, Pevzner SJ, Zhong Q, Sahni N, et al. A proteome-scale map of the human interactome network. Cell. 2014;159:1212–26.

61. Seok J, Xu W, Gao H, Davis RW, Xiao W. JETTA: junction and exon toolkits for transcriptome analysis. Bioinformatics. 2012;28:1274–5.

62. Wang H, Zou J, Zhao B, Johannsen E, Ashworth T, Wong H, et al. Genome-wide analysis reveals conserved and divergent features of Notch1/RBPJ binding in human and murine T-lymphoblastic leukemia cells. Proc Natl Acad Sci U S A. 2011;108:14908–13.

Chapter 7

CASTOR BEAN ORGANELLE GENOME SEQUENCING AND WORLDWIDE GENETIC DIVERSITY ANALYSIS

Maximo Rivarola[1]., Jeffrey T. Foster[2], Agnes P. Chan[3], Amber L. Williams[4], Danny W. Rice[5], Xinyue Liu[1], Admasu Melake-Berhan[3], Heather Huot Creasy[1], Daniela Puiu[3], M. J. Rosovitz[3], Hoda M. Khouri[1], Stephen M. Beckstrom-Sternberg[2,6], Gerard J. Allan[4], Paul Keim[2], Jacques Ravel[1,7], Pablo D. Rabinowicz[1,3,8]

[1] Institute for Genome Sciences, University of Maryland School of Medicine, Baltimore, Maryland, United States of America

[2] Center for Microbial Genetics and Genomics, Northern Arizona University, Flagstaff, Arizona, United States of America

[3] J. Craig Venter Institute, Rockville, Maryland, United States of America

[4]Department of Biological Sciences, Environmental Genetics and Genomics Laboratory, Northern Arizona University, Flagstaff, Arizona, United States of America

[5] Department of Biology, Indiana University, Bloomington, Indiana, United States of America

[6] Pathogen Genomics Division, Translational Genomics Research Institute, Phoenix, Arizona, United States of America

[7] Department of Microbiology and Immunology, University of Maryland School of Medicine, Baltimore, Maryland, United States of America

[8] Department of Biochemistry and Molecular Biology, University of Maryland School of Medicine, Baltimore, Maryland, United States of America

ABSTRACT

Castor bean is an important oil-producing plant in the Euphorbiaceae family. Its high-quality oil contains up to 90% of the unusual fatty acid ricinoleate, which has many industrial and medical applications. Castor bean seeds also contain ricin, a highly toxic Type 2 ribosome-inactivating protein, which has gained relevance in recent years due to biosafety concerns. In order to gain knowledge on global genetic diversity in castor bean and to ultimately help the development of breeding and forensic tools, we carried out an extensive chloroplast sequence diversity analysis. Taking advantage of the recently published genome sequence of castor bean, we assembled the chloroplast and mitochondrion genomes extracting selected reads from the available whole genome shotgun reads. Using the chloroplast reference genome we used the methylation filtration technique to readily obtain draft genome sequences of 7 geographically and genetically diverse castor bean accessions. These sequence data were used to identify single nucleotide polymorphism markers and phylogenetic analysis resulted in the identification of two major clades that were not apparent in previous population genetic studies using genetic markers derived from nuclear DNA. Two distinct sub-clades could be defined within each major clade and large-scale genotyping of castor bean populations worldwide confirmed previously observed low levels of genetic diversity and showed a broad geographic distribution of each sub-clade.

INTRODUCTION

Castor bean (*Ricinus communis*) is an oilseed crop in the Euphorbiaceae family, which includes 245 genera comprising 6,300 species [1]. Other important members of this family include tropical crops such as cassava (*Manihot esculenta*) and rubber tree (*Hevea brasiliensis*), as well as ornamental poinsettias (*Euphorbia pulcherrima*), the invasive weed leafy spurge (*Euphorbia esula*), and the parasitic *Rafflesia* that has the largest flower in the plant kingdom[2].

Castor bean is cultivated in tropical and subtropical areas of the world for oil production as well as an ornamental. The world annual castor oil production in 2004–5 was around 0.5 million tons[3] and the U.S. is among the world largest importers of castor oil and its derivatives [4], [5]. Castor oil contains 90% of the unusual fatty acid ricinoleate (12-hydroxy-oleate), whose hydroxy group has unique chemical and physical properties that make castor oil a vital industrial raw material for numerous products [6], including high-quality lubricants, paints, coatings, plastics, soaps, medications for skin affections, and cosmetics. Castor oil can also be used as a lubricity additive to replace sulfur-based lubricant components in petroleum diesel helping to reduce sulfur emissions [7]. Moreover, Brazil, a highly advanced country in the use of ethanol as motor fuel, is starting to develop the industry of castor oil as a biodiesel component[8].

Castor bean seeds also contain ricin, a highly toxic water-soluble protein. Ricin is a Type 2 ribosome-inactivating enzyme [9] that can be extracted from seeds through a relatively simple process, raising concerns that it could be used as a bioweapon. This, added to its importance as an oil crop, poses a need to gain knowledge of the castor bean plant and its genetic variation in order to help the development of breeding and forensic tools. To this end, organelle genomic sequences can be used to develop robust markers for genetic and phylogenetic studies due to their high level of conservation.

The chloroplast genome contains genes that code for structural and functional components of the organelle. In land plants, chloroplast genomes are contained in a circular molecule of DNA ranging between 115 and 165 kb in size [10]. It contains two copies of a duplicated region arranged as inverted repeats (IR), separated by a large single-copy (LSC) sequence and a small single-copy (SSC) sequence [11], [12].

Complete chloroplast genomic sequences can be obtained using a shotgun strategy [13]. This approach involves DNA purification from isolated organelles [14], mechanical shearing of the DNA, and cloning of random fragments for sequencing [15], [16]. Cloned restriction fragments from purified chloroplast DNA have also been used to complete chloroplast genome sequencing [17], [18]. If a reference chloroplast genome from a related species is available, primers can be designed to amplify multiple PCR fragments to cover and assemble the whole chloroplast genome [19], [20]. Recently, the rolling circle amplification technique [21] has been used to amplify chloroplast DNA and generate the necessary quantities of DNA for standard Sanger capillary sequencing [22], [23] or next-generation sequencing [24]. In addition, chloroplast genomes are often assembled as byproducts of plant genome sequencing projects. Complete chloroplast genome sequences have been obtained using genomic bacterial artificial chromosome (BAC) libraries. In these cases, the libraries are screened by hybridization or PCR amplification of a conserved chloroplast sequence to identify large-insert BAC clones that contain chloroplast DNA. A few BAC clones that cover the complete chloroplast genome are then sequenced by a shotgun strategy [25].

Here we report the compete castor bean chloroplast and mitochondrion genome sequences generated from a whole genome shotgun (WGS) sequencing project of the cultivar (cv.) Hale[26] along with draft chloroplast genome sequences obtained by methylation filtration (MF) [27]from additional castor bean cultivars. MF is used for selectively cloning and sequencing hypomethylated (*i.e.* low-copy) sequences from plant genomes, reducing the recovery of repetitive elements that are usually methylated [28], [29], [30]. As the chloroplast DNA is not methylated [31], [32], MF libraries are typically constructed using nuclear DNA to reduce the recovery of mitochondrion and, mainly, chloroplast DNA clones, which are preferentially selected in MF libraries [33]. Therefore, we used a MF approach to sequence the chloroplast genomes of seven additional castor bean accessions in order to identify polymorphisms for phylogenetic and population genetic studies.

Castor bean has low genetic diversity based on AFLP and SSR studies [34], and recent single nucleotide polymorphism (SNP) analysis in a worldwide collection of accessions identified five main groups of castor bean showing a mixture of genotypes in most geographical regions studied [35]. The chloroplast SNPs analysis described here, contributes to the characterization of the genetic diversity and global castor bean population structure.

RESULTS AND DISCUSSION

Assembly of the Castor Bean Cv. Hale Organelle Genomes

Organelle genomic sequences were extracted from the castor bean cultivar Hale sequence database [26]. The chloroplast genome was assembled using the Celera Assembler [36] using short (3.5–9 kbp) and long (40 kbp) paired-end Sanger reads that showed high similarity to other dicot chloroplast genomes. The 40 kbp insert size fosmid reads allowed us to resolve the proper assembly and orientation of the large inverted repeat, common in angiosperm chloroplasts. This assembly was used to identify additional chloroplast sequences from a larger pool of castor bean sequences that were used to close the gaps in the draft assembly. The resulting closed circular molecule contained 163,161 bp with a near-perfect inverted repeat of 27,347 bp.

Annotation of the finished chloroplast molecule predicted 129 genes, including 38 tRNA genes, 8 ribosomal RNA genes, and 83 protein-coding genes (Figure 1). Each of the 27 kbp inverted repeats contains 17 genes. Fourteen genes contain one intron and 3 genes contain 2 introns. Cassava is the closest relative of castor bean whose chloroplast genome has been sequenced and annotated [37]. Sequence alignment of these two genomes showed a 93% sequence identity over approximately 150 of the 161 kbp of the castor bean chloroplast genome. The gene content of both chloroplasts is identical, and the presence of the conserved *atpF* group II intron in castor bean, which is absent in cassava [37], was confirmed. At the amino acid level, an average of 91% identity was observed between the castor bean and cassava protein-coding genes.

The large (LSC, 89,652 bp) and small (SSC, 18,817 bp) single copy regions are separated by two inverted repeats (IRa and IRb, 27,347 bp each), which are colored in grey. The outer most whorl (4) contains genes transcribed clockwise, and the genes on whorl 3 are transcribed counter-clockwise. The inner-most whorl (1) shows the alignment of a nucleotide blast of *R. communis* against all cassava (*M. esculenta*) tRNAs and rRNAs from the plastid genome. Whorl 2 shows the alignment of a nucleotide to protein blast (BLASTX) of *R. communis* against all proteins of the cassava plastid genome (*M. esculenta*). * Split genes or genes with introns. ** Trans-spliced gene*rps12*.

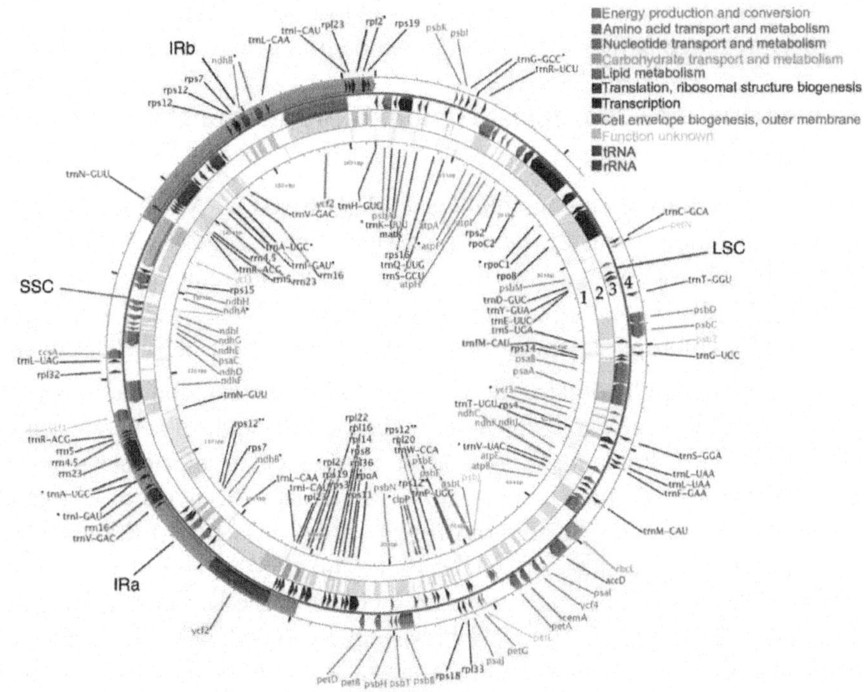

Figure 1: Map of the castor bean chloroplast.

Sequence reads belonging to the mitochondrion genome were extracted using a custom script and assembled using the CAP3 assembler [38]. The assembled genome represents a consensus of the multiple possible genomic and subgenomic configurations that can exist in vivo [39]. The high confidence consensus mitochondrial genome spans 502,773 bp. Due to its larger size and lower abundance, the mitochondrial genome is less suitable for marker development in multiple cultivars than the chloroplast counterpart. Therefore, we did not further utilize the mitochondrion genome for genetic diversity analysis.

The mitochondrial genomes typically code for a few RNA genes as well as some ribosomal proteins and subunits of the oxidative phosphorylation complexes. The castor bean mitochondrial genome contains 37 protein-coding and 3 ribosomal RNA genes, commonly found in angiosperms (Figure S1). Horizontal gene transfer to the nuclear genome often results in variability regarding presence or absence of genes in the mitochondrial genome. In castor bean, the genes coding for ribosomal proteins L2 and S14, often found in plant mitochondrial genomes [40], have been transferred to the nuclear genome, while the gene coding for ribosomal protein L10, which has been recently found in the nuclear or mitochondrial genomes in different angiosperms [41], [42], has been retained in the castor bean mitochondrial genome.

Sequencing the Chloroplast Genomes of Additional Castor Bean Accessions

Based on diversity groupings previously determined by AFLP analysis [34], we selected seven castor bean accessions representing genetically and geographically diverse varieties that originated in different parts of the world (Ethiopia, India, U.S. Virgin Islands, Puerto Rico, El Salvador, Greece, and Mexico) for chloroplast genome sequencing and subsequent SNP discovery.

In order to rapidly obtain chloroplast sequence information we constructed a MF library for each accession and sequenced both ends of several thousand clones per library (Table 1). To estimate the enrichment in chloroplast sequences obtained by MF of leaf genomic DNA from castor bean we also prepared standard shotgun genomic libraries for the accessions from Ethiopia and El Salvador. While the shotgun sequences produced 10% and 9% chloroplast sequences in these two accessions, respectively, the proportion of plastid sequences went up to 29% and 24%, respectively, (nearly a 3-fold enrichment) in the MF libraries. On average, the proportion of chloroplast sequences in the seven accessions' MF libraries was 35% with a maximum of 58% in the Greek accession (Figure 2).

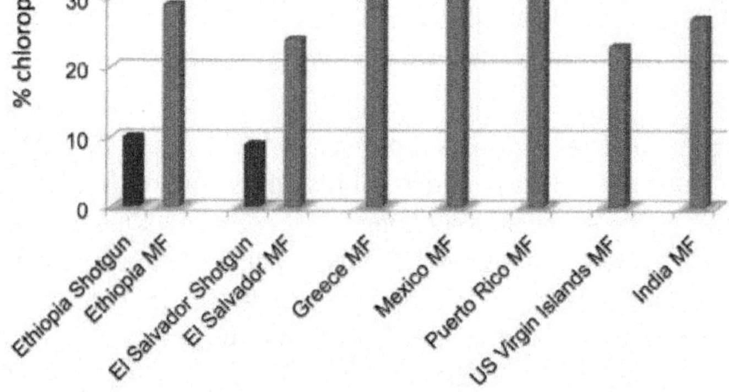

Figure 2: Chloroplast read enrichment in methylation filtration libraries.

Percentage of chloroplast reads in each shotgun (blue) or MF (green) library. Available shotgun reads for cultivars from Ethiopia and El Salvador are shown to highlight the enrichment in chloroplast sequences in MF libraries.

Table 1: MF and WGS sequencing of 7 additional castor bean accessions

Origin	Castor bean accession	Number of shotgun reads	% chloroplast shotgun reads	Number of MF reads	% chloroplast MF reads	Number of contigs
Ethiopia	PI 193851	9,216	10	2,304	29	2
El Salvador	PI 197048	192	9	7,104	24	21
Greece	PI 280219	N/A	N/A	4,032	58	3
Mexico	PI 255238	N/A	N/A	6,144	45	3
Puerto Rico	PI 209132	N/A	N/A	5,376	42	8
US Virgin Islands	PI 209326	N/A	N/A	6,912	23	15
India	PI 173946	N/A	N/A	5,376	27	38

doi:10.1371/journal.pone.0021743.t001

The MF (or the combination of MF and shotgun) reads were assembled for each accession using the AMOS-Cmp assembler, which uses a reference sequence to guide the assembly[43]. We used the closed chloroplast genome that we obtained from the cv. Hale as reference and generated a draft assembly for the chloroplast genome of each cultivar. In this way we were able to order and orient all contigs in each genome, producing a single scaffold for each one. In general, we obtained fairly complete assemblies that would allow SNP identification. Thus, we did not attempt to do any manual work to finish the sequences. Nevertheless, alignment of all the reads in each genome to the reference showed that most of the gaps in the assemblies could potentially be covered by existing sequence reads after manual analysis (Figure 3). In order to obtain genotypic information for some SNPs that fell within gaps in some of the genomes, we performed additional assemblies of the same sequences with the Celera assembler, which uses a different algorithm [36]. In this way we managed to close some gaps, increasing the chances of identifying SNPs with genotypic information in all seven cultivars.

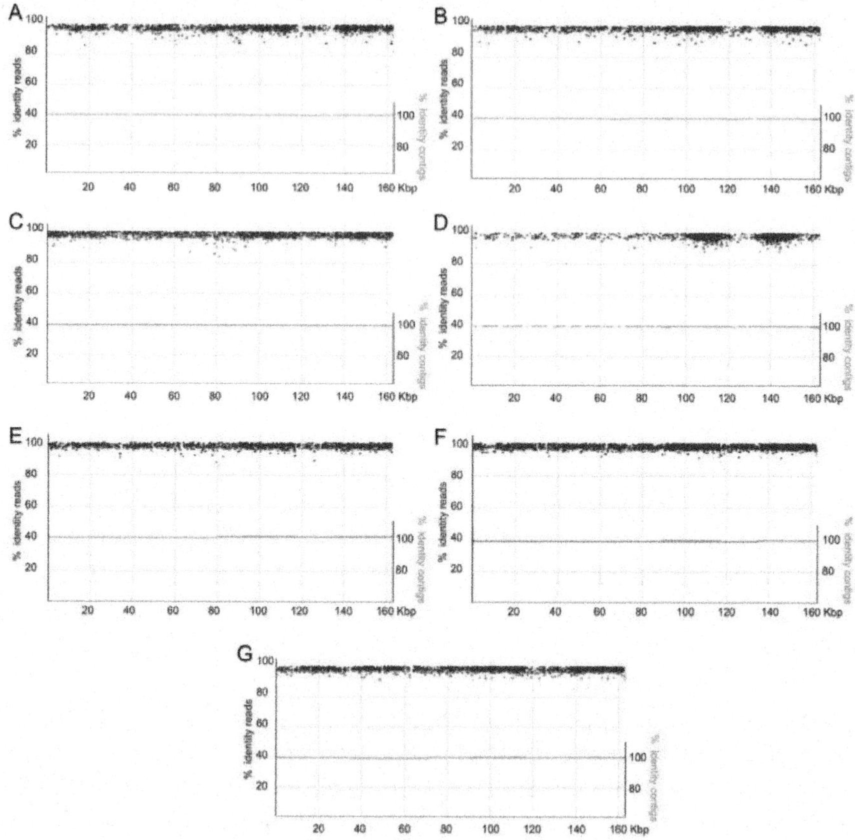

Figure 3: Draft assemblies of the chloroplast genomes of 7 castor bean accessions.

Assembled contigs corresponding to the chloroplast genomes of the different accessions are shown in orange, mapped to the reference v. Hale chloroplast genome (x-axis). Individual reads are shown in blue, aligned to the same reference genome. Left y-axis corresponds to the percent identity of each read relative to the reference cv. Hale sequence. Right y-axis shows the percent identity of contig relative to the Hale sequence. Each panel corresponds to a different cultivar: A) US Virgin Islands, B) El Salvador, C) Puerto Rico, D) India, E) Ethiopia, F) Greece, and G) Mexico.

Chloroplast DNA sequences are useful for phylogenetics and population genetics [44], [45], as well as cultivar identification and forensic analyses [46]. Our results show that MF is a cost-effective way to obtain sequence information of plant organelle genomes given that it does not require time-consuming plastid preparations, library screenings, or amplifications. With this approach, we were able to identify SNPs that allowed us to conduct a worldwide analysis of the castor bean accessions as described below.

SNP Identification and Phylogenetic Analysis

We searched for candidate SNPs in pair-wise alignments between each of the seven castor bean chloroplast genome assemblies and that of cv. Hale. Using custom bioinformatics analyses and stringent criteria (see Methods), 112 SNPs were identified (Table S1). One of them showed a triallelic nature so it was discarded and 29 were not covered in all seven chloroplast genome assemblies. Of these 29, SNP#111 was covered in all genome assemblies except that of the Indian cv. However, because the Greek cv. showed a rare allele for this SNP and was specific to this cultivar (see below) we genotyped this SNP using a TaqMan assay in the Indian cv. to verify that the Greek cv. allele was not present in any of the other genomes. As a result we selected a set of 83 SNPs for which there is high quality sequence data for all cultivars (including SNP#111, Figure 4A) and we used this SNP set to construct a chloroplast-based phylogeny (Figure 4B). Two main groups were evident among the seven castor bean cultivars and we called them clades A and B, which are separated by 69 of the 83 high-quality SNPs.

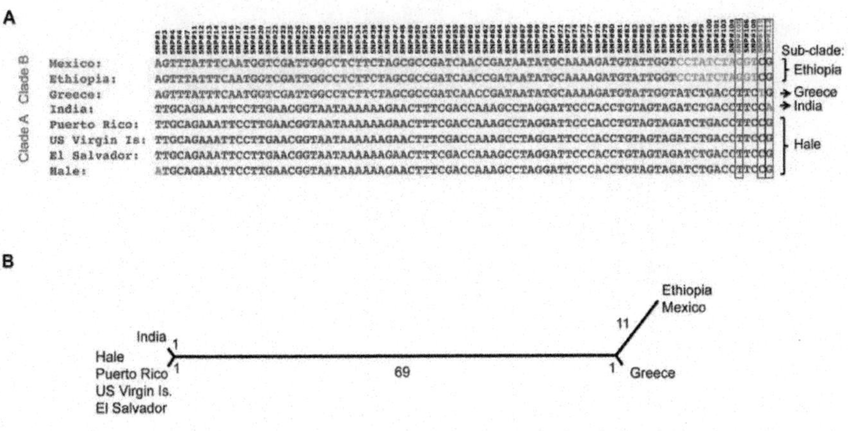

Figure 4: SNP identification and phylogeny obtained from genomic sequencing of 8 castor bean accessions.

A) 83 high quality SNPs identified in the chloroplast genomes of the 8 sequenced chloroplast accessions. The two major clades are shown in red and blue and SNPs that differentiate different sub-clades are shown in orange or green. B) Phylogeny of the 8 castor bean accessions based on the identified SNPs. Two major clades, “A” and “B”, are shown separated by 69 SNPs. Members of each sub-clade are indicated in each branch of the phylogeny.

Members of clade A are highly similar, with only SNP#112 unique to the Indian accession (designating the India sub-clade). The accessions from Puerto Rico, Virgin Islands, and El Salvador were monomorphic for all SNPs and differed from cv. Hale only at SNP#1 (Figure 4A). However, because no assay could be designed for this SNP, these four cultivars are hereafter referred to as the Hale sub-clade. For clade B, the Ethiopian and Mexican accessions did not show polymorphisms for any of the SNPs analyzed and we refer to these accessions as the Ethiopia sub-clade as defined by SNP#105. SNP#111 allowed the identification of the Greek cv. genotype (Greece sub-clade). Thus, the eight accessions with chloroplast sequences were divisible into four sub-clades (Hale, India, Greece, and Ethiopia). We designed TaqMan assays specific to these sub-clades in order to distinguish individuals belonging to each sub-clade among 894 castor bean samples from accessions collected around the world. Overall, genotyping of the worldwide population with these assays showed that limited chloroplast genetic diversity exists throughout the world in castor bean, which is consistent with the low genetic variation observed for nuclear genomic sequences [34], [35].

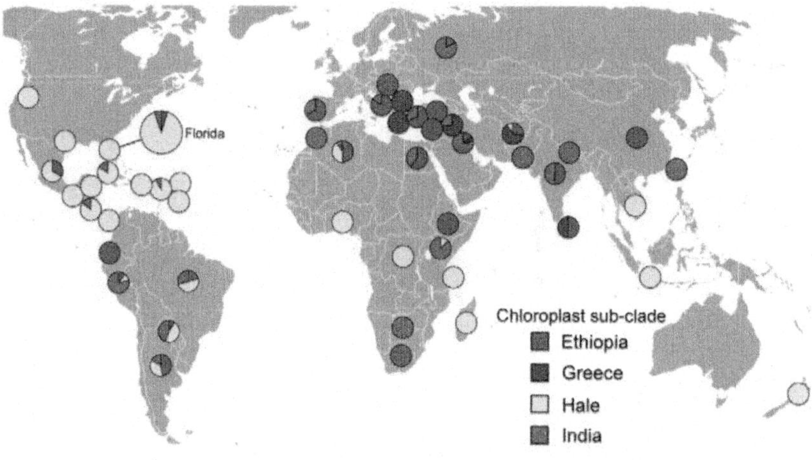

Figure 5: Worldwide distribution of chloroplast genotypes of castor bean from a collection of 894 accessions.

Nevertheless, we could determine that each sub-clade was broadly spread geographically (Figure 5). Within clade A, the Hale sub-clade genotype was distributed throughout the world, dominating in collections from North and Central America, the Caribbean, and parts of South America, Africa, and Asia (Table 2). Members of the India sub-clade, within clade A as well, were also found widespread on every continent sampled. Members of the clade B were found less frequently than those of clade A, with the Ethiopia sub-clade being more widely distributed while the Greece sub-clade was mostly limited to the Middle East and Sri Lanka. The geographic origin of the different accessions could not always be correlated with its genotype. For example, the Mexican and Ethiopian cvs. showed no polymorphisms for any of the SNPs we tested, although the majority of samples from Mexico fell in the Hale sub-clade. A similar scenario was found in clade A, within which there is very little variation but included the Indian cv. together with mostly American varieties. Distribution of chloroplast genotypes corresponding to the sub-clades shown in Figure 4, based on the origin of each accession. The pie chart corresponding to Florida is expanded to reflect the larger number of samples (n=272) that came from that state, relative to other parts of the world.

Table 2: Country or state of origin and the chloroplast sub-clade for 894 *Ricinus communis* samples collected worldwide.

	Number of samples per sub-clade			
Origin	**Ethiopia**	**Greece**	**Hale**	**India**
Afghanistan	3	7	1	
Algeria	5		4	1
Argentina	25		19	9
Bahamas			8	
Benin			8	
Botswana				11
Brazil	9		22	13
Cambodia			8	
China	5			
Costa Rica			5	1
Cuba			17	3
Ecuador	4			
Egypt	3			2
El Salvador			7	
Ethiopia	4			
Florida	13		244	15
Greece		3		
Guatemala			8	
Hale			5	
Hungary		4		
India	52		3	49

Indonesia			8	
Iran	5	3	1	24
Israel	5			
Jamaica			11	1
Jordan	5			
Kenya			1	7
Madagascar			8	
Mexico	3		6	
Morocco				5
Nepal				8
New Zealand			1	
Oregon			3	
Pakistan				6
Panama			8	
Paraguay	1		5	5
Peru	3		3	23
Portugal	4			2
Puerto Rico			8	
Russia	1			5
Serbia	5			1
South Africa	4			
Sri Lanka	1	1		
Syria		8		3
Taiwan				2
Texas			1	
Turkey	40			18
Uruguay			11	
US Virgin Islands			8	
Yugoslavia	5			
Zaire			5	
Zanzibar			2	
Total	205	26	449	214

doi:10.1371/journal.pone.0021743.t002

The two main phylogenetic groups obtained based on chloroplast DNA sequence variation argue for two ancient and differentiated populations that merged to generate the current global populations. In fact, two robust ancestral populations appear to have been involved in the creation of modern hybrid *Ricinus* populations —perhaps even different subspecies. This division into two groups was much less apparent from previous analysis of nuclear DNA sequence variation data [34], [35], likely because of extensive recombination. This has not happened in the cytoplasmic DNAs due to the lack of recombination. It is somewhat fortuitous that the two main groups still exist since cytoplasmic DNA would tend to rapidly go to fixation as one type or the other, especially in small populations.

We compared the nuclear SNP data where five main groups where encountered worldwide [35]with the four chloroplast sub-clades. A strong relationship occurred between groupings based on nuclear and chloroplast SNPs, particularly for the Hale cultivar genotype. Because 451 of the samples that we analyzed in this study had also been analyzed with nuclear SNPs [35], we could compare the grouping results from both types of SNP markers for exactly the same samples. For this subset of samples, 96.1% (171 of 178) of the Hale chloroplast sub-clade genotypes were also grouped together when analyzed with nuclear SNPs. This correspondence was not complete for comparisons between other groups and their most likely population assignment: India 68% (34 of 50), Ethiopia 57% (59 of 86), and Greece 8% (11 of 137).

Global patterns of genetic variation in castor bean have likely been influenced by agricultural and horticultural practices. The pattern of introduction of *R. communis* from multiple chloroplast groupings in many countries is consistent with genetic patterns from nuclear SNPs [35]. In some areas however, such as North America, this is not evident from chloroplast DNA probably because of the limited diversity found in the chloroplast genome. We caution that this low level of genetic diversity may be related to the fact that our samples were limited primarily to those available from the U.S. Department of Agriculture (USDA) germplasm collection, which may not represent the totality of the world's geographic and genetic diversity for castor bean.

MATERIALS AND METHODS

Plant Material

Castor bean accessions were obtained from USDA's National Plant Germplasm System, Agricultural Research Center in Griffin, Georgia and some of the accessions for the worldwide population analysis were also obtained from commercial sources or collected in the wild in the state of Florida, USA. The accessions selected for chloroplast genome sequencing were: Ethiopian cv.: PI193851, Greek cv.: PI280219, Mexican cv.: PI255238, Puerto Rican cv.: PI209132, Indian cv.: PI173946, Salvadorian cv.: PI197048, and cv. from U.S. Virgin Islands: PI209326. Plants were grown in normal greenhouse conditions and leaf samples were collected and dried before DNA extraction as described [34]. Total DNA was prepared using DNeasy kits (Qiagen) following the manufacturer's instructions.

Organelle Genome Sequencing, Assembly and Annotation

Assembly of the castor bean cv. Hale chloroplast genome was carried out by selecting chloroplast reads from the database of whole genome shotgun Sanger reads generated for genome assembly [26]. This database includes paired-end sequence reads from multiple plasmid and fosmid libraries. From this database, we extracted a set of approximately 7,000 3–4 kbp-insert paired-end plasmid reads and selected 765 reads that could be aligned at high stringency against available dicot chloroplast genome sequences from GenBank using sequence similarity searches. These reads were assembled using the Celera Assembler [36]. Subsequently, additional chloroplast reads from fosmid clones were added to the assembly to allow building and orienting the 27 kbp inverted repeats. We then performed standard finishing work to close sequence gaps and resolve ambiguities [47].

Because the sequence and structure of angiosperm mitochondrial genomes are poorly conserved and include unique sequences, it is relatively difficult to identify all reads corresponding to the mitochondrial genome when mixed with a nuclear genome WGS project. To address this difficulty, we developed a method for extending initially assembled reads by iteratively comparing all reads to the ends of the extending assemblies. By examining the read depths of the resulting assemblies, mitochondrial sequences could be separated from those of nuclear and chloroplast origin. The resulting reads were assembled using CAP3 [38].

The chloroplast genomes of the seven additional castor bean cultivars were sequenced using MF libraries. Briefly, castor bean total DNA was purified from leaves and was randomly sheared by nebulization, end-repaired with consecutive BAL31 nuclease and T4 DNA polymerase treatments, and 1.5 to 3 kb fragments were eluted from a 1% low-melting-point agarose gel after electrophoresis. After ligation to BstXI adaptors, DNA was purified by three rounds of gel electrophoresis to remove excess adaptors, and the fragments were ligated into the vector pHOS2 (a modified pBR322 vector) linearized with BstXI. The pHOS2 plasmid contains two BstXI cloning sites immediately flanked by sequencing primer binding sites. The ligation reactions were introduced by electroporation into *E. coli* strain GC10 (*McrBC*$^-$) for regular shotgun libraries or strain DH5α (*McrBC*$^+$) for MF libraries.

Draft assemblies of the genomes of the seven castor bean cultivars sequenced were conducted using AMOS-Cmp [43] using the closed cv. Hale genome sequence as reference. Additional assemblies of the same sequence data were performed with the Celera assembler [36] in order to cover some of the gaps that contained candidate SNPs.

Chloroplast genome annotation was carried out using the software tool DOGMA (Dual Organellar GenoMe Annotator) [48] followed by manual inspection of intron-containing genes. Comparative analysis of castor bean and cassava chloroplast genomes was carried out using CGView [49]. The mitochondrion genome was annotated using comparative analysis to identify mitochondrial genes, taking into account exons and introns. Intergenic regions were further annotated as possible using sequence similarity searches against GenBank.

The finished and annotated castor bean chloroplast and mitochondrion genome sequences from the cv. Hale can be found under GenBank accession numbers JF937588 and HQ874649, respectively. The draft sequence assemblies of the 7 castor bean accessions chloroplast genomes can be found under GenBank accession numbers JF940518 (Ethiopia), JF940520 (India), JF940515 (U.S. Virgin Islands), JF940521 (Puerto Rico), JF940517 (El Salvador), JF940519 (Greece), and JF940516 (Mexico).

SNP Discovery

In order to call candidate SNPs in pair-wise comparisons of the seven castor bean chloroplast genomes versus the Hale cv. genome we applied the following criteria: reads must be longer than 500 bp, the polymorphic base should be covered by at least 3 reads and should not be located in the first or last 30 bp of any read, the Phred quality score of the polymorphic base and mean scores of the 5 bases upstream and 5 bases downstream should be greater than 30, and the polymorphic base should be covered in all cultivars meeting the previous criteria.

Genotyping Assay Development and Screening

We developed three Real-Time TaqMan (Applied Biosystems) minor groove binding (MGB) assays for the four castor bean sub-clades. Primers and probes were designed using Primer Express for TaqMan MGB assays for Allelic

Discrimination version 3.0 (Applied Biosystems). Each assay had dual-probes targeting the specific SNP (Table 3). Although a single SNP separates the Hale cultivar from the cultivars from Puerto Rico, U.S. Virgin Islands, and El Salvador, sequence composition prevented the development of a reliable assay, so these closely related cultivars were grouped together.

Table 3: Sequence for TaqMan primers and probes for defining castor bean chloroplast sub-clades

SNP	Sub-clade	Forward primer sequence	Reverse primer sequence	Sub-clade specific probe*	Alternate probe*
105	Ethiopia	GACATTCCGTCTTCTGAAACCAA	CTAACTATAGTGCAAAGTCGCATCTCTT	TTTCTATATGCC<u>G</u>ATTATGG	TTTCTATATGCC<u>T</u>ATTATG
111	Greece	GATTGATTGGCTGATGTTTCAAAA	AAAGAAACGTCTGTATTCAGCTACAAAG	ATACCCAAAG<u>T</u>TCC	ATACCCAAAG<u>C</u>TCCCA
112	India	TGTCAGGCTATTGTTCTCCTGTTC	GGGAGTCCATCATGTAATCAAAAGA	CTAAAAGTAATG<u>A</u>AGTAAGAC	AAGTAATG<u>G</u>AGTAAGACATC

*SNP-state is underlined in each probe.
doi:10.1371/journal.pone.0021743.t003

We ran the genotyping assays on an ABI Prism 7900HT Sequence Detection System (Applied Biosystems). Each 10 μl PCR mixture contained 1× AB TaqMan Universal Master mix, 0.9 μM each primer, and 0.25 μM each probe. We added 0.2 U of Platinum Taq (Invitrogen) per reaction to increase the efficiency of the amplification. Each assay was run at standard conditions: 2 min inactivation at 50°C, a 10 min hot start at 95°C, followed by 40 cycles of 15 sec denaturation and 1 min annealing at 60°C. Amplification and allelic discrimination plots were visualized using SDS version 2.3 software. We screened 894 samples from 49 countries and 3 US states against all three SNP assays.

ACKNOWLEDGMENTS

We thank the J. Craig Venter Institute's Joint Technology Center and the Informatics group for carrying out sequencing and bioinformatics support, respectively.

AUTHOR CONTRIBUTIONS

Conceived and designed the experiments: JR PK PDR. Performed the experiments: MR JTF APC ALW. Analyzed the data: MR JTF APC DWR SMB-S GJA PDR JR PK. Contributed reagents/materials/analysis tools: AM-B HHC DP MJR HMK XL. Wrote the paper: PDR MR JTF. Edited the manuscript: PK GJA JR DWR

REFERENCES

1. Wurdack KJ, Hoffmann P, Chase MW (2005) Molecular phylogenetic analysis of uniovulate Euphorbiaceae (euphorbiaceae sensu stricto)

Using plastid rbcl and trnl-f dna sequences. American Journal of Botany 92: 1397–1420.

2. Davis CC, Latvis M, Nickrent DL, Wurdack KJ, Baum DA (2007) Floral gigantism in Rafflesiaceae. Science 315: 1812.
3. Gunstone F (2005) Fatty acid production for human consumption. Inform 16: 736–737.
4. Roetheli J, Glaser L, Brigham R (1991) Castor: Assessing the feasibility of US production. Washington, DC: USDA-CSRS, Office of Agricultural Materials.
5. DeVries J, Toenniessen G (2001) Securing the Harvest. Biotechnology, Breeding and Seed Systems for African Crops. New York: CABI Publishers.
6. Brigham R (1993) Castor: Return of an old crop. In: Simon JJaJE, editor. New crops. New York, USA: Wiley & Sons.
7. Goodrum JW, Geller DP (2005) Influence of fatty acid methyl esters from hydroxylated vegetable oils on diesel fuel lubricity. Bioresour Technol 96: 851–855.
8. da Silva Nde L, Maciel MR, Batistella CB, Filho RM (2006) Optimization of biodiesel production from castor oil. Appl Biochem Biotechnol 129–132: 405–414.
9. Endo Y, Mitsui K, Motizuki M, Tsurugi K (1987) The mechanism of action of ricin and related toxic lectins on eukaryotic ribosomes. The site and the characteristics of the modification in 28 S ribosomal RNA caused by the toxins. J Biol Chem 262: 5908–5912.
10. Palmer M (1991) Plastid chromosomes: structure and evolution. In: Vasil IKaB L, editor. The Molecular Biology of Plastids. San Diego: Academic Press. pp. 5–53.
11. Sugiura M, Hirose T, Sugita M (1998) Evolution and mechanism of translation in chloroplasts. Annu Rev Genet 32: 437–459.
12. Sugiura M (1992) The chloroplast genome. Plant Mol Biol 19: 149–168.
13. Anderson S (1981) Shotgun DNA sequencing using cloned DNase I-generated fragments. Nucleic Acids Res 9: 3015–3027.
14. Sandbrink JM, Vellekoop P, Vanham R, Vanbrederode J (1989) A method for evolutionary studies on RFLP of chloroplast DNA, applicable to a range of plant species. Biochem Syst Ecol 17: 45–49.
15. Steane DA (2005) Complete nucleotide sequence of the chloroplast

genome from the Tasmanian blue gum, Eucalyptus globulus (Myrtaceae). DNA Res 12: 215–220.

16. Chumley TW, Palmer JD, Mower JP, Fourcade HM, Calie PJ, et al. (2006) The complete chloroplast genome sequence of Pelargonium×hortorum: organization and evolution of the largest and most highly rearranged chloroplast genome of land plants. Mol Biol Evol 23: 2175–2190.
17. Maier RM, Neckermann K, Igloi GL, Kossel H (1995) Complete sequence of the maize chloroplast genome: gene content, hotspots of divergence and fine tuning of genetic information by transcript editing. J Mol Biol 251: 614–628.
18. Kim KJ, Lee HL (2004) Complete chloroplast genome sequences from Korean ginseng (Panax schinseng Nees) and comparative analysis of sequence evolution among 17 vascular plants. DNA Res 11: 247–261.
19. Shahid Masood M, Nishikawa T, Fukuoka S, Njenga PK, Tsudzuki T, et al. (2004) The complete nucleotide sequence of wild rice (Oryza nivara) chloroplast genome: first genome wide comparative sequence analysis of wild and cultivated rice. Gene 340: 133–139.
20. Chung HJ, Jung JD, Park HW, Kim JH, Cha HW, et al. (2006) The complete chloroplast genome sequences of Solanum tuberosum and comparative analysis with Solanaceae species identified the presence of a 241-bp deletion in cultivated potato chloroplast DNA sequence. Plant Cell Rep 25: 1369–1379.
21. Dean FB, Nelson JR, Giesler TL, Lasken RS (2001) Rapid amplification of plasmid and phage DNA using Phi 29 DNA polymerase and multiply-primed rolling circle amplification. Genome Res 11: 1095–1099.
22. Lee SB, Kaittanis C, Jansen RK, Hostetler JB, Tallon LJ, et al. (2006) The complete chloroplast genome sequence of Gossypium hirsutum: organization and phylogenetic relationships to other angiosperms. BMC Genomics 7: 61.
23. Bausher MG, Singh ND, Lee SB, Jansen RK, Daniell H (2006) The complete chloroplast genome sequence of Citrus sinensis (L.) Osbeck var 'Ridge Pineapple': organization and phylogenetic relationships to other angiosperms. BMC Plant Biol 6: 21.
24. Moore MJ, Dhingra A, Soltis PS, Shaw R, Farmerie WG, et al. (2006) Rapid and accurate pyrosequencing of angiosperm plastid genomes. BMC Plant Biol 6: 17.
25. Saski C, Lee SB, Daniell H, Wood TC, Tomkins J, et al. (2005) Complete chloroplast genome sequence of Gycine max and comparative analyses with other legume genomes. Plant Mol Biol 59: 309–322.

26. Chan AP, Crabtree J, Zhao Q, Lorenzi H, Orvis J, et al. (2010) Draft genome sequence of the oilseed species Ricinus communis. Nat Biotechnol 28: 951–956.
27. Rabinowicz PD, Schutz K, Dedhia N, Yordan C, Parnell LD, et al. (1999) Differential methylation of genes and retrotransposons facilitates shotgun sequencing of the maize genome. Nat Genet 23: 305–308.
28. Whitelaw CA, Barbazuk WB, Pertea G, Chan AP, Cheung F, et al. (2003) Enrichment of gene-coding sequences in maize by genome filtration. Science 302: 2118–2120.
29. Palmer LE, Rabinowicz PD, O'Shaughnessy A, Balija V, Nascimento L, et al. (2003) Maize genome sequencing by Methylation Filtration. Science. in press.
30. Bedell JA, Budiman MA, Nunberg A, Citek RW, Robbins D, et al. (2005) Sorghum genome sequencing by methylation filtration. PLoS Biol 3: e13.
31. Fojtova M, Kovarik A, Matyasek R (2001) Cytosine methylation of plastid genome in higher plants. Fact or artefact? Plant Sci 160: 585–593.
32. McCullough AJ, Kangasjarvi J, Gengenbach BG, Jones RJ (1992) Plastid DNA in Developing Maize Endosperm : Genome Structure, Methylation, and Transcript Accumulation Patterns. Plant Physiol 100: 958–964.
33. Rabinowicz PD (2007) Plant genomic sequencing using gene-enriched libraries. Chem Rev 107: 3377–3390.
34. Allan G, Williams A, Rabinowicz PD, Chan AP, Ravel J, et al. (2008) Worldwide genotyping of castor bean germplasm (*Ricinus communis* L.) using AFLPs and SSRs. Genet Resour Crop Evol 55: 365–378.
35. Foster JT, Allan GJ, Chan AP, Rabinowicz PD, Ravel J, et al. (2010) Single nucleotide polymorphisms for assessing genetic diversity in castor bean (Ricinus communis). BMC Plant Biol 10: 13.
36. Myers EW, Sutton GG, Delcher AL, Dew IM, Fasulo DP, et al. (2000) A whole-genome assembly of Drosophila. Science 287: 2196–2204.
37. Daniell H, Wurdack KJ, Kanagaraj A, Lee SB, Saski C, et al. (2008) The complete nucleotide sequence of the cassava (Manihot esculenta) chloroplast genome and the evolution of atpF in Malpighiales: RNA editing and multiple losses of a group II intron. Theor Appl Genet 116: 723–737.
38. Huang X, Madan A (1999) CAP3: A DNA sequence assembly program. Genome Res 9: 868–877.
39. Woloszynska M (2010) Heteroplasmy and stoichiometric complexity of

plant mitochondrial genomes–though this be madness, yet there›s method in›t. J Exp Bot 61: 657–671.

40. Adams KL, Qiu YL, Stoutemyer M, Palmer JD (2002) Punctuated evolution of mitochondrial gene content: high and variable rates of mitochondrial gene loss and transfer to the nucleus during angiosperm evolution. Proc Natl Acad Sci U S A 99: 9905–9912.
41. Mower JP, Bonen L (2009) Ribosomal protein L10 is encoded in the mitochondrial genome of many land plants and green algae. BMC Evol Biol 9: 265.
42. Kubo N, Arimura S (2010) Discovery of the rpl10 gene in diverse plant mitochondrial genomes and its probable replacement by the nuclear gene for chloroplast RPL10 in two lineages of angiosperms. DNA Res 17: 1–9.
43. Pop M, Phillippy A, Delcher AL, Salzberg SL (2004) Comparative genome assembly. Brief Bioinform 5: 237–248.
44. Newton AC, Allnutt TR, Gillies AC, Lowe AJ, Ennos RA (1999) Molecular phylogeography, intraspecific variation and the conservation of tree species. Trends Ecol Evol 14: 140–145.
45. Provan J, Powell W, Hollingsworth PM (2001) Chloroplast microsatellites: new tools for studies in plant ecology and evolution. Trends Ecol Evol 16: 142–147.
46. Sucher NJ, Carles MC (2008) Genome-based approaches to the authentication of medicinal plants. Planta Med 74: 603–623.
47. Birren B, Green ED, Klapholz S, Myers RM, Roskams J (1997) Genome Analysis. A Laboratory Manual. Analyzing DNA. Cold Spring Harbor, New York: Cold Spring Harbor Laboratory Press.
48. Wyman SK, Jansen RK, Boore JL (2004) Automatic annotation of organellar genomes with DOGMA. Bioinformatics 20: 3252–3255.
49. Grant JR, Stothard P (2008) The CGView Server: a comparative genomics tool for circular genomes. Nucleic Acids Res 36: W181–184.

Chapter 8

ANALYSIS OF EAST ASIA GENETIC SUBSTRUCTURE USING GENOME-WIDE SNP ARRAYS

Chao Tian[1] , Roman Kosoy[1] , Annette Lee[2] , Michael Ransom[1] , John W. Belmont[3] , Peter K. Gregersen[2] , Michael F. Seldin[1]

[1] Rowe Program in Human Genetics, Departments of Biochemistry and Medicine, University of California Davis, Davis, California, United States of America

[2] The Robert S. Boas Center for Genomics and Human Genetics, Feinstein Institute for Medical Research, North Shore LIJ Health System, Manhasset, New York, United States of America

[3] Department of Molecular and Human Genetics, Baylor College of Medicine, Houston, Texas, United States of America

ABSTRACT

Accounting for population genetic substructure is important in reducing type 1 errors in genetic studies of complex disease. As efforts to understand complex genetic disease are expanded to different continental populations the understanding of genetic substructure within these continents will be useful in design and execution of association tests. In this study, population differentiation (Fst) and Principal Components Analyses (PCA) are examined using >200 K genotypes from multiple populations of East Asian ancestry. The population groups included those from the Human Genome Diversity Panel [Cambodian, Yi, Daur, Mongolian, Lahu, Dai, Hezhen, Miaozu, Naxi, Oroqen, She, Tu, Tujia, Naxi, Xibo, and Yakut], HapMap [Han Chinese (CHB) and Japanese (JPT)], and East Asian or East Asian American subjects of Vietnamese, Korean, Filipino and Chinese ancestry. Paired Fst (Wei and Cockerham) showed close relationships between CHB and several large East Asian population groups (CHB/Korean, 0.0019; CHB/JPT, 00651; CHB/Vietnamese, 0.0065) with larger separation with Filipino (CHB/Filipino, 0.014). Low levels of differentiation were also observed between Dai and Vietnamese (0.0045) and between Vietnamese and Cambodian (0.0062).

Similarly, small Fst's were observed among different presumed Han Chinese populations originating in different regions of mainland of China and Taiwan (Fst's <0.0025 with CHB). For PCA, the first two PC's showed a pattern of relationships that closely followed the geographic distribution of the different East Asian populations. PCA showed substructure both between different East Asian groups and within the Han Chinese population. These studies have also identified a subset of East Asian substructure ancestry informative markers (EASTASAIMS) that may be useful for future complex genetic disease association studies in reducing type 1 errors and in identifying homogeneous groups that may increase the power of such studies.

INTRODUCTION

Analysis of population genetic substructure has been enhanced by the ability to perform large genome array studies. The differences and patterns of variation within continental populations are useful for several reasons including recapitulating population migration and origins of ethnic groups, forensic identification, and for defining and applying an understanding of allele frequency variation to genetic association studies. Recent studies by several groups including our own have examined European population substructure [1]–[4]. Importantly, these studies have shown that discerning and accounting for differences in substructure can improve error rates in association studies [5]. With the availability of East Asian (EAS) SNP genotypes, we undertook the current study to perform similar studies for this sub-continental region that contains the largest contribution to the world's population. East Asian population genetic structure is particularly important since multiple genetic studies of complex disease are currently underway including studies of autoimmune diseases in Korean, Chinese and Japanese populations[6]–[11]. An understanding of the relationship among these different populations and ascertaining ancestry informative markers (AIMs) that can discern East Asian substructure will undoubtedly facilitate accurate interpretation of such studies[5].

This study combines high density SNP array genotypes from studies of EAS population groups within the Human Genome Diversity Panel (HGDP) [12], [13] with those of several additional population groups of EAS ancestry. The use of high density SNP genotypes containing over 200 K common autosomal genotypes allows a more comprehensive analyses than those previously performed using limited number of autosomal genotypes. It also complements studies of mitochondrial and Y chromosome haplogroups as well as classical markers that provide important information with respect to part of the history of particular EAS ethnic groups[14]–[20]. Our study expands on previous analyses using HGDP population groups [13] by examining additional parameters of population structure/diversity and by including many additional samples including those from several of the most populous EAS groups (Korean, Filipino and Vietnamese) and Chinese American participants of diverse origin. We apply the genotypic information to identify a set of SNPs that may be useful in the design and execution of association studies.

RESULTS

Population Differentiation Between East Asian Populations

To examine similarities and differences in population differentiation among EAS populations paired Fst values were determined between 19 EAS population groups that were typed with genome-wide SNP arrays (see **Methods)**. The studies included samples derived from HapMap[21], [22], HGDP[13], samples collected in Korea and East Asian American participants (see**Methods**). The Fst values were obtained using three random non-overlapping sets of 3500 SNPs distributed over the autosomal genome (minimum of 50 kb distance between SNPs). This approach was taken to limit potential bias from SNPs in close linkage disequilibrium and to measure of variability of Fst. The small differences in these independent samplings (mean SD=0.0015; median SD=0.0013) indicate that this approach resulted in good estimations of paired Fst values. Relatively large Fst values were evident between many of the relatively small ethnic groups within China (Table 1 and see Figure 1 for geographical information). In particular, those population groups derived from Mongolia or near by provinces including Oroqen, Hezhen, and Daur show relatively large differences with Han Chinese. Similarly, two of the ethnic groups in the southeastern region of China, Lahu and Dai, also showed large paired Fst values with Han Chinese. With respect to population groups derived from very populous groups, the data indicate that Japanese and Korean were very closely related, as were Korean and Han Chinese but that these groups are much further from the south-east Asian populations (Filipino and Vietnamese). The Han Chinese and Japanese groups showed larger separation than either with Korean, although the paired Fst values were still small relative to Chinese/Filipino Fst. The Fst values also showed a close relationship between the Dai ethnic group in China and the Vietnamese population sample. Each of the groups had large paired Fst values with the Yakut from Siberia with the exception of the Mongolian, Hezhen and Oroqen ethnic groups that derive from north-eastern China or Mongolia. The relative size of the Fst values also generally corresponded to the geographical separation of the EAS population groups (depicted in Figure 1).

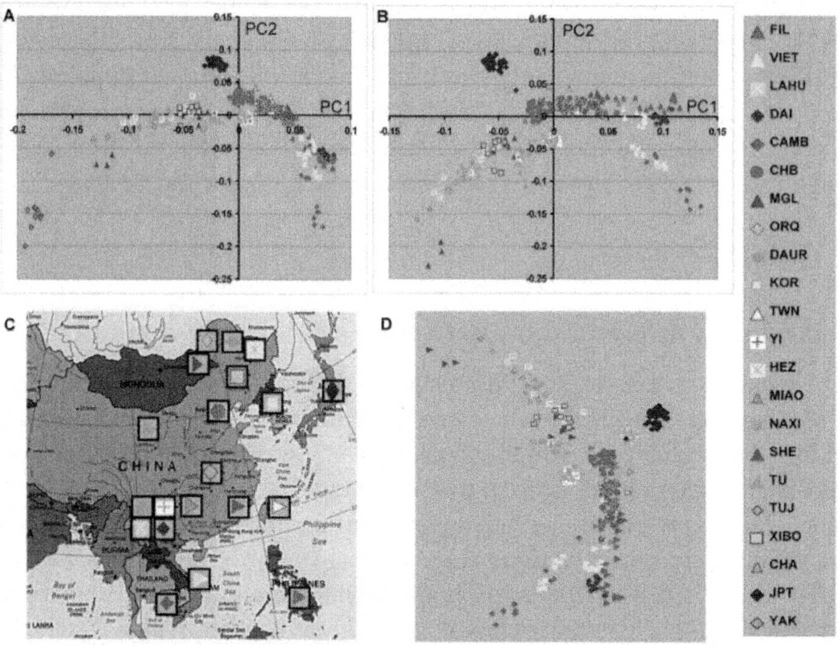

Figure 1: Principal component analyses of substructure in a diverse set of subjects of East Asian descent.

Graphic representation of the first two PCs based on analysis with >200 K SNPs are shown. Color code shows subgroup of subjects for each population group. The subjects included Filipino (FIL), Vietnamese (VIET), Lahu, Dai, Cambodian (CAMB), Han Chinese (CHB), Mongola (MGL), Oroqen (ORQ), Daur, Korean (KOR), Chinese Americans from Taiwan (TWN),Yi, Hezhen (HEZ), Miaozu (MIAO), Naxi, She, Tu, Tujia (TUJ), Xibo, Chinese Americans (CHA), Japanese (JPT), and Yakut (YAK). A, Analyses including the Yakut population group. B, Analysis without Yakut is shown. C, Approximate geographic origin of population group is depicted on a map of East Asia (downloaded from University of Texas Library website). The positions of the HGDP population groups are based on the collection site information[12] and the other population groups were placed based on self-identified country or region of origin. [Note: Yakut are not shown on the map since this population is from Siberia and is a considerable distance north of the depicted region.] D, Shows rotated results of PC1 and PC2 to assist illustration of geographic correspondence of ethnic group locations.

Table 1: Paired Fst values and Standard Deviations for EAS Population Groups[a]

	CHB	KOR	FIL	VIET	CAMB	YAK	YI	DUAR	MGL	LAHU	DAI	HEZH	MIAO	NAXI	OROQ	SHE	TU	TUJ	XIBO
JPT[b]	0.0085+/−0.0003																		
KOR	0.0019+/−0.0002	0.0028+/−0.0003																	
FIL	0.0140+/−0.0007	0.0204+/−0.0004	0.0182+/−0.0015																
VIET	0.0065+/−0.0009	0.0146+/−0.0014	0.0106+/−0.0002	0.0112+/−0.0007															
CAMB	0.0136+/−0.0011	0.0210+/−0.0009	0.0191+/−0.0021	0.0158+/−0.0006	0.0062+/−0.0011														
YAK	0.0289+/−0.0009	0.0297+/−0.0020	0.0279+/−0.0010	0.0430+/−0.0002	0.0376+/−0.0008	0.0377+/−0.0023													
YI	0.0059+/−0.0008	0.0127+/−0.0002	0.0083+/−0.0006	0.0207+/−0.0015	0.0126+/−0.0012	0.0156+/−0.0021	0.0296+/−0.0016												
DUAR	0.0088+/−0.0023	0.0103+/−0.0016	0.0072+/−0.0023	0.0262+/−0.0017	0.0185+/−0.0028	0.0227+/−0.0017	0.0163+/−0.0010	0.0143+/−0.0021											
MGL	0.0049+/−0.0010	0.0082+/−0.0014	0.0052+/−0.0008	0.0212+/−0.0009	0.0139+/−0.0025	0.0161+/−0.0020	0.0141+/−0.0013	0.0085+/−0.0013	0.0081+/−0.0096										
LAHU	0.0223+/−0.0016	0.0295+/−0.0015	0.0271+/−0.0021	0.0300+/−0.0006	0.0219+/−0.0010	0.0237+/−0.0029	0.0511+/−0.0022	0.0261+/−0.0016	0.0346+/−0.0016	0.0280+/−0.0030									
DAI	0.0110+/−0.0004	0.0195+/−0.0007	0.0171+/−0.0023	0.0119+/−0.0020	0.0045+/−0.0008	0.0079+/−0.0020	0.0429+/−0.0015	0.0152+/−0.0018	0.0226+/−0.0006	0.0179+/−0.0016	0.0229+/−0.0016								
HEZH	0.0084+/−0.0011	0.0104+/−0.0016	0.0069+/−0.0010	0.0264+/−0.0009	0.0191+/−0.0001	0.0230+/−0.0009	0.0195+/−0.0032	0.0130+/−0.0010	0.0036+/−0.0007	0.0042+/−0.0005	0.0343+/−0.0037	0.0252+/−0.0016							
MIAO	0.0059+/−0.0010	0.0141+/−0.0003	0.0100+/−0.0001	0.0152+/−0.0010	0.0068+/−0.0008	0.0129+/−0.0005	0.0355+/−0.0019	0.0105+/−0.0008	0.0166+/−0.0017	0.0128+/−0.0017	0.0240+/−0.0032	0.0097+/−0.0014	0.0166+/−0.0004						
NAXI	0.0100+/−0.0014	0.0177+/−0.0021	0.0120+/−0.0010	0.0255+/−0.0025	0.0172+/−0.0004	0.0209+/−0.0008	0.0356+/−0.0023	0.0078+/−0.0013	0.0183+/−0.0030	0.0133+/−0.0033	0.0310+/−0.0007	0.0214+/−0.0016	0.0185+/−0.0010	0.0157+/−0.0019					
OROQ	0.0164+/−0.0011	0.0175+/−0.0003	0.0146+/−0.0007	0.0334+/−0.0027	0.0273+/−0.0038	0.0308+/−0.0021	0.0177+/−0.0006	0.0213+/−0.0014	0.0055+/−0.0014	0.0095+/−0.0008	0.0415+/−0.0010	0.0320+/−0.0028	0.0086+/−0.0013	0.0247+/−0.0028	0.0246+/−0.0032				
SHE	0.0090+/−0.0006	0.0168+/−0.0012	0.0123+/−0.0014	0.0187+/−0.0006	0.0110+/−0.0018	0.0181+/−0.0025	0.0418+/−0.0008	0.0159+/−0.0008	0.0202+/−0.0025	0.0160+/−0.0005	0.0303+/−0.0010	0.0148+/−0.0005	0.0220+/−0.0021	0.0117+/−0.0011	0.0211+/−0.0020	0.0293+/−0.0034			
TU	0.0035+/−0.0008	0.0093+/−0.0010	0.0055+/−0.0003	0.0191+/−0.0009	0.0120+/−0.0010	0.0145+/−0.0014	0.0243+/−0.0011	0.0043+/−0.0014	0.0078+/−0.0017	0.0045+/−0.0030	0.0261+/−0.0020	0.0163+/−0.0022	0.0101+/−0.0021	0.0086+/−0.0012	0.0098+/−0.0007	0.0173+/−0.0006	0.0145+/−0.0015		
TUJ	0.0003+/−0.0002	0.0083+/−0.0006	0.0038+/−0.0011	0.0120+/−0.0013	0.0037+/−0.0025	0.0122+/−0.0007	0.0318+/−0.0027	0.0054+/−0.0012	0.0113+/−0.0007	0.0065+/−0.0038	0.0195+/−0.0027	0.0076+/−0.0019	0.0119+/−0.0015	0.0034+/−0.0014	0.0087+/−0.0007	0.0196+/−0.0026	0.0076+/−0.0011	0.0032+/−0.0005	
XIBO	0.0034+/−0.0014	0.0076+/−0.0012	0.0033+/−0.0011	0.0197+/−0.0004	0.0120+/−0.0020	0.0167+/−0.0004	0.0185+/−0.0012	0.0096+/−0.0012	0.0045+/−0.0003	0.0017+/−0.0015	0.0286+/−0.0009	0.0182+/−0.0025	0.0052+/−0.0017	0.0114+/−0.0028	0.0139+/−0.0026	0.0099+/−0.0032	0.0148+/−0.0011	0.0035+/−0.0006	0.0055+/−0.0009

[a]Paired Fst values are the mean+/−S.D determined from three nonoverlapping sets of 3500 SNPs using the Weir and Cockerham algorithm (see **Methods**).
[b]The EAS population groups included Japanese from Tokyo (JPT) and Chinese from Beijing (CHB) both from HapMap data, Korean (KOR), Filipino (FIL), Vietnamese (VIET) from new typing studies and the following HGDP groups: Cambodian (CAMB), Yakut (YAK), Yi, Daur, Mongolian (MGL), Lahu, Dai, , Hezhen (HEZH), , Miaozu (MIAO), Naxi, Oroqen (OROQ), She, Tu, Tujia (TU), and Xibo. The geographic positions of these populations are provided in Figure 1 with the exception of the Yakut that derive from northern Siberia.

doi:10.1371/journal.pone.0003862.t001

Fis values were also determined for each of the population sample and did not indicate a strong inbreeding component for any of the tested sample groups (**Supplemental Table S1**).

The different Chinese subjects derived from different regions of origin were also examined. For each of the Chinese American groups with self reported origin from North China, South China and Taiwan the paired Fst values with the Han Chinese from Beijing was small (<0.0025) (**Supplemental Table S2**).

Principal Component Analyses Using >200 K SNPs Show Substructure Relationships

To further explore the relationship among EAS population groups and examine population substructure PCA was performed using the genotype results from a set of >200 K SNPs. Analyses were done with and without the inclusion of the Yakut population thought to originate in central Asia, since PCA results are influenced by the inclusion or exclusion of different population groups and we were interested in the relationship between EAS and central Asian populations. The first two principal components in these analyses display the largest genotype variation (**Table 2**) and are graphically depicted in **Figure 1.** Inclusion of the Yakut group showed a possible cline in PC1/PC2 that extends from the current Siberian location of the Yakut to the northern East Asian population groups (**Figure 1A**). Interestingly, the position of the different population groups shows a remarkable correspondence with the geographic origin of each group. This is more clearly suggested when the Yakut population is excluded (**Figure 1B**) and is best illustrated by comparing these geographic locations with rotated PCA results (**Figure 1C and D**). Additional, PCA analyses including the central Asian Uygur and Hazara population groups were also performed but these did not show a clear relationship with the EAS (Supplemental **Figure S1**).

Table 2: Evaluation of Principal Components Analyses in East Asian Populations using 200 K SNPs

PC	All EAS Population Groups[a]			Five Population Groups		
	% Eigen[b]	SHT[c]	K-W Test[d]	% Eigen	SHT	K-W Test
1	17.9%	0.969+/−0.030	2.39E-41	18.0%	0.985+/−0.001	2.66E-26
2	12.0%	0.950+/−0.017	1.01E-40	12.0%	0.951+/−0.010	4.43E-24
3	10.6%	0.798+/−0.109	9.81E-39	9.2%	0.774+/−0.045	1.74E-14
4	10.0%	0.690+/−0.198	2.56E-36	8.8%	0.301+/−0.127	9.32E-01
5	8.9%	0.738+/−0.139	1.43E-31	8.8%	0.011+/−0.013	2.54E-01
6	8.4%	0.481+/−0.055	1.05E-28	8.7%	0.051+/−0.041	4.71E-02
7	8.2%	0.177+/−0.028	6.39E-23	8.7%	0.038+/−0.041	1.79E-01
8	8.0%	0.129+/−0.162	4.97E-10	8.6%	0.069+/−0.032	1.89E-01
9	8.0%	0.033+/−0.016	6.09E-05	8.6%	0.016+/−0.011	5.67E-01
10	7.9%	0.006+/−0.004	4.88E-02	8.6%	0.005+/−0.007	2.00E-01
PC	CHB, KOR, JPT			"Chinese" Groups Alone		
	% Eigen	SHT	K-W Test	% Eigen	SHT	K-W Test
1	15.2%	0.982+/−0.002	6.75E-22	11.5%	0.685+/−0.049	5.39E-06
2	9.8%	0.616+/−0.081	3.26E-08	10.0%	0.059+/−0.060	6.02E-01
3	9.5%	0.003+/−0.003	1.55E-01	10.0%	0.120+/−0.019	3.07E-01
4	9.5%	0.036+/−0.032	1.87E-01	9.9%	0.098+/−0.065	6.90E-01
5	9.5%	0.038+/−0.032	4.98E-01	9.9%	0.014+/−0.018	2.68E-01
6	9.3%	0.053+/−0.045	7.27E-01	9.8%	0.051+/−0.067	1.83E-01
7	9.3%	0.037+/−0.008	2.50E-02	9.8%	0.069+/−0.084	3.77E-01
8	9.3%	0.024+/−0.013	1.18E-01	9.7%	0.113+/−0.073	9.36E-01
9	9.3%	0.035+/−0.041	1.26E-01	9.7%	0.040+/−0.063	4.70E-01
10	9.3%	0.014+/−0.010	4.65E-02	9.7%	0.018+/−0.018	3.05E-01

[a]EAS population groups included each of the populations indicated in Figure 1.
[b]The % eigenvalue (Eigen) is the percentage of the total variance in the first ten PCs.
[c]The Spearman-Brown split half reliability test (SHT)[39] r^2 is the mean+/−SD from the adjusted correlations between: 1) every other chromosomes; 2) half chromosomes (first half each chromosome and second half each chromosome); and 3) first half genome and second half genome (see **Methods**). These correlations were determined after PCA of each individual set.
[d]The Kruskal-Wallis test [23], a nonparametric alternative to the ANOVA was used to examine the statistical significance of the difference in PC scores among subject groups pre-assigned based on self-identification.
doi:10.1371/journal.pone.0003862.t002

The PCA results for PC1 and PC2 are generally consistent with the relative paired Fst values with respect to the distance separation among the different population groups. For example the position of the Korean group approximately midway between the HapMap CHB and JPT groups both graphically (**Figure 1**) and as discussed above for paired Fst values. It is also consistent with the closer relationship between the Dai ethnic group and the Vietnamese subjects. However, the first two PCs do not show the full relationships among the population groups. For example the Lahu ethnic group appears to be closely related to the Cambodian ethnic group (**Figure 1**), although the paired Fst value is relatively large (Table 1). Examination of additional PCs shows the large difference between the Lahu and Cambodian ethnic groups in PCs 3, 4 and 5 (**Figure 2**). Using both the Kruskal-Wallis test [23], a nonparametric alternative to the ANOVA, and a split half reliability test (see Methods) substructure was present in multiple principal components (**Table 2**). Substantial population substructure can be observed by the nonrandom grouping of population groups that extends through PC7.

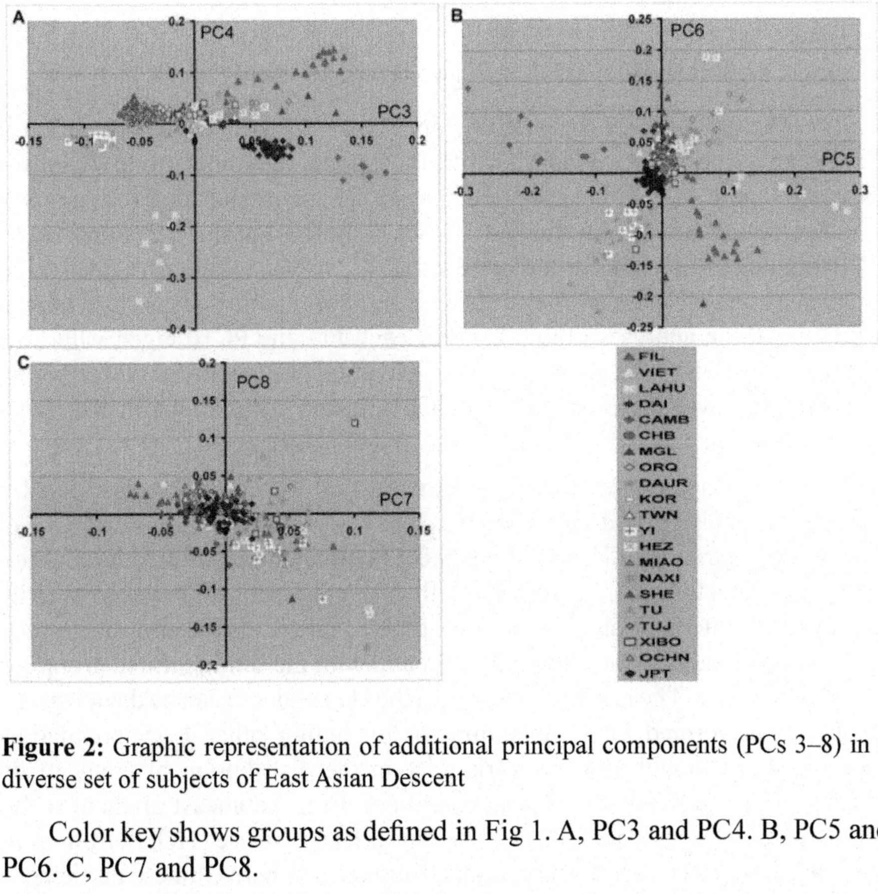

Figure 2: Graphic representation of additional principal components (PCs 3–8) in a diverse set of subjects of East Asian Descent

Color key shows groups as defined in Fig 1. A, PC3 and PC4. B, PC5 and PC6. C, PC7 and PC8.

For the entire EAS population groups studied, the majority of substructure variation defined by PCA appears to be within the first 4 PCs (**Table 2**). The eigenvalues plateau after PC4 with only small differences observed in subsequent PCs (**Figure 3a**). The proportion of the sum of the eigenvalues above this plateau provides a measure of the relative amount of substructure variation defined by each PC (**Figure 3b**). For the total EAS group, >90% of the substructure is defined in the first four PCs by this measurement. For the group of the five populations representing the most populous ethnic groups studied the first two PCs account for 90% of the variation above the plateau.

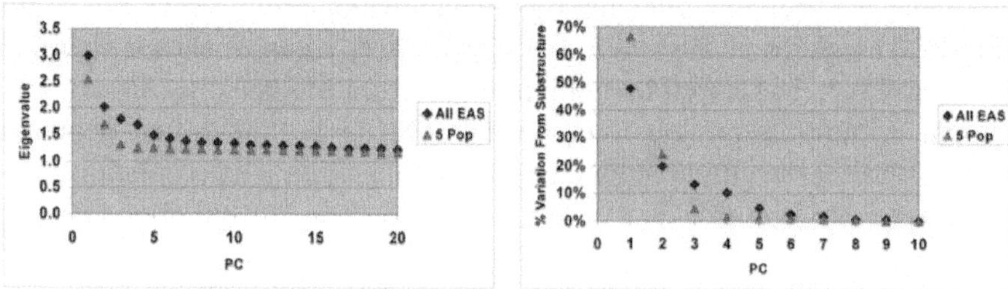

Figure 3: Eigenvalue distribution for principal components.

A, The eigenvalues for each PC are shown for both the entire group of EAS (excluding Yakut), and for the five most populous ethnic groups (Chinese, Korean, Japanese, Filipino and Vietnamese). B, The proportion of the adjusted eigenvalue for each PC for the first 10 PCs is shown. For this measurement the PC10 eigenvalue for each group was used as the baseline. [Note: the eigenvalues plateau as shown in panel A and there is no discernable substructure beyond PC10 for these analyses (Table 2)]. For each PC, the PC10 eigen value was subtracted to determine an "adjusted" eigenvalue. The % substructure variation measurement was the proportion of each adjusted eigenvalue divided by the sum of the adjusted eigenvalues (PC1 through PC10).

Similar analyses were also performed using population sets restricted to the more closely related Han Chinese, Japanese, and Korean groups, as well as a group restricted to Han Chinese and Chinese Americans (**Table 2**). These results as expected indicated substantially less substructure. However, even the subject set limited to Han Chinese and Chinese Americans showed substructure in PC1 using the split half reliability test and with the self identified groupings (ANOVA result). The relationship among the Han Chinese can be demonstrated in PCAs performed either including or excluding other EAS populations (**Figure 4**). Although there is variability in the distribution of many of the self-identified groups there was a general northwest/southeast gradient within these Chinese participants. In PC1 the North Han Chinese (HGDP from north central China[12]) were most separated from the southern Chinese participants including the Chinese American participants from Taiwan or with self-reported southern China origin.

A, Results from PCA performed together with EAS populations. B, PCA performed using only Chinese and Chinese American participants. The color coded population groups included the HapMap Han Chinese from Beijing (CHB), HGDP Han Chinese (HAN), HGDP North Han Chinese (HAN_N), Chinese American North (CHAN), Chinese American South (CHAS), Chinese American Central (CHAC), Taiwan Chinese American (TWN), Korean (KOR), and Hezhen (HEZ).

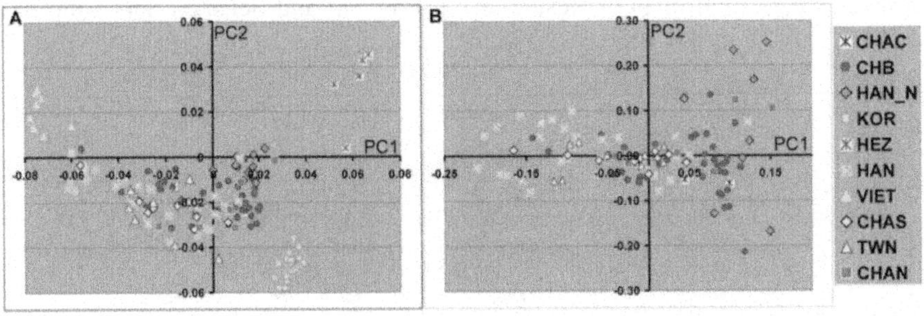

Figure 4: PCA analyses of Han Chinese and Chinese American population groups.

Informativeness of Smaller Sets of SNPS for Large East Asian Population Groups

We next examined the ability of smaller sets of SNPs to define population genetic structure in EAS populations. Random sets of 20 K, 5 K and 1 K SNPs were used to examine substructure in the combined population set and a subset of subjects from the most populous EAS groups (Han Chinese, Japanese, Korean, Filipino and Vietnamese). Correlation values (r^2) were calculated comparing these SNP subsets with the 200 K SNP set. These results, summarized in Table 3, showed that the 20 K random SNP set and 5 K random SNP set corresponded closely with the >200 K SNP set for the first 4 PCs, with decreased correlations observed for the 1 K random SNP set. The relatively poor performance of the 1 K random sets was more pronounced when more closely related population groups were considered e.g. Japanese and Korean for PC1, 20 K/200 K r^2=0.82+/−0.12 (mean+/−SD), 5 K/200 K r^2=0.69+/−0.03, and 1 K/200 K r^2=0.28+/−0.06. These results suggest that random sets of 5 K SNPs may be necessary for resolving and adjusting for substructure in these EAS populations (see discussion).

Table 3: Correlation of PCA Results using Random and Selected Sets with 200 K SNPs

	All EAS Groups[a]			
	PC1	PC2	PC3	PC4
20 K random[b]	0.992+/−0.005	0.977+/−0.002	0.854+/−0.139	0.851+/−0.137
5 K random	0.956+/−0.005	0.888+/−0.021	0.725+/−0.086	0.705+/−0.081
1 K random	0.813+/−0.007	0.514+/−0.019	0.228+/−0.047	0.125+/−0.052
	Five Population Groups (CHB, JPT, KOR, FIL, VIET)			
	PC1	PC2	PC3	PC4
20 K random	0.991+/−0.002	0.961+/−0.005	0.897+/−0.073	0.754+/−0.198
5 K random	0.961+/−0.004	0.862+/−0.019	0.616+/−0.141	0.708+/−0.02
1 K random	0.814+/−0.024	0.419+/−0.074	0.192+/−0.165	0.218+/−0.074
	Test Population Group[c]			
	PC1	PC2		
3 K random	0.862+/−0.035	0.446+/−0.096		
3.0 K AIMs	0.953	0.848		
1.5 K AIMs	0.939	0.819		
750 AIMs	0.886	0.579		

[a]Includes all EAS population groups (see Methods).
[b]Summary of analyses is provided for correlations of three independent random marker sets for each random marker group. For each random group the correlation with the full array set (>200 K SNPs) and is expressed as the mean r^2+/−S.D.
[c]The tester population panel consisted of 20 Chinese, 20 Japanese, 4 Korean, 3 Filipino, 1 Vietnamese, 10 Dai and 10 Cambodian. . This test group did not contain any subjects used in the selection of the EAS-AIMs. As with other comparisons the correlation with the full array set (>200 K SNPs) is expressed as the mean r^2+/−S.D. The EAS-AIMs are provided in Supplemental **Table S3**.
doi:10.1371/journal.pone.0003862.t003

East Asian Substructure Ancestry Informative Markers

AIMs that discern population substructure are likely to be useful in candidate gene, chromosomal position based association studies and defining homogeneous subject sets [24]. Since the application of these methods is most applicable to large population groups we restricted our ascertainment to five populations (Han Chinese, Japanese, Korean, Vietnamese and Filipino) (See **Methods**). To access the potential usefulness of these AIMs an independent set of samples was used and compared with the same number of random SNPs. For this assessment we included Cambodian and Dai samples since we had limited samples from the Vietnamese and Filipino populations. 3 K AIMs showed close correlation between the 200 K results for the first two PCs (Table 3). A set of the best 1.5 K AIMs also showed close correlation (**Figure 5** and **Table 3**). A reduced set of 750 AIMs showed a fall-off in correlation but was still equivalent to 3 K random SNPs. None of the AIM sets correlated with PC3 or PC4 ($r2<0.01$, $p<0.05$), however, these PCs distinguished the Dai and Cambodian from the other population groups and these were not included in our AIM selections. Nevertheless, for the common EAS populations these data suggest that the EAS-AIMs (**Table S3**) will be useful for association studies in the majority of EAS and EAS-American populations.

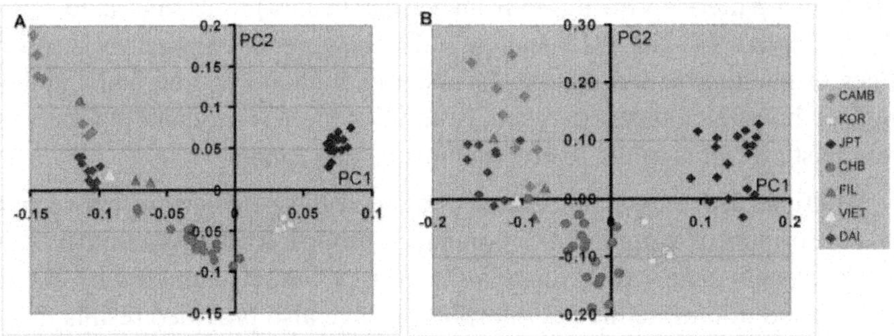

Figure 5: Ability of EAS-AIMs to discern population substructure

A, PCA analysis of tester population samples (see Table 3) using 200 K SNPs. B, PCA analysis of same tester population samples using 1500 EAS-AIMs.

DISCUSSION

The current study extends the definition of EAS population substructure and the relationships among these ethnic groups. The inclusion of participant groups from populous countries in this region with large contributions to the USA population is an important aspect of our study. These population groups complement those included within HapMap studies as well as the HGDP in showing relationships between EAS groups and demonstrating that autosomal genotypes can be used to ascertain membership to various EAS groups. These results emphasize that EAS substructure, similar that previously shown for European substructure, will likely be important for complex disease association studies in defining study participants and reducing type 1 and type 2 error rates.

Our study extends the results of PCA analyses of EAS populations including those of HGDP populations that was recently reported [13]. The graphic representation of the first two PCs showed close correspondence to the historical geographical location and/or sample collection site for most of the EAS population groups. Thus, despite admixture and perhaps uncertain migration patterns, overall the largest component of genotypic variation that is discernable by reducing high order data (all genotypes) to lower order variations (PCs) is consistent with the population geography. This finding supports hypotheses that the relationships among the EAS populations are largely explained by clines formed by demic expansion(s). We speculate that the inclusion of many different related ethnic groups has recapitulated the most common events that separated these ethnic groups. The first PC axis accounting for the largest variation has a north/south orientation. One major part of this pattern forms a line from Siberia (Yakut) to Mongolia to Eastern China (**Figure 1**). The PCA analyses also suggest that at least two separate clines originating or terminating in eastern China at one end and Cambodia and the Philippines at the other end. In addition there is another cline extending from Eastern China to the Korean peninsular and Japan.

Multiple previous studies have examined the relationship between and possible origins of different EAS population groups. Analysis of mitochondrial and Y chromosome haplogroups as well as a limited numbers of classical markers and microsatellite polymorphisms have also provided results that are generally consistent with a north/south orientation of relationships between different EAS population groups [15]–[18]. However, there are exceptions with some studies failing to show this relationship [19], [25]. Summarized by a recent review [26] there are three different postulates regarding the origins of EAS population groups: 1) South East Asian origin [14]–[18], 2) North Asian origin [27] and 3) a combination of northern and southern origin[19], [20]. However, the majority of studies have supported a South-East Asian origin for most EAS populations and include detailed analyses of the age of specific mitochondrial haplogroups, Y chromosome sequences as well as limited marker studies [26]. In contrast, hierarchical trees in the recent HGDP study [13] show branching points consistent with a Yakut derivation. Recent studies using a novel copying model statistical approach appear to suggest an initial northern and southern origin (Cambodians, Mongolians, Xibo, Yi , Tu, Daur, and Naxi receiving large contributions from central-Asian populations) that contribute to Han ancestry[28]. These studies also provide data supporting the derivation of many other EAS groups from a Han expansion (including She, Japanese, Dai, Lahu and Miao). While the current study does not strongly support any of these hypotheses, it does suggest that eastern China is central to the events shaping the population groups in this region.

Multiple additional PCs are necessary to define the overall substructure relationships for the entire group of EAS populations studied as shown in **Figure 2**. However, most of the variation is discerned in the first four PCs for the EAS populations examined and in the first two PCs for the five most populous EAS groups studied. There was no geographic correspondence of the additional PCs and it is unclear whether these additional patterns correspond to individual or multiple different events in the histories of these population groups. Overall the size of the paired Fst values, as expected, showed a strong correlation with the PC eigenvalues summed over the first four PCs (data not shown). Although Fis values do not provide evidence for inbreeding in the current populations, it is unclear whether inbreeding or other factors including bottlenecks during the history of particular EAS ethnic groups may have contributed to the relationships between these populations.

An important aspect of the current study was the identification of EAS-AIM sets. The results show that these AIMs can distinguish the major variation between the populous population groups including Han Chinese, Japanese, Korean, Vietnamese, and Filipino. Additional testing to examine correction for stratification with these population groups was not possible due to limited genotypes currently available. However, by analogy with previous studies in European population groups, these AIMs particular the 1500 EAS-AIM set should be effective in addressing population stratification. The close correspondences of the relative positions in the first two PCs in individual subjects, even within the Han Chinese group, support the potential use of these SNP AIMs. Furthermore, the SHT analysis suggests that studies within the Han Chinese population and Chinese-Americans will benefit from the use of such AIMs in candidate gene studies.

METHODS

Populations Studied

The populations including those from the HGDP, HapMap, the I-control database, a Korean sample set and East Asian Americans. For all but the East Asian American and Korean samples set, genotypes were available from online databases. These included HapMap subjects (44 CHB and 44 JPT) and HGDP subjects (10 Cambodian, 10 Dai, 24 Hazara, 9 Hezhen, 27 Japanese, 10 Miaozu, 7 Naxi, 8 Oroqen, 10 She, 10 Tu, 10 Tujia, 8 Xibo, 13 Yakut and 44 Han Chinese) from the I-ControlDB (www.illumina.com/iControlDB, Illumina, San Diego, CA). Genotypes from other HGDP subjects (10 Daur, 8 Lahu, 9 Mongola, 10 Uygur, 10 Yi,) were from the NIH Laboratory of Neurogenetics (http://neurogenetics.nia.nih.gov/paperdata/public/).

For all EAS American and Korean subjects, blood cell samples were obtained from all individuals, according to protocols and informed-consent procedures approved by institutional review boards, and were labeled with an anonymous code number linked only to demographic information.

The Korean participants were from recruited in Korea (21 subjects). The EAS American samples were individuals born in the respective EAS country and were from Vietnam (22 subjects), Philippines (17 subjects) and different regions of the Peoples Republic of China (23 subjects) and Taiwan (9 subjects). The Filipino American participants included 15 that were recruited as part of the New York Cancer Project (NYCP); a prospective longitudinal study [29] and two recruited in Houston TX. 3 Filipino, 15 Vietnamese and 32 Chinese American samples were recruited in Houston TX. An additional 7 Vietnamese and 3 Korean genotypes were from the I-ControlDB. Of the Chinese American participants (CHA), 28 also indicated their general origin from regions within China (6 north, 10 south, 3 central and 9 subjects Taiwan).

Genotyping

Genotyping was performed using a 300 K Illumina array according to the Illumina Infinium 2 assay manual (Illumina, San Diego), as previously described [30].

Data Filters

SNPs and individual samples with less than 90% complete genotyping information from any data set were excluded from analyses. SNPs that showed extreme deviation from Hardy Weinberg equilibrium ($p<0.00001$) in individual population groups were also excluded from analysis. These filters resulted in a total of 215 K autosomal SNPs that were used for these studies. In addition, for samples from nonHGDP origin individuals with evidence of >10% contribution from other continents were excluded from further study. This was either performed prior to Illumina array genotyping for the Filipino, Vietnamese and CHA subjects using 128 continental AIMS [31]. Samples were also filtered for possible cryptic relationships using the PLINK program [32].

Statistical Analyses

F_{st} and F_{is} was determined using Genetix software[33] that applies the Weir and Cockerham algorithm[34]. A measure of informativeness for each SNP (I_n) was determined using an algorithm previously described [35]. Hardy-Weinberg equilibrium was determined using HelixTree 5.0.2 software (Golden Helix, Bozeman, MT, USA).

Population structure was examined using STRUCTURE v2.1 [36], [37] using parameters and AIMs previously described [31]. This analysis was performed to exclude individuals with evidence of substantial continental admixture from Europe, Africa or the American continent (see Data Filters).

PCA was performed using the EIGENSTRAT statistical package[38]. All analyses were performed after deleting the MHC region on chromosome 6 since regions of high linkage disequilibrium can overly influence PCA results. The Kruskal-Wallis test [23], a nonparametric alternative to the ANOVA was used to examine the statistical significance of the difference in PC scores among subject groups pre-assigned based on self-identification.

The split half reliability test can determine whether independent (non-overlapping) SNP sets provide the same or different results. The split half reliability test was adjusted by the Spearman-Brown formula [39] and was performed three times using 1) alternate chromosomes, 2) alternate half chromosomes, and 3) half genome SNP sets. These sets were chosen to eliminate any dependency in each test between the two half data sets based on linkage disequilibrium.

Selection of EAS-AIMs

Genotypes from 32 Han Chinese (CHA and CHB), 36 Japanese (JPT), 19 Korean, 21 Filipino and 14 Vietnamese were used for SNP selection. An initial set of 3000 EAS substructure AIMs (EAS-AIMs) were based on either I_n values or using SNP scores from PCA. The best performance using a testing panel was observed using a set of SNPs selected using I_n values from a combination of 1) all five population groups (top 600 SNPs), 2) Chinese and Japanese (top 1200 SNPs), and 3) Chinese and Filipino (top 1200 SNPs). The best performance of a 1500 SNP set and a 750 SNP set were observed using a combination of 500 or 250 from each of these three groups. The testing panel consisted of 20 Chinese, 20 Japanese, 4 Korean, 3 Filipino, 1 Vietnamese, 10 Dai and 10 Cambodian. None of the samples in the testing panel overlapped with the ascertainment samples. The Dai and Cambodian samples were included since there were limited numbers of samples available from the Vietnamese group. The performance of the EAS-AIMs was evaluated using correlations in PC1 and PC2 with the >200 K SNP set.

ACKNOWLEDGMENTS

We thank Stephen Johnson and Robert Lundsten for informatics support on the New York Cancer Project samples. We also thank Anthony Liew and Houman Khalili for expert assistance with genotyping. We thank the volunteers from the different populations for donating blood samples.

AUTHOR CONTRIBUTIONS

Conceived and designed the experiments: MFS. Performed the experiments: CT AL PG MFS. Analyzed the data: CT RK MFS. Contributed reagents/materials/analysis tools: CT AL MR JWB PG MFS. Wrote the paper: CT RK JWB PG MFS

REFERENCES

1. Seldin MF, Shigeta R, Villoslada P, Selmi C, Tuomilehto J, et al. (2006) European Population Substructure: Clustering of Northern and Southern Populations. PLoS Genetics 2: 1339–1351.

2. Bauchet M, McEvoy B, Pearson LN, Quillen EE, Sarkisian T, et al. (2007) Measuring European population stratification with microarray genotype data. Am J Hum Genet 80: 948–956.

3. Price AL, Butler J, Patterson N, Capelli C, Pascali VL, et al. (2008) Discerning the ancestry of European Americans in genetic association studies. PLoS Genet 4: e236.

4. Tian C, Plenge RM, Ransom M, Lee A, Villoslada P, et al. (2008) Analysis and application of European genetic substructure using 300 K SNP information. PLoS Genet 4: e4.

5. Tian C, Gregersen PK, Seldin MF (2008) Accounting for ancestry: population substructure and genome-wide association studies. Hum Mol Genet 17: R143–150.

6. Suzuki A, Yamada R, Chang X, Tokuhiro S, Sawada T, et al. (2003) Functional haplotypes of PADI4, encoding citrullinating enzyme peptidylarginine deiminase 4, are associated with rheumatoid arthritis. Nat Genet 34: 395–402.

7. Tokuhiro S, Yamada R, Chang X, Suzuki A, Kochi Y, et al. (2003) An intronic SNP in a RUNX1 binding site of SLC22A4, encoding an organic cation transporter, is associated with rheumatoid arthritis. Nat Genet 35: 341–348.

8. Lee HS, Remmers EF, Le JM, Kastner DL, Bae SC, et al. (2007) Association of STAT4 with rheumatoid arthritis in the Korean population. Mol Med 13: 455–460.

9. Lin YJ, Wan L, Lee CC, Huang CM, Tsai Y, et al. (2007) Disease association of the interleukin-18 promoter polymorphisms in Taiwan Chinese systemic lupus erythematosus patients. Genes Immun 8: 302–307.

10. Shin HD, Sung YK, Choi CB, Lee SO, Lee HW, et al. (2007) Replication of the genetic effects of IFN regulatory factor 5 (IRF5) on systemic lupus erythematosus in a Korean population. Arthritis Res Ther 9: R32.

11. Kobayashi S, Ikari K, Kaneko H, Kochi Y, Yamamoto K, et al. (2008) Association of STAT4 with susceptibility to rheumatoid arthritis and systemic lupus erythematosus in the Japanese population. Arthritis Rheum 58: 1940–1946.

12. Cavalli-Sforza LL (2005) The Human Genome Diversity Project: past, present and future. Nat Rev Genet 6: 333–340.

13. Li JZ, Absher DM, Tang H, Southwick AM, Casto AM, et al. (2008) Worldwide human relationships inferred from genome-wide patterns of variation. Science 319: 1100–1104.

14. Ballinger SW, Schurr TG, Torroni A, Gan YY, Hodge JA, et al. (1992) Southeast Asian mitochondrial DNA analysis reveals genetic continuity of ancient mongoloid migrations. Genetics 130: 139–152.

15. Chu JY, Huang W, Kuang SQ, Wang JM, Xu JJ, et al. (1998) Genetic relationship of populations in China. Proc Natl Acad Sci U S A 95: 11763–11768.

16. Su B, Xiao J, Underhill P, Deka R, Zhang W, et al. (1999) Y-Chromosome evidence for a northward migration of modern humans into Eastern Asia during the last Ice Age. Am J Hum Genet 65: 1718–1724.

17. Yao YG, Kong QP, Bandelt HJ, Kivisild T, Zhang YP (2002) Phylogeographic differentiation of mitochondrial DNA in Han Chinese. Am J Hum Genet 70: 635–651.

18. Shi H, Dong YL, Wen B, Xiao CJ, Underhill PA, et al. (2005) Y-chromosome evidence of southern origin of the East Asian-specific haplogroup O3-M122. Am J Hum Genet 77: 408–419.

19. Karafet T, Xu L, Du R, Wang W, Feng S, et al. (2001) Paternal population history of East Asia: sources, patterns, and microevolutionary processes. Am J Hum Genet 69: 615–628.

20. Cavalli-Sforza LL, Menozzi P, Piazza A (1996) The history and geography of human genes. Princeton, , N.J.: Princeton University Press.

21. 2003) (2003) The International HapMap Project. Nature 426: 789–796.

22. Altshuler D, Brooks LD, Chakravarti A, Collins FS, Daly MJ, et al. (2005) A haplotype map of the human genome. Nature 437: 1299–1320.

23. Hollander M, Wolfe DA (1973) Nonparametric statistical inference. New York: John Wiley and Sons. pp. 115–120.

24. Seldin MF, Price AL (2008) Application of ancestry informative markers to association studies in European Americans. PLoS Genet 4: e5.

25. Ding YC, Wooding S, Harpending HC, Chi HC, Li HP, et al. (2000) Population structure and history in East Asia. Proc Natl Acad Sci U S A 97: 14003–14006.

26. Zhang F, Su B, Zhang YP, Jin L (2007) Genetic studies of human diversity in East Asia. Philos Trans R Soc Lond B Biol Sci 362: 987–995.

27. Nei M, Roychoudhury AK (1993) Evolutionary relationships of human populations on a global scale. Mol Biol Evol 10: 927–943.

28. Hellenthal G, Auton A, Falush D (2008) Inferring human colonization history using a copying model. PLoS Genet 4: e1000078.

29. Mitchell MK, Gregersen PK, Johnson S, Parsons R, Vlahov D (2004) The New York Cancer project: rationale, organization, design, and baseline characteristics. J Urban Health 81: 301–310.
30. Duerr RH, Taylor KD, Brant SR, Rioux JD, Silverberg MS, et al. (2006) A genome-wide association study identifies IL23R as an inflammatory bowel disease gene. Science 314: 1461–1463.
31. Kosoy R, Nassir R, Tian C, White PA, Butler LM, et al. (In press) Ancestry informative marker sets for determining continental origin and admixture proportions in common populations in America. Human Mutation.
32. Purcell S, Neale B, Todd-Brown K, Thomas L, Ferreira MA, et al. (2007) PLINK: a tool set for whole-genome association and population-based linkage analyses. Am J Hum Genet 81: 559–575.
33. Belkhir K, Borsa P, Chikhi L, Raufaste N, Bonhomme F (2001) GENETIX, software under WindowsTM for the genetic of populations. 4.02 ed. Montpellier, France: Laboratory Genome, Populations, Interactions CNRS UMR 5000, University of Montpellier II.
34. Weir B, Cockerham C (1984) Estimating F-statistics for the analysis of population structure. Evolution 38: 1358–1370.
35. Rosenberg NA, Li LM, Ward R, Pritchard JK (2003) Informativeness of genetic markers for inference of ancestry. Am J Hum Genet 73: 1402–1422.
36. Pritchard JK, Stephens M, Donnelly P (2000) Inference of population structure using multilocus genotype data. Genetics 155: 945–959.
37. Falush D, Stephens M, Pritchard JK (2003) Inference of population structure using multilocus genotype data: linked loci and correlated allele frequencies. Genetics 164: 1567–1587.
38. Price AL, Patterson NJ, Plenge RM, Weinblatt ME, Shadick NA, et al. (2006) Principal components analysis corrects for stratification in genome-wide association studies. Nat Genet 38: 904–909.
39. Spearman C (1910) Correlation calculated with faulty data. British Journal of Psychology 3: 271–295.

Chapter 9

ASSEMBLY OF INFLAMMATION-RELATED GENES FOR PATHWAYFOCUSED GENETIC ANALYSIS

Matthew J. Loza[1] , Charles E. McCall[2] , Liwu Li[3] , William B. Isaacs[4,5], Jianfeng Xu6, Bao-Li Chang[7]

[1] Center for Human Genomics, Department of Internal Medicine, Wake Forest University School of Medicine, Winston-Salem, North Carolina, United States of America

[2]Department of Internal Medicine, Wake Forest University School of Medicine, Winston-Salem, North Carolina, United States of America

[3]Department of Biology, Virginia Polytechnic Institute and State University, Blacksburg, Virginia, United States of America

[4]Department of Urology, Johns Hopkins University Medical Institutions, Baltimore, Maryland, United States of America

[5]Department of Oncology, Johns Hopkins University Medical Institutions, Baltimore, Maryland, United States of America

[6] Center for Human Genomics, Department of Epidemiology and Prevention, Wake Forest University School of Medicine, Winston-Salem, North Carolina, United States of America

[7] Center for Human Genomics, Department of Pediatric Medicine, Wake Forest University School of Medicine, Winston-Salem, North Carolina, United States of America

ABSTRACT

Recent identifications of associations between novel variants in inflammation-related genes and several common diseases emphasize the need for systematic evaluations of these genes in disease susceptibility. Considering that many genes are involved in the complex inflammation responses and many genetic variants in these genes have the potential to alter the functions and expression of these genes, we assembled a list of key inflammation-related genes to facilitate the identification of genetic associations of diseases with an inflammation-related etiology. We first reviewed various phases of inflammation responses, including the development of immune cells, sensing of danger, influx of cells to sites of insult, activation and functional responses of immune and non-immune cells, and resolution of the immune response. Assisted by the Ingenuity Pathway Analysis, we then identified 17 functional sub-pathways that are involved in one or multiple phases. This organization would greatly increase the chance of detecting gene-gene interactions by hierarchical clustering of genes with their functional closeness in a pathway. Finally, as an example application, we have developed tagging single nucleotide polymorphism (tSNP) arrays for populations of European and African descent to capture all the common variants of these key inflammation-related genes. Assays of these tSNPs have been designed and assembled into two Affymetrix ParAllele customized chips, one each for European (12,011 SNPs) and African (21,542 SNPs) populations. These tSNPs have greater coverage for these inflammation-related genes compared to the existing genome-wide arrays, particularly in the African population. These tSNP arrays can facilitate systematic evaluation of inflammation pathways in disease susceptibility. For additional applications, other genotyping platforms could also be employed. For existing genome-wide association data, this list of key inflammation-related genes and associated subpathways can facilitate comprehensive inflammation pathway- focused association analyses.

INTRODUCTION

Inflammation is an essential component of immune-mediated protection against pathogens and tissue damage. Immune responses are also responsible for the unfavorable rejection of tissue/organ transplants, hypersensitivity reactions (e.g., atopy, anaphylaxis, contact hypersensitivity, delayed-type hypersensitivity), and septic shock. Aberrant or unchecked immune responses may lead to a state of chronic inflammation [1]–[3]. This may occur when the immune response: 1) is activated in the absence of 'danger' signals; 2) fails to fully turn-off (resolve) after elimination of the danger; and 3) fails to completely clear the danger stimulus. Factors that may influence the initiation, activity, and resolution of immune responses include health (physical and emotional), age, diet, medications, and genetic predisposition.

Inflammation may also be a contributing factor for some diseases. The role of chronic inflammation in a wide variety of diseases is well-appreciated, including rheumatoid arthritis and other autoimmune disorders [4], cardiovascular disease [5]–[7], gastrointestinal disorders [8],[9], and a number of cancers [10]–[14]. Perhaps the best evidence for the importance of chronic inflammation in disease is the efficacy of NSAIDs in reducing the risk or severity of these disorders [15]. There is mounting evidence that dietary factors that may influence inflammation, such as the balance of omega-3 vs. omega-6 polyunsaturated fatty acids (PUFAs), have an impact on disease risk and progression [16]. Genetic studies also provide evidence that variant alleles of genes associated with inflammatory pathways impact the risk of disease initiation, progression, and severity (see **Table 1**). The role of inflammation as a mediator of disease is currently receiving extensive attention, resulting in the National Institute of Allergy and Immunologic Diseases (NIAID) plans for an NIH Roadmap initiative with the overarching theme: "Inflammation as a Common Mechanism of Disease" (http://nihroadmap.nih.gov/inflammation/index.asp).

Table 1: Confirmed associations of genetic variants in inflammation-associated genes and disease

Disease	Gene	Encoded protein	Variant	Odds ratio [a]	p-value	Confirmation method
Age-related macular degeneration [44]	*CFH*	Complement factor H	rs1061170	3.40 [b]	$<1\times10^{-5}$	Case-control/meta-analysis
Atopic asthma [32]	*IL4R*	IL-4 receptor alpha	rs1801275	1.79	3×10^{-9}	Meta-analysis of 7 study populations
Atopic asthma [21]	*TNF*	TNF-alpha	-308 G/A	1.46	1×10^{-4}	Meta-analysis of 15 study populations
Crohn's disease [30]	*CARD15*	Nod2	1007fsinsC	4.3	7×10^{-28}	Meta-analysis of 16 study populations
Breast cancer [33]	*CASP8*	Caspase-8	rs1045485	0.90	0.016	Analysis of 3 study populations (6351 cases/5708 controls)
Breast cancer [33]	*TGFB1*	TGF-beta 1	rs1982073	1.08	0.0088	Analysis of 3 study populations (6863 cases/5587 controls)
Breast cancer [34]	*TNF*	TNF-alpha	rs361525	1.18	0.008	Analysis of two independent study populations
Graves' disease [24]	*PTPN22*	Lymphoid-specific phosphatase	C1858T	1.61	$<1\times10^{-5}$	Meta-analysis of 3 study populations
Inflammatory bowel disease [31]	*IL23R*	IL-23 receptor beta	rs11209026	0.26	5×10^{-9}	Genome-wide screen (raw p-value)
				0.45	8×10^{-4}	Case-control replication
				~0.5 [c]	1.3×10^{-10}	Family-based TDT replication
Psoriatic arthritis [23]	*TNF*	TNF-alpha	-238 G/A	2.29	2×10^{-4}	Meta-analysis of 8 study populations
Rheumatoid arthritis [24]	*PTPN22*	Lymphoid-specific phosphatase	C1858T (R620W)	1.68	$<1\times10^{-5}$	Meta-analysis of 12 study populations
Systemic lupus erythamatosus [25]	*IRF5*	Interferon response factor 5	rs2004640	1.47	4.2×10^{-21}	Case-control/meta-analysis+replication in family-based
Systemic lupus erythamatosus [24]	*PTPN22*	Lymphoid-specific phosphatase	C1858T (R620W)	1.49	$<1\times10^{-5}$	Meta-analysis of 5 study populations
Systemic lupus erythamatosus [22]	*TNF*	TNF-alpha	-308 G/A	2.1	<0.001	Meta-analysis of 10 study populations of European descent
Type 1 diabetes [29]	*CTLA4*	CTLA-4	rs3087243	1.18 [c]	5.6×10^{-6}	Family-based TDT
Type 1 diabetes [28]	*CTLA4*	CTLA-4	rs3087243	1.17 [c]	6×10^{-4}	Family-based TDT
				1.21	1.3×10^{-7}	Case-control
Type 1 diabetes [27]	*CTLA4*	CTLA-4	rs3087243	1.20	3.7×10^{-10}	Case-control
Type 1 diabetes [26]	*IFIH1*	Mda-5, Helicard	rs1990760	0.86	1.42×10^{-10}	Genome-wide, validated in case-control+family-based
Type 1 diabetes [24]	*PTPN22*	Lymphoid-specific phosphatase	C1858T (R620W)	1.85	$<1\times10^{-5}$	Meta-analysis of 6 study populations

[a]Odds ratio for allele test (multiplicative model), unless otherwise indicated.
[b]Odds ratio for dominant model
[c]Risk ratio from family-based transmission disequilibrium test (TDT).
doi:10.1371/journal.pone.0001035.t001

Numerous genetic linkage and case-control association studies have implicated genetic variations in genes important in immunity and inflammation and inflammatory diseases. Single missense heritable mutations can be the sole or major determinant for inflammatory diseases, such as Familial Cold Autoinflammatory Syndrome (missense mutations in exon 3 of *Cias1*account for all cases) [17]–[19] and Familial Mediterranean Fever (*MEFV*, five founder missense mutations, when homozygous, account for 74% of cases) [20]. For many complex inflammatory diseases, polymorphisms in inflammatory genes are more likely to act as modifiers for disease susceptibility rather than sole determinants. Recent meta-analyses report modest associations between single nucleotide polymorphisms (SNPs) in *TNF* (encoding TNF-α) and increased risk for asthma [21], system lupus erythamatosus (SLE) [22], and psoriatic arthritis [23]. A variant of *PTPN22* (encoding a lymphoid-specific protein tyrosine phosphatase) is modestly associated with multiple autoimmune diseases (rheumatoid arthritis, SLE, type 1 diabetes, and Graves' disease) [24]. Associations of *IRF5* (interferon regulator factor 5) genetic variants and increased SLE risk have been highly replicated [25]. *IFIH1*, encoding an innate immunity viral mRNA detector (early type I IFN-β responsive gene, Helicard), is strongly associated with type 1 diabetes risk [26]. Association of a small risk for type 1 diabetes and*CTLA4* has been consistently replicated [27]–[29]. An insertion polymorphism in *CARD15*, the gene encoding the microbial nucleotide detector Nod2, is a major risk factor for Crohn's disease[30]. An association between SNP's in *IL23R* (IL-23 receptor β-chain) and increased risk of inflammatory bowel disease has been reported for a genome-wide association study and confirmed in three independent populations [31]. A recent meta-analysis demonstrated that an*IL4R* (IL-4 receptor alpha chain) variant modestly increases risk for atopic asthma [32]. A description of these studies and reported associations is provided in **Table 1**. This list is not intended to be comprehensive but rather serves as an example of representative genetic variations which have been well-established to be associated with common inflammatory disorders.

In addition to inflammatory/autoimmune diseases, polymorphisms in inflammation-associated genes may also contribute to risk for diseases in which inflammatory/immune-disorders are not the primary characteristic. There is evidence from the Breast Cancer Association Consortium that *CASP8* (caspase 8) and *TGFB1* (TGF-β1) variants impart risk, albeit low penetrance, for breast cancer [33]. Analysis of two independent populations suggest that a rare polymorphism in *TNF* may also be a low-penetrance risk factor for breast cancer [34]. Several groups report associations between various *PTGS2* (COX-2) genetic variants and colorectal cancer risk[35]–[39]. There is also evidence to suggest that a low frequency *COX2* variant allele decreases prostate cancer risk [40], particularly in subjects who frequently eat fish [41]. Interestingly, in a preliminary study using MegAllele™ I&I panel, including 9,275 SNPs in 1,086 genes involved in immunity and inflammation, more SNPs were found to be significantly associated with prostate cancer risk than expected by chance, which suggests multiple genetic variants in this pathway impart modest risk to prostate cancer [42].

Because genetic associations of disease and genetic variations in inflammatory genes are often relatively modest, it is likely that polymorphisms in multiple inflammatory genes cooperate in an additive or synergistic manner to impact disease risk. Pathway analyses may help to reveal gene-gene interactions or risks imparted independently from other genes in the pathway. The advantages of performing analyses at pathway levels are illustrated by Dinu et al. [43]. Associations between *CFH* (complement factor H) and age-related macular degeneration have been replicated in numerous studies [44]. To test whether genetic variations in the multiple complement pathway genes impact macular degeneration risk, Dinu et al. analyzed the existing genome-wide association study data of Klein et al. [45], restricting the analyses to genes in the complement pathway in subjects carrying the *CFH* risk allele. Significant associations were detected for a *C7* and *MBL2* variants and severity of macular degeneration in the context of the complement pathway analysis and the *CHF* risk allele, and these associations would not have been significant in a genome-wide analysis of the data.

RESULTS

Choosing Genes

Various aspects of immunity contribute to the development of an overall inflammatory immune response. These phases include the development of immune cells, sensing of danger, influx of cells to sites of insult, activation and functional responses of immune and non-immune cells, and resolution of the immune response. To broadly cover most aspects of inflammatory responses, the various *phases of immune responses* were considered in choosing genes for the SNP array panel, outlined in **Table 2**. A schematic representation of most of these phases is presented in **Fig. 1**.

Depicted is a schematic representation of an immune response to a generic pathogenic insult. The phases of immune responses (described in Table 2) are shown in bold. Additional aspects not shown are the involvement of secondary lymphoid tissues for initial T cell and B cell activation by dendritic cells that migrate from the site of inflammation to lymph nodes and other secondary lymphoid structures. The resolution of immune responses, immunological memory, and homeostasis are also not depicted.

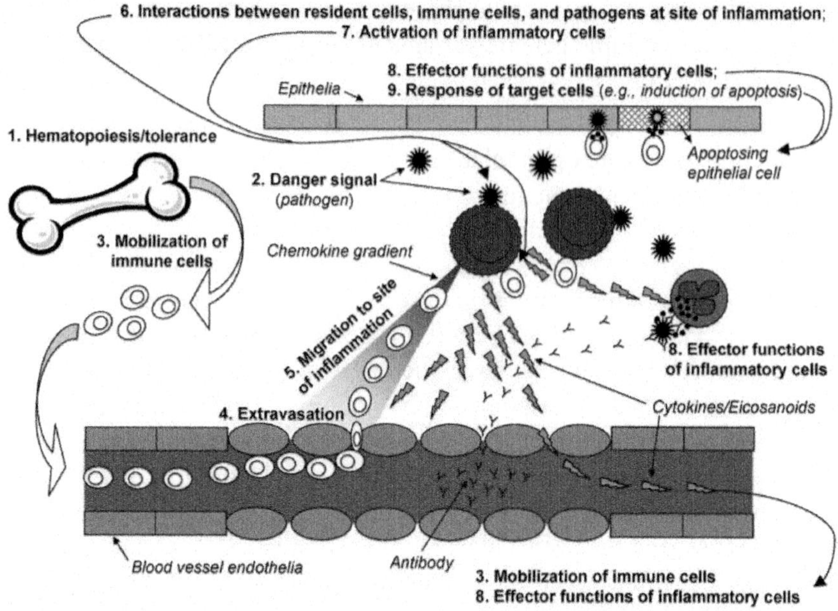

Figure 1: Development of an immune response.

doi:10.1371/journal.pone.0001035.g001

Table 2: Phases of immune response

Phase of immune response	Description
Hematopoiesis/homeosta-sis/tolerance	The generation and differentiation of immune cells and maintenance of their number in circulation and tissues; prevention of self-reactivity.
Danger signal	Innate recognition of and response to pathogenic foreign substances or stress.
Mobilization of immune cells	Systemic soluble mediators informing immune cells in circulation and lymphoid tissues of danger.
Extravasation	The process of circulating immune cells crossing from blood into peripheral tissues and secondary lymphoid tissues.
Migration to site of inflammation	The process of immune cells, after extravasation, reaching the site of inflammatory insult, including chemoattraction, adhesion to substrates, and degradation of extracellular matrix.
Interactions between resident cells, immune cells, and pathogens at site of inflammation	Interactions between resident cells, immune cells, and pathogens at site of inflammation–how infiltrating cells interact with the resident inflammatory cells, non-immune cells (e.g., epithelia), pathogens, and other infiltrating cells, that leads to activation of effector functions.
Activation of inflammatory cells	The signaling pathways and transcription factors stimulated by activating, co-stimulatory, and inhibitory receptors that leads to activation, proliferation, differentiation, and survival of responding immune cells.
Effector functions of inflammatory cells	The factors produced/released by immune cells in attempt to resolve the pathogenic insults, including release of cytotoxic/cytostatic mediators and mediators to enhance or fine-tune the immune response.
Response of target cells	The pathways in non-immune cells (e.g., epithelia) activated in response to the effector functions of immune cells.
Resolution of immune response vs. chronic inflammation	The pathways that lead to the downregulation of immune responses and inflammation after the pathogenic insult is cleared; the factors maintaining late-phase immune responses when the insult is not totally resolved.

doi:10.1371/journal.pone.0001035.t002

Priority was given first to genes of known function in inflammatory responses (in both immune and non-immune cells), and then to genes expressed in immune cells with function implied by homology to other genes but exact function not clear. Ubiquitously expressed genes required for the normal function of most cell types of diverse origin were given lower priority.

However, special emphasis was placed on genes at nodes for signaling to and from multiple pathways, most notably genes in NF-κB, MAPK, and PI3K signaling pathways.

Pathways were built using Ingenuity Pathways Analysis, as described in *Methods* section, using both pre-defined 'canonical pathways' and custom-built pathways based on our own queries for genes/pathways not included in the canonical pathways.

Multiple functional pathways are involved each of the immune response phases, and each functional pathway may contribute to several of the immune response phases. For example (**Fig. 2**), the response of a macrophage responding to a *danger signal* during a gram-negative bacterial infection involves innate pathogen recognition of LPS by the TLR4 complex. TLR4 transduces signals via NF-κB, MAPK, and PI3K signaling pathways, stimulating synthesis of eicosanoids and cytokines to signal other cells of the danger. LPS may also stimulate expression of stress-induced proteins, such as MIC-A and MIC-B (ligands for natural killer cell activating receptors) and T cell co-stimulatory molecules (B7 family proteins).

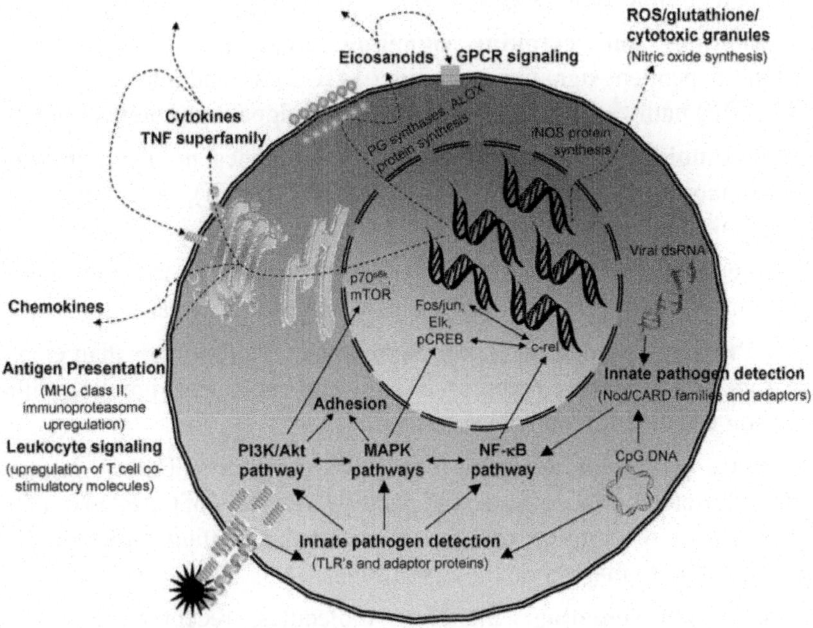

Figure 2: Inflammation subpathways involved in the response to danger signal.

The concerted action of multiple functional subpathways in the initial response of a macrophage to bacteria or virus is depicted. Solid arrows indicate signaling events and dashed arrows stimulated production of proteins and other inflammatory mediators (including autocrine/paracrine responses of the macrophage to the released molecules).

doi:10.1371/journal.pone.0001035.g002

Because immune response phases utilize multiple functional pathways and these pathways are overlapping among phases, the genes chosen for the SNP array panel were assigned to one of the following *functional pathways*:

Adhesion-Extravasation-Migration: adhesion molecules; chemoattractants and chemoattractant receptors; cytoskeletal rearrangement signaling, motility proteins.

Apoptosis signaling: death receptors and ligands and extrinsic apoptosis pathway signaling; mitochondrial-dependent, intrinsic apoptosis pathway signaling; cellular stress signaling.

Calcium signaling: NF-ATs; calcineurins; calcium/calmodulin-dependent kinases.

Complement cascade: components of classical, alternative, and lectin-dependent complement pathways.

Cytokines and cytokine signaling: cytokines; cytokine receptors; cytokine-dependent signaling, including JAK-STAT and interferon-regulatory factor (IRF) pathways; suppressor of cytokine signaling proteins (SOCS).

Eicosanoid synthesis and receptors: enzymes involved in synthesis of prostanoids, leukotrienes, hepoxylins (12-HETE), and lipoxins from arachidonic acid; prostanoid and leukotriene receptors.

Glucocorticoid/PPAR signaling: nuclear receptors for glucocorticoids; steroid-interacting proteins; PPARs and associated proteins.

G-Protein Coupled Receptor Signaling: GPCRs (other than eicosanoid receptors and chemokine receptors); G-protein-dependent signaling pathways, including cAMP-PKA and phospholipase B2.

Innate pathogen detection: Toll-like receptors (TLRs); intracellular nucleotide detectors (e.g., Nod1 and putative members of CARD/Nod family); peptidoglycan recognition proteins; associated signaling molecules linking these detectors to major signaling pathways.

Leukocyte signaling: Signaling molecules, receptors, and adaptors important for regulation of leukocyte activation beyond major signaling pathways (i.e., MAPK, PI3K/Akt, NF-κB, GPCR, and cytokine signaling pathways); including, but not limited to, T cell receptor (TCR) and B cell receptor (BCR) signaling components, B7 family, phosphatases, Foxp3, immunoglobulin receptors, leukocyte inhibitory receptor (CD85) family, scavenger receptors.

MAPK signaling: p38 stress-activated protein kinase, p42/p44 extracellular-regulated kinase (Erk), and Jun kinase (Jnk) signaling pathways.

Natural Killer Cell Signaling: Natural killer (NK) cell-specific activating and inhibitory receptors; adaptors and signaling molecules for transducing NK cell-specific receptor signaling.

NF-kB signaling: Molecules for regulation of NF-κB activation, including adaptors linking. other pathways to and from the NF-κB pathway (e.g., from MAPK, pathogen-detection, and TNF-α signaling pathways).

Antigen presentation: Major Histocompatibility Complex (MHC) molecules and associated proteins; proteins involved in uptake, processing, and loading of peptides on MHC molecules.

PI3K/AKT Signaling: Molecules involved in regulation of PI3K-dependent signaling, including adaptors linking other pathways to and from the PI3K pathway.

ROS/glutathione/cytotoxic granules: Molecules involved in the generation and response to leukocyte-derived cytotoxic agents (reactive oxygen species (ROS), nitric oxide, cytotoxic granules of granulocytes and natural killer cells), including contents of granulocyte and NK cell cytotoxic granules; glutathione peroxidases; peroxiredoxins; catalase; proteinases; superoxide dismutase.

TNF superfamily and signaling: Receptors and ligands of the TNF-α superfamily; adaptors and signaling molecules involved in transducing signals from receptor stimulation to other major signaling pathways (e.g., MAPK and NF-κB pathways).

Examples of the functional subpathways and types of genes chosen for the different phases of immune responses are presented in **Table 3** (proteins encoded by the genes are provided). Most of these immune response phases also utilize common signaling pathways, including elements of MAPK, NF-κB, PI3K/Akt, GCPR, cytokine, and leukocyte signaling pathways. An example of the integration of multiple functional subpathways for one phase of an immune response (danger signal) is depicted in **Fig. 2**. The numbers of genes chosen in each subpathway are listed in **Table 4**, and the complete list of 1027 candidate genes and their primary subpathway (and secondary subpathway for a subset of genes) is provided.

Table 3. Pathways and proteins associated with immune response phases

Phase of immune response	Examples of pathways, proteins, and inflammatory mediators involved in immune response phases
Hematopoiesis/homeostasis/ tolerance	hematopoietic cytokines (M-,G-,GM-CSF;IL-4,-5,-7,-13), stromal factors (c-kit, SCF, Flt3L), regulatory T cell function (Foxp3)
Danger signal	innate pathogen recognition receptors (TLRs, CARDs/NODs, peptidoglycan recognition proteins), scavenger receptors (MSR1), endothelins, adenosine receptors, complement, stress-induced responses (MIC-A,-B), eicosanoid synthesis genes, cytokines, antigen presentation genes
Mobilization of immune cells	systemic inflammatory mediators (IL-1β, IL-6, TNF-a), chemokines, eicosanoids, GPCR signaling (eicosanoids, histamines)
Extravasation	adhesion molecules (integrins, -CAMs), chemokines, vasodilators (eicosanoids/GPCR), cytoskeletal rearrangement singaling molecules (Vav, VASP, MENA), non-muscle myosins
Migration to site of inflammation	adhesion molecules (integrins, -CAMs, maxtrix receptors), chemokines, matrix proteases (MMPs), cytoskeletal rearrangement singaling molecules (Vav,VASP,MENA), focal adhesion proteins (Vav,ROCK), non-muscle myosins
Interactions between resident cells, immune cells, and pathogens at site of inflammation	adhesion molecules, innate detectors of pathogens (TLRs, CARDs/NODs), Fc receptors (FcgRI,II,III; FceRI,II), stress-induced ligands (MIC-A,-B), NK cell-activating receptors, cytokines and receptors, other activating receptors (TCR, BCR complexes; growth factor receptors); co-stimulatory receptors (B7 family, CD2 family), inhibitory receptors (KIRs, LIRs/ILTs), phagocytosis/antigen presentation (XBOX genes, CIITA, TAP, immunoproteasome, HLA molecules)
Activation of inflammatory cells	MAPK pathways (Erk, p38, Jnk), PI3K/Akt signaling, NF-kB signaling, cytokine signaling (JAK/STAT/Tyk, NFIL3, NFIL6, IRFs), GPCR signaling (PKA, PLCb, phosphodiesterases, CREB, Pyk2, Rap1, Src), adaptor signaling proteins (TRAFs, IRAKs, MyD88, DAP10, DAP12, ZAP70, Syk, LAT, SLP76, MyD88, CD3ζ, FcεRγ)
Effector functions of inflammatory cells	cytokines (IFN-γ, IFN-α, TNF-α superfamily, CSFs, interleukins), death receptor ligands (FasL, TRAIL, TNF-a), eicosanoids (prostaglandins, thromboxane, prostacyclin, leukotrienes), cytotoxic mediators (glutathiones/PHOX/ reactive oxygen species, RNS, perforin/granzymes), antibody production, acute phase/fever response (C-reactive protein, factor P)
Response of target cells	cytokine receptors, GPCRs, death receptors, apoptosis signaling, adhesion molecules, growth factor receptors
Resolution of immune response vs. chronic inflammation	apoptosis (death receptor and mitochondrial pathways), TGF-β, IL-10, Foxp3, prostaglandins, phosphatases, inhibitors of cytokine signaling (SOCS, A20/TNFAIP3)

doi:10.1371/journal.pone.0001035.t003

Table 4: Primary subpathways in inflammation panel

Subpathway	Number of genes in subpathway	Number of SNPs in subpathway
Adhesion-Extravasation-Migration	142	1385
Apoptosis Signaling	68	682
Calcium Signaling	14	409
Complement Cascase	40	419
Cytokine signaling	172	1598
Eicosanoid Signaling	39	374
Glucocorticoid/PPAR signaling	21	230
G-Protein Coupled Receptor Signaling	42	1125
Innate pathogen detection	50	457
Leukocyte signaling	121	1743
MAPK signaling	118	1949
Natural Killer Cell Signaling	31	259
NF-kB signaling	33	297
Phagocytosis-Ag presentation	39	286
PI3K/AKT Signaling	37	307
ROS/Glutathione/Cytotoxic granules	22	162
TNF Superfamily Signaling	38	328

doi:10.1371/journal.pone.0001035.t004

Example Application: WFINFLAM tSNP Panel

Because the components comprising inflammation are very numerous and interacting across many pathways, without strong *a priori* evidence it is difficult to choose a handful of candidate genes to fully cover the potential genetic risk factors contributing to the inflammatory component of a particular disease. For this reason, panels of single nucleotide polymorphisms (SNPs) in an array of inflammation-related genes broadly covering most aspects of immunity and inflammation based on our assembled list will be critical in objectively evaluating the impact of genetic variations in inflammation-related genes on an inflammation-dependent outcome.

There is a commercially available product, Affymetrix GeneChip® Human Immune and Inflammation 9K SNP panel, that attempts to serve the purpose. This application -specific panel contains ~9,000 SNPs to cover ~1,000 immunity- and inflammation- related genes (http://www.affymetrix.com/support/technical/datasheets/humanimmune_9k_snp_datasheet.pdf). However, the rationale for choosing the ~1,000 immunity and inflammation genes in this panel is not clear, and the coverage of these genes, regardless of their biological functions, may not be sufficient. In order to investigate the impact of genetic variations in a broad array of inflammation-related genes on disease risk, we created two tagging SNP (tSNP) panels, one each for populations with either Caucasian or African ancestries, for Affimetrix ParAllele genotyping chips. These tSNP panels were designed to capture majority of the genetic variations in these 1027 inflammation candidate genes.

tSNPs for the 1027 inflammation-associated candidate genes were chosen based on a pair-wise r^2 threshold of 0.8 and MAF ≥5% using data in the HapMap Phase II database (HapMap Data Release 21a/phaseII; http://hapmap.org/index.html.en). tSNPs were chosen separately for the CEU (representing European ancestry) and YRI populations (representing African ancestry). In order to accommodate as many relevant inflammation genes as possible, less stringent criteria with r^2 threshold >0.5 were employed for intronic regions (excluding 5kb in the beginning and end of these big introns) greater than 50kb in certain large genes with >100 tSNPs. There were seven genes in this category, and the intronic regions for which a less stringent r^2 threshold was applied are listed in . For genes without genotype information from HapMap, additional SNPs in six inflammation genes were chosen to be included based on information from other resources, as described in *Methods* section.

The resulting inflammation tSNP panels, WFINFLAM-CEU for Caucasians and WFINFLAM-YRI for African descent, include 12,011 SNPs and 21,542 SNPs respectively in 1027 inflammation-associated candidate genes. There is an average of 11.7 and 21 SNPs in each candidate gene in WFINFLAM-CEU and WFINFLAM-YRI panels, respectively. **Table 4** briefly summarizes the numbers of genes and tSNPs included in each subpathway. The annotation table for this panel, including the list of 1027 genes, their chromosomal positions, their associated primary and secondary subpathways, and the number of tSNPs chosen for each gene can be found (and also our website: http://www1.wfubmc.edu/Genomics/PublicationsandData/). Additionally, the annotation file for the SNPs included in the WFINFLAM-CEU and WFINFLAM-YRI panels, including the chromosomal locations, their associated genes and the sub-pathways, can also be found (and our website: http://www1.wfubmc.edu/Genomics/PublicationsandData/). The coverage for the 1027 inflammation candidate genes in CEU is better compared to other widely used genome-wide association panels. For the coverage of the 1027 inflammation-associated candidate genes, 90.4% of the genes have 90% or more SNPs within these genes that can be captured by $r^2 \geq 0.8$ in CEU using WFINFLAM-CEU panel, compared to 78.9% for the Illumina HumanHap 550 genome-wide panel and 45.8% for the Affymetrix 500k genome-wide panel. For populations with African ancestry, the coverage for the 1027 inflammation candidate genes in YRI is greatly improved compared to other widely used genome-wide association panels. For the coverage of the 1027 inflammation-associated candidate genes, 88.2% of the genes have 90% or more SNPs within these genes that can be captured by $r^2 \geq 0.8$ in YRI using WFINFLAM-YRI panel, compared to 27.1% for the Illumina HumanHap 650k genome-wide panel (which was designed to capture more genetic information from YRI population), and 12.5% for the Affymetrix 500k genome-wide panel. The coverage for all these genotyping panels is detailed in .

DISCUSSION

Various components and complex interactions comprise immune and inflammation responses, and numerous genes are involved in this complex network. With a thorough review of various aspects of inflammatory immune responses, and a systematic search for gene-gene interactions using Ingenuity Pathway Analysis, we have provided a comprehensive list of inflammation-associated genes and subpathways for genetic association studies.

Genome-wide association studies have been a very popular approach to test the association between disease phenotypes and genetic variations. However, we believe there are still several advantages for a pathway-focused study. First of all, compared to whole-genome analyses, restricting analyses to SNPs in a specific pathway reduces the number of multiple tests performed in the analysis of a study population, thereby reducing the probability of false positive associations and increasing the effective power of the study.

This kind of study design is particularly effective when inflammation plays an important role in disease etiology and the goal of the studies is to delineate genetic variations in inflammation pathway to disease risk and/or progression. A related second advantage of restricted pathway analysis is in study design. A large proportion of investigators may not have access to the very large number of subjects and multiple confirmation populations needed to overcome false positive associations due to multiple testing in genome-wide association studies. Studies restricted to a pathway analysis permit the use of study populations that are not large enough for use in whole-genome association studies. When target diseases are not prevalent and inflammation is obviously involved in disease etiology, researchers will gain the most out of an inflammation pathway-specific study design. Although some genes not related to inflammation found in whole-genome panels may impart some risk to inflammation-associated diseases, associated genetic variants would be anticipated to be concentrated in a panel of SNPs in inflammation-associated genes. Therefore, the drawback of potentially missing associated non-inflammation genes is offset by the increased probability of detecting true associations in an inflammation-restricted panel. Thirdly, pathway analysis is far less expensive to perform than whole-genome analysis, especially considering the cost for second, and/or third stage confirmation studies needed to follow-up the significant results from an initial screening in order to rule out false positive associations. Lastly, the functional subpathways are also pre-defined with available biological information. This refined information provides investigators with the opportunity to test gene-gene interactions within subpathways in which synergistic interactions are more likely to be concentrated. Additionally, the interplays between subpathways are also clearly defined to enable investigators to test biologically feasible interactions between subpathways.

However, the results from this manuscript also have potential utility for investigators who have more interests in surveying the whole genome. For whole-genome analyses where there is a prior hypothesis for inflammation being associated with the outcome, the inflammation pathway and subpathways defined in this manuscript may provide a framework for testing whether SNPs in the inflammation pathway or subpathways as a whole are overrepresented for significant associations to the outcome. Although pathway networks can be constructed for whole-genome analyses, such networks should be designed *a priori* before beginning the study.

The WFINFLAM and WFINFLAM-YRI SNP array panels for inflammation-associated genes provide a powerful tool for analyzing the contribution of genetic variation in diseases that have inflammatory components. Although whole-genome SNP panels are currently available that include almost all of the genes included in the WFINFLAM panels, the coverage of SNPs in genes included in the WFINFLAM array is superior to the coverage of currently available in whole-genome arrays, especially for populations with African ancestry background. For researchers who would like to use other genotyping platforms, the inflammatory gene list provided here would be a good starting point for designing genotyping assays for other platforms. Additionally, alternate approaches, other than r^2 based method, for choosing SNPs based on the inflammatory gene list provided here could also be considered. For example, researchers may specifically focus on "high-prior" polymorphisms that are known to be functional or have been previously linked to the specific diseases under study, alone or in combination with the tSNPs provided in the WFINFLAM panel. This approach may be more efficient and powerful than the r^2 based method alone, especially if the targeted "high-prior" polymorphisms are causal and their linkage to the nearby tSNPs is incomplete.

In addition, precaution may be warranted for tSNP panels designed based on the HapMap project. The transferability of the LD patterns between populations studied in HapMap project and other study populations may need to be validated. The transferability of HapMap-based selection of tSNPs using the reference CEU population to several other diverse populations of European ancestry has been demonstrated to be almost as effective for overall SNP coverage in selected genomic regions or randomly selected SNPs in the respective populations [46]–[50]. However, the coverage with tSNPs based on HapMap CEU data for a small subset of specific genes or SNPs may be lower in certain populations despite overall similar coverage [51], particularly for isolated indigenous populations [52]. Other than the data provided by the HapMap project, we are not aware of complete sequence variant information available for validation of the transferability of tSNP panels that were designed based on HapMap data to other study populations.

In summary, pathway analysis of inflammation-associated genes is a powerful approach for determining genetic risk factors for both inflammatory diseases and other diseases that may have an under-appreciated modest inflammatory component, such as cancers. The inflammation pathway gene list and functionally-defined subpathways provide useful tools for assessing the impact of genetic variations in inflammation pathways on disease risk, in situations where either pathway-focused studies or genome-wide analyses are employed.

METHODS

Selection of Genes

Networks of genes involved in the regulation of the phases of immune responses (described in*Results* and **Table 2**) were built using Ingenuity Pathways Analysis (Ingenuity Systems:www.analysis.ingenuity.com). Both pre-defined 'canonical pathways' and custom-built pathways based on our own queries for genes/pathways not included in the canonical pathways were used to establish these networks. Canonical pathways included: actin cytoskeleton signaling; antigen presentation pathway; apoptosis signaling; B cell receptor signaling; calcium signaling; cAMP signaling; chemokine signaling; complement and coagulation cascade; death receptor signaling; ERK/MAPK signaling; FcEpsilon receptor signaling; G-coupled protein receptor signaling; GM-CSF signaling; IGF-1 signaling; IL-10 signaling; IL-2 signaling; IL-4 signaling; IL-6 signaling; integrin signaling; interferon signaling; JAK/STAT signaling; leukocyte extravasation signaling; natural killer cell signaling; NF-kB signaling; nitric oxide signaling; notch signaling; p38 MAPK signaling; PI3K/Akt signaling; PPAR signaling; protein ubiquination pathway; PTEN signaling; SPK/JNK signaling; T cell receptor signaling; TGF-β signaling; Toll-like receptor signaling.

These Networks Were Then Arranged Into Inflammation Subpathways By:

combining several networks together (e.g., ERK/MAPK, p38 MAPK, and SAPK/JNK canonical pathways into the 'MAPK signaling' inflammation subpathway; IL-2-, IL-4-, IL-6-, IL-10-, Interferon-, GM-CSF-, IGF-1-, JAK/STAT6-, and TGF-β- signaling canonical pathways into 'cytokine signaling' inflammation subpathway; actin cytoskeleton-, chemokine-, integrin-, and leukocyte extravasation- signaling canonincal pathways into the 'adhesion-extravasation-migration' inflammation subpathway; etc.);

adding additional genes to bridge networks within a subpathway and to include appropriate genes not included in the canonical pathways (e.g., additional cytokines and their receptors were added to the 'cytokine signaling' inflammation subpathway; additional integrins and chemokines/chemoattractant molecules were added to the 'adhesion-extravasation-migration' inflammation subpathway; CD antigens expressed by leukocytes not already included were considered for addition to several subpathways; other missing genes/pathways considered important by the panel of investigators, such as the scavenger receptor network for 'leukocyte signaling' inflammation subpathway and Nod1/CARD family networks for 'innate pathogen detection' inflammation subpathway; etc.);

trimming the networks of genes with low priority for inclusion in the inflammation panel. Genes with lower priority include: genes not expressed in immune cells or not directly involved in cells responding to inflammation, including non-immune cells (e.g., skeletal muscle-specific myosins in the actin cytoskeleton signaling canonical pathway; calsequestrins expressed mainly in various muscle cells, in the calcium signaling canonical pathway; GH1 and GHR, growth hormone expressed in pituitary gland and its receptor, and NGFB and NGFR, nerve growth factor and its receptor, in the NF-κB canonical pathway; etc.; MAPK8IP1, specific for pancreatic cell function, in the PI3K/AKT canonical pathway; etc.); genes with unknown function, though genes with high homology to known inflammatory mediators were considered (e.g., bcl-2 family homologs, IL-1β family homologs). Special emphasis was placed on genes at nodes for signaling to and from multiple pathways, most notably genes in NF-κB, MAPK, and PI3K signaling pathways.

Choosing Tagging SNPs

Tagging SNPs (tSNPs) for candidate genes were chosen using Tagger server (http://www.broad.mit.edu/mpg/tagger/server.html). The target sequences included genomic regions containing the entire candidate genes, 5kb before transcription start site, and 2kb after the transcription end site, based on annotation in NCBI Build 35. When candidate regions for two or more genes overlapped, the combined genomic regions were used for choosing tSNPs. Two separate lists of non-synonymous coding SNPs for the CEPH population (CEU) and for the Yuruba population (YRI) were downloaded from HapMap (rel21a_NCBI_Build35) and forced in as tagging SNPs for ancestry-specific panels. Pair-wise r^2 threshold of 0.8 and MAF≥5% were used. For genes without genotype information from HapMap, SNPs in Affymetrix 500k array, as well as SNPs with frequency data from the Innate Immunity PGA (IIPGA) (http://innateimmunity.net/), were manually chosen to be included in the list based on allele frequencies and inter-SNP distance (the chromosomal positions are based on IIPGA annotation because some of these genes/SNPs were ambiguously mapped in build 35). Some tSNPs were dropped out from the final list because they did not pass the Affymetrix design review due to the following reasons: non-biallelic SNPs, existing SNPs or ambiguous bases too close to the SNP of interest, SNPs exceeding T_m ranges, or SNPs failing BLAST searches. In the attempt to fill in the gaps due to tSNPs failing design criteria, tSNPs for genes with dropped-out tSNPs were chosen again by forcing out the tSNPs failing design review and forcing in the remaining tSNPs when running Tagger. In total, there are 12,011 SNPs in WFINFLAM panel for Caucasians and 21,542 SNPs in WFINFLAM-YRI for African descents in 1027 inflammation-associated candidate genes that passed the Affymetrix design review.

Estimation of genomic coverage by tagging SNPs

We used LdCompare (Hao 2006;http://www.affymetrix.com/support/developer/tools/devnettools.affx) to estimate the coverage of the 1027 inflammation-associated candidate genes with both WFINFLAM panel for Caucasians and WFINFLAM-YRI panel for African descents. We have also computed the coverage of several commercial genotyping arrays, including Affymetrix Mapping 500K, Illumina HumanHap 550 and HumanHap 650, for these 1027 genes. Single-marker coverage (pairwise r^2) and multiple-marker coverage (multiple-marker r^2) were computed and combined to estimate the coverage for the genomic regions containing the entire candidate genes, 5kb before transcription start site, and 2kb after the transcription end site. The summary for the coverage for these genes in all these genotyping panels are detailed in .

AUTHOR CONTRIBUTIONS

Conceived and designed the experiments: JX BC ML. Analyzed the data: BC ML. Wrote the paper: BC ML. Other: Contributed to compilation of data: JX WI LL CM ML BC.

REFERENCES

1. Lawrence T, Gilroy DW (2007) Chronic inflammation: a failure of resolution? Int J Exp Pathol 88: 85–94.
2. Han J, Ulevitch RJ (2005) Limiting inflammatory responses during activation of innate immunity. Nat Immunol 6: 1198–1205.
3. Forrester JS, Bick-Forrester J (2005) Persistence of inflammatory cytokines cause a spectrum of chronic progressive diseases: implications for therapy. Med Hypotheses 65: 227–231.
4. Goronzy JJ, Weyand CM (2005) Rheumatoid arthritis. Immunol Rev 204: 55–73.
5. Ross R (1999) Atherosclerosis–an inflammatory disease. N Engl J Med 340: 115–126.
6. Willerson JT, Ridker PM (2004) Inflammation as a cardiovascular risk factor. Circulation 109: II2–10.

7. Ridker PM, Cushman M, Stampfer MJ, Tracy RP, Hennekens CH (1997) Inflammation, aspirin, and the risk of cardiovascular disease in apparently healthy men. N Engl J Med 336: 973–979.
8. Neuman MG (2007) Immune dysfunction in inflammatory bowel disease. Transl Res 149: 173–186.
9. James SP (2005) Prototypic disorders of gastrointestinal mucosal immune function: Celiac disease and Crohn's disease. J Allergy Clin Immunol 115: 25–30.
10. Coussens LM, Werb Z (2002) Inflammation and cancer. Nature 420: 860–867.
11. Ernst PB, Gold BD (2000) The disease spectrum of Helicobacter pylori: the immunopathogenesis of gastroduodenal ulcer and gastric cancer. Annu Rev Microbiol 54: 615–640.
12. Karin M, Cao Y, Greten FR, Li ZW (2002) NF-kappaB in cancer: from innocent bystander to major culprit. Nat Rev Cancer 2: 301–310.
13. Kuper H, Adami HO, Trichopoulos D (2000) Infections as a major preventable cause of human cancer. J Intern Med 248: 171–183.
14. Shacter E, Weitzman SA (2002) Chronic inflammation and cancer. Oncology (Williston Park) 16: 217–26, 229.
15. Ulrich CM, Bigler J, Potter JD (2006) Non-steroidal anti-inflammatory drugs for cancer prevention: promise, perils and pharmacogenetics. Nat Rev Cancer 6: 130–140.
16. Calder PC (2006) Polyunsaturated fatty acids and inflammation. Prostaglandins Leukot Essent Fatty Acids 75: 197–202.
17. Aganna E, Martinon F, Hawkins PN, Ross JB, Swan DC, et al. (2002) Association of mutations in the NALP3/CIAS1/PYPAF1 gene with a broad phenotype including recurrent fever, cold sensitivity, sensorineural deafness, and AA amyloidosis. Arthritis Rheum 46: 2445–2452.
18. Dode C, Le Du N, Cuisset L, Letourneur F, Berthelot JM, et al. (2002) New mutations of CIAS1 that are responsible for Muckle-Wells syndrome and familial cold urticaria: a novel mutation underlies both syndromes. Am J Hum Genet 70: 1498–1506.
19. Hoffman HM, Mueller JL, Broide DH, Wanderer AA, Kolodner RD (2001) Mutation of a new gene encoding a putative pyrin-like protein causes familial cold autoinflammatory syndrome and Muckle-Wells syndrome. Nat Genet 29: 301–305.
20. Touitou I (2001) The spectrum of Familial Mediterranean Fever (FMF) mutations. Eur J Hum Genet 9: 473–483.

21. Aoki T, Hirota T, Tamari M, Ichikawa K, Takeda K, et al. (2006) An association between asthma and TNF-308G/A polymorphism: meta-analysis. J Hum Genet 51: 677–685.

22. Lee YH, Harley JB, Nath SK (2006) Meta-analysis of TNF-alpha promoter -308 A/G polymorphism and SLE susceptibility. Eur J Hum Genet 14: 364–371.

23. Rahman P, Siannis F, Butt C, Farewell V, Peddle L, et al. (2006) TNFalpha polymorphisms and risk of psoriatic arthritis. Ann Rheum Dis 65: 919–923.

24. Lee YH, Rho YH, Choi SJ, Ji JD, Song GG, et al. (2007) The PTPN22 C1858T functional polymorphism and autoimmune diseases–a meta-analysis. Rheumatology (Oxford) 46: 49–56.

25. Graham RR, Kozyrev SV, Baechler EC, Reddy MV, Plenge RM, et al. (2006) A common haplotype of interferon regulatory factor 5 (IRF5) regulates splicing and expression and is associated with increased risk of systemic lupus erythematosus. Nat Genet 38: 550–555.

26. Smyth DJ, Cooper JD, Bailey R, Field S, Burren O, et al. (2006) A genome-wide association study of nonsynonymous SNPs identifies a type 1 diabetes locus in the interferon-induced helicase (IFIH1) region. Nat Genet 38: 617–619.

27. Howson JM, Dunger DB, Nutland S, Stevens H, Wicker LS, et al. (2007) A type 1 diabetes subgroup with a female bias is characterised by failure in tolerance to thyroid peroxidase at an early age and a strong association with the cytotoxic T-lymphocyte-associated antigen-4 gene. Diabetologia 50: 741–746.

28. Payne F, Cooper JD, Walker NM, Lam AC, Smink LJ, et al. (2007) Interaction analysis of the CBLB and CTLA4 genes in type 1 diabetes. J Leukoc Biol 81: 581–583.

29. Ueda H, Howson JM, Esposito L, Heward J, Snook H, et al. (2003) Association of the T-cell regulatory gene CTLA4 with susceptibility to autoimmune disease. Nature 423: 506–511.

30. Oostenbrug LE, Nolte IM, Oosterom E, van der SG, te Meerman GJ, et al. (2006) CARD15 in inflammatory bowel disease and Crohn's disease phenotypes: an association study and pooled analysis. Dig Liver Dis 38: 834–845.

31. Duerr RH, Taylor KD, Brant SR, Rioux JD, Silverberg MS, et al. (2006) A genome-wide association study identifies IL23R as an inflammatory bowel disease gene. Science 314: 1461–1463.

32. Loza MJ, Chang B (2007) Association between Q551R IL4R genetic variants and atopic asthma risk demonstrated by meta-analysis. J Allergy Clin Immunol In press.

33. Breast Cancer Association Consortium (2006) Commonly studied single-nucleotide polymorphisms and breast cancer: results from the Breast Cancer Association Consortium. J Natl Cancer Inst 98: 1382–1396.

34. Gaudet MM, Egan KM, Lissowska J, Newcomb PA, Brinton LA, et al. (2007) Genetic variation in tumor necrosis factor and lymphotoxin-alpha (TNF-LTA) and breast cancer risk. Hum Genet.

35. Cox DG, Pontes C, Guino E, Navarro M, Osorio A, et al. (2004) Polymorphisms in prostaglandin synthase 2/cyclooxygenase 2 (PTGS2/COX2) and risk of colorectal cancer. Br J Cancer 91: 339–343.

36. Koh WP, Yuan JM, van den BD, Lee HP, Yu MC (2004) Interaction between cyclooxygenase-2 gene polymorphism and dietary n-6 polyunsaturated fatty acids on colon cancer risk: the Singapore Chinese Health Study. Br J Cancer 90: 1760–1764.

37. Lin HJ, Lakkides KM, Keku TO, Reddy ST, Louie AD, et al. (2002) Prostaglandin H synthase 2 variant (Val511Ala) in African Americans may reduce the risk for colorectal neoplasia. Cancer Epidemiol Biomarkers Prev 11: 1305–1315.

38. Siezen CL, van Leeuwen AI, Kram NR, Luken ME, van Kranen HJ, et al. (2005) Colorectal adenoma risk is modified by the interplay between polymorphisms in arachidonic acid pathway genes and fish consumption. Carcinogenesis 26: 449–457.

39. Siezen CL, Bueno-de-Mesquita HB, Peeters PH, Kram NR, van Doeselaar M, et al. (2006) Polymorphisms in the genes involved in the arachidonic acid-pathway, fish consumption and the risk of colorectal cancer. Int J Cancer 119: 297–303.

40. Shahedi K, Lindstrom S, Zheng SL, Wiklund F, Adolfsson J, et al. (2006) Genetic variation in the COX-2 gene and the association with prostate cancer risk. Int J Cancer 119: 668–672.

41. Hedelin M, Chang ET, Wiklund F, Bellocco R, Klint A, et al. (2007) Association of frequent consumption of fatty fish with prostate cancer risk is modified by COX-2 polymorphism. Int J Cancer 120: 398–405.

42. Zheng SL, Liu W, Wiklund F, Dimitrov L, Balter K, et al. (2006) A comprehensive association study for genes in inflammation pathway provides support for their roles in prostate cancer risk in the CAPS study. Prostate 66: 1556–1564.

43. Dinu V, Miller PL, Zhao H (2007) Evidence for association between multiple complement pathway genes and AMD. Genet Epidemiol.
44. Conley YP, Jakobsdottir J, Mah T, Weeks DE, Klein R, et al. (2006) CFH, ELOVL4, PLEKHA1 and LOC387715 genes and susceptibility to age-related maculopathy: AREDS and CHS cohorts and meta-analyses. Hum Mol Genet 15: 3206–3218.
45. Klein RJ, Zeiss C, Chew EY, Tsai JY, Sackler RS, et al. (2005) Complement factor H polymorphism in age-related macular degeneration. Science 308: 385–389.
46. Willer CJ, Scott LJ, Bonnycastle LL, Jackson AU, Chines P, et al. (2006) Tag SNP selection for Finnish individuals based on the CEPH Utah HapMap database. Genet Epidemiol 30: 180–190.
47. De Bakker PI, Graham RR, Altshuler D, Henderson BE, Haiman CA (2006) Transferability of tag SNPs to capture common genetic variation in DNA repair genes across multiple populations. Pac Symp Biocomput 478–486.
48. Montpetit A, Nelis M, Laflamme P, Magi R, Ke X, et al. (2006) An evaluation of the performance of tag SNPs derived from HapMap in a Caucasian population. PLoS Genet 2: e27.
49. Tenesa A, Dunlop MG (2006) Validity of tagging SNPs across populations for association studies. Eur J Hum Genet 14: 357–363.
50. Xu Z, Kaplan NL, Taylor JA (2007) Tag SNP selection for candidate gene association studies using HapMap and gene resequencing data. Eur J Hum Genet.
51. Mueller JC, Lohmussaar E, Magi R, Remm M, Bettecken T, et al. (2005) Linkage disequilibrium patterns and tagSNP transferability among European populations. Am J Hum Genet 76: 387–398.
52. Johansson A, Vavruch-Nilsson V, Cox DR, Frazer KA, Gyllensten U (2007) Evaluation of the SNP tagging approach in an independent population sample-array-based SNP discovery in Sami. Hum Genet 122: 141–150.

Chapter 10

ANALYSIS OF GENE EXPRESSION DATA USING BICLUSTERING ALGORITHMS

Fadhl M. Al-Akwaa[1]

[1]Biomedical Eng. Dept., Univ. of Science & Technology, Sana'a, Yemen

INTRODUCTION

One of the main research areas of bioinformatics is functional genomics; which focuses on the interactions and functions of each gene and its products (mRNA, protein) through the whole genome (the entire genetics sequences encoded in the DNA and responsible for the hereditary information). In order to identify the functions of certain gene, we should able to capture the gene expressions which describe how the genetic information converted to a functional gene product through the transcription and translation processes. Functional genomics uses microarray technology to measure the genes expressions levels under certain conditions and environmental limitations. In the last few years, microarray has become a central tool in biological research. Consequently, the corresponding data analysis becomes one of the important work disciplines in bioinformatics. The analysis of microarray data poses a large number of exploratory statistical aspects including clustering and biclustering algorithms, which help to identify similar patterns in gene expression data and group genes and conditions in to subsets that share biological significance.

What is Clustering?

A large number of clustering definitions can be found in the literature. The simplest definition is shared among all and includes one fundamental concept: the grouping together of similar data items into clusters[1].

Clustering is an important explorative statistical analysis of gene expression data. It aims to identify and group genes that exhibit similar expression patterns over several conditions and also group the conditions based on the expression profiles across set of genes. The successful clustering approach should guarantee two criteria which are homogeneity high similarity between elements in the same cluster, and separation – low similarity between elements from different clusters. When homogeneity and separation are precisely defined, those are two opposing objectives: The better the homogeneity the poorer the separation, and vice versa [2]. Several algorithmic techniques were previously used for clustering gene expression data, including hierarchical clustering [3], self organizing maps [4], and graph theoretic approaches [5].

K-Means

K-means is a classical clustering algorithm [6] invented in 1956 to classify or to group objects (genes) based on attributes or features (experimental conditions) into K number of groups (clusters). K is positive integer number and assumed to be known.

K-means computational approach starts by placing K points into the space represented by the objects that are being clustered. These points represent initial group centroids. We can take any random objects as the initial centroids or the first K objects in sequence can also be used as the initial centroids. Then the K means algorithm will do the four steps below until convergence:

- Determine the centroids coordinate.
- Determine the distance of each object to the centroids using the Euclidean distance.
- Group the objects based on minimum distance.
- Iterate the above steps till no object moves its assigned group.

Each iteration of k-means modifies the current partition by checking all possible modifications of the solution, in which one element is moved to another cluster. This is done by reducing the sum of distances between objects and the centers of their clusters. This procedure is repeated until no further improvement is achieved (No object move the group) and all the objects are grouped into the final required number of clusters.

A disadvantage of K-means algorithm could be perceived in the need to specify the number of clusters K as a parameter value prior to running the algorithm. In cases where there is no expectation about K, user has to make trails with several values of K or use external techniques to guess the no of clusters may be exist.

Hierarchical Clustering (Hcl)

Hierarchical clustering does not partition the genes into subsets. Instead it builds a down-top hierarchy of clusters using agglomerative methods or top - down hierarchy of clusters using divisive methods. The traditional graphical representation of this hierarchy is called dendrogram tree. The divisive method begins at the root and starts to breaks up clusters whose having low similarity. Whereas, the Agglomerative method begins at the leaves of the tree and starts with an initial partition into single element clusters and successively merges clusters until all elements belong to the same cluster [3]. (SeeFigure 1) The agglomerative method is widely used than the divisive one which is not generally available, and rarely has been applied. The idea of the agglomerative method can be summarized as following: Given a set of N items (genes in our case) to be clustered, and an N*N distance (or similarity) matrix [7],

- Assign each item to a cluster, so you have N clusters, each containing just one item.
- Find the closest (most similar) pair of clusters and merge them into a single cluster.
- Compute distances (similarities) between the new cluster and each of the old clusters.
- Repeat steps 2 and 3 until all items are clustered into a single cluster of size N.

In Step 3, distance or similarity measurements between the merged clusters and all the other clusters can be calculated in one of three schemes: single-linkage, complete linkage and average-linkage.

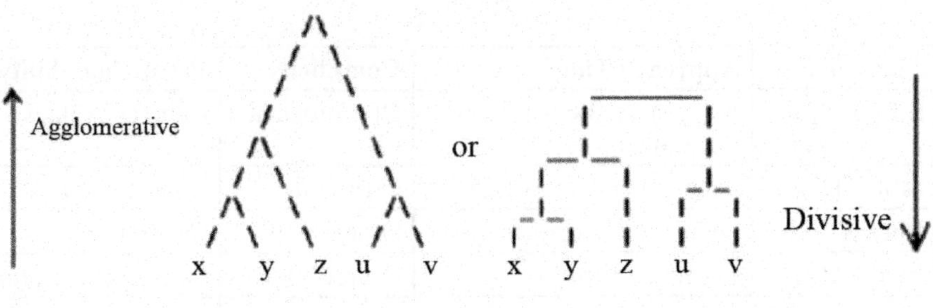

Figure 1: HCL: Agglomerative and Divisive Methods.

Biclustering

Traditional clustering approaches such as k-means and hierarchical clustering put each gene in exactly one cluster based on the assumption that all genes behave similarly in all conditions. However, recent understanding of cellular processes shows that it is possible for subset of genes to be co expressed under certain experimental conditions, and at the same time; to behave almost independently under other conditions. From this context, a new two mode clustering approach called biclustering or co-clustering has been introduced to group the genes and conditions in both dimensions simultaneously.

This allows finding subgroups of genes that show the same response under a subset of conditions, not all conditions. Also, genes may participate in more than one function, resulting in one regulation pattern in one context and a different pattern in another.

Example, if a cellular process is only active under specific conditions and there is a gene participates in multiple pathways that are differentially regulated, one would expect this gene to be included in more than one cluster; and this cannot be achieved by traditional clustering techniques.

Many biclustering methods exist in the literature [8]. Table 1 summarized some of promising biclustering algorithms developed during the last ten years. In brief, we described some of these algorithms according to their prediction strength, their promising results, to what they extend in the community, whether an implementation was available, and the feedback from their authors to explain some ambiguous issues.

Cheng And Church (CC)

CC algorithm[18] is considered to be the first real biclustering implementation after the primary idea has been introduced by Hartigan [19] in 1972.

Table 1: Biclustering Algorithms Comparison

Algorithm	**Approach Time**	**Complicity**	**Prediction ability**
Bivisu [9]	Exhaustive Bicluster Enumeration	$O(m^2nlogm^)a$	Coherent values
MSBE [10]	Greedy Iterative Search	$O((n + m)^2)$	Coherent values
Bimax[11]	Divide-and-Conquer	$O(nm\beta log\beta)$	Coherent values
ROBA [12]	Matrix algebra	O(nmLN)	Coherent Evolution
x-motif [13]	Greedy Iterative Search	$n^{mO(log(1/\alpha)/log(1/\beta)})$	Coherent Evolution
SAMBA [14]	Exhaustive Bicluster Enumeration	O(n2)	Coherent Evolution
OPSM [15]	Greedy Iterative Search	$O(nm^3I)$	Coherent Evolution
Plaid[16]	Distribution Parameter Identification	XX^Xb	Coherent values
ISA [17]	Iterative Signature Algorithm	XXX	Coherent values
CC [18]	Greedy Iterative Search	O((n + m)nm)	Coherent values

CC defines a bicluster as a subset of rows and a subset of columns with a high similarity. The proposed similarity score is called mean squared residue (H) and it is used to measure the coherence of the rows and columns in the single bicluster. Given the gene expression data matrix A = (X;Y); a bicluster is defined as a uniform submatrix (I;J) having a low mean squared residue score as following:

The CC Mean Squared Residue:

$$H(I,J) = \frac{1}{\|I\| \, \|J\|} \sum\nolimits_{i \in I, j \in J} \left(a_{ij} - a_{iJ} - a_{Ij} + a_{IJ}\right)^2 \quad (1)$$

- Where: a_{ij} is gene expression level at row i and column j, a_{iJ} is the mean of row i, a_{Ij} is the mean of column j, a_{IJ} is the overall mean. CC algorithm will identify the submatrix as a bicluster if the score is below a level alpha which is a user input parameter to control the quality of the output biclusters. Generally; CC algorithm performs the following major steps:
- Delete rows and columns with a score larger than alpha.
- Adding rows or columns until alpha level is reached.

Iterate these steps until a maximum number of biclusters is reached or no bicluster is found [18].

Iterative Signature Algorithm (ISA)

The ISA algorithm [17, 20] is a novel method for the biclustering analysis of large-scale expression data. It is an efficient algorithm based on the iterative application of the signature algorithm presented in [17]. ISA considers a bicluster to be a transcription module which can be defined as a set of coexpressed genes together with the associated set of regulating conditions (Figure 2). Starting with an initial set of genes, all samples (conditions) are scored with respect to this gene set and those samples are chosen for which the score exceeds a certain threshold (usually defined by the user). In the same way, all genes are scored regarding the selected samples and a new set of genes is selected based on another user-defined threshold. The entire procedure is repeated until the set of genes and the set of samples converge and do not change anymore.

Multiple biclusters can be discovered by running the ISA algorithm on several initial gene sets. This approach requires identification of a reference gene set which needs to be carefully selected for good quality results. In the absence of pre-specified reference gene set, random set of genes is selected at the cost of results quality[17].

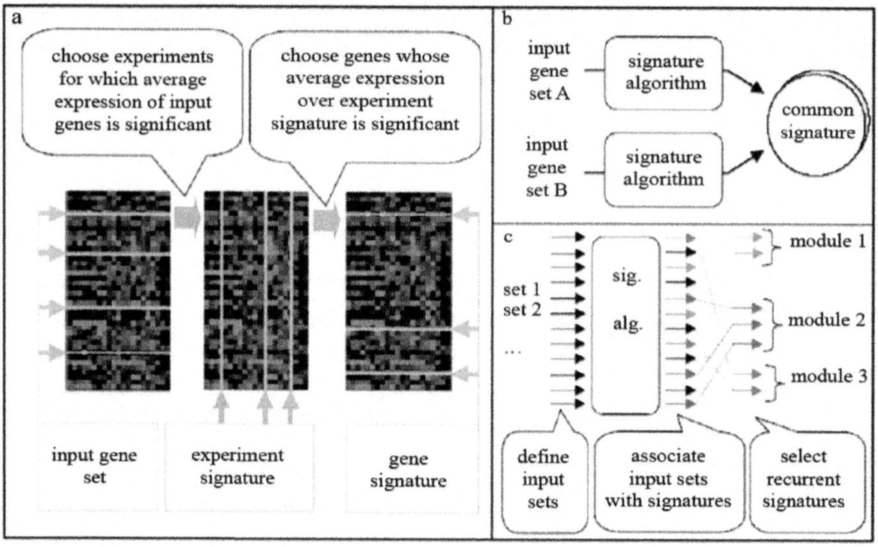

Figure 2: The recurrence signature method. a, The signature algorithm. b, Recurrence as a reliability measure. The signature algorithm is applied to distinct input sets containing different subsets of the postulated transcription module. If the different input sets give rise to the same module, it is considered reliable. c, General application of the recurrent signature method. Copyright © [17].

Biclusters Inclusion Maximal (Bimax)

Bimax[11] is a simple binary model and new fast divide-and-conquer algorithm used to cluster the gene expression data. It is presented in 2006 by Computer Engineering and Networks Laboratory ETH Zurich, Switzerland. Bimax discretized the gene expression data matrix and convert it into a binary matrix by identifying a threshold, so transcription levels (genes expression values) above this threshold become ones and transcription levels below become zeros (or vice versa). Then, it searches for all possible biclusters that contain only ones. This can be done by iterating these steps:

Rearrange the rows and columns to concentrate ones in the upper right of the matrix.

Divide the matrix into two sub matrices.

Whenever in one of the submatrices only ones are found, this sub matrix is returned.

Order Preserving Submatrix(Opsm)

The order-preserving submatrix (OPSM) algorithm [15] is a probabilistic model introduced to discover a subset of genes identically ordered among a subset of conditions. It focuses on the coherence of the relative order of the conditions rather than the coherence of actual expression levels. In other words, the expression values of the genes within a bicluster induce an identical linear ordering across the selected conditions. Accordingly, the authors define a bicluster as a subset of rows whose values induce a linear order across a subset of the columns. The time complexity of this model is $O(nm^3I)$ where n andmare the number of rows and columns of the input gene expression matrix respectively and I is the number of biclusters. A disadvantage of OPSM algorithm is that it takes long time for high dimensional datasets. And this is because its time complexity is cubic with regards to the number of columns (dimensions) of the input matrix [15].

Maximum Similarity Bicluster(Msbe)

MSBE Biclustering algorithm [10] is a novel polynomial time algorithm to find an optimal biclusters with the maximum similarity. The idea behind this algorithm is to find subset of genes that are related to a reference gene. The reference gene is known in advance. MSBE algorithm uses the similarity score for a sub-matrix to find the similar expressions in the microarray datasets. And the threshold of the average similarity score is a user input parameter in order to allow the user to control the quality of the biclustering results.

Clustering or Biclustering

Clustering algorithms [21-23] have been used to analyze gene expression data, on the basis that genes showing similar expression patterns can be assumed to be co-regulated or part of the same regulatory pathway. Unfortunately, this is not always true. Two limitations obstruct the use of clustering algorithms with microarray data. First, all conditions are given equal weights in the computation of gene similarity; in fact, most conditions do not contribute information but instead increase the amount of background noise. Second, each gene is assigned to a single cluster, whereas in fact genes may participate in several functions and should thus be included in several clusters[24].

A new modified clustering approach to uncovering processes that are active over some but not all samples has emerged, which is called biclustering. A bicluster is defined as a subset of genes that exhibit compatible expression patterns over a subset of conditions [11].

During the last ten years, many biclustering algorithms have been proposed (see [8] for a survey), but the important questions are: which algorithm is better? And do some algorithms have advantages over others?

Recently Kevin *et al.* [25]proposed a semantic web algorithm to recommend the best algorithm based on user inputs like: is the dataset contain outliers, is it allowed to get overlapped clusters and the time to retrieve the biclusters.

Generally, comparing different biclustering algorithms is not straightforward as they differ in strategies, approaches, time complicity, number of parameters and prediction ability. In addition, they are strongly influenced by user selected parameter values. For these reasons, the quality of biclustering results is often considered more important than the required computation time. Although there are some analytical comparative studies to evaluate the traditional clustering algorithms[21-23], for biclustering; no such extensive comparison exist even after initial trails have been taken [11]. In the end, Biological merit is the main criterion for evaluation and comparison between the various biclustering methods.

In this chapter we attempt to develope a comparative tool (Bicat-Plus) which is showen in Figure 3 that includes the biological comparative methodology and to be as an extension to the BicAT program[26].

The Goal of BicAT-Plus is to enable researchers and biologists to compare between the different biclustering methods based on set of biological merits and draw conclusion on the biological meaning of the results. In addition, BicAT-Plus help researchers in comparing and evaluating the algorithms results multiple times according to the user selected parameter values as well as the required biological perspective on various datasets.

BicAT-Plus has many features, which could be summarized in the following:-

Algorithms required to be compared could be selected from the biclustering list (left list) to the compared list (right list). External biclustering results for other algorithms could be included in the comparison process. In addition, the organism model, selectable significance level, and GO category should be selected. Finally, Comparison criteria have to be selected based on the user biological metric.

User could perform biclusters functional analysis using the three Gene Ontology (GO) categories (biological process, molecular function and cellular component) (Figure3 with label number 1).

User could evaluate the quality of each biclustering algorithm results after applying the GO functional analysis and display the percentage of the enriched biclusters at different P-values (Figure3 with label number 2).

User could compare between the different biclustering algorithms according to the percentage of the functionally enriched biclusters at the required significance levels, the selected GO category and with certain filtration criteria for the GO terms. (Figure3 with label number 3).

User could evaluate and compare the results of external biclustering algorithms. This gives the BicAT-plus the advantage to be a generic tool that does not depend on the employed methods only. For example, it can be used to evaluate the quality of the new algorithms introduced to the field and compare against the existing ones. (Figure 3 with label number 4).

User could display the results using graphical and statistical charts visualizations in multiple modes (2D and 3D).

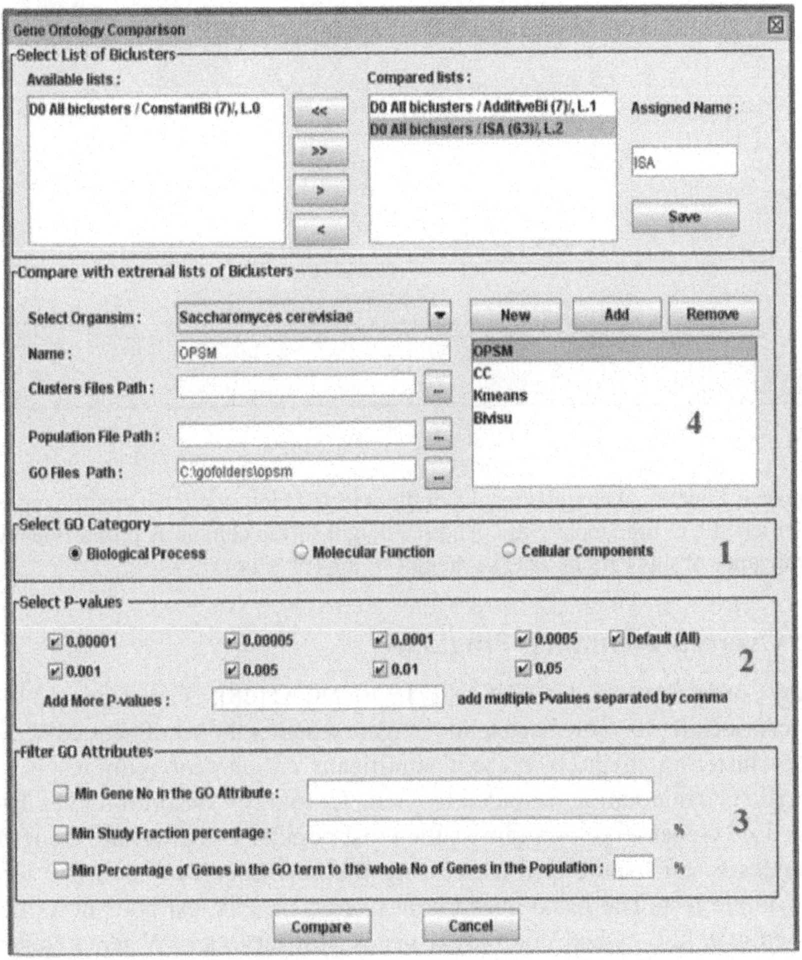

Figure 3: BicAT-Plus Comparison Panel.

MATERIALS AND METHODS

Before using the BicAT-Plus, Active Perl version 5.10 and Java Runtime Environment (JRE) version 6 are required to be installed on your machine. BicAT-Plus has been tested and show good performance on a PC machine with the following configurations: CPU: Pentium 4, 1.5 GHZ, RAM: 2.0 GB, Platform: windows XP professional with SP2.

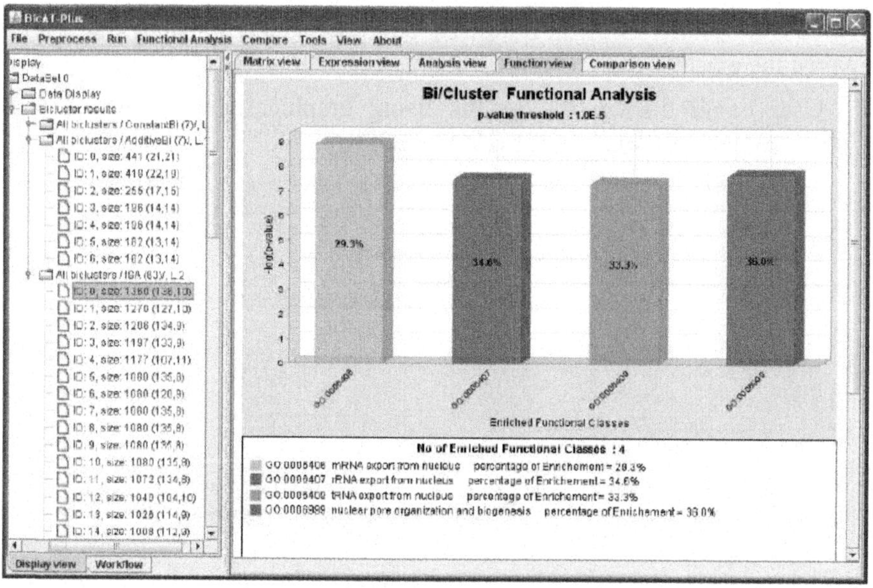

Figure 4: Functional analysis results of the selected bicluster. Each column represents an enriched GO functional class. And the height of the column is proportional to the significance of this enrichment (i.e. height = -log (p-value).

Go Overrepresentation Programs

Many programs like: BINGO[27], FUNCAT[28], GeneMerge[29] and FuncAssociate[30] were used to investigate whether the set of genes discovered by biclustering methods present significant enrichment with respect to a specific GO annotation provided by Gene Ontology Consortium [31]. BicAT-Plus used GeneMerge program as the most popular GO program. GeneMerge provides a statistical test for assessing the enrichment of each GO term in the sample test. The basic question answered by this test is as described by Steven *et al.*[27] "when sampling *X* genes (test set) out of *N* genes (reference set, either a graph or an annotation), what is the probability that *x* or more of these genes belong to a functional category *C* shared by *n* of the *N* genes in the reference set? The hypergeometric test, in which sampling occurs without replacement, answers this question in the form of P-value. It›s counterpart with replacement, the binomial test, which provides only an approximate P-value, but requires less calculation time.»

Comparative Methodologies

BicAT-Plus provides reasonable methods for comparing the results of different biclustering algorithms by:

Identifying the percentage of enriched or overrepresented biclusters with one or more GO term per multiple significance level for each algorithm. A bicluster is said to be significantly overrepresented (enriched) with a functional category if the P-value of this functional category is lower than the preset threshold. The results are displayed using a histogram for all the algorithms compared at the different preset significance levels, and the algorithm that gives the highest proportion of enriched biclusters for all significance levels is considered the optimum because it effectively groups the genes sharing similar functions in the same bicluster.

Identifying The Percentage Of Annotated Genes Per Each Enriched Bicluster

Estimating the predictive power of algorithms to recover interesting patterns. Genes whose transcription is responsive to a variety of stresses have been implicated in a general Yeast response to stress (awkward). Other gene expression responses appear to be specific to particular environmental conditions. BicAT-Plus compares biclustering methods on the basis of their capacity to recover known patterns in experimental data sets. For example, Gasch et al.[32] measure changes in transcript levels over time responding to a panel of environmental changes, so it was expected to find biclusters enriched with one of response to stress (GO:0006950), Gene Ontology categories such as response to heat (GO:0009408), response to cold (GO:0009409) and response to glucose starvation(GO:0042149). The details of this comparison strategy are described in the results and in Table 3.

Comparison Process Steps

The following process diagram shown in Fig 5 summarizes the required steps by the user to compare between the different algorithms using the BicAT-plus:

Download BicAT-Plus from (*www.bioinformatics.org/bicat-plus/*).

Load Gene Expression Data to BicAT-Plus then run the selected five prominent biclustering methods with setting parameters as shown in Table 2.

Run GO comparison tool in the BicAT-Plus and add the available biclustering algorithms to the compared list as shown in Fig 1.

- Select the available GO category e.g. biological process, molecular function and cellular components.
- Select the P-values e.g. 0.00001, 0.0001, 0.01, 0.005, and 0.05.
- Press compare button.
- Press comparison menu, Functional enrichment and select 2D or 3D charts.

Table 2: Default Parameter settings of the compared bi/clustering methods The definitions of these parameters are listed in their original publications [9, 15, 17-18, 20] respectively

Bi/clustering Algorithm	Parameter settings
ISA	$t_g = 2.0$, $t_c = 2.0$, seeds = 500
CC	$\delta = 0.5$, $\alpha = 1.2$, $M = 100$
OPSM	1 = 100
BiVisu	$E = 0.82$, $N_r = 10$, $N_c = 5$, $P_o = 25$
K-means	K=100

RESULTS & DISCUSSION

The above comparison steps is performed on the gene expression data of *S. cerevisiae* provided by Gasch [32]. The dataset contains 2993 genes and 173 conditions of diverse environmental transitions such as temperature shocks, amino acid starvation, and nitrogen source depletion. This dataset is freely available from Stanford University website [33]. For each biclustering algorithm, we used the default parameters as authors recommend in their corresponding publications. See Table 2.

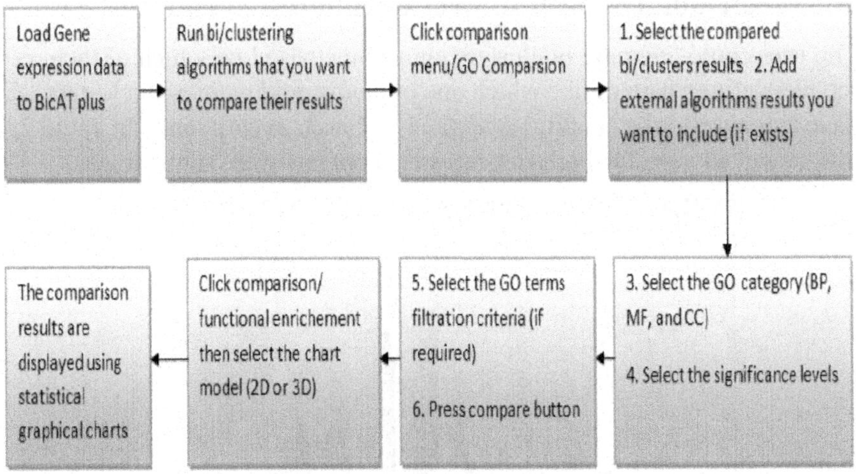

Figure 5: BicAT-Plus Comparison process steps

The Percentage of Enriched Function

After applying the above steps on Gasch data[32], BicAT-plus produce the histogram shown in Fig 6. Investigating this figure, we observed that OPSM algorithm gave a high portion of functionally enriched biclusters at all significance levels (from 85% to 100 %). Next to OPSM, ISA show relatively high portions of enriched biclusters.

In order to evaluate the ability of the algorithms to group the maximum number of genes whose expression patterns are similar and sharing the same GO category, we use the filtration criteria developed in the comparative tool by neglecting those bi/clusters which have study fraction less than 25%. The study fraction of a GO term is the fraction of genes in the study set (bicluster) with this term.

$$Studyfractionof aGOterm = \frac{NoofgenessharingtheGOterminabicluster}{totalnumberofgenesinthisbicluster} \times 100 \quad (2)$$

Figure 7 shows that OPSM and ISA have highly enriched biclusters/ clusters that have large number of genes per each GO category. On the other hand, Bivisu biclusters are strongly affected by this filtration and they contain a lower number of genes per each category. This filtration will help in identifying the powerful and most reliable algorithms which are able to group maximum numbers of genes sharing same functions in one bicluster.

The Predictability Power to Recover Interested Pattern

The user could compare bi/clusters algorithms based on which of them could recover defined pattern like which one of them could recover bi/clusters which have response to the conditions applied in Gasch experiments. In Table 2, the difference between the biclusters/clusters contents were summarized.

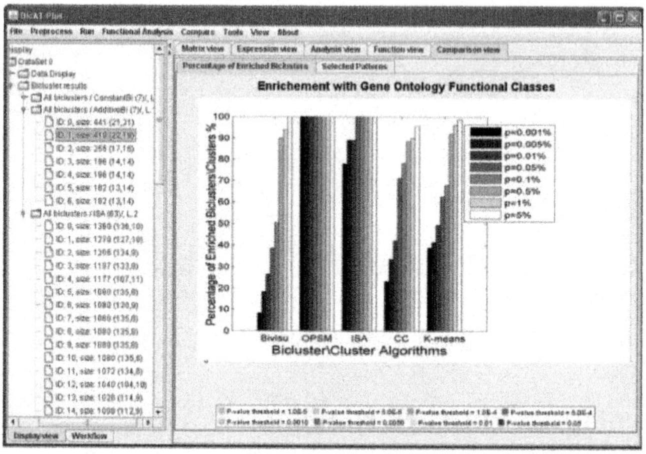

Figure 6: Percentage of biclusters significantly enriched by GO Biological Process category (*S. cerevisiae*) for the five selected biclustering methods and K-means at different significance levels p.

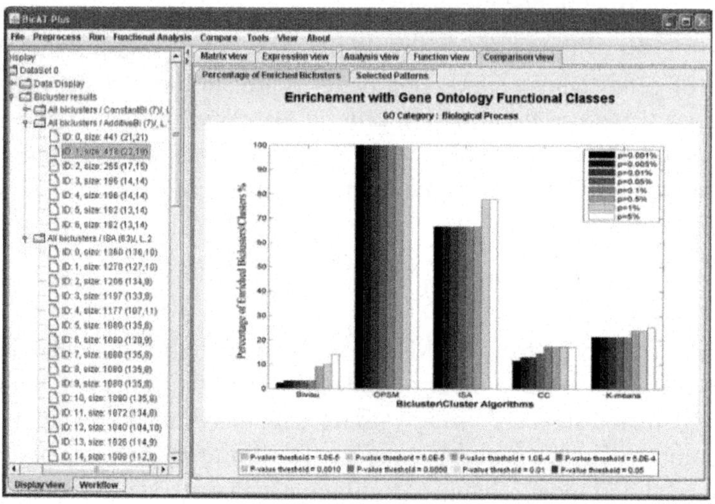

Figure 7: Percentage of significantly enriched biclusters by GO Biological Process category by setting the allowed minimum number of genes per each GO category to 10 and the study fraction to large than 50%.

Although OPSM show high percentage level of enriched biclusters (as shown in Fig 6, 7), its biclusters do not contain any genes within any GO category response to Gasch experiments. The k-means and Bivisu cluster/bicluster results distinguished a unique GO category, which is GO: 0000304 (response to singlet oxygen), and GO: 0042542 (response to hydrogen peroxide) The powerful usage of these bicluster algorithms is significantly appeared in GO: 0006995 "cellular response to nitrogen starvation" where these algorithms were able to discover 4 out of 5 annotated genes without any prior biological information or on desk experiments.

Table 3: Gene Ontology category per number of annotated genes of the Bicluster/cluster algorithm results for the experimental condition on Gasch Experiments[32]

GO Term / (number of annotated genes)	**K-means**	CC	**ISA**	**Bivisu**	**OPSM**
GO:0042493 Response to drug / (118)	4	5	7	6	0
GO:0006970 response to osmotic stress / (83)	3	5	6	3	0
GO:0006979 response to oxidative stress / (79)	2	7	11	0	0
GO:0046686 response to cadmium ion / (102)	2	3	2	2	0
GO:0043330 response to exogenous dsRNA / (7)	2	3	2	2	0
GO:0046685 response to arsenic / (77)	2	0	2	2	0
GO:0006950 response to stress / (532)	9	11	16	2	0
GO:0009408 response to heat / (24)	3	0	2	2	0
GO:0009409 response to cold / (7)	0	0	2	0	0
GO:0009267 cellular response to starvation / (44)	0	2	0	0	0
GO:0006995 cellular response to nitrogen starvation / (5)	4	4	4	0	0
GO:0042149 cellular response to glucose starvation / (5)	0	2	0	0	0

GO:0009651 response to salt stress / (15)	2	7	0	0	0
GO:0042542 response to hydrogen peroxide /(5)	0	0	0	2	0
GO:0006974 response to DNA damage stimulus / (240)	0	22	0	3	0
GO:0000304 response to singlet oxygen / (4)	2	0	0	0	0

CONCLUSION

We have introduced the BicAT-Plus with reasonable comparative methodology based on the Gene Ontology. To the best of our knowledge such an automatic comparison tool of the various biclustering algorithms has not been available in the literature. BicAT-Plus is an open source tool written in java swing and it has a well structured design that can be extended easily to employ more comparative methodologies that help biologists to extract the best results of each algorithm and interpret these results to useful biological meaning.

In other words, the algorithms that show good quality of results (per the dataset) can be used to provide a simple means of gaining leads to the functions of many genes for which information is not available currently (unannotated genes).

Using BicAT-Plus, we can identify the highly enriched biclusters of the whole compared algorithms. This might be quite helpful in solving the dimensionality reduction problem of the Gene Regulatory Network construction from the gene expression data. This problem originates from the relatively few time points (conditions or samples) with respect to the large number of genes in the microarray dataset.

Finally there are several aspects of this research that worth further investigation, according to the Studies carried out so far and also introducing new ideas for consideration

Enrich the BicAT-Plus with more comparative methodologies beside GO. For example, KEGG and promoter analysis by identifying the transcription factors for the clustered genes.

Extend the BicAT-Plus to provide users with multiple export options for the interested enriched biclusters.

Embed the BicAT-Plus as a plug-in in the Cytoscape platform[34] which is open source bioinformatics software for visualizing molecular interaction networks and biological pathways and integrating these networks with annotations, gene expression profiles and other state data. Thus, very promising challenge is to get use of the highly enriched biclusters identified by the BicAT-Plus in solving these integrated networks in the Cytoscape.

REFERENCES

1. G. A. Fung, Overview. Comprehensive, Basic. of, Algorithms. Clustering, Citeseer 20012001137
2. R. Sharan, R. Elkon, R. Shamir, analysis. Cluster, applications. its, gene. to, data. expression, Ernst Schering Res Found Workshop 2002200283108
3. M. B. Eisen, P. T. Spellman, P. O. Brown, D. Botstein, analysis. Cluster, of. display, expression. genome-wide, patterns, Proceedings of the National Academy of Sciences of the United States of America199814863 EOF
4. P. Tamayo, J. Mesirov, Q. Zhu, S. Kitareewan, E. Dmitrovsky, E. S. Lander, T. R. Golub, patterns. Interpreting, gene. of, with. expression, maps. self-organizing, Methods, to. application, dierentiation. hematopoietic, In: Proceedings of the National Academy of Sciences of the United States of America,: 199929072912
5. Sharan RSaR: Click: a clustering algorithm for gene Expression analysis. In: Proceedings of the 8th International Conference on Intelligent Systems for Molecular Biology: 2000307316
6. S. Tavazoie, Campbell. Hughes, R. J. Cho, G. M. Church, determination. Systematic, genetic. of, architecture. network, Nature Genetics1999281 EOF5 EOF
7. S. Johnson, clustering. Hierarchical, Psychometrika. schemes, 1967
8. Madeira SC, Oliveira AL: Biclustering algorithms for biological data analysis: a survey.IEEE/ACM Trans Comput Biol Bioinform 200424 EOF45 EOF
9. Cheng KO, Law NF, Siu WC, Lau TH: BiVisu: software tool for bicluster detection and visualization.Bioinformatics20072342 EOF2344 EOF
10. X. Liu, L. Wang, the. Computing, similarity. maximum, of. bi-clusters, expression. gene, data, Bioinformatics200750 EOF56 EOF
11. A. Prelic, S. Bleuler, P. Zimmermann, A. Wille, P. Buhlmann, W. Gruissem, L. Hennig, L. Thiele, E. A. Zitzler, comparison. Systematic, of. evaluation, methods. biclustering, gene. for, data. expression, Bioinformatics20061122 EOF1129 EOF
12. A. Tchagang, A. Twefik, biclustering. Robust, . R. O. B. A. algorithm, D. N. A. for, data. microarray, analysis, In: IEEE/SP 13thWorkshop on Statistical Signal Processing. 20052005984989

13. T. M. S. K. Murali, conserved. Extracting, expression. gene, from. motifs, expression. gene, data, In: Pac Symp Biocomput. 200320037788

14. A. Tanay, R. S. M. Kupiec, R. Shamir, . Revealing, modularity, in. organization, yeast. the, network. molecular, integrated. by, of. analysis, heterogeneous. highly, data. genomewide, In: Proceedings of the National Academy of Sciences of the United States of America: 200429812986

15. A. Ben-Dor, B. Chor, R. Karp, Z. Yakhini, local. Discovering, in. structure, expression. gene, the. data, submatrix. order-preserving, problem, Journal of Computational Biology2003373 EOF384 EOF

16. H. Wang, W. W. J. Yang, P. S. Yu, . Clustering, Pattern. by, the. p. Similarity, Algorithm. S. I. G. M. O. Cluster, SIGMOD 2002

17. J. Ihmels, G. Friedlander, S. Bergmann, O. Sarig, Y. Ziv, N. Barkai, modular. Revealing, in. organization, yeast. the, network. transcriptional, Nature Genetics2002370 EOF7 EOF

18. Y. Cheng, G. M. Church, of. Biclustering, data. expression, Proceedings of 8th International Conference on Intelligent Systems for Molecular Biology 2000200093103

19. J. Hartigan, Clustering. Direct, a. of, matrix. data, Journal of the American Statistical Association1972123 EOF

20. J. Ihmels, S. Bergmann, N. Barkai, transcription. Defining, using. modules, gene. large-scale, data. expression, Bioinformatics20041993 EOF2003 EOF

21. S. Tavazoie, J. Hughes, M. Campbell, R. Cho, G. Church, determination. Systematic, genetic. of, architecture. network, Nature Genetics1999281 EOF5 EOF

22. R. Guthke, U. Moller, M. Hoffmann, F. Thies, S. Topfer, network. Dynamic, from. reconstruction, expression. gene, applied. data, immune. to, during. response, infection. bacterial, Bioinformatics20051626 EOF1634 EOF

23. P. D'haeseleer, S. Liang, R. Somogyi, network. Genetic, from. inference, clustering. co-expression, reverse. to, engineering, Bioinformatics 2000707 EOF26 EOF

24. D. Reiss, N. Baliga, R. Bonneau, biclustering. Integrated, heterogeneous. of, datasets. genome-wide, the. for, of. inference, regulatory. global, B. M. networks, BMC Bioinformatics2006280 EOF

25. K. Yip, Q. Ya, Peishen, Martin. Schultz, David. W. Cheung, Kei. Cheung-Hoi, Biosphere. A. Sem, Web. Semantic, to. Approach, Microarray.

Recommending, Services. Clustering, In: The Pacific Symposium on Biocomputing. 20062006188199

26. S. Barkow, S. Bleuler, A. Prelic, P. Zimmermann, E. Zitzler, A. T. a. Bic, analysis. biclustering, toolbox, Bioinformatics20061282 EOF1283 EOF
27. S. Maere, K. Heymans, M. Kuiper, N. G. O. a. Bi, plugin. Cytoscape, assess. to, of. overrepresentation, Ontology. Gene, in. categories, Networks. Biological, Bioinformatics 2005
28. A. Ruepp, A. Zollner, D. Maier, K. Albermann, J. Hani, M. Mokrejs, I. Tetko, U. Guldener, G. Mannhaupt, M. Munsterkotter, et al.Fun. The, a. Cat, annotation. functional, for. scheme, classification. systematic, proteins. of, whole. from, genomes, Nucl Acids Res 2004
29. Castillo-Davis CI, Hartl DL: GeneMerge- post-genomic analysis, data mining, and hypothesis testing. Bioinformatics2003
30. G. F. Berriz, O. D. King, B. Bryant, C. Sander, F. P. Roth, gene. Characterizing, with. sets, Associate. Func, Bioinformatics20032502 EOF2504 EOF
31. M. Ashburner, C. Ball, J. Blake, D. Botstein, H. Butler, J. Cherry, A. Davis, K. Dolinski, S. Dwight, J. Eppig, et al.ontology. Gene, for. tool, unification. the, biology. of, The Gene Ontology Consortium. Nat Genet 2000
32. A. P. Gasch, P. T. Spellman, C. M. Kao, O. Carmel-Harel, M. B. Eisen, G. Storz, D. Botstein, P. O. Brown, Expression. Genomic, in. Programs, Response. the, Yeast. of, to. Cells, Changes. Environmental, Mol Biol Cell 20004241 EOF57 EOF
33. http://genome-www.stanford.edu/yeast/_stress
34. P. Shannon, A. Markiel, O. Ozier, N. Baliga, J. Wang, D. Ramage, N. Amin, B. Schwikowski, T. Ideker, a. Cytoscape, environment. software, integrated. for, of. models, interaction. biomolecular, networks, Genome Res 2003

Chapter 11

GENOME SEQUENCE AND GENETIC DIVERSITY OF THE COMMON CARP, CYPRINUS CARPIO

Peng Xu[1,10], Xiaofeng Zhang[2,10], Xumin Wang[3,10], Jiongtang Li[1,10], Guiming Liu[3,10], Youyi Kuang[2,10], Jian Xu[1,10], Xianhu Zheng[2,10], Lufeng Ren[3], Guoliang Wang[3], Yan Zhang[1], Linhe Huo[3], Zixia Zhao[1], Dingchen Cao[2], Cuiyun Lu[2], Chao Li[2], Yi Zhou[4], Zhanjiang Liu[1,5], Zhonghua Fan[3], Guangle Shan[3], Xingang Li[3], Shuangxiu Wu[3], Lipu Song[3], Guangyuan Hou[1], Yanliang Jiang[1], Zsigmond Jeney[6], Dan Yu[3], Li Wang[3], Changjun Shao[3], Lai Song[3], Jing Sun[3], Peifeng Ji[1], Jian Wang[1], Qiang Li[1], Liming Xu[1], Fanyue Sun[5], Jianxin Feng[7], Chenghui Wang[8], Shaolin Wang[9], Baosen Wang[1], Yan Li[1], Yaping Zhu[1], Wei Xue[1], Lan Zhao[1], Jintu Wang[1], Ying Gu[2], Weihua Lv[2], Kejing Wu[3], Jingfa Xiao[3], Jiayan Wu[3], Zhang Zhang[3], Jun Yu[3] & Xiaowen Sun[1,2]

[1]Centre for Applied Aquatic Genomics, Chinese Academy of Fishery Sciences, Beijing, China

[2]Heilongjiang River Fisheries Research Institute, Chinese Academy of Fishery Sciences, Harbin, China

[3]Chinese Academy of Sciences Key Laboratory of Genome Sciences and Information, Beijing Institute of Genomics, Chinese Academy of Sciences, Beijing, China

[4]Stem Cell Program, Division of Hematology and Oncology, Boston Children's Hospital and Dana-Farber Cancer Institute, Harvard Medical School, Boston, Massachusetts, USA

[5]Fish Molecular Genetics and Biotechnology Laboratory, Department of Fisheries and Allied Aquacultures, Auburn University, Auburn, Alabama, USA

[6]Research Institute for Fisheries, Aquaculture and Irrigation, Szarvas, Hungary

[7]Henan Academy of Fishery Science, Zhengzhou, China

[8]College of Fisheries and Life Science, Shanghai Ocean University, Shanghai, China

[9]Department of Psychiatry and Neurobiology Science, University of Virginia, Charlottesville, Virginia, USA

[10]These authors contributed equally to this work. Correspondence should be addressed to X.S

ABSTRACT

The common carp, *Cyprinus carpio*, is one of the most important cyprinid species and globally accounts for 10% of freshwater aquaculture production. Here we present a draft genome of domesticated *C. carpio* (strain Songpu), whose current assembly contains 52,610 protein-coding genes and approximately 92.3% coverage of its paleotetraploidized genome ($2n = 100$). The latest round of whole-genome duplication has been estimated to have occurred approximately 8.2 million years ago. Genome resequencing of 33 representative individuals from worldwide populations demonstrates a single origin for *C. carpio* in 2 subspecies (*C. carpio Haematopterus* and *C. carpio carpio*). Integrative genomic and transcriptomic analyses were used to identify loci potentially associated with traits including scaling patterns and skin color. In combination with the high-resolution genetic map, the draft genome paves the way for better molecular studies and improved genome-assisted breeding of *C. carpio* and other closely related species.

INTRODUCTION

Carp (cyprinids) contribute over 20 million metric tons to fish production worldwide and account for approximately 40% of total global aquaculture production and 70% of total freshwater aquaculture production. They have emerged as the most economically important teleost family. In comparison to other major aquaculture species, such as salmon and shrimp, carp are recognized as an ecofriendly fish because most are omnivorous filter-feeders and thus consume much less fish meal and fish oil. As one of the dominant cyprinid species, *C. carpio* (the common carp) is cultured in over 100 countries worldwide and accounts for up to 10% (over 3 million metric tons) of global annual freshwater aquaculture production[1, 2]. In addition to its value as a food source, *C. carpio* is also an important ornamental fish species. One of its variants, koi, is the most popular outdoor ornamental fish because of its distinctive color and scale patterns.

Most teleosts have undergone a teleost-specific genome duplication (TSGD) and contain 24 to 25 chromosomes in their haploid genome. The haploid genome of *C. carpio* has 50 chromosomes[3], and molecular evidence suggests that an additional whole-genome duplication (WGD) event tetraploidized the genome[4, 5, 6, 7]. Although cytogenetic evidence of the allotetraploidization of *C. carpio* has suggested that 50 bivalents rather than 25 quadrivalents are formed during meiosis[6], genome-scale validation is of great importance. Owing to its economic value in aquaculture, *C. carpio* has been intensively studied in terms of its physiology, development, immunology, disease resistance, selective breeding and transgenic manipulation. In addition, it is also considered an alternative vertebrate fish model to zebrafish (*Danio rerio*). A variety of *C. carpio* genome resources have been developed over the past decade, including a large number of genetic markers[8, 9], genetic maps[10, 11, 12, 13], a BAC-based physical map[14, 15], a large number of ESTs[16, 17, 18] and cDNA microarrays[19]. Recently, a comparative exomic study of *C. carpio* and *D. rerio* has been reported, providing additional genome resourcing data for the research community[20].

Using a whole-genome shotgun strategy and combining data from several next-generation sequencing platforms, we have produced a high-quality genome assembly for *C. carpio* (strain Songpu) and completed the genomic resequencing of 33 *C. carpio* accessions that represent major domesticated strains and populations. In addition to comparative and evolutionary studies of*C. carpio* and its closely related species using the genome sequences, we also demonstrate the genetic basis of phenotypic traits on scale patterning and body color determination, on the basis of data from two distinct domesticated strains (Songpu and Hebao). This study on the *C. carpio*genome provides a valuable resource for the molecular-guided breeding and genetic improvement of the common carp.

RESULTS

Sequencing and Assembly

We prepared genomic DNA from a homozygous double-haploid clonal line from the domesticated strain Songpu, which has a documented breeding history. We performed whole-genome shotgun sequencing on three next-generation sequencing platforms, Roche 454, Illumina and SOLiD, using both single-end and paired-end or mate-pair libraries of various insert size ranging from 250 bp to 8 kb (Supplementary Table 1). Our contig assembly was based on single-end pyrosequencing data (CABOG, Celera Assembler with Best Overlap Graph), and the scaffold assembly was based on paired-end and mate-pair sequences from different sequencing platforms and 29,046 paired BAC-end sequences[14, 15]. After gap filling, the contig and scaffold N50 lengths reached 68.4 kb and 1.0 Mb, respectively. The total length of all scaffolds was 1.69 Gb (Table 1). We estimated the genome size to be 1.83 Gb on the basis of *K*-mer analysis, which is consistent with estimates based on cytogenetic methods[3, 21] (Supplementary Fig. 1). Thus, the scaffolds covered at least 92.3% of the genome (90.2% if we excluded sequence gaps of 40 Mb in length) and 90% of the assembly containing 2,503 large scaffolds, for a total length of 1.53 Gb.

Table 1: Summary of genome assembly

Genome assembly		N50 (size/number)	N90 (size/number)	Total length
	Contigs	68.4 kb/7,171	16.5 kb/25,070	1.65 Gb
	Scaffolds	1.0 Mb/491	96.6 kb/2,503	1.69 Gb
	Chromosomes	50 chromosomes (from 1,456 scaffolds)		875 Mb
Noncoding RNAs		**Copies**		**Length**
	rRNAs	1,012		100.2 kb
	miRNAs	914		20.4 kb
	tRNAs	3,622		268.1 kb
TEs		**Total length**		**Percent of genome**
	Total	529.4 Mb		31.23
	DNA transposons	297.2 Mb		17.53
	Retroelements	169.3 Mb		9.99
Protein-coding genes	**Total number**	**Annotated**		**Unannotated**
	52,610	47,795		4,815

To validate the genome assembly, we mapped all paired-end and mate-pair reads from different sequencing platforms to the assembly and found that an average of 80.3% of the reads (78.1% of Illumina reads, 74.6% of Life Technologies SOLiD reads, 98% of Roche 454 reads and 98.8% of BAC-end reads) could be mapped (Supplementary Fig. 2 and Supplementary Table 2). To assess the accuracy of the assembly, we aligned our assembly to an assembled BAC and five large scaffolds from previously published genome sequences[20]; the result demonstrates high consistency between the two data sets (Supplementary Fig. 3). To assess gene coverage, we mapped assembled transcriptome sequences, including publically available ESTs and new mRNA reads from multiple tissues, to the assembly. The effort yielded ~88.8% coverage of these transcripts by nucleotide sequence similarity (Supplementary Table 3); of all of the mapped genes, 90% were common among the sequenced teleosts (Supplementary Table 4). Owing to the multiple rounds of GWD, *C. carpio* genes are rich in paralogs, which are thought to interfere with assembly. Therefore, we mapped 19 duplicated genes that were shared among teleosts to the genome and found that 16 of the gene pairs mapped to distinct locations whereas the other 3 collapsed into single genes, given the high similarity among the paralogs (Supplementary Table 5).

Genetic Map and Markers

Our attempt to anchor the genome assembly onto a newly updated high-resolution genetic map, constructed using a genetic mapping panel of 107 full siblings produced from the cross of a Songpu pair (Supplementary Note), succeeded in placing a total of 3,470 high-quality SNPs and 773 microsatellite markers (Supplementary Table 6), which clustered into 50 linkage groups (Supplementary Fig. 4) and covered a genetic distance of 3,946.7 cM. The physical coverage of these linkage groups contained 1,456 of the longest scaffolds (~875 Mb in total length and 16.7 Mb of gaps) (Fig. 1 and Table 1), and the ratio of the median genetic distance to the physical distance was 0.2 Mb/cM (Supplementary Figs. 5 and 6).

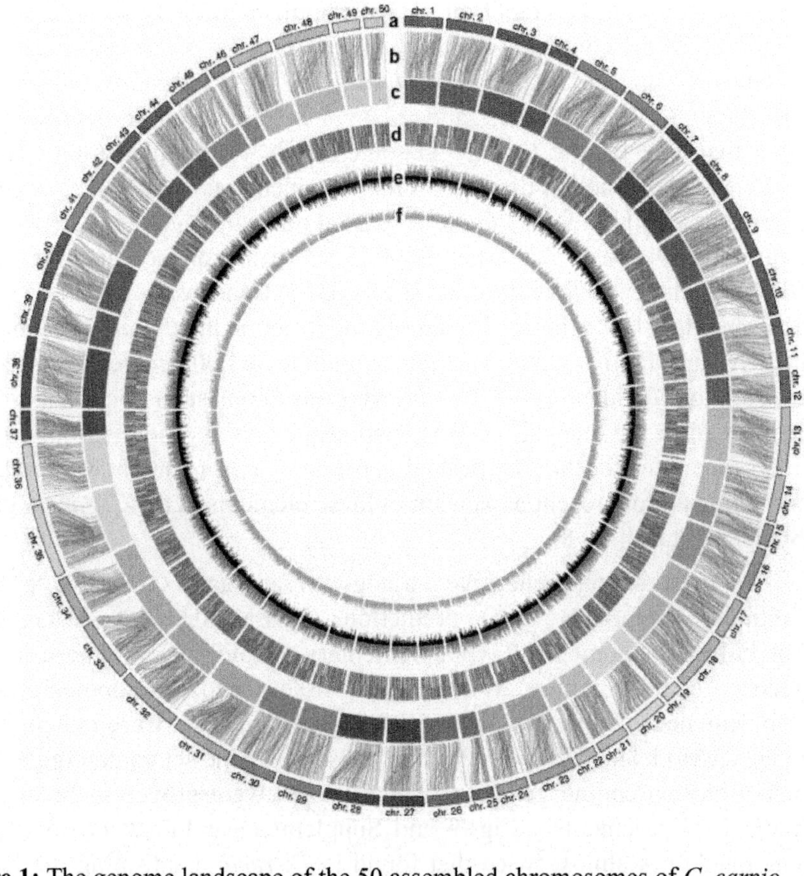

Figure 1: The genome landscape of the 50 assembled chromosomes of *C. carpio*.

(a) The genetic linkage map. **(b)** Anchors between the genetic markers and the assembled scaffolds. **(c)** Assembled chromosomes. **(d)** Gene distribution on each chromosome; red lines indicate genes on the plus strand, and blue lines indicate genes on the minus strand. **(e)** GC content within a 10-kb sliding window; blue indicates GC content that is higher than average, and black indicates GC content that is lower than average. **(f)** Repeat content within a 10-kb sliding window.

Genome Characterization

The *C. carpio* genome has a GC content of 37.0%, slightly higher than that of *D. rerio* but much lower than that of other sequenced teleost genomes (Supplementary Fig. 7 and Supplementary Note). To identify transposable elements (TEs)[22], we constructed a *C. carpio*–specific repeat database. We found that 529 Mb of the assembled contigs (31.3% of the genome assembly) could be attributed to TEs (Table 1 and Supplementary Tables 7 and 8). This proportion of the content is higher than that for most of the sequenced teleost genomes (7.1% in *Takifugu rubripes*[23], 5.7% in*Tetraodon nigroviridis*[24], 30.68% in *Oryzias latipes*[25] and 13.48% in *Gasterosteus aculeatus*[26]) but lower than that for the *D. rerio* genome (59.78%) (Supplementary Table 9). Of the TEs, the fraction of class I TEs (retroelements) was 9.99% of the total genome assembly (4.90% long interspersed nuclear elements (LINEs), 4.35% long terminal repeats (LTRs) and 0.47% short interspersed nuclear elements (SINEs)), whereas that of the class II TEs (DNA transposons) was 17.53%. The most abundant DNA transposon family identified in the *C. carpio* genome was the hAT superfamily, which had approximately 463,000 copies and accounted for 33% of all identified DNA transposons (consistent with our previous findings from the analysis of BAC-end sequences[15]). The distribution of the divergence rates for the TEs peaked at 6% in *C. carpio* and at 8% in *D. rerio*, suggesting a more recent expansion of these elements in the *C. carpio* genome (Supplementary Fig. 8).

We used a comprehensive strategy to annotate *C. carpio* genes by combining *ab initio* gene prediction (FGENESH and AUGUSTUS), protein-based homology (Supplementary Note) and transcript-based evidence (transcriptomes from multiple tissues and developmental stages) (Supplementary Table 10). All predicted gene structures were integrated with EVidenceModeler (EVM)[27] to yield a consensus gene set containing a total of 52,610 protein-coding genes, of which 91.4% were proven to be expressed (Table 1, Supplementary Fig. 9 and Supplementary Tables 11and 12). This gene number is almost twice that found in *D. rerio*, confirming the fact that the tetraploid genome retained a large portion of its gene duplicates after the latest WGD. The average gene and coding sequence lengths were 12,145 bp and 1,487 bp, respectively, and *C. carpio*genes had an average of 7.48 exons per gene (Supplementary Fig. 10 and Supplementary Table 13). In addition, the non-protein-coding genes included 1,012 rRNA, 3,622 tRNA and 914 microRNA (miRNA) genes (Table 1 and Supplementary Table 14).

Genome Evolution

C. carpio has 100 chromosomes, approximately twice as many as are found in other cyprinid fish species. Many studies have corroborated the occurrence of either TSGD or the third round of WGD in most ray-finned fishes[28, 29, 30, 31, 32] and have predicted that TSGD has facilitated the evolutionary radiation and phenotypic diversification of the teleost fishes[29, 31]. The *C. carpio*genome is believed to have undergone an additional round of genome duplication (4R) and to have thus tetraploidized[4, 5, 6, 33]. We have identified approximately 50 chromosome bivalents rather than quadrivalents in the meiotic nuclei of *C. carpio*, suggesting that it is not a true tetraploid species according to karyotyping. The tetraploidy observed in *C. carpio* seems to result from allotetraploidization (species hybridization) rather than autotetraploidization (genome doubling)[6]. We aligned 52,610 high-confidence gene models to the 50 *C. carpio* chromosomes and the *D. rerio* genome ($n = 25$ chromosomes) and identified 8,002 orthologous gene pairs with a clear two-to-one orthologous relationship between the two species, respectively (Fig. 2a). The major obscure synteny found on the long arm of *D. rerio* chromosome 4 is actually in accordance with a recent report that highlighted unique features of this region[34]: the region shows little orthology with other sequenced teleost genomes and harbors zebrafish-specific gene duplication and a high-density small nuclear RNA (snRNA) cluster that accounts for 53.2% of all snRNAs in the genome. This region most likely emerged in *D. rerio* after *Danio-Cyprinus* divergence. In addition, we also observed a number of minor chromosome rearrangements on the carp chromosomes, including on the long arm of chromosome 8 (showing weakened orthology with *D. rerio* chromosome 4) and the region containing orthologs with *D. rerio* chromosome 17. We also identified 2,114 best-match reciprocal paralogous gene pairs and built ohnologous blocks on 25 paired chromosomes. A circular representation of ohnolog pairs clearly demonstrates their one-to-one syntenic relationship (Fig. 2b), consistent with previous observations for genome tetraploidization.

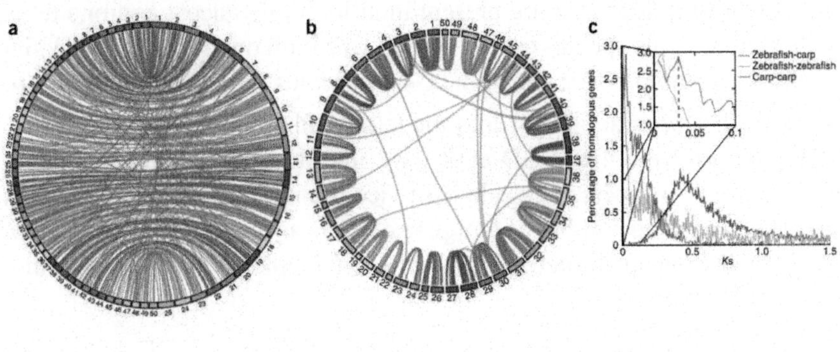

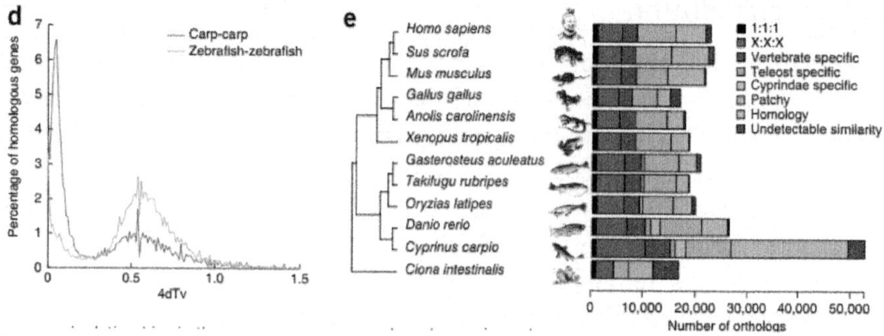

Figure 2: Comparative analyses of the *C. carpio* **genome.**

(**a**) Maps of the 50 common carp chromosomes and of the 25 zebrafish chromosomes based on the positions of 8,002 orthologous pairs demonstrate highly conserved synteny for the 2 species and the tetraploidization of the common carp genome. (**b**) Schematic of major interchromosomal relationships in the common carp genome based on reciprocal best-match relationships. Chromosomes are represented as colored blocks. The positions of duplicated genes on the chromosomes are linked by gray lines. (**c**) The distribution of the synonymous substitution rates (*K*s) of homologous gene groups for intraspecies and interspecies comparisons. The peak (*K*s = 0.03) of the paralogous *K*s distribution (inset) indicates the recent WGD. (**d**) The third and fourth genome duplications in the common carp genome were identified by 4dTv analyses. (**e**) Comparison of the gene repertoire of the *Ciona* species with those of five teleost and six tetrapod genomes, ranging from the highly conserved chordate genes (the black fraction on the left) to species-specific genes (the rust red fraction on the right). "1:1:1" indicates universal single-copy genes, and "X:X:X" indicates any other orthologous group (missing in one species), where X means one or more orthologs per species. "Patchy" indicates the existence of other orthologs that are present in at least one teleost and one tetrapod genome. The species tree on the left was built on the basis of 941 single-copy orthologs, for which the *Ciona* species served as the outgroup.

To further provide insight into the tetraploid nature of the genome at the gene level after the 4R WGD event, we investigated the *hox* gene clusters in *C. carpio*. This species has almost twice the number of *hox* clusters as *D. rerio*[35] and the same number of *hox* gene clusters as the Atlantic salmon (*Salmo salar*)[36], which is an autotetraploid species[37] (Supplementary Figs. 11 and 12, andSupplementary Note).

To determine the date of the *C. carpio* WGD event (4R), we used a total of 5,783 gene families and calculated their synonymous substitution rates (*K*s values) (Fig. 2c and Supplementary Note). On the basis of a *K*s rate of 3.51 $\times 10^{-9}$ substitutions per synonymous site per year[5] and the obtained*K*s value of 0.03, we estimated that the latest WGD (4R) happened 8.2 million years ago, a date more recent than the predictions suggested in previous reports[5, 7]. The carp-zebrafish paralogous genes displayed a distinct peak (*K*s = 0.45) that corresponded to a divergence time of 128 million years ago. In combination with the duplication time and divergence time predictions, these data suggest that the latest WGD event (4R) occurred long after *C. carpio* and *D. rerio* split. Similarly, an analysis of fourfold synonymous third-codon transversion (4dTv) provided additional evidence for an extra round of WGD (Fig. 2d). *C. carpio* and *D. rerio* had a peak in common (4dTv = 0.58), which corresponds to the TSGD event (3R). An extra 4dTv peak within the *C. carpio* paralogous genes (4dTv = 0.1) corresponds to the latest carp-specific WGD (4R).

The annotated gene models of the *C. carpio* genome are substantially better than those of other completely sequenced fish genomes. To understand the evolutionary relationship of the 52,610 gene models with those of other vertebrates, we performed systematic cross-species comparative analysis and classified the genes according to their similarities. We first used five teleosts (*C. carpio, D. rerio, T. rubripes, O. latipes* and *G. aculeatus*), six tetrapods (*Homo sapiens, Mus musculus, Sus scrofa, Gallus gallus, Anolis carolinensis* and *Xenopus tropicalis*) and *Ciona intestinalis* (outgroup) for the comparison (Fig. 2e). We identified 941 single-copy orthologs that were conserved among all investigated species, which only accounted for 1.8% of the predicted gene models of *C. carpio*. Second, we constructed the species phylogeny using a maximum-likelihood approach with multiple alignments of single-copy orthologs. The remaining gene models (98.2%) were more complex and included many-to-many orthologs (28.0%), non-uniformly occurring, patchy orthologs (11.0%) and undetectable models (6.1%). Third, the predicted *C. carpio* gene models corresponded to orthologous genes in *D. rerio* (8,002 orthologous genes), including 2,037 (3.9%) cyprinid-specific gene models. Fourth, we also identified 3,212 species-specific gene models of *C. carpio* that did not have any homologs in the 10 other vertebrates and the *Ciona* species examined. This number is higher than the number of species-specific genes in the *D. rerio* genome, suggesting that a significant number of novel genes were generated in *C. carpio* after the divergence of *C. carpio* and *D. rerio*, likely owing to the latest WGD and independent gene evolution (Supplementary Table 15 and Supplementary Note).

Genetic Diversity

C. carpio, as a genetically diverse and successful species, has adapted to various environments across a broad ecological spectrum in Eurasia and has been domesticated for more than 2,000 years. This species has been bred into numerous strains and local populations, producing distinct phenotypic changes in its growth rate, temperature and hypoxia tolerance, body color, scale pattern and body shape, which are partially attributable to genome diversity due to its two WGD (3R and 4R) events[31]. To investigate its genetic variation, we selected 33 representative *C. carpio*accessions for genome resequencing, which included 13 accessions of 4 wild populations from the Danube River, the Yellow River, the Heilongjiang (Amur) River and the Chattahoochee River and 20 accessions of 6 domesticated strains from Asia and Europe (including Songpu, Xingguo red, Oujiang color, Hebao, Szarvas 22 and koi) (Supplementary Table 16). With a total of 4,176 million paired-end reads (101-bp read length, 417.6 Gb in total length and 229-fold coverage of the genome; Supplementary Table 17), we identified 18,949,596 candidate SNPs and 1,694,102 small insertion-deletions (indels) (Supplementary Table 18).

To investigate the divergence of the representative *C. carpio* accessions from diverse geographical habitats and domestic histories, we constructed phylogenetic trees on the basis of the sequence variations (Fig. 3a and Supplementary Fig. 13). It was obvious that the European and Asian accessions formed two distinct clades. One of the strains, Songpu, was also grouped into the European clade as it was bred from mirror carp originally introduced from Europe in the 1950s. Our principal-component analysis (PCA) yielded a similar result (Fig. 3b), showing Asian accessions as a tight cluster that was separate from the European accessions. We further analyzed the population structure using the Bayesian clustering program STRUCTURE[38]. Because the values of ln likelihood were distinctively high for the models $K =$ 3 and 4 (Supplementary Fig. 14), we show the clusters of $K = 3$ and 4 in Figure 3c. Almost all the accessions either had common ancestry or showed a single origin of the Eurasian population, and the results agree with the hypothesis that modern *C. carpio* evolved from the Caspian Sea ancestor and spread into Europe and the eastern mainland of Asia[39]. There were no uniform patterns covering all the populations, with the exception of two extremely isolated wild populations, Heilongjiang and Oujiang; in other words, extensive genetic admixture has been occurring in both the wild and domesticated *C. carpio* populations. For instance, Songpu carried admixture from the Asian population ($K = 3$), a finding supported by the recent history of introgression after its introduction to China. We also observed that the US accessions separated into both the European and Asian clades, and the trend indicates multiple introductions to North America from both Europe and Asia. This observation is also supported by our PCA and population structure analyses.

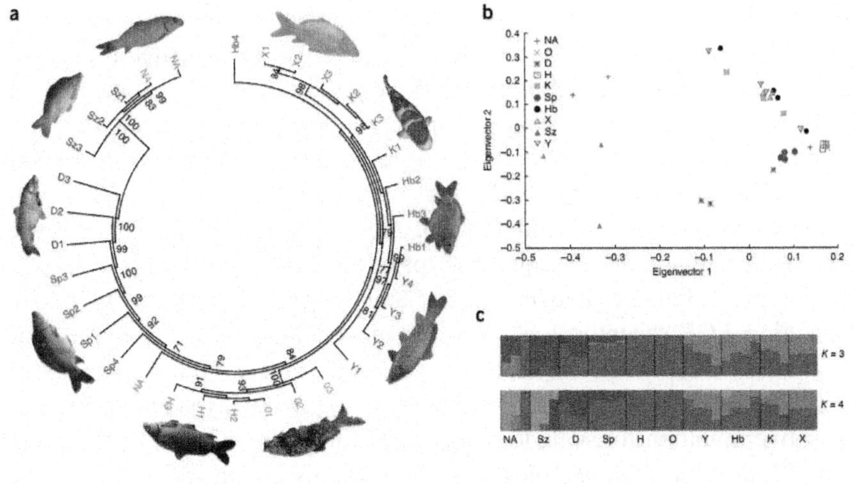

Figure 3: Genetic analysis of ten *C. carpio* **strains using genome resequencing data.**

(**a**) A maximum-likelihood phylogenetic tree of ten strains of common carp generated on the basis of SNPs. Strain abbreviations: Sp, Songpu; D, Danube; Sz, Szarvas; NA, North American; Y, Yellow River; H, Heilongjiang; O, Oujiang color; Hb, Hebao; X, Xingguo; K, koi. (**b**) PCA of *C. carpio* strains. (**c**) The population structure of common carp strains. Each color represents one ancestral population; each strain is represented by a vertical bar, and the length of each colored segment in each vertical bar represents the proportion contributed by ancestral populations. $K = 3$ and 4 was used for analysis with the highest ln value.

We performed a further genetic diversity scan comparing the Hebao and Songpu genomes to identify highly different genomic regions. Hebao is one of the typical strains derived from East Asian subspecies (*C. carpio haematopterus*), whereas Songpu is the strain derived from mirror carp of European subspecies (*C. carpio carpio*) (Fig. 4a). We predict that these two varieties retain substantial genetic differences, given their distinct body shapes, scale morphogenesis and patterns, and skin color phenotypes. We identified a total of 205 genome regions with the highest (top 1% of $\varpi_{Hebao}/\varpi_{Songpu}$) genetic diversity (12.67 Mb in length) containing 326 candidate genes (Supplementary Tables 19 and 20). Gene Ontology (GO) and Kyoto Encyclopedia of Genes and Genomes (KEGG) analysis indicated that a significant portion of these candidate genes were associated with epithelial morphogenesis, hair follicle morphogenesis, pigmentation and immune response, including adherens junction signaling, signaling by Rho family GTPases, tight junction signaling, prolactin signaling, fibroblast growth factor (FGF) signaling, interleukin (IL)-6 signaling and other functional pathways (Fig. 4b and Supplementary Table 21). The results are consistent with the phenotype differences in scale pattern and skin color observed for Hebao and Songpu. We investigated the candidate genes and detected 82 genes and 106 genes that harbored nonsynonymous SNPs in the Songpu and Hebao genomes, respectively (Supplementary Tables 22 and 23). We also identified two Songpu genes (*fgfr1a1* and *lrrc72*) and two Hebao genes (*zpld1*and *nlk*) that harbored deletions in coding regions (Supplementary Figs. 15 and 16). All these instances of sequence diversity altering protein-coding sequences provide candidate loci for assessing phenotypic differences between Hebao and Songpu. Notably, the *fgfr1a1* gene (encoding FGF receptor 1 a1) on chromosome 34 of the Songpu genome contained a 306-bp specific deletion in intron 10 (228-bp deletion) and exon 11 (78-bp deletion) (Fig. 4c). The deletion had previously been reported as the causative mutation for scale loss and reduction in the mirror carp[40]. Extensive investigation on large samples from four strains confirmed that the deletion was only found in Songpu (Supplementary Fig. 16).

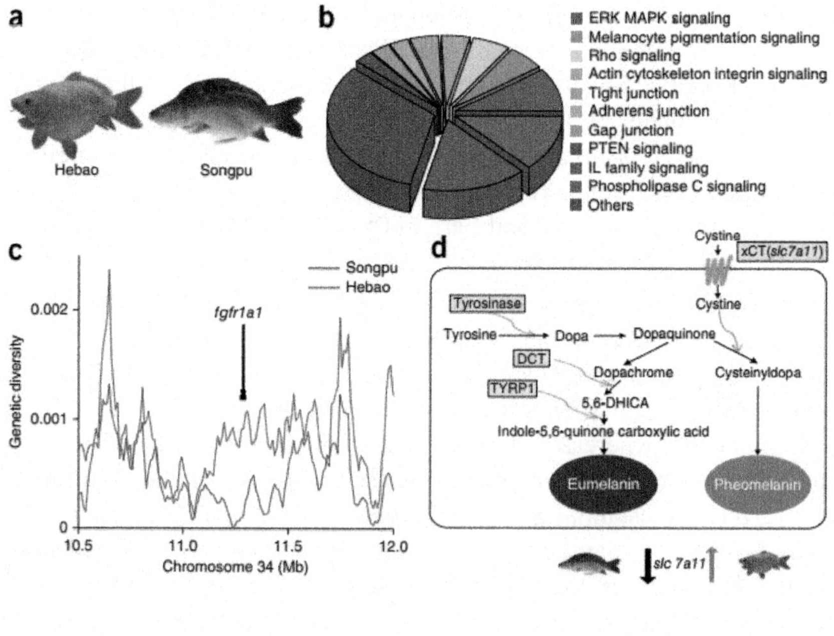

Figure 4: Comparison of the diversity distributions of two *C. carpio* **strains.**

(**a**) Typical Songpu and Hebao carp. (**b**) The distribution of genes involved in KEGG pathways in the regions with the top 1% of $\varpi_{Hebao}/\varpi_{Songpu}$.

(**c**) The diversity (ϖ) distribution on chromosome 34 for Songpu and Hebao showed a 78-bp deletion in exon 11 of the *fgfr1a1* gene in Songpu; this deletion results in a reduced-scale phenotype. (**d**) The pheomelanin synthesis pathway. Transcriptome analysis showed that the *slc7a11* gene is significantly more upregulated in the skin of Hebao than that of Songpu. *slc7a11*encodes the transmembrane cystine/glutamate exchanger (xCT), which transports cystine into melanocytes to synthesize pheomelanin (yellow to red pigment).

Comparative Analysis of the Skin Transcriptome

To further elucidate the differences in the scale pattern and skin color of Hebao and Songpu, we performed a comparative analysis of the skin transcriptome in both strains using a deep RNA sequencing (RNA-seq) approach. We identified 894 differentially expressed transcripts, including 567 upregulated genes in Hebao and 327 upregulated genes in Songpu. The experiment was validated with quantitative RT-PCR (qRT-PCR) on selected genes (Supplementary Fig. 17). Further analysis showed distinct expression patterns in Hebao and Songpu for many genes associated with the Wnt/β-catenin signaling pathway (Supplementary Table 24), which is an essential pathway in initiating hair follicle formation[41]. Both mammalian hair and teleost scales are skin appendages, and their formation involves similar developmental pathways. We inferred that the gene expression differences in these two different carp populations were correlated with the reduced-scale phenotype in Songpu and the full-scale phenotype in Hebao.

We also observed a difference in the expression of the *slc7a11* gene (encoding solute carrier family 7 member 11), the plasma membrane cystine/glutamate exchanger (xCT) that transports cystine into melanocytes. In the melanogenesis pathway, tyrosine is oxidized to form dopaquinone, which is then intracellularly catalyzed to become eumelanin (brown to black pigment) through polymerization and oxidation reactions. However, cystine and dopaquinone can switch off the eumelanin synthesis pathway and promote the synthesis of pheomelanin (yellow to red pigment)[42,43]. *slc7a11* was expressed at a higher level in Hebao than in Songpu, suggesting that more cystine is transported into the pigment cells in Hebao, resulting in the preponderant synthesis of pheomelanin in the skin of Hebao while eumelanin synthesis is suppressed. Higher pheomelanin accumulation in the pigment cells gives Hebao its red skin appearance (Fig. 4d). However, the genetic basis for differences in *slc7a11* expression remains unclear, and further investigation will be necessary to understand the overall role of *slc7a11* in color variation.

DISCUSSION

As one of the most representative carp species, *C. carpio* had a value and global production in 2011 of $5.31 billion and 3.73 million tons, respectively (FAO statistics; see URLs), and the importance of *C. carpio* has been increasing over the past decade. The species is also widely cultured as an ornamental fish because of its various color and scale patterns. We sequenced and assembled the *C. carpio* genome from the genome of a gynogenetic individual using multiple next-generation sequencing platforms and a hybrid assembly strategy. The draft genome provides an important genomic resource to study the genetic basis of economically important traits in carp and to facilitate genome-based genetic breeding technologies in common carp aquaculture. The draft genome also provides insight into the latest WGD event of allotetraploidization that occurred approximately 8.2 million years ago, doubling the chromosome number and gene content of *C. carpio*. The whole-genome resequencing of selected accessions also offers a glimpse into the phylogenetic relationship and population structure of major global accessions of the *C. carpio*population. Comparison of the genomic diversity of two distinct strains, Songpu and Hebao, coupled with additional transcriptomic studies, has allowed us to identify genetic loci and to determine the molecular basis of scale patterns and skin colors, providing a foundation for further studies using comprehensive approaches to completely define the mechanisms underlying these phenotypes. Thus, the draft genome assembly presented here provides a valuable resource for genetic, genomic and biological studies of *C. carpio* and for improving the aquaculturally important traits of farmed *C. carpio* and other key cyprinid species in aquaculture.

METHODS

Ethics Statement.

This study was approved by the Animal Care and Use Committee of the Centre for Applied Aquatic Genomics at the Chinese Academy of Fishery Sciences.

Genome Sequence and Assembly.

A gynogenetic Songpu *C. carpio* was selected as the genomic DNA source for whole-genome sequencing. We constructed 21 shotgun libraries and an 8K mate-pair library according to Roche 454 standard operating procedures. The 22 libraries were sequenced on a Roche 454 genome sequencer using GS FLX Titanium chemistry. We also constructed six paired-end libraries by following the Illumina procedure. Paired-end sequencing of each library was performed on an Illumina HiSeq 2000 instrument to produce the raw data. We then filtered out low-quality and short reads to obtain a set of usable reads. Another library with an 8-kb jumping distance was generated and was sequenced on the SOLiD platform. The published 65,720 clean BAC-end sequences were collected for genome assembly. We assembled the Roche 454 read data set and the Sanger BAC-end sequences into contigs using the Celera assembler[44]. Reads from the Illumina libraries, the SOLiD libraries and the 8,000 Roche 454 mate-pair libraries were aligned to the genomic sequences, and paired-end relationships between the reads were used to construct scaffolds. BAC-end sequences were mapped to the scaffolds and were used for further scaffolding. Finally, we used the paired-end information from the short paired-end reads to fill the gaps between the scaffolds with Gapcloser[45].

Linkage Mapping and Map Integration

Microsatellite and SNP markers were used for genotyping analysis and linkage map construction. A tailed primer protocol was used to amplify microsatellite alleles[46, 47]. PCR products were analyzed on a 3130xl Genetic Analyzer. Restriction site–associated DNA (RAD) technology[48] was used to develop polymorphic SNP markers. JoinMap4.0 software was used to perform the linkage analysis. Linkage between markers was examined by estimating the logarithm of odds (LOD) scores for the recombination rate, and map distances were calculated using the Kosambi mapping function. We then used BLAT (with alignment length coverage of >70%) to align the molecular markers to scaffolds. We linked the scaffolds onto chromosomes with a string of 100 Ns representing the gap between 2 adjacent scaffolds on the basis of a high-resolution genetic map.

Repeat Analysis.

Both homology-based and *de novo* prediction analyses were used to identify the repeat content in the carp genome. For the homology-based analysis, we used Repbase (version 20120418) to perform a TE search with RepeatMasker (3.3.0) and the WuBlast search engine. For the *de novo*prediction analysis, we used RepeatModeler to construct a TE library. Elements within the library were then classified using a homologous search with Repbase and a Support Vector Machine (SVM) method (TEClass).

Gene Prediction and Functional Annotation.

We used three approaches for gene prediction: *ab initio* gene prediction, sequence homology–based prediction and expression evidence–based prediction. Briefly, two *ab initio* prediction software programs, AUGUSTUS[49] and FGENESH[50], were used to predict genes in the repeat-masked genome sequences. Gene model parameters for the programs were trained from long genes and known teleost genes processed by PASA[51]. Sequence homology–based gene prediction included both raw and precise alignments. First, protein sequences from the NCBI non-redundant (nr) database and 68 species sequences in Ensembl (version 68)[52] were collected to build a database. Assembled genome sequences were aligned to their corresponding protein sequences in the database using BLASTX. Identified homologous proteins were selected and then aligned to the genome with TBLASTN. Adjacent and overlapping matches were merged using Perl scripts, building the longest protein for each genomic sequence region. Each target region in the genome was then extended by 10 kb from both ends of the aligned region to cover potential UTRs. Protein sequences were then aligned to those genome fragments by Genewise[53]. Transcriptome reads were generated using the Roche Genome Sequencer FLX (previously released data; available from the Sequence Read Archive (SRA) under accessions SRA009366 and SRA050545) and the Illumina HiSeq 2000 (Supplementary Note). Reads were mapped to genomic sequences by TopHat[54], and Cufflinks[55] was used to produce transcript assemblies. For a gene locus with several alternatively spliced transcripts generated by Cufflinks, the transcript with the longest exon length was chosen. All evidence was merged to form a comprehensive consensus gene set using EVM[27]. PASA was used to update the EVM consensus predictions by adding UTR annotations. To obtain gene function annotations, BLAST searches were conducted against the NCBI nr, SwissProt and TrEMBL protein databases, and homologs were called with E values of $<1 \times 10^{-5}$. The functional classification of GO categories was performed using the InterProScan[56] program. Pathway analysis was performed using the KEGG[57] annotation service, the KEGG Automatic Annotation Server (KAAS)[58].

Comparative Genomic Analysis

Protein-coding genes and coding DNA sequences from 11 species (*D. rerio*, *G. aculeatus*, *O. latipes*, *T. rubripes*, *T. nigroviridis*, *X. tropicalis*, *A. carolinensis*, *H. sapiens*, *M. musculus*, *S. scrofa*, *G. gallus* and *C. intestinalis*) were downloaded from Ensembl (version 68)[52]. For genes with alternatively spliced variants, only the longest transcript was selected. Any genes encoding proteins of fewer than 30 amino acids were discarded. The OrthoMCL pipeline[59] was used to define gene families in the common ancestor of the species. All-against-all similarities were performed using BLASTP, with an E-value cutoff of 1×10^{-5}. The well-aligned regions of each gene family, aligned using MUSCLE[60], were extracted with Gblocks[61]. Phylogenetic analysis of the superalignments was performed using a maximum-likelihood method implemented in PhyML[62] with the Jones-Taylor-Thornton (JTT) model. *C. intestinalis* was selected as the outgroup. We used MCScanX[63]to identify syntenic blocks for *C. carpio* and *D. rerio*, with the gap size set to 15 genes and at least 5 syntenic genes. Circos[64] was used for visualization.

To detect the conserved synteny blocks generated by the fourth round of genome duplication, we identified the reciprocal best-match paralogs from the above all-against-all BLASTP comparisons. Two chromosome regions with the gap size set to 15 genes and at least 5 syntenic genes were considered to have been duplicated.

Genome Evolutionary Analysis

We used two methods to detect genome duplication signatures. All-against-all BLASTP comparisons (E value $< 1 \times 10^{-5}$) were used to identify pairs of homologous genes. For each homologous gene pair in *C. carpio*, the synonymous site divergence value (*K*s) was calculated using the CodeML program (run mode −2) from the PAML package[65]. The distributions of *K*s values for *D. rerio* paralogous pairs and pairs between *D. rerio* and the common carp were analyzed using the same pipeline. We calculated the 4dTv values of paralogous pairs within species and of orthologous pairs between species to give the distribution of the 4dTv value to estimate the speciation and WGD event that occurred during evolutionary history.

Whole-Genome Resequencing and Phylogenetic Analysis

The 10 strains of *C. carpio*, consisting of 33 individuals, were randomly collected across Europe, North America and China. Danube River carp and Szarvas 22 were collected from the carp live gene bank of the Research Institute for Fisheries, Aquaculture and Irrigation of Hungary (HAKI). North American carp were collected from the Chattahoochee River in the United States. All other strains or wild populations were collected from China, including Songpu carp from the Heilongjiang Fishery Research Institute; Yellow River carp from the Henan Academy of Fishery Sciences; Heilongjiang River carp from Fuyuan county, Heilongjiang province; Hebao carp from Wuyuan county, Jiangxi province; Xingguo red carp from Xingguo county, Jiangxi province; Oujiang color carp from Longquan county, Zhejiang province; and koi from the breeding population of the Beijing Fishery Research Institute. Fin chips and blood samples were collected, and DNA was extracted using the DNeasy Blood and Tissue kit (Qiagen). Genome resequencing was conducted using the Illumina HiSeq 2000 platform. Paired-end reads from each accession were aligned to the reference genome using the Burrows-Wheeler Aligner

(BWA)[66]. After mapping, SNPs were identified on the basis of the mpileup files generated with SAMtools[67]. The filtering threshold was set to require a read depth of ≥10 and a quality score of ≥20. Genotypes supported by at least two reads and with a minor allele frequency of ≥0.1 were assigned to each genomic position. We performed all-against-all BLASTP for genes in 5 teleosts (*C. carpio, D. rerio, T. rubripes, O. latipes* and *G. aculeatus*) to determine the similarity for each gene pair and to identify single-copy genes, obtaining 8,375 homozygous SNPs from 7,709 single-copy genes. A maximum-likelihood tree was constructed with PhyML[68] and displayed with MEGA[69]. PCA was performed with EIGENSOFT[70], and homozygous SNPs were used to investigate the population structure using STRUCTURE[39]with 2,000 iterations and 2–8 clusters (*K*). The result of the structure matrix was plotted using DISTRUCT[71] software.

Genome Diversity Analysis and Comparison

We calculated the ϖ distribution for each linkage group using a sliding window method with Tajima›s *D* test in Variscan[72] software. The window width was set to 50 kb, and the stepwise distance was 10 kb. ϖ values were compared, and the ratios were sorted. Using the ratio values, we identified the regions with the 1% highest and lowest diversity, and annotated genes were analyzed. Putative SNPs and deletions in the coding regions were identified by mapping the RNA-seq reads to annotated reference genes using BWA and SAMtools. PCR was performed on the deletion regions to verify the identified gene deletions. PCR products were analyzed via electrophoresis on a 2% agarose gel.

RNA Sequencing Analysis

RNA was extracted from the skin tissues of 18 Hebao and 18 Songpu individuals and pooled for each strain. RNA-seq reads were generated using the Illumina HiSeq 2000 platform (Supplementary Note). Reads with a low quality score and a read length of less than 10 bp were removed. All cleaned reads were mapped to the assembled reference with Bowtie[73]. Then, RSEM (RNA-Seq by Expectation Maximization)[74323 (2011] was used to estimate and quantify gene and isoform abundance. Gene expression was measured in fragments per kilobase of exon per million fragments mapped (FPKM)[55]. Finally, edgeR[75] was used to normalize the expression levels in both strains to identify the differentially expressed transcripts by pairwise comparisons. For qRT-PCR validation, total RNA was isolated and purified from all of the samples using the RNeasy kit (Qiagen) and was quantified using a NanoDrop and a Bioanalyzer 2100 (Agilent Technologies). qRT-PCR was performed on the ABI PRISM 7500 Real-Time PCR System with three replicates using the QuantiTect SYBR Green PCR kit

(Qiagen). The *actb* gene was used as the internal reference. Primer information is provided in Supplementary Table 25. Two-sided *t* tests were used to compare expression levels. GO annotation of the genes was performed on the basis of orthologous relationships with the gene set of *D. rerio*. Pathway analysis was performed using Ingenuity Pathway Analysis (IPA) tools (Ingenuity Systems).

REFERENCES

1. FAO Fisheries and Aquaculture Department. The State of World Fisheries and Aquaculture 2006 (Food and Agriculture Organization of the United Nations, Rome, 2007).
2. Bostock, J. et al. Aquaculture: global status and trends. Phil. Trans. R. Soc. B 365,2897–2912 (2010).
3. Hinegardner, R. & Rosen, D.E. Cellular DNA content and the evolution of teleostean fishes.Am. Nat. 106, 621–644 (1972).
4. Wang, J.T., Li, J.T., Zhang, X.F. & Sun, X.W. Transcriptome analysis reveals the time of the fourth round of genome duplication in common carp (Cyprinus carpio). BMC Genomics 13,96 (2012).
5. David, L., Blum, S., Feldman, M.W., Lavi, U. & Hillel, J. Recent duplication of the common carp (Cyprinus carpio L.) genome as revealed by analyses of microsatellite loci. Mol. Biol. Evol. 20, 1425–1434 (2003).
6. Ohno, S., Muramoto, J., Christian, L. & Atkin, N.B. Diploid-tetraploid relationship among old-world members of the fish family Cyprinidae. Chromosoma 23, 1–9 (1967).
7. Larhammar, D. & Risinger, C. Molecular genetic aspects of tetraploidy in the common carpCyprinus carpio. Mol. Phylogenet. Evol. 3, 59–68 (1994).
8. Ji, P. et al. High throughput mining and characterization of microsatellites from common carp genome. Int. J. Mol. Sci. 13, 9798–9807 (2012).
9. Xu, J. et al. Genome-wide SNP discovery from transcriptome of four common carp strains.PLoS ONE 7, e48140 (2012).
10. Zhang, X. et al. A consensus linkage map provides insights on genome character and evolution in common carp (Cyprinus carpio L.). Mar. Biotechnol. (NY) 15, 275–312 (2013).
11. Zheng, X. et al. A genetic linkage map and comparative genome analysis of common carp (Cyprinus carpio L.) using microsatellites and SNPs. Mol. Genet. Genomics 286, 261–277(2011).
12. Zhang, Y. et al. Genetic linkage mapping and analysis of muscle fiber–related QTLs in common carp (Cyprinus carpio L.). Mar. Biotechnol.

(NY) 13, 376–392 (2011).

13. Sun, X.W. & Liang, L.Q. A genetic linkage map of common carp (Cyprinus carpio L.) and mapping of a locus associated with cold tolerance. Aquaculture 238, 165–172 (2004).
14. Li, Y. et al. Construction and characterization of the BAC library for common carp Cyprinus carpio L. and establishment of microsynteny with zebrafish Danio rerio. Mar. Biotechnol. (NY) 13, 706–712 (2011).
15. Xu, P. et al. Genomic insight into the common carp (Cyprinus carpio) genome by sequencing analysis of BAC-end sequences. BMC Genomics 12, 188 (2011).
16. Christoffels, A., Bartfai, R., Srinivasan, H., Komen, H. & Orban, L. Comparative genomics in cyprinids: common carp ESTs help the annotation of the zebrafish genome. BMC Bioinformatics 7 (suppl. 5), S2 (2006).
17. Zhang, Y. et al. Identification of common carp innate immune genes with whole-genome sequencing and RNA-Seq data. J. Integr. Bioinform. 8, 169 (2011).
18. Ji, P. et al. Characterization of common carp transcriptome: sequencing, de novo assembly, annotation and comparative genomics. PLoS ONE 7, e35152 (2012).
19. Williams, D.R. et al. Genomic resources and microarrays for the common carp Cyprinus carpio L. J. Fish Biol. 72, 2095–2117 (2008).
20. Henkel, C.V. et al. Comparison of the exomes of common carp (Cyprinus carpio) and zebrafish (Danio rerio). Zebrafish 9, 59–67 (2012).
21. Ojima, Y. & Yamamoto, K. Cellular DNA contents of fishes determined by flow cytometry. La Kromosomo II 57, 1871–1888 (1990).
22. Kidwell, M.G. & Lisch, D.R. Transposable elements and host genome evolution. Trends Ecol. Evol. 15, 95–99 (2000).
23. Aparicio, S. et al. Whole-genome shotgun assembly and analysis of the genome of Fugu rubripes. Science 297, 1301–1310 (2002).
24. Van de Peer, Y. Tetraodon genome confirms Takifugu findings: most fish are ancient polyploids. Genome Biol. 5, 250 (2004).
25. Kasahara, M. et al. The medaka draft genome and insights into vertebrate genome evolution. Nature 447, 714–719 (2007).
26. Jones, F.C. et al. The genomic basis of adaptive evolution in threespine sticklebacks. Nature484, 55–61 (2012).

27. Haas, B.J. et al. Automated eukaryotic gene structure annotation using EVidenceModeler and the Program to Assemble Spliced Alignments. Genome Biol. 9, R7 (2008).

28. Meyer, A. & Van de Peer, Y. From 2R to 3R: evidence for a fish-specific genome duplication (FSGD). Bioessays 27, 937–945 (2005).

29. Hoegg, S., Brinkmann, H., Taylor, J.S. & Meyer, A. Phylogenetic timing of the fish-specific genome duplication correlates with the diversification of teleost fish. J. Mol. Evol. 59,190–203 (2004).

30. Santini, F., Harmon, L., Carnevale, G. & Alfaro, M. Did genome duplication drive the origin of teleosts? A comparative study of diversification in ray-finned fishes. BMC Evol. Biol. 9, 194(2009).

31. Crow, K.D. & Wagner, G.P. Proceedings of the SMBE Tri-National Young Investigators Workshop 2005. What is the role of genome duplication in the evolution of complexity and diversity? Mol. Biol. Evol. 23, 887–892 (2006).

32. Jaillon, O. et al. Genome duplication in the teleost fish Tetraodon nigroviridis reveals the early vertebrate proto-karyotype. Nature 431, 946–957 (2004).

33. Zhang, Y. et al. Genome evolution trend of common carp (Cyprinus carpio L.) as revealed by the analysis of microsatellite loci in a gynogentic family. J. Genet. Genomics 35, 97–103(2008).

34. Howe, K. et al. The zebrafish reference genome sequence and its relationship to the human genome. Nature 496, 498–503 (2013).

35. Amores, A. et al. Zebrafish hox clusters and vertebrate genome evolution. Science 282,1711–1714 (1998).

36. Mungpakdee, S. et al. Differential evolution of the 13 Atlantic salmon Hox clusters. Mol. Biol. Evol. 25, 1333–1343 (2008).

37. Davidson, W.S. et al. Sequencing the genome of the Atlantic salmon (Salmo salar). Genome Biol. 11, 403 (2010).

38. Pritchard, J.K., Stephens, M. & Donnelly, P. Inference of population structure using multilocus genotype data. Genetics 155, 945–959 (2000).

39. Balon, E.K. Origin and domestication of the wild carp, Cyprinus carpio: from Roman gourmets to the swimming flowers. Aquaculture 129, 3–48 (1995).

40. Rohner, N. et al. Duplication of fgfr1 permits Fgf signaling to serve as a target for selection during domestication. Curr. Biol. 19, 1642–1647 (2009).

41. Millar, S.E. Molecular mechanisms regulating hair follicle development. J. Invest. Dermatol.118, 216–225 (2002).

42. Hoekstra, H.E. Genetics, development and evolution of adaptive pigmentation in vertebrates.Heredity 97, 222–234 (2006).
43. Ito, S. & Wakamatsu, K. Human hair melanins: what we have learned and have not learned from mouse coat color pigmentation. Pigment Cell Melanoma Res. 24, 63–74 (2011).
44. Miller, J.R. et al. Aggressive assembly of pyrosequencing reads with mates. Bioinformatics24, 2818–2824 (2008).
45. Li, R. et al. De novo assembly of human genomes with massively parallel short read sequencing. Genome Res. 20, 265–272 (2010).
46. Neilan, B.A., Wilton, A.N. & Jacobs, D. A universal procedure for primer labelling of amplicons. Nucleic Acids Res. 25, 2938–2939 (1997).
47. Schuelke, M. An economic method for the fluorescent labeling of PCR fragments. Nat. Biotechnol. 18, 233–234 (2000).
48. Sun, X. et al. SLAF-seq: an efficient method of large-scale de novo SNP discovery and genotyping using high-throughput sequencing. PLoS ONE 8, e58700 (2013).
49. Stanke, M., Schoffmann, O., Morgenstern, B. & Waack, S. Gene prediction in eukaryotes with a generalized hidden Markov model that uses hints from external sources. BMC Bioinformatics 7, 62 (2006).
50. Salamov, A.A. & Solovyev, V.V. Ab initio gene finding in Drosophila genomic DNA. Genome Res. 10, 516–522 (2000).
51. Campbell, M.A., Haas, B.J., Hamilton, J.P., Mount, S.M. & Buell, C.R. Comprehensive analysis of alternative splicing in rice and comparative analyses with Arabidopsis. BMC Genomics 7, 327 (2006).
52. Flicek, P. et al. Ensembl 2013. Nucleic Acids Res. 41, D48–D55 (2013).
53. Birney, E., Clamp, M. & Durbin, R. GeneWise and Genomewise. Genome Res. 14, 988–995(2004).
54. Trapnell, C., Pachter, L. & Salzberg, S.L. TopHat: discovering splice junctions with RNA-Seq. Bioinformatics 25, 1105–1111 (2009).
55. Trapnell, C. et al. Transcript assembly and quantification by RNA-Seq reveals unannotated transcripts and isoform switching during cell differentiation. Nat. Biotechnol. 28, 511–515(2010).
56. Zdobnov, E.M. & Apweiler, R. InterProScan—an integration platform for the signature-recognition methods in InterPro. Bioinformatics 17, 847–848 (2001).
57. Ogata, H. et al. KEGG: Kyoto Encyclopedia of Genes and Genomes. Nucleic Acids Res. 27,29–34 (1999).

58. Moriya, Y., Itoh, M., Okuda, S., Yoshizawa, A.C. & Kanehisa, M. KAAS: an automatic genome annotation and pathway reconstruction server. Nucleic Acids Res. 35, W182–W185(2007).

59. Li, L., Stoeckert, C.J. Jr. & Roos, D.S. OrthoMCL: identification of ortholog groups for eukaryotic genomes. Genome Res. 13, 2178–2189 (2003).

60. Edgar, R.C. MUSCLE: multiple sequence alignment with high accuracy and high throughput.Nucleic Acids Res. 32, 1792–1797 (2004).

61. Talavera, G. & Castresana, J. Improvement of phylogenies after removing divergent and ambiguously aligned blocks from protein sequence alignments. Syst. Biol. 56, 564–577(2007).

62. Guindon, S., Lethiec, F., Duroux, P. & Gascuel, O. PHYML Online—a web server for fast maximum likelihood–based phylogenetic inference. Nucleic Acids Res. 33, W557–W559(2005).

63. Wang, Y. et al. MCScanX: a toolkit for detection and evolutionary analysis of gene synteny and collinearity. Nucleic Acids Res. 40, e49 (2012).

64. Krzywinski, M. et al. Circos: an information aesthetic for comparative genomics. Genome Res. 19, 1639–1645 (2009).

65. Yang, Z. PAML 4: phylogenetic analysis by maximum likelihood. Mol. Biol. Evol. 24,1586–1591 (2007).

66. Li, H. & Durbin, R. Fast and accurate short read alignment with Burrows-Wheeler transform.Bioinformatics 25, 1754–1760 (2009).

67. Li, H. et al. The sequence alignment/map format and SAMtools. Bioinformatics 25,2078–2079 (2009).

68. Guindon, S. & Gascuel, O. A simple, fast, and accurate algorithm to estimate large phylogenies by maximum likelihood. Syst. Biol. 52, 696–704 (2003).

69. Tamura, K. et al. MEGA5: molecular evolutionary genetics analysis using maximum likelihood, evolutionary distance, and maximum parsimony methods. Mol. Biol. Evol. 28,2731–2739 (2011).

70. Price, A.L. et al. Principal components analysis corrects for stratification in genome-wide association studies. Nat. Genet. 38, 904–909 (2006).

71. Rosenberg, N.A. DISTRUCT: a program for the graphical display of population structure.Mol. Ecol. Notes 4, 137–138 (2004).

72. Vilella, A.J., Blanco-Garcia, A., Hutter, S. & Rozas, J. VariScan: analysis of evolutionary patterns from large-scale DNA sequence polymorphism data. Bioinformatics 21, 2791–2793(2005).

73. Langmead, B. & Salzberg, S.L. Fast gapped-read alignment with Bowtie 2. Nat. Methods 9,357–359 (2012).

74. Li, B. & Dewey, C.N. RSEM: accurate transcript quantification from RNA-Seq data with or without a reference genome. BMC Bioinformatics 12, 323 (2011).

75. Robinson, M.D., McCarthy, D.J. & Smyth, G.K. edgeR: a Bioconductor package for differential expression analysis of digital gene expression data. Bioinformatics 26, 139–140

Chapter 12

COMPARATIVE GENOMIC ANALYSIS OF SOYBEAN FLOWERING GENES

Chol-Hee Jung, Chui E. Wong Mohan B. Singh , Prem L. Bhalla

Plant Molecular Biology and Biotechnology Laboratory, ARC Centre of Excellence for Integrative Legume Research, Melbourne School of Land and Environment, The University of Melbourne, Parkville, Victoria, Australia

ABSTRACT

Flowering is an important agronomic trait that determines crop yield. Soybean is a major oilseed legume crop used for human and animal feed. Legumes have unique vegetative and floral complexities. Our understanding of the molecular basis of flower initiation and development in legumes is limited. Here, we address this by using a computational approach to examine flowering regulatory genes in the soybean genome in comparison to the most studied model plant, Arabidopsis. For this comparison, a genome-wide analysis of orthologue groups was performed, followed by an *in silico* gene expression analysis of the identified soybean flowering genes. Phylogenetic analyses of the gene families highlighted the evolutionary relationships among these candidates. Our study identified key flowering genes in soybean and indicates that the vernalisation and the ambient-temperature pathways seem to be the most variant in soybean. A comparison of the orthologue groups containing flowering genes indicated that, on average, each Arabidopsis flowering gene has 2-3 orthologous copies in soybean. Our analysis highlighted that the *CDF3*, *VRN1*, *SVP*, *AP3* and *PIF3* genes are paralogue-rich genes in soybean. Furthermore, the genome mapping of the soybean flowering genes showed that these genes are scattered randomly across the genome. A paralogue comparison indicated that the soybean genes comprising the largest orthologue group are clustered in a 1.4 Mb region on chromosome 16 of soybean. Furthermore, a comparison with the undomesticated soybean (*Glycine soja*) revealed that there are hundreds of SNPs that are associated with putative soybean flowering genes and that there are structural variants that may affect the genes of the light-signalling and ambient-temperature pathways in soybean. Our study provides a framework for the soybean flowering pathway and insights into the relationship and evolution of flowering genes between a short-day soybean and the long-day plant, Arabidopsis.

INTRODUCTION

Plants switch to the reproductive phase of development when environmental and endogenous factors are the most favourable for reproductive success and seed production. This proper timing is the result of elaborate regulatory networks that coordinate the external stimuli with endogenous cues, inducing the expression of genes that initiate the floral transition at the shoot apical meristem (SAM).

Much of our current understanding of the floral initiation process is derived from studies usingArabidopsis thaliana as the model system. More than 180 Arabidopsis genes have been identified that play a role in regulating flowering time, and these genes have been organised into six major pathways (reviewed by Fornara et al. [1]). Although the photoperiod and vernalisation pathways monitor seasonal changes in day length or temperature and, hence, initiate flowering in response to exposure to long days or prolonged cold temperatures, the ambient temperature pathway coordinates the response to daily growth temperatures. The autonomous pathway together with those involving age or gibberellin constitutes the rest of the floral pathways, which function more independently of external stimuli. These pathways are integrated by downstream target genes including LEAFY (LFY), FLOWERING LOCUS T (FT) and SUPPRESSOR OF CONSTANS1 (SOC1), with their resulting outcomes conveyed to floral meristem identity genes such as APETALA1 (AP1) at the SAM that triggers the flowering process [2], [3].

Flowering is one of the most important agronomic traits influencing crop yield. There is thus a great necessity for research that examines the molecular control of this fundamental process in important crop species. This knowledge is critical for the breeding of climate change resilient crop varieties. Soybean, a major food crop, is also a member of the large and diverse legume family, which has the unique capability of forming nitrogen-fixing symbioses with soil microorganisms and has thus been used as part of sustainable agricultural practices for thousands of years. Soybean is distributed broadly across latitudes and is cultivated as different maturity groups, with each having a narrow range of latitudinal adaptation. Unlike Arabidopsis, soybean can undergo a reversion of flowering when plants are shifted from flowering inductive to non-inductive conditions [4]. In addition, soybean also follows a floral developmental plan that is distinct from that of Arabidopsis [5]. Therefore, an understanding of the molecular mechanisms underlying these soybean traits is of fundamental and practical interest. Recent studies have begun to shed light on the molecular adaptation of different soybean cultivars to a wide range of photoperiodic conditions [6], [7]. These studies have highlighted similarities as well as differences in the roles of flowering time genes between soybean and Arabidopsis. The blue light receptor CRYPTOCHROME2 (CRY2) regulates photoperiodic flowering in Arabidopsis; however, in soybean, GmCRY1a but not GmCRY2a is the major regulator of photoperiodic flowering. On the other hand, Kong et al. (2010) revealed that, although soybean contains several FT homologues, the dynamic expression of only two of them is responsible for the adaptation of soybean to diverse photoperiodic environments [6]. Nevertheless, it is still unclear if similar downstream target genes are activated in soybean as in Arabidopsis or how the floral pathway is modified to generate the outputs that reconcile the differences in floral development between the two species.

In view of the recent availability of the soybean genome sequence, we have undertaken a genome-wide analysis for the identification of all soybean orthologues for the corresponding Arabidopsis genes, particularly those involved in flowering. As a paleopolyploid, soybean contains duplicate copies of most genes, and these duplicates may have undergone sub- or neo-functionalisation. We identified 491 putative soybean flowering regulatory genes that are scattered randomly throughout the genome, and then we performed phylogenetic analyses of these gene families to acquire an understanding of the evolutionary relationships among these candidate genes. The identified putative soybean flowering genes were further subjected to anin silico gene expression analysis using two independent transcriptome datasets [8], [9]. Although the distributions of the soybean genes in the paralogue-rich groups are not correlated with the recently duplicated regions in the genome, soybean chromosome 16, especially in the ~1.4 Mb region around 34–35 Mb, is highly enriched for genes within paralogue-rich groups. Our study provides an essential genomic resource for functional analyses of the soybean flowering pathway, facilitating future research and efforts into breeding robust high-yielding crop varieties.

RESULTS

Identification of Soybean Homologues of Arabidopsis Genes

The most recent genome annotation of soybean lists 46,367 genes with high confidence from the current draft genome sequence of soybean [10]. The current soybean annotation (G.max 1.09) identifies the closest Arabidopsis homologue of nearly all of the predicted soybean genes. However, soybean genes are associated with only 55% of the total Arabidopsis genes in the TAIR9 annotation. Thus, we combined the information from the TAIR9 annotation together with the orthologue-based method, which clusters soybean genes and Arabidopsis genes independently into pre-defined orthologue groups (OGs) in the OrthoMCL database (release 5.0) [11]. Then, we matched soybean and Arabidopsis genes under the same OGs as putative orthologues (see Methods). This combined analysis for homologue identification connected 20,730 Arabidopsis genes in 11,344 OGs to 45,175 soybean genes (Dataset S1).

Soybean Flowering-related Genes

In this study, we focus on the 183 Arabidopsis genes that are known to take part in flowering regulatory pathways from previous studies [9], [12], [13], [14], [15], [16], [17], [18], [19], [20],[21], [22]. The orthologue identification analysis found 491 soybean genes that are putative flowering genes (Figure 1 and Table S1). The majority of the Arabidopsis flowering genes have putative soybean orthologues (163 out of 183). However, the soybean orthologues for 20 Arabidopsis flowering genes are not identified by the orthologue-based method used in this study. These Arabidopsis genes include TARGET OF EARLY ACTIVATION TAGGED 2 (TOE2), TOE3 and the vernalisation-insensitive genes VERNALIZATION5/VIN3-LIKE 2(VEL2), and VEL3 (Table S1).

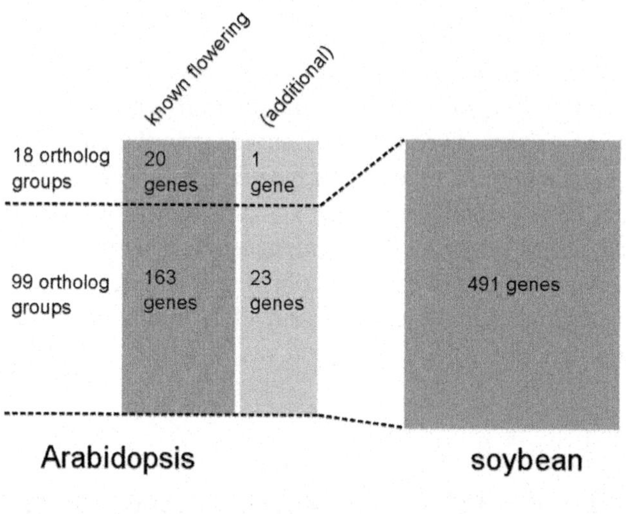

Figure 1: Number of orthologue groups and genes related to flowering pathways.183 Arabidopsis flowering genes are classified into 117 orthologue groups along with 24 additional genes. Function of these additional genes has not been investigated so far. Out of the 117 OGs, 99 OGs contain 491 soybean genes that are putative soybean flowering genes. The numbers of genes are indicated within each box.

Nevertheless, this lack of orthologue identification does not necessarily mean that these 20 Arabidopsis flowering genes are absent in soybean, as they still have similar soybean genes based on a direct BLAST analysis. In addition, the orthologue-based method identified 24 additional Arabidopsis genes that are grouped into the same OGs as known flowering genes (Table S1) but have not been investigated for their role in floral initiation. doi:10.1371/journal.pone.0038250.g001

Subsequent analyses of phylogenetic trees generated from the multiple sequence alignments of the soybean and Arabidopsis genes within each OG estimated that 322 genes are located in the same clades as the Arabidopsis genes known to be involved in flowering pathways (seeMaterials and Methods), indicating that they are likely the true orthologues of their corresponding Arabidopsis flowering genes. The simplified pathway diagram, which contains most of the Arabidopsis flowering-time genes and their putative soybean orthologues, is shown in Figure 2.

Arabidopsis flowering pathway depicted by Higgins et al. (2010) [12] was adapted for this study. Arabidopsis gene symbols are shown in upper box of each node along with the corresponding OG ID and the total number of soybean genes in the same OG. Only the soybean genes that are closer to the Arabidopsis flowering genes in each OG are listed in each node. Arabidopsis genes in grey shades are those that are not assigned with putative soybean orthologues. However, based on BLAST analysis these genes still have homologous soybean genes, the best matching soybean gene in BLAST analysis is shown below these genes. Dashed-lines were used for the arrows or T-bars that involve grey shades. Soybean genes marked with * are those that share the same clade with Arabidopsis genes but have not been investigated so far as flowering genes. OG IDs starting with 'OG5_AT' are arbitrarily generated in this study and do not exist in OrthoMCL 5.0 database. Other conventions are same as those used in the Figure 1A by Higgins et al. (2010) [12]. Arrows show promoting effects, T-bars show repressing effects. Environmental cues are shown as lower case letters in square boxes; 'v' is extended cold (vernalization); 'ld' is long days; 'sd' is short days; 'am' is ambient (non-vernalizing) temperature. Genes are shown in italics and proteins in non-italics in ovals. 'Pfr' indicates P_{fr} phytochrome signaling. Arabidopsis genes assigned to specific pathways are color-coded (photoperiod pathway in green, vernalization in blue and autonomous pathway in purple). Flowering pathway integrators are shown in red. Triple headed arrows indicate activation by red or blue light.

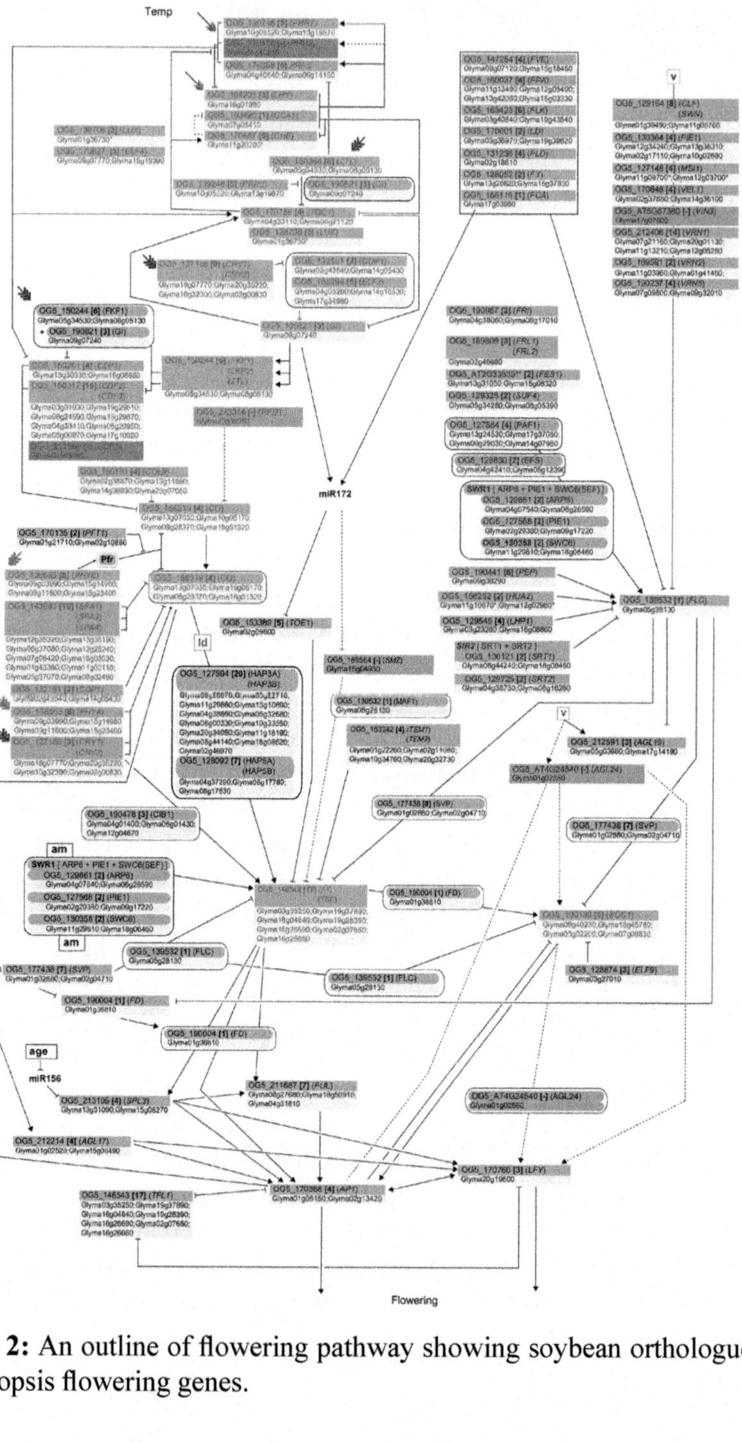

Figure 2: An outline of flowering pathway showing soybean orthologues for Arabidopsis flowering genes.

Expression of Soybean Homologues of Arabidopsis Flowering Genes

The transcriptional activities of the putative soybean floral regulatory genes were examined to gather further evidence for their involvement in flowering. To this end, we utilised the soybean gene expression data from two recent transcriptome analyses [8], [9]. We found that the expression of most of the putative soybean floral genes (449 out of 491 genes; 91.4%) is supported by these two datasets (Figure S1 and Dataset S2). Furthermore, the vast majority of the expressed putative soybean flowering genes, 403 out of 449, exhibited transcriptional activities in flowers (Dataset S2), among which 19 genes are preferentially or specifically expressed in flowers [9]. These 19 genes are spread across 10 OGs, which contain the Arabidopsis MADS box genes, including AP1, PISTILLATA (PI) and SEPALLATA1 (SEP1) (Table S2).

Key Pathways and Gene Families for Flowering

Among the genes for flowering pathways, the key players are those that are involved in the light-signalling pathway, the vernalisation pathway, the autonomous pathway and the ambient temperature pathway, along with genes for meristem identity and flowering pathway integrators[12]. In Arabidopsis, 120 genes, which are grouped into 69 OGs, are known to be key players in flowering [12]. Among these 120 Arabidopsis genes, 112 of them are found in 62 OGs having 314 putative soybean orthologues, (Table 1). Table 1 shows the number of OGs associated with the key pathways or groups of genes for flowering. As one gene can take part in two or more different pathways, one OG can participate in multiple pathways or groups. The 237 soybean genes that are orthologous to key flowering genes of Arabidopsis are scattered throughout the genome rather than clustered within certain regions (Figure 3).

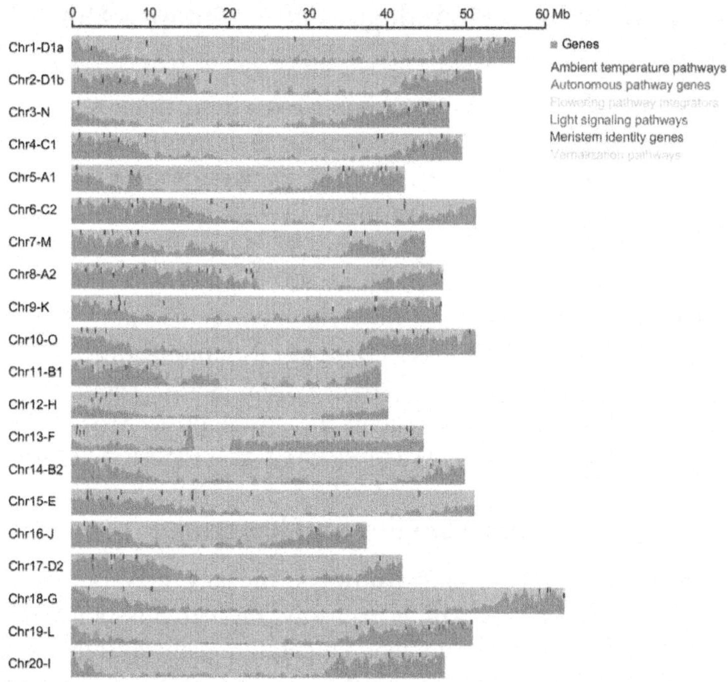

Figure 3: Genomic loci of soybean genes homologous to Arabidopsis floral regulatory genes.

Soybean genes in Table 1 that are homologous to Arabidopsis floral regulatory genes are indicated with solid bars of different colours while the purple shade represents gene density. These soybean genes are randomly spread over the whole genome. Gene density depiction is adapted and modified from Figure 1 by Schmutz et al. (2010) [10].

doi:10.1371/journal.pone.0038250.g003

Table 1: Key flowering pathways and the numbers of associated genes in Arabidopsis, soybean, Medicago, A. lyrata **and** Brachypodium

Pathway	OGs	Arabidopsis[^]	Soybean[*]	Medicago	*A lyrata*	Brachypodium
Light signaling	25	48/(53)	121/(115)	47	59	44
Vernalization	23	32/(36)	81/(71)	46	33	25
Autonomous	16	17/(23)	49/(46)	33	24	23
Ambient temperature	8	16/(19)	38/(30)	28	18	9
Meristem identity	5	7/(7)	19/(18)	6	7	3
Flowering pathway integrators	11	36/(39)	82/(69)	32	40	33

As one gene can take part in two or more different pathways, one OG can participate in multiple pathways or groups.
[^]Numbers within parentheses indicate the total number of Arabidopsis genes belonging to the OGs associated with the given pathways.
[*]Numbers within parentheses indicate the number of soybean genes showing transcriptional activity detected by using either of the two recent transcriptome datasets [8,9].
doi:10.1371/journal.pone.0038250.t001

Light signalling pathways.

Light is one of the main environmental regulators of flowering in plants. Plants sense the time of day and season of year by monitoring the light environment through light signalling pathways[23]. Soybean is a facultative short-day crop, but soybean cultivars also belong to different maturity groups depending upon their photoperiod sensitivity. This strong latitudinal cline is also observed in its undomesticated wild relative, Glycine soja (G. soja). In Arabidopsis, photoperiod pathway genes together with photoreceptor genes and circadian clock components take part in light signalling pathways. The number of known Arabidopsis flowering genes involved in these pathways is 48, which are clustered into 25 OGs. However, these OGs contain 53 Arabidopsis genes in total, suggesting that the additional 5 genes may also be involved in floral initiation (Table 1). In total, 121 soybean genes are identified as putative orthologues of 48 Arabidopsis flowering genes in 25 OGs (Table 1). The multiple sequence alignments followed by phylogenetic tree analyses for the Arabidopsis and soybean gene sequences in each of the 25 OGs revealed that 66 of the soybean genes are more closely located to their corresponding Arabidopsis genes than other soybean genes in the same OGs (Dataset S3). Furthermore, anin silico gene expression analysis of the identified soybean flowering genes determined that 115 of the 121 soybean orthologues are expressed, including 109 genes expressed in flowers [8],[9] (Figure S1 and Dataset S2).

The key Arabidopsis genes involved in the light signalling pathway include the CONSTANS(CO), PHYTOCHROME (PHY) and CRYPTOCHROME (CRY), CIRCADIAN CLOCK ASSOCIATED 1 (CCA1), LATE ELONGATED HYPOCOTYL (LHY) and PSEUDO-RESPONSE REGULATOR 1 [PRR1, also called TIMING OF CAB EXPRESSION 1 (TOC1)] genes. CO, along with CONSTANS-LIKE 1 (COL1) and CONSTANS-LIKE 2 (COL2), are contained in OG5_156319, which also contains four soybean genes as soybean orthologues (Glyma08g28370, Glyma13g07030, Glyma18g51320 and Glyma19g05170) (See Table S1). All four soybean-orthologue candidates of Arabidopsis CO are expressed in tested tissues/developmental stages in the two recent transcriptome datasets [8], [9], but only two candidates are expressed in flowers (Dataset S2). The CRY genes CRY1 and CRY2 are grouped into OG5_127186, which contains nine soybean genes (Table S1). The UV REPAIR DEFECTIVE 3 (UVR3) gene is also grouped into OG5_127186. In the phylogenetic tree of genes contained in OG5_127186, CRY1, CRY2 and UVR3 are all located in the same clade, along with 5 soybean genes (Figure 4A). Among these soybean genes, Glyma08g22400 is the closest orthologue of Arabidopsis UVR3, while Glyma18g07770,

Glyma20g35220, Glyma10g32390 and Glyma02g00830 are closer toCRY2 (Figure 4A). Phylogenetic trees inFigure 4 include putative orthologues in Arabidopsis lyrata (A. lyrata), Medicago truncatula(Medicago) as well as a monocot Brachypodium distachyon (Brachypodium). All three Brachypodium genes and one Medicago genes clustered in the same OG are also found in theCRY1 clade, leaving four soybean genes and one Medicago gene in separate clades, indicating that these may have diverged functions (Figure 4A). Five PHY genes of Arabidopsis (PHYA,PHYB, PHYC, PHYD and PHYE) have eight soybean orthologue candidates, which are contained within OG5_136555 (Table S1). All of these soybean genes, except for Glyma15g23400, are expressed in flowers in one or both of the two transcriptome gene expression analyses integrated in this study [8], [9]. The MYB-transcription factor genes CCA1and LHY are among the key circadian clock components in Arabidopsis and are regulated byTOC1 (also known as PRR1) [24]. CCA1 has a single soybean gene orthologue candidate (Glyma07g05410), while LHY and TOC1 have three and four soybean orthologue candidates, respectively (Table S1). All of the putative soybean orthologues of CCA1, LHY and TOC1 are expressed in the samples tested, including flowers, when analysed for their in silico gene expression [8], [9]. The GIGANTEA (GI) gene in OG5_190821 is a part of the evening loop in Arabidopsis and performs different functions through its interactions with other genes, including the FLAVIN-BINDING, KELCH REPEAT, F BOX 1 (FKF1), LOV KELCH PROTEIN 2 (LKP2) and ZEITLUPE (ZTL) genes contained within OG5_150244, which contains six soybean genes in total (Table S1) [12]. Higgins et al. (2011) reported that GI is a highly conserved single copy gene in Arabidopsis, rice, Brachypodium and barley [12], but it has three orthologous soybean genes (Glyma09g07240, Glyma10g36600, Glyma20g30980) (Table S1).

Arabidopsis genes that are known as flowering genes are shown in red along with their closest soybean homologues in the corresponding phylogenetic trees, which are most likely to be orthologues of Arabidopsis flowering genes. Numbers in each branch indicate the confidence value calculated from 1000 times of bootstrapping. Putative orthologues of A. lyrata, Medicago and Brachypodium are also included in the phylogenetic trees. Gene names starting with 'AT', 'Glyma', 'Alyr_' and 'Bradi' are for genes of A. thaliana, soybean, A. lyrata and Brachypodium, respectively. Medicago gene names start with either 'Medtr' or 'AC'. To clarify relationships between different nodes diverse tree formats for OGs are used. The Arabidopsis genes marked with '*' have not been identified and investigated as flowering genes. Phylogenetic tree for OG5_127186 (A), OG5_139532 (B), OG5_163423 (C), OG5_131236 (D), OG5_147254 (E), OG5_146543 and OG5_126706 (G) is shown.

doi:10.1371/journal.pone.0038250.g004

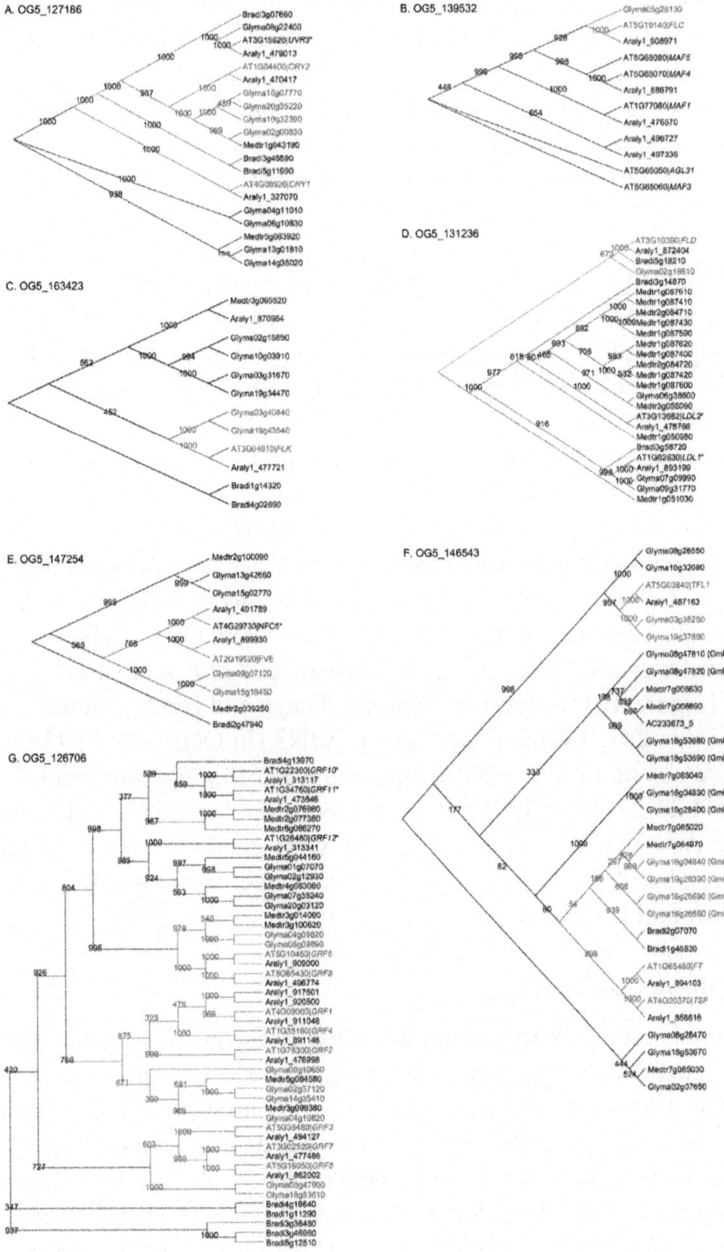

Figure 4: Phylogenetic relationship between soybean and Arabidopsis genes in each orthologue group.

Vernalisation pathway.

Vernalisation involves plants that require prolonged periods of low temperature to initiate flowering. The vernalisation pathway in Arabidopsis involves 32 genes clustered into 23 OGs (Table 1). Among these, 30 Arabidopsis genes in 21 OGs have 81 soybean orthologue candidates (Table 1), of these 81 genes, 71 show evidence of transcription (Dataset S2). However, the orthologous counterparts of the Arabidopsis VERNALISATION INSENSITIVE 3(VIN3) gene in OG5_AT5G57380 and AGAMOUS-LIKE24 (AGL24) gene in OG5_AT4G24540 were not identified in soybean (Table S1) by this method. Nonetheless, a BLAST analysis suggests the potential existence of their soybean orthologues (see below and the Discussion). Among the OGs containing Arabidopsis genes associated with the vernalisation pathway, the ratio of the number of soybean genes to that of Arabidopsis genes is highest in OG5_212406, in which the ratio is 14 soybean genes to 1 Arabidopsis gene, REDUCED VERNALISATION RESPONSE 1 (VRN1) (Table S1). In contrast, the six Arabidopsis genes in OG5_139532, which includes a MADS-box transcription factor gene, FLOWERING LOCUS C (FLC), that negatively regulates flowering [25], share only one soybean gene as a putative orthologue (Glyma05g28130), resulting in the lowest soybean-to-Arabidopsis gene count ratio among the vernalisation-related OGs (Table S1). In the phylogenetic tree of OG5_139532, Glyma05g28130 is most closely related to FLC (Figure 4B). Interestingly, no Medicago and Brachypodium genes are found in this OG. As mentioned above, VIN3 (in OG5_AT5G57380), which is a repressor of FLC in cold temperatures [26], and a flowering promoter gene, AGL24[27], [28], [29] [reviewed by Alexandre and Hennig (2008) [30]] in OG5_AT4G24540, are not assigned with putative soybean orthologues (Table S1), but share closely related soybean genes with other flowering genes (see below and Discussion).

Autonomous pathway.

Autonomous pathways in plants are activated in response to endogenous changes that are independent from the environmental cues leading to flowering [31]. There are 17 genes, grouped into 16 OGs, involved in the Arabidopsis autonomous pathway (Table 1). Each OG has a single Arabidopsis gene that is known to be functional during floral initiation, except for OG5_129164, which contains two Arabidopsis flowering genes: CURLY LEAF (CLF) andSWINGER (SWN) (Table S1). Three other OGs (OG5_147254, OG5_127148 and OG5_131236) also include one or two additional Arabidopsis genes, raising the total number of Arabidopsis genes in the autonomous pathway-related OGs to 23 (Table 1). The total number of orthologous soybean genes to the 17 Arabidopsis genes (or 23 if the additional genes are included) is 49, of which 46 genes are transcriptionally active (Table 1). OG5_163423 has six soybean genes that are orthologous to AT3G04610 [FLOWERING LOCUS KH DOMAIN(FLK)], a repressor of FLC expression [32], which is the highest soybean-to-Arabidopsis gene count ratio among the OGs for autonomous pathways. The subsequent phylogenetic tree analyses revealed that only two soybean genes (Glyma03g40840 and Glyma19g43540) are located in the same clade with Arabidopsis FLK, indicating that they are likely true orthologues of FLK (Figure 4C). Similarly, in OG5_131236 and OG5_147254, only one (Glyma02g18610) and two (Glyma15g18450 and Glyma09g07120) soybean genes, respectively, are found and thus are also likely to be true orthologues of their Arabidopsis counterparts involved in autonomous pathways (Figure 4D,E). OG5_131236 has three Arabidopsis genes, includingFLOWERING LOCUS D (FLD), which down-regulates FLC and has Glyma02g18610 as its closest orthologue according to the phylogenetic tree (Figure 4D), and FVE [also known asMULTICOPY SUPPRESSOR OF IRA1 4 (MSI4)] in OG5_147254, which also down-regulatesFLC and has Glyma15g18450 and Glyma09g07120 as its closest orthologues (Figure 4E). In comparison, four soybean orthologue candidates of Arabidopsis FPA, which has a redundant role with FLD, FVE, and LD [33], are equally distant from their Arabidopsis counterpart (data not shown). Because the minimum number of sequences for the generation of a phylogenetic tree is four, we are unable to generate phylogenetic trees for four OGs (OG5_128052, OG5_155119, OG5_169591 and OG5_170601) (Table S1). Therefore, all of the soybean genes in these OGs are regarded as the closest homologues of the Arabidopsis genes contained in the corresponding groups. Each of VEL2, VEL3 and VIN3 are grouped into a singleton OG and are not assigned orthologous counterparts in soybean (Table S1) but do have homologous genes in soybean according to the direct BLAST analysis (see below andDiscussion).

Ambient temperature pathway.

Plants respond to ambient temperature changes to modulate their flowering times [34]. The ambient temperature pathway in Arabidopsis involves 16 genes that are clustered into 8 OGs that have 38 soybean genes in total (Table 1). Three OGs (OG5_131236, OG5_147254 and OG5_155119) are also involved in autonomous pathways, and the Arabidopsis genes contained in OG5_139532, OG5_129661 and OG5_177438 are also involved in its vernalisation pathway (Table S1). In most of the OGs related to the ambient temperature pathway, the numbers of soybean genes are greater than those of

Arabidopsis genes; however, the opposite findings are observed in the cases for OG5_139532 and OG5_190004. OG5_139532 contains six Arabidopsis genes (including FLC) that are orthologous to only one soybean gene, Glyma05g28130 (see above and Figure 4B). Similarly, Glyma01g36810 is the only soybean orthologue of the Arabidopsis genes AT4G35900 (FD) and AT2G17770 (FDP) in OG5_190004, which encode for the basic leucine zipper (bZIP) domain protein and positively regulate flowering [35]. Arabidopsis AT4G16280 (FCA) in OG5_155119 has one putative soybean orthologue (Glyma17g03960) (Table S1).

Meristem identity genes.

Meristem identity genes are activated by upstream pathways and initiate floral development by triggering the transition of the apical meristem from the vegetative phase to the reproductive phase [12]. Seven Arabidopsis genes, including FD, LFY, SQUAMOSA PROMOTER BINDING PROTEIN-LIKE 3 (SPL3), AP1 and AGL8 [also known as FRUITFULL (FUL)], are involved in this role and are clustered in five OGs (Table 1 and Table S1). The total number of soybean genes clustered within these OGs is 19, 18 of which were expressed in either the transcriptome data [8] or [9] (Table 1). SPL3 in OG5_213105, which positively regulates FT, AP1 and LFY in Arabidopsis [36], has four putative soybean orthologues (Glyma07g31880, Glyma13g24590, Glyma13g31090 and Glyma15g08270). Arabidopsis LFY (AT5G61850) in OG5_170760 has three orthologous counterparts in soybean (Glyma04g37900, Glyma06g17170 and Glyma20g19600) (Table S1). AP1 and CAULIFLOWER (CAL), which are important in initiating flowering, are grouped into OG5_170388, which contains four soybean genes (Table S1).AGL8 (or FUL) in OG5_211687, which is also important for the initiation of flowering [37], [38], has 7 putative soybean orthologues (Table S1).

Flowering pathway integrators.

Genes of flowering pathway integrators integrate signals from several related pathways and determine the exact timing of flowering [2], [3]. In this study, 36 Arabidopsis flowering pathway integrator genes, including FT, LFY, FLC, CO and SOC1, were grouped into 11 OGs (Table 1and Table S1). Among the 82 soybean genes grouped into these same 11 OGs, 69 genes demonstrated expression in at least one of the recent transcriptome datasets of SoyBase [8] or Libault et al. [9] (Table 1). As FT, FLC, LFY and CO are also involved in other flowering pathways, their relationships with soybean counterparts are described above. OG5_169591, which contains the REDUCED VERNALIZATION RESPONSE 2 (VRN2) gene, includes two soybean genes (Glyma01g41460 and Glyma11g03960). OG5_146543 consists of three Arabidopsis genes [FT, TWIN SISTER of FT (TFT) and TERMINAL FLOWER 1 (TFL1)] and 17 soybean genes. These 17 soybean genes include all 10 genes identified as soybean counterparts of Arabidopsis FT in two recent studies [6], [39], validating our approach for the identification of orthologues. The phylogenetic tree for OG5_146543 is also in agreement with that produced by Hecht and colleagues, which sub-classed 10 soybean FT homologues intoFTa, FTb and FTc (Figure 4F) [39]. Eleven members of the NUCLEAR FACTOR Y transcription factors separate into three OGs (OG5_127594, OG5_128092 and OG5_152404) and have 34 soybean homologues (Table S1). The SUPPRESSOR OF OVEREXPRESSION OF CO 1 (SOC1) in OG5_190130 and EARLY FLOWERING 9 (ELF9) in OG5_128874 have four and three putative soybean orthologues, respectively (Glyma03g02200, Glyma07g08830, Glyma09g40230 and Glyma18g45780 for SOC1; Glyma03g27010, Glyma10g31450 and Glyma20g36110 for ELF9) (Table S1). OG5_126706 contains 11 Arabidopsis genes that encode 14-3-3 proteins that are involved in various processes, including signal transduction[40]; 8 of these 11 genes have demonstrated proven roles as flowering pathway integrators in previous studies (Table 1). In soybean, 12 genes are found in the same OG as the putative orthologues of these 11 Arabidopsis genes. In the phylogenetic tree analysis, the GENERAL REGULATORY FACTOR 10 (GRF10), GRF11 and GRF12 genes branch out on their own along with four soybean genes (Figure 4G). Interestingly, these have not yet been investigated for their role as flowering genes even though they group together with eight other 14-3-3 protein genes (that are known as flowering genes) in OG5_126706.

Comparison with Other Species

As Medicago truncatula is another important model species of the legume family that has also been extensively sequenced and studied, we applied the same methods used for the identification of orthologues of Arabidopsis genes to the annotated protein sequences of Medicago (Mt 3.5 annotation) (Dataset S1). Along with Medicago, we also applied the orthologue identification method to Arabidopsis lyrata (A. lyrata) and Brachypodium distachyon(Brachypodium). The numbers of putative orthologues of Arabidopsis genes involved in key flowering pathways in each species are shown in Table 1. Although the number of Medicago genes in the Mt3.5 annotation (version 3) is similar to that of soybean genes in G.max 1.09, smaller numbers of Medicago genes were grouped into the same OGs that contain key flowering genes of Arabidopsis. In particular, the numbers of putative Medicago orthologues of Arabidopsis genes involved in the light signalling pathways, meristem identity and coding for flowering pathway integrators were less than half of the number of soybean genes in these same pathways (Table 1). The number of Brachypodium and A. lyrata genes in each of OGs is also comparable with that of A. thaliana, even though the genome size varies from ~130 Mb (A. thaliana) to ~270Mb (Brachypodium) [41]. (Table 1).

Comparison with Glycine soja

Recently, the draft genome sequence of Glycine soja (G. soja), wild soybean, was released[42]. As Kim et al. identified structural variations, such as large deletions, inversions and insertions, between G. max and G. soja genomes by mapping short reads of G. soja against the G. max genome sequence [42], we determined how many of the G. max genes involved in flowering pathways harbour such structural variations. Of the 1,538 genes in G. max associated with large deletions when promoter regions (the 1 kb region upstream) are taken into account, 10 genes are involved in flowering, 7 of which are involved in key flowering pathways. Similarly, 11 out of the 689 genes associated with large insertions are involved in flowering, but only 4 of them are associated with key flowering pathways (Table S3).Kim et al. also compared the single nucleotide polymorphisms (SNPs) in G. soja to those in the G. max genome [42]. A total of 4,187 SNPs were found within the genic regions of 405 G. soja genes that are counterparts of putative flowering genes of G. max. The number of flowering genes in G. soja that contain SNPs increases to 458 when the 1 kb upstream is included as the promoter region. However, a substantial number of these flowering genes, 182 genes (39.8% of 458 genes), have SNPs only in non-protein coding regions and/or promoters. For example, among the 17 G. soja genes corresponding to G. max genes that are homologous to the Arabidopsis FT, TSF and TFL1(OG5_146543) genes, only five have SNPs in CDSs: Glyma08g47820, Glyma16g04840, Glyma03g35250, Glyma09g26550 and Glyma16g32080. Of these, Glyma08g47820 and Glyma16g04840 were named as GmFT6 and GmFT3a, respectively, by Kong et al. [6]. However, neither GmFT6 nor GmFT3a expression is detected in either of the transcriptome datasets [8], [9].

A subsequent analysis of the proportions of genes containing structural variations or SNPs indicated that these mutations are not particularly enriched or depleted in flowering genes (data not shown).

Soybean Genes with More or Less Paralogues than in Arabidopsis

Soybean flowering genes with more or less paralogues than in Arabidopsis. We subsequently focused on the soybean OGs that are potentially involved in flowering pathways and have significantly more or less paralogues than the corresponding Arabidopsis genes (paralogue-rich and paralogue-less groups, respectively). As soybean underwent additional rounds of whole genome duplication events compared to Arabidopsis, the ratio of soybean gene counts against Arabidopsis gene counts per group is 2.5 on average, which suggests that each OG has 2-3 times more soybean genes than Arabidopsis genes (Figure 5). However, in 8 OGs, the number of soybean genes is far greater than that of Arabidopsis genes (Table 2). The Arabidopsis genes in these groups include PIF3, CDF2/3, EEL, AREB3, AP2,AP3, SVP, AGL8 (or FUL) and VRN1 (Table 2). The majority of the soybean genes in the paralogue-rich groups are transcriptionally active according to the two recent transcriptome datasets for soybean [8], [9] (Dataset S2). However, the hierarchical clustering of the expression profiles of the 82 soybean genes in these 8 paralogue-rich OGs is not in agreement with the gene grouping based on the sequence similarity (i.e., the grouping of OGs), suggesting that the paralogues have diverged in terms of function (Figure 6).

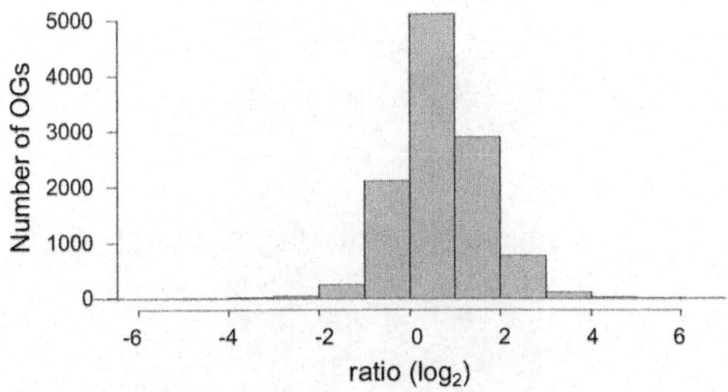

Figure 5: Ratio of soybean gene count to Arabidopsis gene count per OG.

On average, the number of soybean genes per OG is 2-3 times of Arabidopsis genes, but 300 OGs have far more soybean genes (more than 2-standard deviation from average), hence designated as paralogue-rich OGs. doi:10.1371/journal.pone.0038250.g005.

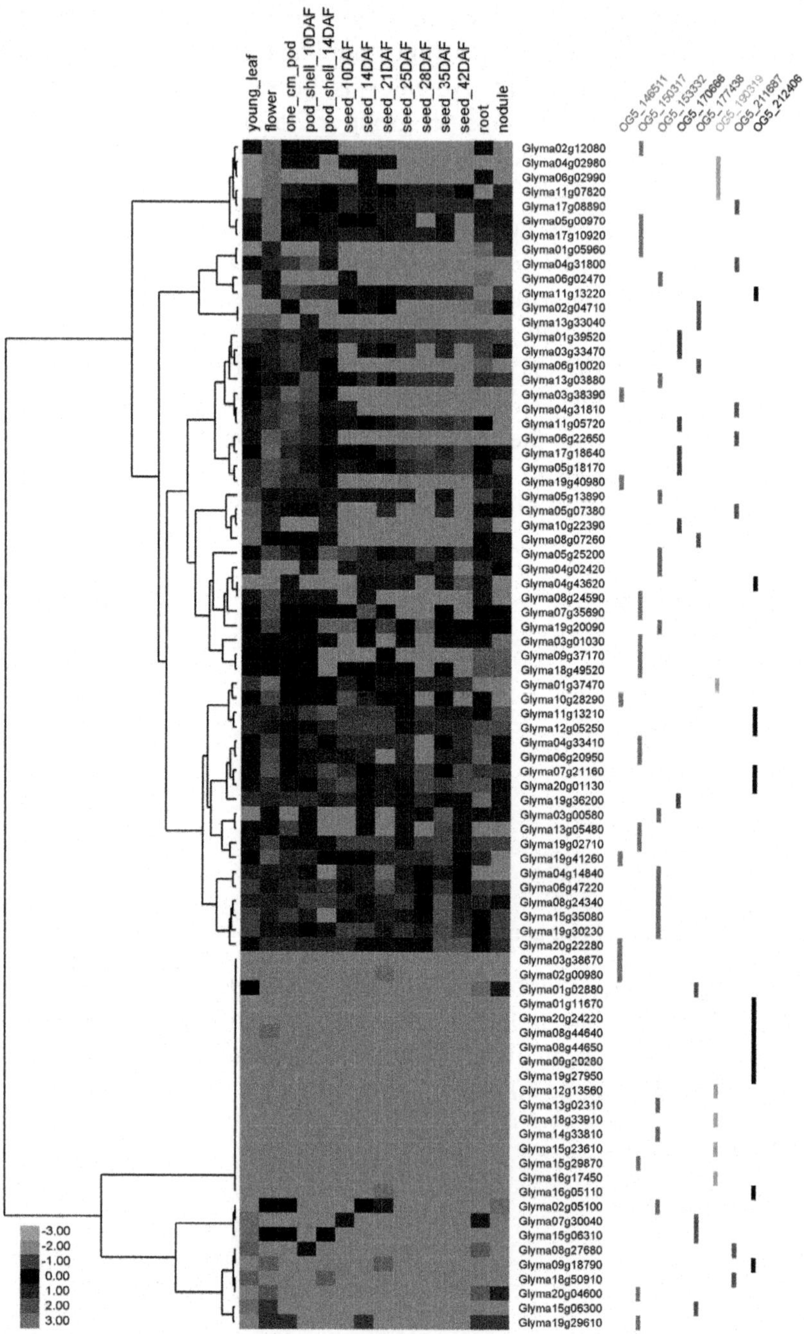

Figure 6: Expression profiles of soybean genes in eight paralogue-rich orthologue groups.

The hierarchical clustering of the expression profiles for 82 soybean genes shows that genes in the same group do not always have similar expression patterns, indicating functional divergence among paralogues. The expression data were extracted from the soybean transcriptome data in SoyBase [8]. DAF: Days After Flowering.doi:10.1371/journal.pone.0038250.g006

Table 2: List of paralogue-rich OGs containing Arabidopsis flowering genes

OrthoMCL ID/soybean gene	Arabidopsis gene/description	PFAM	Panther
OG5_139532/ Glyma05g28130	AT1G77080: *MAF1 (MADS AFFECTING FLOWERING 1)*; transcription factor	K-box region	MADS BOX PROTEIN
	AT5G10140: *FLC (FLOWERING LOCUS C)*; specific transcriptional repressor/transcription factor	K-box region	MADS BOX PROTEIN
	AT5G65050: *AGL31 (AGAMOUS LIKE MADS-BOX PROTEIN 31)*; transcription factor	K-box region	MADS BOX PROTEIN
	AT5G65060: *MAF3 (MADS AFFECTING FLOWERING 3)*; transcription factor	K-box region	MADS BOX PROTEIN
	AT5G65070: *MAF4 (MADS AFFECTING FLOWERING 4)*; transcription factor	K-box region	MADS BOX PROTEIN
	AT5G65080: *MAF5 (MADS AFFECTING FLOWERING 5)*; transcription factor	K-box region	MADS BOX PROTEIN
OG5_189849/ Glyma20g39140	AT1G50680: AP2 domain-containing transcription factor, putative	B3 DNA binding domain	-
	AT1G51120: AP2 domain-containing transcription factor, putative	B3 DNA binding domain	-
OG5_190004/ [illegible]	AT2G17770: *ATBZIP27*; transcription factor	-	-

In contrast to the results for the paralogue-rich OGs, the soybean gene counts are smaller than the Arabidopsis gene counts for the three paralogue-less OGs (OG5_139532, OG5_189849 and OG5_190004) (Table 3). Furthermore, there are 20 Arabidopsis flowering genes in 18 OGs that are not assigned putative soybean orthologues by our method (Table S1). Among these genes are Arabidopsis VEL2, VEL3 and AGL24. VEL2 and VEL3 belong to a small gene family of plant homeodomain (PHD) finger-containing proteins that coordinate flowering through epigenetic regulation [43], [44], while AGL24 is one of the MADS-box genes found to promote flowering by integrating flowering signals from several floral pathways [27], [28], [29]. However, BLAST searches of these Arabidopsis flowering genes against all of the annotated soybean genes reveal that they do have homologous soybean genes. In spite of this, the soybean genes that best match these Arabidopsis genes are putative soybean orthologues of other Arabidopsis genes (Table 4). Table 4 includes Arabidopsis genes in OGs that do not contain any other soybean gene members besides the best-matching soybean gene in the BLAST results. All of the soybean genes bearing a sequence similarity with Arabidopsis genes in OGs lacking other soybean gene members (with BLAST e-values less than 1e-10) are provided in the Dataset S4.

Table 3: List of OGs for flowering genes containing less number of paralogues in soybean.

OrthoMCL ID/soybean gene	Arabidopsis gene/description	PFAM	Panther
OG5_139532/ Glyma05g28130	AT1G77080: *MAF1 (MADS AFFECTING FLOWERING 1)*; transcription factor	K-box region	MADS BOX PROTEIN
	AT5G10140: *FLC (FLOWERING LOCUS C)*; specific transcriptional repressor/transcription factor	K-box region	MADS BOX PROTEIN
	AT5G65050: *AGL31 (AGAMOUS LIKE MADS-BOX PROTEIN 31)*; transcription factor	K-box region	MADS BOX PROTEIN
	AT5G65060: *MAF3 (MADS AFFECTING FLOWERING 3)*; transcription factor	K-box region	MADS BOX PROTEIN
	AT5G65070: *MAF4 (MADS AFFECTING FLOWERING 4)*; transcription factor	K-box region	MADS BOX PROTEIN
	AT5G65080: *MAF5 (MADS AFFECTING FLOWERING 5)*; transcription factor	K-box region	MADS BOX PROTEIN
OG5_189849/ Glyma20g39140	AT1G50680: AP2 domain-containing transcription factor, putative	B3 DNA binding domain	-
	AT1G51120: AP2 domain-containing transcription factor, putative	B3 DNA binding domain	-
OG5_190004/ Glyma01g36810	AT2G17770: *ATBZIP27*; transcription factor	-	-
	AT4G35900: *FD*; DNA binding/protein binding/transcription activator/transcription factor	-	CYCLIC-AMP-DEPENDENT TRANSCRIPTION FACTOR ATF 6

doi:10.1371/journal.pone.0038250.t003

Table 4: Arabidopsis flowering genes in OGs with no soybean genes members and their best BLAST-hit soybean genes

Arabidopsis genes	Symbol/annotation	Best BLAST-hit soybean gene	OG ID for soybean gene	Arabidopsis genes in OG
AT5G67180	AP2 domain-containing transcription factor, putative	Glyma17g18640	OG5_170666	*AP2*
AT2G35670	*FERTILIZATION INDEPENDENT SEED 2 (FIS2)*	Glyma01g41460	OG5_169591	*VRN2*
AT5G27220	protein transport protein-related	Glyma05g21790	OG5_170932*	AT5G48385
AT5G62040	*BFT (brother of FT and TFL1 protein)*	Glyma16g32080	OG5_146543	*FT;TSF;TFL1*
AT2G46790	*ARABIDOPSIS PSEUDO-RESPONSE REGULATOR 9 (APRR9)*	Glyma04g40640	OG5_178368	*APRR5*
AT4G34000	*ABSCISIC ACID RESPONSIVE ELEMENTS-BINDING FACTOR 3 (ABF3)*	Glyma02g14880	OG5_144915	*ABF1;ABF2;ABF4*
AT2G39250	*SCHNARCHZAPFEN (SNZ)*	Glyma15g04930	OG5_153389	*TOE1*
AT3G54990	*SCHLAFMUTZE (SMZ)*	Glyma15g04930	OG5_153389	*TOE1*
AT1G26790	Dof-type zinc finger domain-containing protein	Glyma18g49520	OG5_150317	*CDF3;CDF2*
AT1G69570	Dof-type zinc finger domain-containing protein	Glyma01g05960	OG5_150317	*CDF3;CDF2*
AT2G24790	*CONSTANS-LIKE 3 (COL3)*	Glyma06g06300	OG5_144994	*ATCOL5;ATCOL4*
AT2G47700	zinc finger (C3HC4-type RING finger) family protein	Glyma20g38050	OG5_178422?	AT3G05545
AT2G18870	*VERNALIZATION5/VIN3-LIKE (VEL3)*	Glyma07g09800	OG5_190237	*VRN5*
AT2G18880	*VERNALIZATION5/VIN3-LIKE (VEL2)*	Glyma17g07000	OG5_170648	*VEL1*
AT3G30260	*AGAMOUS-LIKE 79 (AGL79)*	Glyma16g13070	OG5_170388	*AP1;CAL*
AT4G16810	VEFS-Box of polycomb protein	Glyma01g41460	OG5_169591	*VRN2*
AT4G24540	*AGAMOUS-LIKE 24 (AGL24)*	Glyma01g02880	OG5_177438	*SVP*
AT5G42910	basic leucine zipper transcription factor (*BZIP15*)	Glyma04g04170	OG5_144915	*ABF1;ABF2;ABF4*
AT5G57380	*VERNALIZATION INSENSITIVE 3 (VIN3)*	Glyma17g07000	OG5_170648	*VEL1*
AT5G60120	*TARGET OF EARLY ACTIVATION TAGGED (EAT) 2 (TOE2)*	Glyma12g07800	OG5_153389	*TOE1*

*Not among the OGs containing Arabidopsis flowering genes.

doi:10.1371/journal.pone.0038250.t004

Genomic distribution of genes for paralogue-rich groups and one genomic region harbouring numerous homologues of specific Arabidopsis genes.

After we observed that some of the Arabidopsis flowering genes have more or less copies in soybean than average, we particularly expanded the investigation of paralogue-rich groups to the genomic scale. The 300 paralogue-rich OGs that have significantly more numbers of soybean genes (more than 2 standard deviations above the average) contain 4,236 soybean genes, which are spread evenly across the genome. On average, these genes comprise 6–16% of the total number of genes on each chromosome, and their fraction does not have any obvious correlation with the percentage of recently duplicated segments collected from the PHYTOZOME website (www.phytozome.net) (Pearson's Correlation Coefficient: -0.08) (Figure 7A). However, soybean chromosome 16 (Chr16) is exceptionally enriched for genes belonging to paralogue-rich OGs, containing 292 genes, which is 16.2% of the total of genes on Chr16. In particular, 76 of these 292 genes are condensed within a 1.4 Mb region on Chr16 (Figure 7B). These 76 genes are associated with only two OGs: OG5_134835 and OG5_170470. The numbers of total paralogues of these OGs in soybean are 115 and 20, respectively, while Arabidopsis has only one (AT2G34930) and two genes (AT2G44290 and AT2G44300) in OG5_134835 and OG5_170470, respectively. AT2G34930 in OG5_134835 is known as a disease-resistance family protein and is involved in a defence response to fungus and signal transduction (TAIR, www.arabidopsis.org/servlets/TairObject?id=32293&type=locus).

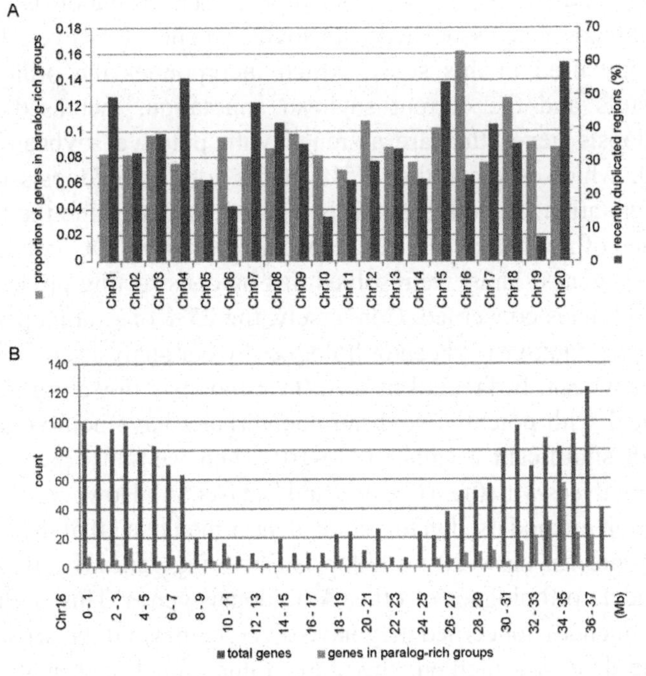

Figure 7: Distribution of genes in paralogue-rich groups in whole genome (A) and chromosome 16 (B).(A) Paralogue-rich orthologue groups and the percentage of recently duplicated segments per chromosome seems to have no apparent correlation.

(B) In chromosome 16, a large number of the genes in paralogue-rich groups, mostly from OG5_134835 or OG5_170470, are condensed in the region around 34-35Mb (see the main text).

doi:10.1371/journal.pone.0038250.g007

Two Arabidopsis genes in OG5_170470 (AT2G44290 and AT2G44300) are annotated as protease inhibitor/seed storage/lipid transfer protein (LTP) family proteins (AT2G44290: TAIR,www.arabidopsis.org/servlets/TairObject?id=33037&type=locus; AT2G44300: TAIR,www.arabidopsis.org/servlets/TairObject?id=33039&type=locus).

DISCUSSION

In this study, we identified potential soybean orthologues of most of the Arabidopsis genes with an emphasis on those that are involved in flowering pathways by associating soybean and Arabidopsis genes based on the current soybean annotation (G.max 1.09) in conjunction with the pre-defined Arabidopsis OGs in the OrthoMCL database [11]. Although the current soybean annotation assigned a closest Arabidopsis gene to nearly all of the predicted soybean genes, only 55% of the total Arabidopsis genes in the TAIR9 annotation are associated with soybean genes. However, the combined methodology used in this study, which incorporates the orthologue-group based results and the current soybean annotation, increased the number of Arabidopsis genes that are assigned with putative soybean orthologues to 20,730, which is more than 75% of the total Arabidopsis genes in the TAIR9 annotation. Nevertheless, it is difficult to determine the true soybean orthologues of Arabidopsis genes, especially when an OG contains multiple Arabidopsis genes. Thus, we resolved this by constructing phylogenetic trees for the candidates concerned. Conversely, the 25% of Arabidopsis genes that failed to have putative soybean orthologues by our analysis are not necessarily absent in soybean. In fact, all of the 20 Arabidopsis flowering genes that are not assigned with potential soybean counterparts have homologous soybean genes with significant e-values (<1e-10) when we subsequently performed BLASTP analyses with them (Table 4 and Dataset S4). However, the orthologue identification method used in this study determined that all of the top-matching soybean genes for these 20 Arabidopsis flowering genes (by BLASTP) were also candidate orthologues of other Arabidopsis genes. This finding suggests that the sequences concerned may have diverged beyond the sensitivity of our orthologue detection method; therefore, future functional analyses of these genes are necessary to confirm their orthology. In addition, it should be also noted that the false positive and false negative rates of OrthoMCL algorithm are 0.17 and 0.06, respectively, even though the OrthoMCL is among the best performing orthology identification tools [45]. Thus, failure of the orthologue identification for a subset of Arabidopsis genes may also due to the error in OrthoMCL-DB.

Using an approach that associates a group of soybean genes with a group of Arabidopsis genes that has the same OG ID also enabled us to investigate which soybean genes have a statistically higher or lower number of paralogues (or copies) in comparison to their Arabidopsis counterparts. For example, the analysis of paralogue-rich and paralogue-less soybean genes can be expanded to whole genes. Based on the near log-normal distribution of soybean-Arabidopsis gene count ratio per OG (mean: 1.03 and standard-deviation: 0.88 in $\log_2$), the number of soybean genes are much higher for 300 OGs (7 time or more than Arabidopsis genes) and smaller for 304 OGs (half or less than Arabidopsis genes) (Figure 5). A preliminary Gene Ontology analysis on the Arabidopsis genes in paralogue-rich and paralogue-less OGs (i.e., those that have far more and less copies in soybean than average, respectively) showed interesting features. Three hundred and ninety nine (399) Arabidopsis genes in 300 paralogue-rich OGs are enriched for 'response to auxin stimulus (GO:0009733)', 'defense response (GO:0006952)', 'response to wounding (GO:0009611)' and 'lipid transport (GO:0006869)' (Table S4). While the 1,946 Arabidopsis genes that have much less copies in soybean (paralogue-less OGs) are also enriched for 'defense response (GO:0006952)', they are mainly enriched for other GO terms, such as 'intracellular signaling cascade (GO:0007242)' (Table S4). The outcome raises questions as to whether the differential accumulation of gene copies between soybean and Arabidopsis is a possible evolutionary innovation that distinguishes the two species from one another.

It is known that duplicated genes (i.e., genes in the same OG) do not necessarily retain the same functions. In our analysis, we found paralogues with diverged expression patterns as well as some more highly related paralogues located within the same clades in the phylogenetic tree and exhibiting similar expression patterns and, hence, more likely to retain similar functions (Figure 6 and Figure 8). Intriguingly, there exist some OGs that consist of soybean genes that are all non-tandem duplicates but with similar expression patterns (for example OG5_150317 inFigure 8). This begs the question as to how the soybean genome keeps copies of the same genes with presumably the same function in different places in the genome, especially because we observed no correlation between the fraction of genes in paralogue-rich OGs in each chromosome and the percentage of recently duplicated segments in the respective chromosomes (Figure 7A).

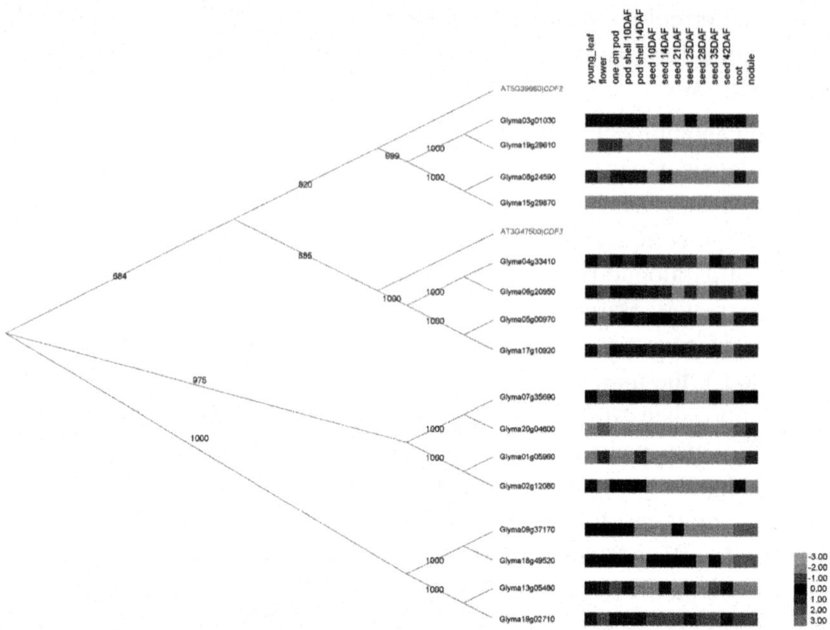

Figure 8: Expression patterns of soybean genes in OG5_150317.While the expression patterns of the 16 soybean genes in OG5_150317 were spread in different clusters as shown in Figure 6, but those that are in the same clade of the phylogenetic tree tend to have similar expression profiles. The expression data were extracted from the soybean transcriptome data in SoyBase [8]. DAF: Days After Flowering.

doi:10.1371/journal.pone.0038250.g008.

Soybean is a short-day species that does not require vernalisation to induce flowering [46]. It is, therefore, intriguing to observe that some of the Arabidopsis genes involved in the vernalisation pathway (e.g., FLC, VRN1 and VRN2) are represented in soybean. A previous study failed to identify any FLC genes, a key regulator of the vernalisation pathway, in Medicago truncatula, soybean, or Lotus japonicus [18], which can be attributed to the more incomplete genome sequence at the time; however, our analysis grouped Arabidopsis FLC along with 5 other Arabidopsis genes and one soybean gene (Glyma05 g28130) into the same OG (OG5_139532), in which the soybean gene and FLC have the closest relationship (Figure 4B). VRN1 forms part of a chromatin-modifying polycomb that is involved in the methylation of histone 3 lysine 9 (H3K9) and histone 3 lysine 27 (H3K27) and, hence, the repression of FLCexpression [47], [48]. VRN1 belongs to a paralogue-rich OG that has far more soybean orthologue candidates than the majority of Arabidopsis genes, although other Arabidopsis genes that mediate vernalisation responsiveness, such as VIN3, are not assigned any putative soybean orthologues. SVP in OG5_177438, which encodes a MADS box transcription factor and is a negative regulator of flowering in Arabidopsis [49], has 8 putative soybean orthologues, while its close homologue AGL24 is not assigned a candidate soybean orthologue. It is likely that the soybean genes in OG5_177438 may have acquired and replaced the function of AGL24 because the most homologous soybean gene of Arabidopsis AGL24 is one of the putative soybean orthologues of Arabidopsis SVP (Table 4).

Two microRNAs (miRNAs), miR156 and miR172, are involved in the regulation of SQUAMOSA PROMOTER BINDING PROTEIN LIKE family genes and APETALA2-LIKE (AP2-like) transcription factors in Arabidopsis [50] and are conserved in soybean [51]. Zhang et al. also concluded that the SPL family genes and AP2-like transcription factors are among the predicted target genes of miR156 and miR172, respectively, in soybean [51]. However, whether or not miR172 plays a similar role in soybean flowering still needs to be investigated.

The functional analysis of genes through reverse genetics approaches is more complicated for soybean than Arabidopsis. Computational analyses such as our study can therefore pinpoint the putative soybean orthologues of Arabidopsis genes with known functions. Indeed, putative soybean orthologues of FT and TFL1 were first identified via computational analysis [18], [52]. This study determines all of the soybean genes that are putative orthologues of Arabidopsis genes by first grouping Arabidopsis and soybean genes into pre-defined orthologue groups and then associating genes in the same group from each species. This method determined not only the inter-species relationship of genes between soybean and Arabidopsis but also the intra-species relationships of genes in terms of their sequence similarities. Subsequently, the most probable soybean orthologues for Arabidopsis genes, especially those involved in key flowering pathways, were inferred through phylogenetic tree analyses. These inferences were also strengthened by referring to publicly available transcriptome datasets for the expression profile of the soybean genes. As more soybean transcriptome data becomes available in the future, it is expected that the putative soybean orthologues of Arabidopsis flowering genes can be more precisely compared by combining the data on their sequence similarities and expression patterns. Additionally, our methods found that 24 Arabidopsis genes, which had not been previously investigated for their roles in flowering, belong to OGs with known Arabidopsis flowering genes. Although sequence similarity does not always indicate functional similarity, these Arabidopsis genes may well be involved in the initiation of flowering. In summary, our study has identified numerous floral regulatory candidate genes in soybean.

Further studies of the genes identified here will provide a new perspective on the molecular processes underlying the floral transition process in soybean.

Materials and Methods

Sequence Data

The protein sequences of annotated soybean genes (G. max 1.09) were downloaded from the PHYTOZOME website (www.phytozome.net). The protein sequences of annotated Arabidopsis genes (TAIR9 release) were downloaded from The Arabidopsis Information Resource (TAIR) website (www.arabidopsis.org). The protein sequences of Medicago genes (Mt3.5 version 3) were downloaded from J. Craig Venter Institute website (www.jcvi.org). For those that have splicing variants, the longest isoforms were selected. Peptide sequences of Brachypodium distachyon were downloaded from PHYTOZOME website, then the longest peptide sequence for each locus was extracted for the analysis [41]. Peptide sequence of filtered Arabidopsis lyrata gene model which best represent each locus is downloaded from PHYTOZOME website[53].

Soybean Homologue Identification and Grouping Orthologous Genes

The current soybean annotation information (G.max 1.09) associates 44,818 out of 46,367 soybean genes to 15,113 Arabidopsis genes. In order to identify soybean orthologues of more Arabidopsis genes, we assigned orthologue group (OG) IDs pre-defined in OrthoMCL database (release 5.0) [11] to Arabidopsis genes. OrthoMCL-DB is a list of OG IDs and the genes under same OG IDs from multiple species. OrthoMCL algorithm examines the all-versus-all BLAST search result and use the Markov Clustering algorithm to find interspecies homologues (orthologues) and intraspecies homologues (paralogues) [11]. Soybean genes that have homologous Arabidopsis genes in the current annotation information were given the same OG ID with the corresponding Arabidopsis genes, and those that do not were assigned OG IDs by OrthoMCL. An OG IDs was assigned to each gene, if possible, via the web-based tool in the OrthoMCL website (http://www.orthomcl.org), which considers the BLAST-hit quality of the input protein sequence to the protein sequences in the OrthoMCL-DB to find the OG containing the closest protein to the input. For the 317 Arabidopsis genes that have homologous soybean genes according to the current soybean annotation but do not have OG IDs assigned by OrthoMCL, arbitrary OG IDs were given, which are 'OG5_' followed by the Arabidopsis gene locus name (e.g., OG5_AT2G33835 for an Arabidopsis gene AT2G33835). Among these, 9 groups have Arabidopsis flowering genes, which are AT2G33835, AT2G18880, AT2G18870, AT3G30260, AT4G16810, AT4G24540, AT5G57380, AT5G42910 and AT5G60120. Then, for each OG, the soybean gene members were regarded as putative soybean orthologues of Arabidopsis gene members.

Transcriptional Activity and Gene Expression Data Analysis

The expression profiles of the 491 putative soybean orthologues of Arabidopsis flowering genes were extracted from SoyBase [8] and the integrated transcriptome atlas of soybean generated by Libault et al. (2010) [9]. For the SoyBase data, a transcriptionally active gene is a gene that has at least two or more sequence reads at one or more of the tested tissues/developmental stages. For the dataset of Libault et al. (2010) [9] any gene that has normalized read counts greater than 0 at least in one tissue/developmental stage is regarded as a transcriptionally active gene. The hierarchical clustering of the expression patterns was performed by CLUSTER 3.0 using the Z-scores of the normalized read counts (bonsai.ims.u-tokyo.ac.jp/~mdehoon/software/cluster), and the clustered results were visualized by Java Treeview (http://rana.lbl.gov/downloads/TreeView/TreeView_vers_1_60.exe).

Multiple Alignment and Phylogenetic Tree Generation

Multiple sequence alignment was carried out by MUSCLE 3.8.31 [54], and phylogenetic trees were generated by CLUSTALW 2.0.12 [55] with the bootstrap option. Dendroscope 2.7.4 [56]was used for the graphical representation of the phylogenetic trees.

Identification of Closer Soybean Orthologues of Arabidopsis Flowering Genes within OGs

The phylogenetic tree information for between all Arabidopsis and soybean genes within the same OG in Newick format was parsed to decide which soybean genes are closer homologues to Arabidopsis genes involved in flowering pathways using in-house Python scripts. Any soybean gene located in the same clade with Arabidopsis flowering genes is regarded as a soybean homologue involved in flowering pathways, unless the soybean gene is equally close or closer to another Arabidopsis gene that are not explicitly involved in flowering pathways. For the small OGs (less than 4 sequences in total), all soybean genes are regarded as close orthologues of corresponding Arabidopsis flowering genes if the OG has only Arabidopsis flowering genes.

AUTHOR CONTRIBUTIONS

Conceived and designed the experiments: CJ PLB MBS. Performed the experiments: CJ. Analyzed the data: CJ. Contributed reagents/materials/ analysis tools: CJ. Wrote the paper: CJ CEW PLB MBS.

REFERENCES

1. Fornara F, de Montaigu A, Coupland G (2010) SnapShot: Control of flowering in Arabidopsis. Cell 141: 550, 550 e551–552:
2. Parcy F (2005) Flowering: a time for integration. Int J Dev Biol 49: 585–593.
3. Simpson GG, Dean C (2002) Arabidopsis, the Rosetta stone of flowering time? Science 296: 285–289.
4. Washburn CF, Thomas JF (2000) Reversion of flowering in Glycine Max (Fabaceae). Am J Bot 87: 1425–1438.
5. Tucker SC (2003) Floral development in legumes. Plant Physiol 131: 911–926.
6. .Kong F, Liu B, Xia Z, Sato S, Kim BM, et al. (2010) Two coordinately regulated homologs of LOWERING LOCUS T are involved in the control of photoperiodic flowering in soybean. Plant Physiol 154: 1220–1231.
7. Zhang Q, Li H, Li R, Hu R, Fan C, et al. (2008) Association of the circadian rhythmic expression of GmCRY1a with a latitudinal cline in photoperiodic flowering of soybean. Proc Natl Acad Sci U S A 105: 21028–21033.
8. .Severin AJ, Woody JL, Bolon YT, Joseph B, Diers BW, et al. (2010) RNA-Seq Atlas of Glycine max: A guide to the soybean transcriptome. BMC Plant Biol 10: 160.
9. .Libault M, Farmer A, Joshi T, Takahashi K, Langley RJ, et al. (2010) An integrated transcriptome atlas of the crop model Glycine max, and its use in comparative analyses in plants. Plant J 63: 86–99.
10. .Schmutz J, Cannon SB, Schlueter J, Ma J, Mitros T, et al. (2010) Genome sequence of the palaeopolyploid soybean. Nature 463: 178–183.
11. Li L, Stoeckert CJ Jr, Roos DS (2003) OrthoMCL: identification of ortholog groups for eukaryotic genomes. Genome Res 13: 2178–2189.
12. .Higgins JA, Bailey PC, Laurie DA (2010) Comparative genomics of flowering time pathways using Brachypodium distachyon as a model for the temperate grasses. PLoS One 5: e10065.
13. .Barton MK, Poethig RS (1993) Formation of the shoot apical meristem

in Arabidopsis thaliana - an analysis of development in the wild-type and in the shoot meristemless mutant. Development 119: 823–831.

14. Endrizzi K, Moussian B, Haecker A, Levin JZ, Laux T (1996) The SHOOT MERISTEMLESS gene is required for maintenance of undifferentiated cells in Arabidopsis shoot and floral meristems and acts at a different regulatory level than the meristem genes WUSCHEL and ZWILLE. Plant J 10: 967–979.
15. Laux T, Mayer KF, Berger J, Jurgens G (1996) The WUSCHEL gene is required for shoot and floral meristem integrity in Arabidopsis. Development 122: 87–96.
16. .Mayer KF, Schoof H, Haecker A, Lenhard M, Jurgens G, et al. (1998) Role of WUSCHEL in regulating stem cell fate in the Arabidopsis shoot meristem. Cell 95: 805–815.
17. .Robles P, Pelaz S (2005) Flower and fruit development in Arabidopsis thaliana. Int J Dev Biol 49: 633–643.
18. .Hecht V, Foucher F, Ferrandiz C, Macknight R, Navarro C, et al. (2005) Conservation of Arabidopsis flowering genes in model legumes. Plant Physiol 137: 1420–1434.
19. .Koo SC, Bracko O, Park MS, Schwab R, Chun HJ, et al. (2010) Control of lateral organ development and flowering time by the Arabidopsis thaliana MADS-box Gene AGAMOUS-LIKE6. Plant J 62: 807–816.
20. .Nakaminami K, Hill K, Perry SE, Sentoku N, Long JA, et al. (2009) Arabidopsis cold shock domain proteins: relationships to floral and silique development. J Exp Bot 60: 1047–1062.
21. .Chen M, Ni M (2006) RFI2, a RING-domain zinc finger protein, negatively regulates CONSTANS expression and photoperiodic flowering. Plant J 46: 823–833.
22. .Han P, Garcia-Ponce B, Fonseca-Salazar G, Alvarez-Buylla ER, Yu H (2008) AGAMOUS-LIKE 17, a novel flowering promoter, acts in a FT-independent photoperiod pathway. Plant J 55: 253–265.
23. Liscum E, Hodgson DW, Campbell TJ (2003) Blue light signaling through the cryptochromes and phototropins. So that's what the blues is all about. Plant Physiol 133: 1429–1436.
24. Alabadi D, Oyama T, Yanovsky MJ, Harmon FG, Mas P, et al. (2001) Reciprocal regulation between TOC1 and LHY/CCA1 within the Arabidopsis circadian clock. Science 293: 880–883.
25. Michaels SD, Amasino RM (1999) FLOWERING LOCUS C encodes a novel MADS domain protein that acts as a repressor of flowering. Plant

Cell 11: 949–956.

26. Wood CC, Robertson M, Tanner G, Peacock WJ, Dennis ES, et al. (2006) The Arabidopsis thaliana vernalization response requires a polycomb-like protein complex that also includes VERNALIZATION INSENSITIVE 3. Proc Natl Acad Sci U S A 103: 14631–14636.

27. Yu H, Xu Y, Tan EL, Kumar PP (2002) AGAMOUS-LIKE 24, a dosage-dependent mediator of the flowering signals. Proc Natl Acad Sci U S A 99: 16336–16341.

28. Michaels SD, Ditta G, Gustafson-Brown C, Pelaz S, Yanofsky M, et al. (2003) AGL24 acts as a promoter of flowering in Arabidopsis and is positively regulated by vernalization. Plant J 33: 867–874.

29. Liu C, Chen H, Er HL, Soo HM, Kumar PP, et al. (2008) Direct interaction of AGL24 and SOC1 integrates flowering signals in Arabidopsis. Development 135: 1481–1491.

30. Alexandre CM, Hennig L (2008) FLC or not FLC: the other side of vernalization. J Exp Bot 59: 1127–1135.

31. Amasino R (2010) Seasonal and developmental timing of flowering. Plant J 61: 1001–1013.

32. Mockler TC, Yu X, Shalitin D, Parikh D, Michael TP, et al. (2004) Regulation of flowering time in Arabidopsis by K homology domain proteins. Proc Natl Acad Sci U S A 101: 12759–12764.

33. Veley KM, Michaels SD (2008) Functional redundancy and new roles for genes of the autonomous floral-promotion pathway. Plant Physiol 147: 682–695.

34. Lee JH, Yoo SJ, Park SH, Hwang I, Lee JS, et al. (2007) Role of SVP in the control of flowering time by ambient temperature in Arabidopsis. Genes Dev 21: 397–402.

35. Abe M, Kobayashi Y, Yamamoto S, Daimon Y, Yamaguchi A, et al. (2005) FD, a bZIP protein mediating signals from the floral pathway integrator FT at the shoot apex. Science 309: 1052–1056.

36. Yamaguchi A, Wu MF, Yang L, Wu G, Poethig RS, et al. (2009) The microRNA-regulated SBP-Box transcription factor SPL3 is a direct upstream activator of LEAFY, FRUITFULL, and APETALA1. Dev Cell 17: 268–278.

37. Gu Q, Ferrandiz C, Yanofsky MF, Martienssen R (1998) The FRUITFULL MADS-box gene mediates cell differentiation during Arabidopsis fruit development. Development 125: 1509–1517.

38. Mandel MA, Yanofsky MF (1995) The Arabidopsis AGL8 MADS box

gene is expressed in inflorescence meristems and is negatively regulated by APETALA1. Plant Cell 7: 1763–1771.

39. .Hecht V, Laurie RE, Vander Schoor JK, Ridge S, Knowles CL, et al. (2011) The Pea GIGAS Gene Is a FLOWERING LOCUS T Homolog Necessary for Graft-Transmissible Specification of Flowering but Not for Responsiveness to Photoperiod. Plant Cell 23: 147–161.

40. Ferl RJ, Manak MS, Reyes MF (2002) The 14–3-3s. Genome Biol 3: REVIEWS3010.1–7.

41. International_Brachypodium_Initiative (2010) Genome sequencing and analysis of the model grass Brachypodium distachyon. Nature 463: 763–768.

42. Kim MY, Lee S, Van K, Kim TH, Jeong SC, et al. (2010) Whole-genome sequencing and intensive analysis of the undomesticated soybean (Glycine soja Sieb. and Zucc.) genome. Proc Natl Acad Sci U S A 107: 22032–7.

43. Greb T, Mylne JS, Crevillen P, Geraldo N, An H, et al. (2007) The PHD finger protein VRN5 functions in the epigenetic silencing of Arabidopsis FLC. Curr Biol 17: 73–78.

44. Sung S, Schmitz RJ, Amasino RM (2006) A PHD finger protein involved in both the vernalization and photoperiod pathways in Arabidopsis. Genes Dev 20: 3244–3248.

45. Chen F, Mackey AJ, Vermunt JK, Roos DS (2007) Assessing performance of orthology detection strategies applied to eukaryotic genomes. PLoS One 2: e383.

46. .Summerfield RJ, Roberts EH (1985) Glycine max; Halevy AH, editor. Boca Raton, Florida: CRC Press. pp. 139–148.

47. Levy YY, Mesnage S, Mylne JS, Gendall AR, Dean C (2002) Multiple roles of Arabidopsis VRN1 in vernalization and flowering time control. Science 297: 243–246.

48. Sung S, Amasino RM (2004) Vernalization in Arabidopsis thaliana is mediated by the PHD finger protein VIN3. Nature 427: 159–164.

49. Hartmann U, Hohmann S, Nettesheim K, Wisman E, Saedler H, et al. (2000) Molecular cloning of SVP: a negative regulator of the floral transition in Arabidopsis. Plant J 21: 351–360.

50. Wu G, Park MY, Conway SR, Wang JW, Weigel D, et al. (2009) The sequential action of miR156 and miR172 regulates developmental timing in Arabidopsis. Cell 138: 750–759.

51. Zhang B, Pan X, Stellwag EJ (2008) Identification of soybean microRNAs

and their targets. Planta 229: 161–182.

52. Tian Z, Wang X, Lee R, Li Y, Specht JE, et al. (2010) Artificial selection for determinate growth habit in soybean. Proc Natl Acad Sci U S A 107: 8563–8568.

53. Hu TT, Pattyn P, Bakker EG, Cao J, Cheng JF, et al. (2011) The Arabidopsis lyrata genome sequence and the basis of rapid genome size change. Nat Genet 43: 476–481.

54. Edgar RC (2004) MUSCLE: multiple sequence alignment with high accuracy and high throughput. Nucleic Acids Res 32: 1792–1797.

55. Larkin MA, Blackshields G, Brown NP, Chenna R, McGettigan PA, et al. (2007) Clustal W and Clustal X version 2.0. Bioinformatics 23: 2947–2948.

56. Huson DH, Richter DC, Rausch C, Dezulian T, Franz M, et al. (2007) Dendroscope: An interactive viewer for large phylogenetic trees. BMC Bioinformatics 8: 460.

Chapter 13

GENOME-WIDE ANALYSIS OF THE NADK GENE FAMILY IN PLANTS

Wen-Yan Li, Xiang Wang, Ri Li, Wen-Qiang Li, Kun-Ming Chen

State Key Laboratory of Crop Stress Biology in Arid Areas, College of Life Sciences, Northwest A&F University, Yangling, Shaanxi, China

ABSTRACT

Background

NAD(H) kinase (NADK) is the key enzyme that catalyzes *de novo* synthesis of NADP(H) from NAD(H) for NADP(H)-based metabolic pathways. In plants, NADKs form functional subfamilies. Studies of these families in *Arabidopsis thaliana* indicate that they have undergone considerable evolutionary selection; however, the detailed evolutionary history and functions of the various NADKs in plants are not clearly understood.

Principal Findings

We performed a comparative genomic analysis that identified 74 NADK gene homologs from 24 species representing the eight major plant lineages within the supergroup Plantae: glaucophytes, rhodophytes, chlorophytes, bryophytes, lycophytes, gymnosperms, monocots and eudicots. Phylogenetic and structural analysis classified these NADK genes into four well-conserved subfamilies with considerable variety in the domain organization and gene structure among subfamily members. In addition to the typical NAD_kinase domain, additional domains, such as adenylate kinase, dual-specificity phosphatase, and protein tyrosine phosphatase catalytic domains, were found in subfamily II. Interestingly, NADKs in subfamily III exhibited low sequence similarity (~30%) in the kinase domain within the subfamily and with the other subfamilies. These observations suggest that gene fusion and exon shuffling may have occurred after gene duplication, leading to specific domain organization seen in subfamilies II and III, respectively. Further analysis of the exon/intron structures showed that single intron loss and gain had occurred, yielding the diversified gene structures, during the process of structural evolution of NADK family genes. Finally, both available global microarray data analysis and qRT-RCR experiments revealed that the NADK genes in *Arabidopsis* and *Oryza sativa* show different expression patterns in different developmental stages and under several different abiotic/biotic stresses and hormone treatments, underscoring the functional diversity and functional divergence of the NADK family in plants.

Conclusions

These findings will facilitate further studies of the NADK family and provide valuable information for functional validation of this family in plants.

INTRODUCTION

NAD(H) and NADP(H) are crucial coenzymes and play important and distinguishable roles in all living organisms [1]. NAD(H) is primarily involved in catabolic reactions, whereas NADP(H) participates in anabolic reactions, such as NADP(H)-dependent reductive anabolic pathways, signal transduction, and defense against oxidative stress [1], [2]. Hence, regulation of the intracellular balance of NAD(H) and NADP(H) is critical. NAD(H) kinase (NADK) is the key enzyme in the *de novo* biosynthesis of NADP(H), catalyzing the transfer of a phosphoryl group from ATP to NAD(H), and thus plays an important role in the regulation of intracellular NAD(H)/NADP(H) balance for NADP(H)-based metabolic pathways.

NADK genes have been found in nearly all living organisms, including Archaea, eubacteria and eukaryotes, except for the intracellular parasite *Chlamydia trachomatis* [3]. NADK genes have been cloned from a wide variety of species, including Archaea (*Methanococcus jannaschii* [4]), eubacteria (*Mycobacterium tuberculosis* [5]), *Escherichia coli* [6], yeast (*Saccharomyces cerevisiae* [7], [8]), humans (*Homo sapiens* [9]) and plants (*Arabidopsis thaliana* [10], [11],[12]). Moreover, the number of NADKs in different organisms varies, with most prokaryotic organisms, such as Archaea and eubacteria, having only one NADK, whereas most eukaryotic organisms, such as yeast and plants, have several NADKs. For example, *Euglena gracilis* [13]has two NADKs and *S. cerevisiae* [14], [15] and *Arabidopsis* have three [10], [11], [16].

The enzymatic properties of natural or recombinant NADKs from several organisms, including their substrate specificity and structural properties, especially of the active site, have been well characterized [2], [9], [17], [18], [19].

All characterized NADKs are homomultimers and exhibit species-dependent phosphoryl donor and acceptor specificity. Hence, these enzymes are classified into two groups: ATP-NAD kinases and inorganic polyphosphate poly(P)/ ATP-NAD kinases, based on the phosphoryl donor specificity [2]. For example, NADKs from gram-positive bacteria (*M. tuberculosis*) and Archaea (*M. jannaschii*) utilize both ATP and poly(P) as phosphoryl donors and they phosphorylate NAD^+/NADH, whereas NADKs from gram-negative bacteria (*E. coli*) and eukaryotes (*S. cerevisiae*, *Arabidopsis* and human) utilize ATP but not poly(P), and phosphorylate NAD^+ and NAD^+/NADH, respectively [2]. In addition, analysis of the sequences and crystal structures of NADKs identified three highly conserved and functionally important motifs within the NADK family (a GGDG motif, an NE/D motif and a Gly-rich motif) and a possible phosphate transfer mechanism [20], [21], [22], [23], [24].

The biological significance and physiological functions of NADKs have been characterized from studies of several organisms. For example, mutation of the single NADK gene in *M. tuberculosis* and *Salmonella enterica* is lethal [3], [4]. Similarly, mutation of all three NADK genes in *S. cerevisiae* (*utr1/yef1/pos5*) or two of the three genes (*utr1/pos5*) also causes lethality, whereas the respective NADK gene mutations are not lethal in *Arabidopsis* [11], [16],[25], [26]. In addition, studies of the physiological functions of NADKs in *S. cerevisiae* and humans revealed that NADKs play a major role in protecting living cells against oxidative stress[8], [15], [26], [27] because NADPH is vital in the intracellular anti-oxidative defense system of most organisms [28].

In plants, NADKs are involved in regulating redox balance, biotic and abiotic stress responses and various developmental processes. Notably, NADK was the first protein in plants demonstrated to be regulated by the calcium-sensing protein calmodulin (CaM) [29], and plant NADKs are divided into CaM-independent and CaM-regulated isoforms. CaM-dependent NADK is essential for survival of plants under difficult conditions and for protecting plants against invading pathogens by helping to provide reductants for the NADPH-dependent oxidative burst [29], [30], [31]. For example, CaM-dependent NADK activity, but not CaM-independent NADK activity, increases under cold stress in green bean [32], and decreased in response to high salinity and drought in tomato [33] and wheat [34]. In *Arabidopsis*, three genes encoding NADK (*NADK1/NADK2/NADK3*) have been identified and their physiological functions have been characterized [2], [35]. *AtNADK1* is located in the cytosol and expressed mainly in roots, *AtNADK2* is located in the chloroplasts and expressed mainly in leaves, whereas *AtNADK3* is located in the peroxisome and is strongly expressed in reproductive tissues, such as the stigma, pollen and carpel vasculature [10], [11], [36]. The *AtNADK1*-deficient mutant exhibits sensitivity to γ-irradiation and paraquat-induced oxidative stress [16]. The *AtNADK2*-deletion mutant displays hypersensitivity to environmental stresses that induce oxidative stress, such as UVB-irradiation, drought, heat shock and high salinity [11]. Similarly, the *AtNADK3*-null mutant shows hypersensitivity to oxidative stress, including methyl viologen, high salinity and osmotic shock [10]. Moreover, AtNADK2 plays a vital role in chlorophyll synthesis and protects chloroplasts against oxidative damage [11]. In addition, plants growth and fertility are affected in the NADK-deficient mutants [2], [11], [37].

Although these studies in bacteria, yeast, human and *Arabidopsis* have led to an understanding of the biochemical properties and physiological functions of NADKs, there has been no systematic study of the evolution and functional divergence of the NADK gene family, especially in Plantae. Here, we performed a comprehensive analysis of the NADK gene family in 24 species, representing the eight major plant lineages within the supergroup Plantae. Phylogenetic analysis was performed to delineate the evolutionary history of the NADK family in Plantae, and exon/intron structure analysis was performed to gain insight into the possible mechanisms of the structural diversity of NADK gene family. Finally, the tissue-specificity and inducibility of NADK gene expression in *Arabidopsis* and rice (*Oryza sativa*) were characterized by examining publicly available microarray datasets and by qRT-PCR experiments. The results obtained here will broaden our understanding of the roles of plant NADKs and provide a framework for further functional investigations of these genes in plants.

RESULTS

Identification of NADK family members in plants

To comprehensively investigate and characterize the NADK gene family in plants, 24 species representing the eight major plant lineages within the supergroup Plantae, were selected for analysis (Figure 1). A hidden Markov model (HMM) search was performed with the obtained sequences and 74 NADK homologs were identified (Figure 1, Table S1). Except for the gymnosperm *Picea sitchensis*, for which the full genome sequence was not yet available, two or more NADK genes were identified in each genome of the selected species (Figure 1). Most of the aquatic algae, including glaucophytes, rhodophytes and chlorophytes, contained two NADK genes per genome and only three species in the chlorophytes, *Micromonas pusilla* RCC299,*Chlamydomonas reinhardtii* and *Volvox carteri*, carried three NADK genes per genome (Figure 1). In contrast, three or more NADK genes were found in all land plants, including bryophytes, lycophytes, monocots and eudicots (Figure 1). In addition, only one NADK gene was identified per genome in the cyanophytes (*Acaryochloris marina* MBIC11017 and *Prochlorococcus marinus* MIT 9301) and bacteria (*E. coli* K-12), outgroup (Figure 1).

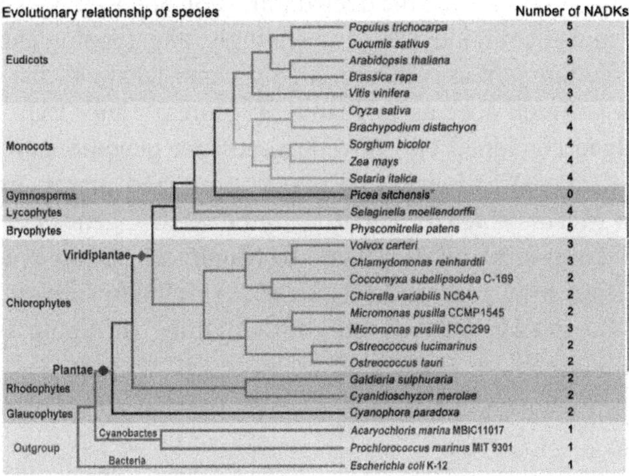

Figure 1: Systematic evolutionary relationships of 24 species among eight lineages within the supergroup Plantae.The numbers of NADK homologs in each species are listed next to the tree. *, the genome sequencing of *Picea sitchensis* is not complete.

doi:10.1371/journal.pone.0101051.g001.

The Pfam and SMART databases were used to analysis the functional domains of the identified NADK candidates. All of the putative NADKs possess a typical NAD_kinase domain (Pfam accession number PF01513) and some also contain other functional domains such as an adenylate kinase domain (ADK; PF00406), a dual-specificity phosphatase, catalytic domain (DSPc; PF00782), and/or a protein tyrosine phosphatase, catalytic domain/ DSPc domain (PTPc/DSPc; SMART accession number SM000012). Only one NADK candidate, CpNADK2, which did not contain a complete NAD_kinase domain, was excluded from the following analysis.

Phylogenetic analysis and classification of the NADK family

To explore the phylogenetic relationships among NADK family members in plants, we first generated a rooted maximum-likelihood phylogenetic tree with the 73 NADKs from the 24 species (Figure 2A), which was inferred from the amino acid sequences of their NAD_kinase domains (Figure S1). Furthermore, phylogenetic trees reconstructed by the neighbor joining, minimum evolution and maximum parsimony methods showed very similar topologies to the maximum-likelihood tree (data not shown). Using *EcNADK1*, *AmNADK1* and *PmNADK1* from bacteria and cyanophytes as the outgroup, all NADK homologs in plants can be classified into four well-conserved subfamilies (I–IV; Figure 2A) with high statistical support, according to the topology and the deep duplication nodes of NADK paralogues in the maximum-likelihood tree. Interestingly, the topological relationship of members within subfamilies was highly consistent with the evolutionary relationships between species in Plantae (Figures 1 and 2A). The majority of aquatic algae contained two NADK genes per genome and grouped into subfamilies II and IV, whereas the majority of land plants, except for the gymnosperm *Picea sitchensis*, carried ≥ three NADK genes per genome and were clustered into subfamilies I, II and III (Figure 2A). Moreover, only one of the NADK genes from green algae (*CrNADK3*) fell into subfamily I and only two (*VcNADK3* and *MpNADK3*) fell into subfamily III (Figure 2A).

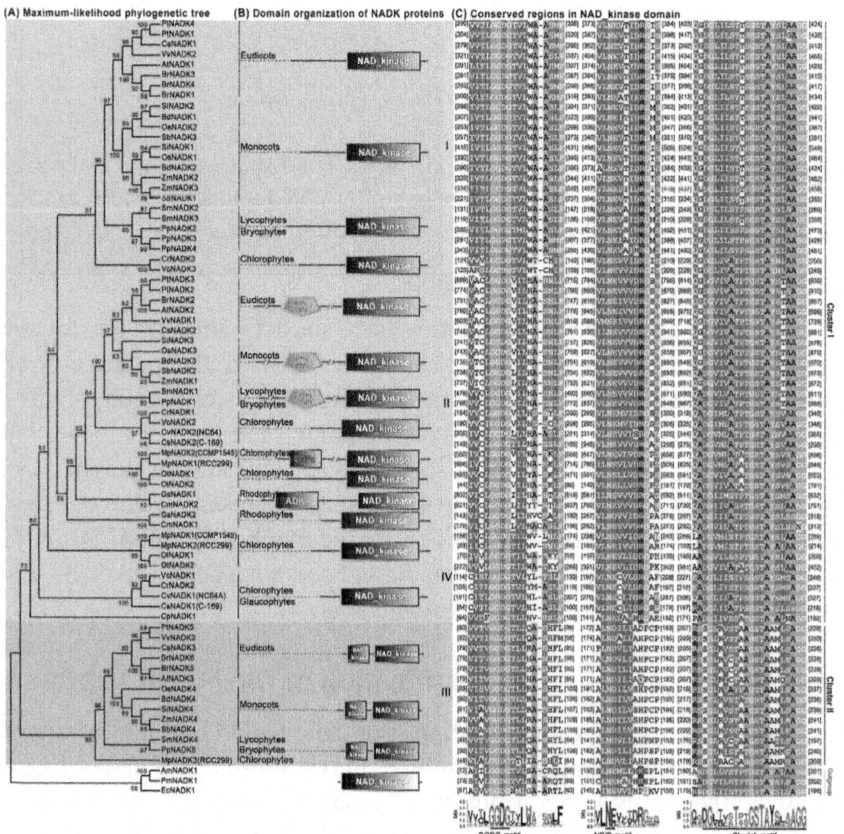

Figure 2: Phylogenetic relationships and domain organization of NADK genes in plants.(**A**) The rooted maximum-likelihood phylogenetic tree of NADK family mem-

bers was inferred from the amino acid sequence alignment of the NAD_kinase domain. Numbers above the nodes represent bootstrap values from 1000 replications. (**B**) Domain organization of the NADKs. (**C**) Amino acid sequence alignment of conserved motifs within the NAD_kinase domain.

doi:10.1371/journal.pone.0101051.g002

Examination of the chromosomal locations of the NADK family genes in the genomes of algae (*Ostreococcus lucimarinus*, *Ostreococcus tauri*, *C. reinhardtii* and *V. carteri*), bryophytes (*Physcomitrella patens*), lycophytes (*Selaginella moellendorffii*), monocots (*Brachypodium distachyon*, rice, *Sorghum bicolor* and *Zea mays*) and eudicots (*Arabidopsis*, *Brassica rapa*,*Populus trichocarpa* and *Vitis vinifera*) showed that NADK genes are randomly distributed (Figure S3). Moreover, a search for NADK paralogs using the Plant Genome Duplication Database (PGDD; http://chibba.agtec.uga.edu/duplication/) [38] revealed eight paralogous gene pairs in *P. patens*, rice, *Z. mays*, *B. rapa* and *P. trichocarpa* (Figure S4A), but none in the other species.

To further explore association of positive selection with duplication and divergence of NADK family genes, the rate of non-synonymous substitution (Ka), synonymous substitution (Ks) and the Ka/Ks ratios were calculated for the eight paralogous gene pairs and used to estimate duplication and divergence times. The Ka/Ks ratios varied from 0.18 to 0.39 among the five different species (Figure S4B). The fact that the Ka/Ks ratios were <1 suggests that the NADK family genes have undergone strong negative selection pressure, and the duplication event was estimated to have occurred ~13.8–67.6 million years ago. Divergence of *P. patens*, rice, *Z. mays*, *P. trichocarpa* and *B. rapa* was estimated to have occurred 67.6, 61.5, 13.8, 20.8 and 23.1–33.1 million years ago, respectively (Figure S4B).

Further analysis of the functional domains showed that domain organization of NADKs in the different subfamilies varied considerably (Figure 2B). All of the identified NADKs contained a typical NAD_kinase domain at the C terminus, but the proteins in subfamily II also carried the additional N-terminal catalytic domains noted earlier (DSPc, PTPc/DSPc or ADK; Figure 2B). Interestingly, the NAD_kinase domain of proteins in subfamily III was not only divided into two parts, but also exhibited low sequence similarity (~30%) between subfamily III members and the other subfamilies (Figures 2C and S1). In addition, based on the NAD_kinase domain sequence similarity, the NADK gene subfamilies could be further divided into two clusters: cluster I (subfamilies I, II and IV) and cluster II (subfamily III).

More detailed analysis of the NAD_kinase domains of the 73 NADKs revealed three highly conserved and functionally important motifs: a GGDG motif, an NE/D motif and a Gly-rich motif (Figures 2C and S1). The GGDG motif is involved in ATP-binding, whereas the NE/D and Gly-rich motifs are involved in NAD(H) binding [2], [20], [21], [23], [24].

Structure analysis of NADK family genes

Intron position is generally very well-conserved in orthologous genes over long evolutionary time intervals, whereas exon/intron structure is slightly less, but sufficiently, conserved in paralogous genes to reveal evolutionary relationships between introns [39], [40], [41]. To investigate the gene structural diversity and possible mechanisms for the structural evolution of NADK homologs in green plants (Viridiplantae, excluding glaucophytes and rhodophytes), we analyzed the exon/intron organization in the coding sequence. Overall, there was considerable diversity in the number of introns (0–11) and the length of introns (50–7452 bp) in the NADK family genes (Figure 3). Interestingly, NADK family members within the same subfamily shared similar gene structure in terms of intron number, exon length, and/or intron phases, with the exception of subfamily IV (Figure 3). For instance, NADK genes in subfamily I had 6–11 introns, 48% (12/25) and 84% (21/25) of which contained 10 and 9–11 introns, respectively. We also investigated intron phases with respect to codons in thc NADK genes. The intron phases were remarkably well conserved among subfamily members, whereas the intron arrangement and phases were strikingly distinct between subfamilies (Figure 3B).

To further explore intron loss or gain within the NADK family genes, we next examined the exon/intron organization of paralogs and orthologs in the land plants (Figure 4A). This analysis revealed that single intron loss and gain likely occurred during the structural evolution of NADK family genes in land plants. For example, The paralogous genes *PtNADK1/4* and *SmNADK2/3*showed conserved exon/intron structure in terms of the number of introns and exon length, whereas a single intron appears to have been lost during the evolution of the *BrNADK1/4* and*PpNADK3/4* paralogs (Figures 3B and 4A). By contrast, a single intron gain occurred between the paralogs *PtNADK2* and *3* (Figures 3B and 4A). Among the orthologous NADK genes in land plants, the majority of subfamilies I, II and III members contained 10, 7 and 4 conserved common introns, respectively (Figure 4A), whereas some subfamily I and II NADK orthologs contained fewer introns. For example, *OsNADK1*, *SiNADK1*, *BdNADK1/2*, *ZmNADK2/3*(subfamily I) have only nine conserved common introns. Similar intron loses were seen with*BrNADK1/3/4*, *SbNADK3* and *SmNADK2/3* (subfamily I), *PpNADK1* (subfamily II), and*SmNADK4* (subfamily III; Figure 4A). Single intron gain was also observed in the NADK genes of land plants. For instance, *OsNADK1*, *SiNADK1*, and *PpNADK2/3/4* in subfamily I and*PtNADK3* in subfamily II appear to have gained an intron, whereas *VvNADK1* in subfamily II appears to have gained two introns (Figure 4A).

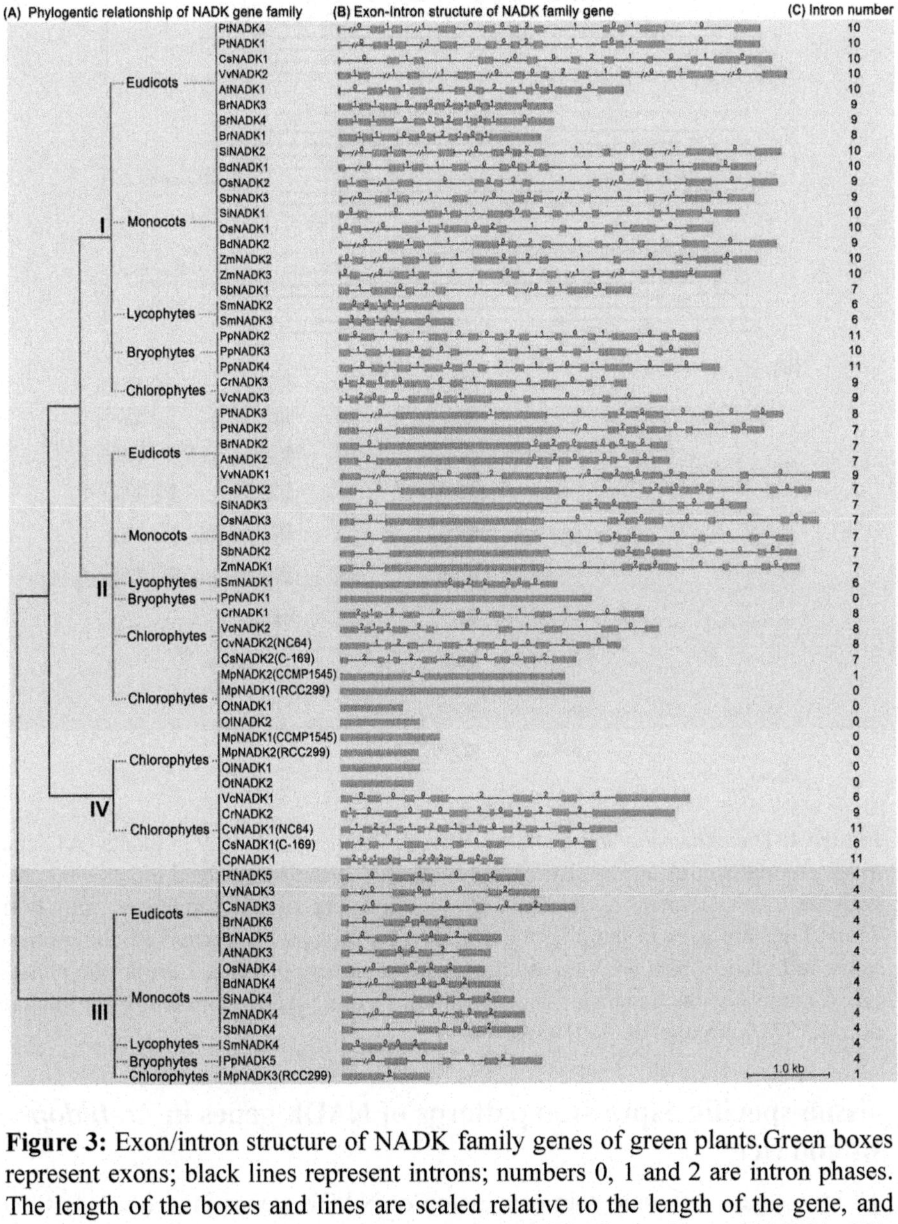

Figure 3: Exon/intron structure of NADK family genes of green plants.Green boxes represent exons; black lines represent introns; numbers 0, 1 and 2 are intron phases. The length of the boxes and lines are scaled relative to the length of the gene, and longer introns are denoted by a double slash.doi:10.1371/journal.pone.0101051.g003.

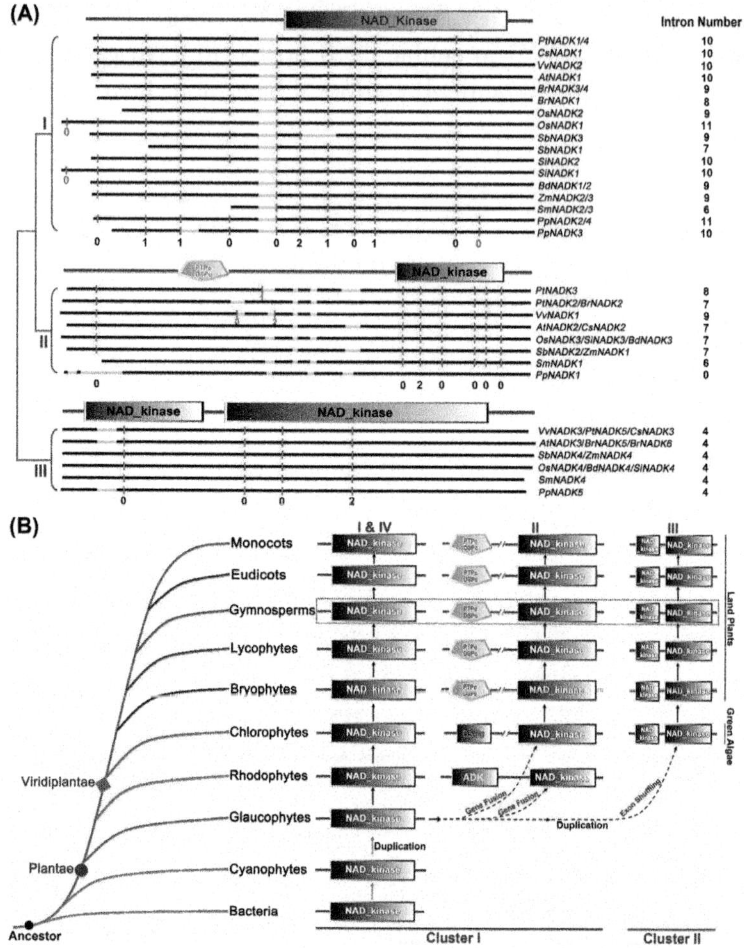

Figure 4: The expansion and evolution of the NADK gene family in plants (**A**) Schematic comparison of intron distribution in NADK orthologs of land plants generated with the CIWOG software. Black horizontal lines are aligned sequences; gray horizontal lines are gaps in the alignment; gray vertical bars are conserved common introns; red vertical bars are gained introns. The numbers 0, 1 and 2 are intron phases. (**B**) A model for the expansion and evolution of the NADK gene family in Plantae. doi:10.1371/journal.pone.0101051.g004.

Tissue-specific expression patterns of NADK genes in *Arabidopsis* and rice

To investigate differences in expression of NADK genes in *Arabidopsis* and rice during plant development, we first analyzed the expression profiles of AtNADK and OsNADK genes available in expressed sequence tag (EST) databases from vegetative and reproductive development stages (Figures 5A and 5B and Tables S3 and S4, respectively). Varying levels of ESTs for each of the AtNADK and OsNADK genes were found the database, indicating that all of the NADK genes in *Arabidopsis* and rice are expressed, and that they are likely differential expressed in different tissues or developmental stages. We then used *Arabidopsis* (ATH1, 22 k array) and rice (Os 51 k array) microarray data in Genevestigator to analyze the expression patterns of the NADK genes in 10 and 9 developmental stages/tissues, respectively (Figure 5C, D and S5). Essentially identical developmental expression profiles for the AtNADK and OsNADK genes were also obtained with data from the *Arabidopsis* and rice eFP browsers in the Bio-Analytic Resource (http://bar.utoronto.ca/welcome.htm) database [42]. *AtNADK3* was not included in the comparison of tissue/developmental expression patterns because it was not included in the microarray data. The combined expression patterns of *AtNADKs* and *OsNADKs*from the EST and microarray data analyses are summarized in Figure 5E.

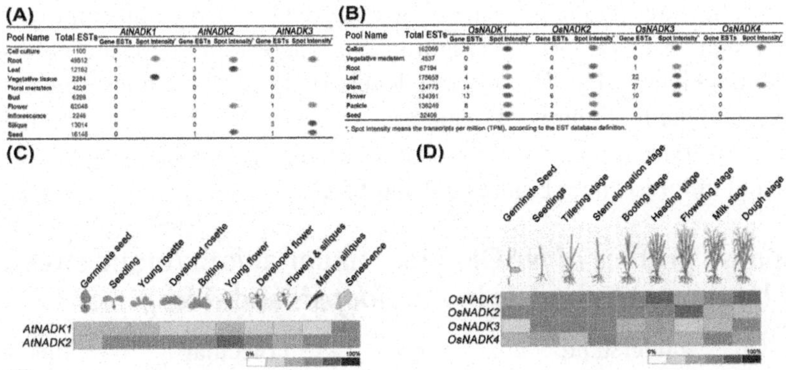

(E)

Gene Name	EST counts	FL_cDNA	Germinate seed	Seedling	Root	Leave	Stem	Flower	Inflorescence	Silique	Seed	Callus
AtNADK1	19	AY383545 AK318804	•	•	••	•	•	•	•	•	•	NA
AtNADK2	30	AF337912 AY099794	•	••	••	•••	••	•••	••	••	•	NA
AtNADK3	25	BT022078 BT033136	•	•	•	•	•	•	•	••	○	NA
OsNADK1	86	AK099730 AK061785	•	••	••	••	••	••	••	NA	••	••
OsNADK2	30	AK289181 AK243031	••	••	••	••	••	••	••	NA	•	••
OsNADK3	69	AK065215 AK101259 AK060751	•	••	••	••	••	••	••	NA	••	••
OsNADK4	10	AK073719	•	•	•	•	••	•	•	NA	•	•

•, weak expression; ••, moderate expression; •••, strong expression; ○, without expression; NA, not available.

Figure 5: Developmental expression patterns of NADK family genes in*Arabidopsis* and rice.Expression profiles of (**A**) *AtNADKs* and (**B**) *OsNADKs* inferred from public EST data. Expression profiles of (**C**) *AtNADKs* (except *AtNADK3*) and (**D**) *OsNADKs* in different developmental stages obtained from microarray data reported in Genevestigator. Results are shown as heat maps in white/gray/red (low to high) that reflect the percent of expression. **(E)** Summary of the relative expression patterns of AtNADK and OsNADK genes inferred from the combined EST (A, B; Tables S3 and S4) and microarray (C, D) profiles.doi:10.1371/journal.pone.0101051.g005.

Overall, all of the AtNADK and OsNADK genes were expressed during the vegetative and reproductive development stages, and they displayed strong tissue specificity. In *Arabidopsis,AtNADK2* showed higher expression in most tissues and developmental stages than *AtNADK1*(Figures 5C and S5). The highest *AtNADK2* expression is seen in leaves and flowers (Figures 5A, C and S5), consistent with Waller et al. (2010) [36]; whereas *AtNADK1* expression is highest in the dry seed and senescence stages (Figures 5C and S5). *AtNADK3* appears to be mainly expressed in the reproductive tissues such as flowers and siliques (Figure 5A), also consistent with the findings of Waller et al. (2010) [36]. In rice, the *OsNADK1*, *OsNADK2* and*OsNADK3* transcripts tend to accumulate to higher levels than the *OsNADK4* transcript in most tissues/developmental stages (Figures 5B, D and S5), but their individual expression patterns differ. For instance, during the germinate seed stage, *OsNADK1* transcript levels are low compared to *OsNADK2* (Figure 5D), whereas the opposite is observed in callus (Figure 5B). The highest levels of *OsNADK3* transcript are seen in leaf and stem (Figures 5D and S5); whereas relatively low levels of the Os*NADK4* transcript are seen in most tissues/developmental stages except for stem and inflorescence (Figures 5D and S5).

Response profiles of NADK genes under abiotic/biotic stresses and hormone treatments in *Arabidopsis* and rice

To understand the molecular mechanism of *AtNADKs* and *OsNADKs* transcriptional regulation under abiotic/biotic stresses and hormone treatments, we first identified potential *cis*-elements in the promoter regions of each NADK gene using the PlantCARE program (Tables 1 and S2). We found that the promoter regions of AtNADK and OsNADK genes contain response elements for several abiotic/biotic stresses, such as low temperature, heat, drought (MYB binding sites), anaerobic conditions, and pathogens (TC-rich repeats, W box, GCC box, Box S, Box-W1 and EIRE) (Tables 1 and S2). In addition, the AtNADK and OsNADK gene promoters contain *cis*-elements for responding to several hormones, such as auxin, gibberellin, abscisic acid (ABA), ethylene, salicylic acid (SA) and methyl jasmonic acid (MeJA) (Tables 1 and S2). Further analysis showed that all of the *AtNADK* and *OsNADK* promoters contain anaerobic- and ABA-responsive elements; and the majority of them also carry heat stress, drought (MYB binding sites) and MeJA response elements (Tables 1 and S2).

Table 1: Abiotic/biotic stress and hormone response elements in *AtNADK* and*Os-NADK* promoters doi:10.1371/journal.pone.0101051.t001

Cis-elements	Core sequences	Functions of the cis-elements	*OsNADK1*	*OsNADK2*	*OsNADK3*	*OsNADK4*	*AtNADK1*	*AtNADK2*	*AtNADK3*
Abiotic/biotic stress									
LTR	CCGAAA	low-temperature responsiveness		√		√			
HSE	AAAAAATTTC	heat-stress responsiveness	√	√	√	√	√		√
MBS	TAACTG	MYB binding site, drought inducibility	√	√	√	√			√
ARE	TGGTTT	anaerobic induction	√	√	√	√	√	√	√
TC-rich repeats	ATTCTCTAAC	defense and stress responsiveness		√	√			√	√
W box	TTGACC	wound and pathogen responsiveness			√		√		
Box S	AGCCACC	wound and pathogen responsiveness	√						
GCC box	AGCCGCC	wound and pathogen responsiveness				√			
EIRE	TTCGACC	Elicitor responsiveness						√	
Box-W1	TTGACC	fungal elicitor responsiveness			√		√		√
Hormone response elements									
AuxRR-core	GGTCCAT	auxin responsiveness			√				
TGA-element	AACGAC	Auxin responsiveness					√	√	
GARE-motif	AAACAGA	Gibberellin responsiveness		√			√		
ABRE	CACGTG	abscisic acid responsiveness	√	√	√	√	√	√	√
ERE	ATTTCAAA	Ethylene responsiveness	√	√	√		√		
TCA-element	GAGAAGAATA	salicylic acid responsiveness					√		√
CGTCA-motif	CGTCA	MeJA responsiveness	√			√	√	√	√

[a]the *cis*-elements were identified with the PlantCARE program (http://bioinformatics.psb.ugent.be/webtools/plantcare/html/) using the sequences 1500 bp upstream from the transcription start site of each NADK gene. The "√" means the NADK gene contains this *cis*-element in the promoter region.
doi:10.1371/journal.pone.0101051.t001

To further demonstrate that the expression of AtNADK and OsNADK genes is induced by abiotic and biotic stress, we again examined *Arabidopsis* and rice microarray data in the Genevestigator database, as well as by qRT-PCR experiments. As shown in Figures 6, 7 andS6, the expression of AtNADK and OsNADK genes is induced to varying degrees by abiotic stress such as cold, heat, drought (PEG), salt (NaCl) and oxidative (methyl viologen, MV), as well as by biotic stress such as the pathogens *Botrytis cinerea*, *Blumeria graminis*,*Pseudomonas syringae*, *Agrobacterium tumefaciens*, *Mycosphaerella graminicola*,*Magnaporthe oryzae*, and *Magnaporthe grisea*. In *Arabidopsis*, *AtNADK1* is up-regulated in whole plants under heat and MV stresses, whereas it is down-regulated and does not change under PEG and cold treatment, respectively (Figure 6). By contrast, *AtNADK2* is down-regulated or does not change under these conditions (Figure 6). *AtNADK3* is up-regulated under cold stress, while it is down-regulated under other abiotic stresses including heat, PEG, NaCl, and MV (Figure 6). Additionally, both *AtNADK1* and *AtNADK2* can be up-regulated by some biotic stresses, although with different expression profiles under the same stress (Figure S6). In rice, *OsNADK1/2* are up-regulated in shoot and root under cold stress, in root under heat stress and in shoot under MV stress, respectively; while they are slightly down-regulated or does not change under these conditions (Figure 7). *OsNADK3* shows up-regulated under PEG in shoot and down-regulated or does not change under other abiotic stresses such as cold, heat, NaCl and MV (Figure 7). Similarly, *OsNADK4* is up-regulated in root under cold and heat and down-regulated or does not change under other abiotic stresses including PEG, NaCl, and MV (Figure 7). Additionally, *OsNADK1–4* can be induced by biotic stresses. For instance,*OsNADK1* is up-regulated in callus after infection by *A. tumefaciens* and root after infection by*M. oryzae*, whereas they are down-regulated in root and leaf after infection by *M. graminicola*and *M. grisea*, respectively (Figure S6).

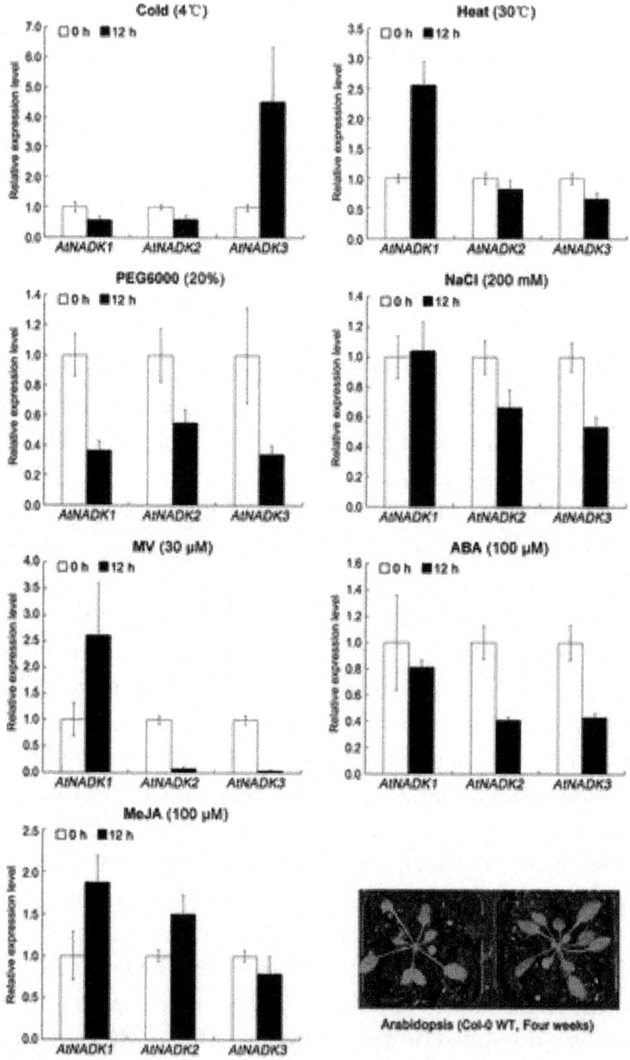

Figure 6: Expression patterns of NADK family genes in *Arabidopsis* under abiotic stress and hormone treatments.Expression levels of *AtNADK1–3* assayed by qRT-PCR under cold (4°C), heat (30°C), drought (20% PEG6000), salt (200 mM NaCl), oxida-

tive (30 μM MV) stresses and MeJA (100 μM), ABA (100 μM) hormone treatments. Data are means ± SD (n=3) and are representative of similar results from three independent experiments.doi:10.1371/journal.pone.0101051.g006.

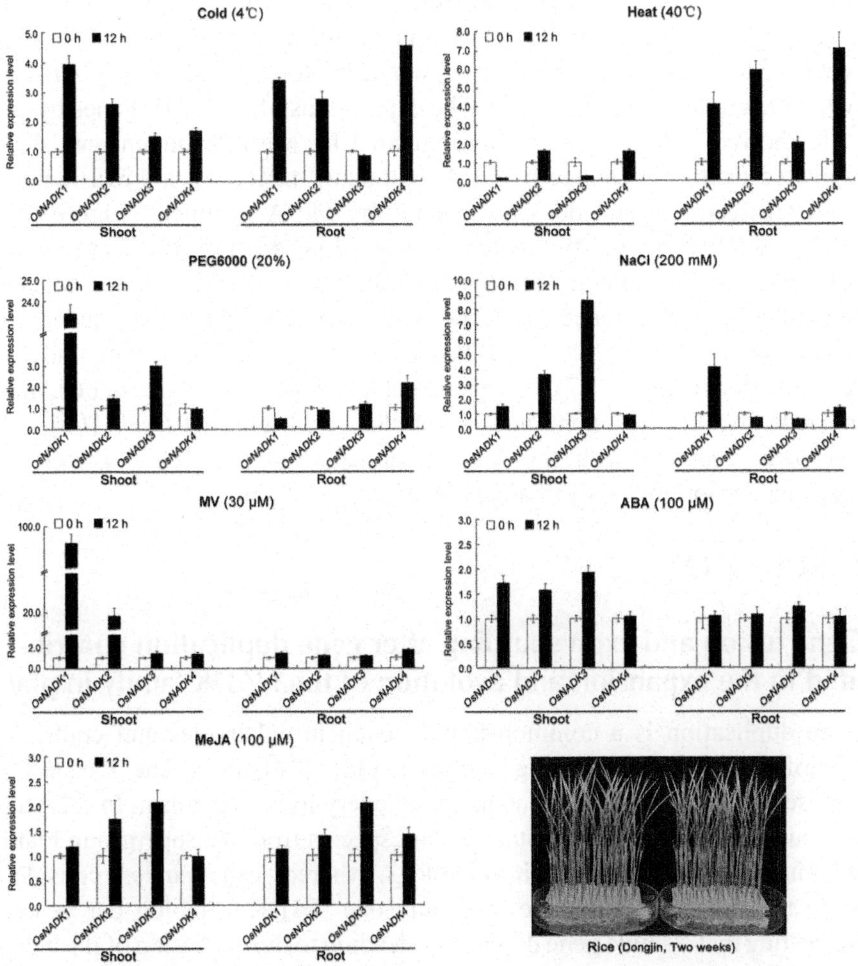

Figure 7: Expression patterns of NADK family genes in rice under abiotic stress and hormone treatments.Expression levels of *OsNADK1–4* assayed by qRT-PCR under cold (4°C), heat (30°C), drought (20% PEG6000), salt (200 mM NaCl), oxidative (30 μM MV) stresses and MeJA (100 μM), ABA (100 μM) hormone treatments. Data are means ± SD (n=3) and are representative of similar results from three independent experiments.doi:10.1371/journal.pone.0101051.g007.

As several hormone response elements were identified in the promoter regions of the AtNADK and OsNADK genes, we also examined their expression profiles under phytohormone treatments in the public *Arabidopsis* and rice microarray data (Figure S7), as well as by qRT-PCR experiments (Figures 6 and 7). *AtNADK1* and *AtNADK2* are differentially expressed in seedlings and cells treated with phytohormones, such as the auxins indole-3-acetic acid (IAA) and naphthaleneacetic acid (NAA), zeatin, SA, ABA, and ABA + SA (Figure S7). Interestingly,*AtNADK1* is up-regulated by ABA, ABA + SA, and ABA + MeJA treatment; whereas *AtNADK2*is down-regulation by ABA and ABA + SA treatment (Figure S7). The expressions of AtNADK genes by qRT-PCR showed *AtNADK1–2* are up-regulated by MeJA treatment and down-regulated or do not change by ABA treatment; *AtNADK3* is down-regulated by ABA treatment and does not change by MeJA treatment (Figure 6). In rice, *OsNADK1–4* are differentially expressed in seedlings and leaves treated with IAA, NAA, zeatin, gibberellin, kinetin, ABA, SA and jasmonic acid (JA) (Figure S7). For instance, *OsNADK3* is up-regulated in roots under*trans*-zeatin treatment, whereas *OsNADK4* is down-regulated in leaves under the same treatment (Figure S7). It is also notable that *OsNADK1–3* can be induced in shoot by ABA treatment and do not change in root, but *OsNADK4* is not obviously changed with the same treatment. All *OsNADK1–4* were up-regulated or not obviously changed by MeJA treatment.

DISCUSSION

Gene fusion and exon shuffling after gene duplication contributed to the expansion and evolution of the NADK family in plants

Gene duplication is a common phenomenon in eukaryotes and contributes to biological diversity during evolution [43], [44], [45]. The NADKs are represented by at least one gene in nearly every living organism. In this study, we found that the 24 representative plant species from the supergroup Plantae examined contain 2–6 NADK homologs, whereas cyanophytes, considered to be the ancestor of Plantae, had only one NADK homolog per genome, suggesting that a single gene duplication leading to the expansion of the NADK gene family, occurred during the divergence of ancestral cyanophytes from Plantae. The scattered distribution of NADK family genes on chromosomes (Figure S3) and the eight paralogous NADK gene pairs found in five land plants (Figure S4) suggest that segmental duplications may have been involved in the expansion of the NADK gene family and caused differences in the number of NADK genes within subfamilies and species of land plants after divergence from aquatic algae.

Gene fusion and exon shuffling after gene duplication are mechanisms that can enhance the functional divergence of duplicated genes by creating additional domains or rearranging the original functional domains [46], [47], [48], [49].

In this study, phylogenetic analysis, together with the domain organizations and gene structures of NADK family genes, showed distinct evolutionary differences among subfamilies (Figure 2). Therefore, a model was constructed to account for the expansion and evolution of the NADK gene family in plants (Figure 4B). In this model, all NADK family members originated from a common ancestor, which contained only the typical NAD_kinase domain and existed in all living organisms from prokaryotic bacteria to eukaryotic angiosperms; cyanophytes were considered to be the ancestor of Plantae, and at least one gene duplication occurred to yield two NADK genes in eukaryotic glaucophytes (Figures 1 and 4B). Moreover, during the evolutionary diversification from glaucophytes to rhodophytes and Viridiplantae, gene fusion events occurred and additional catalytic domains (ADK, DSPc or PTPc/DSPc) were acquired in the N-terminus, leading to the diversified domain organization seen in subfamily II (Figures 2B and 4B). In addition, exon shuffling after gene duplication contributed to the formation of the bifurcated NAD_kinase domain in subfamily III (Figures 2B and 4B).

Single intron loss and gain lead to the diversified gene structures

Gene structural diversity within gene families is another evolutionary mechanism that promotes variability, and intron loss or gain is important in generating structural diversity and complexity[41], [50]. Analyzing the exon/intron structures of NADK genes, we found that the number of introns and intron phases among subfamily members are remarkably conserved, whereas the intron arrangement and intron phases are strikingly distinct between subfamilies (Figures 3 and4). Further analysis of the orthologous and paralogous genes in land plants (Figure 4) suggests that single intron loss as well as intron gain likely occurred and contributed to the diversification of gene structure, and consequent functional diversity and divergence, during the evolution of the NADK family in plants.

NADKs are involved in plant responses to several abiotic and biotic stresses

The three NADK genes of *Arabidopsis* belong to three different subfamilies according to our phylogenetic analysis (Figures 1 and 2). The three *AtNADKs* are known to have anti-oxidative functions [10], [11], [12], [16], [36], but they have distinct mechanisms by which they facilitate plant resistance to oxidative stress. AtNADK1 (subfamily I) is a cytosolic enzyme that indirectly provides cytosolic NADP for plasma membrane NADPH oxidases, which are the key producers of reactive oxygen species under both normal and stress conditions in plants [36], [51], [52],[53]. AtNADK2 (subfamily II) is a chloroplastic enzyme that plays a vital role in chlorophyll synthesis and chloroplast protection against oxidative damage by regulating plastidic NADP-biosynthesis [11], [36]. AtNADK3 (subfamily III) is a peroxisomal enzyme that plays a prominent anti-oxidation role by providing the peroxisomal reductant NADPH [10], [36]. NADK genes in subfamily IV were only present in aquatic algae and grouped with *S. cerevisiae NADK1* (Pos5) in our broader phylogenetic analysis (Figure S2). Pos5 is located in the mitochondria and its deletion causes slow growth, sensitivity to oxidative stress, such as paraquat, hyperoxia, H_2O_2, and Cu^{2+}, and biosynthesis deficiency on iron-sulfur clusters [25], [54], [55]. These observations suggest that plant NADKs in subfamily IV may also localize to mitochondria and protect against oxidative damage by regulating mitochondrial NADP-biosynthesis.

We found that plant NADKs are also involved in responding to several abiotic/biotic stresses, including cold, heat, drought, salt, oxidative and pathogens. *Arabidopsis NADK1* expression is up-regulated by H_2O_2, irradiation and the bacterial pathogen *P. syringae* pv. *tomato*; and the*AtNADK1* deficient mutant is sensitive to γ-irradiation and paraquat-induced oxidative stress[16]. The *AtNADK2* deletion mutant is also hypersensitive to environmental stresses that trigger oxidative stress, such as UVB-irradiation, drought, heat, and high salinity [11]; similarly*AtNADK3* transcription can be induced by MV, high salinity and osmotic shock [10], [36]. In this study, we found that *AtNADK1* and *OsNADK1/2* belongs to subfamily I and can be induced to varying degrees by cold, drought (PEG), salt (NaCl), oxidative (MV) and pathogens (Figures 6,7, S6 and S7). We also found that *AtNADK2*, which belongs to subfamily II, is up-regulated by pathogens such as *P. syringae* pv. *phaseolicola* and *Hyaloperonospora arabidopsidis* (Figure S6); *OsNADK3*, which also belongs to subfamily II, is up-regulated by *Nilaparvata lugens* and down-regulated under anaerobic conditions (Figure S6). *OsNADK4* belongs to subfamily III, and is up-regulated under anaerobic conditions, *N. lugens* and heat in root, and slightly up-regulated by drought or PEG in root (Figures 7 and S6). These observations, together with our analysis of *cis*-elements (Table 1), suggest that NADK genes in subfamilies I, II and III play an important role in plant responses to invading pathogens and abiotic stresses. The functions of NADKs in subfamily IV in abiotic/biotic-stress responses remains to be examined.

NADKs may participate in regulation of hormone signaling

AtNADK3 transcription can be slightly induced by ABA and the mutant is hypersensitive to ABA[10]. In this study, we found that all *AtNADK1–3* showed down-regulated or slightly down-regulated in response to ABA (Figure

6), while *OsNADK1–3* were up-regulated by ABA treatment (Figure 7). It's important to note that response profiles of NADK genes by qRT-PCR are not completely consisted with microarray data or previous studies, because of the different experimental material, processing and analyzing methods used in this study. Moreover, seven AtNADK and OsNADK genes (*AtNADK1–3* and *OsNADK1–4*) contain *cis*-acting ABA responsive elements (Table 1), suggesting that NADKs probably participate in regulation of ABA signaling in plants.

JA and its derivative MeJA are important signaling molecules in plant responses to many abiotic and biotic stresses such as wounding and pathogens [56], [57]. Moreover, JA biosynthesis occurs in peroxisomes [57], [58]. AtNADK3 is also a peroxisomal enzyme and may phosphorylate the NADH derived from β-oxidation to yield NADPH needed for anti-oxidant defense [36]. *AtNADK1/2* and *OsNADK1–4* were up-regulated or slightly up-regulated under MeJA treatment (Figures 6 and 7). Moreover, several of these genes (*AtNADK1/2/3* and*OsNADK1/4*) contain CGTCA-MeJA-responsiveness motifs in their promoter regions (Table 1). These observations suggest that NADKs may also participate in regulation of JA or MeJA signaling in plants.

It should be pointed out that although transcription factors and *cis*-regulatory elements play important roles in regulating gene expression, the relationship is not always direct and the inducibility of expression is often affected by many factors and on multiple levels [59], [60]. For example, ABA-responsive *cis*-regulatory elements were identified in the promoter regions of all seven NADK genes in *Arabidopsis* and rice, but only *OsNADK1–3* were up-regulated or slightly up-regulated by ABA treatment (Figures 7 and S7). Moreover, considering the limitations of the PlantCARE program in predicting *cis*-regulatory elements, further experiments are necessary to verify the relationship between the inducible expression profiles of NADK genes and the *cis*-elements within their promoters.

MATERIALS AND METHODS

Data retrieval and identification of NADK genes

The protein sequences of 24 completely or partially sequenced plant genomes representing the eight major plant lineages were retrieved from public databases. All of the protein sequences were the most current non-redundant sequences from the following sources (Table S1): the glaucophyte *Cyanophora paradoxa* from the *Cyanophora* Genome Project (http://cyanophora.

rutgers.edu/cyanophora/home.php) [61]; the rhodophytes *Cyanidioschyzon merolae* and *Galdieria sulphuraria* from the *Cyanidioschyzon* Genome Project (http://merolae.biol.s.u-tokyo.ac.jp/) [62] and the *Galdieria* Genome Project (http://genomics.msu.edu/galdieria/) [63], respectively; the chlorophytes *O. tauri* (version 2.0)[64], *O. lucimarinus* (version 2.0) [65], *M. pusilla* strain RCC299 (version 3.0) and strain CCMP1545 (version 3.0) [66], *Chlorella variabilis* NC64A (version 1.0) [67], *Coccomyxa subellipsoidea* C-169 (version 2.0) [68], *C. reinhardtii* (version 4.0) [69] and *V. carteri* (version 2.0) and the bryophyte *P. patens* (version 3.0) [70] and lycophyte *S. moellendorffii* (version 1.1)[71] from the Joint Genome Institute (JGI, http://genome.jgi-psf.org/); for the gymnosperm *P. sitchensis* [72] from NCBI (http://www.ncbi.nlm.nih.gov/, partial sequences only because of its genome is only partially sequenced); the monocots *B. distachyon* (version 1.2) [73], *Setaria italica* (version 2.1) [74] and *S. bicolor* (version 2.1) [75] from JGI, rice (version 7.0) and *Z. mays* (version 5.6) from the Institute for Genomic Research Rice Genome Annotation Project (http://rice.plantbiology.msu.edu/index.shtml) [76] and MaizeSequence (http://www.maizesequence.org/index.html) [77], respectively; the eudicots *P. trichocarpa*(version 2.2) [78] and *Cucumis sativus* (version 1.0) [79] from JGI, *Arabidopsis* (version 10.0) from the Arabidopsis Information Resource (http://www.arabidopsis.org/) [80], *B. rapa* (version 1.5) from the *Brassica* Database (http://brassicadb.org/brad/) [81] and *V. vinifera* (version 1.0) from GenoScope (http://www.genoscope.cns.fr/externe/GenomeBrowser/Vitis/) [82]. All of the above protein sequences were integrated into a local protein database for the subsequent identification of NADK homologs.

To identify the NADK genes and their homologs in the Plantae supergroup, HMMER v3.0 [83],[84] was used to perform an HMM search against the local protein database, using the family-specific NAD_kinase domain (PF01513) HMM profile obtained from the Pfam database [85],[86]. The HMM search was performed with the default parameters and an E-value cutoff of $1e^{-5}$. If a candidate gene had multiple alternative splice variants, the longest variant was used to represent the candidate protein. The Pfam and SMART [87], [88] databases were employed to detect conserved domains in the candidate proteins, and the search results were refined manually to eliminate partial NADK domains and other potential false positives.

Sequence alignment and phylogenetic analysis

The NAD_kinase domain sequences of the candidate proteins were aligned using the MUSCLE v3.8 program with the default parameters [89], [90] and the alignments were manually edited using the BioEdit v7.0 program [91]. The rooted maximum-likelihood tree was inferred from the resulting alignments using the Phylip v3.68 package [92] under the γ-corrected Jones–Taylor–Thornton model [93] with default parameters, and the reliability of interior branches was assessed with 1000 bootstrap resamplings. In addition, the MEGA v5.0 program [94] was also used to reconstruct the phylogenetic trees by the neighbor joining, minimal evolution and maximum parsimony methods, and to display the phylogenetic trees.

The rates of non-synonymous substitution (Ka) and synonymous substitution (Ks) were estimated for the orthologous and paralogous gene pairs of NADK family genes using the Codeml program in PAML v4.3 [95] interface tool of PAL2NAL [96], based on the aligned amino acid sequences and the corresponding nucleotide sequences. The duplication and divergence times of each gene pairs were estimated from the Ks of λ substitutions per synonymous site per year as $T=Ks/2\lambda$ ($\lambda=6.5\times10^{-9}$) [97], [98].

Exon/intron structure and conserved motif analysis

The exon/intron structures of individual NADK genes were obtained through the Gene Structure Display Server (http://gsds.cbi.pku.edu.cn) [99] by aligning the coding or cDNA sequences with their corresponding genomic DNA sequences from Phytozome v9.1 (http://www.phytozome.net/) [100]. To illustrate the evolution of introns, gene models were inspected for annotation of introns, and exon/intron boundaries were manually checked. For a subset of genes, predictions pertaining to the types of introns were independently checked using Common Introns Within Orthologous Genes software (http://ciwog.gdcb.iastate.edu/)[101].

The conserved functional motifs within NAD_kinase domains were identified using the MEME v4.9 program (http://meme.sdsc.edu) [102] with the default parameters, and the sequence logos (graphical representations) of these motifs or domains were generated with the WebLogo v3.3 server (http://weblogo.threeplusone.com/create.cgi) [103] based on the results of protein sequence alignments.

Cis-regulatory elements and expression profile analysis

The 1500 bp upstream of the transcription start site of all NADK genes in *Arabidopsis* and rice were obtained from Phytozome v9.1 (http://www.phytozome.net/), and the *cis*-regulatory elements were identified using the PlantCARE program (http://bioinformatics.psb.ugent.be/webtools/plantcare/html/) [104], [105]. ATH1 22 k and Os 51 k microarray data in the Genevestigator V3 database were used to analyze the tissue-specific and inducible expression profiles of NADK genes in *Arabidopsis* and rice [106], respectively. In addition, EST profiles of each NADK gene in *Arabidopsis* and rice were also analyzed by BLASTN search against the corresponding NCBI (http://www.ncbi.nlm.nih.gov/) EST database, with the coding sequences of the individual NADK gene as query. The BLASTN searches were performed with the following criteria: E-value$<1e^{-10}$ and nucleotide identities >95% over 150 bp.

Plant materials, treatments and quantitative real-time PCR (qRT-PCR) analysis

The *Arabidopsis* (Col-0) seeds were pretreated at 4°C for 3 days and directly sown in Sunshine MVP potting soil, and then grown in growth chambers under 16 h light/8 h dark at 22±1°C; while rice (*Oryza sativa* ssp. *japonica* cv. Dongjin) seeds were pretreated germination at 26±1°C for 4 days and transplanted to 1/2 Hogland nutrient solution, and then grown in growth chambers under 16 h light/8 h dark at 26±1°C. Following two (rice) or four (*Arabidopsis*) weeks of growth, the seedlings were grown under 4°C or 30°C (*Arabidopsis*)/40°C (rice) for 12 h for cold and heat treatments, respectively; submerged in 200 mM NaCl or 20% PEG6000 solutions for 12 h for drought and salt treatments, respectively. For oxidative stress and hormone treatments, solutions of 30 μM MV, 100 μM MeJA and ABA were separately sprayed on seedlings for 12 h. All the MV, MeJA and ABA used for treatments were purchased from Sigma-Aldrich. Samples were collected following treatment and immediately frozen at −80°C.

Total RNA was extracted by using Trizol reagent (Takara, Japan) according to the manufacturer's instructions and treated with RNase-free DNase I (Invitrogen, USA) for 15 min to remove any DNA contamination. RNA concentration and quality were verified by using the NanoDrop 1000 Spectrophotometers (Thermo, USA), and cDNAs were synthesized by using oligo $d(T)_{18}$ reverse primer from 5 μg of total RNA in a total volume of 20 μL by using EasyScript First-Strand cDNA Synthesis SuperMix (TransGen Biotech, China). qRT-PCR reactions were carried out in 96-well (20 μL) format by using the FastStart Essential DNA Green Master (Roche, Switzerland), and were performed in an CFX96 Touch Real-Time PCR Detection System (BIO-RAD, USA). The reactions were repeated three times and the quantitative analysis used the $2^{-\Delta\Delta CT}$ method. All gene-specific primers were designed to avoid the conserved region and span introns or cross an exon-exon junction. The detailed primer sequences are shown in Table S5. The *AtTub6* (AT5G12250) and *OsActin1* (accession ID KC140126) were chosen as the internal control in *Arabidopsis* and rice, respectively.

REFERENCES

1. Ying W (2008) NAD+/NADH and NADP+/NADPH in cellular functions and cell death: regulation and biological consequences. Antioxid Redox Signal 10: 179–206. doi: 10.1089/ars.2007.1672
2. .Kawai S, Murata K (2008) Structure and function of NAD kinase and NADP phosphatase: key enzymes that regulate the intracellular balance of NAD(H) and NADP(H). Biosci Biotechnol Biochem 72: 919–930. doi: 10.1271/bbb.70738
3. .Grose JH, Joss L, Velick SF, Roth JR (2006) Evidence that feedback inhibition of NAD kinase controls responses to oxidative stress. Proc Natl Acad Sci U S A 103: 7601–7606. doi: 10.1073/pnas.0602494103
4. .Kawai S, Fukuda C, Mukai T, Murata K (2005) MJ0917 in archaeon *Methanococcus jannaschii* is a novel NADP phosphatase/NAD kinase. J Biol Chem 280: 39200–39207. doi: 10.1074/jbc.m506426200
5. .Kawai S, Mori S, Mukai T, Suzuki S, Yamada T, et al. (2000) Inorganic Polyphosphate/ATP-NAD kinase of Micrococcus flavus and *Mycobacterium tuberculosis* H37Rv. Biochem Biophys Res Commun 276: 57–63. doi: .1006/bbrc.2000.3433
6. .Kawai S, Mori S, Mukai T, Hashimoto W, Murata K (2001) Molecular characterization of*Escherichia coli* NAD kinase. Eur J Biochem 268: 4359–4365. doi: 10.1046/j.1432-1327.2001.02358.x
7. .Kawai S, Suzuki S, Mori S, Murata K (2001) Molecular cloning and identification of UTR1 of a yeast *Saccharomyces cerevisiae* as a gene encoding an NAD kinase. FEMS Microbiol Lett 200: 181–184. doi: 10.1111/j.1574-6968.2001.tb10712.x
8. Outten CE, Culotta VC (2003) A novel NADH kinase is the mitochondrial source of NADPH in *Saccharomyces cerevisiae*. EMBO J 22: 2015–2024. doi: 10.1093/emboj/cdg211
9. **9.**Lerner F, Niere M, Ludwig A, Ziegler M (2001) Structural and functional characterization of human NAD kinase. Biochem Biophys Res Commun 288: 69–74. doi: 10.1006/bbrc.2001.5735
10. Chai MF, Wei PC, Chen QJ, An R, Chen J, et al. (2006) NADK3, a novel cytoplasmic source of NADPH, is required under conditions of oxidative stress and modulates abscisic acid responses in *Arabidopsis*. Plant J 47: 665–674. doi: 10.1111/j.1365-313x.2006.02816.x
11. .Chai MF, Chen QJ, An R, Chen YM, Chen J, et al. (2005) NADK2, n *Arabidopsis*chloroplastic NAD kinase, plays a vital role in both

chlorophyll synthesis and chloroplast protection. Plant Mol Biol 59: 553–564. doi: 10.1007/s11103-005-6802-y

12. .Turner WL, Waller JC, Vanderbeld B, Snedden WA (2004) Cloning and characterization of two NAD kinases from *Arabidopsis*. identification of a calmodulin binding isoform. Plant Physiol 135: 1243–1255. doi: 10.1104/pp.104.040428

13. Stephan C, Renard M, Montrichard F (2000) Evidence for the existence of two soluble NAD(+) kinase isoenzymes in *Euglena gracilis* Z. Int J Biochem Cell Biol. 32: 855–863. doi: 10.1016/s1357-2725(00)00032-7

14. Li YF, Shi F (2006) Partial rescue of pos5 mutants by YEF1 and UTR1 genes in*Saccharomyces cerevisiae*. Acta Biochim Biophys Sin (Shanghai) 38: 293–298. doi: 10.1111/j.1745-7270.2006.00162.x

15. .Shi F, Kawai S, Mori S, Kono E, Murata K (2005) Identification of ATP-NADH kinase isozymes and their contribution to supply of NADP(H) in *Saccharomyces cerevisiae*. FEBS J 272: 3337–3349. doi: 10.1111/j.1742-4658.2005.04749.x

16. Berrin JG, Pierrugues O, Brutesco C, Alonso B, Montillet JL, et al. (2005) Stress induces the expression of AtNADK-1, a gene encoding a NAD(H) kinase in *Arabidopsis thaliana*. Mol Genet Genomics 273: 10–19. doi: 10.1007/s00438-005-1113-1

17. Oganesyan V, Huang C, Adams PD, Jancarik J, Yokota HA, et al. (2005) Structure of a NAD kinase from *Thermotoga maritima* at 2.3 A resolution. Acta Crystallogr Sect F Struct Biol Cryst Commun 61: 640–646. doi: 10.1107/s1744309105019780

18. Mori S, Kawai S, Mikami B, Murata K (2001) Crystallization and preliminary X-ray analysis of NAD kinase from *Mycobacterium tuberculosis* H37Rv. Acta Crystallogr D Biol Crystallogr 57: 1319–1320. doi: 10.1107/s0907444901011362

19. Ando T, Ohashi K, Ochiai A, Mikami B, Kawai S, et al. (2011) Structural Determinants of Discrimination of NAD+ from NADH in Yeast Mitochondrial NADH Kinase Pos5. J Biol Chem 286: 29984–29992. doi: 10.1074/jbc.m111.249011

20. Raffaelli N, Finaurini L, Mazzola F, Pucci L, Sorci L, et al. (2004) Characterization of*Mycobacterium tuberculosis* NAD kinase: functional analysis of the full-length enzyme by site-directed mutagenesis. Biochemistry 43: 7610–7617. doi: 10.1021/bi049650w

21. Garavaglia S, Raffaelli N, Finaurini L, Magni G, Rizzi M (2004) A novel fold revealed by*Mycobacterium tuberculosis* NAD kinase, a key allosteric enzyme in NADP biosynthesis. J Biol Chem 279: 40980–40986. doi:

10.1074/jbc.m406586200

22. Labesse G, Douguet D, Assairi L, Gilles AM (2002) Diacylglyceride kinases, sphingosine kinases and NAD kinases: distant relatives of 6-phosphofructokinases. Trends Biochem Sci 27: 273–275. doi: 10.1016/s0968-0004(02)02093-5
23. Mori S, Yamasaki M, Maruyama Y, Momma K, Kawai S, et al. (2005) NAD-binding mode and the significance of intersubunit contact revealed by the crystal structure of*Mycobacterium tuberculosis* NAD kinase-NAD complex. Biochem Biophys Res Commun 327: 500–508. doi: 10.1016/j.bbrc.2004.11.163
24. Liu J, Lou Y, Yokota H, Adams PD, Kim R, et al. (2005) Crystal structures of an NAD kinase from *Archaeoglobus fulgidus* in complex with ATP, NAD, or NADP. J Mol Biol 354: 289–303. doi: 10.1016/j.jmb.2005.09.026
25. Shianna KV, Marchuk DA, Strand MK (2006) Genomic characterization of POS5, the*Saccharomyces cerevisiae* mitochondrial NADH kinase. Mitochondrion 6: 94–101. doi: 10.1016/j.mito.2006.02.003
26. Bieganowski P, Seidle HF, Wojcik M, Brenner C (2006) Synthetic lethal and biochemical analyses of NAD and NADH kinases in *Saccharomyces cerevisiae*establish separation of cellular functions. J Biol Chem 281: 22439–22445. doi: 10.1074/jbc.m513919200
27. Pollak N, Niere M, Ziegler M (2007) NAD kinase levels control the NADPH concentration in human cells. J Biol Chem 282: 33562–33571. doi: 10.1074/jbc.m704442200
28. Singh R, Mailloux RJ, Puiseux-Dao S, Appanna VD (2007) Oxidative stress evokes a metabolic adaptation that favors increased NADPH synthesis and decreased NADH production in *Pseudomonas fluorescens*. J Bacteriol 189: 6665–6675. doi: 10.1128/jb.00555-07
29. Anderson JM, Charbonneau H, Jones HP, McCann RO, Cormier MJ (1980) Characterization of the plant nicotinamide adenine dinucleotide kinase activator protein and its identification as calmodulin. Biochemistry 19: 3113–3120. doi: 10.1021/bi00554a043
30. Harding SA, Oh SH, Roberts DM (1997) Transgenic tobacco expressing a foreign calmodulin gene shows an enhanced production of active oxygen species. EMBO J 16: 1137–1144. doi: 10.1093/emboj/16.6.1137
31. Zhou CY, Wu GL, Duan ZQ, Wu LL, Gao YS, et al. (2010) H_2O_2-NOX system: an important mechanism for developmental regulation and stress response in plants. Chinese Bulletin of Botany 45: 615–631.

32. Ruiz JM, Sanchez E, Garcia PC, Lopez-Lefebre LR, Rivero RM, et al. (2002) Proline metabolism and NAD kinase activity in green bean plants subjected to cold-shock. Phytochemistry 59: 473–478. doi: 10.1016/s0031-9422(01)00481-2

33. Delumeau O, Paven M-CM-L, Montrichard F, Laval-Martin DL (2000) Effects of short-term NaCl stress on calmodulin transcript levels and calmodulin-dependent NAD kinase activity in two species of tomato. Plant, Cell & Environment 23: 329–336. doi: 10.1046/j.1365-3040.2000.00545.x

34. Zagdanska B (1990) NAD kinase activity in wheat leaves under water deficit. Acta Biochim Pol 1990 37(3): 385–9.

35. Shi F, Li Y, Wang X (2009) Molecular properties, functions, and potential applications of NAD kinases. Acta Biochim Biophys Sin (Shanghai) 41: 352–361. doi: 10.1093/abbs/gmp029

36. Waller JC, Dhanoa PK, Schumann U, Mullen RT, Snedden WA (2010) Subcellular and tissue localization of NAD kinases from *Arabidopsis*: compartmentalization of de novo NADP biosynthesis. Planta 231: 305–317. doi: 10.1007/s00425-009-1047-7

37. Wu LL, Zhou CY, Gao YS, Cong YX, Chen KM, et al. (2011) Cloning and genetic transformation of *OsNADK3* gene in rice. Journal of Nuclear Agricultural Sciences 25: 0863–0870.

38. Lee TH, Tang H, Wang X, Paterson AH (2013) PGDD: a database of gene and genome duplication in plants. Nucleic Acids Res 41: D1152–1158. doi: 10.1093/nar/gks1104

39. **39.**Hardison RC (1996) A brief history of hemoglobins: plant, animal, protist, and bacteria. Proc Natl Acad Sci U S A 93: 5675–5679. doi: 10.1073/pnas.93.12.5675

40. olf YI, Sorokin AV, Mirkin BG, Koonin EV (2003) Remarkable interkingdom conservation of intron positions and massive, lineage-specific intron loss and gain in eukaryotic evolution. Curr Biol 13: 1512–1517. doi: 10.1016/s0960-9822(03)00558-x

41. Li W, Liu B, Yu L, Feng D, Wang H, et al. (2009) Phylogenetic analysis, structural evolution and functional divergence of the 12-oxo-phytodienoate acid reductase gene family in plants. BMC Evol Biol 9: 90. doi: 10.1186/1471-2148-9-90

42. Winter D, Vinegar B, Nahal H, Ammar R, Wilson GV, et al. (2007) An "Electronic Fluorescent Pictograph" browser for exploring and analyzing large-scale biological data sets. PLoS One 2: e718. doi: 10.1371/journal.pone.0000718

43. Bowers JE, Chapman BA, Rong J, Paterson AH (2003) Unravelling angiosperm genome evolution by phylogenetic analysis of chromosomal duplication events. Nature 422: 433–438. doi: 10.1038/nature01521

44. Van de Peer Y, Fawcett JA, Proost S, Sterck L, Vandepoele K (2009) The flowering world: a tale of duplications. Trends Plant Sci 14: 680–688. doi: 10.1016/j.tplants.2009.09.001

45. Magadum S, Banerjee U, Murugan P, Gangapur D, Ravikesavan R (2013) Gene duplication as a major force in evolution. J Genet 92: 155–161. doi: 10.1007/s12041-013-0212-8

46. Kolkman JA, Stemmer WP (2001) Directed evolution of proteins by exon shuffling. Nat Biotechnol 19: 423–428. doi: 10.1038/88084

47. Jones CD, Begun DJ (2005) Parallel evolution of chimeric fusion genes. Proc Natl Acad Sci U S A 102: 11373–11378. doi: 10.1073/pnas.0503528102

48. Kaessmann H (2010) Origins, evolution, and phenotypic impact of new genes. Genome Res 20: 1313–1326. doi: 10.1101/gr.101386.109

49. Morgante M, Brunner S, Pea G, Fengler K, Zuccolo A, et al. (2005) Gene duplication and exon shuffling by helitron-like transposons generate intraspecies diversity in maize. Nat Genet 37: 997–1002. doi: 10.1038/ng1615

50. Zhang Z, Kishino H (2004) Genomic background predicts the fate of duplicated genes: evidence from the yeast genome. Genetics 166: 1995–1999. doi: 10.1534/genetics.166.4.1995

51. Wang GF, Li WQ, Li WY, Wu GL, Zhou CY, et al. (2013) Characterization of Rice NADPH Oxidase Genes and Their Expression under Various Environmental Conditions. Int J Mol Sci 14: 9440–9458. doi: 10.3390/ijms14059440

52. Sagi M, Fluhr R (2006) Production of reactive oxygen species by plant NADPH oxidases. Plant Physiol 141: 336–340. doi: 10.1104/pp.106.078089

53. Foreman J, Demidchik V, Bothwell JH, Mylona P, Miedema H, et al. (2003) Reactive oxygen species produced by NADPH oxidase regulate plant cell growth. Nature 422: 442–446. doi: 10.1038/nature01485

54. Strand MK, Stuart GR, Longley MJ, Graziewicz MA, Dominick OC, et al. (2003) POS5 gene of *Saccharomyces cerevisiae* encodes a mitochondrial NADH kinase required for stability of mitochondrial DNA. Eukaryot Cell 2: 809–820. doi: 10.1128/ec.2.4.809-820.2003

55. Pain J, Balamurali MM, Dancis A, Pain D (2010) Mitochondrial NADH

kinase, Pos5p, is required for efficient iron-sulfur cluster biogenesis in *Saccharomyces cerevisiae*. J Biol Chem 285: 39409–39424. doi: 10.1074/jbc.m110.178947

56. Chehab EW, Kaspi R, Savchenko T, Rowe H, Negre-Zakharov F, et al. (2008) Distinct roles of jasmonates and aldehydes in plant-defense responses. PLoS One 3: e1904. doi: 10.1371/journal.pone.0001904
57. Wasternack C, Hause B (2013) Jasmonates: biosynthesis, perception, signal transduction and action in plant stress response, growth and development. An update to the 2007 review in Annals of Botany. Ann Bot 111: 1021–1058. doi: 10.1093/aob/mct067
58. Schaller A, Stintzi A (2009) Enzymes in jasmonate biosynthesis - structure, function, regulation. Phytochemistry 70: 1532–1538. doi: 10.1016/j.phytochem.2009.07.032
59.
60. **59.**Ma Q, Chirn GW, Szustakowski JD, Bakhtiarova A, Kosinski PA, et al. (2008) Uncovering mechanisms of transcriptional regulations by systematic mining of *cis*regulatory elements with gene expression profiles. BioData Min 1: 4. doi: 10.1186/1756-0381-1-4
61. .Jackson RJ, Hellen CU, Pestova TV (2010) The mechanism of eukaryotic translation initiation and principles of its regulation. Nat Rev Mol Cell Biol 11: 113–127. doi: 10.1038/nrm2838
62. .Price DC, Chan CX, Yoon HS, Yang EC, Qiu H, et al. (2012) *Cyanophora paradoxa*genome elucidates origin of photosynthesis in algae and plants. Science 335: 843–847. doi: 10.1126/science.1213561
63. .Matsuzaki M, Misumi O, Shin IT, Maruyama S, Takahara M, et al. (2004) Genome sequence of the ultrasmall unicellular red alga *Cyanidioschyzon merolae* 10D. Nature 428: 653–657. doi: 10.1038/nature02398
64. Barbier G, Oesterhelt C, Larson MD, Halgren RG, Wilkerson C, et al. (2005) Comparative genomics of two closely related unicellular thermo-acidophilic red algae,*Galdieria sulphuraria* and *Cyanidioschyzon merolae*, reveals the molecular basis of the metabolic flexibility of *Galdieria sulphuraria* and significant differences in carbohydrate metabolism of both algae. Plant Physiol 137: 460–474. doi: 10.1104/pp.104.051169
65. .Derelle E, Ferraz C, Rombauts S, Rouze P, Worden AZ, et al. (2006) Genome analysis of the smallest free-living eukaryote *Ostreococcus tauri* unveils many unique features. Proc Natl Acad Sci U S A 103: 11647–11652. doi: 10.1073/pnas.0604795103

66. .Palenik B, Grimwood J, Aerts A, Rouze P, Salamov A, et al. (2007) The tiny eukaryote*Ostreococcus* provides genomic insights into the paradox of plankton speciation. Proc Natl Acad Sci U S A 104: 7705–7710. doi: 10.1073/pnas.0611046104

67. .Worden AZ, Lee JH, Mock T, Rouze P, Simmons MP, et al. (2009) Green evolution and dynamic adaptations revealed by genomes of the marine picoeukaryotes*Micromonas*. Science 324: 268–272. doi: 10.1126/science.1167222

68. Blanc G, Duncan G, Agarkova I, Borodovsky M, Gurnon J, et al. (2010) The *Chlorella variabilis* NC64A genome reveals adaptation to photosymbiosis, coevolution with viruses, and cryptic sex. Plant Cell 22: 2943–2955. doi: 10.1105/tpc.110.076406

69. .Blanc G, Agarkova I, Grimwood J, Kuo A, Brueggeman A, et al. (2012) The genome of the polar eukaryotic microalga *Coccomyxa subellipsoidea* reveals traits of cold adaptation. Genome Biol 13: R39. doi: 10.1186/gb-2012-13-5-r39

70. .Merchant SS, Prochnik SE, Vallon O, Harris EH, Karpowicz SJ, et al. (2007) The*Chlamydomonas* genome reveals the evolution of key animal and plant functions. Science 318: 245–250. doi: 10.1126/science.1143609

71. Rensing SA, Lang D, Zimmer AD, Terry A, Salamov A, et al. (2008) The*Physcomitrella* genome reveals evolutionary insights into the conquest of land by plants. Science 319: 64–69. doi: 10.1126/science.1150646

72. .Banks JA, Nishiyama T, Hasebe M, Bowman JL, Gribskov M, et al. (2011) The*Selaginella* genome identifies genetic changes associated with the evolution of vascular plants. Science 332: 960–963. doi: 10.1126/science.1203810

73. .Ralph SG, Chun HJ, Kolosova N, Cooper D, Oddy C, et al. (2008) A conifer genomics resource of 200,000 spruce (*Picea spp.*) ESTs and 6,464 high-quality, sequence-finished full-length cDNAs for Sitka spruce (*Picea sitchensis*). BMC Genomics 9: 484. doi: 10.1186/1471-2164-9-484

74. International-Brachypodium-Initiative (2010) Genome sequencing and analysis of the model grass *Brachypodium distachyon*. Nature 463: 763–768. doi: 10.3410/f.2203965.1814064

75. Zhang G, Liu X, Quan Z, Cheng S, Xu X, et al. (2012) Genome sequence of foxtail millet (*Setaria italica*) provides insights into grass evolution and biofuel potential. Nat Biotechnol 30: 549–554. doi: 10.1038/nbt.2195

76. Paterson AH, Bowers JE, Bruggmann R, Dubchak I, Grimwood J, et al. (2009) The*Sorghum bicolor* genome and the diversification of grasses.

Nature 457: 551–556. doi: 10.1038/nature07723

77. Goff SA, Ricke D, Lan TH, Presting G, Wang R, et al. (2002) A draft sequence of the rice genome (*Oryza sativa* L. *ssp. japonica*). Science 296: 92–100. doi: 10.1126/science.1068275
78. Schnable PS, Ware D, Fulton RS, Stein JC, Wei F, et al. (2009) The B73 maize genome: complexity, diversity, and dynamics. Science 326: 1112–1115.
79. Tuskan GA, Difazio S, Jansson S, Bohlmann J, Grigoriev I, et al. (2006) The genome of black cottonwood, *Populus trichocarpa* (Torr. & Gray). Science 313: 1596–1604. doi: 10.1126/science.1128691
80. Huang S, Li R, Zhang Z, Li L, Gu X, et al. (2009) The genome of the cucumber,*Cucumis sativus* L. Nat Genet. 41: 1275–1281. doi: 10.1038/ng.475
81. Arabidopsis-Genome-Initiative (2000) Analysis of the genome sequence of the flowering plant *Arabidopsis thaliana*. Nature 408: 796–815. doi: 10.1038/35048692
82. Wang X, Wang H, Wang J, Sun R, Wu J, et al. (2011) The genome of the mesopolyploid crop species *Brassica rapa*. Nat Genet 43: 1035–1039. doi: 10.1038/ng.919
83. Jaillon O, Aury JM, Noel B, Policriti A, Clepet C, et al. (2007) The grapevine genome sequence suggests ancestral hexaploidization in major angiosperm phyla. Nature 449: 463–467. doi: 10.1038/nature06148
84. Zhang Z, Wood WI (2003) A profile hidden Markov model for signal peptides generated by HMMER. Bioinformatics 19: 307–308. doi: 10.1093/bioinformatics/19.2.307
85. Finn RD, Clements J, Eddy SR (2011) HMMER web server: interactive sequence similarity searching. Nucleic Acids Res 39: W29–37. doi: 10.1093/nar/gkr367
86. Finn RD, Mistry J, Schuster-Bockler B, Griffiths-Jones S, Hollich V, et al. (2006) Pfam: clans, web tools and services. Nucleic Acids Res 34: D247–251. doi: 10.1093/nar/gkj149
87. Punta M, Coggill PC, Eberhardt RY, Mistry J, Tate J, et al. (2012) The Pfam protein families database. Nucleic Acids Res 40: D290–301. doi: 10.1093/nar/gkr1065
88. Schultz J, Milpetz F, Bork P, Ponting CP (1998) SMART, a simple modular architecture research tool: identification of signaling domains. Proc Natl Acad Sci U S A 95: 5857–5864. doi: 10.1073/pnas.95.11.5857
89. Letunic I, Doerks T, Bork P (2012) SMART 7: recent updates to the

protein domain annotation resource. Nucleic Acids Res 40: D302–305. doi: 10.1093/nar/gkr931

90. Edgar RC (2004) MUSCLE: multiple sequence alignment with high accuracy and high throughput. Nucleic Acids Res 32: 1792–1797. doi: 10.1093/nar/gkh340

91. Edgar RC (2004) MUSCLE: a multiple sequence alignment method with reduced time and space complexity. BMC Bioinformatics 5: 113.

92. .Hall TA (1999) BioEdit: a user-friendly biological sequence alignment editor and analysis program for Windows 95/98/NT. Nucleic Acids Symposium Series 41: 95–98.

93. Krawetz S, Retief J (1999) Phylogenetic Analysis Using PHYLIP. Bioinformatics Methods and Protocols: Humana Press. 243–258.

94. Jones DT, Taylor WR, Thornton JM (1992) The rapid generation of mutation data matrices from protein sequences. Comput Appl Biosci 8: 275–282. doi: 10.1093/bioinformatics/8.3.275

95. Tamura K, Peterson D, Peterson N, Stecher G, Nei M, et al. (2011) MEGA5: molecular evolutionary genetics analysis using maximum likelihood, evolutionary distance, and maximum parsimony methods. Mol Biol Evol 28: 2731–2739. doi: 10.1093/molbev/msr121

96. Yang Z (2007) PAML 4: phylogenetic analysis by maximum likelihood. Mol Biol Evol 24: 1586–1591. doi: 10.1093/molbev/msm088

97. Suyama M, Torrents D, Bork P (2006) PAL2NAL: robust conversion of protein sequence alignments into the corresponding codon alignments. Nucleic Acids Res 34: W609–612. doi: 10.1093/nar/gkl315

98. Puranik S, Sahu PP, Mandal SN, B VS, Parida SK, et al. (2013) Comprehensive genome-wide survey, genomic constitution and expression profiling of the NAC transcription factor family in foxtail millet (*Setaria italica* L.). PLoS One 8: e64594. doi: 10.1371/journal.pone.0064594

99. Lynch M, Conery JS (2000) The evolutionary fate and consequences of duplicate genes. Science 290: 1151–1155. doi: 10.1126/science.290.5494.1151

100. .Guo AY, Zhu QH, Chen X, Luo JC (2007) GSDS: a gene structure display server. Yi Chuan 29(8): 1023–6. doi: 10.1360/yc-007-1023

101. .Goodstein DM, Shu S, Howson R, Neupane R, Hayes RD, et al. (2012) Phytozome: a comparative platform for green plant genomics. Nucleic Acids Res 40: D1178–1186. doi: 10.1093/nar/gkr944

102. Wilkerson MD, Ru Y, Brendel VP (2009) Common introns within

orthologous genes: software and application to plants. Briefings in Bioinformatics 10: 631–644. doi: 10.1093/bib/bbp051

103. Bailey TL, Boden M, Buske FA, Frith M, Grant CE, et al. (2009) MEME SUITE: tools for motif discovery and searching. Nucleic Acids Res 37: W202–208. doi: 10.1093/nar/gkp335

104. Crooks GE, Hon G, Chandonia JM, Brenner SE (2004) WebLogo: a sequence logo generator. Genome Res 14: 1188–1190. doi: 10.1101/gr.849004

105. Lescot M, Dehais P, Thijs G, Marchal K, Moreau Y, et al. (2002) PlantCARE, a database of plant *cis*-acting regulatory elements and a portal to tools for in silico analysis of promoter sequences. Nucleic Acids Res 30: 325–327. doi: 10.1093/nar/30.1.325

106. Rombauts S, Dehais P, Van Montagu M, Rouze P (1999) PlantCARE, a plant *cis*-acting regulatory element database. Nucleic Acids Res 27: 295–296. doi: 10.1093/nar/27.1.295

107. Hruz T, Laule O, Szabo G, Wessendorp F, Bleuler S, et al. (2008) Genevestigator v3: a reference expression database for the meta-analysis of transcriptomes. Adv Bioinformatics 2008: 420747. doi: 10.1155/2008/420747

CITATION

CHAPTER 1

Ni J, Hu S, Zhang J, Tang W, Lu W, Zhang C (2015) A Preliminary Genetic Analysis of Complement 3 Gene and Schizophrenia. PLoS ONE 10(8): e0136372. doi:10.1371/journal.pone.0136372.

CHAPTER 2

Montaner D, Dopazo J (2010) Multidimensional Gene Set Analysis of Genomic Data. PLoS ONE 5(4): e10348. doi:10.1371/journal.pone.0010348.

CHAPTER 3

Liu W, Fang L, Li M, Li S, Guo S, Luo R, et al. (2012) Comparative Genomics of Mycoplasma: Analysis of Conserved Essential Genes and Diversity of the Pan-Genome. PLoS ONE 7(4): e35698. doi:10.1371/journal.pone.0035698

CHAPTER 4

Kim Y, Park T (2015) Robust Gene-Gene Interaction Analysis in Genome Wide Association Studies. PLoS ONE 10(8): e0135016. doi:10.1371/journal.pone.0135016.

CHAPTER 5

Pingzhao Hu and Andrew D Paterson, Dynamic pathway analysis of genes associated with blood pressure using whole genome sequence data, DOI: 10.1186/1753-6561-8-S1-S106.

CHAPTER 6

Peng Jiang, Hongfang Wang, Wei Li, Chongzhi Zang, Bo Li, Yinling J. Wong, Cliff Meyer, Jun S. Liu, Jon C. Aster and X. Shirley Liu, Network analysis of gene essentiality in functional genomics experiments, DOI: 10.1186/s13059-015-0808-9.

CHAPTER 7

Rivarola M, Foster JT, Chan AP, Williams AL, Rice DW, Liu X, et al. (2011) Castor Bean Organelle Genome Sequencing and Worldwide Genetic Diversity Analysis. PLoS ONE 6(7): e21743. doi:10.1371/journal.pone.0021743.

CHAPTER 8

Tian C, Kosoy R, Lee A, Ransom M, Belmont JW, Gregersen PK, et al. (2008) Analysis of East Asia Genetic Substructure Using Genome-Wide SNP Arrays. PLoS ONE 3(12): e3862. doi:10.1371/journal.ponc.0003862.

CHAPTER 9

Loza MJ, McCall CE, Li L, Isaacs WB, Xu J, Chang B-L (2007) Assembly of Inflammation-Related Genes for Pathway-Focused Genetic Analysis. PLoS ONE 2(10): e1035. doi:10.1371/journal.pone.0001035.

CHAPTER 10

Fadhl M. Al-Akwaa (2012). Analysis of Gene Expression Data Using Biclustering Algorithms, Functional Genomics, Dr. Germana Meroni (Ed.), ISBN: 978-953-51-0727-9, InTech, DOI: 10.5772/48150.

CHAPTER 11

Peng Xu, Xiaofeng Zhang, Xumin Wang, Jiongtang Li, Guiming Liu, Youyi Kuang, Jian Xu, Xianhu Zheng, Lufeng Ren, Guoliang Wang, Yan Zhang, Linhe Huo, Zixia Zhao, Dingchen Cao, Cuiyun Lu, Chao Li, Yi Zhou, Zhanjiang Liu, Zhonghua Fan, Guangle Shan, Xingang Li, Shuangxiu Wu, Lipu Song, Guangyuan Hou, Yanliang Jiang, Zsigmond Jeney, Dan Yu, Li Wang, Changjun Shao, Lai Song, Jing Sun, Peifeng Ji, Jian Wang, Qiang Li, Liming Xu, Fanyue Sun, Jianxin Feng, Chenghui Wang, Shaolin Wang, Baosen Wang, Yan Li, Yaping Zhu, Wei Xue, Lan Zhao, Jintu Wang, Ying Gu, Weihua Lv, Kejing Wu, Jingfa Xiao, Jiayan Wu, Zhang Zhang, Jun Yu & Xiaowen Sun Show fewer authors, Genome sequence and genetic diversity of the common carp, Cyprinus carpio, doi:10.1038/ng.3098

CHAPTER 12

Chol-Hee Jung , Chui E. Wong , Mohan B. Singh , Prem L. Bhalla; Comparative Genomic Analysis of Soybean Flowering Genes; DOI: 10.1371/journal.pone.0038250

CHAPTER 13

Li W-Y, Wang X, Li R, Li W-Q, Chen K-M (2014) Genome-Wide Analysis of the NADK Gene Family in Plants. PLoS ONE 9(6): e101051. doi:10.1371/journal.pone.0101051

INDEX

Encyclopaedia of MODERN GENETICS ANALYSIS INTEGRATING GENES AND GENOMES

Encyclopaedia of MODERN GENETICS ANALYSIS INTEGRATING GENES AND GENOMES

Volume II: Modern Genetic Analysis

Contributors

Jianliang Ni, Weidong Mao., and Tao He et al.

AURIS
Reference

www.aurisreference.com

Encyclopaedia of Modern Genetics Analysis Integrating Genes and Genomes

Volume II: Modern Genetic Analysis

Contributors: Jianliang Ni, Weidong Mao., and Tao He al.

Published by Auris Reference Limited

www.aurisreference.com

United Kingdom

Printed in 2017 for Sale in the Indian Subcontinent

Modern Genetic Analysis

ISBN: 978-1-78154-772-4 (Set of 3 Volumes)

British Library Cataloguing in Publication Data
A CIP record for this book is available from the British Library

Printed in the United Kingdom

Exclusively distributed by CBS Publishers & Distributors Pvt. Ltd.

Sales & Distribution Rights only for India, Pakistan, Bangladesh, Sri Lanka, Nepal and Bhutan.This book is not to be sold outside these territories.

Contents

List of Abbreviations

AJCPD	Abnormal junction of the choledochopancreatic ducts
APC	Adenomatous polyposis coli
ASO	Allele-specific oligonucleotide
ABC	Approximate Bayesian Computation
BMI	Body mass index
CNS-GCTs	Central nervous system germ cell tumors
CGH	Comparative genomic hybridization
CMLM	Compressed mixed linear model
ERV	Endogenous retrovirus
EGDC	Equine Genetic Diversity Consortium
GIF	Genealogical Index of Familiality
GRF	General random forest
GWAS	Genome-wide association studies
GBS	Genotyping-by-sequencing
HVR-I	Hypervariable region I
IBD	Inflammatory bowel disease
IACUC	Institutional Animal Care and Use Committee
IALHA I	nternational Andalusian and Lusitano Horse Association Registry
LD	Linkage disequilibria
LTR	Long terminal repeat
MSY	Male specific euchromatic
MTA	Material Transfer Agreement
MPWD	Mean number of pairwise differences
MSI	Microsatellite instability
MAF	Minor allele frequency
mtDNA	Mitochondrial DNA
MAGIC	Multiparent advanced generation intercrosses
MSC	Multi-SNPs combination
NHANES	National Health and Nutrition Examination Survey
NAM	Nested association mapping
ORF	Optimum random forest
PCA	Principal components analysis
PAR	Pseudoautosomal region
PODs	Pseudo-observed datasets
QTLs	Quantitative trait loci
ROC	Receiver operating characteristic
RIL	Recombinant inbred line
RGE	Relative genotypic effect
rCRS	Revised Cambridge Reference Sequence
SSvR	Seedling survival rates

SSLP	Simple sequence length polymorphism
SSR	Simple-sequence repeat
SNP	Single nucleotide polymorphism
SNACs	Sinonasal adenocarcinomas
SDS	Sodium dodecyl sulfate
SAP	Sorghum association panel
SVM	Support Vector Machine Algorithm
TWHF	Tennessee Walking Horse Foundation
TRH	Thyrotropin-releasing hormone
UPDB	Utah Population Data Base
WGS	Whole Genome Shotgun

List of Contributors

Weidong Mao
Department of Mathematics & Computer Science Virginia State University Petersburg, VA 23806, USA

Jeonghwa Lee
Department of Computer Science Shippensburg University Shippensburg, PA 17257, USA

Anil V Parwani M.D., Ph.D.
Department of Pathology, The Johns Hopkins Hospital, Baltimore, Maryland

Joseph Geradts M.D.
Department of Pathology and Laboratory Medicine, Roswell Park Cancer Institute, Buffalo, New York

Eric Caspers M.D.
The Academic Medical Center, Amsterdam, The Netherlands

G Johan Offerhaus M.D.
The Academic Medical Center, Amsterdam, The Netherlands

Charles J Yeo M.D.
Department of Surgery, The Johns Hopkins Hospital, Baltimore, Maryland
Department of Oncology, The Johns Hopkins Hospital, Baltimore, Maryland

John L Cameron M.D.
Department of Surgery, The Johns Hopkins Hospital, Baltimore, Maryland

David S Klimstra M.D.
Department of Pathology, Memorial Sloan-Kettering Cancer Center, New York

Anirban Maitra M.D.
Department of Pathology, The Johns Hopkins Hospital, Baltimore, Maryland

Ralph H Hruban M.D.
Department of Pathology, The Johns Hopkins Hospital, Baltimore, Maryland
Department of Oncology, The Johns Hopkins Hospital, Baltimore, Maryland

Pedram Argani M.D.
Department of Pathology, The Johns Hopkins Hospital, Baltimore, Maryland

Takashi Kuramoto
Institute of Laboratory Animals, Graduate School of Medicine, Kyoto University, Sakyo-ku, Kyoto, Japan

Satoshi Nakanishi
Institute of Laboratory Animals, Graduate School of Medicine, Kyoto University, Sakyo-ku, Kyoto, Japan

Masako Ochiai
National Cancer Center Research Institute, Chuo-ku, Tokyo, Japan

Hitoshi Nakagama
National Cancer Center Research Institute, Chuo-ku, Tokyo, Japan

Birger Voigt
Institute of Laboratory Animals, Graduate School of Medicine, Kyoto University, Sakyo-ku, Kyoto, Japan

Tadao Serikawa
Institute of Laboratory Animals, Graduate School of Medicine, Kyoto University, Sakyo-ku, Kyoto, Japan

Dominik T Schneider
Clinic of Paediatric Oncology, Haematology and Immunology, Heinrich-Heine-University, Düsseldorf, Germany

Susanne Zahn
Clinic of Paediatric Oncology, Haematology and Immunology, Heinrich-Heine-University, Düsseldorf, Germany

Sonja Sievers
Clinic of Paediatric Oncology, Haematology and Immunology, Heinrich-Heine-University, Düsseldorf, Germany
Max Planck Institute of Molecular Physiology, Dortmund, Germany

Katayoun Alemazkour
Clinic of Paediatric Oncology, Haematology and Immunology, Heinrich-Heine-University, Düsseldorf, Germany

Guido Reifenberger
Department of Neuropathology, Heinrich-Heine-University, Düsseldorf, Germany

Otmar D Wiestler
German Brain Tumor Reference Center Institute of Neuropathology, University of Bonn, Bonn, Germany

Gabriele Calaminus
Clinic of Paediatric Oncology, Haematology and Immunology, Heinrich-Heine-University, Düsseldorf, Germany

Ulrich Göbel
Clinic of Paediatric Oncology, Haematology and Immunology, Heinrich-Heine-University, Düsseldorf, Germany

Elizabeth J Perlman
Department of Pathology, Children's Memorial Hospital, Chicago, IL, USA

Sue S Yom
Department of Radiation Oncology, The University of Texas MD Anderson Cancer Center, Houston, TX, USA

Asif Rashid
Department of Pathology, The University of Texas MD Anderson Cancer Center, Houston, TX, USA

David I Rosenthal
Department of Radiation Oncology, The University of Texas MD Anderson Cancer Center, Houston, TX, USA

Danielle D Elliott
Department of Pathology, The University of Texas MD Anderson Cancer Center, Houston, TX, USA

Ehab Y Hanna
Department of Head and Neck Surgery, The University of Texas MD Anderson Cancer Center, Houston, TX, USA

Randal S Weber
Department of Head and Neck Surgery, The University of Texas MD Anderson Cancer Center, Houston, TX, USA

Adel K El-Naggar
Department of Pathology, The University of Texas MD Anderson Cancer Center, Houston, TX, USA
Department of Head and Neck Surgery, The University of Texas MD Anderson Cancer Center, Houston, TX, USA

Rui Sousa-Neves
Department of Biology, Case Western Reserve University, Cleveland, Ohio, United States of America

Alexandre Rosas
Departamento de Fı´sica, CCEN, Universidade Federal da Paraı´ba, Joa˜o Pessoa, Paraı´ba, Brazil

Qiuying Huang
Hubei Insect Resources Utilization and Sustainable Pest Management Key Laboratory, Huazhong Agricultural University, Wuhan, China

Ganghua Li
Hubei Insect Resources Utilization and Sustainable Pest Management Key Laboratory, Huazhong Agricultural University, Wuhan, China

Claudia Husseneder
Department of Entomology, Louisiana State University Agricultural Center, Baton Rouge, Louisiana, United States of America

Chaoliang Lei
Hubei Insect Resources Utilization and Sustainable Pest Management Key Laboratory, Huazhong Agricultural University, Wuhan, China

Yinghua Pan
Key Laboratory of Crop Heterosis and Utilization, Ministry of Education, Beijing Key Laboratory of Crop Genetic Improvement, China Agricultural University, Beijing, 100193, China
Rice Research Institute, Guangxi Academy of Agricultural Sciences, Nanning, Guangxi, 530005, China

Hongliang Zhang
Key Laboratory of Crop Heterosis and Utilization, Ministry of Education, Beijing Key Laboratory of Crop Genetic Improvement, China Agricultural University, Beijing, 100193, China

Dongling Zhang
Institute of Crop Science, Chinese Academy of Agricultural Sciences, the National Key Facility for Crop Gene Resources and Genetic Improvement, Key Laboratory of Crop Germplasm Resources and Biotechnology, Ministry of Agriculture, Beijing, 100081, China

Jinjie Li
Key Laboratory of Crop Heterosis and Utilization, Ministry of Education, Beijing Key Laboratory of Crop Genetic Improvement, China Agricultural University, Beijing, 100193, China

Haiyan Xiong
Key Laboratory of Crop Heterosis and Utilization, Ministry of Education, Beijing Key Laboratory of Crop Genetic Improvement, China Agricultural University, Beijing, 100193, China

Jianping Yu
Key Laboratory of Crop Heterosis and Utilization, Ministry of Education, Beijing Key Laboratory of Crop Genetic Improvement, China Agricultural University, Beijing, 100193, China

Jilong Li
Key Laboratory of Crop Heterosis and Utilization, Ministry of Education, Beijing Key Laboratory of Crop Genetic Improvement, China Agricultural University, Beijing, 100193, China

Muhammad Abdul Rehman Rashid
Key Laboratory of Crop Heterosis and Utilization, Ministry of Education, Beijing Key Laboratory of Crop Genetic Improvement, China Agricultural University, Beijing, 100193, China

Gangling Li
Key Laboratory of Crop Heterosis and Utilization, Ministry of Education, Beijing Key Laboratory of Crop Genetic Improvement, China Agricultural University, Beijing, 100193, China

Xiaoding Ma
Institute of Crop Science, Chinese Academy of Agricultural Sciences, the National Key Facility for Crop Gene Resources and Genetic Improvement, Key Laboratory of Crop Germplasm Resources and Biotechnology, Ministry of Agriculture, Beijing, 100081, China

Guilan Cao
Institute of Crop Science, Chinese Academy of Agricultural Sciences, the National Key Facility for Crop Gene Resources and Genetic Improvement, Key Laboratory of Crop Germplasm Resources and Biotechnology, Ministry of Agriculture, Beijing, 100081, China

Longzhi Han
Institute of Crop Science, Chinese Academy of Agricultural Sciences, the National Key Facility for Crop Gene Resources and Genetic Improvement, Key Laboratory of Crop Germplasm Resources and Biotechnology, Ministry of Agriculture, Beijing, 100081, China

Zichao Li
Key Laboratory of Crop Heterosis and Utilization, Ministry of Education, Beijing Key Laboratory of Crop Genetic Improvement, China Agricultural University, Beijing, 100193, China

William R. Yates
Laureate Institute for Brain Research, Tulsa, Oklahoma, United States of America
University of Oklahoma College of Medicine, Tulsa, Tulsa, Oklahoma, United States of America

Craig Johnson
Eating Recovery Center, Denver, Colorado, United States of America

Patrick McKee
Laureate Institute for Brain Research, Tulsa, Oklahoma, United States of America
University of Oklahoma College of Medicine, Oklahoma City, Oklahoma, United States of America

Lisa A. Cannon-Albright
Department of Medicine (Genetic Epidemiology), University of Utah School of Medicine, Salt Lake City, Utah, United States of America
George E. Wahlen Department of Veterans Affairs Medical Center, Salt Lake City, Utah, United States of America

Dong Zhang
Plant Genome Mapping Laboratory, University of Georgia, Athens, GA 30602, USA
Institute of Bioinformatics, University of Georgia, Athens, GA 30602, USA

Wenqian Kong
Plant Genome Mapping Laboratory, University of Georgia, Athens, GA 30602, USA
Department of Crop and Soil Sciences, University of Georgia

Jon Robertson
Plant Genome Mapping Laboratory, University of Georgia, Athens, GA 30602, USA

Valorie H Goff
Plant Genome Mapping Laboratory, University of Georgia, Athens, GA 30602, USA

Ethan Epps
Plant Genome Mapping Laboratory, University of Georgia, Athens, GA 30602, USA

Alexandra Kerr
Plant Genome Mapping Laboratory, University of Georgia, Athens, GA 30602, USA

Gabriel Mills
Plant Genome Mapping Laboratory, University of Georgia, Athens, GA 30602, USA

Jay Cromwell
Plant Genome Mapping Laboratory, University of Georgia, Athens, GA 30602, USA

Yelena Lugin
Plant Genome Mapping Laboratory, University of Georgia, Athens, GA 30602, USA

Christine Phillips
Plant Genome Mapping Laboratory, University of Georgia, Athens, GA 30602, USA

Andrew H Paterson
Plant Genome Mapping Laboratory, University of Georgia, Athens, GA 30602, USA
Institute of Bioinformatics, University of Georgia, Athens, GA 30602, USA
Department of Crop and Soil Sciences, University of Georgia
Department of Plant Biology, University of Georgia
Department of Genetics, University of Georgia

Jessica L. Petersen
University of Minnesota, College of Veterinary Medicine, St Paul, Minnesota, United States of America

James R. Mickelson
University of Minnesota, College of Veterinary Medicine, St Paul, Minnesota, United States of America

E. Gus Cothran
Texas A&M University, College of Veterinary Medicine and Biomedical Science, College Station, Texas, United States of America

Lisa S. Andersson
Swedish University of Agricultural Sciences, Department of Animal Breeding and Genetics, Uppsala, Sweden

Jeanette Axelsson
Swedish University of Agricultural Sciences, Department of Animal Breeding and Genetics, Uppsala, Sweden

Ernie Bailey
University of Kentucky, Department of Veterinary Science, Lexington, Kentucky, United States of America

Danika Bannasch
University of California Davis, School of Veterinary Medicine, Davis, California, United States of America

Matthew M. Binns
Equine Analysis, Midway, Kentucky, United States of America

Alexandre S. Borges
University Estadual Paulista, Department of Veterinary Clinical Science, Botucatu-SP, Brazil

Pieter Brama
University College Dublin, School of Veterinary Medicine, Dublin, Ireland

Artur da Ca^mara Machado
University of Azores, Institute for Biotechnology and Bioengineering, Biotechnology Centre of Azores, Angra do Heroı´smo, Portugal

Ottmar Distl
University of Veterinary Medicine Hannover, Institute for Animal Breeding and Genetics, Hannover, Germany

Michela Felicetti
University of Perugia, Faculty of Veterinary Medicine, Perugia, Italy

Laura Fox-Clipsham
Animal Health Trust, Lanwades Park, Newmarket, Suffolk, United Kingdom

Kathryn T. Graves
University of Kentucky, Department of Veterinary Science, Lexington, Kentucky, United States of America

Ge´ rard Gue´ rin
French National Institute for Agricultural Research-Animal Genetics and Integrative Biology Unit, Jouy en Josas, France

Bianca Haase
University of Sydney, Veterinary Science, New South Wales, Australia

Telhisa Hasegawa
Nihon Bioresource College, Koga, Ibaraki, Japan

Karin Hemmann
University of Helsinki, Faculty of Veterinary Medicine, Helsinki, Finland

Emmeline W. Hill
University College Dublin, College of Agriculture, Food Science and Veterinary Medicine, Belfield, Dublin, Ireland

Tosso Leeb
University of Bern, Institute of Genetics, Bern, Switzerland

Gabriella Lindgren
Swedish University of Agricultural Sciences, Department of Animal Breeding and Genetics, Uppsala, Sweden

Hannes Lohi
University of Helsinki, Faculty of Veterinary Medicine, Helsinki, Finland

Maria Susana Lopes
University of Azores, Institute for Biotechnology and Bioengineering, Biotechnology Centre of Azores, Angra do Heroı´smo, Portugal

Beatrice A. McGivney
University College Dublin, College of Agriculture, Food Science and Veterinary Medicine, Belfield, Dublin, Ireland

Sofia Mikko
Swedish University of Agricultural Sciences, Department of Animal Breeding and Genetics, Uppsala, Sweden

Nicholas Orr
Institute of Cancer Research, Breakthrough Breast Cancer Research Centre, London, United Kingdom

M. Cecilia T Penedo
University of California Davis, School of Veterinary Medicine, Davis, California, United States of America

Richard J. Piercy
Royal Veterinary College, Comparative Neuromuscular Diseases Laboratory, London, United Kingdom

Marja Raekallio
University of Helsinki, Faculty of Veterinary Medicine, Helsinki, Finland

Stefan Rieder
Swiss National Stud Farm, Agroscope Liebefeld-Posieux Research Station, Avenches, Switzerland

Knut H. Røed
Norwegian School of Veterinary Science, Department of Basic Sciences and Aquatic Medicine, Oslo, Norway

Maurizio Silvestrelli
University of Perugia, Faculty of Veterinary Medicine, Perugia, Italy

June Swinburne
Animal Health Trust, Lanwades Park, Newmarket, Suffolk, United Kingdom
Animal DNA Diagnostics Ltd, Cambridge, United Kingdom

Teruaki Tozaki
Laboratory of Racing Chemistry, Department of Molecular Genetics, Utsunomiya, Tochigi, Japan

Mark Vaudin
Animal Health Trust, Lanwades Park, Newmarket, Suffolk, United Kingdom

Claire M. Wade
University of Sydney, Veterinary Science, New South Wales, Australia

Molly E. McCue
University of Minnesota, College of Veterinary Medicine, St Paul, Minnesota, United States of America

Danielle A. Badro
The Lebanese American University, Chouran, Beirut, Lebanon

Bouchra Douaihy
The Lebanese American University, Chouran, Beirut, Lebanon

Marc Haber
The Lebanese American University, Chouran, Beirut, Lebanon
Institut de Biologia Evolutiva (CSIC-UPF), Departament de Cie`ncies de la Salut i de la Vida, Universitat Pompeu Fabra, Barcelona, Spain

Sonia C. Youhanna
The Lebanese American University, Chouran, Beirut, Lebanon

Ange´lique Salloum
The Lebanese American University, Chouran, Beirut, Lebanon

Michella Ghassibe-Sabbagh
The Lebanese American University, Chouran, Beirut, Lebanon

Brian Johnsrud
Modern Thought and Literature, Stanford University, Stanford, California, United States of America

Georges Khazen
The Lebanese American University, Chouran, Beirut, Lebanon

Elizabeth Matisoo-Smith
Allan Wilson Centre for Molecular Ecology and Evolution, University of Otago, Dunedin, New Zealand

David F. Soria-Hernanz
Institut de Biologia Evolutiva (CSIC-UPF), Departament de Cie`ncies de la Salut i de la Vida, Universitat Pompeu Fabra, Barcelona, Spain
The Genographic Project, National Geographic Society, Washington, DC, United States of America

R. Spencer Wells
The Genographic Project, National Geographic Society, Washington, DC, United States of America

Chris Tyler-Smith
The Wellcome Trust Sanger Institute, Wellcome Trust Genome Campus, Hinxton, United Kingdom

Daniel E. Platt
Computational Biology Centre, IBM TJ Watson Research Centre, Yorktown Heights, New York, United States of America

Pierre A. Zalloua
The Lebanese American University, Chouran, Beirut, Lebanon
Harvard School of Public Health, Boston, Massachusetts, United States of America

Barbara Wallner
Institute of Animal Breeding and Genetics, Department of Biomedical Sciences, University of Veterinary Medicine Vienna, Vienna, Austria

Claus Vogl
Institute of Animal Breeding and Genetics, Department of Biomedical Sciences, University of Veterinary Medicine Vienna, Vienna, Austria

Priyank Shukla
Institute of Animal Breeding and Genetics, Department of Biomedical Sciences, University of Veterinary Medicine Vienna, Vienna, Austria

Joerg P. Burgstaller
Institute of Animal Breeding and Genetics, Department of Biomedical Sciences, University of Veterinary Medicine Vienna, Vienna, Austria

Thomas Druml
Institute of Animal Breeding and Genetics, Department of Biomedical Sciences, University of Veterinary Medicine Vienna, Vienna, Austria

Gottfried Brem
Institute of Animal Breeding and Genetics, Department of Biomedical Sciences, University of Veterinary Medicine Vienna, Vienna, Austria

Stefania Vai
Dipartimento di Biologia Evoluzionistica, Università di Firenze, 50122 Florence, Italy

Silvia Ghirotto
Dipartimento di Scienze della Vita e Biotecnologie, Università di Ferrara, 44121 Ferrara, Italy

Elena Pilli
Dipartimento di Biologia Evoluzionistica, Università di Firenze, 50122 Florence, Italy

Francesca Tassi
Dipartimento di Scienze della Vita e Biotecnologie, Università di Ferrara, 44121 Ferrara, Italy

Martina Lari
Dipartimento di Biologia Evoluzionistica, Università di Firenze, 50122 Florence, Italy

Ermanno Rizzi
Institute for Biomedical Technologies, National Research Council, 20090 Segrate, Milan, Italy

Laura Matas-Lalueza
Institut de Biologia Evolutiva, CSIC-UPF, Barcelona 08003, Spain

Oscar Ramirez
Institut de Biologia Evolutiva, CSIC-UPF, Barcelona 08003, Spain

Carles Lalueza-Fox
Institut de Biologia Evolutiva, CSIC-UPF, Barcelona 08003, Spain

Alessandro Achilli
Dipartimento di Chimica, Biologia e Biotecnologie, Università di Perugia, 06123 Perugia, Italy

Anna Olivieri
Dipartimento di Biologia e Biotecnologie "L. Spallanzani", Università di Pavia, 27100,Pavia,Italy

Antonio Torroni
Dipartimento di Biologia e Biotecnologie "L. Spallanzani", Università di Pavia, 27100,Pavia,Italy

Hovirag Lancioni
Dipartimento di Chimica, Biologia e Biotecnologie, Università di Perugia, 06123 Perugia, Italy

Caterina Giostra
Dipartimento di Storia, Archeologia e Storia dell'arte, Università Cattolica del Sacro Cuore, 20123 Milano, Italy

Elena Bedini
Anthropozoologica L.B.A. s.n.c., 57123 Livorno, Italy

Luisella Pejrani Baricco
Soprintendenza per i Beni Archeologici del Piemonte, 10122 Turin, Italy

Giuseppe Matullo
Human Genetics Foundation, 10125 Turin, Italy

Cornelia Di Gaetano
Human Genetics Foundation, 10125 Turin, Italy

Alberto Piazza
Human Genetics Foundation, 10125 Turin, Italy

Krishna Veeramah
Department of Ecology and Evolution, State University of New York, Stony Brook, New York 11794– 5245, United States of America

Patrick Geary
School of Historical Studies, Institute for Advanced Study, Princeton, New Jersey 08540, United States of America

David Caramelli
Dipartimento di Biologia Evoluzionistica, Università di Firenze, 50122 Florence, Italy

Guido Barbujani
Dipartimento di Scienze della Vita e Biotecnologie, Università di Ferrara, 44121 Ferrara, Italy

Preface

Encyclopaedia of Modern Genetics Analysis Integrating Genes and Genomes provides an experiment based introduction on modern genetic analysis. Genetic analysis is the overall process of studying and researching in fields of science that involve genetics and molecular biology. The text *Modern Genetic Analysis* focuses on key advances in genetics and cutting-edge experiments and techniques. First chapter uses a combinatorial method to analyze the genetic data for Crohn's disease and search disease-associated factors for given case/control samples. In second chapter, we analyze large series of resected non–small cell and small cell gallbladder carcinomas for Dpc4, pRB, p16, and p53 protein expression and for K-ras gene mutation. The aim of third chapter is to clarify and characterize the origin and nature of albino and hooded mutations using molecular genetic approaches and combine the findings with historical records. Molecular genetic analysis of central nervous system germ cell tumors with compa ative genomic hybridization has been presented in fourth chapter. Fifth chapter focuses on genetic analysis of sinonasal adenocarcinoma phenotypes. In sixth chapter, we investigate the genetic changes in three closely related species of *Drosophila* by making pair-wise comparisons of their genomes. Seventh chapter contributes to better understanding of the variance of genetic structure and reproductive mode in the genus *Reticulitermes.* Genetic analysis of cold tolerance at the germination and booting stages in rice by association mapping has been revealed in eighth chapter. Ninth chapter supports a specific genetic contribution in the risk for the low body mass index (BMI) phenotype in the Utah population database. In tenth chapter, we perform genome-wide association studies (GWAS) to investigate six components of sorghum inflorescence morphology and 6 traits related to plant height, then compare GWAS-based associations to positional evidence from meta-analysis of QTL likelihood intervals. Eleventh chapter describes the use of a genome-wide set of autosomal SNPs and 814 horses from 36 breeds to provide the first detailed description of equine breed diversity. Twelfth chapter highlights that Y-chromosome and MTDNA genetics reveal significant contrasts in affinities of modern Middle Eastern populations with European and African populations. Last chapter explores on the identification of genetic variation on the horse Y chromosome and the tracing of male founder lineages in modern breeds.

Chapter 1

A COMBINATORIAL ANALYSIS OF GENETIC DATA FOR CROHN'S DISEASE

Weidong Mao[1] & Jeonghwa Lee[2]

[1]Department of Mathematics & Computer Science Virginia State University Petersburg, VA 23806, USA.

[2]Department of Computer Science Shippensburg University Shippensburg, PA 17257, USA

ABSTRACT

The both environmental and genetic factors have roles in the development of some diseases. Complex diseases, suchas Crohn's disease or TypeII diabetes, are caused by a combination of environmental factors and mutations in multiple genes. Patients who have been diagnosed with such diseases cannot easily be treated. However, many diseases can be avoided if people at high risk change their living style, one example being their diet. But how can we tell their susceptibility to diseases before symptoms are found and help them make informed decisions about their health? With the development of DNA microarray technique, it is possible to access the human genetic information related to specific diseases. This paper uses a combinatorial method to analyze the genetic data for Crohn's disease and search disease-associated factors for given case/control samples. An optimum random forest based method has been applied to publicly available genotype data on Crohn's disease for association study and achieved a promising result.

INTRODUCTION

Crohn's disease (also known as regional enteritis) is a chronic, episodic, inflammatory condition of the gastrointestinal tract characterized by transmural inflammation (affecting the entire wall of the involved bowel) and skip lesions (areas of inflammation with areas of normal lining in between). Crohn's disease is a type of inflammatory bowel disease (IBD) and can affect any part of the gastrointestinal tract from mouth to anus. As a result, the symptoms of Crohn's disease can vary among affected individuals. The exact cause of Crohn's disease is unknown. However, research shows that the inflammation seen in the people with Crohn's disease involves several factors: the genes the patient has inherited, the immune system itself, and the environment [1]. In other words, genetic factor has been invoked in the pathogenesis of the disease.

Although the Crohn's disease cannot easily be treated, it can be avoided if people at high risk change their living style, such as their diet. But how can we tell the susceptibility of people to the disease before symptoms are found and help them make informed decisions about their health? With the development of DNA microarray technique, it is possible to access the human genetic information related to specific diseases. Assessing the association between DNA variants and disease has been used widely to identify regions of the genome and candidate genes that contribute to disease [2].

99.9% of one individual's DNA sequences are identical to that of another person. Over 80% of this 0.1% difference will be Single Nucleotide Polymorphisms (SNP) and they promise to significantly advance our ability to understand and treat human disease. A SNP is a single base substitution of one nucleotide with another. Each individual has many single nucleotide polymorphisms that together create a unique DNA pattern for that person. It is important to study SNPs because they represent genetic differences among human beings. Genome-wide association studies require knowledge about common genetic variations and the ability to genotype a sufficiently comprehensive set of variants in a large patient sample [3]. High-throughput SNP genotyping technologies make massive genotype data, with a large number of individuals, publicly available. Accessibility of genetic data makes genome-wide association studies for complex diseases possible.

Success stories when dealing with diseases caused by a single SNP or gene, sometimes called monogenic diseases have been reported [4]. However, most complex diseases, such as psychiatric disorders, are characterized by a non-mendelian, multifactorial genetic contribution with a number of susceptible genes interacting with each other [5]. A fundamental issue in the analysis of SNP data is to define the unit of genetic function that influences disease risk. Is it a single SNP, a regulatory motif, an encoded protein subunit, a combination of SNPs in a combination of genes, an interacting protein complex, a metabolic or a physiological pathway [6]? In general, it may be impossible to associate a single SNP or gene with a disease because a disease may be caused by completely different modifications of alternative pathways, and each gene only makes a small contribution. This makes the identification of genetic factors difficult. Multi-SNP interaction analysis is more reliable but it is computationally infeasible. An exhaustive search among multi-SNP combination is computationally infeasible even for a small number of SNPs. Furthermore, there are no reliable tools applicable to large genome ranges that could rule out or confirm association with a disease.

It is important to search for informative SNPs among a huge number of SNPs. These informative SNPs are assumed to be associated with genetic diseases. Tag SNPs generated by the multiple linear regression based method [7] are good informative SNPs, but they are reconstruction-oriented instead of disease-oriented. Although the combinatorial search method [8] for finding disease-associated multi-SNP combinations has a better result, the exhaustive search is still very slow.

Multivariate adaptive regression spline models [9, 10] are used to detect associations between diseases and SNPs with some degree of success. However, the number of selected predictors is limited, and the type of possible interactions must be specified in advance. Multifactor dimensionality reduction methods [11, 12] are developed specifically to find gene-gene interactions among SNPs, but they are not applicable to a large set of SNPs.

Random forest model has been explored in disease association studies [13], but it was applied on simulated case-control data in which the interacting model among SNPs and the number of associated SNPs are specified, thus making the association model simple and the association is relatively easier to detect. For real data, such as Crohn's disease [14], multi-SNP interaction is much more complex , which involves more SNPs.

In Section 2 of this paper, we propose an optimum random forest model for searching the disease-associated multi-SNP combination for given case-control data. In the optimum random forest model, we generate a forest for each variable (e.g. SNP) instead of randomly selecting some variables to grow the classification tree. We can find the best classifier (a combination of SNPs which includes the SNP) for each SNP, and then we may have M classifiers if the length of the genotype is M. We rank classifiers according to their prediction rate, and the SNP with a higher prediction rates is more disease-associated.

The association of multi-SNP combination can be measured by the disease susceptibility prediction rate. In Section 3 we address the disease susceptibility prediction problem [15, 16, 17, 18]. The goal of disease susceptibility prediction is to assess accumulated information targeted to predicting susceptibility to complex diseases with significantly high accuracy and statistical power. The problem is based on the association study we described above. The Disease-associated multi-SNP combination found in association studies can be used to predict the susceptibility to diseases. On the other side, the prediction results can be used to evaluate the accuracy of the association studies. A higher prediction rate means the higher reliability of the association studies.

The proposed method is applied to analyze the genetic data of the Crohn's disease. We find the disease-associated multi-SNP combination and apply it to predict the susceptibility. The accuracy of the prediction is higher than that of all previously known methods. It can be also applied in disease prevention and control in the near future. For example, after training the available case-control genome data, we can find those significant SNPs which are well associated with the disease. When a patient comes, and we obtain his/her genetic data , we don't need to check the who le sequence, but only disease-associated SNP s ins tea d. This will save much money and time for diagnosis and can be done before the onset of diseases. Therefore, treatment could start earlier to prevent or delay the occurrence of the disease.

DISEASE ASSOCIATION SEARCH FOR CROHN'S DISEASE

In this section we first give an overview of the random forest tree and classification tree, then we will describe the genetic model. Next we propose the optimum random forest algorithm to search Tag SNPs.

Classification Trees and Random Forest

In machine learning, a Random Forest is a classifier that consists of many classification trees. Each tree is grown as follows:

1. If the number of cases in the training set is N, sample N cases at random - but with replacement, from the original data. This sample will be the training set for growing the tree.
2. If there are M input variables, a number m << M is specified such that at each node, m variables are selected randomly out of the M and the best split on these m is used to split the node. The value of m is held constant during the forest growing.
3. Each tree is grown to the largest extent possible. There is no pruning [19].

A different bootstrap sample from the original data is used to construct a tree. Therefore, about one-third of the cases are left out of the bootstrap sample and not used in the construction of the tree. Cross-validation is not required because the one-third oob (out-of-bag) data is used to get an unbiased estimate of the classification error as trees are added to the forest. It is also used to get estimates of variable importance. After each tree is built, we compute the proximities of each terminal node.

In every classification tree in the forest, put down the oob samples and make prediction the classification of the oob samples. In such way we can compute the importance score for variables in each tree based on the number of votes cast for the correct class. All variables can be ranked and those important variables can be found in this way.

Random forest is a sophisticated method in data mining to solve classification problems, and it can be used efficiently in disease association studies to find most disease-associated variables such as SNPs that may be responsible for diseases.

Genetic Model

Recent work has suggested that SNPs in human population are not inherited independently; rather, sets of adjacent SNPs are present on alleles in a block pattern, so called haplotype. Many haplotype blocks in human have been transmitted through many generations without recombination. This means although a block may contain many SNPs, it takes only a few SNPs to identify or to tag each haplotype in the block. A genome-wide haplotype would comprise half of a diploid genome, including one allele from each allelic gene pair. The genotype is the descriptor of the genome which is the set of physical DNA molecules inherited from the organism's parents. A pair of haplotype consists of a genotype.

SNPs are bi-allelic and can be referred as 0 for majority allele and 1, otherwise. If alleles on both haplotypes are the same, then the corresponding genotype is homogeneous, and can be represented as 0 or 1. If the two alleles on the two haplotypes are different, the genotype is heterozygous, represented as 2.

In Figure 1, there are four chromosomes, we assume the first two chromosomes belong to one person and the other two chromosomes belong to another person. We can find on most sites the four chromosomes are identical, but on some sites they are different, nucleotides on these sites are SNP. The haplotype is the concatenation of SNPs and a genotype is composed of two haplotypes.

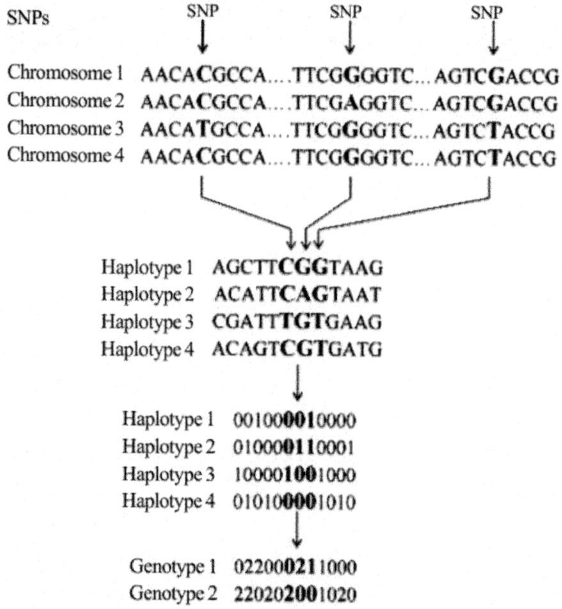

Figure 1: SNP, haplotype and genotype.

The case-control sample populations consist of N individuals who are represented in genotype with M SNPs. Each SNP attains one of the three values 0, 1, or 2. The sample G is an (0, 1, 2)-valued N x M matrix, where each row corresponds to an individual, each column corresponds to a SNP.

The sample G has 2 classes, case and control, and M variables, and each of them represents a SNP. To construct a classification tree, we split the sample S into 3 child sub-samples, depending on the value (0, 1, 2) of the variable (SNP) on the splitting site (loci). In fact we can construct a binary tree (split sample according to homozygous or heterozygous), but there is no way to tell the difference between major allele (1) and minor allele (0). In order to distinguish them we split the sample into 3 sub-samples instead of 2. We grow the tree to the largest possible extent. The construction of the classification tree for case-control sample is illustrated in Figure 2. In the first level, we split the sample (30 genotypes, 14 cases and 16 controls) into 3 sub-samples (17, 8, 5) at loci 5 (the 5th SNP). In the second level, the first sub-sample splits at loci 9 and the second sub-sample splits at loci 7. No splitting is required for the third sub-sample because it is a terminal node with only one class. In the third level, the only split node splits at loci 3. The relationship of a leaf to the tree on which it grows can be described by the hierarchy of splits of branches (starting from the trunk) leading to the last branch from which the leaf hangs. The collection of split site is a Multi-SNPs combination (MSC), which can be viewed as a classification tree. In this example, MSC = {5, 9, 7, 3}and m = 4, which is a collection of 4 SNPs, represented as their loci.

Searching for Disease Associated MultiSNPs

To fully understand the basis of complex diseases, it is important to identify the critical genetic factors involved, which is a combination of multiple SNPs. For a given sample G, S is the set of all SNPs (denoted by loci) for the sample, and a multi-SNPs combination (MSC) is a subset of S. In disease associations, we need to find a MSC which consists of a combination of SNPs that are well associated with the disease. To find such MSC, we need first rank all SNPs according to their association degree (measured as weight) with diseases. Based on the sorting, we can find the n most disease associated SNPs for a given threshold n.

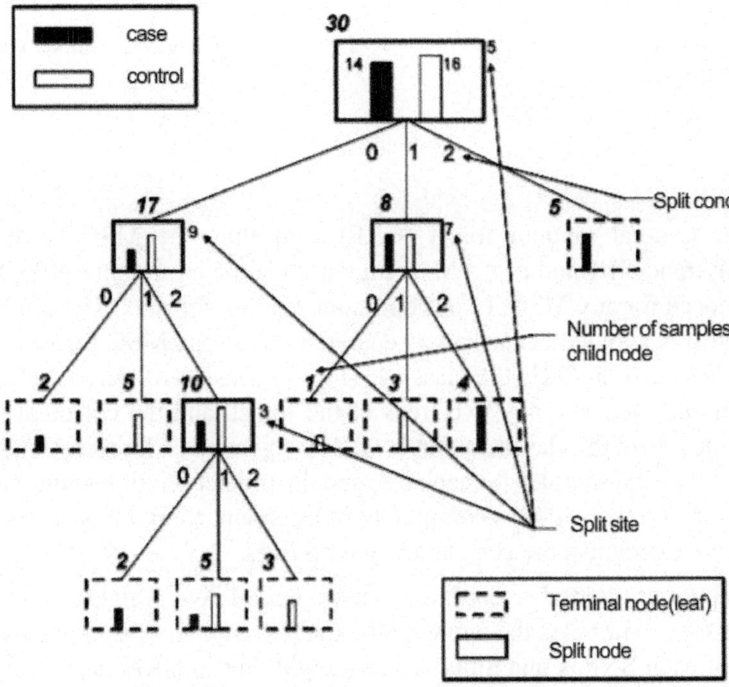

Figure 2: Classification tree for case-control sample.

Although there are many statistical methods to detect the most disease associated SNPs, such as odds ratio or risk rates, the result is not satisfactory. We decide to use the random forest to find them.

Optimum Random Forest

We randomly generate a group of MSCs for each SNP. The size of the MSC should be much less than the size of set S (m << M). Each MSC can be represented as a tree and all trees make the forest F. All trees (or MSCs) of the forest F_i(i=1, 2, ..., M) must include the ith SNP and the other (m-1) SNPs can be randomly chosen from S except the ith SNP. In this way, the M forests cover allSNPs in S.

We grow a classification tree for every MSC in each forest F_i. We run all the testing samples down these trees to get the classifier for each sample in the training set, then we can get a classification rate for each tree in F_i. The MSC_i is the representative for the forest F_i and the MSC_i has the highest classification rate among all trees in F_i. Each member (SNP) of the MSC_i] is assigned a weight w_{ij} (j ∈ MSC) based on the classification rate. The weights for SNPs in the same MSC are the same. We can find M MSCs for the M forests. If a SNP is not a member of MSC_i, then $w_{i,j} = 0$.

The weight for each SNP W_j (j = 1, 2, ..., M) in M is the sum of weights from all MSCs.

$$W_j = \sum_{i=1}^{M} w_{i,j} \qquad (1)$$

In the general random forest (GRF) algorithm, the MSC is selected completely randomly and m << M. It may miss some important SNPs if they are not chosen for any MSC. In our optimum random forest (ORF) algorithm, this scenario is avoided because we generate at least one MSC for each SNP. On the other hand, in GRF, the classifier (forest) consists of trees where there is a correlation between any two trees in the forest, and the correlation will decrease the rate of the classifier. But in ORF, we generate a forest by randomly choosing MSC and samples for each tree and the prediction for testing samples is in this forest only, which is completely independent from the other trees. In this way, we extinguish the correlation among trees.

All SNPs are sorted according to their cumulative weights. The most disease-associated SNP is the one with the highest weight. The contribution to diseases of each SNP is quantified by its weight, but in GRF there is no way tell the difference of contribution among SNPs. The GRF can only tell the difference among classifiers (trees).

DISEASE SUSCEPTIBILITY PREDICTION

In this section we first describe the input and the output of prediction algorithms and then show how to apply the optimum random forest to the disease susceptibility prediction.

Data sets have n genotypes and each has m SNPs. The input for a prediction algorithm includes:

(G1) Training genotype set $g_i = (g_{ij})$, $i = 0, 1, ..., n$, $j = 1, ... m$, $g_{i,j} \in \{0,1,2\}$

(G2) Disease status $s(g_i) \in \{0,1\}$, indicating if g_i, $i = 0, 1, ..., n$, is in case (1) or in control (0) , and

(G3) Testing genotype g_t without any disease status.

We will refer to the parts (G1-G2) of the input as the training set and to the part (G3) as the test set. The output of prediction algorithms is the disease status of the genotype $s(g_t)$.

We use leave-one-out cross-validation to measure the quality of the algorithm. In the leave-one-out cross-validation, the disease status of each genotype in the data set is predicted while the rest of the data is regarded as the training set.

We describe several universal prediction methods below. These methods are adaptations of general computer-intelligence classifying techniques.

Closest Genotype Neighbor (CN): For the test genotype g_t, find the closest (with respect to Hamming distance) genotype g_i in the training set, and set the status $s(g_t)$ equals to $s(g_i)$.

Support Vector Machine Algorithm (SVM) Support Vector Machine (SVM) is a generation learning system based on recent advances in statistical learning theory. SVMs deliver a state-of-the-art performance in real-world applications and have been used in case/control studies [18, 20]. There are some SVM softwares available and we decide to use libsvm-2.71 [19] with the following radial basis function:

$$exp(-\tau * |u-v|^2)$$

General Random Forest (GRF): We use Leo Breiman and Adele Cutler's original implementation of RF version [19]. This version of RF handles unbalanced data to predict accurately. RF tries to perform a regression on the specified variables to produce the suitable model. RF uses bootstrapping to produce random trees and it has its own cross-validation technique to validate the model for prediction/classification.

Most Reliable 2 SNP Prediction (MR2) [17]: This method chooses a pair of adjacent SNPs (site of s_i and $s_i + 1$) to predict the disease status of the test genotype [g_t] by voting among genotypes from the training set which have the same SNP values as [g.sub.t] at the chosen sites s_i and s_{i+1} . They choose the 2 adjacent SNPs with the highest prediction rate in the training set.

LP-based Prediction Algorithm (LP): This method assumes that certain haplotypes are susceptible to the disease while others are resistant to the disease. The genotype susceptibility is then assumed to be a sum of susceptibilities of its two haplotypes.

We want to assign a positive weight to susceptible haplotypes and a negative weight to resistant haplotypes such that for any control genotype the sum of weights of its haplotypes is negative and for any case genotype it is positive. We would also like to maximize the confidence of our weight assignment which can be measured by the absolute values of the genotype weights. In other words, we would like to maximize the sum of absolute values of weights over all genotypes.

This method is based on a graph X = {H, G}, where the vertices H correspond to distinct haplotypes and the edges G correspond to genotypes connecting its two haplotypes. The density of X is increased by dropping SNPs which do not collapse edges with an opposite status. The linear program assigns weights to haplotypes that, for any non-diseased genotype, the sum of weights of its haplotypes is less than 0.5 and greater than 0.5 otherwise. We maximize the sum of absolute values of weights over all genotypes. The status of the testing genotype is predicted as sum of its endpoints [15].

Optimum Random Forest (ORF): In the training set, the optimum random forest algorithm we described above is used to sort all SNPs, and find out the m most disease associated SNPs for a given threshold m. The m most disease associated SNPs (Tag SNPs) are used to build the optimum random forest to test the left-out sample. In leave-one-out test, since the training set is different after leaving one sample out, we may have different Tag SNPs for different training sets. The m variables (SNPs) are used to grow many different classification trees by permuting the order of the splitting site (Note that the tree {3, 9, 5}is different from the tree {5, 9, 3}). We may use the m Tag SNPs to grow many (say, 500) trees and choose the best tree (classifier) to predict the disease status of the testing genotype. The best tree has the highest average prediction rate (over 1000 trials) in the training set. Then we run the testing genotype down the best tree to get its disease status. The Optimum Random Forest algorithm is illustrated in Figure 3.

RESULTS & DISCUSSION

In this section we first describe the genetic data of the Crohn's disease and then discuss our experimental results.

Data Set

The genetic data is derived from the 616 kilobase region of human Chromosome 5q31 that may contain a genetic variant responsible for Crohn's disease by genotyping 103 SNPs for 129 trios [14]. All offspring belong to the case population, while almost all parents belong to the control population. In the entire data, there are 144 case and 243 control individuals. The missing genotype data and haplotypes have been inferred using the 2SNP phasing method [21].

Measures of Prediction Quality

To measure the quality of prediction methods, we need to measure the deviation between the true disease status and the result of predicted susceptibility, which can be regarded as measurement error. We will present the basic measures used in epidemiology to quantify the accuracy of our methods.

The basic measures are:

Sensitivity: the proportion of persons who have the disease and who are correctly identified as cases.

Specificity: the proportion of people who do not have the disease and who are correctly classified as controls.

Input: Training genotype set $G^{N,M}$, N: the number of samples, M: the number of SNPs
Disease status of $G^{N,M}$, $s^{N,M}$,
The threshold m,
Testing genotype g_t.

Sorting the M SNPs, find the MSC with the m most disease-associated SNPs
For $i = 1$ to 500,
  Permute the order of MSC, generate a tree T_i,
    For $j = 1$ to 1000,
    Randomly generate a bootstrapped sample S_j from G,
    Run S_j down the tree T_i to get the classification tree,
    Predict testing sample G'_j ($G'_j = G - S_j$) to get the prediction rate $p_{i,j}$,
  Compute the average prediction rate $\bar{p}_i$ for T_i,
Find the best tree T_b which has the highest $\bar{p}$,
Run g_t down the best tree T_b to get the disease status.

Output: Disease status of the test genotype $s(g_t)$.

Figure 3: Optimum Random Forest Algorithm.

The definitions of these two measures of validity are illustrated in Table 1.

In this table:

a = True positive, people with the disease who test positive b = False positive, people without the disease who test positive

c = False negative, people with the disease who test negative

d = True negative, people without the disease who test negative

From Table 1, we can compute Sensitivity (accuracy in classification of cases, Specificity (accuracy in classification of controls) and accuracy:

$$Sensitivity = \frac{a}{a+c} \quad (2)$$

$$Specificity = \frac{d}{b+d} \quad (3)$$

$$Accuracy = \frac{a+d}{a+b+c+d} \quad (4)$$

Sensitivity is the ability to correctly detect a disease. Specificity is the ability to avoid calling normal as disease. Accuracy is the percent of the population that are correctly predicted.

Results and Discussion

The normalized weights of 103 SNPs are shown in Figure 4. SNPs with higher weights are more associated with the disease.

In Table 2 we compare the optimum random forest (ORF) method with the other 5 methods we described in Section 3. The best accuracy is achieved by ORF 74.4%. From the results we can find that the ORF has the best result since we select the most disease-associated multi-SNPs to build the random forest for prediction. Because these SNPs are well associated with the disease, the random forest may produce a good classifier to reflect the association.

Table1: Classification contingency table

		True Status	
		+	-
Classified	+	a	b
Status	-	c	d

Table 2: The comparison of the prediction rates of 6 prediction methods

Measures	Prediction Methods					
	CN	SVM	GRF	MR2	LP	ORF
Sensitivity	45.5	20.8	34.0	30.6	37.5	70.1
Specificity	63.3	88.8	85.2	85.2	88.5	76.9
Accuracy	54.6	63.6	66.1	65.5	69.5	74.4

Figure 5 shows the receiver operating characteristics (ROC) curve for 6 methods. A ROC curve represents the tradeoffs between sensitivity and specificity. The ROC curve also illustrates the advantage of ORF over all previous methods.

If the size of MSC is m, and the total number of SNPs is M, to get a good classifier, then m should be much less than M. The prediction rate depends on the size of MSC, as shown in Figure 6. In our experiment, we found that the best size of MSC is 19.

CONCLUSION

In this paper, we discuss the potential of applying random forest on disease association studies. The proposed genetic susceptibility prediction method based on the optimum random forest is shown to have a high prediction rate and the multi-SNPs being selected to build the random forest are well associated with diseases. Actually the cause of complex diseases is the combination of the environmental, genetic factors and some other factors such as infection and races. In our future work we are going to analyze the interactive contribution of these factors for the development of complex diseases. Our next project is going to find the relationship between the genetic factor and race in the development of Type 2 Diabetes. The integrated software will be available soon for public use.

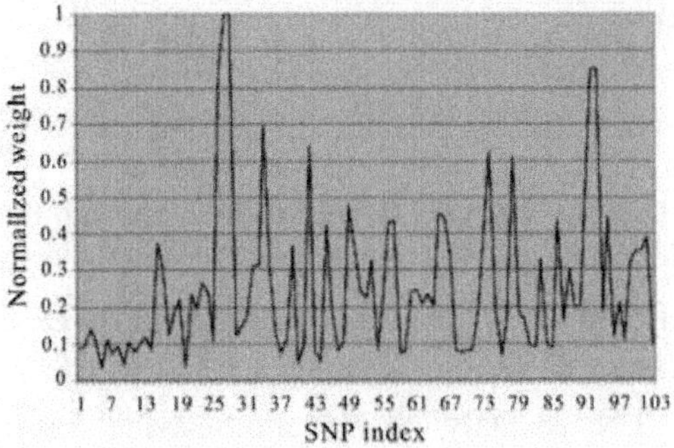

Figure 4: Normalized weights for 103 SNPs.

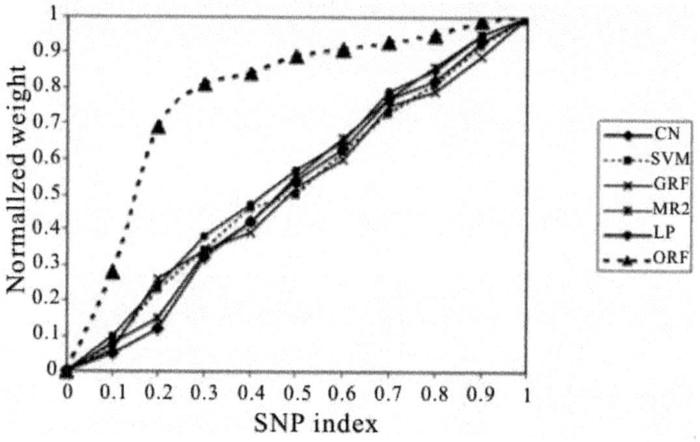

Figure 5: ROC curve for 6 prediction methods.

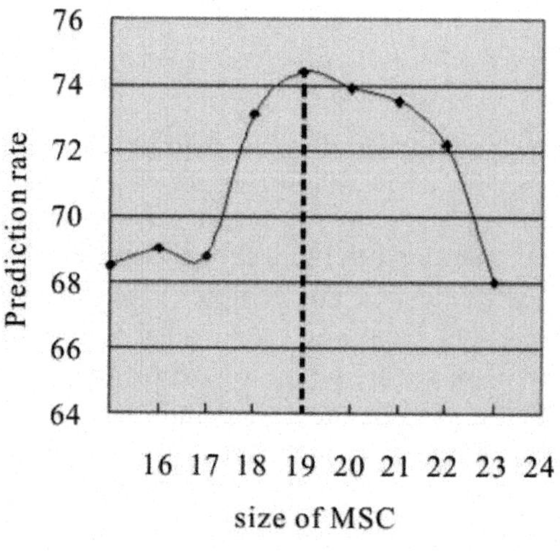

Figure 6: Best MSC size.

REFERENCE

1. National Digestive Diseases Information Clearinghouse (NDDIC), http://digestive.niddk.nih.gov/ddiseases/pubs/crohns.
2. Cardon, L.R. & Bell, J.I. Association Study Designs for Complex Diseases. Nature Reviews: Genetics 2001, 2:91-98.
3. Hirschhorn, J.N. & Daly, M.J. Genome-wide Association Studies for Common Diseases and Complex Diseases. Nature Reviews: Genetics 2005, 6:95-108.

4. Merikangas, KR. & Risch, N. Will the Genomics Revolution Revolutionize Psychiatry. American Journal of Psychiatry, 2003, 160: 625-635.
5. Botstein, D. & Risch, N. Discovering Genotypes Underlying Human Phenotypes: Past Successes for Mendelian Disease, Future Approaches for Complex Disease. Nature Genetics 2003, 33: 228-237.
6. Clark, A.G., Boerwinkle E., Hixson J. & Sing C.F. Determinants of the success of whole-genome association testing. Genome Res. 2005, 15:1463-1467.
7. He, J. & Zelikovsky, A. Tag SNP Selection Based on Multivariate Linear Regression. Proc. of International Conference on Computational Science 2006, LNCS 3992:750-757.
8. Brinza, D., He, J. & Zelikovsky, A. Combinatorial Search Methods for Multi-SNP Disease Association. Proc. of International Conference of the IEEE Engineering in Medicine and Biology 2006, pages 5802-5805.
9. Cook N.R., Zee R.Y. & Ridker P.M. Tree and Spline Based Association Analysis of gene-gene interaction models for ischemic stroke. Stat Med 2004, 23(9):439-453.
10. York T.P. & Eaves L.J. Common Disease Analysis using Multivariate Adaptive Regression Splines (MARS): Genetic Analysis Workshop 12 simulated sequence data. Genetic Epidemiology 2001, 21 (S I):649-654.
11. Ritchie M.D., Hahn L.W., Roodi N., Bailey L.R., Dupont W.D., Parl F.F. & Moore J.H. Multifactor-dimensionality reduction reveals high-order interactions among estrogen-metabolism genes in sporadic breast cancer. Am J Hum Genet. 2001, 69: 138-147.
12. Hahn L.W., Ritchie M.D. & Moore J.H. Multifactor dimensionality reduction software for detecting gene-gene and gene-environment interactions. Bioinformatics 2003, 19:376-382.
13. Lunetta, K., Hayward, L., Segal, J. & Van Eerdewegh, P. Screening Large-scale Association Study Data: Exploiting Interactions Using Random Forests", BMC Genetics 2004, pages 5:32.
14. Daly, M., Rioux, J., Schaffner, S., Hudson, T. & Lander, E. High resolution haplotype structure in the human genome. Nature Genetics 2001, 29:229-232.
15. Mao, W., He, J., Brinza, D. & Zelikovsky, A. A Combinatorial Method for Predicting Genetic Susceptibility to Complex Diseases. Proc. International Conference of the IEEE Engineering In Medicine and Biology Society 2005, pages 224-227.

16. Mao, W., Brinza, D., Hundewale, N., Gremalschi, S. & Zelikovsky, A. Genotype Susceptibility and Integrated Risk Factors for Complex Diseases. Proc. IEEE International Conference on Granular Computing 2006, pages 754-757.

17. Kimmel, G. & Shamir R. A Block-Free Hidden Markov Model for Genotypes and Its Application to Disease Association. J. of Computational Biology 2005, 12(10): 1243-1260.

18. Listgarten, J., Damaraju, S., Poulin B., Cook, L., Dufour, J., Driga, A., Mackey, J., Wishart, D., Greiner,R. & Zanke, B. Predictive Models for Breast Cancer Susceptibility from Multiple Single Nucleotide Polymorphisms. Clinical Cancer Research 2004, 10:2725-2737.

19. Breiman, L. & Cutler, A. http://stat.berkeley.edu/breiman.

20. Waddell, M., Page,D., Zhan, F., Barlogie, B. & Shaughnessy, J., Predicting Cancer Susceptibility from SingleNucleotide Polymorphism Data: A Case Study in Multiple Myeloma. Proc. of the 5th international workshop on Bioinformatics 2005, pages 21-28.

21. Chang, C. and Lin, C. http://www.csie.ntu.edu.tw/libsvm.

22. Brinza, D. & Zelikovsky, A. 2SNP: Scalable Phasing Based on 2-SNP Haplotypes. Bioinformatics 2006, 22(3):371-373.

Chapter 2

IMMUNOHISTOCHEMICAL AND GENETIC ANALYSIS OF NON–SMALL CELL AND SMALL CELL GALLBLADDER CARCINOMA AND THEIR PRECURSOR LESIONS

Anil V Parwani M.D., Ph.D.[1], Joseph Geradts M.D.[4], Eric Caspers M.D.[5], G Johan Offerhaus M.D.[5], Charles J Yeo M.D.[2,3], John L Cameron M.D.[2], David S Klimstra M.D.[6], Anirban Maitra M.D.[1], Ralph H Hruban M.D.[1,3] and Pedram Argani M.D.[1]

[1]Department of Pathology, The Johns Hopkins Hospital, Baltimore, Maryland

[2]Department of Surgery, The Johns Hopkins Hospital, Baltimore, Maryland

[3]Department of Oncology, The Johns Hopkins Hospital, Baltimore, Maryland

[4]Department of Pathology and Laboratory Medicine, Roswell Park Cancer Institute, Buffalo, New York

[5]The Academic Medical Center, Amsterdam, The Netherlands

[6]Department of Pathology, Memorial Sloan-Kettering Cancer Center, New York

ABSTRACT

Gallbladder carcinomas can be highly lethal neoplasms. Relatively little is known about the genetic abnormalities that underlie these tumors, particularly with respect to their timing in neoplastic progression. The authors evaluated 5 noninvasive dysplasias and 33 invasive gallbladder carcinomas (6 small cell carcinomas, 27 non–small cell carcinomas, of which 16 were accompanied by an in situ carcinoma component) for expression of the protein products of the p16, p53, Dpc4, and pRB tumor suppressor genes by immunohistochemistry. Neoplasms were also evaluated for the presence of activating K-ras oncogene mutations. Seventy-five percent of non–small cell gallbladder carcinomas demonstrated loss of p16 expression, whereas 63% accumulated high levels of p53. Loss of Dpc4 and pRB expression was less frequent, seen in 19% and 4% of the neoplasms, respectively. Thirty percent of neoplasms harbored activating K-ras mutations. In contrast, 100% of the small cell carcinomas of the gallbladder demonstrated inactivation of the pRB/p16 pathway; 67% showed loss of pRB expression, and the other 33% lost p16 expression. Eighty-three percent of small cell carcinomas accumulated high levels of p53, whereas loss of Dpc4 expression and activating K-rasmutations were not found. Among 15 evaluable in situ components, 13 harbored the same alterations found in the invasive component. Inactivation of p16 and p53 occur in the majority of non–small cell gallbladder carcinomas. Dpc4 inactivation and K-ras mutations occur in a significant minority of cases. pRB loss is uncommon in non–small cell gallbladder carcinoma, but virtually all small cell carcinomas inactivate the p16/pRB pathway, usually by retinoblastoma protein loss. It is noteworthy that all of these alterations occur at the level of carcinoma in situ.

INTRODUCTION

Carcinoma of the gallbladder is a relatively uncommon, poorly understood, but highly lethal malignancy that tends to present at an advanced stage (1, 2, 3). There are approximately 5000 newly diagnosed cases per year in the United States, with a female predominance. The incidence of gallbladder carcinoma demonstrates marked geographic variation; for example, it is the single largest cause of cancer death for women in Chile but accounts for <0.5% of cancers in women in the United States (1). Much of the geographic variation correlates with the tendency to form gallstones, a recognized risk factor in gallbladder carcinogenesis. Other risk factors include an abnormal junction of the choledochopancreatic ducts (AJCPD); in which the bile duct and pancreatic duct unite well above the sphincter of Oddi, promoting reflux of pancreatic juice into the bile duct), familial adenomatous polyposis, and ulcerative colitis (1, 4). Overall 5-year survival is <5% (1).

The molecular pathogenesis of gallbladder carcinoma remains enigmatic. Loss of heterozygosity (LOH; 5, 6, 7, 8) and cytogenetic (9) studies of non–small cell gallbladder carcinoma have identified several loci of recurrent genetic loss. These loci likely harbor tumor suppressor genes that are inactivated and include chromosome arms 17p, 5q, 9p, 13q, and 18q. Alterations in several specific tumor suppressor genes that map to these loci have been identified (*i.e.*, *p53* at 17p13, *p16* at 9p21, *RB* at 13q), but the data remain cloudy to date as differing studies have yielded differing results (5, 7, 8, 10, 11, 12, 13, 14, 15, 16, 17, 18). Some of the variation in results of immunohistochemical studies may be due to use of different techniques and cutoff points for positivity in different studies. The loci on 18q involved in gallbladder carcinoma have not been identified, with the *DCC* and *DPC4* genes being prime candidates. Finally, mutational activation of the K-*ras* gene has been demonstrated in 5–59% of cases (7, 14, 15, 16, 17, 18), with suggestions that this frequency may be increased in Japanese patients, particularly those with AJCPD. Small cell carcinomas of the gallbladder have been studied even less frequently (19).

In the present study, we analyzed our large series of resected non–small cell and small cell gallbladder carcinomas for Dpc4, pRB, p16, and p53 protein expression and for K-*ras* gene mutation.

METHODS

Case Selection

This study was approved by the Johns Hopkins Institutional Review Board. The computerized files of the Surgical Pathology Division of the Department of Pathology of The Johns Hopkins Hospital were searched over the years 1985–2000 for cases coded as "gallbladder" and "carcinoma." Cases were selected for study on the basis of availability of a paraffin-embedded, formalin-fixed tissue block. An additional case (Case 31) was obtained from Memorial Sloan-Kettering Cancer Center. For each case, a representative formalin-fixed, paraffin-embedded tissue block containing carcinoma and normal tissue was chosen for labeling. For cases in which an *in situ* carcinoma component was noted, a block containing this component was specifically chosen for study.

Cases

We identified five cases of gallbladder dysplasia or carcinoma *in situ*unassociated with invasive carcinoma (Table 1) and identified 33 cases of invasive gallbladder carcinoma for which blocks were available. Patient ages ranged from 37 to 86 years (mean age = 62.8 y; median = 63.5 y), and the male-female ratio was 13:25. Twenty-one of 34 evaluable gallbladders harbored gallstones.

Table 1:- Clinical and Molecular Features of Five Patients with Pure Dysplasia or Carcinoma *In Situ*

Case Number	Age	Sex	Stone	p53	Dpc4	p16	pRB	K-*ras*
1 (LGD)	39	M	Yes	NL	NL	NL	NL	NL
2 (LGD)	37	F	Yes	NL	NL	NL	NL	NL
3 (CIS)	71	M	No	OE	NL	LOSS	NL	MUT D (Val)
4 (CIS)	63	F	Yes	OE	NL	LOSS	NL	NL
5 (CIS)	57	F	Yes	OE	NL	LOSS	NL	NL

NL = normal pattern (intact labeling for p16, Dpc4, and pRB, less than 30% expression of p53, wild type K-*ras*); OE = overexpression; LOSS = loss of expression; LGD = low-grade dysplasia; CIS = carcinoma in situ; MUT = K-*ras* mutation; Stone = gallstone present in specimen.

Two of the pure dysplasia cases were low grade, whereas three were high grade or carcinoma *in situ*. Of the invasive carcinomas, 16 were associated with an *in situ* component; of these, the *in situ* component had a flat component in 10 and was purely papillary in 6.

Twenty-seven of the invasive carcinomas of the gallbladder were non–small cell carcinomas (Cases 1–27); these included 21 adenocarcinomas, 5 adenosquamous carcinomas (defined as showing variable mixture of malignant squamous and glandular components; 1), and 1 purely squamous carcinoma. Of these 27 neoplasms, the primary tumor was evaluated in 24, and a metastasis was the only tissue available in 3. Of these 24 primary tumors, 15 were moderately differentiated (defined as 40–94% of the tumor forming glands; 1), and 9 were poorly differentiated (defined as 5–39% of the tumor forming glands; 1). Six other neoplasms contained a small cell (high-grade neuroendocrine) carcinoma component (Cases 28–33). Two of these six were pure small cell carcinomas (Cases 29, 30), whereas one was predominantly small cell carcinoma with focal squamous differentiation (Case 31). One tumor was associated with flat carcinoma *in situ* (Case 28), whereas two other invasive carcinomas had both small cell carcinoma and adenocarcinoma components along with papillary adenocarcinoma *in situ* (Cases 32, 33).

Several patients› tumors were associated with unusual clinical presentations or pathologic findings. One patient›s gallbladder carcinoma arose within a septate gallbladder (Case 8). Another patient (Case 20) had concurrent colorectal adenocarcinoma and gallbladder carcinoma, whereas another patient›s gallbladder carcinoma extended into the extrahepatic bile ducts, which harbored a carcinoid tumor (Case 13).

Patients tended to present at advanced stage. Using AJCC criteria (20), 29 of 30 patients› primary tumors presented at Stage 3 or above. At least 23 of 30 primary tumor resections were associated with positive surgical margins microscopically, indicating incomplete excision of the neoplasm.

Immunohistochemistry for p53 and Dpc4

Unstained 4-μm sections were cut from the selected paraffin block and deparaffinized by routine techniques. The slides were steamed for 20 minutes in sodium citrate buffer (diluted to 1 × from 10 × heat-induced epitope retrieval buffer; Ventana-Bio Tek solutions, Tucson, AZ). After cooling for 5 minutes, the slides were labeled for 40 minutes at room temperature with either a 1:100 dilution of a monoclonal antibody to Dpc4 (clone B8, Santa Cruz Biotechnology, Santa Cruz, CA) or a 1:250 dilution of a monoclonal antibody to p53 (clone DO-7, DAKO, Carpinteria, CA) using the Bio Tek 1000 automated stainer (Ventana). Labeling was detected by adding biotinylated secondary antibodies, avidin-biotin complex, and 3, 3›-diaminobenzidine. Sections were then counterstained with hematoxylin. Dpc4 and p53 immunolabeling were evaluated jointly by two authors (AVP, PA) using a multiobserver microscope, with agreement on all cases. For p53 labeling, a percentage of positive nuclei was determined. Carcinomas were divided into two groups: normal (<30% nuclear labeling) and positive for high-level accumulation of p53 protein (>30% nuclear labeling). The labeling cutoff was chosen based on previous studies that have demonstrated that this cutoff point correlates best with the status of the *p53* gene in colorectal carcinomas (21). For Dpc4 labeling, any area of uniform cytoplasmic labeling and focal nuclear labeling was considered positive. In statistical analysis, any carcinoma showing even focal nuclear and cytoplasmic labeling was considered positive (expressor), whereas carcinomas demonstrating no expression in a background of intact expression by non-neoplastic cells (desmoplastic stroma, normal peribiliary glands, etc., which served as internal controls) were considered negative (nonexpressors). The rationale for considering carcinomas that labeled only focally as positives is based on the study of Wilentz *et al.* (22), which found that pancreatic tumors with this focal staining pattern proved to have an intact *DPC4* gene.

Immunohistochemistry for p16, pRB

Immunohistochemistry for p16 and pRB was performed in the laboratory of one of the authors (JG) and interpreted by three of the authors (AVP, PA, and JG). Mouse monoclonal anti-RB antibody 3C8 was purchased from QED (San Diego, CA), and mouse monoclonal anti-p16 antibody 16P07 was obtained from LabVision/NeoMarkers (Fremont, CA). Nonspecific mouse IgG was used as a negative antibody control. Standard ABC peroxidase assays were performed as described in detail elsewhere (22, 23, 24, 25, 26). For detection of pRB, deparaffinized sections were incubated with anti-RB antibody 3C8 at 2 µg/mL for 2 hours, after an antigen retrieval step in 0.01 m citrate buffer (95–100° C). For detection of p16, sections were incubated with the anti-p16 monoclonal antibody 16P07 at 1 µg/mL at 4° C overnight, after antigen retrieval in 0.1 m EDTA pH 8.0 (20 min at 95–100° C). The detection reactions for both markers used the Vectastain Elite ABC kit using conditions recommended by the manufacturer. Diaminobenzidine with hematoxylin counterstain was used for color development. Negative controls were labeled under identical conditions. The following external positive controls were used: normal colonic mucosa for pRB and p16 and a p16-positive lung cancer xenograft for p16. In addition, non-neoplastic stromal cells served as internal positive controls for pRB and p16 in every tumor section.

Each case was scored for pRB and p16 reactivity using previously published criteria (19, 23, 24, 25, 26). Briefly, sections were examined for evidence of nuclear labeling above any cytoplasmic background; cytoplasmic labeling itself was disregarded. If there was nuclear labeling in a diffuse or mosaic distribution throughout the lesion, it was considered "positive" (normal) for the respective protein. If all of the neoplastic nuclei failed to label whereas admixed non-neoplastic cells reacted positively, the lesion was scored as "negative" (abnormal). Cases were scored initially by one author (JO), and scoring was subsequently reviewed by two authors (AVP, PA) using a multiobserver microscope with agreement on all cases.

K-*ras* Gene Codon 12 Analysis

For the K-*ras* gene analysis, tumor was microdissected from unstained paraffin-embedded section slides to ensure ≥50% neoplastic cells in the sample. K-*ras*codon 12 analysis was performed with allele-specific oligonucleotide (ASO) hybridization method, as described elsewhere (27). DNA was isolated from microdissected tumor tissue by overnight incubation at 56° C in proteinase K solution, followed by inactivation of proteinase K at 95° C for 10 minutes. DNA was amplified by PCR using K-*ras* codon 12 specific primers (27). PCR products were then digested with *Mva*I, which only recognizes wild-type K-*ras* codon 12. Subsequently, a second-round PCR was performed on both the digested and undigested first-round PCR products. Cell suspensions with mutant–wild-type ratios of 1:100 and 1:1000 were used as positive controls in every PCR procedure. The cell suspensions were made of the human colon cancer cell line SW 480 with a homozygous GGT to GTT mutation at codon 12 of K-*ras* and the human colon cancer cell line HT 29 with wild type K-*ras*. Water was used as a negative control, and placental DNA was used as a control for nonspecific hybridization. After denaturation, the undigested and digested (mutant-enriched) PCR products were spotted onto a nylon membrane and hybridized to each of the K-*ras* codon 12 mutation-specific oligodeoxynucleotides. Final stringency washes were carried out at 63° C, followed by autoradiography. K-*ras* codon 12 mutational analysis was performed twice, in independent experiments.

In all cases, K-*ras* codon 12 mutations identified by ASO hybridization were confirmed by direct fluorescent sequencing using the dideoxy chain termination method (27). Mutant-enriched PCR products (second-round PCR products of ASO-test) were purified using the Qiaquick PCR purification kit (Qiagen Inc.). DNA was sequenced with the DNA sequencing kit, BigDye-Terminator Cycle sequencing Ready Reaction (Applied Biosystems, Warrington, United Kingdom). The reaction products were analyzed on an ABI Prism 3100 Genetic Analyser (Hitachi, Applied Biosystems). Selected wild-type cases were also sequenced.

RESULTS

Pure Dysplasias ($n = 5$)

Neither of the two low-grade dysplasias demonstrated abnormal expression of p53, Dpc4, and p16 or pRB protein, and neither contained a K-*ras* gene mutation. In contrast, all three flat carcinoma *in situ* lesions demonstrated accumulation of p53 protein at high levels and demonstrated loss of p16 expression. pRB expression was intact and diffuse in these three cases, and Dpc4 was detectable in all three (Fig. 1). One of the three carcinoma *in situ* lesions harbored a K-*ras*mutation, involving conversion to valine (Table 1).

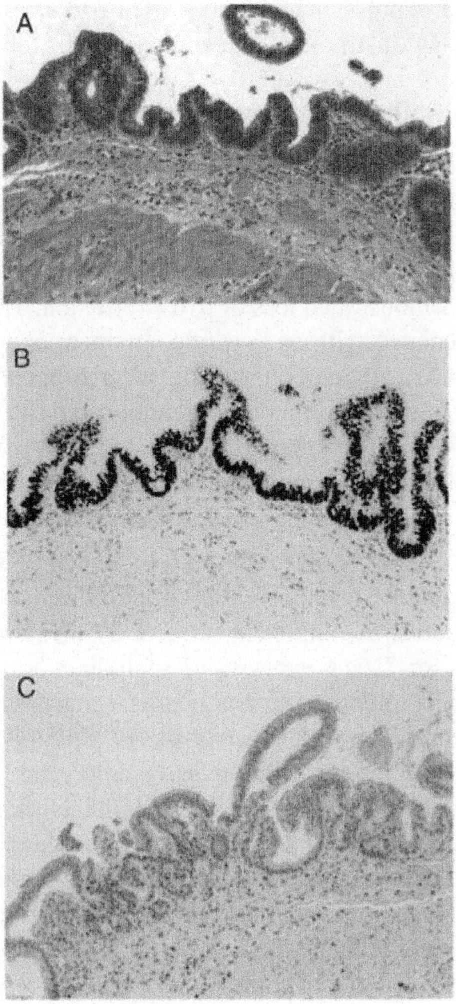

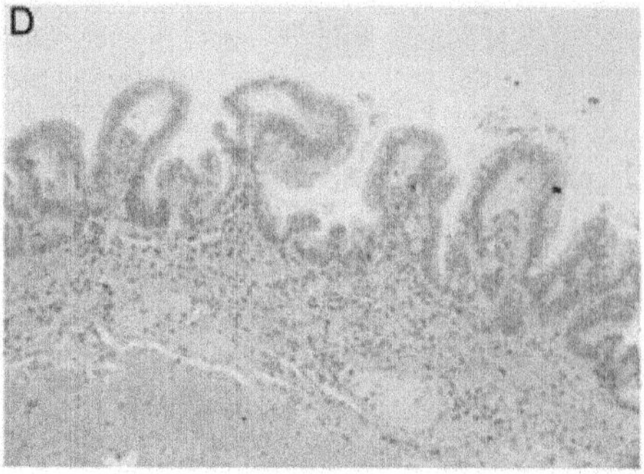

Figure 1: Flat carcinoma *in situ* (A) demonstrating p53 overexpression (B); intact, diffuse pRB expression (C); and loss of p16 expression (D). Note the intact patchy labeling of stromal and endothelial cells for p16.

Invasive Non–Small Cell Carcinomas (*n* = 27)

This group included 14 pure invasive non–small cell carcinomas and 13 invasive non–small cell carcinomas associated with an *in situ* carcinoma component. Loss of p16 expression and abnormal accumulation of p53 were the most common abnormalities noted in this group. Eighteen (75%) of 24 evaluable tumors demonstrated loss of p16 expression. This included 10 of 12 carcinomas unassociated with an *in situ* component and 8 of 12 carcinomas associated with an *in situ* component. The six p16-positive tumors generally displayed a weak to moderate nuclear immunoreactivity in <50% of cells in a mosaic pattern. Of note, in the case (Case 13) with concurrent gallbladder carcinoma and bile duct carcinoid tumor, p16 expression was intact in the carcinoid tumor but was lost in the poorly differentiated gallbladder carcinoma.

Seventeen of 27 carcinomas (63%) accumulated p53 at high levels; this included 8 of 14 pure invasive carcinomas and 9 of 13 invasive carcinomas associated with an *in situ* component. Five of 27 invasive carcinomas (19%) demonstrated loss of Dpc4 expression, including 3 of 14 pure invasive carcinomas and 2 of 13 invasive carcinomas associated with carcinoma *in situ*. Of note, Dpc4 expression was intact in the *in situ* and invasive carcinoma of the gallbladder but absent in the invasive colorectal carcinoma that were concurrently resected (Case 20), supporting the clinical and morphologic impression that these were independent primary neoplasms. Only 1 of 27 non–small cell carcinomas (4%) showed loss of pRB expression; in this tumor the *in situ* component also demonstrated loss of pRB expression. Most tumors with intact pRB expression demonstrated extensive labeling for pRB, particularly those demonstrating loss of p16 expression (Fig. 2).

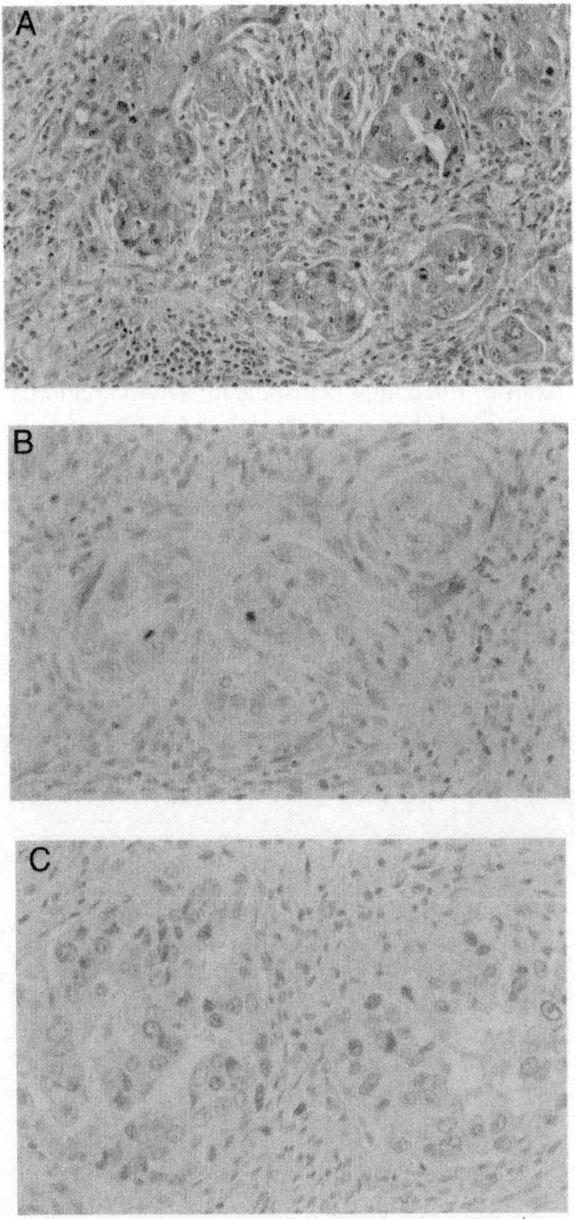

Figure 2: Infiltrating non–small cell carcinoma (A) demonstrating loss of p16 expression (B) and intact, diffuse pRB expression (C). Note the intact labeling of endothelial cells in Figure 2B.

K-*ras* gene mutations were identified in 8 of 27 invasive non–small cell carcinomas (30%). Of the eight invasive adenocarcinomas harboring a

K-*ras*gene mutation, six were primary tumors and two were metastases. Five mutations involved conversion to aspartate, and three, to serine.

There were no significant differences in results for any tested marker among cases classified as moderately *versus* poorly differentiated.

In Situ Carcinoma Components Associated with Invasive Non–Small Cell Carcinoma (*n* = 13)

There was a remarkable concordance between the immunohistochemical labeling of the *in situ* and the invasive non–small cell carcinoma components for p16, pRB, p53, and Dpc4 in the above cases. We identified p16 loss in 7 of 11 (64%) evaluable *in situ* components; the invasive carcinomas corresponding to all of these tumors had shown p16 loss. We identified accumulation of p53 at high levels in 10 of 13 *in situ* lesions (77%) associated with invasive carcinoma; this included all 9 cases in which the invasive component accumulated p53 at high levels. There was one case with discordant findings: in this case, the papillary *in situ* carcinoma component demonstrated high levels of p53, whereas the invasive component showed no p53 expression. We identified loss of Dpc4 expression in both of the *in situ* lesions associated with invasive carcinomas that showed loss of Dpc4 expression (2 of 13 *in situ* lesions overall, 15%; Fig. 3), whereas we found loss of pRB expression in the *in situ* component of the one invasive carcinoma that showed loss of pRB (1 of 13 cases overall, 8%).

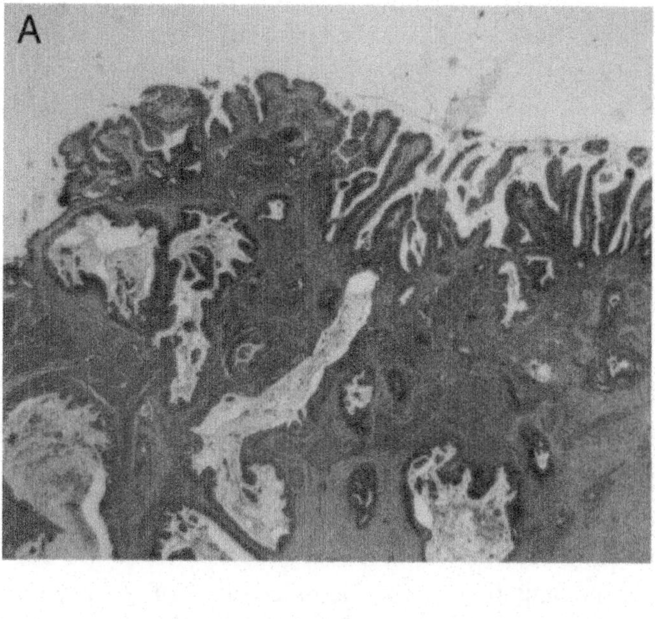

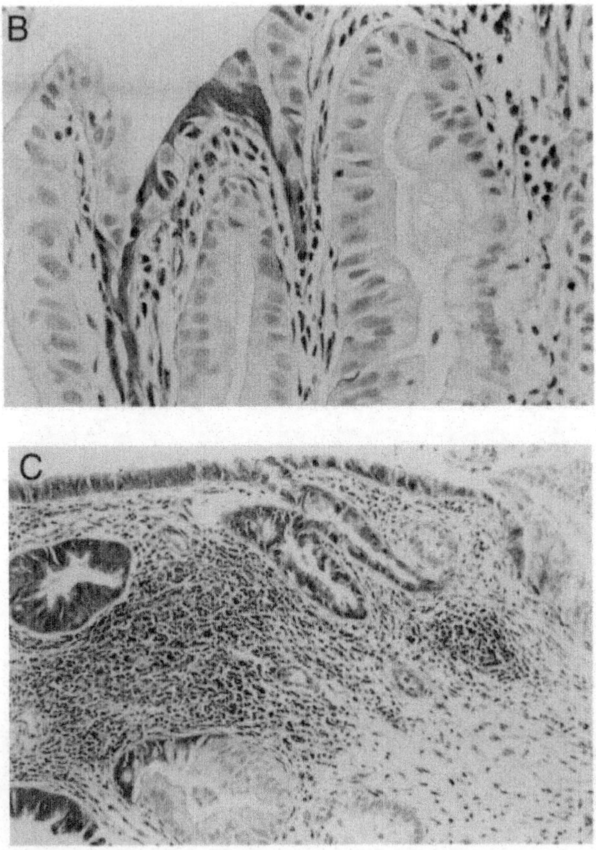

Figure 3: Papillary *in situ* and infiltrating carcinoma (A) demonstrating loss of Dpc4 expression in the *in situ* component (B) and in the invasive component (C). Note the intact labeling of normal stroma and benign gallbladder epithelium (Fig. 3C).

Invasive Small Cell Carcinomas (*n* = 6)

Accumulation of p53 at abnormally high levels was the most common abnormality noted, identified in five of six invasive small cell carcinomas. All six small cell carcinomas demonstrated evidence of inactivation of the pRB/p16 pathway. In contrast to non–small cell carcinomas, loss of pRB expression more common in this tumor type, identified in four of six cases (67%), whereas the other two small cell carcinomas with intact pRB expression demonstrated loss of p16 expression (Fig. 4). Dpc4 protein expression was intact in all six tumors, whereas K-*ras* gene mutations were not identified in any of the three tumors evaluated.

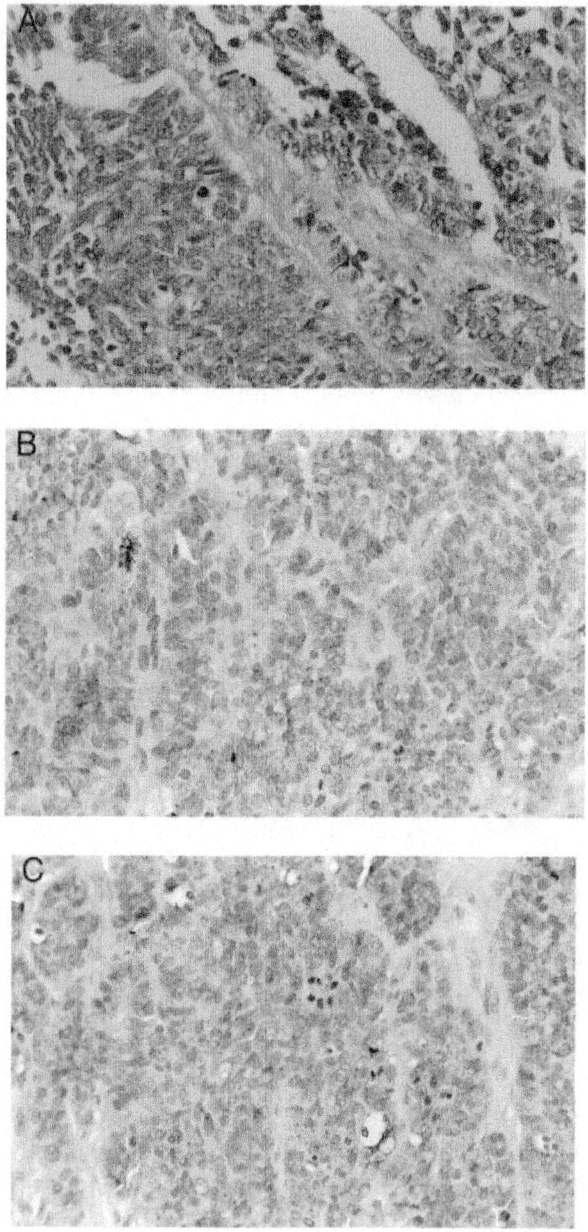

Figure 4: Small cell carcinoma (A) demonstrating loss of pRB expression (B) and intact, diffuse p16 expression (C). Note the intact labeling of endothelial cells for pRB in Figure 4B.

Non–Small Cell Components Associated with Small Cell Carcinomas (*n* = 3)

In two of three cases, the results were concordant. The flat carcinoma *in situ*associated with one small cell carcinoma and the papillary *in situ* and invasive adenocarcinoma associated with another each demonstrated p53 accumulation at high levels and p16 loss like their small cell components. In the other case, only the small cell carcinoma component demonstrated loss of p16 expression; the papillary *in situ* carcinoma and invasive carcinoma with which it was associated expressed p16.

The results of immunohistochemical assays are summarized in Table 2.

Table 2:- Immunolabeling and K-ras Mutation Analysis: Summary

	Dysplasia Only Cases		Invasive Carcinoma Cases: In Situ Component of Non-Small Cell Carcinomas		Invasive Carcinoma Cases: Invasive Component	
	Low Grade	CIS	Flat	Papillary	Non-Small Cell Carcinoma	Small Cell Carcinoma
p53	0/2	3/3	7/9	3/4	17/27 (63%)	5/6 (83%)
p16	0/2	3/3	6/8	1/3	18/24 (75%)	2/5 (40%)
pRB	0/2	0/3	1/9	0/4	1/27 (4%)	4/6 (67%)
Dpc4	0/2	0/3	1/9	1/4	5/27 (19%)	0/6 (0%)
K-*ras*	0/2	1/3	ND	ND	8/27 (30%)	0/3 (0%)

ND = not determined.

* The data are expressed as the number of abnormal results/number of normal and abnormal results. These data do not include the non-small cell carcinoma components associated with the small cell carcinomas.

DISCUSSION

We studied a large series of resected gallbladder carcinomas for expression of the protein products of the *p16, DPC4, RB*, and *p53* tumor suppressor genes and for mutation of the K-*ras* oncogene.

Our immunohistochemical results for p53 generally parallel those previously reported in the literature. A number of prior studies have shown that accumulation of p53 at high levels is frequent in invasive non–small cell carcinomas of the gallbladder and occurs at the level of carcinoma *in situ* (7, 10,11). Although most of these older studies have used primary antibodies other than the D0–7 clone (which is currently used by most laboratories), several have shown good correlation of p53 expression with *p53* gene mutation (10) and LOH at 17p (7). Our study is slightly different from those previously published in that we show a slightly higher frequency of high-level accumulation of p53 in *in situ*carcinoma than in invasive carcinomas. In contrast, most previous studies have shown progressive overexpression of p53 as dysplastic lesions progress to invasion (5, 7). Possible reasons for this difference could include variability in the antibody clone or immunohistochemical technique used, as well as criteria for distinction of low-grade dysplasia from carcinoma *in* *situ*. For example, it is possible that some of the lesions we designated low-grade dysplasia might be termed carcinoma *in situ* by others, which would reduce the frequency of p53 alterations somewhat. Nonetheless, because of the concordance between *in situ*and invasive carcinoma, our results suggest the possibility that gallbladder carcinomas may arise via two pathways, with those arising from flat carcinoma*in situ* overexpressing p53 and with those that do not being less likely to overexpress p53. The high frequency of p53 overexpression that we observed in small cell carcinomas of the gallbladder (83%) is similar to that reported elsewhere (19).

To our knowledge, this is the first study to assess Dpc4 protein expression in non–small cell gallbladder carcinomas. The 18% frequency of loss of protein expression that we observed in non–small cell carcinomas is concordant with results from our previous studies of biliary tract carcinomas. Previously, we showed that loss of Dpc4 expression is as frequent in distal (intrapancreatic) bile duct carcinomas (55%) as it is in pancreatic adenocarcinoma. In contrast, the frequency of Dpc4 loss was lower (overall, 15%) in more proximal (perihilar and intrahepatic) bile duct carcinomas (28). Interestingly, we observed loss of Dpc4 protein in a similar percentage of *in situ* gallbladder carcinomas associated with invasive carcinomas (2 of 13, or 15%), suggesting that *DPC4* inactivation occurs early in these lesions. The 18% frequency of Dpc4 protein loss in these tumors may account for a subset of the approximately 30% frequency of LOH (5, 7) that has been reported at chromosome 18q in prior allelotyping studies of gallbladder carcinomas.

The significant role of *p16* alterations in the pathogenesis of gallbladder carcinomas has emerged in the recent literature. Although studies have shown a rate of mutation that ranges from 30.7 (13) to 80% (12), LOH at the p16 locus on chromosome 9p has been reported to range between 0 and 68% (7, 8, 12). To our knowledge, only two other studies have evaluated p16 protein immunolabeling in gallbladder carcinomas. Shi *et al.* (29) have recently shown loss of p16 expression in 75.7% of gallbladder carcinomas, whereas Kim *et al.*(13) showed loss of labeling in only 23%. Our study is concordant with the results of Shi *et al.* (29) in that we saw frequent loss of p16 protein expression (69%). It is interesting that the two adenocarcinomas associated with small cell carcinomas in this study maintained p16 expression, suggesting that they are biologically different from pure adenocarcinomas.

We found pRB loss to be infrequent in invasive non–small cell gallbladder carcinomas; in fact, most tumors expressed pRB diffusely. Diffuse pRB expression in these tumors likely reflects the sequelae of p16 inactivation, because p16 negatively regulates cellular levels of RB (30). In fact, inverse relationship of p16 and pRB has been demonstrated in several malignancies including bladder carcinomas, lung carcinomas, pancreatic carcinomas, and gliomas (30, 31, 32), and, most recently, gallbladder carcinomas (29). In our series, the inverse relationship was also apparent; no invasive carcinoma demonstrated loss of both p16 and pRB, whereas an inverse relationship (loss of p16 or Rb, but not both) was evident in 25 of 30 tumors. One would not expect inactivation of both of these tumor suppressor genes within the same tumor, because they each function in the same growth-regulatory pathway, and therefore there would be no selective pressure to inactivate both.

In contrast, we found that pRB loss was common (4 of 6 cases) in the small cell carcinomas available for study. Although pRB expression has, to our knowledge, never been evaluated in small cell carcinoma of the gallbladder before, pRB loss is common in pulmonary small cell carcinomas (33) and has been reported in small cell carcinomas in extrapulmonary sites (34, 35). Hence, gallbladder carcinomas appear to fit the paradigm that small cell (high grade) neuroendocrine carcinomas of any site are consistently characterized by loss of pRB expression. It is noteworthy that the two cases that retained pRB showed loss of p16, indicating that all small cell carcinomas of the gallbladder target the p16/pRB pathway.

The role of K-*ras* gene mutation in gallbladder carcinomas is not clear from the literature. Several groups have considered K-*ras* mutations to be "rare and late events" (7) in gallbladder carcinomas. Wistuba *et al.* (36) found K-*ras* gene mutations in only 1 of 21 gallbladder carcinomas, and the mutation was confined to the poorly differentiated area of this tumor. This same group found a higher frequency of K-*ras* gene mutation (25%) in gallbladder adenomas and used this finding to support the idea that adenomas are not the precursor of most invasive carcinomas (36). In contrast, studies from Japan have shown a higher frequency of K-*ras* gene mutation in gallbladder carcinomas and occurrence in precursor lesions. For example, Ajiki *et al.* (15) found K-*ras* gene mutations in 73% of gallbladder dysplasias and 59% of invasive carcinomas. The presence of K-*ras*gene mutations does correlate with the anatomic finding of anomalous junction of the choledochopancreatic ducts (AJCPD) outside of the ampulla of Vater, which is more common in Japan. Hanada *et al.* (10) showed that the rate of K-*ras*mutation was higher in gallbladder carcinomas of patients with AJCPD than in gallbladder carcinomas of patients without AJCPD (50% *versus* 6%, $P < .05$). Our study shows an intermediate rate of mutation in our cohort of 30%. Two of our three metastatic carcinomas harbored K-*ras* gene mutations; however, we demonstrated a mutation in one patient with pure flat carcinoma *in situ*, indicating that this genetic alteration may be an early event in some cases.

In summary, our results with p53, p16, and K-*ras* are consistent with those of previous studies in the literature that have shown a significant rate of alterations in these genes in gallbladder carcinoma. Loss of p16 expression and accumulation of p53 at high levels are the most common abnormalities associated with these carcinomas. Our study is the first to show Dpc4 protein loss in gallbladder carcinoma. We also show that loss of pRB expression is rare in non–small cell gallbladder carcinoma but is common in small cell carcinoma of the gallbladder. Indeed, the p16/pRB pathway was inactivated in all six small cell carcinomas studied. Perhaps most important, using immunohistochemistry, we were able to localize all of these alterations to the stage of carcinoma *in situ*. Indeed, of 15 invasive carcinomas associated with an evaluable *in situ*component, there was complete concordance between the results in 13 cases. Along these lines, one case of pure *in situ* carcinoma in our study contained three distinct alterations: p53 accumulation, p16 loss, and K-*ras* gene mutation.

Hence, carcinoma *in situ* of the gallbladder, although clinically occult, is a genetically advanced lesion. These results suggest that patients with pure carcinoma *in situ* of the gallbladder should receive careful clinical follow-up, particularly if the status of the cystic duct margin is unclear. At the same time, these results highlight our ignorance of mechanisms and markers of progression in this lethal neoplasm.

ACKNOWLEDGEMENTS

Supported by the Margaret Lee Fund for Gallbladder and Bile Duct Cancer research at The Johns Hopkins Hospital.

REFERENCES

1. Albores-Saavedra J, Henson DE, Klimstra DS. *Tumors of the gallbladder, extrahepatic bile ducts, and ampulla of Vater. Atlas of tumor pathology*. 3rd series. Fascicle 27. Washington, D.C.: Armed Forces Institute of Pathology; 2000.
2. Levin B. Gallbladder carcinoma. *Ann Oncol* 1999; 10: S129–S130.
3. Henson DE, Albores-Saavedra J, Corle D. Carcinoma of the gallbladder. Histologic types, stage of disease, grade, and survival rates. *Cancer* 1992;70: 1493–1497.
4. Nugent KP, Spigelman AD, Talbot IC, Phillips RKS. Gallbladder dysplasia in patients with familial adenomatous polyposis. *Br J Surg* 1994; 81: 291–292.

5. Chang HJ, Kim SW, Kim Y-T, Kim WH. Loss of heterozygosity in dysplasia and carcinoma of the gallbladder. *Mod Pathol* 1999; 12: 763–769. |

6. Wistuba II, Albores-Saavedra J. Genetic abnormalities involved in the pathogenesis of gallbladder carcinoma. *J Hepatobiliary Pancreatic Surg*1999; 6: 237–244.

7. Wistuba II, Sugio K, Hung J, Kishimto Y, *et al.* Allele-specific mutations involved in the pathogenesis of endemic gallbladder carcinoma in Chile. *Cancer Res* 1995; 55: 2511–2515.

8. Wistuba II, Tang M, Maitra A, *et al.* Genome-wide allelotyping analysis reveals multiple sites of allelic loss in gallbladder carcinoma. *Cancer Res*2001; 61: 3795–3800.

9. Gorunova L, Parada LA, Limon J, *et al.* Nonrandom chromosomal aberrations and cytogenetic heterogeneity in gallbladder carcinomas. *Genes Chromosomes Cancer* 1999; 26: 312–321.

10. Hanada K, Itoh M, Fujii K, Tsuchida A, Ooishi H, Kajiyama G. K-*ras* and p53 mutations in stage 1 gallbladder carcinoma with an anomalous junction of the pancreaticobiliary duct. *Cancer* 1996; 77: 452–458.

11. Wistuba II, Gadzar AF, Roa I, Albores-Saavedra J. p53 protein expression in gallbladder carcinoma and its precursor lesions. An immunohistochemical study. *Hum Pathol* 1996; 27: 360–365.

12. Yoshida S, Todoroki T, Ichikawa Y, *et al.* Mutations of $p16^{Ink4}$/*CDKN2* and$p15^{Ink4B}$/*MST2* genes in biliary tract cancers. *Cancer Res* 1995; 55: 2756–2760.

13. Kim Y-T, Kim J, Jang YH, *et al.* Genetic alterations in gallbladder adenoma, dysplasia, and carcinoma. *Cancer Lett* 2001; 169: 59–68.

14. Kim SW, Her K-H, Jang J-Y, Kim W-H, Kim Y-T, Park Y-H. K-*ras* oncogene mutation in cancer and precancerous lesions of the gallbladder. *J Surg Oncol* 2000; 75: 246–251.

15. Ajiki T, Fujimori T, Onoyama H, *et al.* K-*ras* gene mutation in gall bladder carcinomas and dysplasia. *Gut* 1996; 38: 426–429.

16. Watanabe M, Asaka M, Tanaka J, Kurosawa M, Kasai M, Miyazaki T. Point mutation of the K-*ras* gene codon 12 in biliary tract tumors. *Gastroenterology* 1994; 107: 1147–1153.

17. Imai M, Hoshi T, Ogawa K. K-*ras* codon 12 mutations in biliary tract tumors detected by polymerase chain reaction denaturing gel electrophoresis.*Cancer* 1994; 73: 2727–2733.

18. Itoi T, Watanabe H, Ajioka Y, *et al.* APC, K-*ras* codon 12 mutations and p53 gene expression in carcinoma and adenoma of the gall-bladder suggest two genetic pathways in gall-bladder carcinogenesis. *Pathol Int* 1996; 46: 333–340.

19. Maitra A, Tascilar M, Hruban RH, Offerhaus GJ, Albores-Saavedra J. Small cell carcinoma of the gallbladder. A clinicopathologic, immunohistochemical, and molecular pathology study of 12 cases. *Am J Surg Pathol* 2001; 25: 595–601.

20. Fleming ID, Cooper JS, Henson DE, *et al.*, editors. *AJCC cancer staging handbook.* Philadelphia, PA: Lippincott-Raven; 1997.

21. Baas IO, Mulder J-WR, Offerhaus GJA, Vogelstein B, Hamilton SR. An evaluation of six antibodies for immunohistochemistry of mutant p53 gene product in archival colorectal neoplasms. *J Pathol* 1994; 172: 5–12.

22. Wilentz RE, Su GH, Dai JL, *et al.* Immunohistochemical labeling for Dpc4 mirrors genetic status in pancreatic adenocarcinomas: a new marker of*DPC4* inactivation. *Am J Pathol* 2000; 156: 37–43.

23. Geradts J, Kratzke RA, Crush-Stanton S, Wen SF, Lincoln CE. Wild-type and mutant retinoblastoma protein in paraffin sections. *Mod Pathol* 1996; 9: 339–347.

24. Geradts J, Hruban RH, Schutte M, Kern SE, Maynard R. Immunohistochemical $p16^{INK4a}$ analysis of archival tumors with deletion, hypermethylation, or mutation of the *CDKN2/MTS1* gene: a comparison of four commercial antibodies. *Appl Immunohistochem Mol Morphol* 2000; 8: 71–79.

25. Maitra A, Roberts H, Weinberg AG, Geradts J. Loss of $p16^{INK4a}$ expression correlates with decreased survival in pediatric osteosarcomas. *Int J Cancer*2001; 95: 34–38.

26. Wilentz RE, Geradts J, Maynard R, *et al.* Inactivation of p16 (INK4A) tumor-suppressor gene in pancreatic duct lesions: loss of intranuclear expression.*Cancer Res* 1998; 58: 4740–4744.

27. Hruban RH, Sturm PDJ, Slebos RJC, *et al.* Can K-*ras* mutations be used to distinguish benign bile duct proliferations from metastases to the liver? A molecular analysis of 101 liver lesions from 93 patients. *Am J Pathol* 1997;151: 943–949.

28. Argani P, Shaukat A, Kaushal M, *et al.* Differing rates of Dpc4 expression and of p53 overexpression among carcinomas of the proximal and distal bile ducts. Evidence for a biologic distinction. *Cancer* 2001; 91: 1322–1341.

29. Shi Y-Z, Hui A-M, Li X, Tayayama T, Makuuchi M. Overexpression of retinoblastoma protein predicts decreased survival and correlates with loss of p16^{INK4} protein in gallbladder carcinomas. *Clin Cancer Res* 2000; 6: 4096–4100.

30. Benedict WF, Lerner SP, Zhou J, Shen X, Tokunaga H, Czerniak B. Level of retinoblastoma protein expression correlates with p16 (MTS-1/INK4A/CDKN2) status in bladder cancer. *Oncogene* 1999; 18: 1197–1203.

31. Sakaguchi M, Fujii Y, Hirabayashi H, *et al.* Inversely correlated expression of p16 and RB protein in non-small cell lung cancers: an immunohistochemical study. *Int J Cancer* 1996; 8: 442–445.

32. Ueki K, Ono Y, Henson JW, Efird JY, von Deimling A, Louis D. CDKN2/p16 or RB alterations occur in the majority of glioblastomas and are inversely correlated. *Cancer Res* 1996; 56: 150–153.

33. Shimizu E, Coxon A, Otterson GA, *et al.* RB protein status and clinical correlation from 171 cell lines representing lung cancer, extrapulmonary small cell carcinoma, and mesothelioma. *Oncogene* 1994; 9: 2441–2448.

34. Herrington CS, Graham D, Southern SA, Bramdev A, Chetty R. Loss of retinoblastoma protein expression is frequent in small cell carcinoma of the uterine cervix and is unrelated to HPV type. *Hum Pathol* 1999; 30: 906–910.

35. Takubu K, Nakamura K, Sawabe M, *et al.* Primary small cell undifferentiated carcinoma of the esophagus. *Hum Pathol* 1999; 30: 216–221.

36. Wistuba II, Miquel JF, Gadzar AF, Albores-Saavedra J. Gallbladder adenomas have molecular abnormalities different from those present in gallbladder carcinomas. *Hum Pathol* 1999; 30: 21–25.

Chapter 3

ORIGINS OF ALBINO AND HOODED RATS: IMPLICATIONS FROM MOLECULAR GENETIC ANALYSIS ACROSS MODERN LABORATORY RAT STRAINS

Takashi Kuramoto[1], Satoshi Nakanishi[1], Masako Ochiai[2], Hitoshi Nakagama[2], Birger Voigt[1], Tadao Serikawa[1]

[1]Institute of Laboratory Animals, Graduate School of Medicine, Kyoto University, Sakyo-ku, Kyoto, Japan

[2]National Cancer Center Research Institute, Chuo-ku, Tokyo, Japan

ABSTRACT

Albino and hooded (or piebald) rats are one of the most frequently used laboratory animals for the past 150 years. Despite this fact, the origin of the albino mutation as well as the genetic basis of the hooded phenotype remained unclear. Recently, the albino mutation has been identified as the Arg299His missense mutation in the Tyrosinase gene and the hooded (*H*) locus has been mapped to the ~460-kb region in which only the *Kit* gene exists. Here, we surveyed 172 laboratory rat strains for the albino mutation and the hooded (*h*) mutation that we identified by positional cloning approach to investigate possible genetic roots and relationships of albino and hooded rats. All of 117 existing laboratory albino rats shared the same albino missense mutation, indicating they had only one single ancestor. Genetic fine mapping followed by *de novo* sequencing of BAC inserts covering the *H* locus revealed that an endogenous retrovirus (ERV) element was inserted into the first intron of the *Kit* gene where the hooded allele maps. A solitary long terminal repeat (LTR) was found at the same position to the ERV insertion in another allele of the *H* locus, which causes the so called Irish (h^i) phenotype. The ERV and the solitary LTR insertions were completely associated with the hooded and Irish coat patterns, respectively, across all colored rat strains examined. Interestingly, all 117 albino rat strains shared the ERV insertion without any exception, which strongly suggests that the albino mutation had originally occurred in hooded rats.

INTRODUCTION

The rat (*Rattus norvegicus*) was the first mammalian species domesticated for scientific research with work dating back to before 1850. Since that time, the rat has a leading role in various research fields, such as physiology, pharmacology, neurosciences, genetics, and medical sciences. Among the rat strains available now, the albino and piebald (hooded) are most common, which is the same situation as in the earliest days when rats were firstly used for scientific research.

Albino rats have played a pioneering role in animal experimentation since its inception. Gregor Mendel reported his famous laws on 'Mendelian inheritance' in 1866. Hugo Crampe was the first scientist to confirm the validity of these laws in animals using some 15,000 white, grey, black, and piebald rats between 1877 and 1885 [1]. These albino rats were in fact the first animals to be domesticated for the purpose of scientific research.

The "hooded" phenotype is also one of the oldest mutations in the rat. Hooded rats have a pattern in which the entire ventral surface is white. Dorsally pigmentation is limited to the head and shoulders (the "hood") and a mid-dorsal stripe extending back to the tip of the tail. In addition to this hooded (*h)*, other modifiers such as Irish and notch alleles are known in the so named *H* locus [2]. The Irish (h^i) causes a white spot on the belly between and behind the front legs. Notch (h^n) rats have a white body with pigmented fur on the sides of the head. The wild-type allele against the *h*, h^i, or h^n is called "self" (*H*). Self animals are pigmented all over.

Despite their undisputed importance to scientific progress, the origin of albino and hooded rats has never been clearly determined. The question remains unanswered, is there only one single spontaneous mutation that has been transferred and bred for more than 100 years into hundreds existing albino rat strains? Or, did several different mutations such as those in mice[3] contribute to the albino phenotype that is now seen in millions of laboratory rodents around the world?

Henry H Donaldson at the Wistar Institute in Philadelphia wrote in 1915: "The Norway rat, *Mus norvegicus* (now *Rattus norvegicus*), is the one mammal easily obtainable both wild and as a domesticated form. This latter is represented by either the albino or the pied rats so common in our laboratories. We do not know whether the common albino variety had a single or multiple origins, or whether the colonies found in Europe are directly related to those now existing here"[4].

A deeper look into the historic records on rodents and a detailed molecular genetic analysis of the genetic patterns of all available coat color variations in rats are the measures of choice for solving these questions. The molecular nature of the albino rat has been identified as Arg299His missense mutation in the rat tyrosinase (*Tyr*) gene [5]. The Arg299His mutation has already been described in humans affected by ocular cutaneous albinism 1A (OCA1A) without any tyrosinase activity and consequential complete lack of pigmentation [6]. The hooded (*H*) locus was mapped to the rat Chr 14 [7]. Recently, Torigoe et al. mapped the *H* locus to the ~460-kb region in which only the *Kit* gene existed but did not found any mutations in the *Kit*coding region of the hooded LEA rats [8].

The aim of this study is to clarify and characterize the origin and nature of albino and hooded mutations using molecular genetic approaches and combine the findings with historical records. The albino related question was addressed by genotyping the *Tyr* gene mutation Arg299His for 172 laboratory strains from different regions of the world. In parallel, we confirmed the mapping of the hooded locus to the *Kit* gene on the rat Chr 14 by genetic studies and subsequently identified genetic variations in the *Kit* gene that are unique for hooded rats.

RESULTS

Genotyping of the Albino Allele

Preliminary sequencing of several albino strains confirmed the Arg299His mutation. Subsequent screening across all 172 strains examined was performed by direct digestion of the PCR product with the specific *Sna*BI restriction endonuclease to genotype the Arg299His mutation in the rat *Tyr* gene. All 55 colored rat strains had the wild-type allele (299Arg) and all of the 117 albino rat strains had the missense mutation (299His) (Table S1).

Fine Mapping of the Rat Hooded (H) Locus

Torigoe et al. mapped the *H* locus between *D14Rat84* and *D14Got40* [8]. Genotyping of the backcross progeny for both SSLP markers revealed six animals carrying the recombinant chromosome between *D14Rat84* and *D14Got40*. By using five SSLP markers, we narrowed the*H* locus down to between *D14Rat13* (position 34,941 kb) and *D14Got40* (position 35,259 kb). Haplotype analysis revealed that the *H* locus spanned from *ENSRNOSNP2799339* (position 34,910 kb) to *ENSRNOSNP2799341* (position 35,033 kb) (Fig. 1). These findings clearly indicated that the *H* locus spanned about 92 kb defined by *D14Rat13* and*ENSRNOSNP2799341* (Fig. 2). This hooded critical region contained the five exons (1 to 5) of the *Kit* gene and the 48.8 kb genomic region upstream of the *Kit* gene.

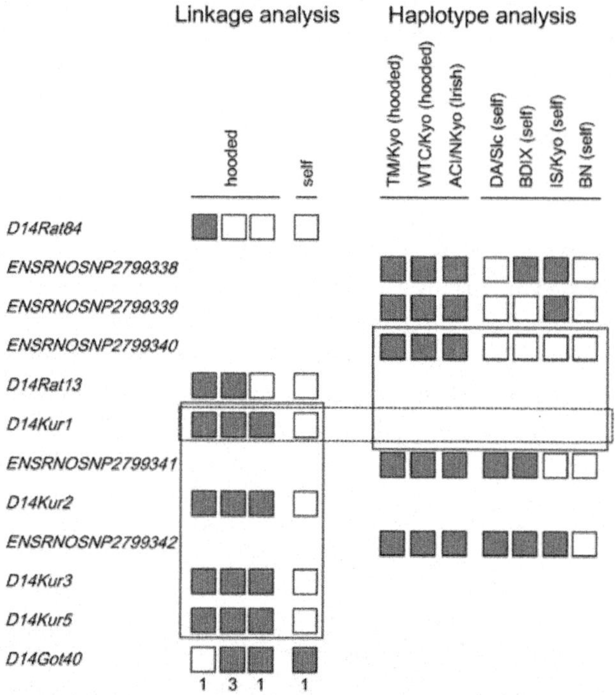

Figure 1: Linkage and haplotype mapping of the hooded (H) locus.

Left: Haplotypes of backcross progeny carrying the recombinant chromosome between D14Rat84 and D14Got40. The grey boxes represent rats homozygous for genetic markers, while the white boxes represent rats heterozygous for genetic markers. The coat pattern phenotypes of the progeny and the number of the progeny for each haplotype are described above and below the haplotypes, respectively. The H locus was narrowed down from D14Rat13 to D14Got40 (boxed with a solid line). Right: SNP haplotypes around the H locus in representative inbred rat strains. Two hooded rat strains and one Irish rat strain shared the identical haplotype of the genomic region between ENSRNOSNP2799338 and ENSRNOSNP2799342. The genomic region that was shared among the four self strains was narrowed down from ENSRNOSNP2799339 to ENSRNOSNP2799341 (boxed with a solid line). The linkage and haplotype analyses demonstrated that the hooded locus was mapped to the region defined by D14Rat13 and ENSRNOSNP2799341 (boxed with a dashed line).

doi:10.1371/journal.pone.0043059.g001

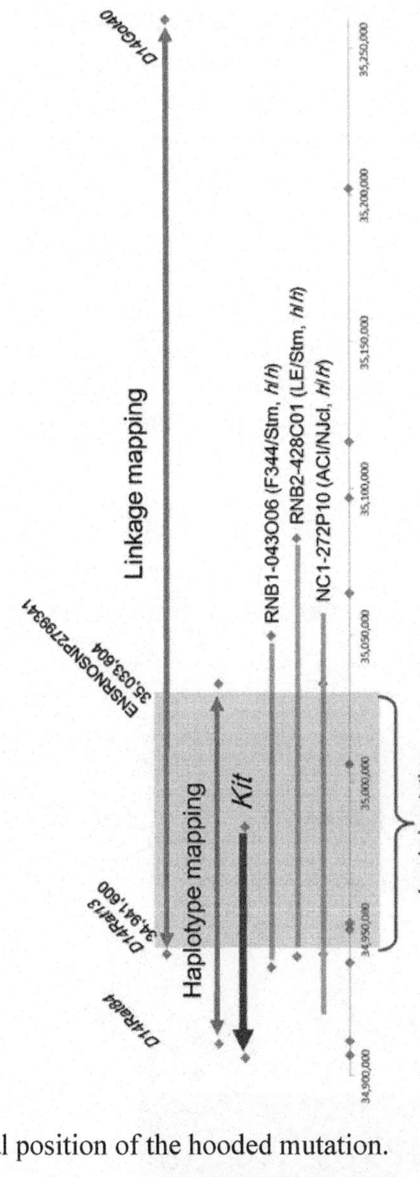

Figure 2: Physical position of the hooded mutation.

The hooded mutation spanned to the 92 kb region defined by D14Rat13 and ENSRNOSNP2799341 (boxed in red). This region contains the upstream region and five exons of the rat Kit gene (black arrow). The hooded critical region was covered by BACs that contain F344/Stm, LE/Stm and ACI/NJcl genomic DNAs. Grey bars represent the regions covered by BACs. Grey squares represent positions of the SSLP or SNP makers examined.

doi:10.1371/journal.pone.0043059.g002

Endogenous Retrovirus Sequences in the Rat Kit Gene

De novo sequencings of the entire inserts of the BAC clones derived from F344/Stm (*c*/*c*, *h*/*h*), and LE/Stm (*C*/*C*, *h*/*h*) rats covering the hooded critical region demonstrated that the 7,098 bp fragment was inserted into intron 1 of the rat *Kit* gene. The position of the insertion was 24,508-bp downstream of exon 1 and 8,054 bp upstream of exon 2 of the *Kit* gene. The 585-bp sequences on the both ends of the insertion showed significant similarities to the LTR of rat ERV (RLTR01_Rn) [9]. The remaining 5,930-bp sequence showed significant similarity to the rat ERV element (RAL_Rn_I) [10]. The ERV insertion was oriented on the minus strand for the*Kit* gene (Fig. 3A). Moreover, the BAC insert containing the Irish allele from the ACI/NJcl rat included the 584 bp LTR element only at the same position of the insertion found in the hooded allele (Fig. 3A).

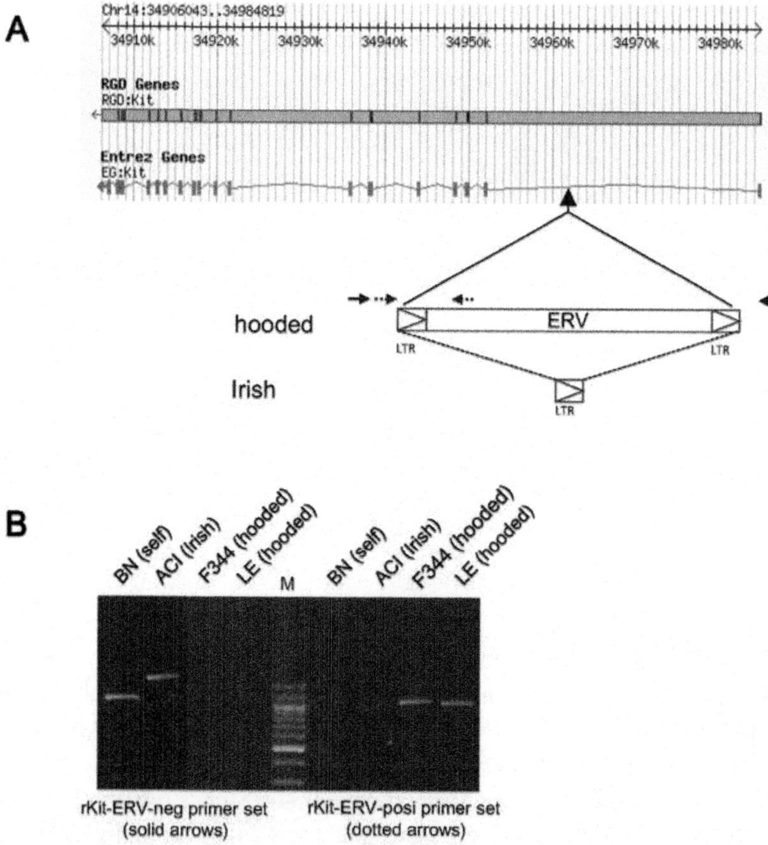

Figure 3: Insertion of the ERV into intron 1 of the rat Kit gene.

A: Position and organization of the ERV insertion found in intron 1 of the rat Kit gene. The insertion was located 8,054 bp upstream of exon 2. The ERV sequence with 2 LTRs on both ends was found in the "hooded" strains. The solitary LTR sequence was found in the "Irish" strains. A set of primers (indicated by solid arrows) was designed to detect the absence of the ERV insertion and the presence of the solitary LTR sequence. A set of primers (indicated by dotted arrows) was designed to detect the presence of the ERV insertion. B (Left): PCR product obtained with the rKit-ERV-neg primer set (solid arrows). A 1,006 bp product was obtained from the BN (self) genome and a 1,590 pb product was obtained from the ACI (Irish) genome. Meanwhile, no products were obtained from the F344 (albino, hooded) genome and the LE (hooded) genome. (Right): PCR products obtained with the rKit-ERV-posi primer set (dotted arrows). Products 997 bp in size were obtained from the F344 genome and the LE genome, while no products were obtained from the BN genome and the ACI genome. M; 100-bp ladder molecular marker.

doi:10.1371/journal.pone.0043059.g003

Genotyping of the 7,098 bp Insertion and the 584 bp Insertion

To find the correlation of the insertions with the hooded or Irish phenotype and survey the prevalence of the insertions among rat strains, we determined the presence of the insertions for 172 rat strains (Fig. 3B). Among 55 colored rat strains, the 18 strains with the self-coat color pattern had no insertion. The 29 strains with the hooded pattern had the ERV (7,098 bp) sequence. Finally, the eight strains with the Irish pattern had the LTR (584 bp) element only. As for the albino rat strains, all of the 117 strains had the ERV (7,098 bp) sequence, indicating the shared uniform haplotype for the albino and hooded mutations (Table S1).

DISCUSSION

The *Kit* gene encodes a cell surface receptor, c-Kit, (molecular weight 145–160 kd) that belongs to the immunoglobulin gene family and carries an intrinsic tyrosine kinase activity in its cytoplasmic portion. The interaction of c-Kit with its ligand, steel factor, leads to receptor dimerization, kinase activation, and tyrosine phosphorylation of cytoplasmic proteins. The c-Kit gene is expressed on melanocytes, gametocytes, mast cells, hematopoietic stem cells, and interstitial cells of Cajal. Thus, the Kit tyrosine kinase membrane receptor is essential for melanogenesis, gametogenesis and hematopoiesis during embryonic development and postnatal life.

c-Kit is expressed in melanoblasts during the melanogenesis starting from the time they leave the neural crest. Expression continues during embryonic development [11]. It is also expressed in the melanocytes of postnatal animals. Thus, mutations of the murine *Kit* gene manifest as dominant white spotting (*W*). In addition to the *Kit* gene, several other genes, such as *Pax3*, *Mitf*, and *Sox10*, could also be involved in the white-spotting phenotype. Mutations in these genes are related to defects of melanocyte development [12]. *W* mutations either alter the coding sequence of the Kit receptor tyrosine kinase, resulting in a receptor with impaired kinase activity, or affect *Kit* expression. *W* mutations that affect *Kit* expression are often located in the regulatory sequence. For example, the $Kit^{W\text{-}57J}$ allele affects the temporal and the spatial pattern of the *Kit* expression, so that $Kit^{W\text{-}57J}/Kit^{W\text{-}57J}$ mice have an irregular band of spotting, lack pigmentation in their feet and tails, and have a head blaze. The $Kit^{W\text{-}57J}$ allele comprises an 80 kb deletion located at the 5′ end of the *Kit*-coding sequence [13]. The $Kit^{W\text{-}bd}$and $Kit^{W\text{-}sh}$ alleles also affect the *Kit* expression pattern during the developmental stage and the both $Kit^{W\text{-}bd}/+$ and $Kit^{W\text{-}sh}/+$ mice show a white band in the trunk region [13], [14]. Both alleles are associated with a genomic inversions located at the 5′ region of the *Kit* gene [13]. These findings indicate that the dysregulation of the *Kit* expression affects the development of melanoblasts and thereby the white spotting appears in mutant mice.

We found the insertion of the ERV sequence in the intron 1 of the rat *Kit* gene. ERV sequences have resulted from both ancient and modern infections of exogenous retroviruses, which have successfully colonized the germ line of their host [15]. Insertion of the ERV disrupts host protein-coding genes or alters gene expression by affecting splicing or by providing novel signals for initiation, regulation or termination of transcription [15]. The ERV insertions of the first intron with antisense orientation lead the small and lower level of transcripts [16], or aberrant splicing [17], [18] in mouse mutants. Thus, we consider that the ERV insertion we found in the hooded allele may provoke the dysregulation of the *Kit*expression and thereby causes the specific hooded pattern.

We also found the solitary LTR sequence in the Irish allele. In mice, it has been known that the internal ERV sequence is deleted via homologous recombination between the 5′ and the 3′ LTRs so that a solitary LTR is left behind [19]. Such deletion reverses the mutant phenotype to the wild-type or, occasionally, attenuates the mutant phenotype. The latter case is found in the reversion of the mouse nonagouti (*a*) to the black-and-tan (a^t) or the white-bellied agouti (A^w). The *a* insertion consists of a 5.5-kb VL30 element that has incorporated 5.5 kb of additional sequence internally; this internal sequence is flanked by 526-bp direct repeats. Homologous recombination utilizing the

526-bp direct repeats generates the a^t allele, containing only the VL30 element with a single internal 526-bp repeat. Homologous recombination utilizing the VL30 LTRs generates the A^w allele, containing only a solitary LV30 LTR [16]. The rat Irish (h^i) causes a white spot on the belly between and behind the front legs. Therefore, we conclude that the solitary LTR in the Irish allele also may have been generated by homologous recombination between the 5′ and 3′ LTRs of the rat ERV that is inserted in the hooded allele and that the Irish allele is a partial revertant of the hooded allele due to the residual solitary LTR.

Coat color variations have historically played important roles in genetic studies to understand the basics of inheritance. Thus, the identification of the causative mutation of the color variations followed by the survey of them for existing strains provides insights into the origin of coat color mutations. Since the coat color variations were noted in the early days of the domestication of the rat, such molecular genetic approach could give new insights on the establishment of laboratory rat strains. In the present study, we focused on the oldest color variations in the rat, albino and hooded.

We collected tissues or DNAs of rat strains from across the world in order to cover all possible albino or hooded mutations that are still available in laboratory rat strains. Thus, it appears very likely that all albino laboratory rat strains developed so far share only one common mutation: 299His in the *Tyr* gene. Most existing albino rat strains are derived from albino strains or stocks of the Wistar Institute or from crosses between Wistar albino rats and other rats including wild rats. The remaining albino rats were not directly derived from the Wistar Institute (Table S1). DON strains, Ihara's rat strains, and TO strains were established in Japan, while the F344 strains and HTX strain were established in the USA, and Yagil's rat strains were established in Israel. They also carry the same *Tyr* mutation. These findings suggest that the heredity of all albino rats can be traced back to one rat with the albino mutation.

Moreover, we found that all albino strains we examined share the 7,098-bp ERV insertion in the*Kit* gene without any exception. According to Donaldson [4], there must have been both albino and hooded stocks before the establishment of the Wistar albino stock. Given the uniform genotypes for the albino and hooded mutations in the albino rat strains, we suggest two possible scenarios for the establishments of the albino and hooded strains. The most likely one is that the albino mutation occurred in a rat of a hooded stock. From that colony, the first albino rats were discovered and were used as founders of albino rats. Some albino rats were introduced into the Wistar Institute and some were used for developments of Wistar-independent albino stocks (Fig. 4). Another scenario is that albino rat strains had been developed independently of hooded stock and subsequent crossing occurred between the albino and the hooded stocks. In the resultant stocks of this cross, albino rats that carried either the hooded (*h*) or self (*H*) allele must be present. Given the evidence provided in this study, albino rats with the particular genotype (*h*/*h*, *c*/*c*) were "by chance" selected and used to establish the albino rat stock. Such newly established albino stock was introduced into the Wistar Institute and some rats were used for the development of the Wistar-independent albino stocks (Fig. 4). Since our survey could not detect any albino rat without the hooded ERV insertion, the second scenario appears very unlikely and is just of hypothetical nature. Therefore, it is very likely that the hooded stock had been developed earlier than the albino strains and the rat albino missense mutation occurred in one hooded rat.

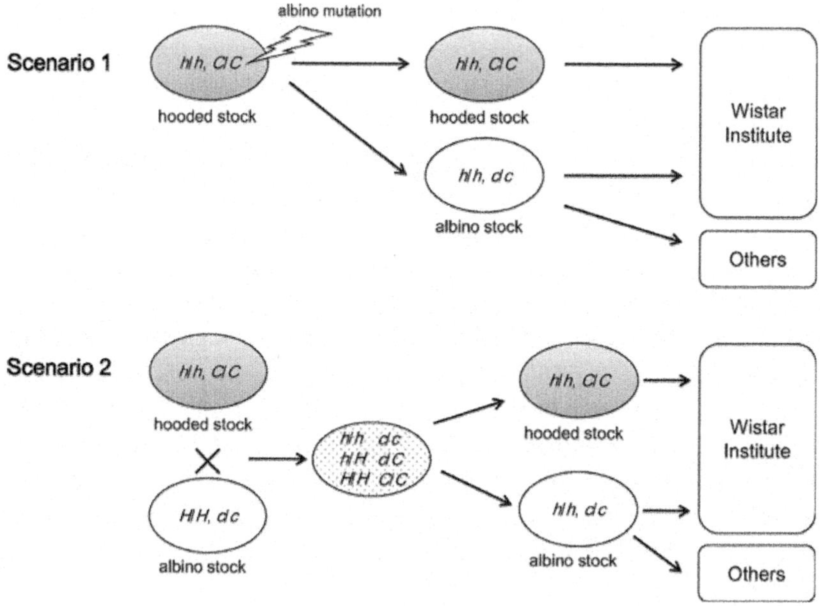

Figure 4: Possible scenarios for the establishment of the albino and hooded rat strains.

Scenario 1: The rat albino (299His) mutation occurred in a rat of the hooded (piebald) rat stock. From the stock, the albino stock was established. Then they were introduced into the Wistar Institute and some albino rats were used for the establishment of the Wistar-independent stocks. Scenario 2: The albino stock that can be traced back to an albino rat that carried the 299His mutation and the hooded (piebald) stock were independently established. Crossing occurred between the albino and the hooded stocks. In the resultant stocks of this cross, albino rats that carried either the hooded (h) or self (H) allele must be present. The albino rats with the particular genotype (h/h, c/c) were "by chance" selected and used to establish the albino rat stock. Such new-established albino stock was introduced into the Wistar Institute and some rats from it were used for the development of the Wistar-independent albino stocks.

doi:10.1371/journal.pone.0043059.g004

MATERIALS AND METHODS

Ethics Statement

Animal experimentation in this study was carried out in strict accordance with the Regulation on Animal Experimentations at Kyoto University. The protocol was approved by the Animal Research Committee of Kyoto University.

Genomic DNAs

Genomic DNAs from 172 rat strains were used. They consist of 55 colored rat strains and 117 albino rat strains (Table S1).

Linkage Analysis

A total of 897 backcross progeny were genotyped for the hooded (*H*) locus by identifying the appearance of the characteristic hooded mark. They consisted of 300 rats from the (ACI × WTC)F1 × WTC cross [20], 207 rats from the (BN × TM)F1 × TM cross, and 390 rats from the (BN × GRY)F1 × GRY cross [21]. All of the rats were genotyped for *D14Rat84* and *D14Got40*and rats that carried the recombinant chromosome between them were used for fine mapping of the *H* locus. Five simple sequence length polymorphism (SSLP) markers (*D14Rat13*,*D14Kur1*, *D14Kur2*, *D14Kur3*, and *D14Kur5*) were used for the fine mapping.

Haplotype Analysis

Two hooded strains (TM/Kyo and WTC/Kyo) and four "self" strains (BDIX, BN/NSlc, DA/Slc, and IS/Kyo) were genotyped for five single nucleotide polymorphism (SNP) markers that were mapped between *D14Rat84* and *D14Got40* [22]. Their names were as follows: ENSRNOSNP2799338 (position: 34,823,877), ENSRNOSNP2799339 (34,910,749), ENSRNOSNP2799340 (34,938,580), ENSRNOSNP2799341 (35,033,604), and ENSRNOSNP2799342 (35,096,682).

BAC Sequencing

RNB1-043O06 BAC containing the hooded allele of the rat strain F344/Stm, RNB2-428C01 BAC containing the hooded allele of the rat strain LE/Stm and NC1-272P10 BAC containing the Irish allele of the rat strain ACI/NJcl, were used for BAC sequencing. High molecular weight BAC DNAs were prepared by using a Large-Construct Kit (QIAGEN) and physically fractionated into sections several hundred base pairs in length by using a Covaris S-series (Covaris, MT, USA). The BAC inserts were sequenced by using a Genome Sequencer FLX system (Roche Diagnostics) and the sequences were assembled by using a GS *De Novo*assembler (Roche Diagnostics).

Genotyping

The genotyping of the Arg299His missense mutation in the *Tyr* gene was performed by the PCR-RFLP analysis [5]. The PCR products amplified with a set of primers (rTyr12&13 TTTCATTCATATGTAAGTCCCTTG and GCTTAGCATTGCAAAACTCACA) were digested with *Sna*BI restriction endonuclease (New England Biolabs). In this assay, the PCR products from the wild-type (299Arg) are digested, while those from the albino-type (299His) are not digested.

To determine the presence of the ERV insertion or the LTR insertion in intron 1 of the *Kit* gene, primer sets were designed. The rKit-ERV-posi (GGCCTGTGAGTGTGAATTTG and GGACGAGCCCCATAAATA) were used to detect the ERV insertion and the rKit-ERV-neg (ACTTAAAGACCACTGAGGACA and AATGCGGAACATCTTTCAA) were used to detect the LTR insertion.

Supporting Information

Table_S1.xlsx

Table S1: Coat color phenotypes and the genotypes for Tyr and Kit genes in laboratory rat strains.

doi:10.1371/journal.pone.0043059.s001

(XLSX)

	1	2	3	4	5	6	7	8	9
1									
2	Table S1. Coat color phenotypes and the genotypes for Tyr and Kit genes in laboratory rat strains								
3									
4	Strain family	Sub strain name	Source1	NBRP No.2	Coat color phenotype	Genotype of the hooded locus	Tyr gene	Kit gene	Origin3
5							missence mutation Arg 299 His	ERV insertion type	

6	ACI and its derived strains								Established from a cross beteen August and COP strains by MR Curtiss and WF Dunning at Columbia University, Institute for Cancer Research, USA in 1926
7		ACI/NHok	NBRP-Rat	33	agouti hooded Irish	hood-ed Irish	Arg	LTR	From the NIH, USA, to Tokyo Biochemical Research Institute, Japan, and to Hokkaido University, Japan in 1967.
8		ACI/NJcl	CLEA	NA	agouti hooded Irish	hood-ed Irish	Arg	LTR	From the NIH, USA, to National Institute of Genetics, Japan, to Central Institute of Experi-mental Animals, Japan in 1980, then to CLEA Japan Inc in 1986.
9		ACI/NKyo	NBRP-Rat	1	agouti hooded Irish	hood-ed Irish	Arg	LTR	From the NIH, USA to Kyoto University, Japan in 1988.
10		ACI-Lystbg/ Kyo	NBRP-Rat	2	beige hooded Irish	hood-ed Irish	Arg	LTR	Derived from spontaneous beige coat colour mutation found in ACI/NKyo at Institute of Laboratory Animals, Kyoto University in 1999, and maintained as a congenic strain.
11		ACI/NMna	NBRP-Rat	203	agouti hooded Irish	hood-ed Irish	Arg	LTR	Derived from ACI/NMs by M Matsushima at Fujita Health University School of Medicine, Japan
12		ACI/NSlc	NBRP-Rat	158	agouti hooded Irish	hood-ed Irish	Arg	LTR	From National Institute of Infectious Disease, Tokyo, Japan to Japan Slc Inc in 1992
13		ACI-Chib/Kyo	NBRP-Rat	465	agouti hooded Irish	hood-ed Irish	Arg	LTR	Derived from small body size mutation found in ENU-mutagenesis using ACI/Nslc, and maintained as congenic line.
14		ACIS/Hok	NBRP-Rat	34	agouti with white face	hood-ed Irish	Arg	LTR	Derived from a spontaneous mutation showing wider white pigmentation found in ACI/Hok in 1981, and maintained by crossing its heterozy-gous rats with wild type rats.

15	Albany rat strain								Albany rat colony derived from an unknown source, supplied by CE Bills at the Mead Johnson Company of Evansville, Indiana, USA, breeding started by A Knudson at the Departement of Biochemistry, the Albany Medical College, New York, USA in 1930. Representatives of the Albany stock were transferred to the NIH by JM Wolfe and AM Wright at Albany Medical CollegeUSA, where inbreeding started.
16		ALB/Hok	NBRP-Rat	35	non-agouti dilute brown	self	Arg	-	To Hokkaido University in 1956 (F8), to National Institute of Genetics, Mishima, Japan in 1958, then to the Center for Experimental Plants & Animals Chromosome Research Unit, Faculty of Science, Hokkaido University, Japan
17	Bio Breeding Laboratory derived rat strains								Established from spontaneous mutants, which develop autoimmune Type 1 diabetes, found in a Wistar rat stock at Bio-Breeding Laboratories, Ontario, Canada
18		BB/OK	NBRP-Rat	260	albino	N.D.	His	LTR-ERV1-LTR	Introduced from the Bio Breeding Laboratories at University of Greifswald, Karlsburg, Germany in 1983, and established as BB subline by I Klötting.
19		BB/WorTky	NBRP-Rat	30	albino	N.D.	His	LTR-ERV1-LTR	Introduced from the Bio Breeding Laboratories to the massachusetts medical school, Worcester in 1977, where inbreeding began. Imported to Tokyo Medical College, Japan, by K Komeda in 1983.
20	Berlin-Druckrey strains								BDI was derived from a pair of pink-eyed yellow rats obtained from Kröning, Göttingen, and the inbred strain was served as progenitor strain for further Berlin-Druckrey strains.
21		BDII	HAN-NOVER	NA	albino	N.D.	His	LTR-ERV1-LTR	Derived from an outbred colony of Wistar rats obatained from the Unilever Research Institute, Rotterdam, The Netherlands.

22		BDIX/NemOda	NBRP-Rat	304	agouti	self	Arg	-	Derived from a cross between BDI and BDVIII. BDVIII was derived from F2 cross between BDII and BDIII with the same origin as BDI.
23		KB/Oda	NBRP-Rat	306	albino	N.D.	His	LTR-ERV1-LTR	A segregating inbred strain for albino locus, derived from a cross between Wistar rat of a breeder Nihon rat Ltd. in Japan and BDI.
24	Brown Norway rat strains								Established from a brown coat color rat found in a pen-bred colony of Wistar Insitute maintained by HD King and P Aptekman. The origin are derived from wild rats captured in the vicinity of Philadelphia. F35 by 1934.
25		BN/1Hok	NBRP-Rat	37	brown	self	Arg	-	BN substrain maintained at the Chromosome Research Unit, Hokkaido University in 1987 (F15).
26		BN/2Hok	NBRP-Rat	38	brown	self	Arg	-	BN substrain maintained at the Chromosome Research Unit, Hokkaido University in 1987 (F16).
27		BN/fMailHok	NBRP-Rat	36	brown	self	Arg	-	Introduced to Hokkaido University through Aichi Colony institute from Kyoto University in 1976, and then transferred from the Laboratory of Experimental Animal Science to the Center for Experimental Plants & Animals at Hokkaido University, Japan in 1982.
28		BN/KtsSlc	NBRP-Rat	165	brown	self	Arg	-	To Japan Slc Inc from Baba at Kitasato University School of Medicine, Japan in 2002.
29		BN/KunKtsSlc	NBRP-Rat	164	brown	self	Arg	-	To Japan Slc Inc from Baba at Kitasato University School of Medicine, Japan in .2001.
30		BN/Seac	KYUDO	NA	brown	self	Arg	-	BN substrain maintained at Seac Yoshitomi Co Ltd.

31		BN/SsNHsd	HAR-LAN	NA	brown	self	Arg	-	To the NIH, USA in 1972 from WK Silvers at F34. Silvers began brother-sister matings with selection for histocompatibility in 1958 from a brown mutation in a stock of wild rats maintained by HD King in a penbred colony.
32		BN/SsNSlc	NBRP-Rat	149	brown	self	Arg	-	To Japan SLC Inc from the NIH in 1994
33	Buffalo rat strains								Derived from Buffalo stock of H Morris, presumably from Wistar commercial stock in 1946.
34		BUF/Mna	NBRP-Rat	200	albino	N.D.	His	LTR-ERV1-LTR	Established from sib mating of Buffalo rat colony transferred from Biological Associates in America, USA. To Department of Pathology, Nagoya City University in 1964. At F18 generation, the rats showed 100% thymoma and were named as BUF/Mna by M Matsuyama at Fujita Health University, Japan in 1977.
35		BUF/NacJcl	CLEA	NA	albino	N.D.	His	LTR-ERV1-LTR	To Japan Slc inc from National Cancer Center Research Institute, Tokyo, Japan.
36	Cohen diabetes rat strains								Derived from the Hebrew University albino rats, and genetitacally selected for blood glucose levels by AM Cohen in Israel at about 1965.
37		CDR/Ygl	YAGIL	NA	albino	N.D.	His	LTR-ERV1-LTR	Cohen diabetic-resistant strain developed by selective inbreeding for a lower blood glucose level.
38		CDS/Ygl	YAGIL	NA	albino	N.D.	His	LTR-ERV1-LTR	Cohen diabetic-sensitive strains developed by selective inbreeding for a higher blood glucose level.

39	Copenhagen rat strains								Original rats were obtained from a breeder in Copenhagen by J Rosenstirn and introduced to MR Curtis at Columbia University Institute for Cancer Research in 1920.Inbreeding began in 1922 by WF Dunning at Crocker Reserach Institute of Columbia University, USA.
40		Iar:Copenhagen Rat	IAR	NA	brown hooded	hood-ed	Arg	LTR-ERV1-LTR	To Institue of Animal Reproduction from the NIH, USA, in 1986.
41	DA strain								Derived from heterogenous stocks of unknown origin, and developed by TT Odell at OakRidge National Laboratory. It may be related to COP. It was named DA because it express the Ag-D blood group allele.
42		DA/Slc	NBRP-Rat	157	agouti	self	Arg	-	To Japan SLC Inc from Kumamoto University School of Medicine, Japan in 1983.
43	Dahl rat strains								Derived from Sprague-Dawley outbred rats, and selected for salt sensitive hypertension by LK Dahl at Brookhaven National Laboratories, New York, USA.
44		DahlS/Seac	KYUDO	NA	albino	N.D.	His	LTR-ERV1-LTR	Salt sensivive hypertension substrain
45		SR/JrNgs	NBRP-Rat	485	albino	N.D.	His	LTR-ERV1-LTR	Control strain of SS/JrNgs
46		SS/JrNgs	NBRP-Rat	484	albino	N.D.	His	LTR-ERV1-LTR	Salt sensivive hypertension substrain
47	DOB strain								Derived from wild rats trapped at goat shed at the Shitara field, Graduate School of Bioagri-cultural Sciences and School of Agricultural Sciences Nagoya University. Japan in 2000.

48		DOB/Oda	NBRP-Rat	307	agouti	self	Arg	-	From S Oda at Nagoya University to Institute of Laboratory Animals, Kyoto University, Japan, in 2007 (F14), where this inbred strain was established in 2009.
49	Donryu rat strains								Derived from albino rats purchased from a local breeder in Ohota city, Gumma, Japan, by R Sato in 1950.
50		ALD/Hyo	NBRP-Rat	576	albino	N.D.	His	LTR-ERV1-LTR	Derived from a male rat showing hepatomegaly and splenomegaly, acid lipase deficiency, in non-inbred Donryu rat found by H Yoshida at Ehime University, Japan in 198, and maintained in heterozygous condition.

Coat color phenotypes and the genotypes for Tyr and Kit genes in laboratory rat strains.

ACKNOWLEDGMENTS

We thank the National Bio Resource Project – Rat (http://www.anim.med.kyoto-u.ac.jp/nbr/) for providing the rat genome DNA. We are grateful to Dr. Dirk Wedekind for providing DNA from rat strains PAR, R33, and BDII, Prof. Yoram Yagil and Dr. Chana Yagil for providing DNA for rat strains CDR/Ygl, CDR/Ygl, SBH/Ygl, and SBN/Ygl, Dr. Pernilla Strindh-Forsgren for providing tissue samples of the M520 rat strain. The RNB1 and RNB2 BAC clones are available from the RIKEN Bio Resource Center (http://dna.brc.riken.jp/index.html). The genome sequences of BAC inserts are available from the DDBJ with accession numbers (AP012486 for NC1-272P10, AP012487 for RNB1-043O06, and AP012488 for RNB2-482C01).

AUTHOR CONTRIBUTIONS

Conceived and designed the experiments: TK TS. Performed the experiments: TK SN. Analyzed the data: TK BV TS. Contributed reagents/materials/analysis tools: MO HN. Wrote the paper: TK BV TS.

REFERENCES

1. Crampe H (1885) Landwirthschaftliche Jahrbücher.
2. Castle WE (1951) Variation in the hooded pattern of rats, and a new allele of hooded. Genetics 36: 254–266.
3. Beermann F, Orlow SJ, Lamoreux ML (2004) The *Tyr* (albino) locus of the laboratory mouse. Mamm Genome 15: 749–758. doi: 10.1007/s00335-004-4002-8
4. Donaldson HH (1915) The Rat data and reference tables.
5. Blaszczyk WM, Arning L, Hoffmann KP, Epplen JT (2005) A Tyrosinase missense mutation causes albinism in the Wistar rat. Pigment Cell Res 18: 144–145. doi: 10.1111/j.1600-0749.2005.00227.x
6. Gershoni-Baruch R, Rosenmann A, Droetto S, Holmes S, Tripathi RK, et al. (1994) Mutations of the tyrosinase gene in patients with oculocutaneous albinism from various ethnic groups in Israel. Am J Hum Genet 54: 586–594.
7. Serikawa T, Kuramoto T, Hilbert P, Mori M, Yamada J, et al. (1992) Rat gene mapping using PCR-analyzed microsatellites. Genetics 131: 701–721.
8. Torigoe D, Ichii O, Dang R, Ohnaka T, Okano S, et al. (2011) High-resolution linkage mapping of the rat hooded locus. J Vet Med Sci 73: 707–710. doi: 10.1292/jvms.10-0529

9. Jurka J (2009) Long terminal repeat of ERV1 enodogenous retrovirus from rat: consensus. Repbase Reports 9: 1214.
10. Nakamuta M, Furuich M, Takahashi K, Suzuki N, Endo H, et al. (1989) Isolation and characterization of a family of rat endogenous retroviral sequences. Virus Genes 3: 69–83. doi: 10.1007/bf00301988
11. Keshet E, Lyman SD, Williams DE, Anderson DM, Jenkins NA, et al. (1991) Embryonic RNA expresson patterns of the c-Kit receptor and its cgnate ligand suggest multiple functional roles in mouse development. EMBO J 10: 2425–2435.
12. Baxter LL, Hou L, Loftus SK, Pavan WJ (2004) Spotlight on spotted mice: a review of white spotting mouse mutants and associated human pigmentation disorders. Pigment Cell Res 17: 215–224. doi: 10.1111/j.1600-0749.2004.00147.x
13. Kluppel M, Nagle DL, Bucan M, Bernstein A (1997) Long-range genomic rearrangements upstream of Kit dysregulate the developmental pattern of Kit expression in W^{57} and W^{banded} mice and interfere with distinct steps in melanocyte development. Development 124: 65–77.
14. Duttlinger R, Manova K, Chu TY, Gyssler C, Zelenetz AD, et al. (1993) W-sash affects positive and negative elements controlling c-kit expression: ectopic c-kit expression at sites of kit-ligand expression affects melanogenesis. Development 118: 705–717.
15. Stocking C, Kozak CA (2008) Murine endogenous retroviruses. Cell Mol Life Sci 65: 3383–3398. doi: 10.1007/s00018-008-8497-0
16. Bultman SJ, Klebig ML, Michaud EJ, Sweet HO, Davisson MT, et al. (1994) Molecular analysis of reverse mutations from nonagouti (*a*) to black-and-tan (a^t) and white-bellied agouti (A^w) reveals alternative forms of agouti transcripts. Genes Dev 8: 481–490. doi: 10.1101/gad.8.4.481
17. Bowes C, Li T, Frankel WN, Danciger M, Coffin JM, et al. (1993) Localization of a retroviral element within the rd gene coding for the beta subunit of cGMP phosphodiesterase. Proc Natl Acad Sci U S A 90: 2955–2959. doi: 10.1073/pnas.90.7.2955
18. Hofmann M, Harris M, Juriloff D, Boehm T (1998) Spontaneous mutations in SELH/Bc mice due to insertions of early transposons: molecular characterization of null alleles at the nude and albino loci. Genomics 52: 107–109. doi: 10.1006/geno.1998.5409
19. Maksakova IA, Romanish MT, Gagnier L, Dunn CA, van de Lagemaat LN, et al. (2006) Retroviral elements and their hosts: insertional mutagenesis in the mouse germ line. PLoS Genet 2: e2. doi: 10.1371/journal.pgen.0020002

20. Kuramoto T, Yokoe M, Hashimoto R, Hiai H, Serikawa T (2011) A rat model of hypohidrotic ectodermal dysplasia carries a missense mutation in the *Edaradd* gene. BMC Genet 12: 91. doi: 10.1186/1471-2156-12-91

21. Tokuda S, Kuramoto T, Tanaka K, Kaneko S, Takeuchi IK, et al. (2007) The ataxic groggy rat has a missense mutation in the P/Q-type voltage-gated Ca^{2+} channel α1A subunit gene and exhibits absence seizures. Brain Res 1133: 168–177. doi: 10.1016/j.brainres.2006.10.086

22. Saar K, Beck A, Bihoreau MT, Birney E, Brocklebank D, et al. (2008) SNP and haplotype mapping for genetic analysis in the rat. Nat Genet 40: 560–566. doi: 10.1038/ng.124

Chapter 4

MOLECULAR GENETIC ANALYSIS OF CENTRAL NERVOUS SYSTEM GERM CELL TUMORS WITH COMPARATIVE GENOMIC HYBRIDIZATION

Dominik T Schneider[1], Susanne Zahn[1], Sonja Sievers[1,2], Katayoun Alemazkour[1], Guido Reifenberger[3], Otmar D Wiestler[4], Gabriele Calaminus[1], Ulrich Göbel[1]and Elizabeth J Perlman[5]

[1]Clinic of Paediatric Oncology, Haematology and Immunology, Heinrich-Heine-University, Düsseldorf, Germany

[2]Max Planck Institute of Molecular Physiology, Dortmund, Germany

[3]Department of Neuropathology, Heinrich-Heine-University, Düsseldorf, Germany

[4]German Brain Tumor Reference Center Institute of Neuropathology, University of Bonn, Bonn, Germany

[5]Department of Pathology, Children's Memorial Hospital, Chicago, IL, USA

ABSTRACT

The limited information available to date regarding the genetic alterations in germ cell tumors of the central nervous system has raised concerns about their biologic relationship to other germ cell tumor entities. We investigated fresh-frozen or archival tumor samples from 19 patients with central nervous system germ cell tumors (CNS-GCTs), including seven germinomas, eight malignant nongerminomatous germ cell tumors and four teratomas, using chromosomal comparative genomic hybridization to determine recurrent chromosomal imbalances. All 15 malignant CNS-GCTs and two of four teratomas showed multiple chromosomal imbalances. Chromosomal gains (median: 4 gains/tumor, range: 0–9 gains/tumor) were observed more frequently than losses (median: 1.6 losses/tumor, range: 0–6 losses/tumor). Gain of 12p, which is considered characteristic for germ cell tumors of the adult testis, was detected in 11 of 19 tumors and 10 of 15 malignant CNS-GCTs. In one tumor, gain of 12p was confined to an amplicon at 12p12, corresponding to the commonly amplified region on 12p. Other common gains were found on chromosome arms 1q and 8q (n=9, each). Among the chromosomal losses, parts of chromosome 11 (n=5), 18 (n=4), and 13 (n=3) were deleted most frequently. Notably, we observed no difference in the genetic profiles of germinomatous and nongerminomatous CNS-GCTs; however, the average number of imbalances was higher in the latter group. A meta-analysis comparing 116 malignant gonadal and extragonadal germ cell tumors revealed that the genomic alterations in CNS-GCTs are virtually indistinguishable from those found in their gonadal or other extragonadal counterparts of the corresponding age group. These data strongly argue in favor of common pathogenetic mechanisms in gonadal and extragonadal germ cell tumors.

During childhood and adolescence, the majority of germ cell tumors arise outside of the gonads, and beyond early childhood, the central nervous system and the mediastinum constitute the most frequent sites of extragonadal germ cell tumors.[1] The vast majority of central nervous system germ cell tumors (CNS-GCTs) develop in the pineal gland or the suprasellar region, and approximately 10% of all CNS-GCTs present as bifocal tumors.[2,3] The extragonadal appearance of germ cell tumors is likely related to errors in germ cell migration during early embryonal development. Accordingly, imprinting studies of gonadal and extragonadal germ cell tumors show loss of the methylation imprint at imprinting control regions, correlating with the methylation status within early stages of primordial germ cell development.[4,5,6] However, the molecular mechanisms that interfere with normal homing of germ cells to the gonadal ridge and that allow for a sustained survival of germ cells in an extragonadal environment remain unclear. Furthermore, the molecular mechanisms of the apparent tropism to the pineal gland and suprasellar region in particular have to be elucidated.

Morphologically, gonadal and extragonadal germ cell tumors are virtually indistinguishable. Apart from spermatocytic seminomas, which may develop in the testis of elderly patients, all histologic entities of germ cell tumors may also arise in the central nervous system. Among malignant germ cell tumors, germinomas are distinguished from nongerminomatous germ cell tumors, which include embryonal carcinoma, yolk sac tumor and choriocarcinoma. In addition, teratomas can be distinguished as a distinct group with a histologically benign appearance. These may include completely differentiated organoid (mature teratoma) or immature structures, the latter resembling immature steps of embryonal and fetal organ development. Characteristically, malignant nongerminomatous germ cell tumors present as tumors with composite histology that include different histologic subentities. It is generally assumed that this phenomenon is related to the cell of origin, the totipotent primordial germ cell, which may differentiate along many different pathways.

Among all germ cell tumors, malignant testicular germ cell tumors of young adults constitute the most intensively studied entity. These are characterized by a specific cytogenetic aberration, the isochromosome 12p (i (12p)), which is considered pathognomonic and can be detected in about 80% of malignant testicular germ cell tumors.[7] In the remaining tumors, gain of chromosomal material of 12p is related to either marker chromosomes or homogeneously staining regions that contain chromosomal material derived from chromosomal band 12p11.2–12.1.[8, 9] Additional chromosomal imbalances have also been reported in testicular germ cell tumors. However, these are less frequent and less consistent as compared to gain of 12p.[10, 11, 12] Notably, testicular germ cell tumors that arise during infancy and childhood show a divergent genetic profile that is characterized by loss of 1p and 6q, and gain of 1q and 20, while gain of 12p11–12 is not detectable.[13, 14, 15, 16] In extragonadal germ cell tumors, similar age-dependent cytogenetic and molecular-genetic patterns have been identified.[13, 17]

There are only limited data regarding cytogenetic and molecular genetic patterns of intracranial germ cell tumors. The largest genetic study published to date includes 15 pineal germ cell tumors analyzed with comparative genomic hybridization (CGH). In this study, only a minority of germinomas showed gain of 12p.[18] Others detected gain of 12p at varying frequencies.[19, 20, 21, 22, 23]Therefore, several authors concluded that gain of 12p, which constitutes the biologic hallmark of germ cell tumors in adolescents and adults, does not play a major role in the biology of CNS-GCTs, thus distinguishing these from their counterparts at gonadal or other extragonadal sites, such as the mediastinum.[18,23] Of note, the genetic profiles of pineal germ cell tumors were different from pineal parenchymal tumors, which frequently showed gain of 12q and 22.[24]

Considering these conflicting data, we have performed chromosomal CGH analysis of 19 tumors from representative histologic groups in order to define characteristic patterns of chromosomal imbalances in CNS-GCTs. In addition, we compared the CGH profiles of CNS-GCTs to those of 101 malignant germ cell tumors from gonadal and different extragonadal sites, including tumors of different age groups and all histologic entities.

MATERIALS AND METHODS

Patients

Fresh-frozen tumor tissue (*n*=14) or formalin-fixed and paraffin-embedded archival tumor tissue (*n*=6) was obtained from the germ cell tumor bank of the Pediatric Oncology Group, the German Brain Tumor Reference Center in Bonn, Germany, the Department of Neuropathology in Düsseldorf, Germany, and the German Children's Tumor Registry in Kiel. In one case, both fresh frozen and paraffin embedded tissue was available for analysis, and both experiments yielded identical profiles. All tumors were classified according to the WHO classification of brain tumors and germ cell tumors.[25, 26] Hematoxylin–Eosin stained sections were evaluated prior to DNA extraction. Only those tumor samples that showed more than 70% viable tumor cells were analyzed. Germinomas with pronounced lymphocytic infiltration comprising greater than 30% of the specimen cellularity were excluded. A brain metastasis of a testicular nonseminomatous germ cell tumor was included for comparison to primary CNS-GCTs. None of the patients had a history of a previous, synchronous or metachronous testicular, ovarian or extracranial extragonadal germ cell tumor. The study was reviewed and approved by the Local Ethics Committee of the Heinrich-Heine University, Düsseldorf, Germany, and the Joint Committee on Clinical Investigations of the Johns-Hopkins Medical Institutions, Baltimore, MD, USA.

DNA Extraction from Fresh-Frozen Tissue

DNA was salt-extracted from twenty 5 μm sections after sodium dodecyl sulfate (SDS)/proteinase K (GibcoBRL, Rockville, MD, USA) digestion for 24 h, and purified with phenol–chloroform–isoamyl alcohol extraction followed by ethanol precipitation using standard protocols.[27] Alternatively, DNA was extracted with the DNeasy Tissue Kit (Qiagen, Hilden, Germany), according to the manufacturer's instructions. Reference DNA was extracted from normal male lymphocytes.

DNA Extraction from Archival Tissue

Five 10 μm sections were deparaffinized with xylene and rehydrated through an alcohol series. For tumor dissection, five to 10 10-μm sections on glass slides were manually dissected after deparaffinization with a razor blade using a stained section as a guide. The tissue was incubated at 55°C for 72 h with SDS/proteinase K (1.5 mg/ml, new enzyme added after 24 and 48 h).[28] DNA was extracted according to the phenol–choloform–isoamyl alcohol method[27] and ethanol-precipitated.

Labeling of Tumor and Reference DNA

Prior to labeling, the presence of high-molecular weight DNA and absence of overwhelming DNA degradation (in archival tissues) were confirmed by 1% agarose gel electrophoresis, and the amount of DNA was determined with UV spectroscopy. Tumor and reference DNA (2 μg) were directly labeled with Spectrum-Green- and Spectrum-Red-dUTP (Vysis, Downers Grove, IL, USA) using nick translation, as previously described.[14, 17, 29] The amount of DNase and

DNA-Polymerase (GibcoBRL) and the reaction time were carefully adjusted in order to obtain labeled DNA fragment lengths of 500–2000 base pairs on a 1% agarose gel.

Comparative Genomic Hybridization

In total, 500 ng of labeled tumor and reference DNA were ethanol coprecipitated with 20 µg of Cot-1 DNA (GibcoBRL). The precipitate was resuspended in hybridization buffer containing 50% formamide, 2 × SSC, and 10% dextran sulfate. The probe was denatured at 75°C for 10 min and partially reannealed at 37°C for 30 min. The probe was hybridized at 37°C for 3 days to methotrexate-synchronized male metaphase cells that previously had been aged for 5 days, stored at -20°C, washed in 2 × SSC, and denatured at 70°C for 2 min in 70% formamide, 2 × SSC, pH 7.0. Posthybridization washes were performed for 5 min each in 2 × SSC at 70°C, 37°C, and in water at room temperature. The chromosomes were counterstained with 0.1 µg/ml 4'6/diamidino-2-phenylindole-2HCl (DAPI) in Antifade (Oncor, Gaithersburg, MD, USA). The experimental protocol has been validated using control samples that have been characterized by conventional cytogenetic analysis.

Microscopic Analysis

Gray-level images were aquired for each fluorescent dye with a charge-coupled device camera on a Zeiss Axioscope epifluorescence microscope, using the Applied Imaging Corporations's dedicated Cytovision soft- and hardware for 'high-resolution CGH' (according to the protected designation by Cytovision Inc., Santa Clara, CA, USA).[30, 31] The chromosomes of at least 15 metaphase cells were identified using reverse DAPI banding. The background fluorescence was subtracted, and the green:red ratio of each entire metaphase was normalized to 1.0. According to the designated high-resolution modus for CGH analysis, average fluorescence ratios and 95% confidence intervals were calculated from at least 12 representative chromosomes and compared to dynamic standard reference intervals,[32] thus increasing the sensitivity and spatial resolution of chromosomal CGH to approximately 5 MB.[33]

RESULTS

Fourteen tumors were from male and five tumors from female patients. Age at diagnosis ranged from a neonate to 25 years, with a median of 11.5 years. This distribution is in accordance with the general epidemiologic pattern of CNS-GCTs, indicating that this panel of CNS-GCTs constitutes a representative group.[1]Among 18 patients with complete clinical data, pineal tumors were most frequent (n=11), followed by the suprasellar region (n=5) and the spine (n=2). In a 3-year-old child (Table 1, patient no. 12), a ventricular recurrence of a primarily suprasellar tumor was analyzed. Nine patients were prospectively enrolled onto the German MAKEI (Maligne Keimzelltumoren) 89 or SIOP CNS GCT 96 protocol, allowing for analysis of clinical follow-up data. The patient with tumor recurrence died as a result of tumor progression, whereas the others are in continuous remission. Table 1 summarizes the documented clinical and pathologic data in relation to the chromosomal gains and losses detected by CGH analysis.

Table 1:- Summary of the clinical, pathologic and genetic data of the 19 patients with CNS-GCTs, sorted by histology and age

#	*Tissue*	*Age*	*Sex*	*Site*	*Histology*	*Status*	*Gains*	*Losses*
1	FFT	Neonate	F	Spinal cord	MT	NED	None	None
2	FFT	6 y	M	Pineal gland	IT	NED	1q, 8, 17q, X	20
3	FFT+FFPET	8 y	M	Pineal gland	IT	NED	1q, 8, 12p, X	9q, 11, 18
4	FFT	20 y	M	Pineal gland	IT	ND	None	None
5	FFPET	10 y	F	Pineal gland	GER, TER	NED	1q, 8, 21	—
6	FFPET	11 y	M	Pineal gland	GER, TER	NED	12p, 17q	—
7	FFPET	12 y	M	Pineal gland	GER	ND	1, 2, 8, 10, 12	4, 13q, 22q
8	FFPET	14 y	M	Pineal gland	GER	NED	1q21–q24, 7q22–qter, amp12p12, 17q24–qter, 21, 22	Y
9	FFT	15 y	F	Suprasellar region	GER	ND	8q11–q21	—
10	FFPET	25 y	M	Suprasellar region	GER	NED	7, 12p	13
11	FFPET	Adult	F	Pineal gland	GER	ND	3p, 12p	1p, 6, 11
12	FFT	3 y	M	Relapse (3rd ventr.)	EC, YST	DOD	1p21–qter, 2pter–2q22, 2q34–qter, 3	11q24–qter, 20q
13	FFPET	9 y	M	Pineal gland	IT, YST	NED	1q, 5q, 8, 10q22–q23, 12p, 16q23–qter, 22	11q14–q22, 18, Y
14	FFPET	10 y	M	Pineal gland	IT, YST	NED	3,12p,14,18,21	—
15	FFPET	11 y	M	Pineal gland	YST, EC, CHC	NED	2q32–qter, 8, 12, amp12p, 14, 21, 22, X, Y	4, 9p, 10, 11
16	FFPET	13 y	F	Suprasellar region	YST	ND	1q21–q32, 2p16–p22, 3p, 8, 10, 12p, 17q, 20q, 21	—
17	FFPET	17 y	M	Suprasellar region	EC, TER	ND	X, Y	3p13–p14, 5q32-qter, 8q24-qter. 10q25–qter, 13, 18q
18	FFPET	20 y	M	Spinal cord	EC, YST	ND	1q, 5	—
19	FFPET	20 y	M	Pineal gland	YST	NED	2, 7, 8q, 12p, 14, 18p, 21, X, Y	8p21–pter, 15, 18q

Four patients presented with pure mature (n=1) or immature (n=3) teratomas, whereas all other patients suffered from malignant germ cell tumors. Seven patients had germinomas, which showed additional teratomatous elements in two patients. Eight patients were diagnosed with malignant nongerminomatous germ cell tumors. In these tumors, yolk sac tumor was the most common histologic type, followed by embryonal carcinoma and teratoma, while choriocarcinoma was detected in one patient only.

Of 19 analyzed tumors, 17 showed chromosomal imbalances (Figure 1). In general, chromosomal gains (mean: 4 gains/tumor, range: 0–9 gains/tumor) were more frequent than losses (mean: 1.6 losses/tumor, range: 0–6 losses/tumor). The most frequent chromosomal imbalance, gain of 12p, was detected in 11 of 19 tumors and 10 of 15 malignant CNS-GCTs. In two of these tumors, the whole chromosome 12 was gained, with one of these two tumors showing a high level gain of 12p. Notably, in one germinoma, gain of 12p was limited to the chromosomal band 12p12 (Figure 2). Gain of 12p was absent in tumors of the three patients who were younger than 8 years of age.

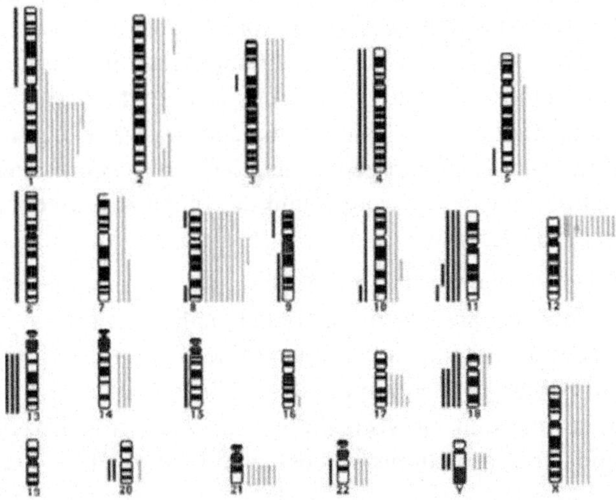

Figure 1: High-resolution CGH analysis of 19 CNS-GCTs: Each line to the right or left side of a chromosome represents a region gained or lost, respectively. Chromosomal amplifications (>1.5:1 ratio) are indicated as thick bars.

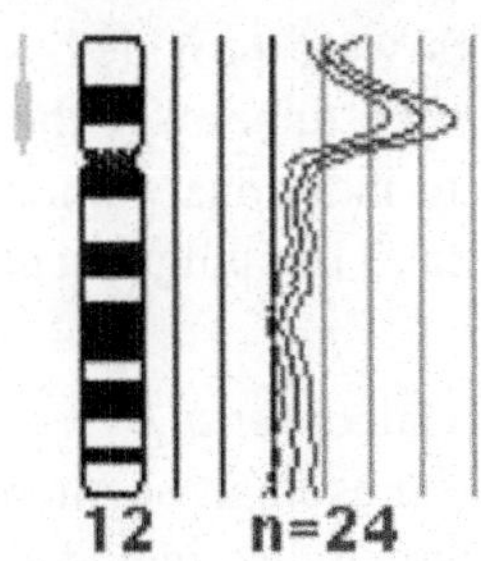

Figure 2: CGH profile of a pineal germinoma showing a circumscribed high-level amplification restricted to chromosomal band 12p12.

The second most common chromosomal gains were found at 1q, with a region of overlap mapping to 1q21–q24, and chromosome 8, with a commonly gained region at 8q11–q21. In six patients, chromosome 21 was gained, and five tumors showed gain of the X chromosome. Due to restricted tumor material and lack of normal reference tissue, no analysis defining the constitutional sex chromosomal status could be performed. Chromosome arms

11q and 18q as well as chromosome 13 were most frequently affected by losses of genetic material. The only tumors with balanced CGH profiles were two pure teratomas, while all malignant germ cell tumors showed chromosomal imbalances. However, the CGH profile of the two imbalanced teratomas resembled those found in malignant CNS-GCTs. Among malignant germ cell tumors, chromosomal imbalances were more frequent in nongerminomatous germ cell tumors (mean: 8.1 imbalances/tumor) than in germinomas (mean: 4.1 imbalances/tumor, Student'*st*-test, P=0.03). Particularly, chromosomal gains were more frequently detected in nongerminomatous germ cell tumors (mean: 5.9 gains/tumor) than in germinomas (mean: 3 gains/tumor, P=0.045), whereas chromosomal losses were found at comparable frequencies (mean: 2.3 *vs* 1.1 losses/tumor, P=0.27).

Otherwise, no significant differences were detected when specific imbalances were compared to the different malignant histologic types. We also did not observe a significant correlation between a specific chromosomal imbalance and patient gender, tumor site or clinical outcome. Last, there was no significant correlation between the tissue type (fresh frozen or paraffin embedded) and chromosomal profiles detected with CGH.

Meta-Analysis of Gonadal and Extragonadal Germ Cell Tumors

We compared the chromosomal patterns observed in the 15 malignant CNS-GCTs of this series to those published in other major CGH studies of malignant gonadal and extragonadal germ cell tumors, in which the chromosomal profiles of individual tumors were provided, sorted by age and tumor site ([14, 17, 18, 29, 34]plus 16 unpublished tumors). Other studies, in which the CGH profile cannot be attributed to individual tumors, were excluded (eg Korn *et al*[11] and Mostert *et al*[34]). In total, 116 malignant germ cell tumors were included in this meta-analysis.

Similar to the study of Chibon *et al*,[35] we divided the genome into 100 datapoints resulting in chromosomal segments of approximately 30–40 Megabases. Since most of the studies included did not specifically indicate chromosomal amplification, we only designated chromosomal gain as +1 and loss as -1, while balanced regions were counted as 0. In Figure 3, all chromosomal gains are illustrated as green bars, and losses as red bars, respectively.

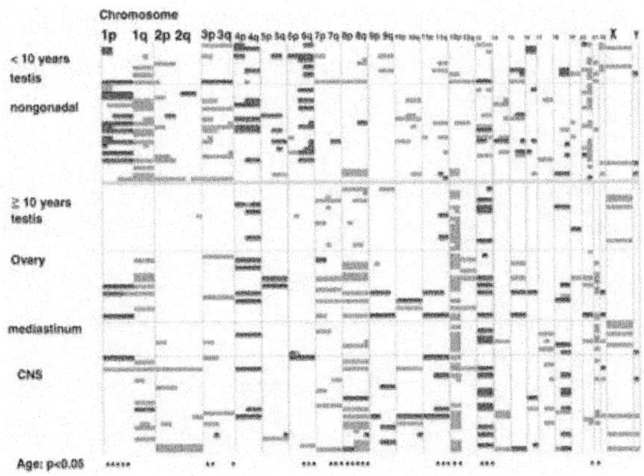

Figure 3: Meta-analysis of 116 malignant gonadal and extragonadal GCTs, separated by age and sorted by site[14, 17, 18, 29, 34] (including this series and 16 unpublished tumors). Each line represents 100 chromosomal data points of a single tumor, with chromosomal gains illustrated as green bars and losses as red bars, respectively. At the bottom, chromosomal regions showing statistically significant differences that correlated with age are indicated by an asterix.

This comparison revealed no detectable difference between seminomas or malignant nonseminomatous germ cell tumors. Instead, significant differences according to site and age are obvious. Loss of 1p is most frequent in sacrococcygeal and mediastinal germ cell tumors of infancy and childhood ($P<0.01$), while gain of 1q can be found in both age groups. Loss of 6q is virtually exclusive to malignant childhood germ cell tumors ($P<0.001$), irrespective of site. Moreover, gain of 20q represents a distinctive imbalance of childhood germ cell tumors ($P<0.001$). In contrast, gain of 12p represents the most important marker of both gonadal and extragonadal malignant germ cell tumors of adolescent and adult patients ($P<0.001$), but is only exceptionally found in children. In addition, gain of 8/8q is also common in this age group ($P<0.01$). Lastly, tumors arising in adolescents and adults are characterized by more frequent gain of 7q, 21q and loss of 13 ($P<0.001$), while gain of 3p is more commonly observed in pediatric tumors/($P<0.05$). Taken together, this meta-analysis of 116 germ cell tumors clearly demonstrates the reproducibility of patterns detected by CGH across different laboratories.

Last, a specific comparison of the 34 CNS-GCTs analyzed by Rickert *et al* and our group with 36 testicular germ cell tumors was performed (Figure 4). This analysis included 18 children younger than 10 years. This analysis indicates that with regard to chromosomal imbalances detectable with CGH, there are no apparent differences according to site. The only detectable difference appears to be a higher percentage of gain of chromosome 8 in CNS-

GCTs ($P<0.05$); however, this difference is not detectable, if age is use as a stratum for comparison, indicating that the chromosomal constitution of testicular and CNS-GCTs is identical.

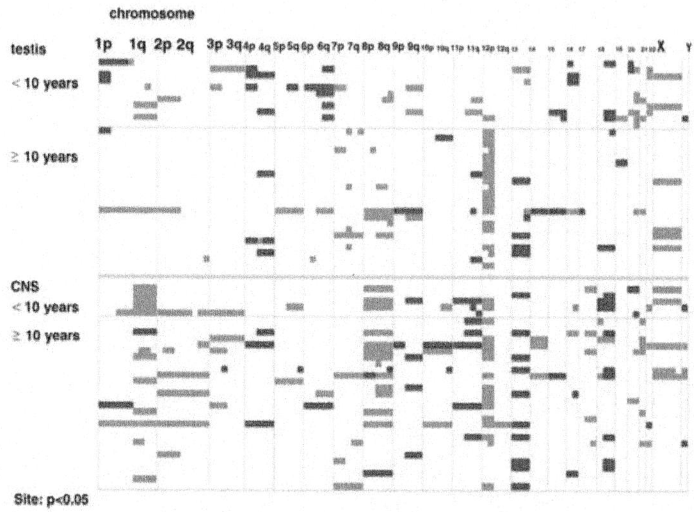

Figure 4: Meta-analysis of 36 testicular and 34 CNS-GCTs, separated by site and sorted by age[15, 18, 34] (including this series and eight unpublished tumors). Each line represents 100 chromosomal data points of a single tumor, with chromosomal gains illustrated as green bars and losses as red bars, respectively. As indicated at the bottom, no chromosomal regions showed statistically significant differences correlated with site.

DISCUSSION

The last 25 years have been characterized by a significant breakthrough in the treatment of malignant germ cell tumors, which can mainly be attributed to the introduction of cisplatin-based chemotherapy and the optimization of multidisciplinary therapeutic strategies.[36] Recently, this success has been translated to the treatment of CNS-germinomas and, to a lesser extent, to nongerminomatous germ cell tumors.[2, 37, 38, 39, 40, 41] To a certain degree, the less favorable prognosis of CNS-GCTs that secrete alpha fetoprotein or human chorionic gonadotrophin (embryonal carcinoma, yolk sac tumor, choriocarcinoma) compared to secreting germ cell tumor at other sites might be related to anatomic problems that interfere with surgical resection. Nevertheless, it remains possible that specific biologic aspects of CNS-GCTs may additionally contribute to the poorer prognosis compared to malignant germ cell tumors at other sites.

In general, it can be assumed that lessons learned in testicular germ cell tumors might also be transferred to CNS-GCTs. This assumption is supported by imprinting studies implicating a common precursor cell for both gonadal and extragonadal germ cell tumors.[5, 6] However, there are several lines of evidence that suggest certain biologic differences between central nervous system and non-CNS-GCTs. Testicular germ cell tumors show a characteristic association with a preinvasive carcinoma *in situ*, the testicular intratubular neoplasia.[7, 31, 42] No such *in situ* lesion has been seen in the brain of patients with CNS-GCT. Furthermore, it has been demonstrated that certain histologic types express markers associated with distinct maturational stages of germ cell development, which presumably are dependent on a specific gonadal environment.[7, 43, 44, 45] Lastly, there are some data suggesting that the chromosomal marker characteristic of testicular germ cell tumors, the isochromosome 12p, plays only a minor role in CNS-GCTs.[18, 22] As a consequence, it must be debated whether the information generated in the more frequent (and more easily assessable) testicular germ cell tumors can indeed be transferred to CNS-GCTs without scrutiny. This concern relates not only to our general understanding of CNS-GCT biology but also to the development of future therapeutic strategies.

In this context, the meta-analysis of germ cell tumors of the different clinical and histologic groups (Figures 3 and 4) provides important information. While age-dependent differences become clearly apparent, there is no obvious correlation with either site or histologic subtype of malignant germ cell tumor.[13, 14, 46] In particular, our meta-analysis unequivocally indicates that there is no apparent difference in the chromosomal constitution of testicular and CNS-GCTs (Figure 4). In accordance with the observation that the vast majority of CNS-germ cell tumors develop in adolescents and adults,[1] CNS-GCTs also display CGH profiles characteristic of malignant germ cell tumors of this age group. In fact, gain of 12p, which in most cases is related to an isochromosome 12p, constitutes the most frequent chromosomal imbalance in our series of CNS-GCT. Furthermore, the additional chromosomal imbalances such as gain of 1q or 8 also occur at comparable frequencies as in other gonadal and nongonadal germ cell tumors of adults (Figures 3 and 4). Although CNS-GCTs show no bimodal age distribution as testicular or mediastinal germ cell tumors,[1] our analysis suggests that also in the CNS, germ cell tumors of infancy and early childhood are genetically distinct from those of adolescence and adulthood, as indicated by the analysis of patients 1 and 12, which show profiles characteristic of teratoma and malignant germ cell tumors of childhood.[47] However, in the patients 2, 3 and 13, who were 6, 8 and 9 years old, profiles characteristic of adolescent patients have been found. Therefore, in contrast to other sites such as the mediastinum, an age cutoff somewhat younger than 10 years might be more appropriate for CNS-GCTs, indicating for modulating local environmental factors. Nevertheless, our observation and our meta-analysis substantiate the notion that age and not site constitutes the primary distinguishing parameter.

In other extragonadal germ cell tumors (in particular mediastinal germ cell tumors), there is a significant association with constitutional Klinefelter syndrome that distinguishes these tumors from testicular germ cell tumors, for which this association is not seen.[48, 49] Furthermore, there have been single reports of CNS-GCTs in the presence of Klinefelter syndrome.[50, 51, 52] In our series of patients, we have no clinical data indicating constitutional Klinefelter syndrome. A previous FISH analysis of six suprasellar teratomas (age 2–25 years) revealed a high frequency of sex chromosomal abnormalities, with a normal constitutional karyotype in one analyzed patient.[53] In accordance to this study, both teratomas with chromosomal imbalances and three malignant nongerminomatous germ cell tumors showed gains of the X chromosome. Unfortunately, we did not have access to extratumoral tissue allowing for an analysis of the constitutional karyotype.

The Role of Gain of 12p in CNS-GCTs

The question has been raised whether those CNS-GCTs that lack gain of 12p constitute a different biologic subtype.[18] In contrast to our series, in which five of six germinomas displayed gain of 12p, Rickert *et al* found gain of 12p in only one out of eight central nervous system germinomas. This statistically significant difference (χ^2-test, P=0.02) could probably be related to methodical differences such as sample selection. Germinomas often show a pronounced lymphocytic infiltration that could significantly dilute the tumor DNA content, and therefore reduce the ability to detect chromosomal imbalances. In accordance with this hypothesis, Rickert *et al* detected no chromosomal imbalances in two of eight germinomas, whereas in our series, all malignant tumors showed genomic alterations. Furthermore, it might be argued whether the high-resolution CGH algorithm might additionally increase the heterogeneity tolerance as it has been postulated for array CGH.[54] In this context, it has to be noted that although the spatial resolution of chromosomal CGH is limited, even with array CGH approaches no discrepant or additional frequently recurrent imbalances have been detected in extracranial germ cell tumors[16] or CNS-GCTs (Chin C Lau, oral presentation on the Second International Symposium in CNS-GCTs in Los Angeles, November 2005[55]).

Furthermore, other issues such as a patient selection bias must also be considered. First, the three youngest patients included in this series (and the youngest patient of the series published by Rickert *et al*) do not show gain of 12p but a CGH profile comparable to that of testicular, sacrococcygeal or mediastinal CGH at this age with imbalances of chromosome 1 and gain of 1q or 20q in malignant germ cell tumors (Figures 1 and 3). Second, three of the tumors that lacked gain of 12p are teratomas. Characteristically, the vast majority of teratomas indeed do not show chromosomal aberrations on cytogenetic, CGH or array CGH analyses.[13, 16, 17, 47]

Nevertheless, it is very important to note that two of the teratomas included in our study (and both immature central nervous system teratomas analyzed by Rickert *et al*) show chromosomal imbalances resembling those of malignant germ cell tumors. This might indicate that in contrast to teratomas in infants, teratomas of adolescents and adults might be related to a maturational differentiation of malignant germ cell tumors with a preservation of genetic aberrations that may result in a potentially malignant clinical presentation (accordingly, patient no. 3 showed tumor cell dissemination in the cerebrospinal fluid). In previous classification systems, this phenomenon had been attributed the term 'malignant teratoma, differentiated'. In accordance, we and other have found significant cytogenetic and CGH aberrations in residual or growing teratomas of patients with malignant nonseminomas (unpublished observation and van Echten *et al*[56]). Therefore, the histologic diagnosis of teratoma within the brain of a postpubertal patient should not result in confidence of benignity.

Considering these exceptions, 10 out of 14 malignant germ cell tumors of patients older than 5 years showed gain of 12p. The Figures 3 and 4 illustrate that this frequency is almost identical to that found in testicular[11, 34] and mediastinal germ cell tumors of adolescents and adults.[17] In addition, our meta-analysis underlines that CNS-GCTs without 12p gain do not show chromosomal patterns distinct from other germ cell tumors.

Restricted Amplification of 12p12

We observed one patient whose tumor demonstrated a circumscribed amplification of 12p12. To date, comparable profiles have only been detected in testicular and ovarian (Riopel *et al*) germ cell tumors but not in extragonadal germ cell tumors. Albrecht *et al*[57] published a cytogenetic analysis of a central nervous system germinoma that showed a homogeneously staining region (HSR), the chromosomal origin of which could not be identified by FISH with painting probes to the chromosomes 6 or 12.

In testicular tumors, gain of chromosomal material at 12p in the absence of an isochromosome 12p is also related to the presence of either an HSR or double minutes. Microarray and FISH analyses helped to delineate the commonly amplified region at 12p12. It is generally assumed that the amplification of tumor genes at this region triggers malignant transformation at the transition from TIN to invasive malignant germ cell tumors.[8, 9, 31] Since in our patient the amplicon is located at the identical chromosomal region as in testicular germ cell tumors, this observation provides additional evidence in favor of a common pathogenesis of testicular and CNS-GCTs.

In conclusion, this CGH data obtained in our present series of 19 CNS-GCTs and their comparison to previous findings in gonadal and extragonadal germ cell tumors clearly indicate that at the resolution level of CGH, CNS-GCTs are virtually indistinguishable from malignant germ cell tumors at other sites presenting in the same age group. This observation strongly argues for a similar genetic constitution and biology of germ cell tumors at different sites.

ACKNOWLEDGEMENTS

This study has been supported by a Max-Eder Grant of the German Cancer Aid (Deutsche Krebshilfe), the Children Cancer Foundation of Maryland and the American Cancer Society, Grant Number RPG-97-113-01-CCE. We thank Dieter Harms for providing tissue samples, Amy E Schuster for technical assistance and Renate Groth and Susanne Koch for data management.

REFERENCES

1. Schneider DT, Calaminus G, Koch S, *et al.* Epidemiological analysis of 1442 children and adolescents registered in the german germ cell tumor protocols. *Pediatr Blood Cancer* 2004;42:169–175.
2. Calaminus G, Bamberg M, Baranzelli MC, *et al.* Intracranial germ cell tumors: a comprehensive update of the European data. *Neuropediatrics* 1994;25:26–32.
3. Calaminus G, Bamberg M, Harms D, *et al.* AFP/beta-HCG secreting CNS germ cell tumors: long-term outcome with respect to initial symptoms and primary tumor resection. Results of the cooperative trial MAKEI 89.*Neuropediatrics* 2005;36:71–77.
4. Bussey KJ, Lawce HJ, Himoe E, *et al.* SNRPN methylation patterns in germ cell tumors as a reflection of primordial germ cell development. *Genes Chromosomes Cancer* 2001;32:342–352.
5. Schneider DT, Schuster AE, Fritsch MK, *et al.* Multipoint imprinting analysis indicates a common precursor cell for gonadal and nongonadal pediatric germ cell tumors. *Cancer Res* 2001;61:7268–7276
6. Sievers S, Alemazkour K, Zahn S, *et al.* IGF2/H19 imprinting analysis with the methylation-sensitive single nucleotide primer extension method in human germ cell tumors reflects their origin from different stages of primordial germ cell development. *Genes Chromosomes Cancer*2005;44:256–264.

7. Oosterhuis JW, Looijenga LH. Testicular germ-cell tumours in a broader perspective. *Nat Rev Cancer* 2005;5:210–222.

8. Zafarana G, Gillis AJ, van Gurp RJ, *et al.* Coamplification of DAD-R, SOX5, and EKI1 in human testicular seminomas, with specific overexpression of DAD-R, correlates with reduced levels of apoptosis and earlier clinical manifestation. *Cancer Res* 2002;62:1822–1831.

9. Zafarana G, Grygalewicz B, Gillis AJ, *et al* 12p-Amplicon structure analysis in testicular germ cell tumors of adolescents and adults by array CGH.*Oncogene* 2003;22:7695–7701.

10. van Echten J, Oosterhuis JW, Looijenga LH, *et al.* No recurrent structural abnormalities apart from i(12p) in primary germ cell tumors of the adult testis. *Genes Chromosomes Cancer* 1995;14:133–144.

11. Korn WM, Oide Weghuis DE, Suijkerbuijk RF, *et al.* Detection of chromosomal DNA gains and losses in testicular germ cell tumors by comparative genomic hybridization. *Genes Chromosomes Cancer*1996;17:78–87.

12. Oosterhuis JW, Looijenga LH, van Echten J, *et al.* Chromosomal constitution and developmental potential of human germ cell tumors and teratomas.*Cancer Genet Cytogenet* 1997;95:96–102.

13. Bussey KJ, Lawce HJ, Olson SB, *et al.* Chromosome abnormalities of eighty-one pediatric germ cell tumors: sex- , age-, site-, and histopathology-related differences – a Children's Cancer Group study. *Genes Chromosomes Cancer* 1999;25:134–146.

14. Perlman EJ, Hu J, Ho D, *et al.* Genetic analysis of childhood endodermal sinus tumors by comparative genomic hybridization. *J Pediatr Hematol Oncol* 2000;22:100–105.

15. Mostert M, Rosenberg C, Stoop H, *et al.* Comparative genomic and *in situ*hybridization of germ cell tumors of the infantile testis. *Lab Invest*2000;80:1055–1064.

16. Veltman I, Veltman J, Janssen I, *et al.* Identification of recurrent chromosomal aberrations in germ cell tumors of neonates and infants using genomewide array-based comparative genomic hybridization. *Genes Chromosomes Cancer* 2005;43:367–376.

17. Schneider DT, Schuster AE, Fritsch MK, *et al.* Genetic analysis of mediastinal nonseminomatous germ cell tumors in children and adolescents. *Genes Chromosomes Cancer* 2002;34:115–125.

18. Rickert CH, Simon R, Bergmann M, *et al.* Comparative genomic hybridization in pineal germ cell tumors. *J Neuropathol Exp Neurol*2000;59:815–821.

19. de Bruin TW, Slater RM, Defferrari R, *et al.* Isochromosome 12p-positive pineal germ cell tumor. *Cancer Res* 1994;54:1542–1544.

20. Losi L, Polito P, Hagemeijer A, *et al.* Intracranial germ cell tumour (embryonal carcinoma with teratoma) with complex karyotype including isochromosome 12p. *Virchows Arch* 1998;433:571–574.

21. Dal Cin P, Dei Tos AP, Qi H, *et al.* Immature teratoma of the pineal gland with isochromosome 12p. *Acta Neuropathol (Berl)* 1998;95:107–110.

22. Lemos JA, Barbieri-Neto J, Casartelli C. Primary intracranial germ cell tumors without an isochromosome 12p. *Cancer Genet Cytogenet*1998;100:124–128.

23. Okada Y, Nishikawa R, Matsutani M, *et al.* Hypomethylated X chromosome gain and rare isochromosome 12p in diverse intracranial germ cell tumors.*J Neuropathol Exp Neurol* 2002;61:531–538.

24. Rickert CH, Simon R, Bergmann M, *et al.* Comparative genomic hybridization in pineal parenchymal tumors. *Genes Chromosomes Cancer*2001;30:99–104.

25. Kleihues P, Burger PC, Scheithauer BW. The new WHO classification of brain tumours. *Brain Pathol* 1993;3:255–268.

26. Mostofi FK, Sobin LH. *Histopathological Typing of Testis Tumors*. World Health Organization: Geneva, 1993.

27. Ausubel FM, Brent R, Kingston RE, *et al.* Preparation and analysis of DNA. In: Ausubel FM, Brent R, Kingston RE, Moore DD, Seidman JG, Smith JA, Struhl K (eds). *Short Protocols in Molecular Biology*. 3 edn. John Wiley & Sons, Inc.: New York, 1997, pp 2-1–2-43.

28. Isola J, DeVries S, Chu L, *et al.* Analysis of changes in DNA sequence copy number by comparative genomic hybridization in archival paraffin-embedded tumor samples [see comments]. *Am J Pathol* 1994;145:1301–1308.

29. Riopel MA, Spellerberg A, Griffin CA, *et al.* Genetic analysis of ovarian germ cell tumors by comparative genomic hybridization. *Cancer Res*1998;58:3105–3110.

30. Ness GO, Lybaek H, Houge G. Usefulness of high-resolution comparative genomic hybridization (CGH) for detecting and characterizing constitutional chromosome abnormalities. *Am J Med Genet* 2002;113:125–136.

31. Ottesen AM, Skakkebaek NE, Lundsteen C, *et al.* High-resolution comparative genomic hybridization detects extra chromosome arm 12p material in most cases of carcinoma *in situ* adjacent to overt germ

cell tumors, but not before the invasive tumor development. *Genes Chromosomes Cancer* 2003;38:117–125.

32. du Manoir S, Schrock E, Bentz M, *et al.* Quantitative analysis of comparative genomic hybridization. *Cytometry* 1995;19:27–41.
33. Kirchhoff M, Rose H, Lundsteen C. High resolution comparative genomic hybridisation in clinical cytogenetics. *J Med Genet* 2001;38:740–744. |
34. Mostert MM, van de PM, Olde WD, *et al.* Comparative genomic hybridization of germ cell tumors of the adult testis: confirmation of karyotypic findings and identification of a 12p-amplicon. *Cancer Genet Cytogenet* 1996;89:146–152.
35. Chibon F, Mariani O, Mairal A, *et al.* The use of clustering software for the classification of comparative genomic hybridization data. an analysis of 109 malignant fibrous histiocytomas. *Cancer Genet Cytogenet* 2003;141:75–78.
36. Einhorn LH, Donohue JP. Chemotherapy for disseminated testicular cancer.*Urol Clin North Am* 1977;4:407–426.
37. Bamberg M, Kortmann RD, Calaminus G, *et al.* Radiation therapy for intracranial germinoma: results of the German cooperative prospective trials MAKEI 83/86/89. *J Clin Oncol* 1999;17:2585–2592.
38. Bouffet E, Baranzelli MC, Patte C, *et al.* Combined treatment modality for intracranial germinomas: results of a multicentre SFOP experience. Societe Francaise d'Oncologie Pediatrique. *Br J Cancer* 1999;79:1199–1204.
39. Balmaceda C, Finlay J. Current advances in the diagnosis and management of intracranial germ cell tumors. *Curr Neurol Neurosci Rep* 2004;4:253–262.
40. Kellie SJ, Boyce H, Dunkel IJ, *et al.* Primary chemotherapy for intracranial nongerminomatous germ cell tumors: results of the second international CNS germ cell study group protocol. *J Clin Oncol* 2004;22:846–853.
41. Calaminus G, Bamberg M, Jürgens H, *et al.* Impact of surgery, chemotherapy and irradiation on long term outcome of intracranial malignant non-germinomatous germ cell tumors: results of the German Cooperative Trial MAKEI 89. *Klin Pädiatr* 2004;216:141–149.
42. Jong JD, Stoop H, Dohle GR, *et al.* Diagnostic value of OCT3/4 for pre-invasive and invasive testicular germ cell tumours. *J Pathol* 2005;206:242–249.

43. Stoop H, van Gurp R, de Krijger R, *et al.* Reactivity of germ cell maturation stage-specific markers in spermatocytic seminoma: diagnostic and etiological implications. *Lab Invest* 2001;81:919–928.

44. Looijenga LH, Stoop H, De Leeuw HP, *et al.* POU5F1 (OCT3/4) identifies cells with pluripotent potential in human germ cell tumors. *Cancer Res*2003;63:2244–2250.

45. Honecker F, Stoop H, de Krijger RR, *et al.* Pathobiological implications of the expression of markers of testicular carcinoma *in situ* by fetal germ cells. *J Pathol* 2004;203:849–857.

46. Schneider DT, Perlman EJ, Harms D, *et al.* Mediastinal germ cell tumors (MGCT) in children and adolescents: age correlates with histological differentiation, genetic profiles and clinical outcome. In: Harnden P, Joffe JK, Jones WG (eds). *Germ Cell Tumours V.* Springer: London, 2002, pp 127–128.

47. Schneider DT, Schuster AE, Fritsch MK, *et al.* Genetic analysis of childhood germ cell tumors with comparative genomic hybridization. *Klin Pädiatr*2001;213:204–211.

48. Hasle H, Mellemgaard A, Nielsen J, *et al.* Cancer incidence in men with Klinefelter syndrome. *Br J Cancer* 1995;71:416–420.

49. Nichols CR, Heerema NA, Palmer C, *et al.* Klinefelter's syndrome associated with mediastinal germ cell neoplasms. *J Clin Oncol*1987;5:1290–1294.

50. Hashimoto M, Hatasa M, Shinoda S, *et al.* Medulla oblongata germinoma in association with Klinefelter syndrome. *Surg Neurol* 1992;37:384–387.

51. Casalone R, Righi R, Granata P, *et al.* Cerebral germ cell tumor and XXY karyotype. *Cancer Genet Cytogenet* 1994;74:25–29.

52. Prall JA, McGavran L, Greffe BS, *et al.* Intracranial malignant germ cell tumor and the Klinefelter syndrome. Case report and review of the literature. *Pediatr Neurosurg* 1995;23:219–224.

53. Yu IT, Griffin CA, Phillips PC, *et al.* Numerical sex chromosomal abnormalities in pineal teratomas by cytogenetic analysis and fluorescence*in situ* hybridization. *Lab Invest* 1995;72:419–423.

54. Garnis C, Coe BP, Lam SL, *et al.* High-resolution array CGH increases heterogeneity tolerance in the analysis of clinical samples. *Genomics*2005;85:790–793.

55. Lau CC. Genomic profiling of intracranial germ cell tumors [abstract]. *Neurooncol* 2005;7:514–515.

56. van Echten J, Sleijfer DT, Wiersema J, *et al.* Cytogenetics of primary testicular nonseminoma, residual mature teratoma, and growing teratoma lesion in individual patients. *Cancer Genet Cytogenet* 1997;96:1–6.

57. Albrecht S, Armstrong DL, Mahoney DH, *et al.* Cytogenetic demonstration of gene amplification in a primary intracranial germ cell tumor. *Genes Chromosomes Cancer* 1993;6:61–63.

Chapter 5

GENETIC ANALYSIS OF SINONASAL ADENOCARCINOMA PHENOTYPES: DISTINCT ALTERATIONS OF HISTOGENETIC SIGNIFICANCE

Sue S Yom[1], Asif Rashid[2], David I Rosenthal[1], Danielle D Elliott[2], Ehab Y Hanna[3], Randal S Weber[3] and Adel K El-Naggar[2,3]

[1]Department of Radiation Oncology, The University of Texas MD Anderson Cancer Center, Houston, TX, USA

[2]Department of Pathology, The University of Texas MD Anderson Cancer Center, Houston, TX, USA

[3]Department of Head and Neck Surgery, The University of Texas MD Anderson Cancer Center, Houston, TX, USA

ABSTRACT

Sinonasal adenocarcinomas, a relatively rare entity, are composed of distinctly different morphologic subtypes with variable biological behavior. To investigate the genetic events associated with their development and clinicopathologic features, we analyzed the alterations in K-ras, APC, β-catenin, hMLH1 and hMSH2 and p53 genes expression in a cohort of 15 primary tumors comprising the two main sinonasal adenocarcinoma subtypes (enteric and seromucinous). The patients consisted of 13 men and two women, who ranged in age from 50 to 87 years. Tumors were predominantly located in the ethmoid sinus. Eight tumors were Enteric-type, and seven were seromucinous type. Nine patients were smokers and four were nonsmokers; and no information was available on two patients. Two of the eight enteric-type, had K-ras mutation at codons 12A and 12B, and one showed microsatellite instability at BAT-25. Two patients with enteric-type tumors had a history of wood-dust exposure, and one had a K-ras mutation at 12A codon as well as p53 overexpression. No patients with the seromucinous type had any genetic abnormalities, except for overexpression of p53 in two tumors. Our results show that (1) a subset of enteric-type sinonasal adenocarcinoma shares certain genetic alterations with colonic adenocarcinomas, (2) the seromucinous-type sinonasal adenocarcinoma lacks alterations and may develop through a different pathway, (3) high p53 expression is associated with aggressive tumor features in both subtypes and (4) the enteric-type runs a more malignant course than the seromucinous counterpart.

Primary sinonasal adenocarcinomas (SNACs), excluding those of salivary origin, are uncommon and represent approximately 10–20% of malignant neoplasms at these locations. Despite their common origin from an ectodermally derived respiratory mucosa, they manifest two distinct phenotypic categories, including the enteric and the seromucinous adenocarcinomas.[1, 2] The underlying mechanism for their histopathologic diversity is unknown. Our group has recently shown that the respiratory epithelium undergoes intestinal epithelial metaplasia (Figure 2a) prior to the development of the enteric-form of sinonasal adenocarcinoma, supporting histogenetic resemblance to primary colonic carcinoma.[3] We, therefore, hypothesize that the enteric-type shares common genetic alterations with primary colonic adenocarcinoma and differs from the seromucinous-type.

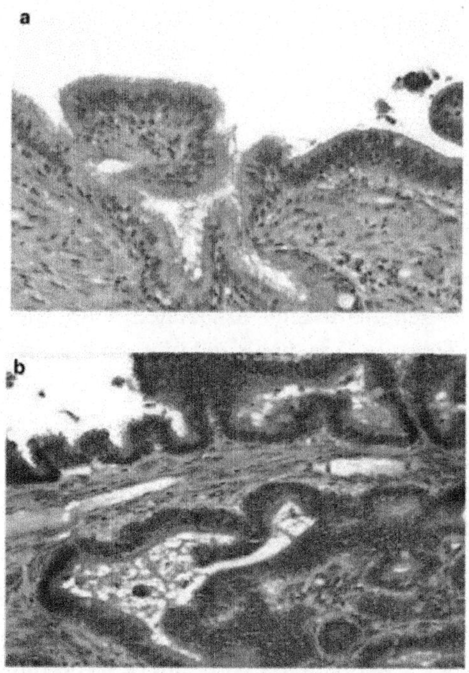

Figure 2: Intestinal metaplasia of the respiratory epithelium (a) and underlying intestinal-type sinonasal adenocarcinoma (b).

Previous investigations have shown that certain genetic alterations at the *adenomatous polyposis coli* (*APC*), β-catenin and *K-ras* genes, and mutations and/or deletions of the *p53* suppressor gene characterize colorectal tumorigenesis.[4, 5, 6, 7, 8, 9, 10, 11, 12, 13, 14] Also, microsatellite instability (MSI) caused by alterations in nucleotide mismatch repair genes, including *hMSH2,hMLH1*, *PMS1*, *PMS2* and *GTBP*, has been reported to be associated with a subset of these tumors.[10, 11, 12] Molecular studies of SNAC are rare and limited to the intestinal type and have reported contradictory results. Wu *et al*[13] in a study of the enteric form, reported a lack of point mutations in the first or second exon of the K-*ras* oncogene, and only two p53 mutations in 11 specimens. However, a recent study by Saber *et al*[14] found four K-*ras* mutations at codons 12 or 13 in 28 patients with sinonasal adenocarcinoma, while only one of 13 enteric-type tumors had K-*ras* mutation.

To examine the molecular events associated with the phenotypic diversity of SNAC, we performed molecular and immunohistochemical analysis on 15 primary tumors of both types.

MATERIALS AND METHODS

Specimens

The surgical pathology database of MD Anderson Cancer Center from 1995 to 2003 was searched and 16 patients were identified who had sinonasal adenocarcinomas with available tissue blocks from curative resection or tumor biopsy. All slides were reviewed to confirm the original diagnosis and to select blocks for this study. Histologic classification was based on criteria of Kleinsasser[15] and Batsakis *et al.*[16] Histological slides were reviewed by two pathologists operating independently of clinical information. Patient information was extracted from medical records. Records were reviewed for age, sex, race, history of exposure to dust or smoking, site and stage of tumor, type of treatment, and follow-up status.

DNA Extraction

DNA was prepared from tissue specimens stored in paraffin-embedded, 4% formalin-fixed blocks. In a few cases for which this isolation technique was insufficient, genomic DNA was extracted by microdissection of hematoxylin- and eosin-stained slides without coverslip. A 27 1/2-gauge needle was used under low magnification ($\times$ 4) in selected areas where the neoplastic cellularity was >50%. Genomic DNA was extracted from the microdissected tissue, as described previously.[17] Briefly, three 5-μm-thick sections were extracted twice in 1000 μl of xylene for 30 min and twice in 1000 μl of 100% ethanol for 3–5 min. The solid residue was dried by evaporation, after which 75 μl of lysis buffer was added (0.25% Tween-20, TE9, 2 mg/ml Proteinase K). The mixture was incubated overnight at 56°C. The Proteinase K was deactivated for 10 min at 100°C. The condensate was separated from the insoluble material by centrifugation.

Molecules Analyses

DNA could not be isolated in one patient with enteric-type. Therefore, seven enteric-type and seven nonseromucinous tumors were analyzed.

Analysis of K-*ras*, APC, and β-catenin gene mutation was carried out as previously described.[5] Immunohistochemistry for microsatellite analysis of markers hMLH1 and hMSH2 was carried out according to the method of Alexander *et al.*[18]

Immunohistochemistry for p53 protein was also performed according to previously published method.[19] Overexpression of p53 protein was considered to be present when more than 50% of the nuclei of tumor cells were strongly stained.

RESULTS

Clinical Data

The clinicopathologic findings of all sinonasal adenocarcinoma cases are summarized in Table 1. Histopathologically, seven tumors were seromucinous-type (Figure 1a) and eight were enteric-type (Figures 1b and 2b). All of the seromucinous adenocarcinomas were of the low-grade category. Six of the eight enteric-type carcinomas were moderately differentiated villo-tubular adenocarcinomas and two were well-differentiated adenocarcinoma with mucinous component. Of the 15 primary tumors, 13 occurred in men and two in women. The ages ranged from 50 to 87 years, with a mean of 67 years. Ten primary tumors originated in the ethmoid sinus, one in the maxillary sinus and two in the nasal cavity; and the primary site was unknown for two cases. Six of eight patients with enteric-type carcinomas and three of seven with seromucinous form had a history of smoking. Two of the eight patients with the enteric-type carcinoma had a history of wood-dust exposure and were also smokers. Initial surgical treatment included five craniofacial resections, four medial maxillectomies, two endoscopic resections, one partial maxillectomy, one transpalatal resection, and two unknown.

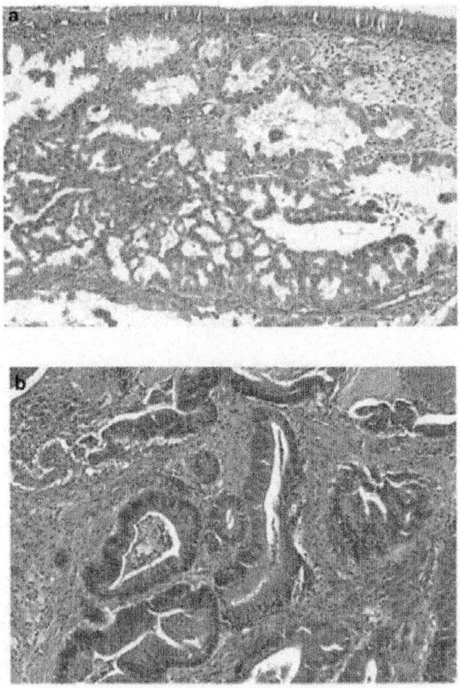

Figure 1: Photomicrographs of seromucinous (a) and intestinal-type (b).

Table 1:- Demographic, clinicopathologic and follow-up data of patients with sinonasal adenocarcinomas

Type	*Age*	*Sex*	*Race*	*Risk factors*	*Site*	*P. Surgery*	*Stage*	*Margins*	*XRT*	*CHX*	*Follow-up*
Enteric-type											
1	73	M	White	Smoking (20 py)	ES	CFR	4	Negative	N/A	N/A	N/A
2	67	M	White	None	ES	ER	2	Positive	Yes	Yes	LR
3	68	M	White	Smoking (20 py), chemicals, asbestos	ES	MM	2	Negative	Yes	No	NED
4	53	M	Black	Wood dust; smoking (20 py)	ES	CFR	4	Negative	No	No	NED
5	76	M	White	Smoking (20 py)	ES	ER	2	Positive	Yes	No	LR
6	71	F	N/A	N/A	N/A	N/A	N/A	N/A	N/A	N/A	N/A
7	67	M	White	Wood dust; smoking (175 py)	ES	PM	2	Positive	Yes	No	LR
8	67	M	White	Smoking (42 py)	ES	CFR	4	Negative	Yes	No	NED
Nonenteric											
9	71	M	White	None	NC	ER	2	Negative	Yes	No	NED
10	68	F	N/A	N/A	N/A	N/A	N/A	N/A	N/A	N/A	N/A
11	69	M	White	Smoking (90 py)	ES	MM	2	Negative	Yes	No	NED
12	67	M	Asian	Smoking (47 py)	NC	MM	2	Positive	Yes	No	NED
13	50	M	White	None	ES	CFR	2	Negative	Yes	No	NED
14	51	M	White	None	MS	CFR	4	Negative	Yes	Yes	DOD
15	87	M	Hispanic	Smoking (10 py)	ES	MM	2	Negative	N/A	N/A	NED

Genetic Alterations

Table 2 presents the molecular findings of sinonasal adenocarcinoma. Two K-*ras*mutations at codon 12 were identified in two of seven enteric-type carcinomas (29%) but in none of the seromucinous carcinomas (Figure 3). One of these tumors manifested associated p53 overexpression and MSI (14%). Two of seven (29%) seromucinous-type tumors had p53 overexpression, and none had any genetic alterations at the genes tested. No mutation at the APC or β-catenin genes was present in any tumor of either type. Alteration in hMLH1 expression was identified in two moderately differentiated enteric-type carcinomas, and one also had alteration of hMSH2. None of the seromucinous cancers showed MSI.

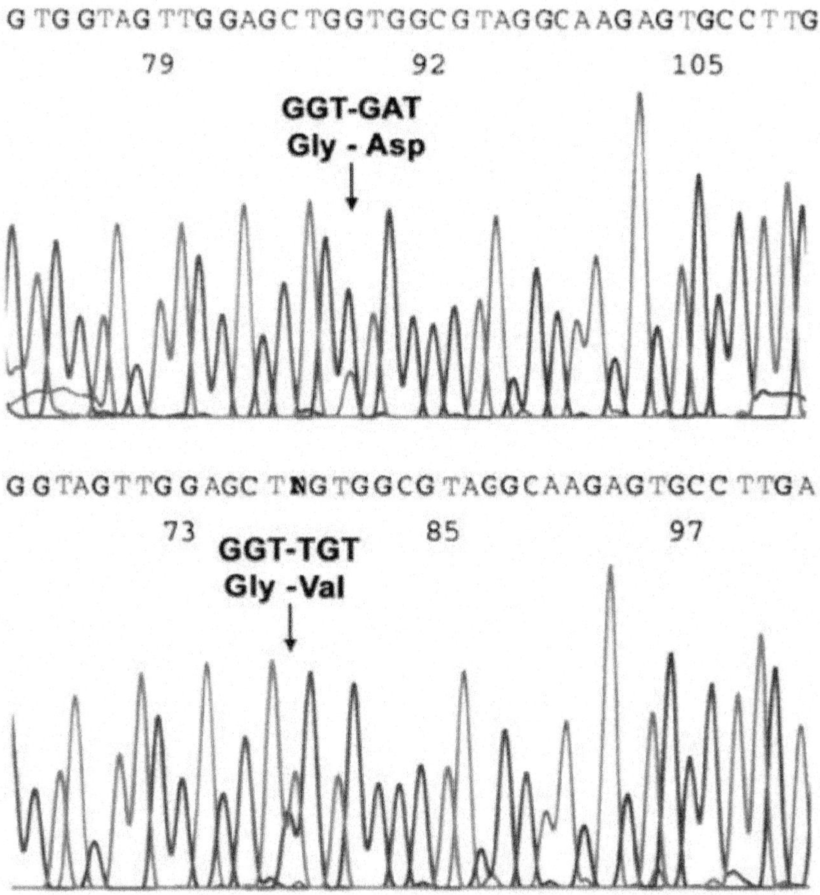

Figure 3: G → A transition mutation of the Ras gene in intestinal-type sinonasal adenocarcinoma.

Table 2:- Molecular analysis of sinonasal adenocarcinomas

Type #	*K-ras*	*APC*	*β-Catenin*	*p53*	*hMLH1*	*hMSH2*	*Bat 25*	*Bat 26*
Enteric								
1	●	○	○	○	○	○	○	○
2	○	○	○	○	○	○	○	○
3	○	○	○	○	○	○	○	○
4	○	○	○	○	○	○	○	○
5	○	○	○	○	○	○	○	○
6	○	○	○	○	○	○	○	○
7	●	○	○	●	○	○	●	○
8	N/A	N/A	N/A	N/A	N/A	N/A	N/A	N/A
Non-enteric								
9	○	○	○	●	○	○	○	○
10	○	○	○	○	○	○	○	○
11	○	N/A	○	○	○	●	○	○
12	○	○	○	○	●	○	○	○
13	○	○	○	○	●	●	○	○
14	○	○	○	●	○	○	N/A	○
15	○	○	○	○	○	○	○	○

●, alteration; ○, no alteration; N/A, not assessed.

DISCUSSION

Our results show K-*ras* mutations in 29% of the enteric-type carcinoma tumors and in none of the seromucinous-subtype. A similar incidence of mutations at this gene has previously been reported in studies of primary colorectal carcinomas.[14]This is further underscored by the detection of an MSI, a feature commonly present in a subset of primary enteric adenocarcinomas, in one of the enteric-type with K-*ras* mutation. Because of the phenotypic similarities between primary colonic and enteric-type sinonasal adenocarcinomas the presence of K-*ras* mutations in both indicates an early association with their development, regardless of the site of origin. In addition evidence that intestinal metaplasia of the sinonasal mucosa precede to the development of the enteric-type lends further credence to this hypothesis.[3]

The lack of other genetic features typically reported in primary colonic adenocarcinoma in our enteric-type tumors, however, suggest that specific events related to their site of origin underlie these differences. We contend that subsequent genetic events, especially in the former, are most likely acquired during the progression as a result of loco-regional and other epidemiological factors.[20, 21, 22, 23, 24, 25, 26, 27] These results, along with the lack of any alteration in the seromucinous subtype, support a distinct pathway for the evolution and progression of the intestinal and primary colonic adenocarcinomas.

In our cohort, only three high-grade tumors overexpressed p53, one of which was enteric in type and also had a K-*ras* mutation. Two patients with these tumors developed local recurrence or metastasis. Studies of colonic adenocarcinoma and a recent study of sinonasal adenocarcinoma have reported a high incidence of p53 abnormalities.[28, 29] Recent studies of p53 in enteric-type adenocarcinoma have associated this finding with sawdust exposure.[13, 28, 29] Similarly an association between dust exposure and epigenetic alterations at the CpG island of certain genes in these tumors has been reported.[28] The underlying factors for the differences between our findings and those of other investigators can be attributed to epidemiologic and/or patient populations variables. Only two patients in our cohort had a history of wood-dust exposure, and it is conceivable that sawdust contributes only to the development of a subset of the enteric-form tumors and that an alternative pathway may play a role in tumors' developing in nondust-exposed patients.

Our findings also suggest that patients with dust exposure develop via a different pathway than those without such history, and that the enteric-form are different molecularly from seromucinous adenocarcinoma. In our cohort, patients with the enteric-form had a more aggressive clinical course, than those with the seromucinous-type. However, because of the small number of tumors examined further studies are needed to determine the effect of tumor differentiation on the clinical behavior of these tumors. Sinonasal adenocarcinomas are generally unresponsive to medical management modalities and definitive surgical excision which achieves local control rates of 50% is the primary treatment of choice.[30] Recently, however, the addition of postoperative radiation following surgical resection have increased the 5-year local control rate to 59%.[31] Since our patient outcomes are consistent with these findings, we, recommend that patients with the enteric-type be treated with surgery and adjuvant radiation therapy.

ACKNOWLEDGEMENTS

This work was supported in part by the Kenneth D Müller Professorship and a National Cancer Institute Specialized Program of Research Excellence grant in head and neck cancer.

REFERENCES

1. Batsakis JG, Rice DH, Solomon AR. The pathology of head and neck tumors: squamous and mucous-gland carcinomas of the nasal cavity, paranasal sinuses, and larynx, part 6. *Head Neck Surg* 1980;2:497–508.
2. Franquemont DW, Fechner RE, Mills SE. Histologic classification of sinonasal–intestinal type adenocarcinoma. *Am J Surg Pathol* 1991;15:368–375.
3. Choi H-R, Sturgis EM, Rashid A, *et al.* Sinonasal adenocarcinoma: evidence for histogenetic divergence of the Enteric and nonenteric phenotypes. *Hum Path* 2003;34:1101–1107.

4. Vogelstein B, Fearon ER, Hamilton SR, *et al.* Genetic alterations during colorectal-tumor development. *N Engl J Med* 1988;319:525–532.

5. Lee JH, Abraham SC, Kim HS, *et al.* Inverse relationship between APC gene mutation in gastric adenomas and development of adenocarcinoma. *Am J Pathol* 2002;161:611–618.

6. Kim IJ, Kang HC, Park JH, *et al.* Determination of tumor aggressiveness in colorectal cancer by K-ras-2 analysis. *Arch Surg* 1993;128:526–531.

7. Finkelstein SD, Sayegh R, Bakker A, *et al.* Prediction of biologic aggressiveness in colorectal cancer by p53/K-ras-2 topographic genotyping.*Mol Diagn* 1996;1:5–28.

8. Cunningham J, Lust JA, Schaid DJ, *et al.* Expression of p53 and 17p allelic loss in colorectal carcinoma. *Cancer Res* 1992;52:1974–1980.

9. Ohue M, Tomita N, Monden T, *et al.* A frequent alteration of p53 gene in carcinoma in adenoma of colon. *Cancer Res* 1994;54:4798–4804.

10. Kim IJ, Kang HC, Park JH, *et al.* Development and applications of a beta-catenin oligonucleotide microarray: beta-catenin mutations are dominantly found in the proximal colon cancers with microsatellite instability. *Clin Cancer Res* 2003;9:2920–2925.

11. Fukushima H, Yamamoto H, Itoh F, *et al.* Frequent alterations of the beta-catenin and TCF-4 genes, but not of the APC gene, in colon cancers with high-frequency microsatellite instability. *J Exp Clin Cancer Res*2001;20:553–559.

12. Fujiwara T, Stolker JM, Watanable T, *et al.* Accumulated clonal genetic alterations in familial and sporadic colorectal carcinomas with widespread instability in microsatellite sequences. *Am J Pathol* 1998;153:1063–1078.

13. Wu TT, Barnes L, Bakker A, *et al.* K-ras-2 and p53 genotyping of intestinal-type adenocarcinoma of the nasal cavity and paranasal sinuses. *Mod Pathol*1996;9:199–204.

14. Saber AT, Nielsen LR, Dictor M, *et al.* K-ras mutations in sinonasal adenocarcinomas in patients occupationally exposed to wood or leather dust. *Cancer Lett* 1998;126:59–65.

15. Kleinsasser O. Terminal tubulus adenocarcinoma of the nasal seromucous glands. A specific entity. *Arch Otorhinolaryngol* 1985;241:183–193.

16. Batsakis JG, Mackay B, Ordonez NG. Enteric-type adenocarcinoma of the nasal cavity. An electron microscopic and immunocytochemical study.*Cancer* 1984;54:855–860.

17. Goelz SE, Hamilton SR, Vogelstein B. Purification of DNA from formaldehyde fixed and paraffin embedded human tissue. *Biochem Biophys Res Commun* 1985;130:118–126.
18. Alexander J, Watanable T, Wu TT, *et al.* Histopathological identification of colon cancer with microsatellite instability. *Am J Pathol* 2001;158:527–535.
19. Baas IO, Mulder J-WR, Offerhaus GJA, *et al.* An evaluation of six antibodies for immunohistochemistry of mutant p53 gene product in archival colorectal neoplasms. *J Pathol* 1994;172:5–12.
20. Cecchi F, Buiatti E, Kriebel D, *et al.* Adenocarcinoma of the nose and paranasal sinuses in shoemakers and woodworkers in the province of Florence, Italy (1963–77). *Br J Ind Med* 1980;37:222–225.
21. Kleinsasser O, Schroeder HG. Adenocarcinomas of the inner nose after exposure to wood dust. Morphological findings and relationships between histopathology and clinical behavior in 79 cases. *Arch Otorhinolaryngol*1988;245:1–5.
22. Wills JH. Nasal carcinoma in woodworkers: a review. *J Occup Med*1982;24:526–530.
23. Klintenberg C, Olofsson J, Hellquist H, *et al.* Adenocarcinoma of the ethmoid sinuses. A review of 28 cases with special reference to wood dust exposure. *Cancer* 1984;54:482–488.
24. Hayes RB, Gerin M, Raatgever JW, *et al.* Wood-related occupations, wood dust exposure, and sinonasal cancer. *Am J Epidemiol* 1986;124:569–577.
25. Leclerc A, Martinez Cortes M, *et al.* Sinonasal cancer and wood dust exposure: results from a case–control study. *Am J Epidemiol*1994;140:340–349.
26. Van den Oever R. Occupational exposure to dust and sinonasal cancer. An analysis of 386 cases reported to the N.C.C.S.F. Cancer Registry. *Acta Otorhinolaryngol Belg* 1996;50:19–24.
27. Lopez JI, Nevado M, Eizaguirre B, *et al.* Intestinal-type adenocarcinoma of the nasal cavity and paranasal sinuses. A clinicopathologic study of 6 cases. *Tumori* 1990;76:250–254.
28. Perrone F, Oggionni M, Birindelli S, *et al.* TP53, p14ARF, p16INK4a and H-ras gene molecular analysis in intestinal-type adenocarcinoma of the nasal cavity and paranasal sinuses. *Int J Cancer* 2003;105:196–203.
29. Bashir AA, Robinson RA, Benda JA, *et al.* Sinonasal adenocarcinoma: immunohistochemical marking and expression of oncoproteins. *Head Neck*2003;25:763–771.

30. Perez P, Dominguez O, Gonzalez S, *et al.* Ras gene mutations in ethmoid sinus adenocarcinoma: prognostic implications. *Cancer* 1999;86:255–264.

31. Claus F, Boterberg T, Ost P, *et al.* Postoperative radiotherapy for adenocarcinoma of the ethmoid sinuses: treatment results for 47 patients. *Int J Radiat Oncol Biol Phys* 2002;54:1089–1094.

Chapter 6

AN ANALYSIS OF GENETIC CHANGES DURING THE DIVERGENCE OF DROSOPHILA SPECIES

Rui Sousa-Neves[1], Alexandre Rosas[2]

[1]Department of Biology, Case Western Reserve University, Cleveland, Ohio, United States of America,

[2]Departamento de Fı´sica, CCEN, Universidade Federal da Paraı´ba, Joa˜o Pessoa, Paraı´ba, Brazil

ABSTRACT

Background

It has been long appreciated that speciation involves changes in body plans and establishes genetic, reproductive, developmental and behavioral incompatibilities between populations. However, little is still known about the genetic components involved in these changes or the sequence and scale of events that lead to the differentiation of species.

Principal Findings

In this paper, we investigated the genetic changes in three closely related species of *Drosophila* by making pair-wise comparisons of their genomes. We focused our analysis on the modern relatives of the alleles likely to be segregating in pre-historic populations at the time or after the ancestor of *D. simulans* became separated from the ancestor of *D. melanogaster*. Some of these genes were previously implicated in the genetics of reproduction and behavior while the biological functions of others are not yet clear.

Conclusions

Together these results identify different classes of genes that might have participated in the beginning of segregation of these species millions of years ago in Africa.

INTRODUCTION

One of the greatest challenges in modern Biology is the identification of genes that operate during evolution and diversification of species. However, since species are already separated and preserved samples of the prehistoric species are scarce, the investigation of the scale and types of genetic changes occurred during and after speciation has been limited.

The problem of speciation has puzzled generations of biologists and is of great significance to a wide variety of fields in biology. For instance, during evolution, mutations modify the architecture of brains and external appearances, novel behaviors appear and reciprocal lethal/sterile genetic systems emerge to block gene flow. The recurrent appearance of these themes across different phyla suggests a conservation of genetic processes. However, little is still known about the scale of the genetic changes that occur during speciation.

To investigate this issue, we chose to use the genetic workhorse *Drosophila melanogaster* and two sequenced sibling species, *D. simulans* and *D. sechellia*. The fact that *D. melanogaster* is a close relative of *D. simulans* and *D. sechellia* and has a myriad of genetic and genomic resources, makes it an ideal model to study evolutionary processes. The latest estimates suggest that *D. simulans* diverged from *D. melanogaster* approximately 5.4 million years ago in Africa, while *D. sechellia* diverged from *D. simulans* 0.5 million years ago in the Indian island of Seychelles [1]. Similar estimates suggest that a fourth species, the incompletely sequenced*D. mauritiana*, diverged from *D. simulans* 0.1–0.3 million years ago in the island of Mauritius.

Anatomically, these sibling species are almost identical, except for the different appearance of the male genitalia in all four species and minor ambiguous morphological features [2]. All four species have a set of 4 chromosomes mostly homosequential, but *D. simulans*, *D. sechellia*and *D. mauritiana* have nearly identical rearrangements that distinguish them from *D. melanogaster* [3]–[6]. These rearrangements suggest that the common ancestor of these species had already diverged chromosomally from *D. melanogaster* between 0.5 and 5.4 Million years ago [3]–[6].

Despite their similarities, the three sibling species (*D. melanogaster*, *D. simulans* and *D. sechellia*) have an intriguing different biology. For instance, the mating between *D. simulans*males and *D. melanogaster* females results in a dramatic larval death of the male offspring that is accompanied by a reduction of the brain and lack of imaginal discs [7], [8], while the surviving adult females are sterile [3], [9]. The reciprocal mating between *D. melanogaster*males and *D. simulans* or *D. sechellia* females is rarely successful and results in embryonic lethality of female and sterility of male offspring [3], [10], [11]. Similar results are obtained with hybrids between *D. sechellia* and *D. melanogaster* and somewhat less extreme phenotypes with *D. sechellia* and *D. simulans* hybrids. In the latter case, both sexes survive, but the male progeny is sterile [3]. Behaviorally, these species exhibit a mating asymmetry and it has been proposed that females of the newest species (i.e. *D. simulans* and *D. sechellia*) reject males of the oldest species archetype (i.e. *D. melanogaster*). In contrast, females of the oldest species accept males of the newest species [3], [12], [13]. Similar mating asymmetries appear in a significant number of closely related species [14].

Together, the facts highlighted above suggest that *D. simulans* and *D. sechellia* are more closely related to each other than to *D. melanogaster*, and quite conceivably share incompatible genes that affect mitotic, embryonic, reproductive, sensory perception and behavioral systems. However, with few notable exceptions, the genes and alleles involved in these processes still remain elusive [15]–[19].

Here we screened the genome for genetic variants that might reveal the changes occurred during or after the divergence of the ancestor of *D. simulans* from the ancestor of *D. melanogaster*. For convenience, *D. melanogaster* is taken as the archetypical or ancestral form as previously suggested [5], [12]. In particular, we searched for alleles with little or no divergence between *D. simulans* and *D. sechellia* that greatly diverged from *D. melanogaster*. These alleles are expected to have appeared at the time or right after the divergence of the ancestor of *D. simulans* from the ancestor of *D. melanogaster*, but before the separation of *D. simulans* from *D. sechellia*. For this reason, we refer to them as *ancestral alleles*. It is noteworthy that ancestral alleles of *D. simulans* and *D. sechellia*, also happen to be fast evolving alleles, when the Melanogaster subgroup is used as a reference. The analysis of the predicted gene products of ancestral alleles reveal which classes of genes might have been involved in the segregation of these species.

RESULTS

Number of Coding Sequences Identified in *D. Simulans* and *D. Sechellia*

The major objective of this search was to quantify and identify alleles that might have been segregating in pre-historic populations of the ancestor of *D. simulans* that were inherited by the descendants *D. simulans* and *D. sechellia*. We expected that ancestral alleles should be informative of the developmental, reproductive and behavioral novelties that distinguish *D. simulans* and *D. sechellia* from *D. melanogaster*.

To begin addressing this issue, we extracted and compared the annotated coding sequences of *D. melanogaster* to the sequence of computationally predicted coding sequences of *D. simulans* and *D. sechellia* (Fig. 1A). A total of 13,740 predicted coding sequences were assembled from *D. simulans* and *D. sechellia* genomes: 2,226 on the X chromosome, 5,355 on the second chromosome, 6,074 on the third chromosome and 85 on the fourth chromosome.

A

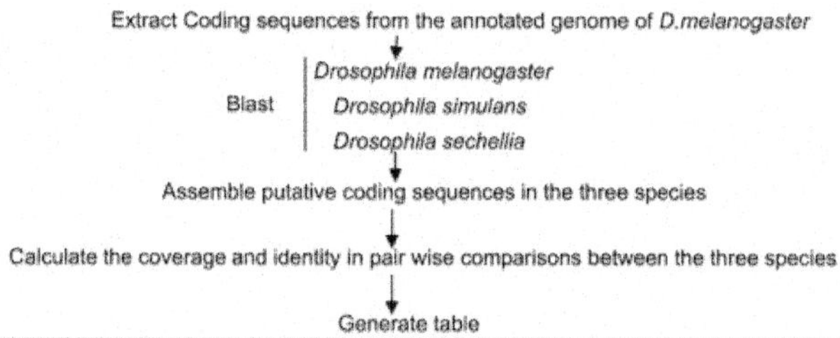

B

Gene #	Gene Name	% not covered in sim	% not covered in sec	Divergence sim sec	Divergence sim-mel	Divergence sec-mel	Control mel-mel
1	CG17636	0.29	0.29	0.58	3.38	3.73	3.55
2	RhoGAP1A	24.74	0	25.51	27.71	3.95	0.26
3	CG17707	0.07	0	1.02	4.12	4.46	0.03
4	SP71	0	0	0.33	2.75	2.89	0.09
5	CG3038	0.19	0.19	0.92	3.69	3.88	0.18
6	CG2995	11.36	0.02	12.41	14.94	3.68	0.08
7	cin	0.06	0	1.11	4.54	4.1	0.11
8	CG13377	0	0	0.6	3.02	3.63	0
9	CG13376	0.17	0.17	5.04	6.89	5.55	5.38
10	ewg	12.07	0	12.81	14.53	3.1	0.23

Select with the same coverage sim-sec

Select only genes with 99% identity or more in the mel-mel

C

Gene #	Gene Name	% not covered in sim	% not covered in sec	Divergence sim sec	Divergence sim-mel	Divergence sec-mel	Control mel-mel
8	CG13377	0	0	0.6	3.02	3.63	0
13	CG12470	0	0	0.34	1.87	2.21	0
15	CG32816	0	0	0.79	6.12	6.21	0
17	ac	0	0	0.83	3.14	3.3	0
18	sc	0	0	0.58	3.27	3.07	0
20	pcl	0	0	1.8	3.76	4.25	0
21	ase	0	0	1.1	2.4	2.81	0
32	CG17778	0	0	1.56	1.04	2.08	0
38	vnd	0	0	0.4	3	2.94	0
43	CG13364	0	0	1.03	3.59	3.59	0

Figure 1: Overview of the data collection and sorting.

A) Exons of coding sequence were extracted from the annotated genome of *D. melanogaster* using Extractor and electronically joined using Analyst to obtain complete coding sequences. These coding sequences were then automatically blasted against the genome of *D. melanogaster*, *D. simulans* and *D. sechellia* with Megablast. Analyst scanned the resulting alignments for the best hits and assembled the coding sequences in the three species from them. B) Analyst also calculated the coverage, the percentage not covered, the divergence in sim-sec, sim-mel, sec-mel as well as the control mel-mel and organized this data in a table. C) To minimize artifacts due to incomplete clone representation in the genomic libraries, the coding sequences were filtered and only genes with the same coverage in *D. simulans* and *D. sechellia* retrieved. To avoid genes truncated by Megablast (i.e. usually genes with small exons), only genes with a mismatch up to 1% in the control mel-mel were retrieved. After these two filters were applied, a new table like the one exemplified in C) was generated for each chromosome.

doi:10.1371/journal.pone.0010485.g001

Identification of High Confidence Genes by Sorting and Filtering Data

The data collected from each chromosome arm was organized in a table, which consists of eight columns with the following information: (1) coding sequence number in *D. melanogaster*; (2) gene name; (3) percentage of bases not covered in *D. simulans* and (4) in *D. sechellia*; (5) divergence between *D. simulans* and *D. sechellia*; (6) divergence between *D. melanogaster*and *D. simulans*; (7) divergence between *D. melanogaster* and *D. sechellia*; and finally, (8) assembly control (i.e. percentage of mismatches between the actual coding sequences of *D. melanogaster* vs. the coding sequences assembled from Blast results in *D. melanogaster*).

To avoid false positives due to truncated fragments in the WGS libraries, we first applied a filter that discards all genes with different coverage in *D. simulans* and *D. sechellia* (Fig. 1C, Columns 3 and 4). The remaining genes were sorted using values of the control Blast mel vs. mel (Fig. 1C, Column 8) in ascending order. In addition, only genes with a mismatch of up to 1% were retrieved. Our control of automatic assembly of coding sequences assured that only high quality coding sequences assembled from Blast alignments (i.e. 99% match or greater) were analyzed. After filtering the data using the criteria above, 8,416 reliable coding sequences corresponding to 61% of the total number of coding sequences extracted from *D. melanogaster*were obtained: 1,039 on the X; 3,407 on the second; 3,951 on the third and 19 on the fourth chromosome.

Identification of Genes that vary the least between *D. Simulans* and *D. Sechellia* and the most in *D. Melanogaster*

To identify genes that diverged the least in the pair sim-sec and the most in the pairs sim-mel and sec-mel (ancestral alleles), we employed two strategies. The first strategy selects genes in the pair sim-sec that diverged less than the average plus the standard deviation of all genes in the same chromosome, and in addition that also diverged more than the average plus the standard deviation in sim-mel and sec-mel pairs. The second method, which will be explained in more detail in the next section, is based on the observation that most of the genes in *D. simulans*/*D. sechellia* diverge linearly from *D. melanogaster*, while few very similar genes diverge non-linearly.

Out of the total 8,416 reliable coding sequences selected previously, the first method led to the identification of 517 genes: 67 genes on the X; 112 on the left arm of the second; 106 on the right arm of the second; 88 on the left arm of the third; and 144 on the right arm of the third chromosome (Table 1, Fig. 2). No ancestral alleles were identified on the fourth chromosome due to the fact this chromosome has highly homogeneous divergences (data not shown).

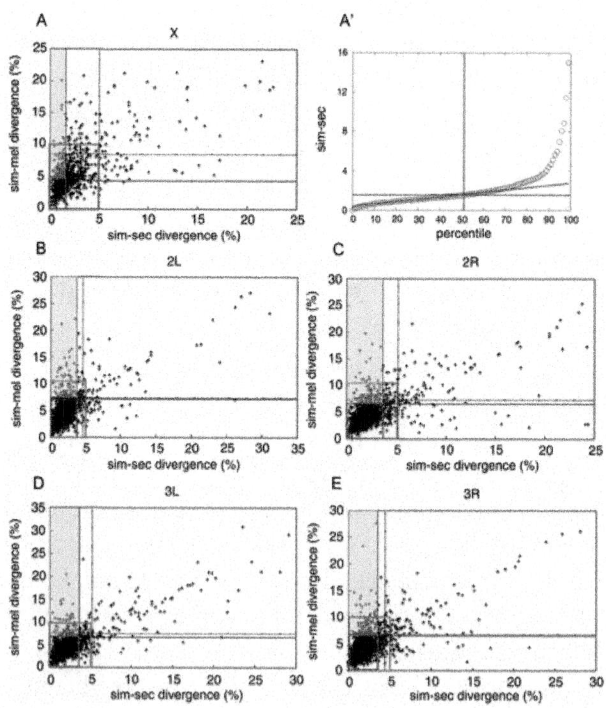

Figure 2: Patterns of divergence along chromosomes and two screening methods.

A, B, C, D and E) Each graph corresponds to a chromosome or chromosome arm (X, 2L, 2R, 3L and 3R), where the genes are ordered from the least divergent to the most divergent in sim-sec (abscissa) and sim-mel (ordinate). The horizontal and vertical dashed lines delimit the averages plus one standard deviation of the divergence between sim-sec (horizontal) and sim-mel (vertical). The upper left quadrants delimit genes found by the method of averages and standard deviations. Note that the divergences of most genes in all 5 graphs are clustered in a quadrant that can be roughly delimited between the abscissa values of 0%–5% and ordinate values 0%–10% (red rectangles). In this quadrant, the genes have a good fit with a linear distribution (P<0.0001). To better delimit the quadrant in which the divergence is linear in each chromosome, the data was divided in percentiles of divergences of sim-sec, sim-mel and sec-mel. (A›) exemplifies the percentiles of the X chromosome. Since each point in these curves represents one percentile, the percentage of genes that diverge linearly is equal to the number of points that can be transected by a straight line. Once this linear interval is defined, the values on the x and y axes become known and can be used to redefine the quadrant of linear divergences (inferior left quadrant in blue). The region where the genes in sim-sec vary the least and the genes in sim-mel vary the most is the adjacent upper quadrant to the left of the point where the horizontal and vertical lines cross (gray quadrant).

doi:10.1371/journal.pone.0010485.g002

	X	2L	2R	3L	3R	Totals
Screening 1	67	112	106	88	144	517
Screening 2	101	74	89	77	98	439
Common	14	73	76	61	96	320

The genes identified by method 1 have divergences inferior to the average plus standard deviation for the chromosomes in which they are located in sim-sec and higher divergences than the average plus standard deviation in sim-mel and sec-mel. Screening 2 selects genes that retained ancestral features and diverge significantly form *D. melanogaster.* Note that screening 2 appears more stringent than screening 1. Note also that a large number of common genes that can be found by both screenings in the autosomes, but considerably less on the X. No ancestral alleles could be found on the fourth chromosome (see text).

doi:10.1371/journal.pone.0010485.t001

Table 1: Number of genes identified by the screening methods 1 and 2.

doi:10.1371/journal.pone.0010485.t001

Genes that Diverge Linearly and Non-Linearly in *D. Simulans* and *D. Sechellia*

In the second strategy to identify ancestral alleles, we searched for patterns of divergence among the 8,416 coding sequences. In this case, the data was sorted in ascending order of identity, and the values of the sim-sec pair were plotted against the sim-mel pair for the X chromosome, 2R and 2L, 3R and 3L chromosome arms (Fig. 2). Since the divergences of sim-mel and sec-mel are approximately the same (data not shown), the graphs of the sec-mel pair were not included in the figure.

The graphs in Figure 2 show that the genes that diverge the least in sim-sec pair, but also diverge the most in sim-mel pair fall between values 0 to 5% of the abscissa. Moreover, it is clear that the majority of genes also fall roughly within the ordinate range of 0 to 10%, and that within this interval there is a linear fit (Fig. 2, $P<0.0001$). We refer this interval to as *initial linear interval* (Fig. 2 A–E, outlined in red). Conversely, for values above 10% in the ordinate there is no significant agreement with a linear fit ($P>0.05$). We refer this interval to as *initial non-linear interval* (Fig. 2 A–E, outside the blue region). These results suggest that within the initial linear interval, the sim-sec genes diverge from sim-mel fairly linearly, while within the initial non-linear interval, this linearity breaks down. Thus, the non-linear interval contains the genes that varied the least in sim-sec and the most in sim-mel.

The results above suggest the existence of at least two gene populations; one large group that changes at a similar pace over generations in *D. simulans* and *D. sechellia* and a smaller group with a high degree of divergences.

Delimiting a Quadrant with Genes that Diverged from *D. Melanogaster*and were Inherited in *D. Simulans* and *D. Sechellia*

In the graphs shown in Fig. 2, we noticed the presence of a linear and a non-linear interval, but it is difficult to determine the boundary between the two intervals. In order to define more precisely this limit, we divided the data of sim-mel, sim-sec and sec-mel in percentiles, as shown in Fig. 2A′ for the X chromosome. Since each point in the graph represents one percentile, the line that transects the largest number of linear points along this curve reveals the percentage of genes that diverge linearly. On the X chromosome, this range corresponds to the 51st percentile. Thus, the linearly distributed genes fall between the abscissa values 0 to 1.56 (sim-sec) and ordinate values 0 to 4.35 (sim-mel) (Fig. 2A′). We also applied the same methodology for the autosomes (data not shown).

We identified a more approximate interval where the genes diverge linearly in the abscissa and ordinate (Fig. 2, the inferior left quadrants in blue), and selected the genes that are in the quadrant above in the ordinate (Fig. 2, highlighted in gray). Using the same methodology for the pairs sim-sec and sec-mel, we identified the common set of genes in both searches (i.e. common genes to the percentiles of sim-sec vs. sim-mel and sim-sec vs. sec-mel).

A total of 439 genes common to *D. simulans* and *D. sechellia* were identified: 101 genes on the X, 74 on 2L, 89 on 2R, 77 on 3L and 98 on 3R (Table 1, Fig. 2). We were unable to perform a similar analysis for the fourth chromosome due to the fact that only 19 genes of high confidence were identified, which precluded the use of percentiles.

Both Screening Strategies Identify a Large Number of Common Genes

When the results of the percentile search are combined with the results of the search of averages and standard deviations, we notice that a significant number of genes are represented in both searches. In particular, 73 (98.6%) genes common to both searches were found on 2L; 76 (85.4%) on 2R, 61(79.2%) on 3L; and 96 (98.0%) on the 3R. Interestingly, on the X chromosome only 14 (13.9%) genes common to both searches were found (Table 1). The relatively low percentage of common genes observed on the X chromosome stems from the fact that the average divergence plus the standard deviation on this chromosome is higher than in the autosomes (Fig. 2, note the different spacing between the dashed line and solid lines on the X with autosomes). This variation results in the exclusion of several genes found by the method of percentiles and at the same time, in the inclusion of others. Thus, when the results of both screening are combined, the number of genes found by the averages method on the X chromosome is inferior to those found by the percentiles method. The main conclusions that can be drawn from these results are that both searches identify a large number of common genes and that the X chromosome genes evolve slightly faster than autosomal genes.

Most Genes Identified are Orthologous

To test whether the 320 genes identified in *D. simulans* and *D. sechellia* by both methods correspond to true orthologues as opposed to paralogues, we blasted the *D. simulans* and *D. sechellia* genes separately against the genome of *D. melanogaster*. Out of these 320 genes, 307 genes from each species matched the chromosomal position and gene used in *D. melanogaster* at the beginning of the screening. Thus, 96% of these genes correspond to homologues, not paralogs. The remaining 13 (4%) were excluded since they correspond to paralogs.

Spatial Distribution of Ancestral Alleles within the Genome and Divergence Hotspots

We next tested whether the ancestral alleles are clustered in specific genomic locations or whether they appear evenly distributed across the genome. To address this issue we plotted their occurrence along the 20 divisions of the major chromosomes, using chromosomal coordinates of *D. melanogaster* [20] (Fig. 3). Since the method of averages and standard deviations produces a distortion that results in fewer common genes on the X chromosome, we used the results obtained from the percentile search to plot the position of these genes. Our data suggest that although these alleles can be found in almost every division of the three major chromosomes, some regions are hotspots for ancestral alleles. These regions were identified by searching for regions that have more ancestral alleles than the average plus 1 standard deviation. In particular, the X chromosome division 1 and division 9 have more ancestral alleles than most divisions on this chromosome. Similarly, three divisions on the left arm of the second chromosome (i.e. 22, 23 and 34) also harbor more ancestral alleles than the average plus one standard deviation for this arm. The right arm of the second chromosome also seems to have three hotspots in divisions 44, 54 and 59, while the distribution of ancestral alleles on the left arm of the third chromosome does not contain prominent hotspots, except perhaps by divisions 61, 64, 68 and 70. Finally, on the right arm of the third chromosome, one prominent hotspot appears at division 82. The significance of this clustering is not yet clear, but we note that some of these hotspots are located nearby known rearrangement breakpoints observed in *D. simulans* such as in divisions 1–2, divisions 21–22, 59–60 and 82 [21].

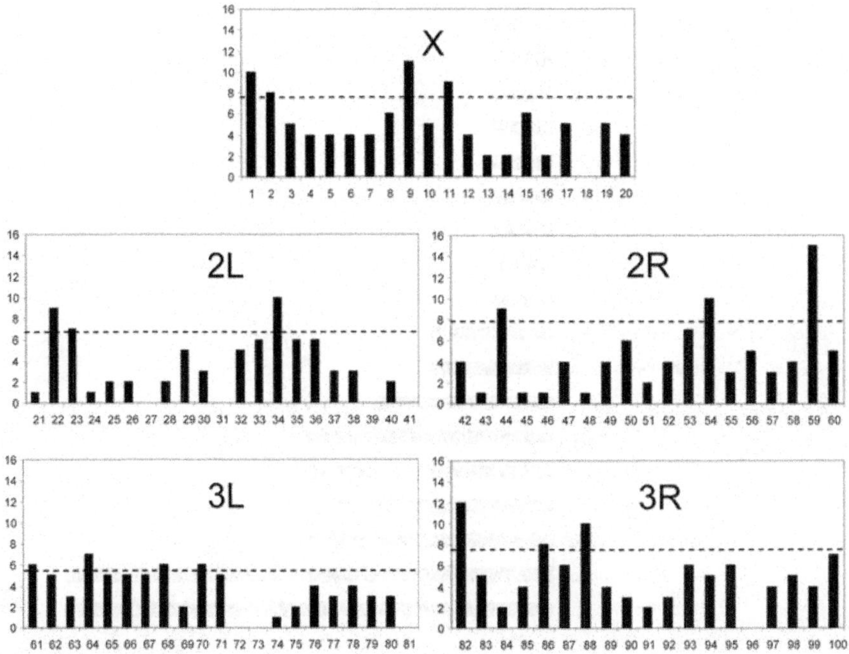

Figure 3: Distribution of ancestral alleles in the three major chromosomes.

The 20 division coordinates used are those of *D. melanogaster*. Almost all divisions have one or more ancestral alleles. The dashed lines indicate the average number of alleles plus one standard deviation. Note that some divisions have a higher density of ancestral alleles than others.

doi:10.1371/journal.pone.0010485.g003

Annotated Biological Functions of Ancestral Alleles

To test whether the genes identified have functions consistent with roles in species differentiation, we cross-referenced them to Gene Ontology (GO). If ancestral alleles participated in the segregation of these species, then we should expect to find biological functions consistent with pre and post-zygotic barriers, such as those that interfere with mating and cause interspecific lethality, sterility and mitotic defects in hybrids.

The GO referencing shows that despite the fact that less than 40% of these genes have either known molecular or biological functions, several ancestral alleles fall in discrete GO functional groups such as hybrid lethality, oogenesis, gamete generation, female meiosis, sperm competition and displacement, chemical perception of taste and olfaction, and immunity (Fig. 4).

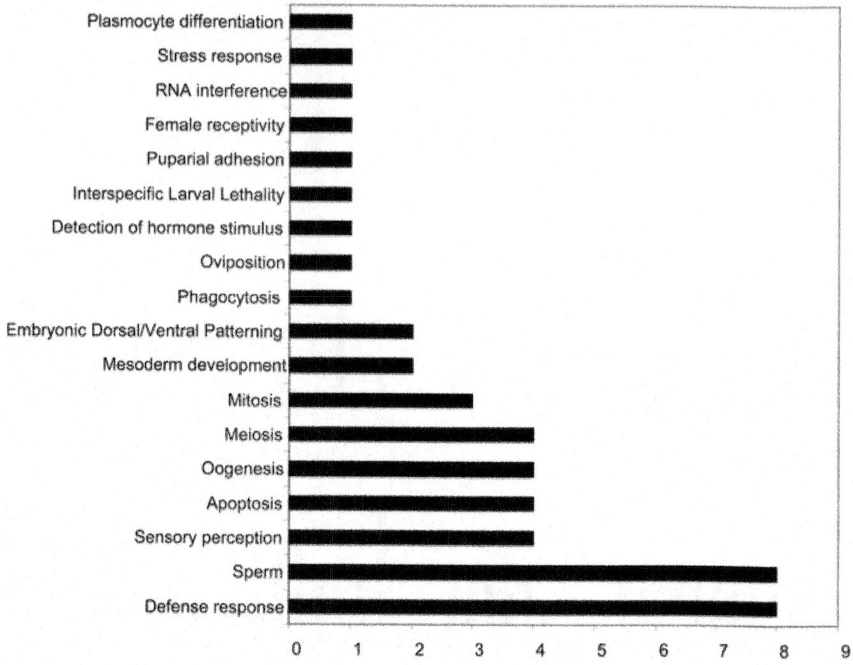

Figure 4: The biological functions of 48 ancestral alleles defined by Gene Ontology.

The graph shows only genes with biological functions assigned by assays, inferred by sequence similarity or phenotype and does not include general biochemical properties such as phosphorylation, transcription initiation, signal transduction, proteolysis among others.

doi:10.1371/journal.pone.0010485.g004

Among known genes that cause zygotic barriers, our search readily identified *Lethal hybrid rescue* (*Lhr*) as an ancestral allele. The wild type allele of *Lhr* in *D. simulans* is responsible for the larval lethality in the sons of *D. simulans* males and *D. melanogaster* females [15], [19]. Similarly, the gene *CG14781* appears as an ancestral allele. *CG14781* has been implicated in mitotic spindle elongation and recently shown to correspond to *mei-38* [22], [23]. Null mutants of *mei-38* cause abnormal meiotic non-disjunction in females, abnormal mitosis and consequently lethality due to aneuploidy [23]. Thus, *mei-38* could be potentially involved in the sex-specific offspring hybrid lethality in females.

Our search also identified a number of genes with functions consistent with the formation of pre-zygotic barriers. For instance, accessory gland proteins such as Acp29AB and Acp98AB appear as ancestral alleles and it has been suggested that Acp29AB confers a resistance to the sperm of one male to be displaced by the sperm of another male, while Acp98AB appears to negatively regulate female receptivity [24], [25]. We also found genes involved in perception of taste such as *Gr59d* and *Gr59f* and odors like the *Odorant binding proteins Obp19a*, *Obp22a*and *Obp47a* [26]. These genes have been implicated in the sensory perception of chemical stimuli [27] and could potentially participate in the formation of pre-mating barriers through species-specific mate recognition.

Finally, our search identified ancestral alleles that potentially have novel and previously unsuspected functions such as immunity (Fig. 4). Together, these results suggest that the segregation of an ancestral population into three distinct species involved changes in reproduction, embryonic development, nervous system development and physiology, and immunity.

DISCUSSION

Genomes and Footprints of Evolution

Extensive circumstantial evidences suggest that the genes that once created a sharp barrier between the ancestor of *D.melanogaster* and its sibling species might share an unusual conservation in *D. simulans* and *D. sechellia*. We tested this hypothesis by comparing their coding sequences and found 439 genes with little divergence in *D. simulans* and *D. sechellia*, but that diverge significantly from *D. melanogaster*.

The ancestral alleles identified in this work possibly record the earliest events in the differentiation of these *Drosophila* lineages that can be detected in extant species. The fact that these genes are very similar in *D. simulans/D. sechellia* but diverged from *D. melanogaster* more than most genes in the genome suggests two possible scenarios. In the first, the high divergence of ancestral alleles was acquired focally in time (i.e. this divergence is the result of one or few events that happened in short periods of time). The second possible scenario is that they were acquired over a longer period of time. (i.e. these genes are more prone to mutations and evolve faster than other genes). There are at least two evidences that favor the first hypothesis, but these are not yet conclusive. If these alleles were more prone to mutations, then we should expect that they would continue diverging at high rates after the separation of *D. simulans* from *D. sechellia*, but we did not observe such continuing divergence in the genome samples currently available. Also, if these alleles were more prone to mutations, then we should expect to observe high rates of polymorphism in *D. melanogaster*and *D. simulans*, which has not been reported in genomic results of different strains sequenced yet. In addition, we found a higher frequency of ancestral alleles near known chromosomal rearrangements, which raises the interesting possibility that these alleles could have been generated at the time those rearrangements appeared.

The High Divergence of X-Chromosome Genes, Recombination and Segregation Patterns

In our search, we analyzed each chromosome separately to test whether there were variations in the rate of divergence among distinct chromosomes. The existence of such differences might provide an insight into the mechanism involved in the generation of ancestral alleles. Our analysis reveals that the average divergence plus the standard deviation for the X chromosome genes is higher than that of autosomal genes. Conversely, the same analysis suggests that the fourth chromosome has a lower average divergence than the other autosomes and the X chromosome. Together, these results show that the chromosome X evolves faster than the other autosomes and suggest that the fourth chromosome evolves slower. Since the rates of mutation and recombination on the X chromosome and in the two large autosomes do not appear significantly different [28](http://flybase.org/maps/chromosomes/maps.html), the discrepancy between the divergences of the X and the remaining chromosomes is intriguing. However, this discrepancy possibly stems from the fact that the X chromosome is the only chromosome that exists in one or two copies (X/X and X/Y) in every generation. The existence of a hemizygote state allows recessive mutations on the X chromosome to be subject to the scrutiny of natural selection at least one generation before and in more individuals per generation, than a similar recessive mutation in an autosome. Thus, even with the same rate of mutation, recessive mutations on the X chromosome are subject to more rounds of selection than mutations in the autosomes, and consequently should have a better chance to become fixed.

The difference in the Mean plus Standard Deviation on the Fourth Chromosome Genes suggest a Possible Mechanism for the Generation of Ancestral Alleles

While analyzing the fourth chromosome, we detected an unusually low divergence in this chromosome. One possible explanation for this low divergence is that this is the only autosome that does not recombine during meiosis. Without recombination, errors acquired due to abnormal crossover are almost inexistent and the possibility of combining in a single chromosome different alleles floating in a population is equally low. Thus, the lack of errors during recombination and the combination of these mutations in a single chromosome could be accountable to some extent for the low levels of generation and accumulation of ancestral alleles on the fourth chromosome. However, since only 19 out of the 85 genes on this chromosome could be analyzed, this hypothesis needs to be more thoroughly tested as new high quality sequences become available for this chromosome.

The Advantages and Limits of the Analysis of Ancestral Alleles

The literature of speciation mechanisms has some examples of cleverly designed experiments to isolate genes required to block gene flow among closely related species. However, despite the fact that these screenings are of great significance and provide invaluable information about the approximate position of genes involved in speciation, researchers often face tremendous challenges to identify them molecularly. A typical example is *Lhr*, a gene identified genetically in 1979, which was only molecularly cloned almost 30 years later. This gene was readily identified as an ancestral allele in our search.

Our search can also potentially simplify the identification of other genes involved in speciation. For instance, Sawamura and cols. (2004) genetically mapped a female sterile mutant from *D. simulans*, presumably involved in the sterility observed in *D. melanogaster*/*D. simulans* hybrids, near the chromosomal division 32. Despite their efforts to narrow the region down to a 170 kb interval containing 20 coding sequences, they could not identify molecularly which of those 20 genes had a major effect on female fertility [29]. Our screening has identified 5 ancestral alleles on subdivision 32, and within the interval identified by Sawamura et al, there are only two ancestral alleles:Vm32E and CG14926. The GO of Vm32E suggests a role in the formation of embryonic vitelline membrane, which is consistent with female sterility, while CG14926 has no defined function but is expressed in male spermatocytes.

Although our analysis can potentially simplify the search and characterization of novel genes involved in speciation, there are some limits to its capabilities. The first one is the quality of the sequenced genomes. For instance, our search failed to identify *Hybrid male rescue*, since this gene does not have the same coverage in *D. simulans* and *D. sechellia* and for this reason was excluded from our analysis. Several other genes in the genome of *D. simulans* and *D. sechellia* also have a poor coverage. We expect that the search of evolutionary genes using the strategy outlined here will be greatly improved as more sequence gaps are filled.

The sequence comparison tools developed in this work can also be used in other types of screenings to identify genes involved in other biological processes unique to each sibling *Drosophila* species. For instance, since our screening was designed to identify ancestral alleles of *D. simulans* and *D. sechellia*, it eliminated genes required for particular specializations in each species. Our screening most likely missed genes that might be necessary for the feeding habits that make *D. simulans* and *D. melanogaster* cosmopolitan and *D. sechellia* restricted to *Morinda*. To identify the genes required for such differences, it would be necessary to screen for highly conserved genes in *D. simulans* and *D. melanogaster*that diverged in *D. sechellia*. Similarly, genes involved in the female choice for males would be expected to be missed by the current screening since females of the three species prefer males of their own species. To identify these genes the search should be directed to fast evolving genes (i.e. genes that are most divergent in the three species). Together, our results identify a relatively small number of genes that can be tested for speciation roles among *D. melanogaster* sibling species.

MATERIALS AND METHODS

Gene Extraction and Searches

Usually genome searches that aim to find variation in coding sequences focus on translations since non-synonym amino acid variation is generally believed to produce phenotypic variation. However, this approach eliminates synonym substitutions that result in protein variation (e.g. mutations in splicing enhancers). For this reason, here we took all nucleotide variation in consideration.

Annotated sequences from the *Drosophila* library NT corresponding to the X (AE014298), 2L (AE014134), 2R (AE013599), 3L (AE014296), 3R (AE014297) and 4 (AE014135) arms or chromosomes were downloaded from the NCBI website and the coding sequences (CDS) were extracted using Extractor, a software developed by us. The extracted genes were then Blasted against the Whole Genome Sequences of *D. melanogaster* (mel-mel), *D. simulans* (mel-sim) and *D. sechellia* (mel-sec) obtained from the Whole Genome Shotgun (WGS) NCBI›s library (14-Mar-2008). This library contains sequences from different strains of *D. simulans* and thus provides samples of gene variation in different populations. We developed another program, the Analyst, to automatically assemble the blast hits of the clone with best coverage using the*D. melanogaster* positions of the annotated coding sequences as a template. The Analyst also reported the coverage of all coding genes in *D. simulans* and *D. sechellia* and calculated the divergence between the pairs mel-sim, mel-sec and sim-sec by using their respective alignments.

Blast Settings and Controls

Several different Blast settings were used in control experiments where annotated coding sequences of Drosophila were blasted against Whole Genome

Sequences of *D. melanogaster*. These controls were used to define the Blast program that most consistently identifies the largest number of complete coding sequences in *D. melanogaster*. Discontinuous Megablast was chosen since it yielded the best reconstruction of the coding sequences in the control mel-mel.

Triangulation of Alleles with the same or Similar Nucleotide Composition in *D. Simulans* and *D. Sechellia* that Diverged from *D. Melanogaster*

Identity values generated by Blast alignments provide information about the percentage of substitutions within a DNA segment, but not about the position of these substitutions. Thus, if a query gene in one species has the same identity of the subjects in two other species and this identity is different from 0, the two subjects may or may not contain mutations in the same position. To identify mutations in the same position, we triangulated the position of these substitutions by coverage and identity in pair wise comparisons between mel-sim, mel-sec and sim-sec. Using this system, genes that diverged significantly from *D. melanogaster* and were inherited by *D. simulans* and *D. sechellia* lineages should appear with the same coverage and an identical or similar identity in the *D. simulans/D. sechellia* comparison, and with equally fewer identities in the *D. melanogaster/D. simulans* and *D. melanogater/D. sechellia*comparisons.

To minimize errors due to the incorrect automatic assembly of the coding sequences in *D. simulans* and *D. sechellia* that could interfere with the evaluation of the divergence (i.e. truncated coding sequences due to the inability of Blast to identify particular exons), the coding sequences of *D. melanogaster* were Blasted against the *D. melanogaster* WGS library and the predicted coding sequences assembled from the Blast hits. The identities of these predicted coding sequences were then compared to the actual coding sequences in the annotated genome and only genes with at least 99% of identity with the annotated coding sequences were included in the analysis. To avoid false-positives due to incomplete clone representation, the data was sorted by coverage in sim and sec, and only genes with the same coverage were selected.

Cross-Referencing to Gene Function

The functional cross-referencing of the genes identified was done using annotated biological functions from flybase (http://flybase.bio.indiana.edu/), as well as descriptions in the literature.

ACKNOWLEDGMENTS

We would like to thank Claudia Mieko Mizutani and Joseph Schinaman for comments on the manuscript.

AUTHOR CONTRIBUTIONS

Conceived and designed the experiments: RSN AR. Performed the experiments: RSN AR. Analyzed the data: RSN AR. Contributed reagents/materials/analysis tools: RSN AR. Wrote the paper: RSN AR.

REFERENCES

1. Tamura K, Subramanian S, Kumar S (2004) Temporal patterns of fruit fly (Drosophila) evolution revealed by mutation clocks. Mol Biol Evol 21: 36–44.
2. Acebes A, Cobb M, Ferveur JF (2003) Species-specific effects of single sensillum ablation on mating position in Drosophila. J Exp Biol 206: 3095–3100.
3. Lachaise D, David JR, Lemeunier F, Tscas L, Ashburner M (1986) The Reproductive Relationships of Drosophila sechellia with D. mauritiana, D. simulans, and D. melanogaster from the Afrotropical Region. Evolution 40: 262–271.
4. Lemeunier F, Ashburner MA (1976) Relationships within the melanogaster species subgroup of the genus Drosophila (Sophophora). II. Phylogenetic relationships between six species based upon polytene chromosome banding sequences. Proc R Soc Lond B Biol Sci 193: 275–294.
5. Masly JP, Jones CD, Noor MA, Locke J, Orr HA (2006) Gene transposition as a cause of hybrid sterility in Drosophila. Science 313: 1448–1450.
6. Podemski L, Ferrer C, Locke J (2001) Whole arm inversions of chromosome 4 in Drosophila species. Chromosoma 110: 305–312.
7. Seiler T, Nöthiger R (1974) Somatic cell genetics applied to species hybrids of Drosophila. Experientia: 30: 709.
8. Sanchez L, Düberdorfer A (1983) Development of Imaginal Discs from Lethal Hybrids between Drosophila melanogaster and Drosophila mauritiana. Roux's Arch Dev Biol 192: 48–50.
9. Orr HA, Madden LD, Coyne JA, Goodwin R, Hawley RS (1997) The developmental genetics of hybrid inviability: a mitotic defect in Drosophila hybrids. Genetics 145: 1031–1040.

10. Sturtevant AH (1929) Contributions to the genetics of Drosophila simulans and Drosophila melanogaster. I. The genetics of Drosophila simulans. Publs Carnegie Instn 399: 1–62.

11. Hadorn E (1961) Zur Anatomie und Phasenspezifität der Letalität von Bastarden zwischen *Drosophila melanogaster* und *D.simulans*. Rev Swisse Zool 68: 197–207.

12. Watanabe TK, Kawanishi M (1979) Mating Preference and the Direction of Evolution in Drosophila. Science 205: 906–907.

13. Tomaru M, Doi M, Higuchi H, Oguma Y (2000) Courtship song recognition in the Drosophila melanogaster complex: heterospecific songs make females receptive in D. melanogaster, but not in D. sechellia. Evolution 54: 1286–1294.

14. Wirtz P (1999) Mother species-father species: unidirectional hybridization in animals with female choice. Anim Behav 58: 1–12.

15. Watanabe TK (1979) A gene that rescues the lethal hybris between Drosophila melanogaster and D.simulans. The Japanese Journal of Genetics 325–331.

16. Hutter P, Roote J, Ashburner M (1990) A genetic basis for the inviability of hybrids between sibling species of Drosophila. Genetics 124: 909–920.

17. Sawamura K, Yamamoto MT, Watanabe TK (1993) Hybrid lethal systems in the Drosophila melanogaster species complex. II. The Zygotic hybrid rescue (Zhr) gene of D. melanogaster. Genetics 133: 307–313.

18. Sawamura K, Taira T, Watanabe TK (1993) Hybrid lethal systems in the Drosophila melanogaster species complex. I. The maternal hybrid rescue (mhr) gene of Drosophila simulans. Genetics 133: 299–305.

19. Brideau NJ, Flores HA, Wang J, Maheshwari S, Wang X, et al. (2006) Two Dobzhansky-Muller genes interact to cause hybrid lethality in Drosophila. Science 314: 1292–1295.

20. Bridges CB (1935) Salivary chromosome maps with a key to the banding of the chromosomes of Drosophila melanogaster. J Hered 26: 60–64.

21. Horton IH (1939) A Comparison of the Salivary Gland Chromosomes of Drosophila Melanogaster and D. Simulans. Genetics 24: 234–243.

22. Baker BS, Carpenter AT (1972) Genetic analysis of sex chromosomal meiotic mutants in Drosophilia melanogaster. Genetics 71: 255–286.

23. Wu C, Singaram V, McKim KS (2008) mei-38 Is required for chromosome segregation during meiosis in drosophila females. Genetics 180: 61–72.

24. Wolfner MF, Harada HA, Bertram MJ, Stelick TJ, Kraus KW, et al. (1997) New genes for male accessory gland proteins in Drosophila melanogaster. Insect Biochem Mol Biol 27: 825–834.
25. Chapman T (2001) Seminal fluid-mediated fitness traits in Drosophila. Heredity 87: 511–521.
26. Clyne PJ, Warr CG, Carlson JR (2000) Candidate taste receptors in Drosophila. Science 287: 1830–1834.
27. Galindo K, Smith DP (2001) A large family of divergent Drosophila odorant-binding proteins expressed in gustatory and olfactory sensilla. Genetics 159: 1059–1072.
28. Haag-Liautard C, Dorris M, Maside X, Macaskill S, Halligan DL, et al. (2007) Direct estimation of per nucleotide and genomic deleterious mutation rates in Drosophila. Nature 445: 82–85.
29. Sawamura K, Karr TL, Yamamoto MT (2004) Genetics of hybrid inviability and sterility in Drosophila: dissection of introgression of D. simulans genes in D. melanogaster genome. Genetica 120: 253–260.

Chapter 7

GENETIC ANALYSIS OF POPULATION STRUCTURE AND REPRODUCTIVE MODE OF THE TERMITE RETICULITERMES CHINENSIS SNYDER

Qiuying Huang[1], Ganghua Li[1], Claudia Husseneder[2], Chaoliang Lei[1]

[1]Hubei Insect Resources Utilization and Sustainable Pest Management Key Laboratory, Huazhong Agricultural University, Wuhan, China

[2]Department of Entomology, Louisiana State University Agricultural Center, Baton Rouge, Louisiana, United States of America

ABSTRACT

The subterranean termite *Reticulitermes chinensis* Snyder is an important pest of trees and buildings in China. Here, we characterized genetic structure and reproductive modes of *R. chinensis* from China for the first time. A total of 1,875 workers from 75 collection sites in Huanggang, Changsha and Chongqing cities were genotyped at eight microsatellite loci. Analysis of genetic clusters showed two subpopulations in Chongqing city. The Huanggang population showed a uniform genetic pattern and was separated from the other populations by the largest genetic distances (F_{ST}: 0.17–0.20). In contrast, smaller genetic distances (F_{ST}: 0.05–0.12) separated Changsha, Chongqing-1 and Chongqing-2 populations. Chongqing-1 was the only population showing a genetic bottleneck. Isolation by distance among colonies in the Huanggang population indicated limited alate dispersal or colony budding. Lack of isolation by distance among colonies within the populations of Changsha, Chongqing-1 and Chongqing-2, suggested long-range dispersal by alates and/or human-mediated transport. Overall, extended family colonies (73.91%) were predominant in all four populations, followed by simple (20.29%), and mixed family colonies (5.80%). Most simple families were headed by inbred related reproductive pairs in the Changsha population, while most simple families in the Chongqing-1 population were headed by outbred unrelated pairs. Simple families in the Huanggang population were a mixture of colonies headed by outbred or inbred reproductive pairs. The sample size of simple families in the Chongqing-2 population was too small to yield significant results. Extended families in all four populations were headed on the average by ≤10 neotenics. Mixed families likely originated from pleometrosis. Presence of heterozygote genotypes showed that all neotenic reproductives collected in addition from five field colonies in Wuhan city were sexually produced, suggesting that these colonies did not undergo parthenogenesis. This study contributes to better understanding of the variance of genetic structure and reproductive mode in the genus *Reticulitermes*.

INTRODUCTION

An increasing number of studies have employed molecular markers, such as microsatellites, to investigate population genetic structure and colony breeding system in termites, which has improved our understanding of evolutionary genetics, phylogeography, invasion biology, patterns and processes of dispersal, and social organization in termites [1], [2]. The analyses of genetic differentiation and patterns of gene flow within and among termite populations promoted understanding of population dynamics and dispersal patterns [3]–[5], and can help clarify the effect of ecological factors and human disturbance on the levels of genetic differentiation and gene flow [6]–[8]. Moreover, the studies on population genetic structure of invasive termites are helpful to identify potential source populations and to assess whether changes in genetic variability and/or breeding system are associated with invasiveness[9]–[11]. Previous studies on colony breeding structure showed that most populations of subterranean termites were composed of different percentages of simple and extended families[4], [12]–[16], while mixed families were generally less common [6], [17]. Additionally, the number of reproductives and the degree of relatedness among them in termite colonies can also be inferred using molecular markers [3],[18]–[22]. Recent studies on the modes of reproduction using microsatellite genotyping found that queens of *Reticulitermes speratus* and*R. virginicus* can use conditional parthenogenesis, where primary queens produce secondary queens by parthenogenesis (asexual queen succession AQS), but use sexual reproduction to produce workers [23], [24]. Studies on the model of conditional use of sexual and asexual reproduction are helpful to understand the advantages and disadvantages of different reproductive modes in termites [25]. However, it is still unknown how widespread AQS is among *Reticulitermes* spp.

The genus *Reticulitermes* has received much attention among the Isoptera when it comes to population and colony genetic studies, but almost all the species used in these studies were from America and Europe. In Asia, only colony genetic structure of *R. speratus* from Japan was studied using a small number of colonies on a limited spatial scale [26], but population genetic structure of this species has not yet been investigated. Since phylogeographic divisions were found among species of *Reticulitermes* termites across different countries even within Europe [27]–[30], it is necessary to thoroughly investigate and compare population genetic structure and colony breeding system of several representative termite species from different places in Asia. In a phylogenetic analysis of the family Rhinotermitidae, Austin *et al.*[31] found that *R. speratus* and *R. chinensis* were close relatives within the genus*Reticulitermes*, belonging to the same clade. Since *R. speratus* is one of the species featuring AQS, knowledge of the evolution of this unusual breeding strategy could be significantly advanced by studying how widespread AQS is within the genus *Reticulitermes* and if the development of AQS is congruent with phylogeny. Thus, we test the hypothesis that colonies of *R. chinensis* undergo AQS as well.

The subterranean termite *R. chinensis* is widely distributed in China, including Beijing, Tianjin, Shanxi and the Yangtze River drainage basin [32]. This termite species is an important pest of forest trees and urban buildings. To date, some basic information on the biology of *R. chinensis* has been obtained through field observations and laboratory experiments, including the growth of incipient colonies, the formation of neotenics, and the time of alate swarms [33],[34]. Since *R. chinensis* has cryptic foraging and nesting habits with hidden royal chambers, it is very difficult to find and collect reproductives. Thus, population genetic structures and colony breeding systems of *R. chinensis* are still poorly understood. The objectives of this study were to use microsatellite genotyping to (1) describe the population genetic structure of*R. chinensis*, (2) infer colony breeding system, (3) and determine whether colonies of *R. chinensis* undergo AQS as it has been suggested for other *Reticulitermes* species.

RESULTS

Colony Assignments and Population Genetic Structure

STRUCTURE analysis showed that the *R. chinensis* colonies from Huanggang, Changsha and Chonqing cities (Figure 1) belong to two major genetic clusters (K=2, DeltaK=1348.8) but the Evanno plot (Figure S1) also shows the presence of a secondary pattern consisting of four genetic clusters indicating that there might be subpopulations in one of the regions (DeltaK=417.5). The 25 samples collected from Huanggang (Figure 1c) belonged to 24 different colonies. No subpopulations were detected in Huanggang (Figure 2), but significant isolation by distance at 500 m spatial scale was found (Figure 3). The 25 samples collected from Changsha (Figure 1d) belonged to 23 colonies.

STRUCTURE indicated the potential presence of multiple subpopulations (2–4 clusters) among the samples from Changsha (Figure 2). However, assignment of most colonies to any one of the clusters was weak (<80%). Clusters were not congruent with the two geographic clusters of sample sites (Figure 1d) and no significant isolation by distance was detected (Figure 3). Therefore, we considered all Changsha colonies to belong to one population in the further analyses. The 25 samples collected from Chongqing (Figure 1e) belonged to 23 colonies. Results generated from STRUCTURE indicated the presence of two distinct genetic clusters in Chongqing, with 13 colonies assigned to the first subpopulation (Chongqing-1) and 9 colonies assigned to a second subpopulation (Chongqing-2). One colony could not be assigned to either of the two subpopulations of Chongqing (Figure 1e, open triangle) but belonged to the genetic cluster representing the Changsha population (Figure 2). The majority (83%) of colonies were strongly (≥80%) assigned to one or the other subpopulation. Genetic separation was supported by 10 private alleles distinguishing subpopulation Chongqing-1 from Chongqing-2 and 15 private alleles distinguishing Chongqing-2 from Chongqing-1. Thus, colonies from the Chongqing-1 and Chongqing-2 subpopulations were treated as separate populations in the following analyses, thereby increasing the total number of populations to four. There was no significant isolation by distance within either subpopulation (Figure 3). Interestingly, there was no obvious relationship between the genetic clusters and geographical location of the colonies (Figure 1e). Evidence was found for a recent genetic bottleneck in the Chongqing-1 subpopulation, but not in Chongqing-2. The Chongqing-1 subpopulation had significant heterozygote excess under the infinite allele model and the two-phase model (P=0.002) and marginal heterozygote excess under the stepwise mutation model (P=0.097).

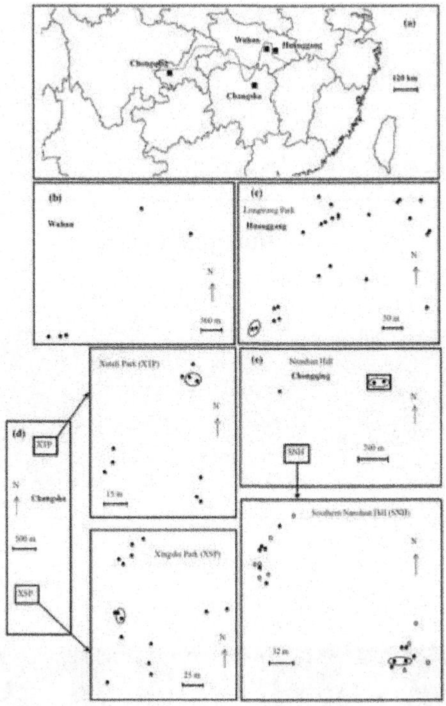

Figure 1: Map of the sample locations for the studies of population structure and reproductive mode.

(a) Overview over the locations of the cities in China where samples were obtained from (Huanggang, Changsha and Chongqing and Wuhan). (b) Sample sites of entire nests including neotenics from Wuhan City, Hubei Province. (c) Sample sites of workers from Longwang Park,Huanggang City, Hubei Province. (d) Sample sites of workers from Xuteli Park and Xingsha Park, Changsha City, Hunan Province. (e) Sample sites of workers from Nanshan Hill, Nanan District, Chongqing City. The dotted line represents the Yangtze River. Genetic analyses revealed two subpopulations in Chongqing city (*filled circle* subpopulation 1; *open circle* subpopulation 2; *Open triangle* representing one colony that could not be assigned to either subpopulation of Chongqing. The loops represent samples assigned to the same colony.

doi:10.1371/journal.pone.0069070.g001

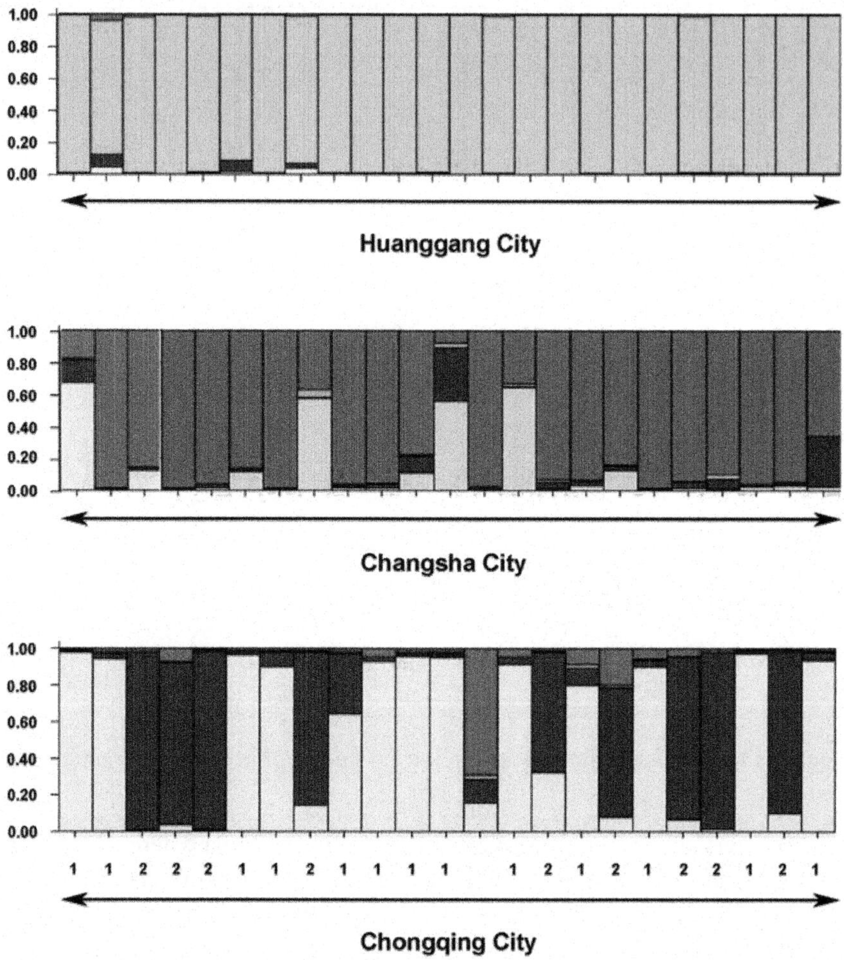

Figure 2: Assignment of individuals (each representing a colony) sampled from three cities to genetic clusters.

Columns represent colonies, each color represent a different genetic cluster defined by STRUCTURE (K=4). The colors in each column represent the likelihood with which a colony is assigned to each genetic cluster. The numbers (1 and 2) in this figure represent the colonies assigned to the Chongqing-1 population and the Chongqing-2 population, respectively.

doi:10.1371/journal.pone.0069070.g002

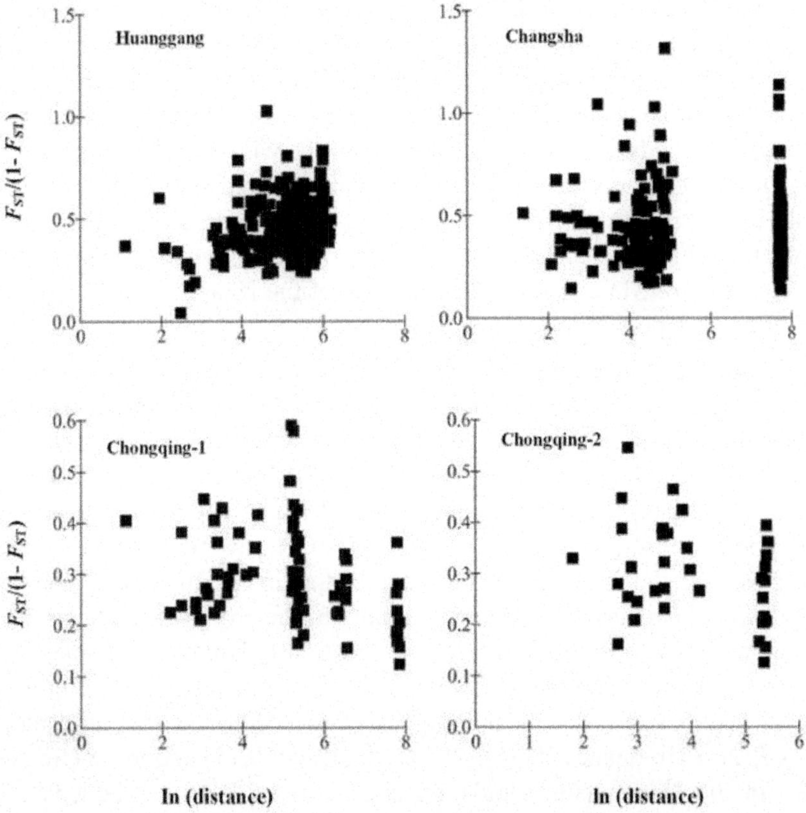

Figure 3: Isolation by distance analysis for the colonies within each population.

The correlation coefficient was significant for the relationship for Huanggang population (r=0.2245, P=0.0014, Mantel test). The correlation coefficients were not significant for either of the relationships (r=0.0130, −0.3284 and −0.3426, P=0.4462, 0.9654 and 0.9015, Mantel test, for Changsha, Chongqing-1 and Chongqing-2 population, respectively).

doi:10.1371/journal.pone.0069070.g003

Permutation tests showed significant genetic differentiation among all four populations (P=0.001, 6000 permutations) confirming the existence of four genetically differentiated groups indicated by STRUCTURE analysis. Genetic distances (F_{ST}) among the four populations of *R. chinensis* ranged from 0.05 to 0.2 (Table S1). The Huangang population was clearly separated from the other populations by large genetic distances ranging from 0.17 to 0.20. Its genetic pattern was uniform showing little contributions to the genetic make-up of the other populations in the dataset. Although the remaining three populations were significantly differentiated from each other according to the permutation tests, the genetic distances ranging from 0.05 to 0.12 between the populations, were rather small in particular between the two Chongqing subpopulations (Table S1). Two colonies from Changsha were assigned to the genetic clusters representing the two Chongqing subpopulations with >80%. One colony from Chongqing could not be assigned to either subpopulation of Chongqing but could be assigned with 70% to the genetic cluster representing the Changsha population (Figure 2). The STRUCTURE results showing two main genetic clusters (Figure S1) confirm the separation of the Huangang population from the other three populations (Figure S2).

Table 1 shows the summary statistics for genetic variability and inbreeding in the populations. Between eight (*RS78*) and 21 (*Rf21-1*) alleles were detected per locus with a mean of 14.38 per locus. The expected heterozygosity ranged from 0.69 to 0.79, and the observed heterozygosity ranged from 0.55 to 0.84 (Table 1). The inbreeding coefficient ranged from −0.1 to 0.21. Overall, the four populations deviated significantly from Hardy–Weinberg equilibrium. There was a deficit of heterozygotes at three loci in the Huanggang population, and at two loci in the Changsha population. This heterozygote deficit at a subset of loci might have been caused by amplification failure of null alleles. Micro-Checker software confirmed deviations from Hardy-Weinberg-Equilibrium due to heterozygote deficit and suggested the occurrence of null alleles at locus *Rf21-1* (frequency: 0.31), *RS15* (frequency: 0.15) and *RS68* (frequency: 0.11) in the Huanggang population and at locus *Rf21-1* (frequency: 0.11) and *RS78* (frequency: 0.28) in the Changsha population. Heterozygote excess was detected across all loci in Chongqing-1 possibly due to a recent genetic bottleneck (see above). No evidence for linkage disequilibrium, large allele dropout or scoring error due to stuttering was detected in any of the four populations.

Table 1: Summary statistics for the four genetically differentiated populations

Population	N	N_A	H_E	H_O	F_{IS}
Huanggang	24	9	0.79	0.65	0.19
Changsha	23	7.5	0.69	0.55	0.21
Chongqing-1	13	5.88	0.76	0.84	−0.1
Chongqing-2	9	5.88	0.75	0.65	0.13

N, number of colonies screened per location; N_A, average number of alleles per locus; H_E, expected heterozygosity; H_O, observed heterozygosity; F_{IS}, inbreeding coefficient.

doi:10.1371/journal.pone.0069070.t001

doi:10.1371/journal.pone.0069070.t001

Colony Genetic Structure and Breeding System

Overall, extended family colonies were predominant (55–83%) in the four populations from Huanggang, Changsha and Chongqing, followed by simple family colonies (17–31%), and mixed family colonies (0–9%). The proportion of extended, simple and mixed family colonies varied among the four populations, with the highest proportion of extended families (83%) in the Huanggang population, the highest proportion of simple families (31%) in the Chongquing-1 population and the most mixed family colonies (9%) in the Chongquing-2 population (Table 2).

Table 2: *F*-statistics and relatedness coefficients for workers from *R. chinensis*colonies from four populations in China and expected for possible breeding systems of subterranean termite colonies previously derived from computer simulations [20], [21]

	F_{IT} (*SE*)	F_{CT} (*SE*)	F_{IC} (*SE*)	*r* (*SE*)
Huanggang				
Simple family colonies (n = 4)	0.14 (0.10)^{a}	0.33 (0.05)	−0.29 (0.06)^{a}	0.58 (0.05)^{a}
Extended family colonies (n = 20)	0.18 (0.08)^{A}	0.30 (0.02)	−0.19 (0.09)^{A}	0.52 (0.02)^{A}
Changsha				
Simple family colonies (n = 4)	0.23 (0.06)^{b}	0.36 (0.05)	−0.20 (0.02)^{b}	0.58 (0.05)^{a}
Extended family colonies (n = 18)	0.14 (0.09)^{C}	0.29 (0.02)	−0.21 (0.11)^{AB}	0.50 (0.03)^{A}
Mixed family colonies (n = 1)				
Chongqing-1				
Simple family colonies (n = 4)	−0.07 (0.06)^{c}	0.19 (0.03)	−0.31 (0.07)^{a}	0.40 (0.07)^{a}
Extended family colonies (n = 8)	0.03 (0.04)^{D}	0.25 (0.03)	−0.28 (0.04)^{C}	0.48 (0.04)^{A}
Mixed family colonies (n = 1)				
Chongqing-2				
Simple family colonies (n = 2)	0.03 (0.10)^{a}	0.29 (0.07)	−0.35 (0.09)^{a}	0.56 (0.10)^{a}
Extended family colonies (n = 5)	0.11 (0.10)^{A}	0.23 (0.02)	−0.16 (0.12)^{AB}	0.42 (0.05)^{A}
Mixed family colonies (n = 2)	−0.06 (0.04)$^{\alpha}$	0.11 (0.03)	−0.19 (0.04)$^{\alpha}$	0.24 (0.06)$^{\alpha}$
Simulated breeding systems				
(A) Simple family colonies with				
(1) outbred unrelated pairs	0^{ac}	0.25	−0.33^{a}	0.5^{a}
(2) inbred related pairs	0.33^{ab}	0.42	−0.14^{ab}	0.62^{a}
(B) Extended family colonies with inbreeding among multiple neotenics				
(1) $N_f = N_m = 1$, $X = 1$	0.26^{AC}	0.65	−0.14^{A}	0.55^{A}
(2) $N_f = 2$, $N_m = 1$, $X = 3$	0.52^{B}	0.59	−0.17^{A}	0.78^{B}
(3) $N_f = N_m = 10$, $X = 1$	0.33^{A}	0.34	−0.01^{A}	0.51^{A}
(4) $N_f = 200$, $N_m = 100$, $X = 3$	0.33^{A}	0.34	0^{B}	0.5^{A}
(C) Mixing between unrelated colonies, $N_f = N_m = 1$, $X = 3$, $p = 0.8$	0.57$^{\beta}$	0.43	0.25$^{\gamma}$	0.55$^{\gamma}$
(D) Mixing between related colonies, $Nf = Nm = 1$, $X = 3$, $p = 0.9$	0.66$^{\beta}$	0.64	0.04$^{\gamma}$	0.77$^{\gamma}$
(E) Pleometrosis				
(1) colonies headed by two queens and one king	0$^{\alpha}$	0.19	-0.23$^{\alpha}$	0.38$^{\alpha}$
(2) colonies headed by two queens and one king, then $N_f = N_m = 10$, $X = 3$	0.27$^{\alpha}$	0.29	-0.03$^{\alpha}$	0.45$^{\alpha}$
(3) colonies headed by five queens and five kings, then $N_f = N_m = 10$, $X = 3$	0.1$^{\alpha}$	0.12	-0.02$^{\alpha}$	0.22$^{\alpha}$

n, number of colonies; *SE*, standard error derived from jackknifing over colonies (or loci if sample size ≤3). One sample *t*-tests were performed using F_{IT}, F_{IC}, and *r* values across individual colonies. Significant differences between empirical values and expected values are indicated by different letters (uppercase for extended families, lowercase for simple families, Greek alphabet lower case for mixed families). For simulated breeding systems, *X* represents generation number of neotenics within a colony; N_f and N_m represent number of replacement females and males per generation respectively; *p* presents mixing proportion between workers from different colonies.

doi:10.1371/journal.pone.0069070.t002

doi:10.1371/journal.pone.0069070.t002

The *F*-statistics and relatedness values estimated from the worker genotypes in each population are shown in Table 2, along with values derived from simulations for different breeding systems computed in previous studies [20], [21], [35]. Mean observed F_{IT} and F_{IC} values of simple family colonies in Changsha were significantly higher than the expected values in outbred unrelated pairs (Table 2, case A1) (all $P<0.05$, one sample *t* test), but not significantly different from the expected F_{IT} and F_{IC} values for inbred related pairs (Table 2, case A2). Thus, most of simple-family colonies in Changsha were headed by monogamous pairs of inbred related reproductives. However, mean observed F_{IT} and F_{IC} values of simple family colonies in Chongqing-1 were significantly and marginally significantly lower than the expected values in inbred related pairs, respectively (Table 2, case A2) ($P<0.01$, one sample *t*test), but not significantly different from the expected F_{IT} and F_{IC} values for outbred unrelated pairs (Table 2, case A1). Therefore, the majority of simple-family colonies in Chongqing-1 were likely headed by monogamous pairs of outbred unrelated reproductives. The empirical values for simple families in Chongqing-2 also matched closest with the assumption of outbred unrelated founder pairs, however, no statistical significances could be achieved due to the small sample size (n=2). In Huanggang, the *F*-statistics and relatedness values of simple family colonies were not significantly different from either of the expected values and the standard errors were large (Table 2, cases A1 and 2). This indicates that some colonies in the Huanggang population might be headed by outbred unrelated pairs and others by inbred related pairs.

The values of F_{IT}, F_{IC}, and *r* of the extended family colonies in the Huanggang population were consistent with those expected for colonies having small numbers (≤10) of neotenic reproductives representing the first generation of replacement reproductives (Table 2, cases B1 and 3), but differed significantly ($P<0.05$, one sample *t* test) from breeding systems assuming three generations of replacement reproductives and/or hundreds of neotenics in at least one of the three genetic parameters (Table 2, cases B2 and 4). In Changsha, the extended family colonies showed F_{IT}, F_{IC}, and *r* values consistent with those expected for colonies headed by a pair of first generation neotenic reproductives (Table 2, case B1), but differed significantly ($P<0.05$, one sample *t* test) from the other breeding systems in at least one of the three genetic parameters (Table 2, cases B2, 3 and 4). In Chongqing-1, the values of F_{IT}, F_{IC}, and *r* for the extended family colonies differed significantly ($P<0.05$, one sample *t*test) from all the four simulated breeding systems in at least one of the three genetic parameters (Table 2, cases B1, 2, 3 and 4), thus no single predominant type (in regard to numbers and generations of neotenics) could be assigned to the extended families in this population. In Chongqing-2, the F_{IT}, F_{IC}, and *r* values in extended family colonies were consistent with the simulated values for colonies headed by the first generation of ≤10 neotenics (Table 2, cases B1, and 3), but differed significantly ($P<0.05$, one sample *t* test) from the other breeding systems with multiple generations and/or high numbers of neotenics in at least one of the three genetic parameters (Table 2, case B2 and 4).

The Changsha and Chongqing-1 populations contained only one mixed family colony each, so$F$-statistics could not be obtained. The F-statistics and relatedness estimates for the two mixed family colonies from Chongqing-2 were consistent with simulated values for pleometrosis (Table 2, case E1, 2 and 3), but F_{IT} differed significantly ($P<0.05$, one sample *t*test) from the breeding systems involving mixing of colonies (Table 2, cases C and D).

Inferring Reproductive Mode of Neotenics' Parents

Of the nine field colonies collected in Wuhan city (Figure 1a) and dissected to obtain the reproductives, four colonies contained single pairs of primary reproductives and were thus simple families. Five colonies (Figure 1b) contained multiple neotenics with a range of 1 to 51 secondary kings and 5 to 74 secondary queens (Table 3). Most of the secondary reproductives developed from nymphs, but there were one male and 5 female ergatoid neotenics (i.e. derived from workers) in colony 5, and two female ergatoid neotenics in colony 2 (Table S2). Neither primary kings nor primary queens were found in addition to the neotenics. We determined that all of the 59 secondary queens and 29 secondary kings genotyped were sexually produced because they were heterozygote at one to eight loci (Table S3). This allows us to test whether these neotenics originated from simple, extended or mixed family colonies. The results showed that the neotenics in colony 1, 3, 4 and 5 had more genotypes than possible for offspring of a monogamous pair or the observed frequencies of the genotypes deviated significantly from those expected in simple families ($P<0.05$, G-test). Thus, the neotenics from these four colonies were offspring of extended family colonies headed by multiple neotenics and as such arose at least two generations after colony foundation (Table 3). However, neotenics in colony 2 had six alleles at locus *Rf21-1*, which suggests that they originated from a mixed family colony (Table 3).

Table 3: Compositions of the five colonies headed by neotenics and reproductive mode of neotenics' parents

Colony	Location	Collection date	No. of secondary queens genotyped/ collected	No. of secondary kings genotyped/ collected	Reproductive mode of neotenics' parents	Breeding system of the colonies of the neotenics' parents
1	Shizi Hill, Wuhan City, China	1 Aug 2011	20/30	3/5	Sexual reproduction	Extended family
2	Shizi Hill, Wuhan City, China	15 Aug 2011	20/74	20/51	Sexual reproduction	Mixed family
3	Shizi Hill, Wuhan City, China	23 Apr 2011	8/8	4/4	Sexual reproduction	Extended family
4	Houshan Hill, Wuhan City, China	20 Apr 2011	6/6	1/1	Sexual reproduction	Extended family
5	Yujia Hill, Wuhan City, China	20 Apr 2011	5/5	1/1	Sexual reproduction	Extended family

doi:10.1371/journal.pone.0069070.t003

doi:10.1371/journal.pone.0069070.t003

DISCUSSION

Large-scale population genetic structure analysis revealed the existence of four populations with significant genetic differentiation in the dataset. The four populations were divided into two main genetic clusters. The population from Huanggang was separated from the other populations by comparatively large genetic distances and showed a uniform genetic pattern, suggesting that there was little contribution of alleles to or from the other populations. The sample sites from Huanggang lie north of the Yangtze River and are thus separated from the sample sites of Changsha and Chongqing, which are located south of the Yangtze River. Thus, geographical barriers, including the Yangtze River in addition to long distances, might limit gene flow between Huanggang and the other three populations. In contrast, the remaining three populations from Changsha, Chongqing-1 and -2 showed some level of gene flow among them. Surprisingly, colonies from Changsha were being assigned to the genetic clusters representing Chongqing populations and *vice versa*, which is indirect evidence for gene flow even across large distances (approx. 640 km). The most likely routes of migrants connecting Changsha and Chongqing are occasional human-mediated movement of infested material from one population to the other through roads and trade network, such as building timbers, ornamental plants, and barges travelling along rivers. Similarly, human transport was considered to be a factor in the gene flow for introduced *R. flavipes* populations between North American and France [9], for *R. urbis* between Balkans and Western Europe [8], and for *Coptotermes formosanus* between Fukue and Kyushu islands [7].

The existence of two populations with slightly different genetic characteristics in the same area in Chongqing was peculiar. The Chongqing-1 population showed signs of a genetic bottleneck, but the Chongqing-2 population was in mutation-drift equilibrium. This difference between the two populations colonizing the same area in Chongqing might be explained by the introduction history. The first introduction to Nanshan Hill in Chongqing took place in the early 1990s when a number of ornamental plants and landscape timbers were transported to Nanshan Hill for constructing Nanshan Arboretum [36]. The second introduction to Nanshan Hill in Chongqing happened in the early 2000s when a lot of building timbers and ornamental plants were transported to construct hotels in Nanshan Hill. To explain the difference between the two populations, we suggest that the earlier introduced population (Chongqing-2) might have contained a larger population size mitigating the initial bottleneck effects and/or enough time since the introduction has passed for the bottleneck to become undetectable, while the more recent introduction (Chongqing-1) still is experiencing the effects of a bottleneck [11]. Relative excess of heterozygosity is a transient effect and is expected to be detectable only for a limited number of generations [37]. Interestingly, bottlenecks could be detected as far back as the 1950's and 60's in another subterranean termite with a known introduction history, *C. formosanus* [11]. Either *R. chinensis* has a higher generation rate accelerating the return to mutation-drift equilibrium, the initial bottlenecks were not as severe as in *C. formosanus* 'intercontinental journey (i.e. the initial founder population size was larger), and/or the small sample size prevented detection of weak bottleneck effects in the Chongqing-1 population [38]. Since there are no conceivable barriers preventing gene flow, we expect the differences between the sympatric populations of Chonqing-1 and -2 to disappear over time.

There was significant isolation by distance among colonies in the Huanggang population, indicating that dispersal by primary reproductives is relatively limited over the spatial scale studied (550 m), either because of short range mating flights by alates and/or frequent colony reproduction by budding. Similarly, significant isolation by distance has been found in two *R. flavipes* populations from France [9]. In contrast, there was no significant isolation by distance among colonies in the Chongqing-1 and Chongqing-2 populations, at the spatial scale studied (about 2.6 km). This lack of isolation by distance suggests that during mating flights reproductives disperse relatively far over the spatial scale studied. Simple-family colonies in Chongqing-1 were headed by monogamous pairs of outbred unrelated reproductives, which further supports the hypothesis of long-range dispersal flights of alates. In addition, the human-mediated introduction and movement of colonies via landscaping and building material and activities might have further masked any isolation by distance imposed by natural dispersal limits.

Similar to the Chongqing populations, the colonies in the Changsha population also did not show isolation by distance at the spatial scale studied (2.4 km). In contrast to the Chongqing colonies, however, simple family colonies in Changsha were headed by inbred related monogamous pairs. This result does not support the existence of long-range mating flights in this population. Alternatively, as suggested for the Chongqing population, human-mediated colony movement may play an important role in population expansion and reducing population viscosity, i.e. isolation by distance, in Changsha.

The four populations of *R. chinensis* in our study consisted of a combination of simple family and extended family colonies. A few mixed family colonies were found in all but the Huanggang population. Overall, extended families with a low number of neotenics and generation turnovers were the dominant breeding system. This finding is consistent with previous studies in *R. flavipes* [9], [13], [21], [39], *Zootermopsis nevadensis nuttingi* [40], *R.*

grassei [35], [41], *R. urbis* [8] and *C. formosanus* in Guangdong, Hunan, Kauai, Maui 1 and Charleston [10], [11], [16]. Moreover, small percentages of mixed family colonies were also detected for *R. flavipes* [13], [39] and *R. grassei* [41]. Proportions of termite colonies with different breeding systems vary across populations and depend on age structure of the colonies, dynamics of colony-colony interactions, food quantity and quality, soil characteristics and disturbance or treatment [4], [14], [20], [21], [42]. There were plenty of 20–30 years old pine trees and camphor trees in the parks and hills of sample collection in Huanggang, Changsha and Chongqing, suggesting that the *R. chinensis* colonies had enough time and food to grow and develop into extended families headed by neotenics [35]. Even more recently established populations, such as the Chongqing populations, showed predominance of extended family colonies, possibly due to disturbance and movement of colonies by human activities, which might lead to splitting colonies, death and replacement of colony founders and an overall increase in extended family colonies in the population as was previously shown in a population of another subterranean termite, *C. formosanus*, which was subjected to disturbance by building and landscaping activities [42].

Simple family colonies in Changsha were headed by inbred related monogamous pairs. This result suggested that dispersal by primary reproductives is limited in the Changsha population as discussed above. Similarly, simple family colonies headed by inbred related monogamous pairs were found in studies of *R. grassei* in southwestern France [35], *Z. n. nevadensis* in North Carolina [40], and *C. formosanus* in Kyushu, Fukue and Hawaii [7], [11], [16]. However, simple family colonies in Chongqing-1 were headed by outbred unrelated monogamous pairs, which suggested that colonies reproduce by relatively long-range mating flights within this population. Similarly, simple family colonies headed by outbred unrelated monogamous pairs were reported in studies of other termite species, including *R. flavipes* in Massachusetts [21]and North Carolina [3], [13], [43], [44], [45], *R. virginicus* in North Carolina [44], and *C. formosanus* in Charleston [10]. Moreover, our results indicated that simple family colonies in Huanggang were a mixture of colonies whose workers were offspring of outbred and unrelated or inbred and related monogamous pairs. A previous study in one *R. grassei* population in southwestern France has also reported that simple family colonies were a mixture of colonies headed by outbred or inbred monogamous pairs [35]. The sample size of only two simple families in the Chongqing-2 population was too small to reject either of the simulated breeding systems with sufficient statistical support.

The strongly negative F_{IC} values for the extended families in Huanggang and Changsha suggest that the mean number of neotenics in our colonies was low (≤10), and that all neotenics were inbred and ultimately the descendants from monogamous pairs of founders. Similarly, extended family colonies were found to be headed by low numbers of inbred neotenics in populations of *R. flavipes* [3], [43], [44], [45], *R. hageni* [45], *R. virginicus* [45], *C. formosanus* in South Carolina, Louisiana, North Carolina, Hawaii, Guangdong and Hunan [10], ,*Z. n. nevadensis* [40], *Z. angusticollis* and *Z. n. nuttingi* [5]. Surprisingly, the extended family colonies in Chongqing-1 were inconsistent with all of thc simulated breeding systems. Similar results have been found in studies of other termite species and were suggested to arise from colonies with multiple reproductives representing a mixture of several breeding systems [35],[40], [43]. Although classified as extended family colonies, the breeding structure estimates for extended family colonies in Chongqing-1 were most similar to values expected for simple family colonies headed by outbred monogamous pairs. One possibility is that these colonies only recently made the transition from simple to extended family by developing neotenics and thus contained still a measurable proportion of offspring of the previous (i.e. simple family) generation [40], [42]. Since colonies of the Chongqing-1 population were probably recently moved by human activities (see above), it is plausible to assume that disturbance might have increased incidental death of the colony founders resulting in recent development of replacement reproductives and thus the transition to extended family colonies [42]. Overall, the results suggested that the extended family colonies in Chongqing-1 contained small numbers of neotenics as shown for the populations of Huanggang and Changsha. The values of F_{IT},F_{IC}, and r for extended families in Chongqing-2 were not significantly different from those for the simulated breeding system based on high numbers of neotenics (Table 2, Case B4), which is possibly due to the low sample size of extended family colonies in this population. Nevertheless, the strongly negative F_{IC} values (−0.16) for the extended families in Chongqing-2 suggested that the number of neotenics in these colonies was low (≤10). Low numbers of neotenics were also found in the extended family colonies of the other three populations (Huanggang, Changsha and Chongqing-1).

Mixed family colonies in Chongqing-2 displayed breeding structure estimates consistent with pleometrosis. Evidence for mixed family colonies originating from pleometrosis was previously found in the lower termite *Z. nevadensis* [40] and the higher termites *Nasutitermes cornigerand* *M. michaelseni* [46], [47]. In pleometrosis, multiple queens and kings (alates) cooperate in colony foundation and expansion during the primary stages of colony development [18], [47]. However, pleometrosis has never been found in any of the other *Reticulitermes* species reported from America and Europe [1]. Some other factors have been suggested to explain mixed family colonies, including colony fusion, invasion of mature colonies by other alates, or sharing foraging galleries by neighboring colonies [40], [43]. For example, mixed family colonies were the result of colony fusions in native and introduced populations of *R. flavipes*[43], [48]. Since the percentage of mixed families was low in *R. chinensis* populations, larger sample sizes are needed to unequivocally determine the origin of mixed family colonies.

In addition to information about breeding systems in *R. chinensis* colonies derived from worker genotypes, we also directly observed number and types of reproductives in nine field colonies collected from Wuhan. Four of the dissected field colonies were simple families headed by a pair of primary kings and queens, and five colonies contained multiple neotenics. Three of the five colonies had low numbers of reproductives (6–12). The range is largely consistent with the results from the genetic analyses of worker genotypes from the populations of Huanggang, Changsha and Chongqing. The percentage of our collected field colonies containing secondary kings in *R. chinensis* was higher than those in *R. speratus* and *R. virginicus* [23], [24]. In *R. speratus,* secondary kings have been found in only 6.67 percent of colonies (2 of 30 colonies;[23]). Secondary kings were not found at all in five colonies of *R. virginicus* [24]. In contrast, each colony of *R. chinensis* in the present study had at least one secondary king and up to 51 secondary kings in colony 2. Almost all the neotenics in *R. speratus* and *R. virginicus*differentiated from nymphs [23], [24], but there were eight ergatoid neotenics in *R. chinensis* in this study. Our results based on the inferred genotypes of the neotenics' parents in each of the five colonies showed that the preceding parental generation in the five colonies also consisted of multiple reproductives. The breeding system of the five colonies headed by the neotenics' parents was in four cases an extended family colony headed by neotenics and in one case a mixed family colony. These results suggest that extended family colonies headed by neotenics and mixed family colonies can continue the colony life cycle by producing the next generation of neotenics in *R. chinensis*.

Our results showed that all the neotenics in the five field colonies of *R. chinensis* were sexually produced, suggesting that these colonies did not undergo conditional parthenogenesis. Based on the classical model of AQS [23], we did not expect the ergatoid neotenics to be produced parthenogenetically since they are derived from workers, which are almost exclusively sexually produced in other *Reticulitermes* species that undergo AQS [24]. We also did not necessarily expect the secondary kings to be parthenogens. Queens usually increase their reproductive output and fitness by producing multiple replacement daughters via parthenogenesis, which at least in *R. speratus* can continue the queen's full genetic contribution throughout generations by producing parthenogenetic daughters themselves. However, none of the nymphoid secondary queens in *R. chinensis* was homozygote at all loci.

In contrast to our findings in *R. chinensis*, the seven colonies of *R. speratus* in Kyoto [23] and the five colonies of *R. virginicus* in North Carolina and Texas [24] have been shown to undergo AQS. A previous phylogenetic analysis of the family Rhinotermitidae found that *R. speratus*and *R. virginicus* belonged to distinctly different clades and were thus not closely related, which prompted the hypothesis that AQS might have arisen early in the evolution of the genus*Reticulitermes*. However, the same study showed that *R. chinensis* is a close relative of *R. speratus* with both species sharing the same clade [31]. Nevertheless their reproductive mode differs since *R. chinensis* apparently does not undergo AQS. Therefore, AQS has either evolved independently in *R. speratus* and *R. virginicus* [24] or *R. chinensis* secondarily lost the ability to undergo AQS.

Clinal variation of ecological factors, such as temperature, moisture, seasonality, availability of wood and soil composition have been shown to influence breeding structure in *Reticulitermes*species [49], however, geographic parthenogenesis has not yet been studied in termites. In facultative parthenogenic plants and animals in general, asexual populations tend to occur at higher latitudes and altitudes and in disturbed habitats [50]. Previous studies in populations of the stick insect, *Clitarchus hookeri* showed that the incidence of parthenogenesis increased at higher latitudes both in the Northern and Southern Hemisphere [51]. The colonies of *R. chinensis* in our study were collected from lower latitudes than the colonies of *R. speratus* and*R. virginicus* (>35°N) that exhibited parthenogenesis, except for one *R. virginicus* colony from South Houston (<30°N) [23], [24]. Thus, it would be interesting to study whether *R. chinensis*colonies located further north and at higher altitudes than the ones in this study display AQS or if parthenogenesis simply does not occur in this species regardless of the local ecological conditions.

MATERIALS AND METHODS

Sampling for Population Genetic Study

Between May and July 2010, samples of *R. chinensis* were collected from fallen trees and stumps from 75 collection sites in three geographic regions of China (Figure 1a): Longwang Park, Huanggang city, Hubei Province (Figure 1c); Xuteli Park and Xingsha Park, Changsha city, Hunan Province (Figure 1d); and Nanshan Hill, Nanan District, Chongqing city (Figure 1e). All the parks and hills of sample collection were inside cities. The geographic distance was 300 km, 650 and 800 km between Huanggang and Changsha, between Changsha and Chongqing, and between Chongqing and Huanggang, respectively. The latitudes ranging from 28°14′20.1″ to 30°27′34.5″ and longitudes ranging from 106°36′53.8″ to 114°52′00.3″ of all sample sites were determined using a GPS localizer (Table S4). More than 30 workers per sample were collected in the field and immediately preserved in 95% ethanol. A few soldiers per sample were also collected in order to identify the species.

None of the parks and hills where samples were collected are privately-owned or protected in any way, and *R. chinensis* is not endangered or protected. Thus, no specific permissions were required to access and sample at these locations.

Microsatellite Genotyping

DNA was extracted from whole worker bodies using the DNeasy Tissue Kit (Beijing Dingguo Changsheng Biotech Co. Ltd). Twenty-five workers from each collection site were genotyped at 8 microsatellite loci (Table S5), identified from *R. flavipes* [52], [53]. PCR amplifications were performed as described in Vargo [52]. PCR products were separated by electrophoresis on 6% polyacrylamide gels run on a LI-COR 4300 DNA analyzer. Alleles were scored using the computer program SAGA^{GT} (LI-COR, Inc.). The software Micro-Checker [54] was used to test loci for null alleles and possible scoring errors derived from large allele dropout and the presence of microsatellite stutter bands.

Colony Identification

Samples from the 25 collection sites in each region were tested for significant genotypic differentiation using log-likelihood *G*-Statistics (FSTAT, [55]). *P* values were obtained by permutations of the multilocus genotypes between each pair of samples and standard Bonferroni corrections were used. If samples significantly differed from each other, they were assigned to different colonies [11], [14].

Population Genetic Structure

Since colony members are close kin, only a single randomly chosen individual per colony was used for population genetic analyses to avoid bias by nonindependent genotypes. The number of genetic clusters in the three geographic regions was assessed on the basis of a comparison of the penalized log likelihoods over independent simulation runs using STRUCTURE 2.3.3 with variable numbers of assumed genetic clusters (*K*) [56]. All simulations were done using the default settings, i.e. the admixture model and correlated allele frequencies, with 100,000 runs in the data collection phase following a burn-in period of 100,000 runs [11]. Three repeats of simulations were run for each model with different numbers of assumed genetic clusters (K=1 to 7). The number of genetic clusters was determined following the Delta K method of Evanno *et al.* [56] using STRUCTURE HARVESTER v.0.6 [57]. According to the methods of Pritchard *et al.* [58], the estimated membership coefficients in K clusters (degree of admixture) of each multilocus genotype (i.e. individual representing a colony) were determined and were graphically displayed in form of a histogram. Colonies from the three geographical regions were probabilistically assigned to the populations for which they showed dominant membership (>80%). As long as considerable structure was found, datasets would be separated into subsets and reanalyzed until the number of subpopulations was determined. Confirmed subpopulations were treated as genetically differentiated populations. Pairwise genetic differentiation among populations at the 1/1000 nominal level was confirmed via permutation tests using multilocus G-statistics with Bonferroni corrections as implemented in FSTAT [55]to obtain the final number of populations in the data set.

To estimate isolation by distance, pairwise F_{ST} values for all pairs of colonies within each population were obtained. The matrix correlation between $F_{ST}/(1-F_{ST})$ and the ln of geographic distance (m) for all pairs of colonies within each population was determined [10]. The significance of the correlation coefficient was estimated using Mantel tests (one sided *p*-values from 10,000 permutations, TFPGA v. 1.3, [59]. In addition, genetic differentiation (F_{ST}) between each pair of populations was computed based on a hierarchical analysis using the program GENETIC DATA ANALYSIS version 1.1 (GDA, [60]). Exact tests for Hardy–Weinberg equilibrium for each locus and linkage disequilibrium between all pairs of loci were performed using the program GDA (GDA, [60]) with 3200 shufflings [7]. To determine whether populations had experienced a recent genetic bottleneck, worker genotypes were tested for heterozygosity excess across loci based on a Wilcoxon sign-rank test under three mutation models (infinite allele model, two-phased model of mutation, and stepwise mutation model) using the program Bottleneck v. 1.2.02 [37].

Colony Genetic Structure and Breeding System

Referring to the previous methods [20]–[21], [43], colonies were classified into one of three family types: simple, extended or mixed families. Colonies would be regarded as simple families if the numbers of genotypes of the workers were compatible with those expected for offspring of a monogamous pair and if the frequencies of the observed genotypes were not different from the expected Mendelian ratios. Colonies would be regarded as extended families if there were more than four genotypes, or if genotype frequencies were significantly different from expected Mendelian ratios (P<0.05, G-test). Mixed family colonies would be characterized by the presence of more than four alleles at one or more loci.

Colony genetic structure was analyzed by F-statistics using the methods of Weir & Cockerham[61] as implemented in FSTAT [55]. Separate analyses were performed for each population to assure that F values ($F_{IT,}$ F_{CT} and F_{IC}) used to infer breeding system were not confounded by higher-level genetic structure [7], [20], [21]. The coefficient of relatedness (r) was determined by the average nest mate relatedness within all colonies of the respective population using the program FSTAT [55]. These empirical values were then compared to expected values resulting from previously compiled computer simulations of possible breeding systems of termites via one-sample t-tests not assuming equal variances (P<0.05), to infer predominant breeding systems in the populations of this study [20], [21], [35].

Inferring Reproductive Mode of Neotenics' Parents

To infer reproductive mode of neotenics' parents, that is, parthenogenesis versus sexual reproduction in *R. chinensis*, we used the methods of Matsuura *et al.* [23] and Vargo *et al.* [24]to collect field colonies and conduct microsatellite analysis. Between July 2010 and September 2011, we collected four field colonies with primary reproductives (Figure S3a) and five field colonies with neotenics (Figure S3b) in Wuhan city (Figure 1a). All reproductives from each colony were immediately preserved in 95% ethanol in a vial for subsequent microsatellite analysis. We discriminated secondary neotenic reproductives from primary reproductives based on the lighter body color and absence of eyes in neotenics. Sex of all reproductives was identified based on the conformation of the caudal sternites under a stereoscope [62]. According to the presence or absence of wing pads, neotenics would be divided into nymphoid and ergatoid neotenics [23].

We genotyped up to 20 male and female neotenics (nymphoid and ergatoid) from five field colonies collected at Shizi Hill (colony 1, 2 and 3), Houshan Hill (colony 4), and Yujia Hill (colony 5) in Wuhan city using the eight microsatellite loci and protocols described above (Table S3). The latitudes ranging from 28°28′49.86″ to 30°32′15.3″ and longitudes ranging from 114°20′58.38″ to 114°24′9.66″ of the five field colonies were also determined using a GPS localizer (Figure 1b). If neotenics were homozygous at all loci, they would be considered to be produced by parthenogenesis. If neotenics were heterozygous at one or more loci, they would be considered to be sexually produced [23], [24]. For sexually produced neotenics, we tested whether these neotenics are offspring of simple, extended or mixed family colonies using the same methods as described for workers above [20]–[21], [43].

ACKNOWLEDGMENTS

We are thankful to Xintian Cai and Bing Zhou for their help in field collection. We thank the technical assistance for genotyping from Double Company. We thank Drs. Xianchun Li, Lane D. Foil and the anonymous reviewers for providing valuable comments on earlier drafts of this manuscript.

AUTHOR CONTRIBUTIONS

Conceived and designed the experiments: QH GL CH CL. Performed the experiments: QH GL. Analyzed the data: QH GL CH CL. Contributed reagents/materials/analysis tools: QH GL CH CL. Wrote the paper: QH GL CH CL. Designed the software used in analysis: QH CH GL CL.

REFERENCES

1. Vargo EL, Husseneder C (2009) Biology of subterranean termites: insights from molecular studies of *Reticulitermes* and *Coptotermes*. Annu Rev Entomol 54: 379–403. doi: 10.1146/annurev.ento.54.110807.090443
2. Vargo EL, Husseneder C (2011) Genetic structure of termite colonies and populations. In *Biology of Termites: A Modern Sythesis* (eds D. E. Bignell, Y. Roisin & N. Lo), 321–348. Dordrecht, The Netherlands: Springer Science+Business Media B.V.
3. Vargo EL (2003) Hierarchical analysis of colony and population genetic structure of the eastern subterranean termite, *Reticulitermes flavipes*, using two classes of molecular markers. Evolution 57: 2805–2818. doi: 10.1111/j.0014-3820.2003.tb01522.x
4. Husseneder C, Messenger MT, Su NY, Grace JK, Vargo EL (2005) Colony social organization and population genetic structure of an introduced population of formosan subterranean termite from New Orleans, Louisiana. J Econ Entomol 98: 1421–1434. doi: 10.1603/0022-0493-98.5.1421
5. Booth W, Brent CS, Calleri DV, Rosengaus RB, Traniello JFA, et al. (2012) Population genetic structure and colony breeding system in dampwood termites (*Zootermopsis angusticollis* and *Z. nevadensis nuttingi*). Insectes Soc 59: 127–137. doi: 10.1007/s00040-011-0198-2
6. Goodisman MAD, Crozier RH (2002) Population and colony genetic structure of the primitive termite *Mastotermes darwiniensis*. Evolution 56: 70–83. doi: 10.1554/0014-3820(2002)056[0070:pacgso]2.0.co;2

7. Vargo EL, Husseneder C, Grace JK (2003) Colony and population genetic structure of the Formosan subterranean termite, *Coptotermes formosanus*, in Japan. Mol Ecol 12: 2599–2608. doi: 10.1046/j.1365-294x.2003.01938.x
8. Leniaud L, Dedeine F, Pichon A, Dupont S, Bagnères AG (2010) Geographical distribution, genetic diversity and social organization of a new European termite,*Reticulitermes urbis* (Isoptera: Rhinotermitidae). Biol Invasions 12: 1389–1402. doi: 10.1007/s10530-009-9555-8
9. Dronnet S, Chapuisat M, Vargo EL, Lohou C, Bagnères AG (2005) Genetic analysis of the breeding system of an invasive subterranean termite, *Reticulitermes santonensis*, in urban and natural habitats. Mol Ecol 14: 1311–1320. doi: 10.1111/j.1365-294x.2005.02508.x
10. Vargo EL, Husseneder C, Woodson D, Waldvogel MG, Grace JK (2006) Genetic analysis of colony and population structure of three introduced populations of the Formosan subterranean termite (Isoptera: Rhinotermitidae) in the continental United States. Environ Entomol 35: 151–166. doi: 10.1603/0046-225x-35.1.151
11. Husseneder C, Simms DM, Delatte JR, Wang C, Grace JK, et al. (2012) Genetic diversity and colony breeding structure in native and introduced ranges of the Formosan subterranean termite, *Coptotermes formosanus*. Biol Invasions 14: 419–437. doi: 10.1007/s10530-011-0087-7
12. Copren KA (2004) The genetic and social structure of the western subterranean termite, *Reticulitermes hesperus* (Isoptera: Rhinotermitidae). PhD Thesis, University of California, Davis, CA.
13. Vargo EL, Carlson JC (2006) Comparative study of breeding systems of sympatric subterranean termites (*Reticulitermes flavipes* and *R.hageni*) in central North Carolina using two classes of molecular genetic markers. Environ Entomol 35: 173–187. doi: 10.1603/0046-225x-35.1.173
14. Husseneder C, Simms DM, Riegel C (2007) Evaluation of treatment success and patterns of reinfestation of the Formosan subterranean termite (Isoptera: Rhinotermitidae). J Econ Entomol 100: 1370–1380. doi: 10.1603/0022-0493(2007)100[1370:eotsap]2.0.co;2
15. Thompson GJ, Lenz M, Crozier RH, Crespi BJ (2007) Molecular-genetic analyses of dispersal and breeding behaviour in the Australian termite *Coptotermes lacteus*: evidence for non-random mating in a swarm-dispersal mating system. Aust J Zool 55: 219–227. doi: 10.1071/zo07023
16. Husseneder C, Powell JE, Grace JK, Vargo EL, Matsuura K (2008) Worker size in the Formosan subterranean termite in relation to colony

breeding structure as inferred from molecular markers. Environ Entomol 37: 400–408. doi: 10.1603/0046-225x(2008)37[400:wsitfs]2.0.co;2

17. DeHeer CJ, Vargo EL (2008) Strong mitochondrial DNA similarity but low relatedness at microsatellite loci among families within fused colonies of the termite *Reticulitermes flavipes*. Insectes Soc 55: 190–199. doi: 10.1007/s00040-008-0999-0
18. Atkinson L, Adams ES (1997) The origins and relatedness of multiple reproductives in colonies of the termite *Nasutitermes corniger*. Proc R Soc Lond B 264: 1131–1136. doi: 10.1098/rspb.1997.0156
19. Husseneder C, Brandl R, Epplen C, Epplen JT, Kaib M (1998) Variation between and within colonies in the termite: morphology, genomic DNA, and behaviour. Mol Ecol 7: 983–990. doi: 10.1046/j.1365-294x.1998.00416.x
20. Thorne BL, Traniello JFA, Adams ES, Bulmer M (1999) Reproductive dynamics and colony structure of subterranean termites of the genus *Reticulitermes* (Isoptera Rhinotermitidae): a review of the evidence from behavioral, ecological, and genetic studies. Ethol Ecol Evol 11: 149–169. doi: 10.1080/08927014.1999.9522833
21. Bulmer M S, Adams ES, Traniello JFA (2001) Variation in colony structure in the subterranean termite *Reticulitermes flavipes*. Behav Ecol Sociobiol 49: 236–243. doi: 10.1007/s002650000304
22. Ross KG (2001) Molecular ecology of social behaviour: analyses of breeding systems and genetic structure. Mol Ecol 10: 265–284. doi: 10.1046/j.1365-294x.2001.01191.x
23. Matsuura K, Vargo EL, Kawatsu K, Labadie PE, Nakano H, et al. (2009) Queen Succession Through Asexual Reproduction in Termites. Science 323: 1687. doi: 10.1126/science.1169702
24. Vargo EL, Labadie PE, Matsuura K (2012) Asexual queen succession in the subterranean termite *Reticulitermes virginicus*. Proc R Soc Lond B 279: 813–819. doi: 10.1098/rspb.2011.1030
25. Matsuura K (2011) Sexual and asexual reproduction in termites. In *Biology of Termites: A Modern Sythesis* (eds D. E. Bignell, Y. Roisin & N. Lo), 255–277. Dordrecht, The Netherlands: Springer Science+Business Media B.V.
26. Hayashi Y, Kitade O, Gonda M, Kondo T, Miyata H, et al. (2005) Diverse colony genetic structures in the Japanese subterranean termite *Reticulitermes speratus*(Isoptera: Rhinotermitidae). Sociobiology 46: 175–184. doi: 10.13102/sociobiology.v60i4.446-452

27. Clément JL, Bagnères AG, Uva P, Wilfert L, Quintana A, et al. (2001) Biosystematics of *Reticulitermes* termites in Europe: morphological, chemical and molecular data. Insectes Soc 48: 202–215. doi: 10.1007/pl00001768

28. Marini M, Mantovani B (2002) Molecular relationships among European samples of*Reticulitermes* (Isoptera, Rhinotermitidae). Mol Phylogenet Evol 22: 454–459. doi: 10.1006/mpev.2001.1068

29. Kutnik M, Uva P, Brinkworth L, Bagnères AG (2004) Phylogeography of two European*Reticulitermes* (Isoptera) species: the Iberian refugium. Mol Ecol 13: 3099–3113. doi: 10.1111/j.1365-294x.2004.02294.x

30. Luchetti A, Marini M, Mantovani B (2005) Mitochondrial evolutionary rate and speciation in termites: data on European Reticulitermes taxa (Isoptera, Rhinotermitidae). Insectes Soc 52: 218–221. doi: 10.1007/s00040-005-0806-5

31. Austin JW, Szalanski AL, Cabrera BJ (2004) Phylogenetic analysis of the subterranean termite family Rhinotermitidae (Isoptera) by using the mitochondrial cytochrome oxidase II gene. Ann Entomol Soc Am 97: 548–555. doi: 10.1603/0013-8746(2004)097[0548:paotst]2.0.co;2

32. Huang FS, Zhu SM, Ping ZM, He XS, Li GX, et al.. (2000) Fauna Sinica: Insecta, Vol. 17: Isoptera. Beijing: Science Press.

33. Pan YZ, Liu YZ, Tang GQ (1990) The establisment and development of colonies in*Reticulitermes chinensis* Snyder. Acta Entomol Sin 33: 200–206.

34. Liu YZ, Tang TY (1994) The colony development and growth by substitute reproductives of *Reticulitermes chinensis*.. Acta Entomol Sin 37: 38–43.

35. DeHeer CJ, Kutnik M, Vargo EL, Bagnères AG (2005) The breeding system and population structure of the termite *Reticulitermes grassei* in Southwestern France. Heredity 95: 408–415. doi: 10.1038/sj.hdy.6800744

36. Yang WT, Zhang JL (2011) Analysis Landscape of Plants in Nanshan Botanical Park. South China Agriculture 5: 1–5.

37. Piry S, Luikart G, Cornuet JM (1999) BOTTLENECK: a computer program for detecting recent reductions in the effective size using allele frequency data. J Hered 90: 502–503. doi: 10.1093/jhered/90.4.502

38. Peery ZM, Kirby R, Reid BN, Stoelting R, Doucet-Bëer E, Robinson S, et al. (2012) Reliability of genetic bottleneck tests for detecting recent population declines. Mol Ecol 21 3403–3418. doi: 10.1111/j.1365-294x.2012.05635.x

39. DeHeer CJ, Kamble ST (2008) Colony genetic organization, fusion and inbreeding in*Reticulitermes flavipes* from the Midwestern U.S. Sociobiology. 51: 307–325.

40. Aldrich BT, Kambhampati S (2007) Population structure and colony composition of two*Zootermopsis nevadensis* subspecies. Heredity 99: 443–451. doi: 10.1038/sj.hdy.6801022

41. Nobre T, Nunes L, Bignell DE (2008) Colony interactions in *Reticulitermes grassei*population assessed by molecular genetic methods. Insectes Soc 55: 66–73. doi: 10.1007/s00040-007-0971-4

42. Aluko GA, Husseneder C (2007) Colony dynamics of the Formosan subterranean termite in a frequently disturbed urban landscape. J Econ Entomol 100: 1037–1046. doi: 10.1603/0022-0493(2007)100[1037:cdotfs]2.0.co;2

43. DeHeer CJ, Vargo EL (2004) Colony genetic organization and colony fusion in the termite *Reticulitermes flavipes* as revealed by foraging patterns over time and space. Mol Ecol 13: 431–441. doi: 10.1046/j.1365-294x.2003.2065.x

44. Vargo EL (2003) Genetic structure of *Reticulitermes flavipes* and *R. virginicus*(Isoptera, Rhinotermitidae) colonies in an urban habitat and tracking of colonies following treatment with hexaflumuron bait. Environ Entomol 32: 1271–1282. doi: 10.1603/0046-225x-32.5.1271

45. Parman V, Vargo EL (2008) Population density, species abundance, and breeding structure of subterranean termite colonies in and around infested houses in central North Carolina. J Econ Entomol 101: 1349–1359. doi: 10.1603/0022-0493(2008)101[1349:pdsaab]2.0.co;2

46. Thorne BL (1984) Polygyny in the Neotropical termite *Nasutitermes corniger*: life history consequences of queen mutualism. Behav Ecol Sociobiol 14: 117–136. doi: 10.1007/bf00291903

47. Hacker M, Kaib M, Bagine R, Epplen J, Brandl R (2005) Unrelated queens coexist in colonies of the termite *Macrotermes michaelseni*. Mol Ecol 14: 1527–1532. doi: 10.1111/j.1365-294x.2005.02507.x

48. Perdereau E, Bagnères AG, Dupont S, Dedeine F (2010) High occurrence of colony fusion in a European population of the American termite *Reticulitermes flavipes*. Insect Soc 57: 393–402. doi: 10.1007/s00040-010-0096-z

49. Vargo EL, Leniaud L, Swoboda LE, Diamond SE, Michael D, et al.. (2013) Clinal variation in colony breeding structure and level of inbreeding in the subterranean termites *Reticulitermes flavipes* and *R. grassei*. Mol Ecol: doi: 10.1111/mec.12166.

50. Glesener RR, Tilman D (1985) Sexuality and the components of environmental uncertainty: Clues from geographic parthenogenesis is terestrial animals. Am Nat 112: 659–673. doi: 10.1086/283308

51. Morgan-Richards M, Trewick SA, Stringer IAN (2010) Geographic parthenogenesis and the common tea-tree stick insect of New Zealand. Mol Ecol 19: 1227–1238. doi: 10.1111/j.1365-294x.2010.04542.x

52. Vargo EL (2000) Polymorphism at trinucleotide microsatellite loci in the subterranean termite *Reticulitermes flavipes*. Mol Ecol 9: 817–820. doi: 10.1046/j.1365-294x.2000.00915.x

53. Dronnet S, Bagnères AG, Juba TR, Vargo EL (2004) Polymorphic microsatellite loci in the European subterranean termite, *Reticulitermes santonensis* Feytaud. Mol Ecol Notes 4: 127–129. doi: 10.1111/j.1471-8286.2004.00600.x

54. Van Oosterhout C, Hutchingson WF, Wills DPM, Shipley P (2004) Micro-Checker: software for identifying and correcting microsatellite data. Mol Ecol Notes 4: 535–538. doi: 10.1111/j.1471-8286.2004.00684.x

55. Goudet J (2001) FSTAT, a program to estimate and test gene diversities and fixation indices (version 2.9.3). Unil website. Available:http://www.unil.ch/izea/softwares/fstat.html. Accessed 2012 Dec 23.

56. Evanno G, Regnaut S, Goudet J (2005) Detecting the number of clusters of individuals using the software structure: a simulation study. Mol Ecol 14: 2611–2620. doi: 10.1111/j.1365-294x.2005.02553.x

57. Earl DA, vonHoldt BM (2012) STRUCTURE HARVESTER: a website and program for visualizing STRUCTURE output and implementing the Evanno method. Conservation Genetics Resources 4: 359–361. doi: 10.1007/s12686-011-9548-7

58. Pritchard JK, Stephens M, Donnelly P (2000) Inference of population structure using multilocus genotype data. Genetics 155: 945–959.

59. Miller MP (1997) Tools for population genetic analyses (TFPGA) 1.3: A Windows program for the analysis of allozyme and molecular population genetic data. *February 2000.* Usu website. Available: http://bioweb.usu.edu/mpmbio. Accessed 2012 Nov 18.

60. Lewis PO, Zaykin D (2000) Genetic Data Analysis: Computer program for the analysis of allelic data, Version 1.0 (d12). Uconn website. Available:http://lewis.eeb.uconn.edu/lewishome/software.html. Accessed 2012 Jun 16.

61. Weir B, Cockerham CC (1984) Estimating F-statistics for the analysis of population structure. Evolution 38: 1358–1370. doi: 10.2307/2408641

62. Roonwal ML (1975) Sex ratio and sexual dimorphism in termites. J Sci Ind Res india 34: 402–416.

Chapter 8

GENETIC ANALYSIS OF COLD TOLERANCE AT THE GERMINATION AND BOOTING STAGES IN RICE BY ASSOCIATION MAPPING

Yinghua Pan[1,3], Hongliang Zhang[1], Dongling Zhang[2], Jinjie Li[1], Haiyan Xiong[1], Jianping Yu[1], Jilong Li[1], Muhammad Abdul Rehman Rashid[1], Gangling Li[1], Xiaoding Ma[2], Guilan Cao[2], Longzhi Han[2], Zichao Li[1]

[1]Key Laboratory of Crop Heterosis and Utilization, Ministry of Education, Beijing Key Laboratory of Crop Genetic Improvement, China Agricultural University, Beijing, 100193, China

[2]Institute of Crop Science, Chinese Academy of Agricultural Sciences, the National Key Facility for Crop Gene Resources and Genetic Improvement, Key Laboratory of Crop Germplasm Resources and Biotechnology, Ministry of Agriculture, Beijing, 100081, China

[3]Rice Research Institute, Guangxi Academy of Agricultural Sciences, Nanning, Guangxi, 530005, China

ABSTRACT

Low temperature affects the rice plants at all stages of growth. It can cause severe seedling injury and male sterility resulting in severe yield losses. Using a mini core collection of 174 Chinese rice accessions and 273 SSR markers we investigated cold tolerance at the germination and booting stages, as well as the underlying genetic bases, by association mapping. Two distinct populations, corresponding to subspecies *indica* and *japonica* showed evident differences in cold tolerance and its genetic basis. Both subspecies were sensitive to cold stress at both growth stages. However, *japonica* was more tolerant than *indica* at all stages as measured by seedling survival and seed setting. There was a low correlation in cold tolerance between the germination and booting stages. Fifty one quantitative trait loci (QTLs) for cold tolerance were dispersed across all 12 chromosomes; 22 detected at the germination stage and 33 at the booting stage. Eight QTLs were identified by at least two of four measures. About 46% of the QTLs represented new loci. The only QTL shared between *indica* and*japonica* for the same measure was *qLTSSvR6-2* for SSvR. This implied a complicated mechanism of old tolerance between the two subspecies. According to the relative genotypic effect (RGE) of each genotype for each QTL, we detected 18 positive genotypes and 21 negative genotypes in *indica*, and 19 positive genotypes and 24 negative genotypes in*japonica*. In general, the negative effects were much stronger than the positive effects in both subspecies. Markers for QTL with positive effects in one subspecies were shown to be effective for selection of cold tolerance in that subspecies, but not in the other subspecies. QTL with strong negative effects on cold tolerance should be avoided during MAS breeding so as to not cancel the effect of favorable QTL at other loci.

INTRODUCTION

Low temperature is one of the major abiotic stresses that threaten the adaptability of rice and its production. Due to the long growing season and frequent low temperature in North China, Korea, Japan and other countries at high latitudes, the growth duration of cultivars that cannot tolerate low temperature has to be shortened and this usually results in low yields. Cold stresses that often occur during flowering and grain filling all over the world [1–2] may also lower grain quality and yield. Cold tolerance of rice is a complex quantitative trait significantly influenced by environment. Thus research on the underlying genetic mechanisms and discovery of genes affecting to cold tolerance in rice will be helpful for developing elite cultivars with strong cold tolerance.

Segregating populations are usually used for mapping and cloning genes for cold tolerance. More than 30 quantitative trait loci (QTLs) for cold tolerance have been mapped on 12 chromosomes (according to www.gramene.com). Most of them were for cold tolerance at the germination stage, and a few were cloned. Using backcross inbred lines of Livorno/Haymasari, the QTL *qLTG3–1* was cloned and proven to contribute to the high germinability under low temperatures [3]. Mapping QTL for cold tolerance at the booting stage is more difficult because of difficulties in phenotyping and the underlying complex biological and genetic mechanisms involved. Eight QTLs related to spikelet fertility under cold stress were mapped on chromosomes 1, 4, 5, 7, 10 and 11 based on NILs derived from KMXBG as donor and Towada as the recipient [4–5]. Saito et al. mapped the QTL *Ctb1* encoding an F-box protein and contributing to normal anther development under cold stress [6].

In contrast to traditional QTL mapping, “association mapping” is an alternative approach for studying complex traits. Association mapping, also known as linkage disequilibrium mapping, utilizes allelic variation in natural populations, and is capable of simultaneously identifying many loci for multiple traits [7]. Association mapping has been employed in rice to identify QTLs or genes related to yield [8–10], heading date [11], disease resistance [12–13] and tolerance to abiotic stresses [14].

Using 174 rice accessions from the mini core collection of cultivated rice and 273 simple-sequence repeat (SSR) markers we detected QTLs associated with cold tolerance during the germination and booting stages. We subsequently screened the accessions with strong cold tolerance and QTL markers showing strong effects, which will be the potential parents and markers in developing cultivars with cold tolerance by molecular marker assisted selection.

MATERIALS AND METHODS

Plant Materials

The research material comprised 174 diverse accessions from the mini core collection of cultivated rice [15]. Among them, 118 were landraces, 56 were modern improved cultivars and 109 were subspecies *indica*, and 65 were *japonica*.

Evaluation of Cold Tolerance at the Germination Stage

Fifty seeds of each accession were placed in a drying oven at 50°C for 48 h to break dormancy, and then soaked in distilled water in a 15 cm Petri dish for 24 h. The pre-soaked seeds were germinated in a growth chamber at a constant temperature of 32°C for 36 h. Germinated seeds with 5 mm coleoptiles were stressed at 5°C for 10 days, and then moved to a greenhouse at 20°C for 10 days to allow seedlings to recover and resume normal growth. Seedling survival rates (SSvR) were then scored as a measure of cold tolerance at the germination stage.

Evaluation of Cold Tolerance at the Booting Stage under Naturally Low Temperature Conditions

One measure of cold tolerance at the booting stage is the seed setting rate under naturally low temperature conditions (SStR-NL). Suitable conditions are available at an experimental farm of Yunnan Academy of Agricultural Sciences, Kunming, where atmospheric temperatures are 15–19°C during the booting stage extending from June to August [16]. Seedlings of each cultivar were grown in a nursery seedbed planted 31 March 2006, and transplanted on 9 May with two replications. Twenty plants of each accession were transplanted in a single row with a row × plant distance of 25 × 15 cm. The field was fertilized with N (120 kg/ha^2) and P_2O_5 (80 kg/ha^2). SStR-NL was scored as the mean seed setting rate of 10 plants from the middle 18 plants in each row.

Evaluation of Cold Tolerance at the Booting Stage under Cold Water Stress

The other two measures of cold tolerance at the booting stage were seed setting rate under cold water (SStR-CW) and relative seed setting rate (RSStR-CW, i.e. ratio of seed setting rate under cold water stress to seed setting rate under normal conditions). Cold tolerance under cold water stress was evaluated in Gongzhuling, Jilin province. Seedlings were grown in a nursery seedbed after planting on 15 April 2006, and transplanted on 25 May with two replications. Twelve plants of each accession were transplanted in a single row with a line × plant distance 25 × 15 cm. The field was fertilized with N (120 kg/ha^2) and P_2O_5 (80 kg/ha^2). On 1 July at panicle initiation, the field was irrigated continuously for 40 days with cold water pumped from a deep well where the water temperature is a constant 19°C. The cold water was maintained at 20 cm depth at the early stage to 30 cm at the late growth stage, such that the young panicles were covered by cold water. Late accessions that could not be harvested from late August to mid-October were enclosed by a plastic greenhouse to avoid the effects of low atmospheric temperatures on grain filling. Another two replications were transplanted and managed under the same conditions, except that they were irrigated with surface water at higher temperatures (daily mean temperature of 22–26°C from 1 July to 10 August). Mean seed setting rates for all accessions under both cold and normal conditions were recorded on the middle 10 plants in each row.

SSR Polymorphism and Genotyping

DNA was extracted from the young leaves using the CTAB method. Two hundred and seventy three simple sequence repeat (SSR) markers randomly distributed across all 12 rice chromosomes were selected for genotyping. The average distance between markers was 835 kb. The minimum distance between adjacent markers was 0.06 kb and the maximum was 4,050 kb; 18% of distances between adjacent markers were less than 150 kb, more than 50% were less than 750 kb, and more than 90% were less than 1,800 kb. PCR conditions, gel electrophoresis of PCR products and genotype scoring methods were as described in Zhang *et al.* [15].

Statistical Analysis and Association Mapping

STRUCTURE 2.2 [17], a model-based program, was utilized to infer population structure with 60 unlinked SSRs from the 273 markers. Ten independent simulations were run for each K (the number of clusters, from 1 to 10) using a burn-in length of 10,000, run length of 100,000 and a model for admixture and correlated allelic frequencies. To determine the *K* value, the LnP(D) value in the STRUCTURE output and Evanno's K between successive *K* [18] were used.

Pair-wise linkage disequilibria between markers were evaluated by TASSEL V4.0 (http://www.maizegenetics.net/tassel) in a total of 174 accessions and in each of two clusters inferred by STRUCTURE (here, the clusters were subspecies, *indica* and *japonica*).

Association analyses between SSR markers and measures of cold tolerance were carried out by TASSEL V4.0. The Q-Q plot of GLM and MLM for four measures of cold tolerance indicated that MLM overestimated the effect of genetic relationship among individuals. The GLM model was therefore used to analyze the association between SSR markers and four measures of cold tolerance for *indica* and *japonica* separately. To control false discovery in association analyses, the significance level α was adjusted upward by α' in a rate $(1-\alpha)^R$ for R-rejected hypotheses. Here, we set α = 0.05, then α' = $0.05(1–0.05)^R$. We denoted markers with p-values lower than α' as significantly associated ones.

To provide information about the markers and germplasm resources that may be used in breeding cultivars with strong cold tolerance, we investigated the genotypic effect of each genotype for each QTL. Due to the distinct difference in cold tolerance between the two subspecies, the relative genotypic effect for the ith genotype in the sth subspecies for the corresponding indicator was denoted as $RGE_{si} = (\bar{x}_{si} - \bar{x}_s)/\bar{x}_s$, where, s represents subspecies (*indica* or *japonica*), $\bar{x}_{si}$ is the average phenotype of ith genotype in the s^{th}subspecies, and $\bar{x}_s$ is the average phenotype of the sth subspecies. Error MS for each marker (MS_E) estimated during association analysis was used to estimate the error MS of RGE as $MS_{RGEsi} = (MS_E / n_i)/\bar{x}_s^2$. Given that MS_E was estimated from a large sample, the z-test was used to test whether the RGE of each genotype was biased from zero. We denoted genotypes significantly larger than zero as positive or cold-tolerant genotypes, and those smaller than zero as negative or cold-sensitive genotypes.

RESULTS

Population Structure and Linkage Disequilibrium

When we ran the STRUCTURE simulation using 60 SSRs, the LnP(D) value increased with *K*from 1 to 10, but showed an evident knee and there was a sharp peak of Evanno's Δ*K* at *K* = 2 (Fig. 1). These results indicated that there were two distinctly divergent populations, corresponding to subspecies *indica* and *japonica*. To survey the influence of population structure on LD, we analyzed LD in the whole population and in each of the subgroups (i.e.*indica* and *japonica*) identified by structure analysis (Fig. 2). At the whole population level, the r^2 within the whole genome was 0.0624±0.0865. For SSR markers with physical distances less than 50 kb, the r^2 was 0.1443±0.2149; for those between 50 kb and 150 kb, $r^2$was 0.1332±0.1829; for those between 150 kb and 500 kb, r^2 was 0.0926±0.1353; for those between 500 kb and 1000 kb, r^2 was 0.0624±0.0765; and for those more than 1,000 kb, r^2 was 0.0605±0.0818. Thus LD for markers with physical distances shorter than 50 kb was not obviously different from those between 50 kb and 150 kb, but decreased dramatically for those further than 150 kb. Compared to the whole population, LDs within *indica* and *japonica* were obviously lower (Fig. 2). For SSR markers with physical distances shorter than 50 kb, mean r^2 were 0.1388±0.2135 and 0.1810±0.3203 for *indica* and *japonica*, respectively; for markers between 50 kb and 150 kb, 0.0761±0.0796 and 0.0760±0.0368; for markers between 150 kb and 500 kb, 0.0331±0.0573 and 0.0392±0.0616; for markers between 500 kb and 1000 kb, 0.0155±0.0148 and 0.0336±0.0505; and for markers more than 1000 kb, 0.0134±0.0165 and 0.0201±0.0261. These results indicated that LDs between close markers (such as physical distances less than 50 kb) in both *indica* and *japonica* did not decrease relative to the whole population, however, the LDs between distant markers especially those more than 150 kb decreased dramatically. Given the above results, we carried out the association analysis within *indica* and *japonica* independently in order to avoid false positive associations; and given that few non-linked markers showed strong LD, we used the GLM model that controls population structure but not kinship for the association analysis within*indica* and *japonica*.

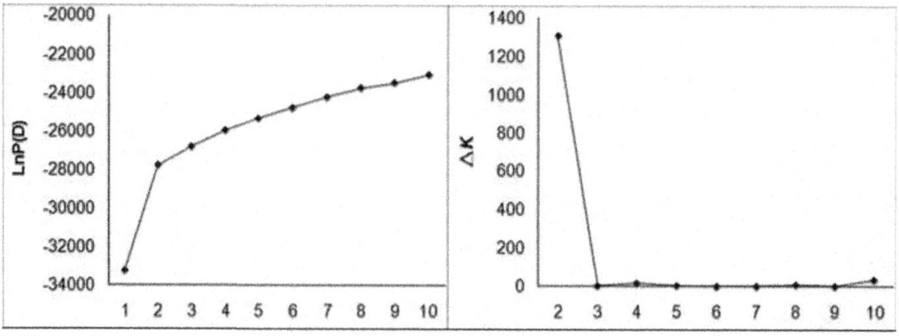

Figure 1: Average LnP(D) and Δ*K* **over 10 repeats of STRUCTURE simulation.**

doi:10.1371/journal.pone.0120590.g001

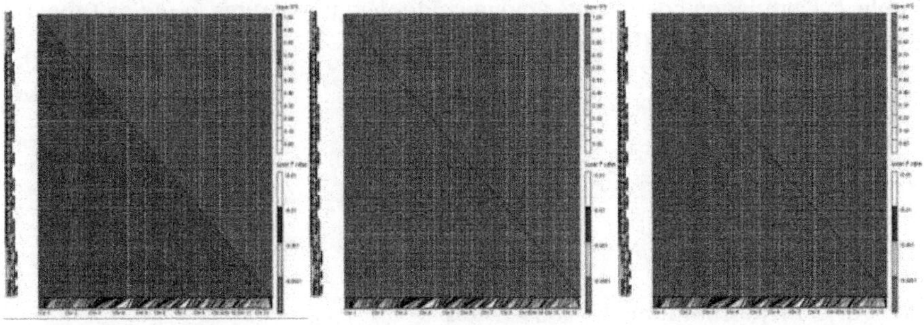

Figure 2: Distribution of LD across 273 SSR loci on 12 linkage groups in the total population (left), *indica* **(center) and** *japonica* **(right).**

doi:10.1371/journal.pone.0120590.g002

Cold Tolerance at the Germination and Booting Stages

All four measures of cold tolerance showed high variation among accessions at both the germination and booting stages, particularly the latter (Table 1). *Japonica* had apparently strong cold tolerance at the germination stage as measured by seedling survival rate (SSvR = 79.11%) under low temperature; this was more than three-fold stronger than *indica* (23.39%). Both *indica* and *japonica* were sensitive to cold stress at the booting stage, as measured by seed setting rate under natural low temperature conditions in Kunming (SStR-NL), seed setting rate under cold water irrigation in Gongzhuling (SStR-CW), and relative seed setting rate under cold water irrigation in Gongzhuling (RSStR-CW). However, *japonica* was more tolerant to cold stress (SStR-NL and RSStR-CW were higher than 25%) than *indica* (SStR-NL and RSStR-CW were lower than 25%). Correlation analyses among the four measures revealed low correlations between cold tolerance at the germination stage and at the booting stage, and low correlations between cold tolerances in different environments (Table 2). Higher correlations in cold tolerance among different stages and environments occurred in the case of *japonica*.

Table 1: Phenotypic variation in four cold-tolerant measures in *indica* and*japonica*

Measure	*Indica*				*Japonica*			
	Mean (%)	Range	SD (%)	CV %	Mean (%)	Range	SD (%)	CV %
SSvR	23.39	1.11–100.00	18.03	77.08	79.11**	1.11–100.00	28.67	36.23
SStR-NL	18.88	0–74.42	21.33	113.01	29.89**	0–80.43	26.41	68.35
SStR-CW	14.71	0–83.49	21.24	144.39	16.87	0–64.31	19.54	115.82
RSStR-CW	14.57	0–89.14	20.53	140.92	25.79*	0–90.47	25.84	100.18

* Significant at $P = 0.05$;
**: significant at $P = 0.01$;
SSvR, survival at low temperature expressed as percentage (%) of surviving plants;
SStR-NL, percentage (%) of filled grains per panicle under natural low temperature conditions in Kunming;
SStR-CW, percentage (%) of filled grains per panicle under cold water irrigation in Gongzhuling;
RSStR-CW, percentage (%) relative seed setting under cold water irrigation in Gongzhuling.

doi:10.1371/journal.pone.0120590.t001

doi:10.1371/journal.pone.0120590.t001

Table 2: Correlation coefficients among cold-tolerant measures in *indica* **(above the diagonal) and** *japonica* (below the diagonal)

Measure	SSvR	SStR-NL	SStR-CW	RSStR-CW
SSvR	1.00	0.16	0.11	0.03
SStR-NL	0.16	1.00	0.00	0.17
SStR-CW	0.17	0.23	1.00	0.98 **
RSStR-CW	0.27	0.25	0.91**	1.00

**: Significant at P = 0.01.

doi:10.1371/journal.pone.0120590.t002

doi:10.1371/journal.pone.0120590.t002

QTLs Associated With Cold Tolerance at the Germination and Booting Stages

Fifty one QTLs related to cold tolerance at the germination and booting stages were detected. They were distributed on all 12 chromosomes, with the maximum number on chromosome 1 and only one on chromosome 9 (Fig. 3). Among them, 22 were detected at the germination stage, 33 at booting, and eight were identified by at least two measures. The genetic contribution of each QTL was 29.20% on average, with a minimum 5.25% (*qLTSSvR3–1* on chromosome 3) and the maximum 59.28% (*qLTSSvR7–1* on chromosome 7). Of the reported QTLs related to cold tolerance in rice (www.gramene.com) more than 54% (21 of 39 QTLs) were detected in the current study (Fig. 3). We identified four loci associated with cold tolerance at the booting stage that were also detected by Cui *et al.* by association mapping [19].

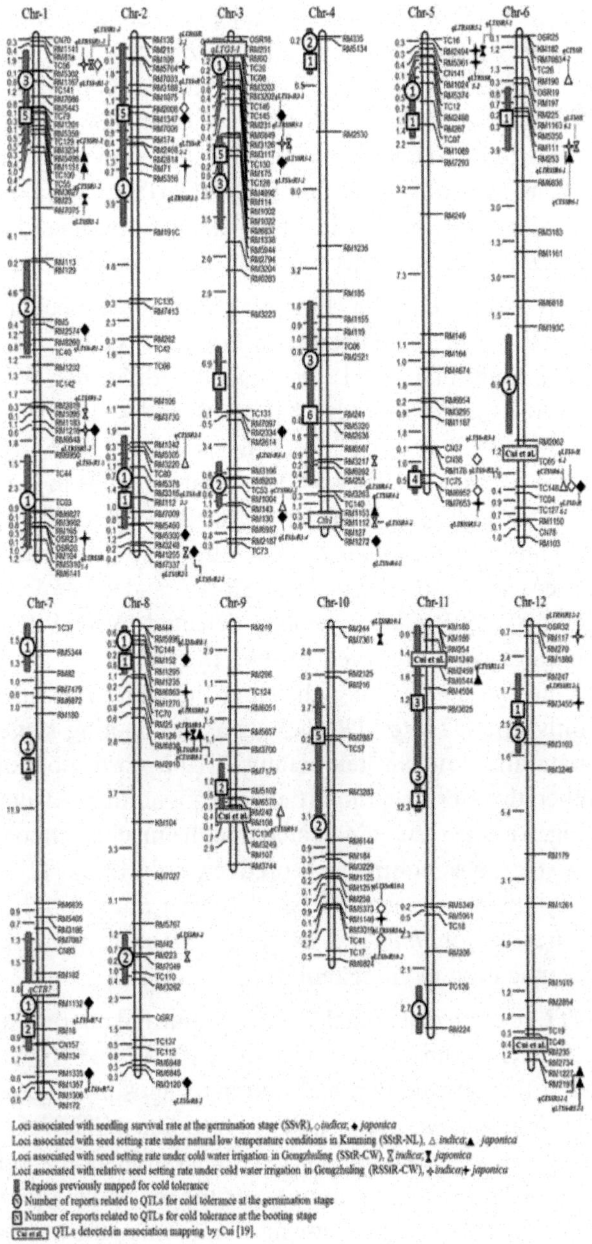

Figure 3: Chromosome maps of QLTs for cold tolerance at the germination and booting stages (the distances between markers are Mb).

doi:10.1371/journal.pone.0120590.g003

Among the 22 QTLs related to cold tolerance at germination as measured by SSvR, 6 were detected only in *indica* with low average contributions to phenotypic variation (CPV = 12.0%), 15 only in *japonica* with high average CPV (36.2%), and *qCTSSR6–3* in both *indica* and*japonica* with average CPV of 13.5%. The results showing that only one QTL was shared between *indica* and *japonica* and that CPV of QTLs in *japonica* were much higher than those in*indica*, implied that the mechanism of cold tolerance at the germination stage in *japonica* was different from that in *indica*.

Of 12 QTLs associated with SStR-NL, five were detected in *indica* with low average CPV (21.8%), and seven in *japonica* with higher average CPV (33.9%) (Fig. 3). Eight of 12 QTLs associated with SStR-CW were identified in *indica* with low average CPV (24.0%), compared with four in *japonica* with high average CPV (35.2%). Six of 15 QTLs associated with RSStR-CW, were in *indica* with low average CPV (24.3%), compared to nine in *japonica* with high average CPV (38.2%). No QTL related to SStR-NL, SStR-CW or RSStR-CW was shared between *indica* and *japonica*. Under the same stress and environmental conditions, three (*qLTRSSR6–1*, *qLTRSSR3–1* and *qLTRSSR1–3*) of eight QTLs identified by SStR-CW were also detected by RSStR-CW in *indica*, and similar findings were observed for two (*qLTRSSR8–1* and *qLTRSSR5–1*) of four QTLs in *japonica*. Although using the same measure (seed setting rate, SStR), only one QTL (*qLTRSSR8–1*) at the booting stage was detected under both naturally low air temperatures (the corresponding measure is SStR-NL) and cold water irrigation (the corresponding measure is SStR-CW) in *japonica* (Fig. 3). The results indicated that the mechanism of cold tolerance at the booting stage was complex. Moreover, only one QTL (*qLTRSSR8–1*) was detected by the same measure (seed setting rate, SStR) under different kinds of cold stress in *japonica*, i.e. naturally low air temperature in Kunming and cold water irrigation in Gongzhuling.

Among QTLs detected at both the germination and booting stages, the same two QTLs (*qLTSSR1–3* and *qCTSSR6–3* with CPV 9.1 and 10.9%, respectively) were detected at the germination and booting stages in *indica*, but there was no similar QTL in *japonica*. That is,*qLTSSR1–3* and *qCTSSR6–3* may provide a common mechanism of cold tolerance at the germination and booting stages to some degree in *indica*, although most of the cold tolerance at both stages may be attributed to different mechanisms. In *japonica* the mechanisms of cold tolerance at the germination and booting stages appear to be more different than in *indica*.

Markers and Germplasm Resources with Strong Cold Tolerance Potentially useful for Breeding

According to the relative genotypic effect (RGE) of each genotype at each QTL, we detected 18 positive genotypes and 21 negative genotypes in *indica*, and 19 positive genotypes and 24 negative genotypes in *japonica*. Generally, the negative effects were much stronger than the positive effects in both subspecies. Using these genotypes as selection markers we independently screened *indica* and *japonica* accessions containing different numbers of positive genotypes (Fig. 4 and 5). For positive genotypes, the cold tolerance as measured by the four measures increased with the increasing number of positive genotypes in *indica*, but not in *japonica* (Fig. 4). Likewise, for positive genotypes detected in *japonica*, the cold tolerance as measured by the four measures generally increased in *japonica*, but not in *indica*, apart from two exceptions for measures SStR_NL (Fig. 5B) and RSStR_CW (Fig. 5D). We checked the accessions with four genotypes positive for SStR_NL (Fig. 5B), among which one accession (Ye Li Cang Hua) had quite low seed setting (18%) and contained a strong negative genotype (RM5496_136), which reduced seed setting by up to 60%. Similarly, among the accessions with three genotypes positive for RSStR_CW (Fig. 5D), two cultivars (Ning Hui 21 and Dan Dong Lu Dao) had low seed setting (7% and 13.5%, respectively), and both contained a strong negative genotype (RM6863_155) that reduced seed setting by as much as 67%.

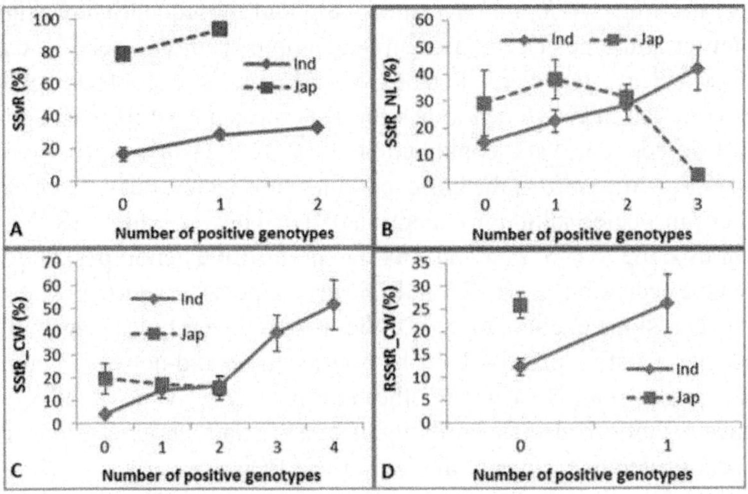

Figure 4: Phenotypes of four measures in *indica* **and** *japonica* **accessions with different numbers of positive QTLs detected in** *indica*.

doi:10.1371/journal.pone.0120590.g004

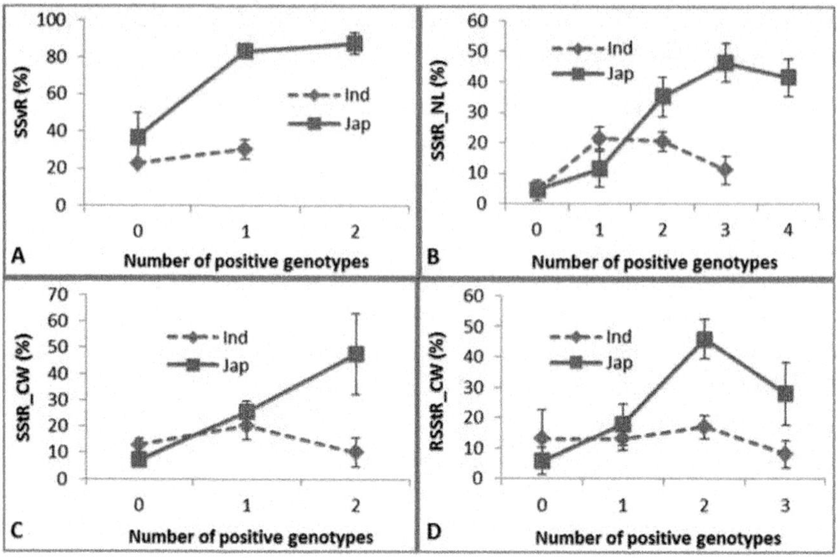

Figure 5: Phenotypes of four measures in *indica* **and** *japonica* **accessions with different numbers of positive QTLs detected in** *japonica.*

doi:10.1371/journal.pone.0120590.g005

In addition to screening accessions using positive genotypes we also screened the top five cultivars showing high cold tolerance according to each measure in *indica* and *japonica*, respectively, and subsequently investigated the distribution of the positive and negative genotypes among these accessions in each subspecies. There were three obvious characteristics for the distribution of the positive and negative genotypes for each measure within each subspecies. Firstly, the top cultivars contained at least one positive genotype of the corresponding measure in the corresponding subspecies, but not necessarily in the other subspecies, although exceptions existed in *indica* for SStR_NL (where accessions Z102, Z146, Z54 and Z64 had no positive genotypes) and RSStR_CW (where accessions Z126, Z5 and Z60 had no positive genotypes). Secondly, the top accessions excluded most of the negative genotypes, especially those with strong negative effects. Thirdly, both positive and negative genotypes in one subspecies may exist in the other subspecies but with no effect on cold tolerance in that subspecies, implying that *indica* and *japonica* do not share the same set of tolerance genes or alleles at most loci.

DISCUSSION

QTLs for Cold Tolerance in *Indica* and *Japonica*

Association analysis is an effective approach to dissect the genetic bases underlying complex traits in plants [9–10], animals [20] and humans [21]. In the present study we used association analysis to identify 51 QTLs related to cold tolerance at the germination and booting stages in rice. Among them, 46.17% were new QTLs, and 27 corresponded to QTLs reported previously. For example, *qLTSSR4–1* associated with SStR-CW was near the cloned *Ctb1* on chromosome 4; *qLTSSvR7–1* associated with SSvR is 1.4 cM from *qCTB7* on chromosome 7; and *qLTRSSR12–1* associated with RSStR-CW is near *qCTS12*. Four loci identified in the present study were detected at the booting stage by Cui *et al.* [17]. Of these, *qLTSSvR6–1* on Chr6 in the current study is near RM528 as reported by Cui *et al.*, *qCTSSR9–1* on Chr9 is near RM160, *qCTSSR11–1* on Chr11 is near RM4B, and *qCTSSR12–1* on Chr12 is near RM235.

Our results clearly indicated that *japonica* has a higher level of tolerance to cold stress than*indica* at both the germination and booting stages. This agrees with results published by others [22–25]. More QTLs were identified in *japonica* (36) than in *indica* (26). Five QTLs were shared among measures in *indica* or *japonica* at the booting stage. Only one QTL (*qCTSSR6–3*) associated with seedling survival rate (SSvR) under low temperatures was shared by *indica*and *japonica*. Reasons for this deserve further discussion.

During the thousands of years of cultivation and utilization under diverse environmental conditions tremendous genetic differentiation has occurred in rice produced under various agro-ecosystems. *Indica* is known to be adapted to tropical environments at low latitudes or altitudes typified by warm climatic conditions, whereas *japonica* cultivars were grown in temperate areas at high latitudes or altitudes with relatively cool conditions [26], at high altitudes in tropical and sub-tropical areas, and in areas where paddy fields are irrigated by cold water. In such areas, water and soil temperatures at sowing are below 15°C [27–28], and thus*japonica* cultivars were preferred due to tolerance to prevailing low temperatures. In double-cropped rice-growing regions, water and soil temperatures at sowing are higher than 15°C but cold air currents from Siberia often occurring in April may cause early rotting of seedlings, resulting in heavy seedling losses and a delayed growing period. *Indica* cultivars grown in such areas are usually characterized by tolerance to short periods of low temperature. Evident correlation has been found between cold stress and altitude or atmospheric temperature. Low temperatures (15–19°C) at booting cause sterile pollen which directly leads to spikelet sterility and ultimately causes serious yield losses [23,29–31]. In Heilongjiang province an increase of 1°C in negative accumulated temperature GDD17 at the inflorescence differentiation stage can result in a 3.51 kg/ha yield loss, and an increase of 1°C in negative accumulated temperature GDD19 at booting stage can cause a 4.95 kg/ha yield loss [32]. *Indica* is adapted to warm areas and is more sensitive to low temperatures. Pollen sterility is observed in *indica* when average temperatures remain below 22°C for three days [33]. The seasonal duration for rice gets progressively shorter due to decreased average temperature and accumulated temperature from southern to northern China. *Japonica* cultivars are mainly grown in the area north of the Yellow River, whereas most *indica* cultivars are grown to the south. This distribution implies that *indica-japonica* differentiation resulted from adaptation to different climatic and geographic environments.

The significant *indica–japonica* difference in cold tolerance was probably enhanced by rare genetic exchanges between the two subspecies. During cultivation and improvement over thousands of years, *indica* and *japonica* became genetically separated with adaptation to different agro-ecosystems. Differentiation between the subspecies was further enforced by fertility barriers that prevented genetic exchange between the two subspecies [34].

Observations of apparent differences in cold tolerance between *indica* and *japonica* may be partially attributed to the use of SSRs originating from noncoding regions. Only 15% of the SSRs used in the present study were from ESTs: this is higher than the proportion (1.5–7.5%) among all SSRs reported in rice [35–36]. Detection of marker-trait associations based on linkage disequilibria (LD) in genetically diverse materials can help to identify QTLs controlling target traits. However, this depends on the chromosome distance between marker and QTL and times between mutational changes in marker and the causal mutation of the target QTL. An association can be detected easily in a population when the mutation of the test marker is near in distance and time to the functional mutation in the target QTL [37]. However, 85% of SSRs in the study were nonfunctional markers and there are few genomic exchanges between the two subspecies as mentioned above. This may partially explain why *indica* and *japonica*share so few QTLs of cold tolerance.

Both Positive and Negative Genotypes should be considered in Breeding by MAS

Association mapping is a proven and efficient way to detect natural variation in a diverse population and provides many candidate genes or genomic regions for marker assisted selection in breeding. For example, Huang *et al.* used a worldwide panel of cultivated rice varieties to map 32 new loci associated with flowering time and 10 grain-related traits, allowing identification of candidate genes for 18 associated loci [10]. Marker assisted selection is an efficient breeding approach to pyramid elite genes in crop improvement; for example, Narayanan *et al.* developed a line combining blast and blight resistance genes Pi-1, Piz-5 and Xa2 using markers [38]. In the present study, we identified 18 and 19 positive genotypes (i.e. cold-tolerant genotypes) in *indica* and *japonica*, respectively, based on 51 QTLs. Our results indicated that the more positive genotypes the selected cultivars contain, the stronger cold tolerance they show. This proved that positive markers can be used to pyramid cold tolerance genes. In addition to those positive genotypes, however, we identified even more negative genotypes (sensitive to cold stress), numbering 21 and 24 in *indica* and *japonica*, respectively. Some cultivars (such as Ning Hui 21 and Dan Dong Lu Dao) are very sensitive to low temperatures due to a single strong negative genotype (such as RM6863_155) even though they may contain up to three positive genotypes. This indicates that certain strong negative genotypes may play the role of the shortest piece of wood as showed in the bucket theory and should be avoided by the breeders lest they cover up the advantage of the positive genotypes.

ACKNOWLEDGMENTS

We thank Professor Robert A McIntosh, University of Sydney, for suggested revisions to the manuscript.

AUTHOR CONTRIBUTIONS

Conceived and designed the experiments: YHP HLZ ZCL. Performed the experiments: XDM GLC YHP HLZ DLZ. Analyzed the data: YHP HLZ DLZ JJL HYX JPY JLL MARR GGL. Contributed reagents/materials/analysis tools: YHP HLZ DLZ JJL XDM GLC LZH. Wrote the paper: YHP HLZ ZCL MARR.

REFERENCES

1. Guirguis K, Gershunov A, Schwartz R, Bennet S. Recent warm and cold daily winter temperature extremes in the Northern Hemisphere. Geophysical Research Letters. 2011 Sep 2. pii: S0094–8276. doi: 10.1029/2011GL0487622.
2. Kodra E, Steinhaeuser K, Auroop R. Persisting cold extremes under 21st-century warming scenarios. Geophysical Research Letters. 2011 Apr 28. pii: S0094–8276. doi: 10.1029/2011GL047103.

3. Fujino K, Sekiguchi H, Matsuda Y, Sugimoto K, Ono K, Yano M. Molecular identification of a major quantitative trait locus, *qLTG3–1*, controlling low-temperature germinability in rice. Proc Natl Acad Sci USA. 2008; 105: 12623–12628. doi: 10.1073/pnas.0805303105. pmid:18719107

4. Xu LM, Zhou L, Zeng YW, Feng MW, Zhang HL, Shen SQ, et al. Identification and mapping of quantitative trait loci for cold tolerance at the booting stage in a *japonica*rice near-isogenic line. Plant Sci. 2008; 174: 340–347. doi: 10.1016/j.plantsci.2007.12.003

5. Zhou L, Zeng YW, Zheng WW, Tang B, SM Yang, Zhang HL, et al. Fine mapping a QTL, qCTB7, for cold tolerance at the booting stage on rice chromosome 7 using a near-isogenic line. Theor Appl Genet. 2010; 121: 893–905. doi: 10.1007/s00122-010-1358-x

6. Saito K, Hayano-Saito Y, Kuroki M, Sato Y. Map-based cloning of the rice cold tolerance gene *Ctb1*. Plant Sci. 2010; 179: 97–102. doi: 10.1016/j.plantsci.2010.04.004

7. Flint-Garcia S, Thuillet AC, Yu JM, Gael-Pressoir G, Romero S, Mitchell SE, et al. Maize association population: a high-resolution platform for quantitative trait locus dissection. Plant J. 2005; 44: 1054–1064. pmid:16359397 doi: 10.1111/j.1365-313x.2005.02591.x

8. Iwata H, Uga Y, Yoshioka Y, Ebana K, Hayashi T, Hayashi T. Bayesian association mapping of multiple quantitative trait loci and its application to the analysis of genetic variation among *Oryza sativa* L. germplasms. Theor Appl Genet. 2007; 114: 1437–1449. pmid:17356864 doi: 10.1007/s00122-007-0529-x

9. Agrama HA, Eizenga GC, Yan W. Association mapping of yield and its components in rice cultivars. Mol Breeding. 2007; 19: 341–356. doi: 10.1007/s11032-006-9066-6

10. Huang X, Wei X, Sang T, Zhao Q, Feng Q, Zhao Y, et al. Genome-wide association studies of 14 agronomic traits in rice landraces. Nature Genetics. 2010; 42: 961–970. doi: 10.1038/ng.695. pmid:20972439

11. Huang XH, Yan Z, Xing HW, Li CY, Wang A, Zhao Q, et al. Genome-wide association study of flowering time and grain yield traits in a worldwide collection of rice germplasm. Nature Genetics. 2012; 44: 32–39. doi: 10.1093/jncimonographs/lgs007. pmid:22623593

12. Li XB, Yan WG, Agrama H, Jia LM, Jackson A, Moldenhauer K, et al. Unraveling the complex trait of harvest index with association mapping in rice (*Oryza sativa* L.). PLoS ONE. 2012; 7: e29350. doi: 10.1371/journal.pone.0029350. pmid:22291889

13. Jia LM, Yan WG, Zhu CS, Agrama HA, Jackson A, Yeater K, et al. Allelic analysis of sheath blight resistance with association mapping in rice. PLoS ONE. 2012; 7: e32703. doi: 10.1371/journal.pone.0032703. pmid:22427867

14. Zhang YM, Zou DT. Association analysis of rice cold tolerance at tillering stage with SSR markers in *japonica* cultivars in Northeast China. Chin J Rice Sci. 2012; 26: 423–430.

15. Zhang DL, Zhang HL, Wang MX, Sun JL, Qi YW, Wang FM, et al. Genetic structure and differentiation of *Oryza sativa* L. in China revealed by microsatellites. Theor Appl Genet. 2009; 199: 1105–1117. doi: 10.1007/s00122-009-1112-4

16. Dai LY, Lin XH, Ye CR, Ise K, Saito K, Kato A, et al. Identification of quantitative trait loci controlling cold tolerance at the reproductive stage in Yunnan landrace of rice, Kunmingxiaobaigu. Breed Sci. 2004; 54: 253–258. doi: 10.1270/jsbbs.54.253

17. Falush D, Stephens M, Pritchard JK. Inference of population structure using multilocus genotype data: linked loci and correlated allele frequencies. Genetics. 2007; 164: 1567–1587.

18. Evanno G, Regnaut S, Goudet J. Detecting the number of clusters of individuals using the software structure: a simulation study. Molecular Ecology. 2005; 14: 2611–2620. pmid:15969739 doi: 10.1111/j.1365-294x.2005.02553.x

19. Cui D, Xu CY, Tang CF, Yang CG, Yu TQ, A XX, et al. Genetic structure and association mapping of cold tolerance in improved *japonica* rice germplasm at the booting stage. Euphytica. 2013; 193: 369–382 doi: 10.1007/s10681-013-0935-x

20. Bolormaa S, Hayes BJ, Savin K, Hawken R, Barendse W, Arthur PF, et al. Genome-wide association studies for feedlot and growth traits in cattle. J Anim Sci. 2011; 6: 1684–1697. doi: 10.2527/jas.2010-3079. pmid:21239664

21. Barrett JC, Hansoul S, Nicolae DL, Cho JH, Duerr RH, Rioux JD, et al. Genome-wide association defines more than 30 distinct susceptibility loci for Crohn›s disease. Nat Genet. 2008; 40: 955–962. doi: 10.1038/ng.175. pmid:18587394

22. Fujino K, Sekiguchi H, Sato T, Kiuchi H, Nonoue Y, Takeuchi Y, et al. Mapping of quantitative trait loci controlling low-temperature germinability in rice (*Oryza sativa* L.). Theor Appl Genet. 2004; 108: 794–799. pmid:14624339 doi: 10.1007/s00122-003-1509-4

23. Andaya VC, Mackill DJ. QTLs conferring cold tolerance at the booting stage of rice using recombinant inbred lines from a *japonica* × *indica* cross. Theor Appl Genet. 2003; 106: 1084–1090. pmid:12671757
24. Tai LM, Du J, Zeng YW. The cluster analysis of peculiarity of *indica* and *japonica* in Simao. Southwest China J Agaric Sci. 2006; 19: 49–53.
25. Zeng YW, Li ZC, Yang ZY. Geographical distribution and cline classification of*indica*/*Japonica* subspecies of Yunnan local rice resources. Acta Agronomica Sin. 2001; 1: 15–20.
26. Chang TT, Oka HI. Genetic variousness in the climatic adaption of rice cultivars. In: Proceedings of the Symposium on Climate and Rice. International Rice Research Institute, Los Banŏs, The Philippines. 1976; 87–111.
27. Fujino K, Sekiguchi H, Sato T, Kiuchi H, Nonoue Y, Takeuchi Y, et al. Mapping of quantitative trait loci controlling low-temperature germinability in rice (*Oryza sativa* L.). Theor Appl Genet. 2004; 108: 794–799. pmid:14624339 doi: 10.1007/s00122-003-1509-4
28. Sthapit BR, Witcombe JR. Inheritance of tolerance to chilling stress in rice during germination and plumule greening. Crop Sci. 1998; 8: 660–665. doi: 10.2135/cropsci1998.0011183x003800030007x
29. Nishiyama I. Male sterility caused by cooling treatment at the young microspore stage in rice plants. XXIII. Another length, pollen number and the difference in susceptibility to coolness among spikelets on the panicle. Jpn J Crop Sci. 1982; 514: 462–469. doi: 10.1626/jcs.51.462
30. Ji SL, Jiang L, Wang YH, Liu SJ, Liu X, Zhai HQ, et al. QTL and epistasis for low temperature germinability in rice. Acta Agron Sin. 2008; 34: 551–556. doi: 10.1016/s1875-2780(08)60021-8
31. Ye C, Fukai S, Godwin I, Reinke R, Snell P, Schiller J, et al. Cold tolerance in rice varieties at different growth stages. Crop Pasture Sci. 2009; 60: 328–338. doi: 10.1071/cp09006
32. Liu XF, Zhang Z, Shuai JB, Wang P, Shi WJ, Chen Y, et al. Effect of chilling injury on rice yield in Heilongjiang province. Acta Geographica Sin. 2010; 67(9): 1223–1232.
33. Wen S, Yao H. Global warming over the period 1961–2008 did not increase high-temperature stress but did reduce low-temperature stress in irrigated rice across China. Agricultural and Forest Meteorology. 2011; 9: 1193–1201. doi: 10.1016/j.agrformet.2011.04.009

34. Kato S, Kosaka H, Hara S. On the affinity of rice varieties as shown by the fertility of hybrid plants. Bull Sci Fac Agric. 1928; 3: 132–147.
35. Thiel T, Michalek W, Varshney RK, Graner A. Exploiting EST databases for the development of cDNA derived microsatellite markers in barley (*Hordeum vulgare* L.). Theor Appl Genet, 2003; 106: 411–422. pmid:12589540
36. Lawson MJ, Zhang L. Distinct patterns of SSR distribution in the Arabidopsis thaliana and rice genomes. Genome Biol. 2006; 7: R14. pmid:16507170
37. Meuwissen THE, Hayes BJ, Goddard ME. Prediction of total genetic value using genome-wide dense marker maps. Genetics. 2001; 157: 1819–1829. pmid:11290733
38. Narayanan NN, Niranjan B, Norman PO, Casiana MV, Gnanamanickam SS, Datta K, et al. Molecular breeding: marker-assisted selection combined with biolistic transformation for blast and bacterial blight resistance in *indica* rice (cv. CO39). Mol Breeding. 2004; 14: 61–71. doi: 10.1023/b:molb.0000037995.63856.2d

Chapter 9

GENETIC ANALYSIS OF LOW BMI PHENOTYPE IN THE UTAH POPULATION DATABASE

William R. Yates[1,2], Craig Johnson[3], Patrick McKee[1,4], Lisa A. Cannon-Albright[5,6]

[1]Laureate Institute for Brain Research, Tulsa, Oklahoma, United States of America

[2]University of Oklahoma College of Medicine, Tulsa, Tulsa, Oklahoma, United States of America

[3]Eating Recovery Center, Denver, Colorado, United States of America

[4]University of Oklahoma College of Medicine, Oklahoma City, Oklahoma, United States of America

[5]Department of Medicine (Genetic Epidemiology), University of Utah School of Medicine, Salt Lake City, Utah, United States of America

[6]George E. Wahlen Department of Veterans Affairs Medical Center, Salt Lake City, Utah, United States of America

ABSTRACT

The low body mass index (BMI) phenotype of less than 18.5 has been linked to medical and psychological morbidity as well as increased mortality risk. Although genetic factors have been shown to influence BMI across the entire BMI, the contribution of genetic factors to the low BMI phenotype is unclear. We hypothesized genetic factors would contribute to risk of a low BMI phenotype. To test this hypothesis, we conducted a genealogy data analysis using height and weight measurements from driver's license data from the Utah Population Data Base. The Genealogical Index of Familiality (GIF) test and relative risk in relatives were used to examine evidence for excess relatedness among individuals with the low BMI phenotype. The overall GIF test for excess relatedness in the low BMI phenotype showed a significant excess over expected (GIF 4.47 for all cases versus 4.10 for controls, overall empirical p-value<0.001). The significant excess relatedness was still observed when close relationships were ignored, supporting a specific genetic contribution rather than only a family environmental effect. This study supports a specific genetic contribution in the risk for the low BMI phenotype. Better understanding of the genetic contribution to low BMI holds promise for weight regulation and potentially for novel strategies in the treatment of leanness and obesity.

INTRODUCTION

Genetic factors increase the risk for a high body mass index (BMI), overweight and obesity[1]. The role of genetic factors in low BMI is less well understood.

Family studies have found family clustering for low BMI [2]–[4]. However, family studies cannot distinguish between genetic and familial factors. Twin studies have estimated the heritability of BMI across the entire BMI range at between 50 and 74% [5]–[6].

Molecular genetic studies have identified a series of candidate genes for low BMI including a thyrotropin-releasing hormone (TRH) receptor polymorphism [7], the Ser23 allele of the serotonin 2C receptor [8], a genetic variant on chromosome 16p11.2 [9] and a copy number variant identified as gremlin1 [10]. Allelic variants of the *FTO* gene linked to obesity risk are infrequently found in thin individuals [11].

Understanding the genetic and environmental contributions to low BMI are important because low BMI has been linked to medical and psychiatric illnesses as well as increased mortality. In females, low BMI during childhood and adolescence increases women›s risk for later endometriosis [12], preterm birth [13], low infant birth weight [14] and increased risk for placental abruption [15]. Infants born to mothers with low BMI have an increased risk for atrial septal defect, genital abnormalities including hypospadias [16].

The psychiatric illness *anorexia nervosa* is defined by a low BMI in association with extreme fear of becoming fat [17].

Mortality rates across BMI categories in many studies display a U-shaped curve with increased death rates for the low BMI as well as those with a high BMI. Mortality rates are higher by an estimated 73% in those with BMI below 18.5 [18]. A Japanese study estimated the mortality risk increased by 78% in those with a BMI<18.5 and increased by 155% in those with a BMI<16 [19]. The exact mechanism for the mortality increase in low BMI populations is unclear. Some of the increase may be due to higher rates of death in severe illness and surgical procedures such as lung transplantation [20].

A study of excess deaths related to being low BMI and high BMI in the United States provides a reference for the relative contribution of low BMI to mortality [21]. High BMI was estimated to contribute to 111,909 deaths in the U.S. in 2000 while low BMI was estimated to be associated with 33,746 deaths. Thus, low BMI is estimated to contribute about three deaths for every ten deaths related to being high BMI.

To further understand the prevalence and genetic contributions of leanness, we examined the low BMI phenotype in the Utah Population Data Base (UPDB). The UPDB provides a strategy to examine the possibility of both environmental and genetic contributions to a phenotype by estimating risk in both close and distant relatives and by testing for excess familial clustering. An observation of excess close relationships alone would not have allowed discrimination between shared genes and shared environment, but the UPDB allows us to consider more distant relationships that are unlikely to represent lifestyle sharing beyond what is expected in the Utah population. We hypothesized that low BMI individuals would demonstrate near and distant familial clustering consistent with a genetic contribution.

METHODS

Ethics Approval

The protocol for this study was approved by the University of Utah Institutional Review Board and the Resource for Genetic and Epidemiologic Research. The Resource for Genetic and Epidemiologic Research is the oversight board for the Utah Population Data Base. All data used in this research study contains no individual identifiers. A waiver of consent was approved for this study due to the lack of individual identifiers for all subjects. Consent requirements were waived for this study since obtaining consent would have unnecessarily identified individuals in the anonymous database. Review of this protocol by the University of Utah Institutional Review Board and the Resource for Genetic and Epidemiologic Research includes a review and approval of consent issues and other ethical aspects of the research.

Utah Population Data Base (UPDB)

The UPDB is a unique computerized database primarily representing the pioneer founders of Utah and their modern day descendants. It includes up to 15 of genealogy data dating back to the original Utah founding pioneers [22], as well as current generations. The genealogy data has been linked to statewide data including driving license (DL) data, births, deaths, the Utah Cancer Registry, and Utah Hospital Discharge Data, among other data sets (www.huntsmancancer.org/groups/ppr). The DL data includes height and weight and is available for over three million Utah drivers.

For the genetic analyses performed here we selected only from those 1,192,768 individuals in the UPDB who have genealogy data for both parents, all four grandparents, and six of their eight great grandparents and whose genealogy connects to the original Utah genealogy, and the 593,704 of these individuals who have Utah Drivers License data. These strict criteria allow for appropriate matching of cases and controls in terms of quality and quantity of genealogical data.

The oversight board for the UPDB encourages collaboration with outside investigators and institutions. Researchers with interest in using the UPDB to test hypotheses may contact the board or one of the authors for information about methods to apply for access.

The UPDB has been successfully used to define familial clustering and genetic influences in a variety of disorders including cancer [23]–[27], coronary artery disease [28], diabetes[29] rotator cuff disease [30], and deaths due to influenza [31] and asthma [32]. The methods used to identify phenotypes, assess familial and genetic effects and identify pedigrees using UPDB data have been described in detail in these studies. The study of high-risk pedigrees identified in the UPDB has led to multiple gene identifications, including*BRCA1* [33], *BRCA2* [34], *CDKN2A* (melanoma) [35]–[36] and *HPC2/ELAC2* [37].

Low BMI Phenotype

The phenotype of adult leanness was established using Utah DL data available for 593,704 individuals (with acceptable genealogy data as described) included in the UPDB. We identified all male and female drivers whose most recent calculated BMI (from height and weight provided) was <18.5. Rates for the low BMI phenotype were calculated by age group and are shown in Table 1, which includes the age group, the number of lean individuals in the age group, the total number of individuals with DL data in the age group, the leanness prevalence and the 95% confidence interval for prevalence by age group, estimated by the method of Clopper and Pearson [38].

Table 1: Prevalence Rates for Low BMI (<18.5) in the UPDB

Age	BMI<18.5	N	Prevalence (%)	95% CI
15–19	6,259	57,010	11.0	10.7, 11.2
20–24	2,847	62,070	4.6	4.4, 4.8
25–29	1,629	63,548	2.5	2.4, 2.7
30–34	922	49.069	1.9	1.8, 2.0
35–39	544	39,353	1.4	1.3, 1.5
40–44	400	36,097	1.1	1.0, 1.2
45–49	311	40,719	0.8	0.7, 0.9
50–54	258	42,998	0.6	0.5, 0.7
55–59	176	38,040	0.4	0.3, 0.5
60–64	135	26,947	0.5	0.4, 0.6
65–69	191	27,076	0.7	0.6, 0.8
70–74	255	26,207	1.0	0.9, 1.1
75–79	354	25,122	1.4	1.3, 1.6
80 or older	569	26,979	2.1	1.9, 2.3

doi:10.1371/journal.pone.0080287.t001

doi:10.1371/journal.pone.0080287.t001

Statistical Analysis

The Genealogical Index of Familiality (GIF) statistic was used to test the hypothesis of excess relatedness among individuals in the low BMI phenotype. The GIF was developed specifically for the UPDB [39]–[40]. Briefly, the GIF measures the average pair-wise relatedness of a set of individuals and compares that measurement to the average pair-wise relatedness expected in the Utah population. The GIF test differs from relative risk (RR) in that it includes analysis of all genetic relationships, both close and distant. The GIF utilizes the Malecot coefficient of kinship to measure pair-wise relatedness. The coefficient is defined as the probability that randomly selected homologous genes from two individuals are identical by descent from a common ancestor [41]. The coefficient is 0.50 for parent/offspring, 0.25 for a sibling pair, 0.125 for an uncle/nephew pair, 0.0625 for a first cousin pair, and so forth. The contribution to the GIF statistic is therefore smaller for individual pairs with greater genetic distance between them; more closely related pairs contribute more.

To evaluate the significance of the GIF test, we estimated the average pair-wise relatedness for 1,000 sets of controls matched to the cases by birth year, sex, and birthplace (Utah or not). These controls are chosen from among the 593,704 individuals with acceptable quality genealogy data who also have DL data. The empirical significance of the GIF test is measured by comparing the case GIF to the distribution of 1,000 control GIF values.

The GIF statistic measures familial clustering, which can be due to genetic (genes related to low BMI phenotype), or to shared familial environmental effects (i.e. familial preference for low calorie diet or rigorous physical exercise), or to a combination of both. In order to better distinguish these effects, we recalculate the case GIF and the control GIFs while ignoring close relationships (first and second degree). If this distant GIF (dGif) test is significant, it provides strong evidence that there is significant distant excess relatedness that is unlikely to be due to shared environment.

The calculation of RR in relatives provides the more traditional mechanism for identifying genetic effects. A genetic contribution to a phenotype is supported when both close and distant relatives have elevated risk. RRs for the low BMI phenotype were estimated for first-, second- and third degree relatives of low BMI individuals as follows. First-degree relatives include parents, siblings and offspring; second-degree relatives are the first-degree relatives of the first-degree relatives (e.g. uncle, grandmother); third-degree relatives are the first-degree relatives of the second-degree relatives (e.g. first cousin, great grandchild), All 593,704 individuals in the UPDB with acceptable quality and quantity genealogy data as described and with DL data were assigned to one of 132 cohorts based on birth year (in five year groups), sex, and birthplace (Utah or not). Cohort-specific rates of low BMI were estimated by dividing the total number of low BMI individuals per cohort by the total number of individuals with DL data per cohort. Expected numbers of low BMI first-degree relatives were estimated by counting the number of first-degree relatives with DL data and genealogy data by cohort (without duplication), multiplying by the rate of low BMI in each cohort, and summing over all cohorts. Observed numbers of low BMI individuals (BMI<18.5) among relatives were counted without duplication. RRs were estimated for each degree of relationship as observed/ expected number of low BMI individuals; 95% confidence intervals for the RR were calculated using the method of Agresti [42]

High-risk Pedigree Identification

It is possible to identify pedigrees in the UPDB with a significant excess of low BMI using the same tools listed above. We first identify all possible related clusters of individuals with low BMI; no cluster is a subset of any other cluster, but individuals can be identified in more than 1 cluster. These clusters represent all sets of related individuals with low BMI descending from a common founder (a pedigree), but they are not necessarily high-risk for low BMI, they can represent chance clusters. To identify which of the clusters are high-risk for low BMI, we apply the internal cohort-specific rates for low BMI we estimated from the UPDB (see above methods for Relative Risks) to all of the descendants of each cluster. We compare the observed number of low BMI cases among the descendants to the expected number of low BMI cases among the descendants to identify those pedigrees with a significant excess of low BMI cases.

RESULTS

Prevalence Rates and Proband Selection

The prevalence of BMI<18.5 in the UPDB individuals with acceptable quality and quantity genealogy are summarized by age group in Table 1. The prevalence of BMI<18.5 ranged from .4% in the 55 to 59 year old age group to 11.0% in those aged 15 to 19 years. The BMI in the sample ranged from a minimum of 12.10 to a maximum of 62.93.

Table 1 illustrates some developmental factors associated with weight: i) stable adult weight often is not reached until midlife; ii) younger individuals have different ranges in BMI as they progress through adolescence and young adulthood, and iii) in older geriatric age groups, rates of low BMI increase with the loss of lean body mass with aging. To minimize phenotypic heterogeneity due to these other factors, we elected to include individuals between the ages of 25 and 64. This age period showed relatively stable low BMI population prevalence rates and occurred outside the periods of early and late adult development. We identified 4,375 individuals between the ages of 25 and 64 years with BMI<18.5 with acceptable quality genealogy data.

Genealogy Index of Famililiality (GIF)

The GIF test for excess relatedness in low BMI was performed on all of the 14,867 low BMI individuals and the results were compared to the average relatedness observed among 1,000 sets of matched controls selected from all individuals who had a Utah driver's license.Table 2 shows the number of cases, the average relatedness of the cases, the mean control relatedness, the empirical p value for the test for excess overall relatedness, and the empirical p value for the test for excess distant relatedness. The overall GIF test for excess relatedness for everyone with low BMI shows a significant excess over expected (p-value<0.001). The distant GIF test, ignoring close relatives (genetic distance<4) also showed a significant excess (p-value=0.031).

Table 2: GIF Test for Low BMI (<18.5) in the UPDB

Group	N	Case GIF	Control GIF	GIF p-value	dGIF p-value
All BMI<18.5	14,867	4.47	4.10	<0.001	0.031
BMI<18.5, ages 25–64	4,375	4.84	4.19	<0.001	<0.001

doi:10.1371/journal.pone.0080287.t002

doi:10.1371/journal.pone.0080287.t002

Table 2 also shows the results of the GIF test for the 4,375 low BMI adults aged 25–64 years of age. Similar results were obtained for this subgroup of lean individuals from which individuals at the extremes of the age distribution were removed to reduce bias. Figure 1shows the contribution to the GIF statistic, separately for cases and controls, by pairwise genetic distance where genetic distance 1=parent/offspring, 2=siblings, 3=avunculars, 4=first cousins, and so forth for the low BMI individuals aged 25–64 years. As seen, the distribution of relatedness for cases is in excess up to genetic distance=4 (e.g. first cousins) and beyond genetic distance=9 (e.g. second-cousins once-removed. Although the pairwise relatedness distributions for cases and matched controls cross at some points, as seen inFigure 1, the pairwise relatedness for cases is significantly elevated over that for matched controls when all genetic distances are considered as determined by distant Gif test (dGif p=0.031).

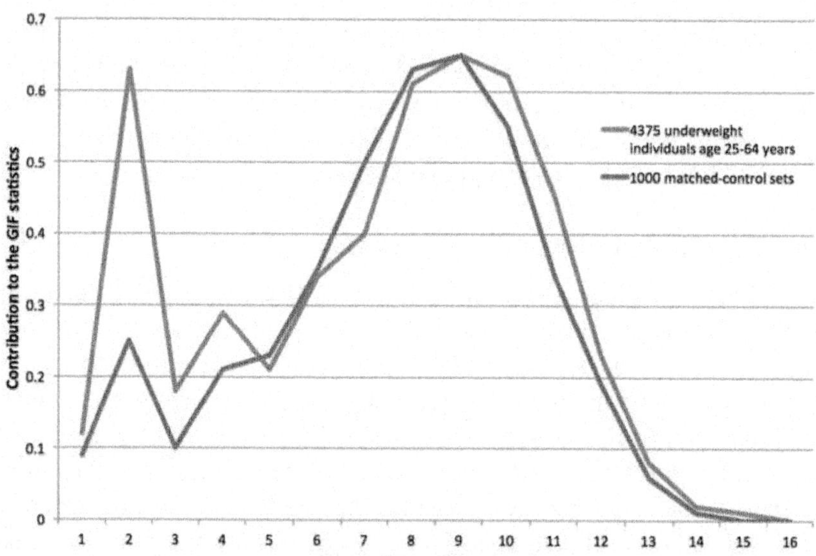

Figure 1: The contribution to the GIF statistic by genetic distance for 4,375 low BMI cases aged 25–64 years old compared to 1,000 sets of matched UPDB controls with BMI data.

Genetic distance between pairs is shown on the x-axis and represents an increasing measure of relatedness (1=parent/offspring; 2=siblings, e.g.; 3=uncle/niece, e.g.; 4=first cousins, e.g.) from close to distant; the most distant relationships noted (genetic distance=16) could represent, for example, two individuals who have a common ancestor 8 generations past. The cumulative contribution to the GIF statistic for each relatedness (as measured by genetic distance) for all pairs identified at that genetic distance is represented on the y-axis. The contribution to the GIF statistic for each larger genetic distance is one-half as large; the contribution for genetic distance 1=½, for genetic distance 2=¼, and so forth. The distribution for controls represents the expected relatedness of a group of individuals just like the cases (ignoring BMI) and is smoother because it is averaged over 1,000 different sets of controls tested. The distribution for cases represents only the analysis of the single set of cases and is more irregular. The peak at genetic distance=2 (e.g. siblings) in comparison with genetic distance=1 (parent/offspring) is seen for both cases and controls and represents that we observe more sib pairs than parent/offspring pairs in our data. A similar peak for cases at genetic distance 4 also indicates that we observed more cousins (same generation) than avunculars, for example.

doi:10.1371/journal.pone.0080287.g001

Relative Risks (RRs)

Estimates of relative risks in relatives of lean adults are shown in Table 3. The table shows the total number of each type of relative with BMI data (# relatives), the observed number of those relatives with BMI<18.5 (obs), and the expected number of relatives with BMI<18.5 (expected) based on birth year, sex, and birthplace cohort-specific rates for low BMI among all UPDB individuals with BMI data.RRs for adult leanness were significantly elevated among first-, second-, and third-degree relatives of lean adults. The smaller number of second degree lean adults observed is not unexpected, given that second degree relatives are primarily in different generations (avunculars, grandparent/child), while first and third-degree relatives occur in the same generation (siblings and cousins, respectively) as well as in different generations (parent/offspring). Since the DL data exist only after 1980, there is a very narrow window that limits observations across generations; however, our results support the GIF results, where the contribution from first- (genetic distance 1 and 2), second- (genetic distance=3) and third-degree relatives (genetic distance=4) was in excess for cases compared to controls. Table 4 shows the RR for leanness among the relatives of those lean Utah adults who were aged 25–65 years. Although the RR estimates are slightly larger, the conclusions are not different.

Table 3: Relative Risk of Low BMI (<18.5) in Relatives of All Lean Utah Adults

Relatives	N	Observed	Expected	p-value	Relative Risk	95% CI
First-degree	54,324	3,565	1,615.5	<0.00001	2.21	2.13, 2.28
Second-degree	96,151	2,236	1,799.7	2.1 e-23	1.24	1.19, 1.30
Third-degree	175,286	5,204	4,546.0	7.7 e-22	1.14	1.11, 1.18
Fourth-degree	317,807	6,897	6,773.4	0.133	1.02	.99, 1.04
Fifth-degree	501,420	12,417	12,271.5	0.191	1.01	.99, 1.03
Sixth-degree	571,643	14,180	14,147.9	0.788	1.00	.99, 1.02
Seventh-degree	588,223	14,796	14,772.6	0.849	1.00	.99, 1.02

doi:10.1371/journal.pone.0080287.t003

doi:10.1371/journal.pone.0080287.t003

Table 4: Relative Risk of Low BMI (<18.5) in Relatives of Lean Utah Adults Ages 25–65 Years

Relatives	N	Observed	Expected	p-value	Relative Risk	95% CI
First-degree	20,551	1,177	458.0	<0.00001	2.57	2.42, 2.72
Second-degree	32,849	1,056	775.4	<0.00001	1.36	1.28, 1.45
Third-degree	63,856	1,668	1,363.8	7.0 e-16	1.22	1.17, 1.28
Fourth-degree	153,915	3,597	3,427.9	0.0039	1.05	1.02, 1.08
Fifth-degree	309,372	6,899	6,774.1	.130	1.02	0.99, 1.04
Sixth-degree	474,053	11,512	11,376.2	.204	1.01	0.99, 1.03
Seventh-degree	562,037	12,984	13,955.3	.809	1.00	0.99, 1.02

doi:10.1371/journal.pone.0080287.t004

doi:10.1371/journal.pone.0080287.t004

Pedigree Identification

We identified all possible clusters of the 4,375 low BMI individuals between 25 and 65 years of age. These clusters merely represent all related sets of individuals with low BMI, they are not necessarily high-risk for BMI. We further evaluate each cluster by testing for an excess of low BMI among all the descendants of the founding pair of the cluster using the low BMI rates estimated from the UPDB. We identified over 4,000 clusters of low BMI relatives, ranging in size from 2 related cases (n=789 clusters) to size 168 (n=1 cluster).

We were able to identify thousands of individual pedigrees that may assist in future molecular genetic studies of the low BMI phenotype. As an example, we have identified 63 pedigrees with a significant excess of individuals with low BMI ($p<0.0001$) with at least 10 cases. An example of one of these pedigrees is shown in Figure 2. As can be observed, Utah driver's license data is only available for the most recent two or three generations of the Utah genealogy; earlier generations remain unknown for the phenotype of interest.

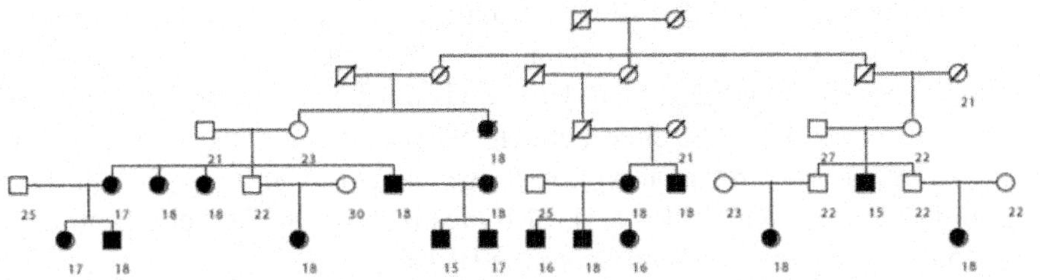

Figure 2: Example UPDB pedigree with a statitstical excess of low bmi (<18.5) individuals.

Individuals with bmi <18.5 are fully shaded and BMI is shown beneath subjects where available.

doi:10.1371/journal.pone.0080287.g002

DISCUSSION

Familial clustering of the low BMI phenotype in the UPDB is consistent with both genetic as well as environmental contributions to low BMI humans. Increased relative risks for the low BMI phenotype in first-, second-, and third-degree relatives suggests that genetic factors contribute to the familial clustering pattern.

The GIF analysis confirms a genetic contribution to low BMI in relatives with an even more distant degree of relatedness. Excess relatedness among close relatives could represent either shared environment or shared genetics, or a combination,. However, the finding of excess relatedness in distant relatives is much more likely to result from shared genes than shared environment.

Our study represents the largest genealogical population genetics study of low BMI to date. There are no comparable low BMI studies using a similar methodology. However, our identification of a genetic contribution to low BMI is consistent with findings in family studies[2]–[4] and in a single adoption study [43].

A primary limitation of this study is the reliability and validity of Department of Motor Vehicle height and weight self-reported measures. There is limited study of the accuracy of self-reported height and weight in DL data. Self-reported weights in overweight and obese individuals might be significantly underestimated due to the social stigma of obesity. In contrast, social stigma issues in reporting an accurate weight in the low BMI may be less than in obesity. Nevertheless, it remains possible that there is some social pressure to overestimate weight among the low BMI.

Utah driver›s license BMI data from the Utah Population Data Base has been compared to BMI data obtained by the CDC Behavioral Risk Factor Surveillance System (BRFSS). This analysis found BMI means generally varied in males by only three percent between the databases with no bias toward over or underestimation across age categories. In younger female age groups (between 25 and 34 years), BMI means from Utah driver›s license data was 5 to 8% lower than means from the BRFSS [44].

We compared rates of low BMI in the UPDB to the National Health and Nutrition Examination Survey (NHANES) to look for evidence of a self-report bias. [45]–[46]. The NHANES includes data from direct measurement of weight and height in a representative sample of individuals in the United States. The overall NHANES estimated prevalence of low BMI (BMI<18.5) is 1.8% in the U.S. adult population.

Prevalence rate estimates of low BMI in the UPDB generally are in close agreement with estimates from the NHANES. Rate estimates for BMI<18.5 by age group for the UPDB and (NHANES) were:20 to 39 years of age, 3.3%, (2.6%), 40 to 59 years of age, 0.9%, (1.2%), 60 to 74 years of age, .9%, (.9%), 75 years of age and older, 2.0%, (1.7%). This agreement supports the validity of the low BMI phenotype in the UPDB. Nevertheless, there is no easy method to directly measure the reliability and validity of the drivers license self-reported weight and height in the UPDB.

Some degree of inaccurate self-report of weight may be present and contribute to variance in our relative risk estimates. However, weight self-report bias would likely lead to underestimation of the relative risk.

Identification of familial relationships in the UPDB is also based on self-report. It is possible, self-reported relationships may differ from biological relationships. However, thousands of UPDB high-risk pedigrees have been genotyped with a very high rate of accuracy between self-reported relationships and genotype.

The UPDB cohorts cross several generations. Environmental dietary and physical exercise patterns likely change across generations and may have influenced a portion of BMI data.

Despite these potential limitations, this study finds excess clustering of the low BMI phenotype among both close and distant relatives supporting a significant genetic contribution. Environmental factors are also likely to contribute to familial clustering of low BMI. Families often share diet, exercise and other lifestyle patterns that contribute to body weight.

Pedigrees in the UPDB with a significant excess of low BMI individuals among the descendants of founder couples have been identified. These pedigrees could prove valuable for additional molecular genetic studies of low BMI. Figure 2 shows an example pedigree with a significant excess of individuals with low BMI. The pedigree founder has 257 descendants with BMI data. Eighteen of these individuals have a BMI<18.5. This is significantly greater than 5.3, the number expected in the pedigree ($p=1.13\ e^{5}$).

Further studies to confirm our results and to explore the molecular genetics of the low BMI phenotype are needed. Such studies may uncover the mechanisms for increased morbidity and mortality in low BMI populations. Additionally, further understanding of the genetic and environmental contributions to low BMI may provide insight for prevention and treatment of both low BMI as well as obesity.

AUTHOR CONTRIBUTIONS

Conceived and designed the experiments: WRY CJ PM LACA. Performed the experiments: LACA. Analyzed the data: LACA. Contributed reagents/materials/analysis tools: LACA. Wrote the paper: WRY.

REFERENCES

1. Day FR, Loos RJ (2011) Developments in obesity genetics in the era of genome-wide association studies. J Nutrigenet Nutrigenomics 24: 222–38. doi: 10.1159/000332158
2. Lazkarzewski PM, Khoury P, Morrison JA, Kelly , Mellies MJ, et al. (1983) Familial obesity and leanness. Int J Obesity 1983 7: 505–27.
3. Magnusson PKE, Rasmussen F (2002) Familial resemblance of body mass index and familial risk of high and low body mass index. A study of young men in Sweden. Int J Obesity 26: 1225–1231. doi: 10.1038/sj.ijo.0802041
4. Borecki IB, Higgins M, Schreiner PJ, Arnett DK, Mayer-Davis E (1998) Evidence for multiple determinants of the body mass index: The National Heart, Lung, and Blood Institute Family Heart Study. Obes Res 6: 107–14. doi: 10.1002/j.1550-8528.1998.tb00323.x
5. Stunkard AJ, Harris JR, Pedersen NL, McClearn GE (1990) The body mass index of twins who have been reared apart. N Engl J Med 322: 1483–7. doi: 10.1056/nejm199005243222102
6. Dubois L, Ohm Kyvik K, Girard M, Tatone-Tokuda F, Perusse D (2012) Genetic and environmental contributions to weight, height, and BMI from birth to19 years of age: an international study of over 12,000 twin pairs. PLoS One 7 (2) e30153. doi: 10.1371/journal.pone.0030153

7. Liu XG, Tan LJ, Lei SF, Liu YJ, Shen H, et al. (2009) Genome-wide association and replication studies identified TRHR as an important gene for lean body mass. Am J Hum Genet 84: 418–23. doi: 10.1016/j.ajhg.2009.02.004

8. Bah J, Westberg L, Baghaei F, Henningsson S, Rosmond R, et al. (2010) Further exploration of the possible influence of polymorphisms in HTR2C and 5HTT on body weight. Metabolism 59: 1156–63. doi: 10.1016/j.metabol.2009.11.007

9. Jacquemont S, Reymond A, Zufferey F, Harewood L, Walters RG, et al. (2011) Mirror extreme phenotypes associated with gene dosage at the chromosome 16p11.2 locus. Nature 478: 97–102.

10. Hai R, Pei YF, Shen H, Zhang L, Liu XG, et al. (2012) Genome-wide association study of copy number variation identified gremlin1 as a candidate gene for lean body mass. J Hum Genet 57: 33–7. doi: 10.1038/jhg.2011.125

11. Hunt SC, Stone S, Xin Y, Scherer CA, Magness CL, et al. (2008) Association of the*FTO* gene with BMI. Obesity (Silver Spring) 16: 902–4. doi: 10.1038/oby.2007.126

12. Vitonis AF, Baer HJ, Hankinson SE, Laufer MR, Missmer SA (2010) A prospective study of body size during childhood and early adulthood and the incidence of endometriosis. Hum Reprod 5: 1325–34. doi: 10.1093/humrep/deq039

13. Khashan AS, Kenny LC (2009) The effects of maternal body mass index on pregnancy outcome. Eur J Epidemiol 24: 697–705. doi: 10.1007/s10654-009-9375-2

14. Kalk P, Guthmann F, Krause K, Relle K, Godes , et al. (2009) Impact of maternal body mass index on neonatal outcome. Eur J Med Res 14: 216–22.

15. Deutsch AB, Lynch O, Alio AP, Salihu HM, Spellacy WN (2010) Increased risk of placental abruption in low BMI women. Am J Perinatol 27: 235–40. doi: 10.1055/s-0029-1239490

16. Salihu HM, Lynch O, Alio AP, Mbah AK, Kornosky JL, et al. (2009) Extreme maternal low BMI and feto-infant morbidity outcomes: a population-based study. J Matern Fetal Neonatal Med 22: 428–34. doi: 10.1080/14767050802385764

17. American Psychiatric Association. (2000) Diagnostic and Statistical Manual of Mental Disorders, 4th ed, text rev. Washington, DC: American Psychiatric Association.

18. Orpana HM, Berthelot JM, Kaplan MS, Feeny DH, McFarland B, et al. (2010) BMI and mortality: results from a national longitudinal study of Canadian adults. Obesity (Silver Spring) 18: 214–8. doi: 10.1038/oby.2009.191
19. Tamakoshi A, Yatsuya H, Lin Y, Tamakoshi K, Kondo T, et al. (2010) JACC Study Group. BMI and all-cause mortality among Japanese older adults: findings from the Japan collaborative cohort study. Obesity (Silver Spring) 18: 362–9. doi: 10.1038/oby.2009.190
20. Lederer DJ, Wilt JS, D›Ovidio F, Bacchetta MD, Shah L, et al. (2009) Obesity and low BMI are associated with an increased risk of death after lung transplantation. Am J Respir Crit Care Med 180: 887–95. doi: 10.1164/rccm.200903-0425oc
21. Flegal KM, Graubard BI, Williamson DF, Gail MH (2005) Excess deaths associated with low BMI, overweight, and obesity. JAMA 293: 1861–7. doi: 10.1001/jama.298.17.2028
22. Skolnick M (1980) The Utah genealogical database: a resource for genetic epidemiology. In: Cairns J, Lyon JL, Skolnick M, editors. Cancer incidence in defined populations. Cold Spring Harbor, NY: Cold Spring Harbor Laboratories pp 285–97
23. Cannon LA, Bishop DT, Skolnick MH (1986) Segregation and linkage analysis of breast cancer in the Dutch and Utah families. Genet Epidemiol Suppl 1: 43–8. doi: 10.1002/gepi.1370030707
24. Cannon-Albright LA, Goldgar DE, Neuhausen S, Gruis NA, Anderson DE, et al. (1994) Localization of the 9p melanoma susceptibility locus (MLM) to a 2-cM region between D9S736 and D9S171. Genomics 23: 265–8. doi: 10.1006/geno.1994.1491
25. Shirts BH, Burt RW, Mulvihill SJ, Cannon-Albright LA (2010) A population-based description of familial clustering of pancreatic cancer. Clin Gastroenterol Hepatol 8: 812–6. doi: 10.1016/j.cgh.2010.05.012
26. Maul JS, Burt RW, Cannon-Albright LA (2007) A familial component to human rectal cancer, independent of colon cancer risk. Clin Gastroenterol Hepatol 5: 1080–4. doi: 10.1016/j.cgh.2007.04.025
27. Teerlink C, Farnham J, Allen-Brady K, Camp NJ, Thomas A, et al. (2012) A unique genome-wide association analysis in extended Utah high-risk pedigrees identifies a novel melanoma risk variant on chromosome arm 10q. Hum Genet 131: 77–85. doi: 10.1007/s00439-011-1048-z
28. Horne BD, Camp NJ, Muhlestein JB, Cannon-Albright LA (2006) Identification of excess clustering of coronary heart diseases among

extended pedigrees in a genealogical population database. Am Heart J 152: 305–11. doi: 10.1016/j.ahj.2005.12.028

29. Weires MB, Tausch B, Haug PJ, Edwards CQ, Wetter T, et al. (2007) Familiality of diabetes mellitus. Exp Clin Endocrinol Diabetes 115: 634–40. doi: 10.1055/s-2007-984443

30. Tashjian RZ, Farnham JM, Albright FS, Teerlink CC, Cannon-Albright LA (2009) Evidence for an inherited predisposition contributing to the risk for rotator cuff disease. J Bone Joint Surg Am 91: 1136–42. doi: 10.2106/jbjs.h.00831

31. Albright FS, Orlando P, Pavia AT, Jackson GG, Cannon Albright LA (2008) Evidence for a heritable predisposition to death due to influenza. J Infect Dis 197: 18–24. doi: 10.1086/524064

32. Teerlink CC, Hegewald MJ, Cannon-Albright LA (2007) A genealogical assessment of heritable predisposition to asthma mortality. Am J Respir Crit Care Med 176: 865–70. doi: 10.1164/rccm.200703-448oc

33. Miki Y, Swensen J, Shattuck-Eidens D, Futreal PA, Harshman K, et al. (1994) A strong candidate for the breast and ovarian cancer susceptibility gene BRCA1. Science 266: 66–71. doi: 10.1126/science.7545954

34. Tavtigian SV, Simard J, Rommens J, Couch F, Shattuck-Eidens D, et al. (1996) The complete BRCA2 gene and mutations in chromosome 13q-linked kindreds. Nat Genet 12: 333–7. doi: 10.1038/ng0396-333

35. Cannon-Albright LA, Goldgar DE, Meyer LJ, Lewis CM, Anderson DE, et al. (1992) Assignment of a locus for familial melanoma, MLM, to chromosome 9p13–p22. Science 258: 1148–52. doi: 10.1126/science.1439824

36. Kamb A, Shattuck-Eidens D, Eeles R, Liu Q, Gruis NA, et al. (1994) Analysis of the p16 gene (*CDKN2*) as a candidate for the chromosome 9p melanoma susceptibility locus. Nat Genet 8: 23–6. doi: 10.1038/ng0994-22

37. Vesprini D, Nam RK, Trachtenberg J, Jewett MA, Tavtigian SV, et al. (2001) HPC2 variants and screen-detected prostate cancer. Am J Hum Genet 68: 912–7. doi: 10.1086/319502

38. Clopper CJ, Pearson ES (1934) The use of confidence or fiducial limits illustrated in the case of the binomial. Biometrika 26: 404–413. doi: 10.1093/biomet/26.4.404

39. Hill JR (1980) A survey of cancer sites by kinship in the Utah Mormon population. In: Cairns J, Lyon JL, Skolnick M, editors. Cancer incidence in defined populations. Cold Spring Harbor, NY: Cold Spring Harbor Laboratories pp 299–318.

40. Cannon-Albright LA (2008) Utah family-based analysis: past, present and future. Hum Hered 65: 209–20. doi: 10.1159/000112368
41. Malecot G (1948) Les mathematiques de l'heredite. Paris: Masson & Cie.
42. Agresti A (1990) Categorical Data Analysis. New York: Wiley.
43. Costanzo PR, Schiffman SS (1989) Thinness—not obesity—has a genetic component. Neurosci Biobehav Rev 13: 55–58. doi: 10.1016/s0149-7634(89)80052-1
44. Centers for Disease Control and Prevention (CDC) (2008) Behavioral Risk Factor Surveillance System Survey Data. Atlanta, Georgia: U.S. Department of Health and Human Services, Centers for Disease Control and Prevention.
45. Centers for Disease Control and Prevention (CDC) (2011) National Center for Health Statistics (NCHS). National Health and Nutrition Examination Survey Data. Hyattsville, MD: U.S. Department of Health and Human Services, Centers for Disease Control and Prevention. Available:http://www.cdc.gov/nchs/nhanes/about_nhanes.htm
46. Center for Disease Control and Prevention (CDC) National Center for Health Statistics (NCHS). Health E-Stat. Prevalence of Low BMI Among Adults: United States, 2003–2006. Available: http://www.cdc.gov/nchs/data/hestat/low BMI/low BMI_adults.htm

Chapter 10

GENETIC ANALYSIS OF INFLORESCENCE AND PLANT HEIGHT COMPONENTS IN SORGHUM (PANICOIDAE) AND COMPARATIVE GENETICS WITH RICE (ORYZOIDAE)

Dong Zhang[1,2], Wenqian Kong[1,3], Jon Robertson[1] , Valorie H Goff[1] , Ethan Epps[1] , Alexandra Kerr[1] , Gabriel Mills[1] , Jay Cromwell[1] , Yelena Lugin[1] , Christine Phillips[1] and Andrew H Paterson[1,2,3,4,5]

[1]Plant Genome Mapping Laboratory, University of Georgia, Athens, GA 30602, USA

[2]Institute of Bioinformatics, University of Georgia, Athens, GA 30602, USA

[3]Department of Crop and Soil Sciences, University of Georgia

[4]Department of Plant Biology, University of Georgia

[5]Department of Genetics, University of Georgia

ABSTRACT

Background

Domestication has played an important role in shaping characteristics of the inflorescence and plant height in cultivated cereals. Taking advantage of meta-analysis of QTLs, phylogenetic analyses in 502 diverse sorghum accessions, GWAS in a sorghum association panel (n = 354) and comparative data, we provide insight into the genetic basis of the domestication traits in sorghum and rice.

Results

We performed genome-wide association studies (GWAS) on 6 traits related to inflorescence morphology and 6 traits related to plant height in sorghum, comparing the genomic regions implicated in these traits by GWAS and QTL mapping, respectively. In a search for signatures of selection, we identify genomic regions that may contribute to sorghum domestication regarding plant height, flowering time and pericarp color. Comparative studies across taxa show functionally conserved 'hotspots' in sorghum and rice for awn presence and pericarp color that do not appear to reflect corresponding single genes but may indicate co-regulated clusters of genes. We also reveal homoeologous regions retaining similar functions for plant height and flowering time since genome duplication an estimated 70 million years ago or more in a common ancestor of cereals. In most such homoeologous QTL pairs, only one QTL interval exhibits strong selection signals in modern sorghum.

Conclusions

Intersections among QTL, GWAS and comparative data advance knowledge of genetic determinants of inflorescence and plant height components in sorghum, and add new dimensions to comparisons between sorghum and rice.

BACKGROUND

The Sorghum genus has recently become an important botanical model for Andropogoneae grasses, by virtue of its relatively small and largely-sequenced genome, a minimum of gene duplication thanks to 70 million years of abstinence from polyploidy, and its close relationship to grasses such as maize, sugarcane and Miscanthus that have much more complex genomes [1]. Cultivated sorghum (*Sorghum bicolor*) ranks fifth in importance among the world's grain crops, is a versatile source of food, fodder, and fuel, and possesses a great diversity of cultivated forms that may reflect its wide range of adaptation [2-4].

The ~30 year history of using linked molecular markers to dissect complex traits in plants has broadly used two complementary approaches. Conventional biparental QTL mapping [5] has been widely used and has provided foundational information that led to some successes in the identification of causal genes in many organisms. However, biparental QTL mapping generally offers relatively coarse resolution that is not sufficient to determine causative genes. Highly saturated recombination maps, multiparent advanced generation intercrosses (MAGIC) [6] or nested association mapping (NAM) [7], offer options to enhance mapping resolution of QTLs. Dramatic increases in genomic data provide rich resources with which to investigate genes and gene functions on a much finer scale than QTL mapping by taking advantage of historical accumulation of recombination events in a gene pool using 'association genetics' [8]. However, association mapping can require extremely high DNA marker densities to thoroughly scan a genome for genes influencing a trait, and complex measures to distinguish between artifacts such as relatedness among genotypes (especially in improved germplasm) and true evidence of functional association between a mutation and a phenotype [8-11]. Although GWAS is able to explore for causative loci on a genome-wide scale, population structure and genetic relatedness may confound associations at causative loci. Some GWAS conducted in rice and *A. thaliana* have suggested that known causative loci showed weaker signals than nearby markers [12]. In contrast, carefully designed crossing schemes in QTL mapping may be more targeted to locate relevant QTLs. Identifying intersections between results from biparental QTL mapping and association genetic data is a potentially powerful means to mitigate constraints associated with each approach, accelerating progress toward identifying specific genes that function in biological processes of relevance to agriculture.

The grass inflorescence, the primary food source for humanity, has been repeatedly selected during domestication [13]. Some well-characterized domestication traits related to the inflorescence include pericarp color of seed, seed shattering, awn length/presence and seed size/yield. Of similarly high and recurring importance in plant domestication and crop improvement are flowering time and plant height, which often show significant genetic correlation with one another [13,14].

In this study, we use GWAS to investigate 6 components of sorghum inflorescence morphology and 6 traits related to plant height, then compare GWAS-based associations to positional evidence from meta-analysis of QTL likelihood intervals. QTL meta-analysis, the comparison of multiple independent QTL studies in different germplasm and environments, is used here to provide a more comprehensive picture of the true genetic control of a trait than analysis of any single population [15], for example revealing plant height and flowering of sorghum to be genetically more complex than had been realized after more than 70 years of investigation [16]. We note that a few classically-identified loci (*dw1-dw4)* had already been compared to GWAS data [4], one of which (*dw3*) had been cloned [17], however here we address additional loci identified only by meta-analysis of QTL likelihood intervals.

Inbreeding organisms have limits to the precision in the association mapping studies [12]. Sorghum is largely inbreeding, which can result in strong LD patterns, and may lead to low genetic resolution in specific local regions along the genome. Hence, we reported the hotspots underlying 12 traits in sorghum, instead of gene candidates. The identified 'hotspots' provide a valuable advance toward the goal of uncovering causative variants for the trait of interest.

It is accepted widely that QTL intervals controlling common traits have non-random correspondence across and within cereals, and GWAS adds a new dimension to the ability to compare the genetic control of common traits in different cereals (and other taxa). For example, in an early comparison, the observed probabilities that seed mass (size) QTLs in sorghum, rice, and maize would correspond so frequently by chance was conservatively estimated as 0.1 to 0.8% [18]. Another early study [14] indicated that 8 of 25 regions affecting flowering of maize fall into 4 homoelogous regions. Numerous studies have shown that some orthologs across taxa have similar functions underlying common phenotypes, but other causative genes have no obvious counterparts that contribute to similar traits even in their close relatives. Thus, genetic correspondence may reflect functionally conserved 'hotspots' existing across taxa, but may or may not be conserved 'genes'. Genes in a pathway exhibit significantly higher genomic clustering than expected by chance in eukaryotes [19]; for example, co-regulated clusters of genes have been implicated in QTLs affecting cotton fiber traits [20].

The common ancestor of rice, maize and sorghum experienced a whole-genome duplication (WGD; named rho) that is still readily discernible in their genomes [21], making it possible to test the hypothesis of convergent evolution across an estimated 70 million years, with the possibility of subfunctionalization among homoeologous regions. A hypothesis worthy of further exploration is that a co-regulated cluster of genes in the cereal common ancestor may have experienced gain/loss and functional divergence of some members in the subsequent 70 my of divergence, with independent domestications conferring additional functional changes in similar locations of different taxa but which are not strictly orthologous.

METHODS

Genotype

The imputed version of ~265,000 published SNPs characterized in 971 worldwide accessions based on genotyping-by-sequencing (GBS) was employed [4]. About 72% of annotated genes contain ≥1 SNP site. A total of 228 of 354 accessions from a US sorghum association panel (SAP) [22] are converted tropical lines that are photoperiod insensitive, early maturing, and short stature phenotypes, produced via crossing exotic lines and modern U.S. cultivars. It has been demonstrated [4] that the population has sufficient power to dissect a trait, such as inflorescence architecture, that was not a target of selection in the sorghum conversion program. We used the 354 accessions from the SAP to perform GWAS. A total of 502 accessions, each characterized to a unique morphological type, were used to analyze the population structure. Wild sorghums [*Sorghum bicolor* ssp. *verticilliflorum* (L.) Moench] (n = 31) of four races, namely *aethiopicum*,*arundinaceum*, *verticilliflorum* and *virgatum*, were used to calculate expected heterozygosity for the wild population, and to compare with the heterozygosity in the 502 accessions of cultivated sorghum.

Phenotype for GWAS

Phenotypic data for the 354 accessions in the SAP from three different growouts was utilized. We used a completely randomized design for all the 354 accessions (unreplicated trial with two observations per plot). On seed grown during 2008 in Lubbock, TX, the average RGB values for pericarp color for each genotype were determined from images of the surfaces of 5 seeds. Using the conversion formula of RGB→CIE-L*ab, RGB was transformed to CIE-L*ab color space. RGB is summarized by three values: R, G and B, and CIE-L*ab is summarized by: L, a and b. From growouts in 2009 (seeds sowed on May 19th) and 2010 (seeds sowed on May 26th), near Watkinsville GA, we took two representative samples (from each plot) per genotype in each year for the following measurements. Awn presence: presence or absence of awn, with 2 for abundant, 1 for occasional, and 0 for absent; base-flag length: measured in cm, from the plant base to the flag leaf; flag-rachis length: measured in cm, from the flag leaf to the top node, with positive values indicating the flag leaf below the rachis and negative values indicating the flag leaf above the rachis; inflorescence length: measured in cm, from the rachis to the top of inflorescence; inflorescence width: measured in cm, at the widest point; nodes: the number of nodes; whorls: the number of whorls; dry inflorescence weight; dry stalk weight; total plant height; and flowering time: the average number of days for the first five heads to flower, from the planting date. For traits measured on two samples, we used the average values per genotype to assess heritability and phenotypic correlations. Since the raw data in all the quantitative traits in our study is distributed in near-normal fashion, we conducted GWAS by using the data without transformation. Considering that data combined across years may weaken association levels for those traits with low heritability, we performed GWAS on data for each year individually.

QTL Mapping

QTL confidence intervals were from two resources. (1) We compiled 1-LOD likelihood intervals, which have been identified to underlie any of the 12 traits from published literature [13,14,18,23-29]. (2) We used a recombinant inbred line (RIL) from an interspecific cross between *S. bicolor* and *S. propinquum* (BTxSP), the widest cross that can be made with *S. bicolor*using conventional techniques, containing 141 loci on 10 linkage groups collectively spanning 773.1 cM to map the confidence intervals for base-flag length, flowering time, nodes and total plant height [29] for 2010 and 2011 data. Phenotypic data were collected with two replications per genotype in both 2010 and 2011. A LOD score of 2.5 was used for QTL detection. The methods for anchoring QTL intervals to the reference genome have been discussed in Zhang et al., 2013 [16]. Briefly, based on colinearity between genetic and physical positions of markers, a QTL region is delineated by two flanking markers nearest to the likelihood peak that have alignment information (BLASTN hits). Genomic positions for intervals are available from Additonal file 2: Table S2.

GWAS

The Compressed Mixed Linear Model (CMLM) involves genetic marker-based kinship matrix modeling of random effects, used jointly with population structure estimated by principal components analysis (PCA) to model fixed effects [30-32]. Since more extensive genetic variations may confer a phylogeny closer to the true one, the total of 265,487 SNPs in the SAP were used to analyze population structure. The compression level and optimal number of principal components that adequately explain population structure were previously determined by the Genomic Association and Prediction Integrated Tool [30-32]. Log quantile–quantile (QQ) P-value plots for 265,487 single-SNP tests of association implied that there were few systematic sources of spurious association using CMLM, noting the close adherence of P values to the null hypothesis over the most of the range.

Significance Threshold

We performed Bonferroni-like multiple testing correction [33] to determine significance thresholds for GWAS. Instead of 265,487 independent tests assumed in the Bonferroni method, the total number of tests was estimated by using the average extent of LD across the genome. On average, LD decays to background levels ($r^2 < 0.1$) within 150 kb in the current GBS data [4]. The effective number of independent tests was defined as LD bins [reference genome size (730 Mb)/average LD extent (150 kb)]. Given 0.05 as the desired experiment wide probability of type I error, a significance cutoff within about an order of magnitude of 10^{-5} was estimated.

Overlap between QTL and Heterozygosity Reduction

Since the hypergeometric probability distribution (sampling without replacement) $p = \frac{\binom{l}{m}\binom{n-l}{s-m}}{\binom{n}{s}}$ can assess the correspondence between QTLs [14,23], we used the hypergeometric probability distribution to evaluate genetic overlap between plant height/flowering time QTLs and significant heterozygosity reduction in wild sorghum. n is the total number of intervals (defined as 30 cM, approximating a QTL likelihood interval) along the whole genome; l is the number of intervals having significant heterozygosity reduction; s is the number of intervals having plant height/flowering time QTL; *m*is the number of intervals having both features (overlapping intervals). The purpose of this test is to show that biparental QTL mapping may capture genetic variations for domestication traits that were evolved from wild sorghum and of importance in the history of sorghum selection. Thus, we used QTL intervals for plant height/flowering time that were only determined by biparental QTL mapping. The regions with the largest 1% of heterozygosity reduction values were selected for testing. We utilized alignment between a high-density genetic recombination map [34] and the sorghum reference genome [1] to unify the genetic positions of QTLs and heterozygosity reduction regions (based on windows of 500 consecutive SNPs).

Reference Genomes

The gene annotations refer to JGI annotation release Sbi1.4 [1] and Michigan State University Rice Genome Annotation Project (MSU-RGAP release 7) [35].

RESULTS AND DISCUSSION

Phylogenetic Relationships of Five Main Sorghum Races

Morris et al used a genome-wide SNP map to explore the population structure of 971 sorghum accessions, and illustrated the differentiation of their geographic origins [4]. However, several intriguing questions remain, for example: (1) what is the most primitive sorghum type?, and (2) how many independent domestications has sorghum experienced?

It is generally accepted that the domestication of sorghum started in Africa. Bicolor, guinea, caudatum, durra and kafir are five main morphological types that are well recognized to represent genetic diversity in the cultivated sorghum. Wild sorghum [*Sorghum bicolor* ssp. *verticilliflorum* (L.) Moench] including four races (namely *aethiopicum*, *arundinaceum*,*verticilliflorum* and *virgatum*), is the progenitor of the cultivated sorghum [*Sorghum bicolor* (L.) Moench] [36,37]. Thc phylogenetic relatedness in sorghum races has been discussed in several studies [38-41], but ambiguous clustering patterns have often been found, part of which may be attributable to the limitations of either low-density markers or small population size. To re-investigate inferences drawn before, we focused on a subset of 502 accessions, in which 471 are each characterized uniquely to one and only one of the five primary cultivated races and 31 are wild types. Both the phylogenetic tree and the PCA plots indicate that bicolor is the most primitive race, based on having close phylogenetic relationship with wild types (Figure 1a and b). The level of population differentiation, fixation index (F_{ST}), was measured between wild types and each of five primary types. Race bicolor (F_{ST} = 0.04) exhibits closer genetic relationship with wild sorghums than any of the other 4 primary races (F_{ST} (guinea-wild) = 0.11, (durra-wild) = 0.20, (kafir-wild) = 0.33, (caudatum-wild) = 0.14), with guinea and caudatum apparently representing early derivatives. A large block of bicolor accessions are intermediate among durra types, potentially consistent with an ancestral relationship (Figure 1b), noting that F_{ST} (0.14) supports small population differentiation between bicolor and durra. Likewise, a large block of guinea accessions clustering within caudatum may suggest another derivation in the history of sorghum selection, also supported by evidence of minimal population differentiation (F_{ST} = 0.17). Races caudatum, durra and kafir show clustering patterns that are substantially distinct from one another, save for occasional single accessions that could be misclassifications (Figure 1b). Pairs of these 3 sorghum races show relatively high levels of population differentiation [F_{ST} (durra-caudatum) = 0.26, (durra-kafir) = 0.46, (caudatum-kafir) = 0.33]. Although the first three components of PCA are able to explain 89.1% of the total genetic variance, ambiguous clustering patterns are still occasionally observed, especially for the two primitive races bicolor and guinea. Both the PCA plot and neighbor-joining tree show quite similar clustering patterns by using the complete SNP set (265,487 SNPs) (Figure 1) and a set including more informative SNPs (missing rate ≤ 50% and MAF ≥ 0.02). The complex history of diffusion and selection in sorghum may confound phylogenetic inference using the clustering patterns. The phylogenetic relationships of sorghum races are still open to questions that may be more accurately addressed with growing genotype data.

a

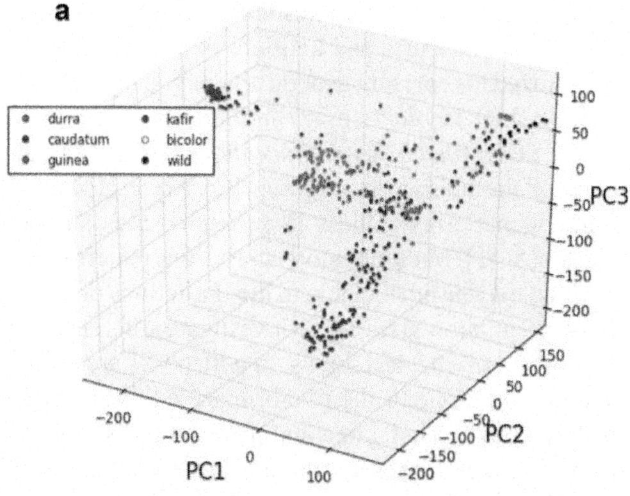

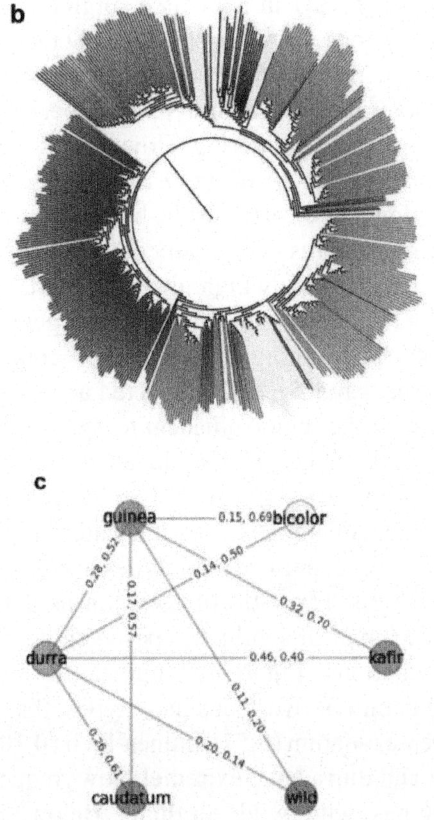

Figure 1: Population structure of 502 worldwide sorghum accessions. 476 belong to the five main cultivated races and 26 are wild types. (a)PCA plots of the first three components for 265,487 SNPs. The five main cultivated races and the wild type are color-coded. (b)Neighbor-joining tree of 502 sorghum accessions. (c) Population differentiations and frequencies of common two-locus haplotypes for 100 SNPs adjacent to the *Sh1* gene [2] for pairs of populations. All the connections for guinea and durra are shown. The F_{ST}values paired with frequencies of common two-locus haplotypes are indicated.

Using representative SNPs across the *Shattering1* (*Sh1*) gene, a key domestication locus [2], a recent discovery revealed four haplotypes, in which three represent non-shattering forms. Specifically, guinea and durra share a common haplotype (SC265); a second haplotype (Tx623) is prevalent in kafir; and a third haplotype (Tx430) is dominant in caudatum. Three non-shattering haplotypes strongly suggest at least three domestication episodes for reduced shattering in the five main sorghum races [2,3]. In general, domestication loci can exhibit very strong LD, because of reduction in genetic divergence. Unusually high genetic divergence at *Sh1* in cultivated sorghum races may explain why the surrounding region does not show stronger LD than the background level. The clustering pattern in the neighbor-joining tree and F_{ST} values support the inference that kafir and caudatum have experienced two independent domestication events for non-shattering. However, is the non-shattering allele derived from a common ancestor of guinea and durra? Generally, we did not observe a close phylogenetic relationship between guinea and durra on the basis of the clustering patterns of the genetic tree (Figure 1b), and the large value of F_{ST} (0.28). This relatively high level of genetic differentiation implies that guinea and durra may have experienced convergent domestication events. The types of haplotypes resulting from selection could be restricted by the limited number of representative SNPs detected across *Sh1*. To overcome this limitation, we examined the region adjacent to *Sh1* to provide a pool enriched by SNPs that are tightly linked to the shattering locus. Since *Sh1* was not genotyped in the current data set, it would be challenging to know the exact number of SNP sites that are linked to *Sh1*. Instead, using 100 SNPs in a 200 kb region adjacent to *Sh1*, we examined haplotypes in consecutive pairs of loci for each sorghum population. The ratio of common haplotypes/total haplotypes was calculated for each pair of sorghum types. Among the 9 possible pairwise comparisons of populations (Figure 1c), guinea-durra only shows a modest frequency (0.52) of common two-locus haplotypes, that is lower than guinea-bicolor (0.69), guinea-caudatum (0.57), guinea-kafir (0.70) and durra-caudatum (0.61). Both guinea and durra have extremely low frequencies (0.20 and 0.14) of common haplotypes with wild sorghum. Below, using two additional domestication genes mapped herein, we further investigate the hypothesis that guinea and durra may have experienced independent domestication events but achieved non-shattering by convergence.

Meta-Analysis of Sorghum QTLs

Biparental QTL mapping is based on the principle that genes and linked DNA markers largely co-segregate during meiosis save for occasional recombination events, thus allowing their analysis in the progeny. The limited number of recombination events captured in progeny of recent crosses may result in QTL likelihood intervals that contain dozens or even hundreds of genes. Further, the environment and parental lines used in a cross can limit the power to accurately estimate the number of QTLs and magnitude of their effects. Using a database that we have recently described [16], we compiled l-LOD QTL likelihood intervals resulting from 11 independent biparental QTL mapping studies to yield a more complete picture of the genetic control of a trait than could be obtained from any individual study.

The pericentromeric region of sorghum chromosome 6 (Sb06) has repeatedly shown evidence of genetic control of plant height, and was thought to harbor two classic dwarfing genes (*dw2* and *dw4*) [4]. The general lack of recombination in this region allowed QTL confidence intervals to cross centromeres and cover broad genomic areas. Two other regions repeatedly associated with plant height are in the euchromatin of Sb07 and Sb09, which were considered to contain *dw3* and *dw1* respectively [4]. An additional 9 nonoverlapping regions in the sorghum genome containing height QTLs show that genetic control of sorghum plant height involves substantially more than the four genes reported in classical studies [42]. Likewise, 14 flowering QTL likelihood intervals published in six studies fall into at least 11 non-overlapping regions (Figure 2), strongly indicating far more than the classically suggested six genes, *Maturity1* (*Ma1*) to *Ma6*, in genetic control of sorghum flowering time [16,42,43]. Flowering time and plant height show significant genetic correlations on chromosomes Sb01, Sb03, Sb04, Sb06, and Sb09, indicating that their inheritance is linked either functionally (pleiotropy) or physically (linkage disequilibrium) [13].

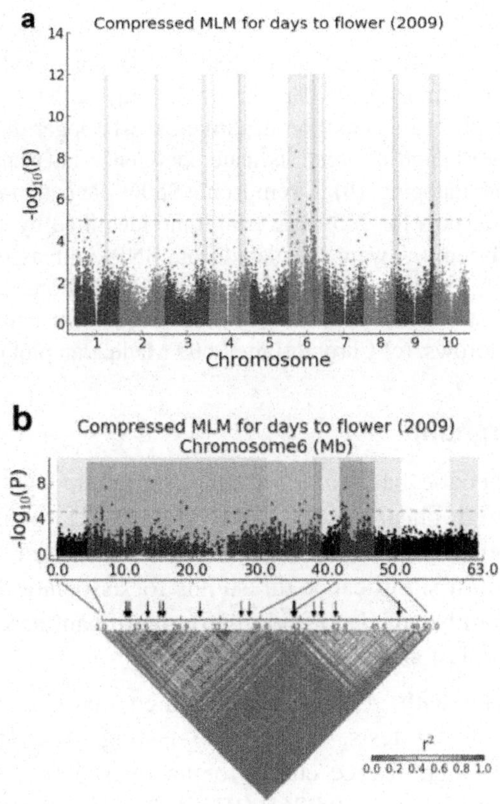

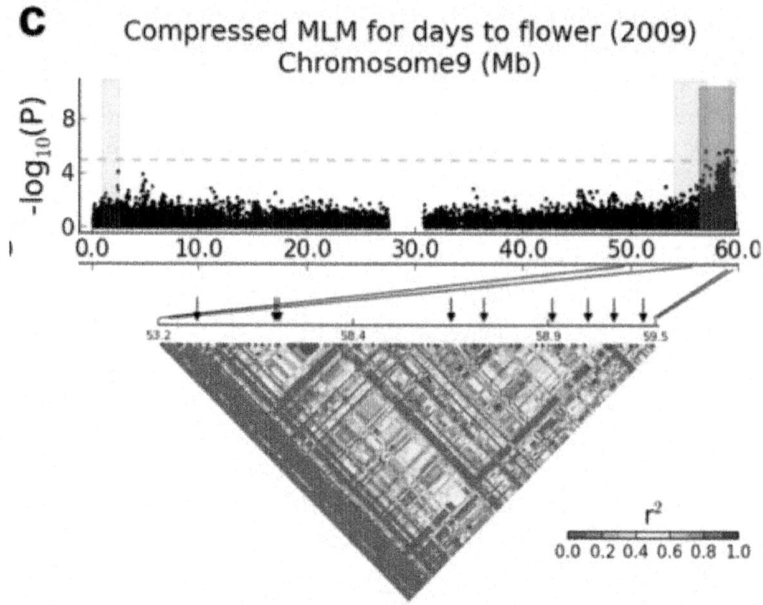

Figure 2: GWAS for flowering time in 2009. (a) Genome-wide Manhattan plot of CMLM. Significance threshold is denoted by the gray dashed line. The 10 sorghum chromosomes are plotted against the negative base-10 logarithm of the association P value. Areas highlighted in green indicate confidence intervals for flowering time determined by QTL mapping. (b) Chromosome Sb06 Manhattan plot of CMLM (top). Red areas show hotspots for 2009 flowering time identified by association mapping. Linkage disequilibrium (r^2 were calculated from SNPs with association $p \leq 0.05$ and missing data $\leq 50\%$) matrices (bottom) are plotted for regions denoted by anchoring lines. Regions of strong LD are shown in red. Significant association markers are denoted by black arrows. (c) Chromosome Sb09 Manhattan plot of CMLM.

GWAS for Sorghum

To dissect the genetic basis of 12 traits in sorghum by GWAS, we used a compressed mixed linear model (CMLM) [30,32] to assess evidence of phenotype-genotype association. Three steps were taken into consideration: (1) determination of significance thresholds for association, (2) identification of linkage disequilibrium (LD) regions for significant association signal and (3) positive control of association.

A major issue with genome scans, which involve many thousands of independent statistical tests, is multiple testing. The Bonferroni method approximates the significance cutoff for an overall (i.e. genome-wide) 5% probability of type I error as $0.05/265{,}487 = 1.89 \times 10^{-7}$ in our studies. However, this method has been criticized for its stringency [44] owing to the fact that genotype at some SNP loci are correlated thus are not independent hypotheses. Sorghum is largely inbreeding, which can result in strong LD patterns along the genome, so that an appropriate significance threshold may be larger than 1.89 × 10^{-7}. Here, we used the quantified average LD information in sorghum to adjust the Bonferroni correction [33] (See details in Methods). A significance cutoff within about an order of magnitude of 10^{-5} is inferred to balance an acceptable false positive rate with sufficient power to detect true associations.

It is also important to determine LD with single SNP association, especially when causative variants are not genotyped (or at least not known). On the basis of pairwise measures of LD (r^2), 'block-like' structures can be visually apparent. It is now well understood that the extent of LD in the pericentromeric region, which experiences relatively little recombination, is greater than in the euchromatin, which experiences more frequent recombination. A long LD block with association signals is most likely to contribute striking features to the 'skyline' of a genome-wide Manhattan plot.

Known genes and biparental QTL intervals that have been identified previously are useful to assess association validity. If knowledge of such candidate genes/intervals is limited in the species of interest, information from closely related species might be utilized, using synteny-based approaches to deduce orthology. In addition to the compilation of QTL likelihood intervals and *Dwarf1*(*Dw1*)-*Dw4* loci [4] in sorghum, sorghum maturity (*Ma*) genes *Ma1* [14] and *Ma6* [45], and genes *yellow seed1* (*y1*) [46,47] and *Tannin1* (*Tan1*) [48] for sorghum pericarp color provide positive controls for GWAS.

Traits related to the Sorghum Inflorescence

We investigated 6 properties of the sorghum inflorescence, including awn presence, pericarp color, dry inflorescence weight, inflorescence length and width, and whorl number.

Using a *S. bicolor* intraspecific map (BTxIS), Hart et al., [24] identified an interval controlling the presence of awns in euchromatin near the 3′ end of chromosome Sb03. The most striking association based on GWAS of awn presence for two years was in the genetically mapped interval (Figure 3a, Figure 4b). Both mapping strategies achieve similar genetic resolution, with intervals spanning ~4.7 megabases (Mb). We found 10 additional significant association hotspots in 2009 and 7 in 2010, none of which are consistent in the two years, which could represent modifiers affected by environment, or false positive associations.

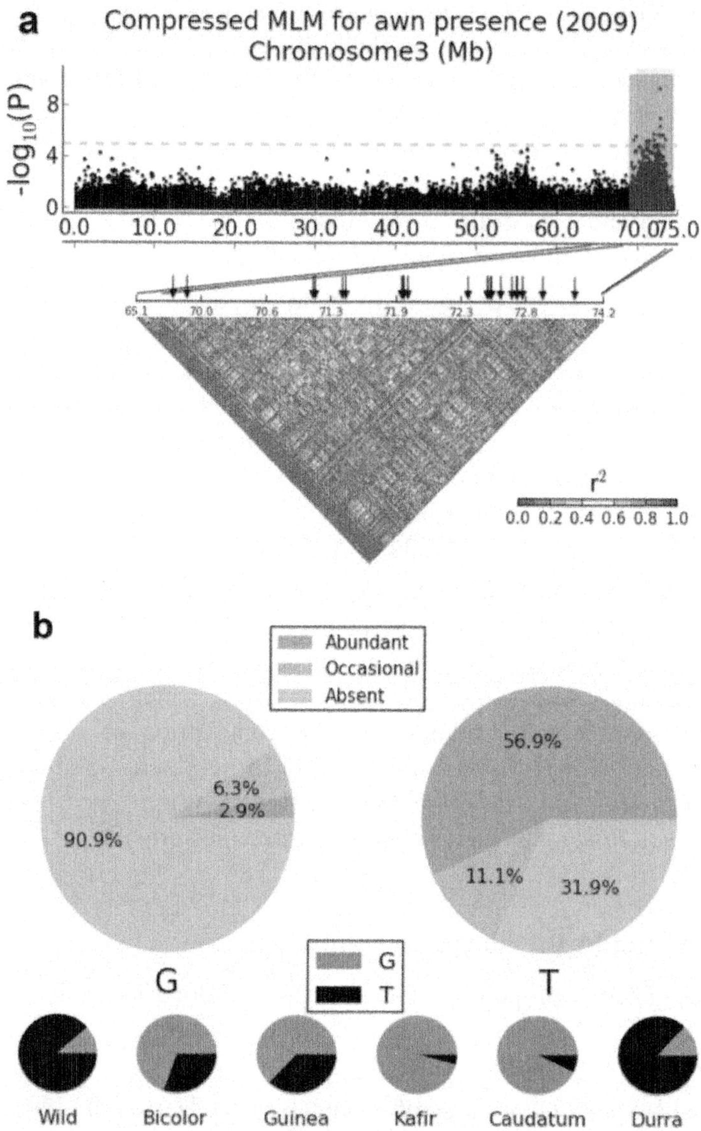

Figure 3: The spectrum of awn presence variation and allele frequencies at locus "S3_72702502" on chromosome Sb03. (a) Chromosome Sb03 Manhattan plot for 2009 awn presence in sorghum is plotted with the hotspot (red area) identified by GWAS, with the prior interval (green area) determined by QTL mapping [24], and with the LD pattern determined by r^2. Significant association markers are denoted by black arrows. (b) Three types of awn classification, which are "abundant", "occasional" and "absent" are color-coded in their frequencies plots for alleles "G" and "T". The allele frequencies are plotted for the five main cultivated races and wild sorghum.

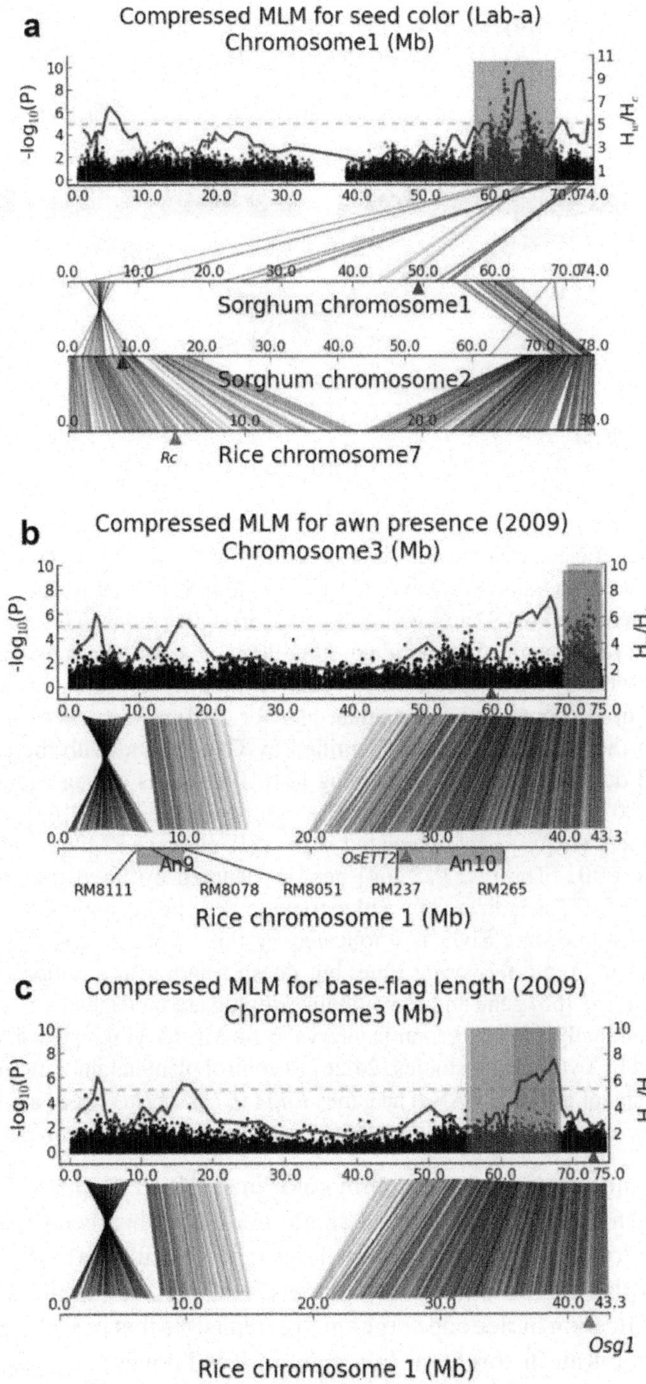
a
Compressed MLM for seed color (Lab-a)
Chromosome1 (Mb)
-log10(P)
Hu/Hc
Sorghum chromosome1
Sorghum chromosome2
Rc
Rice chromosome7
b
Compressed MLM for awn presence (2009)
Chromosome3 (Mb)
-log10(P)
Hu/Hc
An9
OsETT2
An10
RM8111
RM8078
RM8051
RM237
RM265
Rice chromosome 1 (Mb)
c
Compressed MLM for base-flag length (2009)
Chromosome3 (Mb)
-log10(P)
Hu/Hc
Osg1
Rice chromosome 1 (Mb)

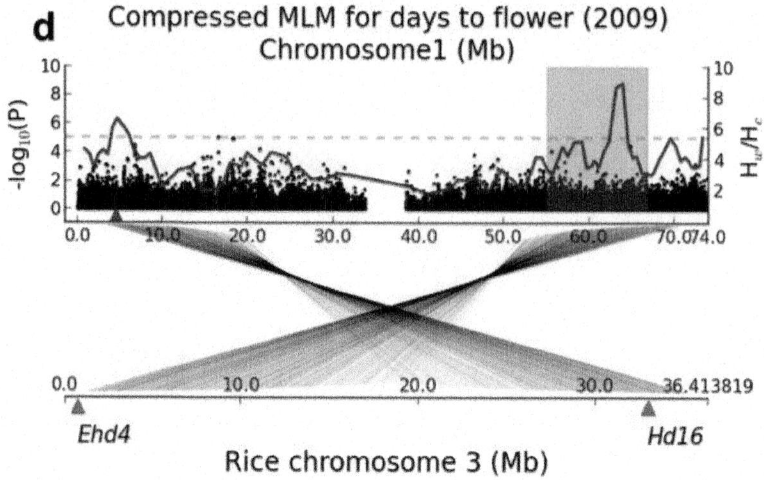

Figure 4: Genetic correspondence across taxa. (a) Chromosome Sb01 Manhattan plot for pericarp color (a value of L*ab model) of sorghum is plotted with the hotspot (red area) identified by GWAS, and with the curve of H_w/H_c ratios. The *Rc* [52] gene is denoted by a red triangle. Two sorghum homologs of *Rc* on chromosomes Sb01 and Sb02 are denoted by blue triangles. Gray connecting lines indicate pairs of duplicated genes. (b) Chromosome Sb03 Manhattan plot for 2010 awn presence in sorghum is plotted with the hotspot (red area) identified by GWAS, and with the prior interval (green area) determined by QTL mapping [24]. Rice awns co-segregated with SSR marker RM8078 tightly linked to An9 on chromosome Os01 [63]. The interval An10 for rice awn [63] was associated with SSR markers RM265 and RM237 on chromosome Os01. The *OsETT2* [64] gene is denoted by a red triangle. Sorghum ortholog of *OsETT2* is indicated by a blue triangle. (c) The genomic interval (~55 Mb-67 Mb) on chromosome Sb03 is implicated by three linkage studies [14,26,28] to affect plant height and flowering time, but doesn't harbor association signal in our GWAS. The *Osg1* [66] gene and its sorghum ortholog are indicated by a red and a blue triangle individually. (d) The genomic interval (~58 Mb-64 Mb) on chromosome Sb01 is implicated by two linkage studies [24,26] to control plant height, but doesn't harbor association signal in our GWAS. The genes *Ehd4* [67] and *Hd16* [68] are indicated by red triangles, and sorghum ortholog of*Hd16* is denoted by a blue triangle.

Inheritance studies of endosperm color in sorghum proposed that the trait was oligogenic [49,50]. To date, a single gene (*Rc*) has been verified to be responsible for pericarp pigmentation of rice grains (Figure 4a), while gene *Rd* is assumed to play a role in spreading pigment [51,52]. In view of the similar levels of gene duplication in rice and sorghum, it is plausible that pericarp color is also an oligogenic trait in sorghum. In sorghum, it is known that the gene *yellow seed1* (*y1*) is required for the production of red phlobaphene pigments in the grain pericarp [46,47], and the *Tan1* gene controls tannin biosynthesis to affect pericarp color [48]. In order to minimize the limitations of artificial descriptions of colors, we applied two commonly used color models, RGB and CIE-L*ab (See the Methods also for the details), to phenotype pericarp color of sorghum seeds. On chromosome Sb01, an association peak (~61.45 Mb) in a hotspot (58 Mb-67 Mb) for red, green and blue in RGB measurement, and the values of 'L' and 'a' in the Lab model, is close to *y1* (~61 Mb). An additional association peak (~61.86 Mb) near *Tannin1* (*Tan1*) (~61.6 Mb), is localized in a hotspot 57.5 Mb-62 Mb on chromosome Sb04 to be associated with red and green in the RGB and 'L' in the Lab.

Dry inflorescence weight, inflorescence length/width, and whorl number, were each very sensitive to the environment, with no colocalized hotspots between the two years for which we have data. The most striking 'skyline' determined by GWAS for dry inflorescence weight is centered at ~57 Mb on chromosome Sb09, which is also associated with plant height and flowering time. For inflorescence length the most striking association hotspot centered at ~45 Mb on chromosome Sb06, a location also connected with plant height and flowering.

To further investigate the inference that guinea and durra may have experienced independent domestication events but achieved non-shattering by convergence, we examined two-locus haplotypes for loci associated with pericarp color and awn presence respectively, both considered to be traits that may be subject to selection. For pericarp color, the association peak (S1_61453639) and 7 linked ($r^2 \geq 0.6$) SNPs in the hotspot (chromosome Sb01: 57,017,032 bp-68,413,103 bp) are used to calculate the ratio of common haplotypes/total haplotypes for each pair of sorghum types (See the "Phylogenetic relationships of five main sorghum races" also for the details). Similarly, we used an association peak (S3_72702502) and 18 linked SNPs in the hotspot (chromosome Sb03: 68,912,931 bp-74,241,979 bp) to calculate two-locus haplotype ratios for awn presence. Compared to other sorghum race pairs, guinea-durra only shows modest frequencies of common two-locus haplotypes for pericarp color (freq = 0.22) and awn presence (freq = 0.26). Our findings further support the hypothesis that independent domestication of guinea and durra involved convergent selection for non-shattering.

Race-Specific Patterns in awn Presence Variation

Geographic origins and domestication history can result in patterns of phenotypic variation among genotypes within a gene pool. We investigated whether awn presence exhibits variation patterns correlated with race-specific alleles for sorghum. The hotspot (chromosome Sb03: 68,912,931 bp-74,241,979 bp) for sorghum awn presence, with the association peak (S3_72702502), is detected by both GWAS and QTL mapping (See "Traits related to the sorghum inflorescence" also). The association peak is located at an intergenic locus, which is flanked by two annotated genes [*Sb03g045420*(chromosome Sb03: 72,681,274 bp-72,687,688 bp) (similar to Hexokinase-3) and *Sb03g045430* (chromosome Sb03: 72,703,668 bp-72,704,913 bp) (similar to Putative uncharacterized protein)], based on published reduced representation sequence [4]. In this scenario, it is probable that the causative genes/loci were not genotyped, and that S3_72702502 and the causative loci shared history of mutation and recombination.

An early study [53] described morphology of panicles for two of the five main sorghum races. We (Figure 3b) show allele distribution at the locus S3_72702502, reflecting alleles 'T' and 'G' associated with awned and awnless sorghum panicles individually, finding correlation between the alleles and the race-specific morphology of awns. Allele 'G' with the dominant frequency in race kafir corresponds to awnless cylindrical-shaped panicles of kafir [53]. Indeed, we found 84% of kafir accessions to have awnless panicles in our phenotypic data. Similarly, allele 'T' with the dominant frequency in race durra, corresponds to bearded and hairy panicles of durra [53]. The major allele in race caudatum is 'G', consistent with the finding that 92% of caudatum accessions for which we have data are awnless. Our findings also suggest that the two most primitive sorghum types, bicolor and guinea, derived more allele 'G' from wild sorghum which is dominated by more ancestral allele 'T'.

Traits related to Plant Height

We conducted GWAS for 6 traits related to sorghum plant height, including total plant height, distance from base to flag leaf (base-flag), distance from rachis to flag leaf (rachis-flag), number of nodes on the main stalk (nodes), dry stalk weight and days to flowering. Strong positive correlation are observed among the phenotypic data of total plant height, base-flag, nodes, dry stalk weight and days to flowering. It is important to remember that 228 accessions in the sorghum association panel (SAP), are converted tropical lines that are photoperiod insensitive, early maturing, and short statured phenotypes, developed by crossing exotic lines and U.S. cultivars [14,22]. Three independent studies [4,14,54] revealed three consistent genomic regions in sorghum that contribute to the introgression of cultivar-specific alleles to exotic sorghum lines. It is also clear that *Dwarf* (*Dw*)*4* and *Dw2* are associated with the introgression region in the heterochromatin of chromosome Sb06, and *Dw3* and *Dw1* are associated with the introgression regions in the 3′ terminal euchromatin of chromosome Sb07 and Sb09 respectively.

In our studies, the *Dw1-Dw4* regions reported by Morris et al. [4] show the most striking signals associated with base-flag leaf length for two years. Likewise, the association results of total plant height are consistent with that of Morris et al [4]. This suggests that environmental variables have relatively little effect on *Dw1-Dw4*.

In addition to *Dw1-Dw4*, we found noteworthy hotspot(s) for plant height located on chromosome Sb04. Our BTxSP mapping population [29] and Shiringani et al., [28] each suggested two overlapping QTL likelihood intervals on Sb04, 57.98 Mb-64.93 Mb and 48.80 Mb-58.58 Mb, to contribute to plant height. Within both intervals, we found multiple SNPs significantly associated with base-flag and dry stalk weight. Because the terminal region of Sb04 holding these QTLs exhibited weak LD, we could not set clear boundaries for the hotspot(s). GWAS of another plant height related trait, dry stalk weight in 2009 data, also shows a clear association 'skyline' in the region of 48.80 Mb-64.93 Mb on Sb04.

Plant height and flowering time show significant genetic correlation. In further agreement, three association loci for flowering time show strong LD with dwarfing genes, and are distributed in the introgression regions. There is some confusion concerning the identities of *Ma1* and *Ma6*, with one group suggesting that *Ma1* is a sorghum ortholog of a Triticeae flowering gene, *PRR37* [55], but which is located very near the published position of *Ma6* [45]. Many of the same authors who report that *PRR37* is *Ma1* have stated under separate cover that *PRR37* is *Ma6* [56]. The LD pattern suggests two haplotype blocks in the region of 6 Mb-46 Mb on chromosome Sb06 (Figure 2b), suggesting that *Ma1* (or *Ma6*) and *Dw4*[4] are tightly linked in the region of 6 Mb-39 Mb, and *Dw2* [4] is linked with *Ma6* (or *Ma1*) in the region of 40 Mb-46 Mb on Sb06. We also identified a haplotype block of 56 Mb-59.5 Mb on Sb09 (Figure 2c), which is strongly associated with plant height and flowering time. Greater mapping resolution is required to pinpoint the causative genetic mutation(s) that affect each trait.

Selective Signatures and Phenotype-Genotype Association

A phenotype-genotype association is no guarantee that the trait or its candidate gene has been historically important or is an adaptation [57]. Reduction in genetic diversity, which can be assessed by heterozygosity, can bear the signature of domestication [4,58,59]. Using quantified genome-wide heterozygosity, low heterozygosity regions in sorghum have been reported [4] but have not shown significant correspondence with identified loci associated with plant height, one of the key domestication traits. Here, we refined analysis of selective signatures from domestication and explored the genomic regions that have been important in sorghum adaptation for inflorescence morphology and plant height. Genetic diversity of the sorghum population was assessed by the ratio of expected heterozygosity in wild sorghum to that in cultivated types (H_w/H_c) across the sorghum genome (Figure 5). Overall, the refined pattern of reduction in genetic diversity is consistent with that measured previously [4]. Among the hotspots determined by association mapping for the 12 traits, the genomic region 58 Mb-64 Mb on chromosome Sb01 for pericarp color shows a strong selection signal based on reduction in heterozygosity ratio (Figure 4a). Such correspondence was also discovered in rice [60], consistent with the observation that pericarp color has profoundly influenced the popularity of cultivated cereal varieties [61].

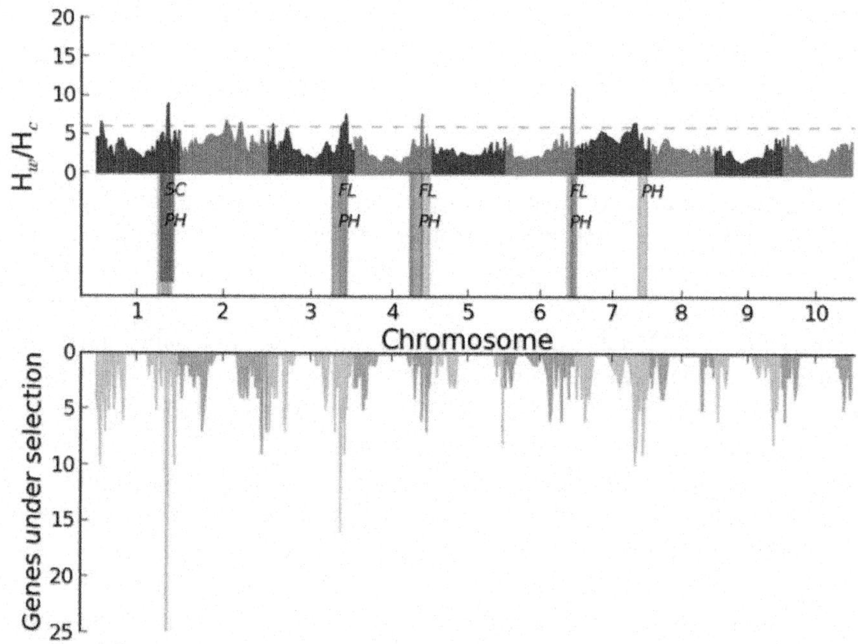

Figure 5: Genome-wide selection signatures. We calculated average ratios (expected heterozygosity of wild type sorghum (H_w) / expected heterozygosity of cultivated sorghum (H_c)) for windows of 500 consecutive SNPs throughout the genome. The gray dashed line is the cutoff for the top 5% of heterozygosity ratios. Green areas indicate confidence intervals determined by QTL mapping to contribute to either flowering time (FL) or plant height (PH), that also have strong selection signals. The hotspot on chromosome Sb01 identified by GWAS for pericarp color (SC) is highlighted by red. The distribution of 725 candidate genes implicated in domestication and/or improvement via gene-based population summary statistics [41] is shown below the heterozygosity ratios.

Some confidence intervals identified by QTL mapping for plant height and flowering time [14,24-28] colocalized non-randomly ($p = 0.231 \times 10^{-6}$) (See the overlapping testing in the "Methods") with regions having striking reductions in heterozygosity ratio (Figure 5). For example, Hart et al., [24] and Ritter et al., [26] each identified a region (55.1 Mb-66.7 Mb) for plant height on chromosome Sb01 which showed strong selection signals. A region of 55 Mb-62 Mb on chromosome Sb06, showing strong selection signals, was identified for plant height by Kebede et al., [25], Srinivas et al., [27] and Shiringani et al., [28], and for flowering time by Shiringani et al., [28]. Additional heterozygosity ratio peaks overlapping with plant height and flowering time intervals are found on Sb03 and Sb04. Most such colocalizing intervals affect plant height and flowering time simultaneously, and do not colocalize with hotspots determined by association mapping.

Using 8 M SNPs characterized in 44 sorghum lines, a recent study [41] implicated 725 candidate genes in sorghum domestication and/or improvement via gene-based population summary statistics. We show (Figure 5) at least two regions having striking reductions in heterozygosity ratio, to also be enriched for candidate genes under selection. One striking case is the 58 Mb-64 Mb interval on chromosome Sb01, in genetic control of pericarp color and plant height [24,26]. The other case is the 62 Mb-64 Mb interval on chromosome Sb03, in genetic control of plant height and flowering time [14,26,28].

Much of the 'missing heritability' from GWAS has been speculated to be attributed to the low/rare frequency of causative mutations, which are unlikely to be detected by most association studies [62]. Some genetic variations with large effect may be hidden by neutral genetic diversity in a natural population. By contrast, biparental QTL mapping uses genetic markers (e.g. SSR and RFLP), and is often derived from parental lines that differ strongly in phenotype such as our *S. bicolor* × *S. propinquum* cross between parents which are separated by 1-2 million years. Sometimes, biparental QTL mapping can be more efficient than GWAS to detect alleles that are rare in populations. QTL mapping and GWAS are thus two complementary approaches to determine "saturation" in terms of QTL discovery. Our findings lead to the prediction that the likelihood intervals on chromosome Sb01, Sb03, Sb04 and Sb06 identified by QTL mapping should bear the signature of selection, and may encode important variants with low/rare frequency for plant height and flowering time.

Genetic Correspondence between Sorghum and Rice

Synteny and colinearity have been well conserved between grass species such as rice, maize, and sorghum since their divergence about 50 Mya [21], enabling us to compare causal loci in corresponding regions across taxa. SSR mapping in rice revealed that two intervals affecting awns are located on chromosome Os01, and designated as *An9* and *An10*, respectively (Figure 4b) [63]. *An10* produced shorter and sparser awns than *An9*, and is thought to be more important for awn presence in rice. Recently, a rice gene, *OsETT2*, which colocalizes with *An10*, has been identified using genetic analyses and RNA-silencing experiments, to be involved in awn formation [64]. Chromosome Sb03 harbors the most significant hotspot for awn presence in sorghum, bounded in a region sharing large-scale homoeology with a region on Os01 containing *An10* and *OsETT2* (Figure 4b). The two hotspots do not appear to have direct correspondence across taxa, making it unclear whether there has been some localized rearrangement, functional divergence of different members of a co-regulated cluster of genes, or some other explanation.

Duplication and subsequent rearrangements can result in chromosomes exhibiting a complex pattern of homoeology across taxa. The *Rc* gene is responsible for conditioning red pericarp in rice [52]. Association mapping shows no noteworthy signal around the sorghum ortholog (*Sb02g006380*) of *Rc*, or for another sorghum gene copy (*Sb01g028230*). Rearrangement following polyploidization can lead to a scrambled patchwork in genome organization [65]. In the case of pericarp color, the correspondence of gene arrangement between the most significant hotspot in sorghum and the rice *Rc*region is still discernible through multiple alignment (Figure 4a). The hotspot on sorghum chromosome Sb01 shares homoeology with three small genomic segments on Sb01, one of which has correspondence with the region enclosing*Sb02g006380* on sorghum chromosome Sb02. Thus, despite discernible parallels in genome organization, corresponding genes do not appear to be responsible for phenotypic variation in seed color of sorghum and rice. This may simply reflect that 40-50 My of rice-sorghum divergence, did involve many genetic changes that were different, as well as some that may have been convergent [18].

Both linkage studies and reductions in heterozygosity ratios (Figure 4c and d) implicate two genomic regions in genetic control of plant height and flowering time, which are located in the 58 Mb-64 Mb on chromosome Sb01 [24,26] and in the 62 Mb-64 Mb on chromosome Sb03 [14,26,28] individually. In rice, gene *Osg1* [66], which is located on chromosome Os01 (Figure 4c) orthologus to sorghum chromosome Sb03, has been known to affect plant height via reduction in its expression; genes *Ehd4* [67] and *Hd16* [68] on chromosome Os03 orthologous to sorghum chromosome Sb01, are characterized by their function in flowering time of rice. We observed that sorghum has experienced gene loss for *Ehd4*, and none of three rice genes shows direct correspondence with the two genomic regions in sorghum. These findings indicate that different genes, rather than orthologs of *Osg1*, *Ehd4* and *Hd16*, were identified by QTL mapping on Sb03 and Sb01.

QTL Correspondence Across Homoeologous Regions within Sorghum

QTLs may share homoeology not only across taxa, but also within taxa. Maize, in which many homoeologous chromosome segments have been identified as a result of lineage-specific genome duplication, is a particularly favorable organism in which to show such QTL correspondence [14]. Only occasional cases have been found in sorghum [14], probably due to much greater antiquity of genome duplication [1].

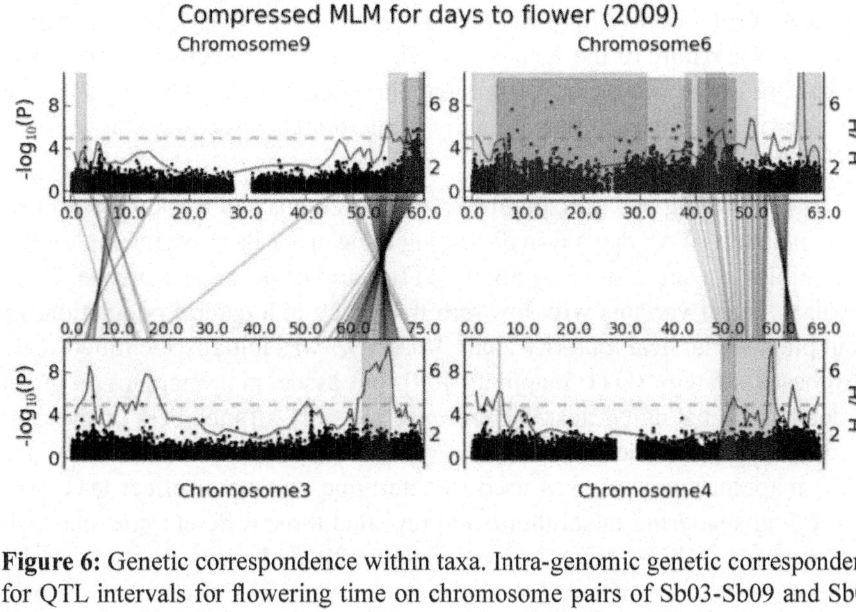

Figure 6: Genetic correspondence within taxa. Intra-genomic genetic correspondence for QTL intervals for flowering time on chromosome pairs of Sb03-Sb09 and Sb04-Sb06.

A total of 4 pairs of homoeologous QTLs for plant height/flowering time are found in our study. For flowering time, we observed two instances of homoeologous QTLs in sorghum (Figure 6). The confidence interval mapped on chromosome Sb03 [28] shares homoeology with the hotspot of 54 Mb-59.6 Mb on chromosome Sb09 which has been identified by QTL mapping [23] and our GWAS. The other correspondence was determined between 48.6 Mb-59.3 Mb on chromosome Sb04 [26,28] and 38 Mb-50.7 Mb on chromosome Sb06 [14,18]. Likewise, we found homoeologous QTLs for plant height on chromosome pairs of Sb03-Sb09 and Sb04-Sb06. The additional cases for flowering time are shown on chromosome pairs of Sb02-Sb07 and

Sb05-Sb08. Such genetic correspondence within sorghum implies that some homoeologous regions created in a genome duplication 70 million years ago [21] may still retain some genes with similar functions. In most instances, we found that one QTL hotspot exhibits much stronger selection signal than its homoeologous counterpart on the basis of genetic diversity reduction in the wild population. Independent episodes of domestication and crop improvement may preferentially select one counterpart, but may or may not select the other.

CONCLUSIONS

Our studies illustrate how GWAS may be used to complement the genetic resolution of causal elements (genes) of quantitative phenotypes that is typically attained from conventional likelihood intervals determined by QTL mapping. The degree of improvement in resolution by GWAS over QTL mapping is related to the nature of the 'genomic environment' surrounding a gene – with substantial improvement in recombinationally-active euchromatin but much less improvement in recombinationally recalcitrant heterochromatin with long LD blocks.

Understanding and utilizing the relative strengths and weaknesses of QTL mapping and GWAS can aid in dissecting the genetic basis of a complex trait. A carefully chosen cross can allow QTL mapping to have better statistical power to detect variants with low/rare frequency in a natural population. For example, the classical maturity locus *Ma3/phyB* was initially identified with a map-based strategy (QTL mapping) [69], but association mapping is unable to detect striking signal near*Ma3* because virtually all members of the panel are wild type (i.e. the mutation is rare). Another case is that a map-based (QTL mapping) strategy was used to determine one major-effect QTL (*sh1*) controlling shattering in sorghum, and revealed three different non-shattering haplotypes widely existing in five major cultivated races that may reflect low/rare frequency of causative polymorphism [2,3]. Hence, association mapping may have reduced power to find phenotype-genotype association for *sh1*. Additionally, complex population structure in unrelated individuals may compromise genetic variation at the true causative loci, resulting in cases of true negative detection in GWAS. More generally, multiple independent domestications may create a scenario under which many QTLs cannot be verified with the current sorghum association panel.

An important fundamental question in gene mapping is why the GWAS approach has revealed so much less variation than anticipated [62]. Even given 'perfect' (i.e. 100% accurate) information about phenotype and genotype, some associations may not be repeatable due to interaction between genotype and environment [57]. We observed that traits related to the inflorescence were more greatly affected by environment than traits related to plant height (Figure 7). The interaction between genotype and environment plays an important role in the variation between individuals in their observable characteristics, even for traits with high heritability.

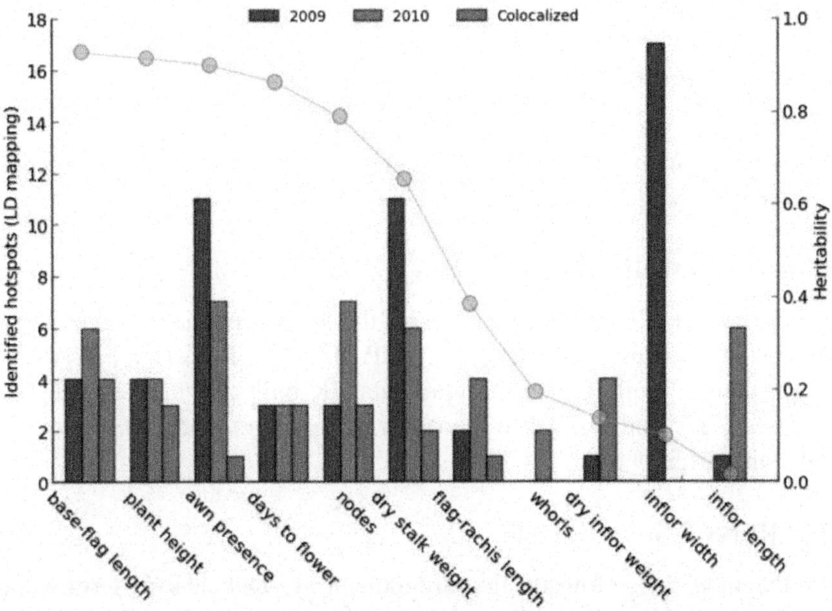

Figure 7: Summary of hotspots identified by GWAS for multi-year trait data. For each trait, numbers of hotspots for 2009, 2010 and both years are indicated with bars. The values of heritability for the traits are indicated with the curve.

Identification of intersections among QTL, GWAS and comparative data advance knowledge of the genetic determinants of variation in sorghum, and provide a finer-scale comparison than was previously possible of the genetics of important traits between sorghum and rice. Non-random correspondence of sorghum GWAS signals to rice QTLs, i.e. between divergent panicoid and oryzoid grasses, adds a new dimension to evidence of the ability to leverage genetic data about the important traits across divergent plants. More variants can be gained through re-sequencing gene candidates. Joint evidence from hotspots and re-sequencing data are likely to be extremely powerful to answer the question of whether such genetic correspondence implicates 'conserved' single genes or co-regulated clusters of genes that exist for the trait of interest after cereal divergence.

Rich SNP data also clarifies the phylogenetic relatedness of sorghum races, improving on several prior studies [38-41]. Based on the high-density genotype in the large population, we showed joint evidence from the striking clustering patterns in population structure and fixation index values, suggesting several hypotheses that might shed light on the unknown ancestry of sorghum races.

DECLARATIONS

Acknowledgements

This work was supported by the United Sorghum Checkoff Program, and the US Department of Energy-US Department of Agriculture Plant Feedstock Program (project grant numbers: 2011-02786, 2012-03304).

Authors' Contributions

AHP conceived of the project and guided the process of analyses. DZ and WK conducted the genome data analyses. AHP, DZ, WK, JR, VHG, EE, AK, GM, JC, YL and CP collected phenotypic data. JR built and curated phenotypic databases. AHP and DZ wrote the paper. All authors read and approved the final manuscript.

REFERENCES

1. Paterson AH, Bowers JE, Bruggmann R, Dubchak I, Grimwood J, Gundlach H, et al. The Sorghum bicolor genome and the diversification of grasses. Nature. 2009;457:551–6.
2. Lin Z, Li X, Shannon LM, Yeh C-T, Wang ML, Bai G, et al. Parallel domestication of the Shattering1 genes in cereals. Nat Genet. 2012;44:720–4.
3. Olsen KM. One gene' s shattering effects. Nat Genet. 2012;44:616–8.
4. Morris GP, Ramu P, Deshpande SP, Hash CT, Shah T, Upadhyaya HD, et al. Population genomic and genome-wide association studies of agroclimatic traits in sorghum. Proc Natl Acad Sci U S A. 2012;110:453–8.
5. Paterson AH, Lander ES, Hewitt JD, Peterson S, Lincoln SE, Tanksley SD. Resolution of quantitative traits into Mendelian factors by using a complete RFLP linkage map. Nature. 1988;335:721–6.
6. Kover PX, Valdar W, Trakalo J, Scarcelli N, Ehrenreich IM, Purugganan MD, et al. A multiparent advanced generation inter-cross to fine-map quantitative traits in Arabidopsis thaliana. PLoS Genet. 2009;5.

7. Buckler ES, Holland JB, Bradbury PJ, Acharya CB, Brown PJ, Browne C, et al. The genetic architecture of maize flowering time. Science. 2009;325:714–8.

8. Yu J, Pressoir G, Briggs WH, Vroh Bi I, Yamasaki M, Doebley JF, et al. A unified mixed-model method for association mapping that accounts for multiple levels of relatedness. Nat Genet. 2006;38:203–8.

9. Thornsberry JM, Goodman MM, Doebley J, Kresovich S, Nielsen D, Buckler ES. Dwarf8 polymorphisms associate with variation in flowering time. Nat Genet. 2001;28:286–9.

10. Yu J, Holland JB, McMullen MD, Buckler ES. Genetic design and statistical power of nested association mapping in maize. Genetics. 2008;178:539–51.

11. Myles S, Peiffer J, Brown PJ, Ersoz ES, Zhang Z, Costich DE, et al. Association mapping: critical considerations shift from genotyping to experimental design. Plant Cell. 2009;21:2194–202.

12. Morrell PL, Buckler ES, Ross-Ibarra J. Crop genomics: advances and applications. Nat Rev Genet. 2011;13:85–96.

13. Brown PJ, Klein PE, Bortiri E, Acharya CB, Rooney WL, Kresovich S. Inheritance of inflorescence architecture in sorghum. Theor Appl Genet. 2006;113:931–42.

14. Lin Y, Keith F, Paterson AH. Comparative analysis of QTLs affecting plant height and maturity across the poaceae, in reference to an interspecific sorghum population. Genetics. 1995;141:391–411.

15. Rong J, Feltus FA, Waghmare VN, Pierce GJ, Chee PW, Draye X, et al. Meta-analysis of polyploid cotton QTL shows unequal contributions of subgenomes to a complex network of genes and gene clusters implicated in lint fiber development. Genetics. 2007;176:2577–88.

16. Zhang D, Guo H, Kim C, Lee T-H, Li J, Robertson J, et al. CSGRqtl, a comparative quantitative trait locus database for Saccharinae grasses. Plant Physiol. 2013;161:594–9.

17. Multani DS, Briggs SP, Chamberlin MA, Blakeslee JJ, Murphy AS, Johal GS. Loss of an MDR transporter in compact stalks of maize br2 and sorghum dw3 mutants. Science. 2003;302:81–4.

18. Paterson AH, Lin YR, Li Z, Schertz KF, Doebley JF, Pinson SR, et al. Convergent domestication of cereal crops by independent mutations at corresponding genetic Loci. Science. 1995;269:1714–8.

19. Lee JM, Sonnhammer ELL. Genomic gene clustering analysis of pathways in eukaryotes. Genome Res. 2003;13:875–82.

20. Paterson AH, Wendel JF, Gundlach H, Guo H, Jenkins J, Jin D, et al. Repeated polyploidization of Gossypium genomes and the evolution of spinnable cotton fibres. Nature. 2012;492:423–7.
21. Paterson AH, Bowers JE, Chapman BA. Ancient polyploidization predating divergence of the cereals, and its consequences for comparative genomics. Proc Natl Acad Sci U S A. 2004;101:9903–8.
22. Casa AM, Pressoir G, Brown PJ, Mitchell SE, Rooney WL, Tuinstra MR, et al. Community resources and strategies for association mapping in sorghum. Crop Sci. 2008;48:30–40.
23. Feltus FA, Hart GE, Schertz KF, Casa AM, Casa AM, Kresovich S, et al. Alignment of genetic maps and QTLs between inter- and intra-specific sorghum populations. Theor Appl Genet. 2006;112:1295–305.
24. Hart GE, Schertz KF, Peng Y, Syed NH. Genetic mapping of sorghum bicolor (L.) Moench QTLs that control variation in tillering and other morphological characters. TAG Theor Appl Genet. 2001;103:1232–42.
25. Kebede H, Subudhi PK, Rosenow DT, Nguyen HT. Quantitative trait loci influencing drought tolerance in grain sorghum (Sorghum bicolor L. Moench). TAG Theor Appl Genet. 2001;103:266–76.
26. Ritter KB, Jordan DR, Chapman SC, Godwin ID, Mace ES, Lynne McIntyre C. Identification of QTL for sugar-related traits in a sweet × grain sorghum (Sorghum bicolor L. Moench) recombinant inbred population. Mol Breed. 2008;22:367–84.
27. Srinivas G, Satish K, Madhusudhana R, Reddy RN, Mohan SM, Seetharama N. Identification of quantitative trait loci for agronomically important traits and their association with genic-microsatellite markers in sorghum. Theor Appl Genet. 2009;118:1439–54.
28. Shiringani AL, Frisch M, Friedt W. Genetic mapping of QTLs for sugar-related traits in a RIL population of Sorghum bicolor L. Moench. Theor Appl Genet. 2010;121:323–36.
29. Kong W, Jin H, Franks CD, Kim C, Bandopadhyay R, Rana MK, et al. Genetic analysis of recombinant inbred lines for Sorghum bicolor × Sorghum propinquum. G3 (Bethesda). 2013;3:101–8.
30. Bradbury PJ, Zhang Z, Kroon DE, Casstevens TM, Ramdoss Y, Buckler ES. TASSEL: software for association mapping of complex traits in diverse samples. Bioinformatics. 2007;23:2633–5.
31. Zhang Z, Ersoz E, Lai C-Q, Todhunter RJ, Tiwari HK, Gore MA, et al. Mixed linear model approach adapted for genome-wide association studies. Nat Genet. 2010;42:355–60.

32. Lipka AE, Tian F, Wang Q, Peiffer J, Li M, Bradbury PJ, et al. GAPIT: genome association and prediction integrated tool. Bioinformatics. 2012;28:2397–9.
33. Matthies I, Malosetti M, Röder MS, Eeuwijk FV. Genome-Wide association mapping for kernel and malting quality traits using historical European barley records. PLoS One. 2014;9(11):e110046.
34. Bowers JE, Abbey C, Anderson S, Chang C, Draye X, Hoppe AH, et al. A High-density genetic recombination map of sequence-tagged sites for sorghum, as a framework for comparative structural and evolutionary genomics of tropical grains and grasses. Genetics. 2003;386:367–86.
35. Kawahara Y, de la Bastide M, Hamilton JP, Kanamori H, McCombie WR, Ouyang S, et al. Improvement of the Oryza sativa Nipponbare reference genome using next generation sequence and optical map data. Rice (N Y). 2013;6:4.
36. Doggett H. Sorghum, 2nd edition. UK: Longman; 1988.
37. Wiersema JH, Leon B. World Economic Plants: A standard Reference. New York: CRC press; 1999.
38. Brown PJ, Myles S, Kresovich S. Genetic support for phenotype-based racial classification in sorghum. Crop Sci. 2011;51:224.
39. Bouchet S, Pot D, Deu M, Rami J-F, Billot C, Perrier X, et al. Genetic structure, linkage disequilibrium and signature of selection in Sorghum: lessons from physically anchored DArT markers. PLoS One. 2012;7:e33470.
40. Billot C, Ramu P, Bouchet S, Chantereau J, Deu M, Gardes L, et al. Massive sorghum collection genotyped with SSR markers to enhance use of global genetic resources. PLoS One. 2013;8:e59714.
41. Mace ES, Tai S, Gilding EK, Li Y, Prentis PJ, Bian L, et al. Whole-genome sequencing reveals untapped genetic potential in Africa's indigenous cereal crop sorghum. Nat Commun. 2013;4.
42. Quinby JR. Sorghum Improvement and the Genetics of Growth. College Station: Texas A&M University Press; 1974.
43. Rooney WL, Aydin S. Genetic control of a photoperiod-sensitive response in Sorghum bicolor (L.) Moench. Crop Sci. 1999;39:397–400.
44. The Wellcome Trust Case Control Consortium. Genome-wide association study of 14,000 cases of seven common diseases and 3,000 shared controls. Nature. 2007;447:661–78. **Central**
45. Brady JA. Sorghum Ma5 and Ma6 Maturity Genes. Ph.D. dissertation. Texas A&M University; 2006.

46. Chopra S, Gevens A, Svabek C, Wood KV, Peterson T, Nicholson RL. Excision of the Candystripe1 transposon from a hyper-mutable Y1-cs allele shows that the sorghumY1 gene controls the biosynthesis of both 3-deoxyanthocyanidin phytoalexins and phlobaphene pigments. Physiol Mol Plant Pathol. 2002;60:321–30.

47. Ibraheem F, Gaffoor I, Chopra S. Flavonoid phytoalexin-dependent resistance to anthracnose leaf blight requires a functional yellow seed1 in Sorghum bicolor. Genetics. 2010;184:915–26.

48. Wu Y, Li X, Xiang W, Zhu C, Lin Z, Wu Y, et al. Presence of tannins in sorghum grains is conditioned by different natural alleles of Tannin1. Proc Natl Acad Sci U S A. 2012;109:10281–6.

49. Worzella WW, Khalidy R, Badawi Y, Daghir S. Inheritance of Beta-carotene in grain sorghum hybrids. Crop Sci. 1965;5:591–2.

50. Gorbet DW, Weibel DE. Inheritance and genetic relationships of six endosperm types in sorghum. Crop Sci. 1972;12:378–82.

51. Kinoshita T. Gene analysis and Linkage in Biology of Rice. In: Tsunoda S, Takahashi N, editors. JSSP. Tokyo: Elsevier; 1984. p. 187–274.

52. Sweeney MT, Thomson MJ, Pfeil BE, Mccouch S. Caught red-handed: Rc encodes a basic helix-loop-helix protein conditioning red pericarp in rice. Plant Cell. 2006;18:283–94.

53. Magness JR, Markle GM, Compton CC. Food and feed crops of the United States. Interregional Research Project IR-4, IR Bul. 1 (Bul. 828 New Jersey Agr. Expt. Sta.); 1971.

54. Thurber CS, Ma JM, Higgins RH, Brown PJ. Retrospective genomic analysis of sorghum adaptation to temperate-zone grain production. Genome Biol. 2013;14:R68.

55. Murphy RL, Klein RR, Morishige DT, Brady JA, Rooney WL, Miller FR, et al. Coincident light and clock regulation of pseudoresponse regulator protein 37 (PRR37) controls photoperiodic flowering in sorghum. Proc Natl Acad Sci U S A. 2011;108:16469–74.

56. Mullet JE, Rooney WL, Klein PE, Morishige DT, Murphy R, Brady JA. Discovery and utilization of sorghum genes (MA5/MA6). Office USP ed., vol. US 8,309793 B2. USA: The Texas A&M University System; 2012.

57. Ross-Ibarra J, Morrell PL, Gaut BS. Plant domestication, a unique opportunity to identify the genetic basis of adaptation. Proc Natl Acad Sci U S A. 2007;104 Suppl:8641–8.

58. Doebley JF, Gaut BS, Smith BD. The molecular genetics of crop domestication. Cell. 2006;127:1309–21.
59. Xu X, Liu X, Ge S, Jensen JD, Hu F, Li X, et al. Resequencing 50 accessions of cultivated and wild rice yields markers for identifying agronomically important genes. Nat Biotechnol. 2012;30:105–11.
60. Huang X, Kurata N, Wei X, Wang Z-X, Wang A, Zhao Q, et al. A map of rice genome variation reveals the origin of cultivated rice. Nature. 2012;490:497–501.
61. Sieglinger JB. Seed-color inheritance in certain grain-sorghum crosses. J Agric Res. 1934;XXVII(1).
62. Manolio TA, Collins FS, Cox NJ, Goldstein DB, Hindorff LA, Hunter DJ, et al. Finding the missing heritability of complex diseases. Nature. 2009;461:747–53.
63. Matsushita S, Kurakazu T, Sobrizal Doi K, Yoshimura A. Mapping of genes for awn in rice using Oryza meridionalis introgression lines. Rice Genet Newsl. 2003;20:17.
64. Toriba T, Hirano H-Y. The DROOPING LEAF and OsETTIN2 genes promote awn development in rice. Plant J. 2014;77:616–26.
65. Shoemaker RC, Polzin K, Labate J, Specht J, Brummer EC, Olson T, et al. Genome duplication in soybean (Glycine subgenus soja). Genetics. 1996;144:329–38.
66. Wan L, Zha W, Cheng X, Liu C, Lv L, Liu C, et al. A rice β-1,3-glucanase gene Osg1 is required for callose degradation in pollen development. Planta. 2011;233:309–23.
67. Gao H, Zheng X-M, Fei G, Chen J, Jin M, Ren Y, et al. Ehd4 encodes a novel and Oryza-genus-specific regulator of photoperiodic flowering in rice. PLoS Genet. 2013;9:e1003281.
68. Hori K, Ogiso-Tanaka E, Matsubara K, Yamanouchi U, Ebana K, Yano M. Hd16, a gene for casein kinase I, is involved in the control of rice flowering time by modulating the day-length response. Plant J. 2013;76:36–46.
69. Childs KL, Miller FR, Cordonnier-Pratt MM, Pratt LH, Morgan PW, Mullet JE. The sorghum photoperiod sensitivity gene, Ma3, encodes a phytochrome B. Plant Physiol. 1997;113:611–9.

Chapter 11

GENETIC DIVERSITY IN THE MODERN HORSE ILLUSTRATED FROM GENOME-WIDE SNP DATA

Jessica L. Petersen[1], James R. Mickelson[1] , E. Gus Cothran[2] , Lisa S. Andersson[3] , Jeanette Axelsson[3] , Ernie Bailey[4] , Danika Bannasch[5] , Matthew M. Binns[6] , Alexandre S. Borges[7] , Pieter Brama[8] , Artur da Caˆmara Machado[9] , Ottmar Distl[10], Michela Felicetti[11], Laura Fox-Clipsham[12], Kathryn T. Graves[4] , Ge´ rard Gue´ rin[13], Bianca Haase[14], Telhisa Hasegawa[15], Karin Hemmann[16], Emmeline W. Hill[17], Tosso Leeb[18], Gabriella Lindgren[3] , Hannes Lohi1[6], Maria Susana Lopes[9] , Beatrice A. McGivney[17,] Sofia Mikko[3] , Nicholas Orr[19], M. Cecilia T Penedo[5] , Richard J. Piercy[20], Marja Raekallio[16], Stefan Rieder[21], Knut H. Røed[22], Maurizio Silvestrelli[11], June Swinburne[12,23], Teruaki Tozaki[24], Mark Vaudin1[2], Claire M. Wade[14], Molly E. McCue[1]

[1] University of Minnesota, College of Veterinary Medicine, St Paul, Minnesota, United States of America

[2] Texas A&M University, College of Veterinary Medicine and Biomedical Science, College Station, Texas, United States of America

[3] Swedish University of Agricultural Sciences, Department of Animal Breeding and Genetics, Uppsala, Sweden

[4] University of Kentucky, Department of Veterinary Science, Lexington, Kentucky, United States of America

[5] University of California Davis, School of Veterinary Medicine, Davis, California, United States of America

[6] Equine Analysis, Midway, Kentucky, United States of America

[7] University Estadual Paulista, Department of Veterinary Clinical Science, Botucatu-SP, Brazil

[8] University College Dublin, School of Veterinary Medicine, Dublin, Ireland

[9] University of Azores, Institute for Biotechnology and Bioengineering, Biotechnology Centre of Azores, Angra do Heroı´smo, Portugal

[10] University of Veterinary Medicine Hannover, Institute for Animal Breeding and Genetics, Hannover, Germany

[11] University of Perugia, Faculty of Veterinary Medicine, Perugia, Italy

[12] Animal Health Trust, Lanwades Park, Newmarket, Suffolk, United Kingdom

[13] French National Institute for Agricultural Research-Animal Genetics and Integrative Biology Unit, Jouy en Josas, France

[14] University of Sydney, Veterinary Science, New South Wales, Australia

[15] Nihon Bioresource College, Koga, Ibaraki, Japan

[16] University of Helsinki, Faculty of Veterinary Medicine, Helsinki, Finland

[17] University College Dublin, College of Agriculture, Food Science and Veterinary Medicine, Belfield, Dublin, Ireland

[18] University of Bern, Institute of Genetics, Bern, Switzerland

[19] Institute of Cancer Research, Breakthrough Breast Cancer Research Centre, London, United Kingdom

[20] Royal Veterinary College, Comparative Neuromuscular Diseases Laboratory, London, United Kingdom

[21] Swiss National Stud Farm, Agroscope Liebefeld-Posieux Research Station, Avenches, Switzerland

[22] Norwegian School of Veterinary Science, Department of Basic Sciences and Aquatic Medicine, Oslo, Norway

[23] Animal DNA Diagnostics Ltd, Cambridge, United Kingdom

[24] Laboratory of Racing Chemistry, Department of Molecular Genetics, Utsunomiya, Tochigi, Japan

ABSTRACT

Horses were domesticated from the Eurasian steppes 5,000–6,000 years ago. Since then, the use of horses for transportation, warfare, and agriculture, as well as selection for desired traits and fitness, has resulted in diverse populations distributed across the world, many of which have become or are in the process of becoming formally organized into closed, breeding populations (breeds). This report describes the use of a genome-wide set of autosomal SNPs and 814 horses from 36 breeds to provide the first detailed description of equine breed diversity. F_{ST} calculations, parsimony, and distance analysis demonstrated relationships among the breeds that largely reflect geographic origins and known breed histories. Low levels of population divergence were observed between breeds that are relatively early on in the process of breed development, and between those with high levels of within-breed diversity, whether due to large population size, ongoing outcrossing, or large within-breed phenotypic diversity. Populations with low within-breed diversity included those which have experienced population bottlenecks, have been under intense selective pressure, or are closed populations with long breed histories. These results provide new insights into the relationships among and the diversity within breeds of horses. In addition these results will facilitate future genome-wide association studies and investigations into genomic targets of selection.

INTRODUCTION

With a world-wide population greater than 58 million [1], and as many as 500 different breeds, horses are economically important and popular animals for agriculture, transportation, and recreation. The diversity of the modern horse has its roots in the process of domestication which began 5,000–6,000 years ago in the Eurasian Steppe [2]–[4]. Unlike other agricultural species such as sheep [5] and pigs [6], [7], archaeological and genetic evidence suggests that multiple horse domestication events occurred across Eurasia [2], [8]–[12]. During the domestication process, it is believed that gene flow continued between domesticated and wild horses [13] as is likely to also have been the case during domestication of cattle [14], [15]. Concurrent gene flow between domestic and wild horses would be expected to allow newly domestic stock to maintain a larger extent of genetic diversity than if domestication occurred in one or few events with limited individuals.

Prior genetic work aimed at understanding horse domestication has shown that a significant proportion of the diversity observed in modern maternal lineages was present at the time of domestication [2], [8], [16]. The question of mitochondrial DNA (mtDNA) diversity was further addressed by recent sequencing of the entire mtDNA genome. These studies estimate that, minimally, 17 to 46 maternal lineages were used in the founding of the modern horse [2], [17]; however, those data were unable to support prior studies suggesting geographic structure among maternal lineages [9], [18]. Recent nuclear DNA analyses have utilized "non-breed" horses sampled across Eurasia to attempt to understand the population history of the horse. These microsatellite-based studies suggest a weak pattern of isolation by distance with higher levels of diversity in, and population expansion originating from Eastern Asia [13], [19]. High diversity as observed by both mtDNA and microsatellites and the absence of strong geographical patterns is likely a result of continued gene flow during domestication, the high mobility of the horse, and its prevalent use for transportation during and after the time of domestication. Interestingly, while significant diversity is observed in maternal lineages, paternal input into modern horse breeds appears to have been extremely limited as shown by a lack of variation at the Y-chromosome [20], [21].

Diversity in the founding population of the domestic horse has since been exploited to develop a wealth of specialized populations or breeds.

While some breeds have been experiencing artificial selection for hundreds of years (e.g. Thoroughbred, Arabian), in general, most modern horse breeds have been developed recently (e.g. Quarter Horse, Paint, Tennessee Walking Horse) and continue to evolve based upon selective pressures for performance and phenotype (Table 1). Horse breeds resulting from these evolutionary processes are generally closed populations consisting of individual animals demonstrating specific phenotypes and/or bloodlines. Each breed is governed by an independent set of regulations dictated by the respective breed association. Not all breeds are closed populations. Some breed registries allow admixture from outside breeds (e.g. Swiss Warmblood, Quarter Horse), and others are defined by phenotype (e.g. Miniature). Finally, some populations that are often referred to as breeds are classified simply by their geographic region of origin and may not be actively maintained by a formal registry (e.g. Mongolian, Tuva) (Table 1). Those breeds that may be free ranging and experience lesser degrees of management may more appropriately be termed "landrace populations." Therefore, genetic characteristics within horse breeds are expected to differ based upon differences in the definition of the breed, the diversity of founding stock, the time since breed establishment, and the selective pressures invoked by breeders. The extent of gene flow not only varies within breed, but among horse breeds, the direction and level of gene flow is influenced by breed restrictions/requirements, and potentially by geographic distance.

Table 1: Populations (breeds) included in the study, region of breed origin and sampling location, notes on population history relevant to diversity statistics, and breed classification based upon use and phenotype.

doi:10.1371/journal.pone.0054997.t001

Breed	Geographic Origin	Region Sampled	Population size (approx)	Population Notes	Classification(s)
Akhal Teke	Turkmenistan	US & Russia	3,500	Pedigree records began-1885, Stud book-1941	Riding horse, endurance
Andalusian	Spain	United States	185,000	US registry formed in 1995 including Pura Raza Española & Lusitano bloodlines	Riding horse, sport
Arabian	Middle East	United States	1 million	Arabian type bred for over 3,500 years; US stud book-1908	Riding horse, endurance
Belgian	Belgium	United States	common	US Association began-1887	Draft
Caspian	Persia	United States	rare	Rediscovered in 1965 with N~50, no breeding records prior; Stud book-1966	Riding and driving pony
Clydesdale	Scotland	US & UK	5,000	Registry formed-1877 in Scotland; Stud book-1879	Draft
Exmoor	Great Britain	United Kingdom	2,000	Exmoor Pony Society-1921	Riding and driving pony
Fell Pony	England	United Kingdom	6,000	Fell Pony Society began in 1922; outcrossed with Dale's pony until 1970s	Light draft pony
Finnhorse	Finland	Finland	19,800	Stud book-1907	Light draft; riding horse; trotting
Florida Cracker	United States	United States	rare	Introduced to US in 1500s; association began-1989 with 31 horses	Riding horse, gaited
Franches-Montagnes	Switzerland	Switzerland	21,000	Official stud book-1921; Current breeding association established-1997	Light draft, riding horse
French Trotter	France	France	common	Population closed-1937 although allows some Standardbred influence	Riding horse, trotting
Hanoverian	Germany	Germany	20,000 (Germany)	Outcrossing allowed	Riding horse
Icelandic	Iceland	Sweden	180,000	Isolated >1,000 years; Federation of Icelandic Horse Association began-1969	Riding horse, gaited
Lusitano	Portugal	Portugal	12,000	Stud book-1967 after split from Spanish Andalusian breed	Riding horse, sport
Mangalarga Paulista	Brazil	Brazil	common	Registry began-1934	Riding horse
Maremmano	Italy	Italy	7,000	Breed identification based upon conformation and inspection	Riding horse
Miniature	United States	United States	185,000	Two US registries founded in 1970s; Maximum height restrictions for registration	Driving pony, extreme small size
Mongolian	Mongolia	Mongolia	2 million	Many types based upon purpose and geography	Riding horse, landrace
Morgan	United States	United States	100,000	Founding sire born in 1789; Registry-1894	Riding and driving horse
New Forest Pony	England	United Kingdom	15,000	Stud book-1910 with a variety of sires; No outcrossing since 1930s	Light draft, riding pony, landrace
North Swedish Horse	Sweden	Sweden	10,000	Breed association-1894; Stud book-1915	Draft
Norwegian Fjord	Norway	Norway	common	Stud book-1909	Riding and light draft
Paint	United States	United States	1 million	Registry-1965; One parent can be Quarter Horse or Thoroughbred	Riding horse, stock horse
Percheron	France	United States	20,000	Stud book-1893	Draft
Peruvian Paso	Peru	United States	25,000	Breed type over 400 years old; Closed population	Riding horse, gaited
Puerto Rican Paso Fino	Puerto Rico	Puerto Rico	250,000	Breed type ~500 years old; Association founded-1972	Riding horse, gaited
Quarter Horse	United States	United States	4 million	Association formed-1940; One parent may be Paint or Thoroughbred	Riding horse, stock horse, racing
Saddlebred	United States	United States	75,000	Breed type founded in late 1700s; Association began-1891	Riding and driving horse, some gaited
Shetland	Scotland	Sweden	common	Stud book-1891	Riding pony
Shire	England	United States	7,000	1st Shire organization-1877 (UK); stud book-1880; US assoc-1885	Draft
Standardbred	United States	Norway	common	Stud book-1871; Some outside trotting bloodlines (French Trotter) allowed	Riding horse, harness racing (trot)
Standardbred	United States	United States		Stud book-1871; Harness racing in early 1800s included pacing horses	Riding horse, harness racing (trot or pace)
Swiss Warmblood	Switzerland	Switzerland	15,000	Stud book-1921; Crossed with European Warmbloods, Thoroughbreds, Arabians	Riding horse, sport
Tenn Walking Horse	United States	United States	500,000	Registry-1935; Blood typing and parentage verification mandated in 1993	Riding horse, gaited
Thoroughbred	England	UK & Ireland	common	Stud book-1791; Closed population	Race horse, riding horse, sport
Thoroughbred	England	United States			Race horse, riding horse, sport
Tuva	Siberia	Russia	30,000	Different types depending on region	Light draft, landrace

doi:10.1371/journal.pone.0054997.t001

Considering modern breeds, unlike mtDNA, nuclear markers can discern breed membership[12]. However, studies of nuclear genetic diversity of modern breeds to date have most commonly focused on a single population of interest, sets of historically related breeds, or breeds within a specific geographic region [22]–[36]. Additionally, these analyses of nuclear genetic diversity in horse breeds are largely based upon microsatellite loci, which do not often permit consolidation of data across studies. Thus, large, across-breed investigations of nuclear diversity in the modern, domestic horse are lacking.

The Equine Genetic Diversity Consortium (EGDC), an international collaboration of the equine scientific community, was established in an effort to quantify nuclear diversity and the relationships within and among horse populations on a genome-wide scale. The development of this consortium has facilitated the collection of samples from 36 breeds for genotyping on the Illumina 50K SNP Beadchip. The breeds included in this report represent many of the most popular breeds in the world as well as divergent phenotypic classes, different geographic regions of derivation, and varying histories of breed origin (Table 1). The standardized SNP genotyping platform permits the compilation of data across breeds at a level never before achieved. Results of this collaboration now allow for the detailed description of diversity and assessment of the effects of genetic isolation, inbreeding, and selection within breeds, and the description of relationships among breeds. These data will also facilitate future across breed genome-wide association studies as well as investigations into genomic targets of selection.

RESULTS

Samples

Of the 38 populations sampled, two breeds were represented by geographically distinct populations: the Thoroughbred was sampled in the both the United States (US) and the United Kingdom and Ireland (UK/Ire), and the Standardbred was sampled in the US as well as in Norway. Eight Standardbred horses sampled from the US were noted to be pacing horses as opposed to the Norwegian and remaining US individuals that were classified as trotters. In addition, the International Andalusian and Lusitano Horse Association Registry (IALHA) in the US maintains one stud book but designates whether the individual was derived from Spanish (Pura Raza Española) or Portuguese (Lusitano) bloodlines, or a combination of both. Of the Andalusian horses collected in the US, five were noted to have Portuguese bloodlines.

Phenotypic classifications of the horse breeds include those characterized by small stature (Miniature Horse, pony breeds), breeds characterized by large stature and/or large muscle mass in proportion to size (draft breeds), light horse or riding breeds, gaited breeds, rare breeds, breeds founded in the past 80 years, and populations that are relatively unmanaged ("landrace"). The number of samples, sampling location, region of breed origin, and a list of primary breed characteristics are found in Table 1.

After pruning of individuals for genotyping quality and relationships (see methods), and keeping a similar number of individuals per breed, 814 of the 1,060 horses remained in the analysis. Of the horses removed, 12 had known pedigree relationships at or more recent to the grandsire/dam level, 44 individuals were removed at random from overrepresented breeds to equalize sample size across breeds, 4 failed to genotype at a rate greater than 0.90, and 186 were removed due to pi hat values (pairwise estimates of identity by descent) above the allowed threshold. Of those last 186 horses that were removed, 122 were from disease studies where relationships were common due to sampling bias.

Within Breed Diversity

Diversity indices were calculated using 10,536 autosomal SNPs that remained after pruning for minor allele frequency (MAF), genotyping rate, and linkage disequilibrium (LD) across breeds (referred to as the primary SNP set). Diversity indices were also calculated using three other SNP sets, resulting from different levels of LD-based pruning (see methods). Individuals noted as outliers in parsimony and cluster analyses (see below) were excluded from within-breed diversity calculations.

Using the primary SNP set, diversity, as measured by expected heterozygosity (H_e), ranged from 0.232 in the Clydesdale, to 0.311 in the Tuva (Table 2). Considering the SNP sets pruned less stringently for LD, the diversity within the Thoroughbred increased in relationship to the other breeds, as did that of nine other breeds. Mean and total heterozygosity increased with increased number of loci and less stringent LD pruning (Table 2). Inbreeding coefficients (F_{IS}) calculated on the primary SNP set showed significant excess homozygosity in 17 populations, which was greatest in the Andalusian (0.065). Three of the four lowest F_{IS} values were found in the Thoroughbred samples (UK/Ire, US, and when considered together) (Table 2).

Table 2: Number of samples (N), effective population size (N_e), individual inbreeding estimates (f), inbreeding coefficient (F_{IS}), and expected heterozygosity (H_e) from four SNP sets pruned based upon varying levels of LD.

							Expected Heterozygosity (He)			
				Individual inbreeding (f)			r2 0.1	R2 0.1	r2 0.2	r2 0.4
Breed	**N**	**Ne**	**FIS**	**Min**	**Max**	**Mean**	**10,536**	**6,028**	**18,539**	**26,171**
Akhal Teke	19	302	0.015*	0.015	0.297	0.101	0.287	0.281	0.303	0.311
Andalusian	18a	329	0.065*	0.028	0.274	0.114	0.296	0.293	0.308	0.312
Arabian	24a	346	0.033*	0.060	0.060	0.060	0.287	0.280	0.302	0.310
Belgian	30b	431	−0.002	0.039	0.166	0.111	0.278	0.276	0.284	0.284
Caspian	18	351	−0.022	−0.033	0.136	0.041	0.294	0.292	0.305	0.308
Clydesdale	24	194	0.004	0.128	0.323	0.261	0.232	0.225	0.238	0.236
Exmoor	24	216	0.034*	0.055	0.556	0.239	0.247	0.242	0.253	0.252
Fell Pony	21	289	0.002	0.069	0.178	0.114	0.278	0.272	0.285	0.285
Finnhorse	27	575	−0.004	0.011	0.100	0.052	0.296	0.296	0.302	0.301
Florida Cracker	7	171	0.026*	0.004	0.359	0.159	0.270	0.263	0.284	0.291
Franches-Montagnes	19a	316	0.003	0.018	0.203	0.095	0.284	0.279	0.297	0.301
French Trotter	17a	233	−0.018	0.064	0.173	0.105	0.275	0.262	0.295	0.307
Hanoverian	15a	269	−0.010	0.002	0.087	0.052	0.294	0.280	0.320	0.335
Icelandic	25c	555	0.006*	0.043	0.234	0.083	0.289	0.288	0.290	0.288
Lusitano	24	391	0.039*	0.008	0.220	0.090	0.296	0.292	0.309	0.315
Maremmano	24	341	−0.012	−0.015	0.109	0.038	0.298	0.287	0.318	0.329
Miniature	21	521	0.005	0.043	0.161	0.075	0.291	0.292	0.296	0.295
Mangalarga Paulista	15	155	−0.011	0.176	0.320	0.242	0.235	0.228	0.246	0.250
Mongolian	19a	751	0.001	−0.034	0.055	0.015	0.309	0.308	0.314	0.314
Morgan	40	448	0.040*	0.003	0.307	0.090	0.296	0.287	0.310	0.317
New Forest Pony	15	474	0.000	−0.022	0.066	0.025	0.304	0.300	0.316	0.319
Norwegian Fjord	21a	335	−0.003	0.053	0.168	0.122	0.274	0.274	0.278	0.277
North Swedish Horse	19	369	0.011*	0.069	0.210	0.133	0.275	0.276	0.279	0.278
Percheron	23	451	0.003	0.043	0.143	0.086	0.287	0.284	0.292	0.293
Peruvian Paso	21	433	0.002	0.008	0.134	0.055	0.298	0.293	0.306	0.310
Puerto Rican Paso Fino	20	321	−0.003	0.004	0.298	0.103	0.280	0.278	0.287	0.290
Paint	25	399	0.006*	−0.013	0.101	0.040	0.302	0.289	0.324	0.337
Quarter Horse	40a	426	0.011*	−0.012	0.144	0.047	0.302	0.290	0.323	0.336
Saddlebred	25d	297	−0.008	0.051	0.145	0.103	0.279	0.268	0.297	0.306
Shetland	27	365	0.032*	0.108	0.370	0.182	0.264	0.268	0.268	0.266
Shire	23	357	0.024*	0.130	0.258	0.187	0.261	0.252	0.268	0.267
Standardbred - Norway	25e	232	−0.004	0.063	0.202	0.130	0.272	0.255	0.289	0.298
Standardbred - US	15	179	0.039*	0.097	0.222	0.153	0.276	0.262	0.293	0.303
Standardbred - all	40	290	0.022*	−0.028	0.323	0.130	0.276	0.260	0.293	0.303
Swiss Warmblood	15a	271	0.005	0.023	0.117	0.059	0.296	0.281	0.322	0.337
Thoroughbred - UK/Ire	19a	143	−0.028	0.089	0.171	0.133	0.264	0.245	0.292	0.309
Thoroughbred - US	17a	163	−0.015	0.093	0.182	0.134	0.267	0.250	0.295	0.313
Thoroughbred - all	36	190	−0.019	0.089	0.182	0.134	0.266	0.248	0.294	0.312
Tuva	15	533	0.016*	−0.028	0.116	0.022	0.311	0.309	0.320	0.322
Tennessee Walking Horse	19	230	0.008*	0.065	0.276	0.148	0.269	0.256	0.284	0.291
Mean	22.3	341	0.007	0.039	0.204	0.107	0.282	0.275	0.295	0.300
Total	814						0.313	0.303	0.329	0.336
Min			−0.028	−0.034	0.055	0.015	0.232	0.225	0.238	0.236
Max			0.005	0.176	0.556	0.261	0.311	0.309	0.324	0.337

[a]Individuals from this breed also included in [41];
[b]20 of these individuals were also reported in [41];
[c]17 of these individuals were also reported in [41];
[d]21 of these individuals were also reported in [41];
[e]19 of these individuals were also reported in [41].
F_{IS} and f were calculated based upon the primary SNP set (10,536 loci). Samples also used in [41] are indicated in the footnotes.
*Indicates significance at α<0.05 determined by 10,000 permutations.
doi:10.1371/journal.pone.0054997.t002

Inbreeding coefficients (f) calculated for each individual based upon observed and expected heterozygosity showed several individuals with significant loss of heterozygosity. The highest individual value of f (0.56) was found in an Exmoor pony. Within breeds, average individual estimates of f were greatest in the Clydesdale, Mangalarga Paulista, and Exmoor while the lowest breed means were found in the landrace populations (Table 2).

Effective population size (N_e), as estimated by LD [37] using an autosomal SNP set pruned within each breed for quality, was lowest (143) in the UK/Ire sample of the Thoroughbred (UK/Ire) but also low in the other racing breeds as well as the Clydesdale (Table 2). Highest values of N_e were observed in the Eurasian landrace populations, the Mongolian (743) and Tuva (533), and also in the Icelandic (555), Finnhorse (575), and Miniature (521). Breed-specific decay of LD essentially mirrors the results of the N_e calculation given the relationship between the statistics.

Parsimony and Principal Component Analyses

With a domestic ass designated as the outgroup, parsimony analysis of 10,066 loci pruned for LD of $R^2=0.2$ (see methods) resulted in generally tight clustering and monophyly of samples within breeds, supported by high bootstrap values (Figure 1). Major clades of the tree show grouping of the Iberian breeds (Lusitano and Andalusian), ponies (Icelandic, Shetland, Miniature), Scandinavian breeds (Finnhorse, North Swedish Horse, Norwegian Fjord), heavy draft horses (Clydesdale, Shire, Belgian, Percheron), breeds recently admixed with and/or partly derived from the Thoroughbred (Paint, Quarter Horse, Maremmano, Swiss Warmblood, Hanoverian), modern US breeds (American Saddlebred (hereafter "Saddlebred" and Tennessee Walking Horse), trotting breeds (Standardbred and French Trotter), and Middle Eastern breeds (Akhal Teke and Arabian). Exceptions to monophyly include the Paint and Quarter Horse as well as the Hanoverian and Swiss Warmblood, which are mixed in clades surrounding the Thoroughbred and Maremmano. In addition, the Clydesdale was placed as a clade within the Shire breed and the Shetlands as a clade within the Miniatures. Strong bootstrap support for monophyly is present within a subset each of Lusitanos (83%), and Andalusians (87%); however the remainder of individuals from these breeds were intermixed. No structure was found within the US sample regarding individual Andalusians noted to have Portuguese bloodlines opposed to those with Spanish bloodlines. The Mongolian and most Tuva horses were grouped together while a subset of the Tuvas fell out as a sister clade to the Caspians. Several individuals were not positioned in the clades that represented the majority of the other individuals in the breed (Figure 1). These include three Shires, two Mongolians, a Caspian, and a Norwegian Fjord. In each instance, the outlier status of these individuals was also supported by cluster analysis (see below).

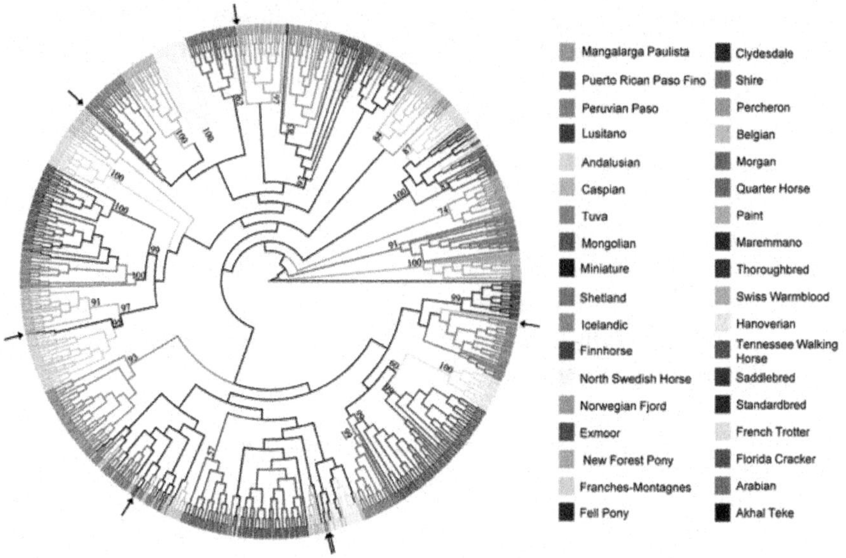

Figure 1: Individual and breed relationships among 814 horses illustrated by parsimony.

Parsimony tree created from 10,066 SNPs and rooted by the domestic ass. Breeds are listed in the legend in order starting from the root and working counterclockwise. Individual outliers with respect to their breeds are noted with arrows. Bootstrap support calculated from 1,000 replicates is shown for major branches when greater than 50%.

doi:10.1371/journal.pone.0054997.g001

Principal component analysis (PCA) also serves to visualize individual relationships within and among breeds. All Thoroughbred samples, regardless of origin, are separated from the others by PC1 and form a cluster at the top of the figure. Intermediate between the Thoroughbred and central cluster of breeds are the Hanoverian, Swiss Warmblood, Paint, and Quarter Horse. The Shetland, Icelandic, and Miniature split from the remainder of samples in PC2, falling out in the lower left corner, and the British drafts anchor the figure at the lower right. While most breeds cluster tightly, several are dispersed across one or both PCs. The Hanoverian, Swiss Warmblood, Paint, and Quarter Horse, as noted above, are extended along PC1, while the Arabian and Franches-Montagnes show similar spreading, also along PC1. The Tuva, Clydesdale, and Shire individuals also are not as tightly clustered as other populations despite the low within breed diversity of the latter two.

Distance Analysis

An unrooted neighbor joining (NJ) tree of Nei's distance [38] was constructed using SNP frequencies within breeds from the 10,536 SNP data set (Figure 2). The relative placement of breeds reflects that seen in the parsimony tree with several exceptions. The Paint, Quarter Horse, Swiss Warmblood, Hanoverian, Maremmano, and Thoroughbred, are found in one large branch of the tree, although the Maremmano is placed outside of the clade containing the aforementioned breeds. The position of the Morgan with the Saddlebred and Tennessee Walking Horse also deviates from parsimony analysis but reflects historic records of relationships among these breeds. The Scandinavian breeds remain in one branch of the clade, which also includes the Shetland and Miniature. Unlike the parsimony cladogram, the Caspian falls in a clade with the other Middle Eastern breeds, the Arabian and Akhal Teke. Finally, the Exmoor, a British breed, is placed with another British breed, the New Forest Pony, rather than with the Scandinavian breeds as in the parsimony analysis. Each branch shows support of over 50%, with many clades being supported by over 99% of the 1,000 bootstrap replicates.

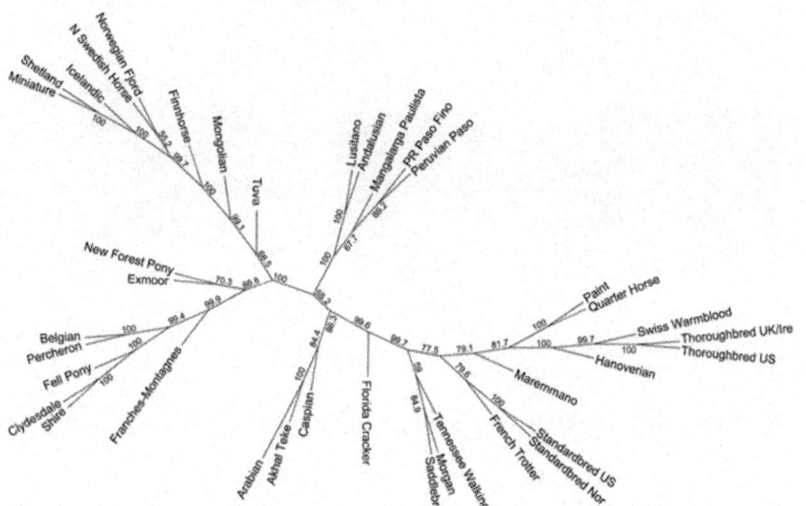

Figure 2: Distance based, neighbor joining tree calculated from SNP frequencies in 38 horse populations.

Majority rule, neighbor joining tree created from 10,536 SNP makers using Nei's genetic distance and allele frequencies within each population. Percent bootstrap support for all branches calculated from 1,000 replicates is shown.

doi:10.1371/journal.pone.0054997.g002

Cluster Analysis

Likelihood scores for runs of various K in Structure showed an increase in overall mean ln P(X|K) until K=35. A clear "true" value of K is not obvious examining the likelihood scores or using the Evanno method [39] (data not shown); however, variance among runs begins to increase with a diminishing increase in likelihood scores after K=29, which is near the peak of the curve. The value of the highest proportion (breed average q-value) of assignment of each breed for each value of K.

The first breeds to have all individuals assign strongly to one cluster are the Thoroughbred and Clydesdale (with Shire) at K=2, followed by the Shetland at K=3; these four breeds do not show signs of admixture at any K value analyzed. Evidence of weak geographic grouping is observed at K =4, which consists of: 1, the Middle Eastern and Iberian breeds (pink); 2, the Thoroughbred and breeds to which it continues to be or was historically crossed (yellow); 3, breeds developed in Scandinavia and Northern Europe (orange); and 4, the British Isles draft breeds (blue) (Figure 3).

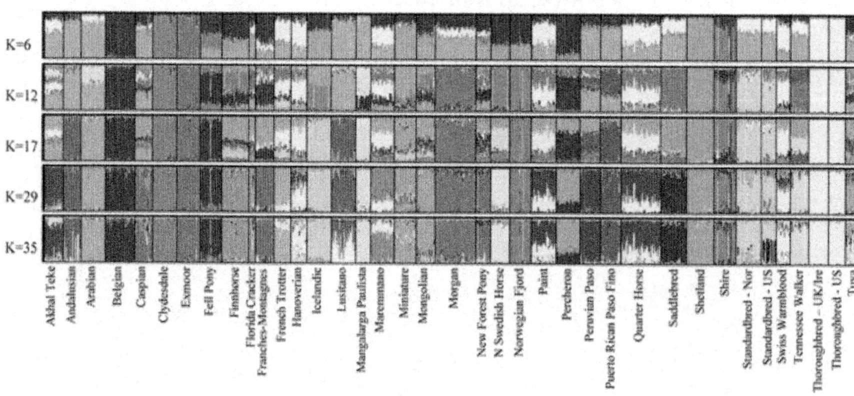

Figure 3: Bayesian clustering output for five values of K in 814 horses of 38 populations.

Structure output for five values of K investigated. Each individual is represented by one vertical line with the proportion of assignment to each cluster shown on the y axis and colored by cluster.

doi:10.1371/journal.pone.0054997.g003

Middle Eastern and Iberian Breeds

As also observed in the NJ tree, clustering of the Iberian and Middle Eastern breeds with the Mangalarga Paulista, Peruvian Paso, and Puerto Rican Paso

Fino (q >0.5) is observed until K =8, after which point the Mangalarga Paulista assigns with q =0.93 to another cluster. The remaining breeds cluster together until K=12, at which time the Middle Eastern breeds (Arabian, Akhal Teke, and Caspian) are assigned to their own cluster, leaving the Iberian breeds clustered with the Peruvian Paso and Puerto Rican Paso Fino. At low values of K (*i.e.* K <6) the Florida Cracker, Saddlebred, Standardbreds, Morgan, and Tennessee Walking Horse fall into the cluster with the Iberian and Middle Eastern breeds with breed mean q >0.5. At K=29, each of these breeds is assigned with q >0.72 to an individual cluster with the exception of the Lusitano and Andalusian, which remaining clustered together.

Thoroughbreds and Thoroughbred Crossed Breeds

Relationships described by the NJ tree among the Thoroughbred, Hanoverian, Swiss Warmblood, Paint, Quarter Horse, and Maremmano are also seen in cluster analysis. Clustering of those breeds with the Thoroughbred is observed throughout the values of K examined although at moderate frequencies (Figure 3). At K =29, the Hanoverian and Swiss Warmblood remain assigned to the cluster defined by the Thoroughbred but with assignment probabilities of 0.51 each. The Quarter Horse and Paint also assign to this cluster with q-values of 0.30 and 0.34, respectively. Neither the Quarter Horse, Paint, Hanoverian, or Swiss Warmblood populations assign to any cluster with q >0.62 at K =29. No evidence of population substructure is observed between the US and UK/Ire Thoroughbreds as also shown by PCA and parsimony analyses.

Scandinavian and Northern European Breeds

As in the NJ and parsimony trees, the Finnhorse, Icelandic, Miniature, North Swedish Horse, Norwegian Fjord, and Shetland are parsed into the same cluster (q-value >0.5) through K =5. However, unlike the NJ tree, at K =4, the highest value of assignment places the Belgian and Percheron into this cluster although with q <0.5 (0.42 and 0.38, respectively). The relationship remains until K =6, at which time the Miniature, Icelandic, and Shetland fall into a different cluster. At K =10, the Icelandic clusters again with the North Swedish Horse and Norwegian Fjord. The Norwegian and United States Standardbred populations, which at K =4 assign with q >0.5 to the cluster containing the Scandinavian breeds, separate from the Scandinavian breeds at K =5. At K =31, substructure appears in the Standardbred samples, which correlates to those individuals identified as pacers and that fall into an individual clade in the parsimony tree. At K =29, the Miniature and Shetland continue to be assigned to the same cluster (q-values =0.55 and 0.95, respectively). The next highest proportions of assignment of the Miniature horse are to the clusters described by the New Forest Pony ($q = 0.20$) and Icelandic ($q = 0.11$). No value of K evaluated eliminated signals of admixture from all populations in the dataset at $K=38$ (the actual number of populations sampled) or any value of K through 45 (data not shown).

British Isles Draft

The Clydesdale and Shire cluster together, and apart from the other breeds beginning at $K = 3$. In addition, the Fell Pony, which is placed within the same clade in the NJ and parsimony trees, and proximal to the Clydesdale and Shire in PCA, shows moderate assignment to this cluster ($0.29 < q < 0.41$) for several values of K from 4 to 14. At $K = 29$, the Shire assigns to the same cluster as the Clydesdale with $q = 0.69$. The individual outliers from the Shire breed also noted in parsimony analysis are evident beginning at $K = 3$. Excluding these outliers, at $K = 29$, the proportion of assignment for the Shires to the cluster with the Clydesdale increases to 0.74.

F_{ST}

All pairwise F_{ST} values calculated between the 37 populations (excluding the Florida Cracker) were significant as tested by 20,000 permutations (Figure 4). The lowest level of differentiation was found between the Paint and Quarter Horse populations ($F_{ST} = 0.002$), while the greatest divergence was observed between the Clydesdale and Mangalarga Paulista ($F_{ST}=0.254$). The two Thoroughbred populations had an F_{ST} value of 0.004, while the two Standardbred populations had 10-fold greater divergence ($F_{ST} = 0.020$) than the minimum observed value in this dataset; this value is similar to that observed between the Lusitano and Andalusian (0.021). An F_{ST} value of 0.006 was identified between the Tuva and Mongolian populations. The global F_{ST} value was 0.100. AMOVA computed on the set of 37 samples (excluding the outliers identified in Structure and the Florida Cracker) showed that 10.03% of the variance was accounted for among populations ($p = < 0.001$), 0.53% of the variance was among individuals within populations ($p=0.19$), and 89.44% of the variation was within individuals ($p<0.001$).

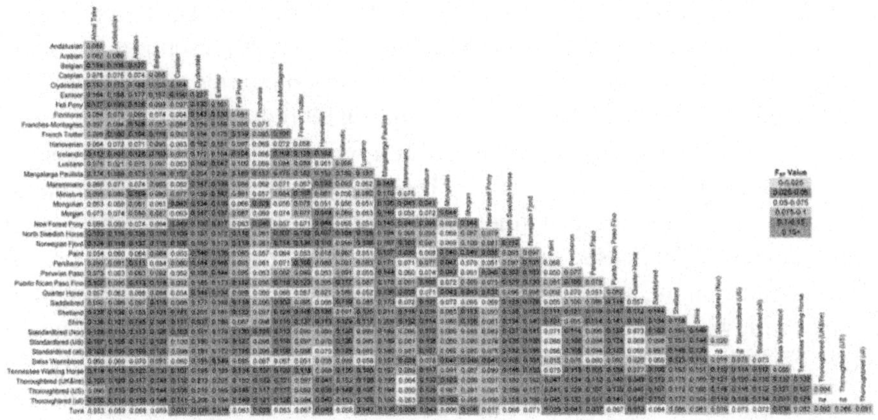

Figure 4: Pairwise F_{ST} **values based upon 10,536 SNPs in 37 horse populations.**

Pairwise F_{ST} values as calculated in Arlequin using 10,536 autosomal SNPs and significance tested using 20,000 permutations. All pairwise values are significantly different from zero. (individual outliers were removed from this analysis).

doi:10.1371/journal.pone.0054997.g004

DISCUSSION

These data are gathered from populations that represent tremendous diversity in phenotype and breed specialization. With breeds sampled across four continents, the resulting relationships observed largely reflect similarities of geographic origin, documented breed histories, and shared phenotypes. In general the highest within breed diversity was observed in breeds that are recently derived, continue to allow introgression of other populations, those that have a large census population size, and landrace populations that experience a lesser degree of controlled breeding. Not surprisingly, low diversity is observed in breeds with small census size, relatively old breeds with closed populations, and those with documented founder effects, whether due to population bottlenecks or selective breeding.

A total of seven individuals were identified by parsimony and cluster analysis as outliers with respect to the breed to which they were assigned. The pedigrees of these individuals were unknown. Because it is possible these horses were unknowingly crossbred or subject to mishandling in the field or laboratory, they were excluded from the within-breed analyses to avoid potential bias in indices of diversity. In addition, the potential impact of SNP ascertainment bias on diversity calculations must be acknowledged. The reference genome is from a Thoroughbred mare [40] and SNP identification was based upon the reference genome and data from seven other horses representing six breeds. Therefore, SNPs are generally derived to identify modern variation within the Thoroughbred as well as between the Thoroughbred and these other breeds. Thus, the SNPs identified may reflect an upward bias in diversity indices in the Thoroughbred and closely related breeds [41]. It seems that ascertainment bias may have particularly influenced the results when considering the data sets that have an increased number of loci resulting from relaxed LD pruning. These results show an increase in the relative diversity of the Thoroughbred, breeds with which the Thoroughbred continues to actively interbreed, and the other SNP discovery breeds, with respect to other breeds in the study. This is opposite of what may be expected given the high levels of genome-wide LD in the Thoroughbred. Without considering SNP ascertainment bias, it is expected that measured diversity would increase in breeds with low LD more quickly than in those with high LD, due to greater independence of markers in the former breeds. These SNPs, derived largely from the Thoroughbred, are apparently detecting a higher proportion of Thoroughbred-specific, rare variants and it appears that as more loci are included, more of these Thoroughbred-based variants are assayed, resulting in the observed increase in variation in the Thoroughbred, Thoroughbred-influenced breeds, and breeds used in SNP ascertainment.

The majority of the analyses were performed using 10,536 SNP markers pruned across breeds for LD of $r^2>0.2$ as well as MAF of 0.05 or above. Even though additional markers could have been used for analysis, many population-level statistics assume independence of loci. The stringent pruning for LD was therefore undertaken to eliminate bias in the test statistics that may result from substantial breed-specific differences in LD [40], [41]. A truncated data set also helped to make calculations, especially cluster analysis, computationally feasible. On the other hand, diversity indices were calculated after pruning the full data set to $r^2=0.2$ and $r^2=0.4$ (using pairwise correlation), and one replicate setting the threshold to $R^2<0.1$ (using the variance inflation factor), to examine the effect of allowing for varying levels of LD and therefore varying numbers of loci (see methods).

Within-breed Diversity

Even considering SNP ascertainment, low diversity as measured by H_e was observed in the Thoroughbred as well as the Standardbred, which both experience high selective pressures and are closed populations. Low diversity was also observed in breeds that have undergone a severe population bottleneck, such as the Exmoor and Clydesdale, and breeds that have small census population sizes, such as the Florida Cracker. Although the Thoroughbred is a large population that is widely distributed on a geographic scale, historic records suggest that one sire is responsible for 95% of the paternal lineages in the breed and as few as 30 females make up 94% of maternal lineages [28]. In addition, the population has been largely closed to outside gene flow since the formation of the first stud book in 1791 [42] and individuals within the breed are subject to selective pressure for racing success; therefore low, within-breed diversity is not at all surprising.

Using LD-based calculations, the estimated N_e for the Thoroughbred was similar to that found in a UK sample [43] and among the lowest of the study set despite the large census population size and geographic distribution of this breed. Individual inbreeding values based upon observed vs. expected homozygosity indicate that individual Thoroughbred horses show signs of inbreeding, with a mean loss of heterozygosity of 16.3%. This value is slightly larger than that found in [28] (13.9%). Using the same SNP array, [44] also showed inbreeding in the Thoroughbred, and specifically an increase in inbreeding over time. The only breeds with higher f values were the Exmoor, Clydesdale, Mangalarga Paulista, and Shire. Despite low individual diversity, F_{IS} values do not show significant inbreeding in either of the Thoroughbred populations as a whole, or in the Norwegian Standardbred although F_{IS} is significant in the US Standardbred population (discussed below).

The Clydesdale and Exmoor, in addition to having high individual estimated coefficients of inbreeding, also show the lowest within-breed diversity observed in the dataset. A lack of diversity in the Clydesdale and another British draft breed, the Shire, is likely a result of a severe population bottleneck observed in most draft breeds with the onset of industrialization and after the conclusion of World War II (WWII) as well as selection for size and color [45], [46]. The Exmoor pony, considered to be one of the purest native breeds of Britain, has been naturally selected for survival in harsh winter conditions on the moors in southwest England[45], [47]. Similar to the draft breeds, the Exmoor population decreased significantly after WWII to approximately 50 individuals, undoubtedly influencing the diversity observed in this study. The effect of low population size and selection is also reflected in extremely high individual estimates of f within some individuals. Finally, the Mangalarga Paulista shows low levels of heterozygosity, and as discussed below, the greatest divergence as measured by pairwise $F_{ST}$of all breeds in the study. While these results could be due to geographic distance between this and other breeds, and/or genetic drift, unfortunately these horses were all sampled from only two farms and likely do not represent the entirety of the diversity present in the breed; therefore we cannot rule out sampling error which would inflate the estimated level of divergence between these individuals and the other breeds and result in a decrease in H_e. However, a lack of diversity in sampling of the breed would not have an effect on estimates of individual inbreeding coefficients, which were among the highest of the entire data set.

Converse to the above examples, high levels of diversity as measured by both H_e and N_e, accompanied by low estimates of inbreeding (f and F_{IS}), are observed in the Mongolian, Tuva, and New Forest Pony. The Mongolian and Tuva are unique in that they represent landrace populations that are less managed than the popular breeds of Western Europe and North America; they occupy a diverse range of habitat, have been selected for meat and milk in addition to use in transportation, and originate in the region where domestication was likely to have occurred. The population of Mongolian horses is large and individuals are phenotypically diverse [48]. In 1985, approximately two million Mongolian horses of four different types were estimated to live within the country [45]. The Tuva is not as numerous as the Mongolian but is similar in its purpose and also has high within-breed phenotypic diversity. In addition, it is suggested that the Tuva has experienced outcrossing in order to increase its size and stamina[45] as may also be the case in the Mongolian [49]. Similarly, the New Forest Pony was historically a free-ranging population in Great Britain, but was crossbred until the 1930's. These traits: old populations, large population size, outcrossing, high phenotypic diversity, and lesser artificial selection/management, result in the high levels of genetic diversity observed. This extent of diversity appears to diminish as populations are restricted by selective pressures into formal breeds.

Other population characteristics are likely the cause of the diversity observed in the Finnhorse, Icelandic, and Miniature. In the case of the Icelandic, the high level of diversity was possibly maintained by a large census population size despite isolation for almost a thousand years and several population bottlenecks due to natural disasters [45]. In the case of the Finnhorse, diversity may be due to within-breed substructure into four sections of the studbook established in 1970: the work horse (draft), trotters, riding horse, and pony [50]. Finally, high diversity in breeds such as the Miniature is likely a result of a diverse founding stock [45], [51], [52]; horses of small size from a variety of geographic regions and bloodlines were utilized in founding the breed, which is defined by phenotype.

All of these three factors, large population size, phenotypic diversity within the breed, and a diversity of founding stock, also lead to the relatively high levels of diversity observed in the Paint and Quarter Horse; in addition, these breeds both allow continued outcrossing between themselves and with the Thoroughbred and have experienced a tremendous population expansion since the formal foundation of the breeds within the past 45–75 years. Due to the relative infancy of these populations, it could be argued that the Paint, Quarter Horse, and other, newly-derived breeds, have not yet had time to undergo the evolutionary processes necessary to be genetically distinct populations as is observed in breeds with longer histories and closed studbooks. However, even with high within-breed diversity and large census population sizes (over 1 million and 4 million worldwide for the Paint and Quarter Horse, respectively), N_e for these breeds account for only a fraction of the census size, demonstrating non-random mating and selection. Outcrossing is also continued in the Swiss Warmblood and Hanoverian breeds, which show similar trends in diversity measures as the Quarter Horse and Paint. The relatively low N_e in these breeds, accompanied by moderate H_e may partially be due to significant crossing with the Thoroughbred, which would contribute long blocks of LD [40],[41], resulting in decreased estimates of N_e.

Of note in breeds such as the Quarter Horse, Lusitano, and Andalusian, is that despite moderate to high relative levels of H_e, and low to moderate estimates of f, F_{IS} values in each breed are significantly positive. Significant F_{IS} was also previously observed in the Iberian breeds using microsatellite markers [53]. While selection and inbreeding may be responsible for significant values of F_{IS} in some of these breeds, another instance in which F_{IS} may be significantly positive is in the presence of subpopulation structure within the sample. Evidence of this in the Lusitano and Andalusian is present in parsimony analysis where individuals of the two breeds fall into one clade, but within that clade are two highly supported branches represented by a subset of each breed. In addition, when forcing high values of K in Structure, such as observed at K =35, Andalusian and Lusitano individuals fall into one of two clusters with q-value >0.5 in a nonbreed-specific manner (data not shown). These results support [54], which showed potential subpopulation structure in the Lusitano via microsatellite analysis. In the Quarter Horse, subpopulation structure is evident through the evaluation of bloodlines and the selection of popular sires for diverse performance classes. This population substructure in the Quarter Horse has also been demonstrated by marked differences in allele frequencies among performance types (cutting, western pleasure, halter, racing, etc.) [55]. A similar instance is found in the US population of the Standardbred, which also has significant excess homozygosity (F_{IS}). Unlike Standardbreds in Europe, which are raced at a trot, those in the US are divergently selected for racing at either the pace or the trot, creating structure within the breed [56].

Finally, several rare populations are included in this dataset. The Caspian is one of the oldest breeds in the Middle East and was thought to be extinct until its recent rediscovery in 1965. The Florida Cracker, a now rare breed, was developed in the United States from feral stock of Iberian descent [57]. The sample size of the Florida Cracker limits the conclusions that can be drawn regarding within-breed diversity. However, the Caspian shows high N_e, H_e, and estimates of f, given its rarity. After rediscovery of the breed, which historically was believed to represent a type of landrace population, [58] describes a three-year survey, which found approximately 50 individuals remaining, noting that many could not be considered "pure." In addition, [33] were unable to show evidence of a recent bottleneck in the Caspian breed. The diversity observed in what are now considered Caspian horses likely stems from high levels of diversity within those individuals that founded the modern population.

Among-breed Diversity

The expectation of homogeneity within breeds due to closed populations and selection is supported by the results of AMOVA, which show significant variation present among populations, but a non-significant proportion of variance within. However, the variation among samples lends information about current and historic relationships. Observed trends include patterning based upon geographic origin and/or phenotypic similarities, and relatively low divergence observed in comparisons that include breeds with high within-breed diversity. In Structure analysis, K =29 was chosen as the most likely value of K; however no single value stood out as the "best" number of clusters to describe these data. Regardless, patterns observed in clustering were also supported by pairwise F_{ST} values, parsimony, PCA, and NJ dendograms.

High Diversity and Low Divergence – Landrace Breeds

The Mongolian and Tuva populations are believed to have been influential in the spread of horses across Asia and Europe [45], [59]; these landrace populations, harboring high levels of within-breed diversity, were found to be similar to one another, with a pairwise F_{ST} value of 0.006. In addition, with the exception of six Tuva individuals that fell into a clade with the Caspians, both parsimony and NJ analyses place the Tuva and Mongolian into the same clade of each tree. Examining all comparisons, low F_{ST} values were observed between the Mongolian and Tuva compared to the other breeds in this study, supporting the potential role of Eurasian horses of similar type in founding modern stocks. This also aligns with high microsatellite diversity observed in Eastern Eurasian "non-breed" (landrace) populations in [19]. On the other hand, breeds with high diversity in general show lower levels of divergence as measured by F_{ST}, while those with low diversity show higher values of F_{ST}. Low divergence in breeds with high diversity is expected as variation within a breed may indicate outcrossing with other populations, and high variation also makes these breeds more likely to share variation with others by chance. In contrast, if a breed has little within-breed variation, it is less likely to share genetic variation with another breed by chance, especially with another breed that is relatively homogeneous itself. As demonstrated in human literature, source populations are expected to contain greater diversity than those populations which they found [60], [61]; this is also suggested in the horse by [11], which showed greater mtDNA diversity in Iberian breeds than the recently founded American breeds. If the argument is made that the low F_{ST} values of Tuvas and Mongolians supports their role in founding modern breeds, the same argument could be made for the Quarter Horse, which also shows low levels of pairwise divergence; however that argument would be unreasonable as the Quarter Horse was developed in only the past century. Therefore, the relative values of F_{ST} are informative, but these F_{ST} values and data, which represent modern breeds generally derived from limited founding stock, and subjected to intense artificial selective pressures, cannot be used independently to elucidate the evolution of the modern horse. Also, as is the case for other analyses, while the relationships observed can shed light on the history of breeds, they cannot distinguish between recent admixture and shared ancestry.

Thoroughbred-influenced Breeds

The Thoroughbred is believed to have founding sires of Arabian, Turk, and Barb ancestry [42], and [28] found that two sires, noted as being Arabian (Godolphin Arabian and Darley Arabian) together contributed to over 20% of the modern population. However, it is likely that the "Arabian" foundation stallions were not Arabians as the breed is known today. It is noted in [62]that the Godolphin Arabian was a Turkoman stallion with partial Arabian blood, while in other work it is suggested he was a Barb [45]. Regardless of the true ancestry of these stallions, restrictions placed upon the export of purebred Arabians during the 16th and 17th centuries, as well as the general use of the term "Arab" for horses of Middle Eastern descent, it is likely other "Arabian" horses with influence on the Thoroughbred breed also had Turkoman, and Barb bloodlines [45], [62]. The pairwise F_{ST} values between the Thoroughbred and the Arabian do not suggest any less divergence than observed between the Thoroughbred and a majority of the other breeds. In addition, at K=29 the Arabian assigns to the Thoroughbred cluster at only 2.3%. If the Arabian did have significant influence on the Thoroughbred breed, there are several possible explanations for why the supposed Arabian influence is not more apparent. The first is related to SNP ascertainment and the bias of SNPs toward modern variation in Thoroughbred. It is possible that the genes derived from the Arabian are at or near fixation in the Thoroughbred, which would reduce the chance that these SNPs, and the variation described within them are present in the dataset. Another possibility is that the current Arabian sample, taken from the United States, may not reflect the Arabian lineage(s) that were influential in the founding of the Thoroughbred. Finally, as noted above and also suggested elsewhere [63], it may simply be that Arabian bloodlines were not as instrumental in the Thoroughbred breed as once thought or that the initial Arabian influence (and genes) have been selected against or lost to drift during the development of the modern Thoroughbred racehorse.

Within the Thoroughbred itself, divergence between the US and European samples had a significant F_{ST} of 0.004, similar to that observed between the Hanoverian and Swiss Warmblood (0.008) and Mongolian and Tuva (0.006), but larger than the minimally observed value seen between the Paint and the Quarter Horse (0.002). Although artificial insemination is prohibited in the Thoroughbred and would be anticipated to limit gene flow to some extent, the founder effect in the original European Thoroughbred by few high-impact sires and dams, accompanied by shared selective pressures, relatively recent importation of the breed to the United States, and ongoing shipment of horses between continents are likely contributing to the lack of geographic population structure identified by parsimony and cluster analyses.

While within-breed diversity of the Thoroughbred was relatively low, and a notable relationship with the Arabian was not observed, among-breed analysis shows a clear influence of the Thoroughbred on many other breeds. Placed with the Thoroughbred in parsimony analysis are the Hanoverian, Maremmano, and Swiss Warmblood. The Maremmano, an Italian breed, shows a q-value of assignment to the Thoroughbred clade of 0.26 at K =29. This is not surprising given reports that Thoroughbreds contributed over 13% of the maternal lineages to the stallion lines within the stud book [64]. Low differentiation of the Maremmano compared to the Hanoverian was also reported in [29], which is logical given similar influence of the Thoroughbred on the Hanoverian. This and the allowed crossing of Thoroughbreds into the Swiss Warmblood population is reflected in minimal measures of divergence between these samples. The continued influence of the Thoroughbred on the Paint and Quarter Horse is also reflected both by low F_{ST} values as well as greater than 30% assignment of the Paint and Quarter Horse to the Thoroughbred cluster. As each of these breeds experience continued gene flow from the Thoroughbred, and had Thoroughbred founding stock, these results are not unexpected. While outcrossing of these breeds is allowed (with restrictions), it is likely that even in cases of breeds with closed stud books, some outcrossing, intentional, unintentional, and/or undocumented, has occurred; it has been demonstrated that historical pedigrees, while helpful, are not always accurate [65]. Issues of outcrossing and individual identification can now be more easily addressed using genetic testing and have the potential to assist managers in decisions regarding breeding and registration.

Finally, the Standardbred samples, which also represent two continents, had a significant pairwise F_{ST} value of 0.020, five-fold greater than that observed between the two Thoroughbred samples. This comparison may reflect geographic structure, the influence of French Trotter bloodlines in the European sample, and selection for pacing horses in the US population. Within the horses included in this study, eight were noted to be pacers. These pacers all fall within one clade of the parsimony tree, supported by a bootstrap value of 98%. The limited sample size does not allow a thorough comparison of the pacing vs. trotting Standardbreds, however significant genetic differentiation between horses of the two racing groups has been reported [56].

Middle Eastern and Iberian Breeds

The Middle Eastern breeds, Arabian, Akhal Teke, and Caspian, were placed into a single clade of the NJ tree, with the Arabian and Akhal Teke in their own, highly supported clade. This relationship was supported by low values of K in Structure, which placed the Iberian and Middle Eastern breeds into the same cluster until K =12. However, parsimony analysis did not support this relationship between the Middle Eastern and Iberian breeds as the Arabian and Akhal Teke individuals were placed into one clade, apart from the Iberian samples and from the Caspians.

In Europe, the Iberian breeds (Lusitano and Andalusian) have only recently been distinguished from one another depending upon the region in which they are bred and divergent selective pressures. In the US, horses of each breed are occasionally interbred [51]. The close relationship between the Andalusian and Lusitano samples in this study is reflected in the minimum F_{ST} value observed in either breed, 0.021. The parsimony tree and PCA also shows that individuals cannot necessarily be distinguished from one another regardless of whether they were sampled in the US or Portugal. Of the two clades that appear to suggest population substructure, one includes only Portuguese Lusitanos while the other includes only US samples, although two are of Portuguese ancestry.

Iberian and Gaited Breeds

A horse that is considered to be "gaited" naturally moves in a means other than the traditional walk, trot, canter, and gallop. Alternative gaits in horses are distinguished from traditional gaits by their unique footfall pattern and/or rhythm. The genetic basis of gait has recently been investigated and suggests that all modern gaited breeds share a common ancestor as supported by a shared, extended haplotype spanning a variant significantly associated with the ability to pace [66]; [67]. There is a great deal of historical evidence that the shared ancestry of gaited breeds traces back to Iberian bloodlines, in particular to the Spanish Jenette [45], [51]. Influence of the Iberian breeds on modern gaited breeds is seen in early clustering in Structure analysis as well as in the NJ tree where the Puerto Rican Paso Fino and Peruvian Paso are placed on the same branch as the Andalusian and Lusitanos. In addition to Iberian lines, the Narragansett Pacer is often named as instrumental in the founding of American breeds that may gait including the Saddlebred, Standardbred, and Tennessee Walking Horse [45]. Within those breeds, the Tennessee Walking Horse was documented to be greatly influenced by the Saddlebred, Standardbred, and Morgan [68], [69]; and the Standardbred itself had influence from the Thoroughbred and Morgan (among others) [70]. While we do not have samples of the now extinct Narragansett Pacer, our data set does support a close relationship between the Tennessee Walking Horse and Saddlebred, as observed in the NJ and parsimony analysis as well as with the Morgan and Standardbred. At low values of K in cluster analysis, the Tennessee Walking Horse and Saddlebred cluster strongly. In NJ analysis, the Florida Cracker, which has many individuals that demonstrate the ability to gait, is found intermediate to the Iberian and the modern US gaited breeds. Interestingly, the Icelandic, a four- or five-gaited breed, does not show any significant affinity to the other gaited breeds although they share the recently identified major locus that appears to be essential to the ability to perform alternate gaits [66]; [67]. It thus seems that the genetic variant associated with the gait phenotype arose well before the separation of breeds. Instead of clustering with the other gaited breeds, the Icelandic clusters with the Shetland through K=16 and also is within a highly supported branch of the NJ tree with the Shetland and Miniature. Finally, the influence of the Shetland on the development of the Miniature is observed at all values of K as well as in the parsimony tree where both breeds occupy the same clade.

Drafts

The Shire and Clydesdale populations share assignment to the same cluster throughout Structure runs. The similarity between these breeds is also seen in a lack of monophyly in the parsimony tree, the sharing of a branch of the NJ tree, positioning in PCA, and a pairwise F_{ST} value of 0.037. The Fell Pony, a British breed, falls out as sister taxa to the British draft horses, the Shire and Clydesdales. However F_{ST} values show that divergence between the Fell Pony and either the Clydesdale or Shire is not significantly less than seen with most other populations. The other branch of the "draft" clade of the NJ tree contains the breeds from the European mainland, the Belgian, Percheron, and Franches-Montagnes; each of these breeds shows monophyly in parsimony analysis. In addition, similarities among draft and light draft breeds are reflected in cluster analyses at K =6, which show the populations from the European mainland and Scandinavian Peninsula (Belgian, Finnhorse, Franches-Montagnes, North Swedish Horse, Norwegian Fjord, and Percheron) assign to one cluster with q >0.5. The grouping of the Scandinavian breeds is similar to that previously reported [31]. Geographic relationships are also suggested by the two British breeds, the New Forest Pony and Exmoor that fall just basal to the draft clade in the NJ tree.

Summary

This data set resulting from a large international collaboration represents the first study in the horse to provide an extensive overview of nuclear genetic diversity within, and relationships among a diverse sample of breeds and landrace populations. These data are now available for use in subsequent studies of population-level relationships and provide a baseline for monitoring changes in breed diversity. With high mtDNA diversity but limited paternal input during domestication, this increased understanding of nuclear diversity within the horse will allow for the identification of genomic regions of importance to breed derivation and will be instrumental in guiding across-breed gene discovery projects.

METHODS

Ethics Statement

DNA sampling was limited to the collection of blood by jugular venipuncture performed by a licensed veterinarian or from hairs pulled from the mane or tail by the horse owner or researcher. All animal work was conducted in accordance with and approval from the international and national governing bodies at the institutions in which samples were collected (the University of Minnesota Institutional Animal Care and Use Committee (IACUC); the University of Kentucky IACUC; the University College Dublin, Animal Research Ethics Committee; Swiss Law on Animal Protection and Welfare; the Ethical Board of the University of Helsinki; the Animal Health Trust Clinical Research Ethics Committee; Norwegian Animal Research Authority; UK Home Office License; and the Lower Saxon state veterinary office).

Samples and Genotyping

Tissue samples and previously collected genotypes from 1,060 horses were obtained from members of the EGDC or were obtained by our laboratory. 814 samples representing the 38 populations included in this study were selected from the EDGC sample collection with the goal of obtaining as random of a sample as possible and to minimize close relationships among individuals. In some cases, genotypes were available from breeds collected for genome-wide association studies (GWAS). In all cases, when pedigree information was available, no relationships were allowed at or more recent to the grandsire/dam level. If no pedigrees were available, once genotyping was performed, individuals were removed from the analyses to reduce genome sharing as measured by autosomal estimates of identity by descent (pi hat) values in PLINK [71] greater than 0.3 (after pruning for MAF>0.05). In samples that were obtained as a result of GWAS, "control" individuals were preferentially chosen for inclusion in these analyses. When necessary, DNA isolation from hair roots took place using a modification of the Puregene (Qiagen) protocol for DNA purification from tissue. Modifications include the addition of 750 µl of isopropanol rather than 300, increasing the precipitation spin time to 15 m at 4°C, and washing the pellet twice. Approximately 1 µg of DNA was used for SNP genotyping using the Illumina SNP50 Beadchip according to the manufacturer's protocol. All genotype calls were extracted from the raw intensity data using GenomeStudio (Illumina) with the minimum gencall score threshold of 0.15. The raw intensity scores were available for all populations with the exception of the Lusitano and Maremmano.

Data Pruning

SNP discovery was conducted using horses from seven breeds (Akhal Teke, Andalusian, Arabian, Icelandic, Quarter Horse, Standardbred, Thoroughbred) as well as the reference genome of a Thoroughbred mare [41]. To eliminate ascertainment bias as much as possible, horses from the discovery breeds were removed from the dataset, which was then pruned to exclude SNPs with MAF less than 0.05. All horses were then replaced and those SNPs removed from all analyses. In this new, complete data set, SNP markers that failed to genotype in at least 99% of the individuals and SNPs that had a MAF of 0.05 or less across all samples were removed as well as SNPs on ECAX. SNPs that were in LD across breeds were also removed; files used for basic diversity indices were pruned for $r^2<0.1$ in PLINK [71] considering 100 SNP windows and moving 25 SNPs per set (–indep-pairwise 100 25 1.11). Allowing for additional LD, data sets were also created for r2<0.2 and 0.4. An additional data set, used for Structure analysis was pruned for $R^2<0.1$ in Plink (–indep;

R=multiple correlation coefficient), which is similar to the above method but instead of analyzing pairwise relationships of SNPs as in the former method, uses a multiple regression approach upon the SNPs in the analysis window. Files were converted for usage between analyses programs using PLINK, perl script, CONVERT [72] and/or PGDSpider 2.0.1.4 [73].

Within-breed Diversity

Expected heterozygosity (H_e) and AMOVA were calculated in Arlequin3.5 [74] on all four data sets. AMOVA was conducted on the primary data set with breeds designated as populations and excluding the Florida Cracker due to small sample size. Analyses were also conducted grouping the two Thoroughbred and Standardbred samples together by breed. F_{IS} was calculated and significance tested on the primary data set, with 10,000 permutations of the data in Genetix [75]. Individual inbreeding coefficients (f) were calculated in PLINK based upon loss of heterozygosity (–het).

Among Breed Relationships

Pairwise F_{ST} values were calculated on the primary, 10,536 SNP dataset in Arlequin3.5 [74]using Reynolds' distance [76] with significance tested using 20,000 permutations.

A neighbor joining (NJ) cladogram was built using breed allele frequencies calculated from the primary SNP set using the packages seqboot, gendist, neighbor, and consense and Nei's genetic distance [38] in PHYLIP ver3.69 [77]. Bootstrap support from 1,000 iterations of the data was used to assess support for the resulting majority rule consensus cladogram.

A parsimony cladogram was constructed using 10,066 SNP markers pruned from the original data set using the MAF and genotyping rate criteria as above and allowing for $R^2<0.2$ (–indep-pairwise). The domestic ass was included as an outgroup for traditional and new-technology searches in TNT [78].

Principal component analysis (PCA) was conducted in snpStats in R (http://cran.r-project.org) on the full SNP set consisting of all 814 individuals and 38,755 autosomal SNPs (pruned only for MAF and genotyping rate).

Cluster Analysis

Clustering of breeds into genetic groups was examined using the program Structure 2.3.3 [79],[80] assuming K = 1 to 45. The Structure algorithm included the admixture model and correlated allele frequencies. Three iterations of each K value were conducted with 35,000 MCMC repetitions (15,000 burn-in). The convergence of Structure runs was evaluated by equilibrium of alpha and likelihood scores. The value of K most suitable to explain the diversity in these data was predicted by the highest mean estimated ln P(X|K) while minimizing variance and also making biological sense [80], [81]. The replicates from each run of K were input into CLUMPP [82] and the average cluster membership calculated using the LargeK Greedy algorithm. Output from CLUMPP was visualized in Distruct [83].

Effective Population Size

To estimate effective population size (N_e), the full set of 54,602 SNP markers was pruned within each population to remove those with MAF <0.01 and genotyping rate of <0.05. Pairwise r^2 values between remaining SNPs were calculated in Haploview [84], for each population considering intermarker distances from 0 to 4 Mb in 50 kb increments. Values of N_e were calculated using the method of [37], which includes a correction for small sample size and the assumption that 1 Mb =1 cM.

Data Access

All SNP genotype data are available at the NAGPR Community Data Repository (animalgenome.org) for the purpose of reconstructing the analyses. The only exception is the data collected from the Tennessee Walking Horse, which, under agreement from the granting agency (to the University of Minnesota from the Foundation for the Advancement of the Tennessee Walking Show Horse (FAST) and the Tennessee Walking Horse Foundation (TWHF)), is only available under a Material Transfer Agreement (MTA) between interested individuals and the University of Minnesota.

ACKNOWLEDGMENTS

The computational support provided by Rob Schaefer and Aaron Rendahl is gratefully acknowledged. Finally, the EGDC thanks the Havemeyer Foundation for their continued support of equine genomics research.

AUTHOR CONTRIBUTIONS

Conceived and designed the experiments: JLP JRM MEM. Performed the experiments: JLP MEM. Analyzed the data: JLP. Contributed reagents/materials/analysis tools: JLP JRM EGC LSA JA EB DB MMB ASB PB ACM OD MF LFC KTG GG BH TH KH EWH TL GL HL MSL BAM SM NO MCTP RJP MR SR KHR MS JS TT MV CMW MEM. Wrote the paper: JLP JRM EGC MEM.

REFERENCES

1. .FAOSTAT (2010). Available: http://www.faostat.fao.org.Accessed 2012 Jun 5.
2. .Lippold S, Matzke NJ, Reissmann M, Hofreiter M (2011) Whole mitochondrial genome sequencing of domestic horses reveals incorporation of extensive wild horse diversity during domestication. BMC Evol Biol 11: 328. doi: 10.1186/1471-2148-11-328
3. .Ludwig A, Pruvost M, Reissmann M, Benecke N, Brockmann GA, et al. (2009) Coat color variation at the beginning of horse domestication. Science 324: 485. doi: 10.1126/science.1172750
4. .Outram AK, Stear NA, Bendrey R, Olsen S, Kasparov A, et al. (2009) The earliest horse harnessing and milking. Science 323: 1332–1335. doi: 10.1126/science.1168594
5. .Pedrosa S, Uzun M, Arranz JJ, Gutierrez-Gill B, Primitivo FS, et al. (2005) Evidence of three maternal lineages in near eastern sheep supporting multiple domestication events. P R Soc B 272: 2211–2217. doi: 10.1098/rspb.2005.3204
6. .Larson G, Dobney K, Albarella U, Fang M, Matisoo-Smith E, et al. (2005) Worldwide phylogeography of wild boar reveals multiple centers of pig domestication. Science 307: 1618–1621. doi: 10.1126/science.1106927
7. Wu GS, Yao YG, Qu KX, Ding ZL, Li H, et al. (2007) Population phylogenomic analysis of mitochondrial DNA in wild boars and domestic pigs revealed multiple domestication events in East Asia. Genome Biol 8: R245. doi: 10.1186/gb-2007-8-11-r245
8. .Cieslak M, Pruvost M, Benecke N, Hofreiter M, Morales A, et al. (2010) Origin and history of mitochondrial DNA lineages in domestic horses. PLOS One 5: e15311. doi: 10.1371/journal.pone.0015311
9. .Jansen T, Forster P, Levine MA, Oelke H, Hurles M, et al. (2002) Mitochondrial DNA and the origins of the domestic horse. PNAS 99: 10905–10910. doi: 10.1073/pnas.152330099
10. .Lei CZ, Su R, Bower MA, Edwards CJ, Wang XB, et al. (2009) Multiple maternal origins of native modern and ancient horse populations in China. Anim Genet 40: 933–944. doi: 10.1111/j.1365-2052.2009.01950.x
11. .Lira J, Linderholm A, Olaria C, Brandstrom Durling M, Gilbert MT, et al. (2010) Ancient DNA reveals traces of Iberian Neolithic and Bronze Age lineages in modern Iberian horses. Mol Ecol 19: 64–78. doi: 10.1111/j.1365-294x.2009.04430.x

12. .Vila C, Leonard JA, Gotherstrom A, Marklund S, Sandberg K, et al. (2001) Widespread origins of domestic horse lineages. Science 291: 474–477. doi: 10.1126/science.291.5503.474

13. .Warmuth V, Eriksson A, Bower MA, Barker G, Barrett E, et al. (2012) Reconstructing the origin and spread of horse domestication in the Eurasian steppe. PNAS 109: 8202–8206. doi: 10.1073/pnas.1111122109

14. .Beja-Pereira A, Caramelli D, Lalueza-Fox C, Vernesi C, Ferrand N, et al. (2006) The origin of European cattle: evidence from modern and ancient DNA. PNAS 103: 8113–8118. doi: 10.1073/pnas.0509210103

15. .Gotherstrom A, Anderung C, Hellborg L, Elburg R, Smith C, et al. (2005) Cattle domestication in the Near East was followed by hybridization with aurochs bulls in Europe. P R Soc B 272: 2345–2350. doi: 10.1098/rspb.2005.3243

16. .Keyser-Tracqui C, Blandin-Frappin P, Francfort HP, Ricaut FX, Lepetz S, et al. (2005) Mitochondrial DNA analysis of horses recovered from a frozen tomb (Berel site, Kazakhstan, 3rd Century BC). Anim Genet 36: 203–209. doi: 10.1111/j.1365-2052.2005.01316.x

17. .Achilli A, Olivieri A, Soares P, Lancioni H, Kashani BH, et al. (2012) Mitochondrial genomes from modern horses reveal the major haplogroups that underwent domestication. PNAS 109: 2449–2454. doi: 10.1073/pnas.1111637109

18. .McGahern A, Bower MA, Edwards CJ, Brophy PO, Sulimova G, et al. (2006) Evidence for biogeographic patterning of mitochondrial DNA sequences in Eastern horse populations. Anim Genet 37: 494–497. doi: 10.1111/j.1365-2052.2006.01495.x

19. .Warmuth V, Manica A, Eriksson A, Barker G, Bower M (2012) Autosomal genetic diversity in non-breed horses from eastern Eurasia provides insights into historical population movements. Anim Genet. In press. doi:10.1111/j.1365–2052.2012.02371.x.

20. .Lindgren G, Backstrom N, Swinburne J, Hellborg L, Einarsson A, et al. (2004) Limited number of patrilines in horse domestication. Nat Genet 36: 335–336. doi: 10.1038/ng1326

21. .Ling Y, Ma Y, Guan W, Cheng Y, Wang Y, et al. (2010) Identification of y chromosome genetic variations in Chinese indigenous horse breeds. J Hered 101: 639–643. doi: 10.1093/jhered/esq047

22. .Aberle KS, Hamann H, Drogemuller C, Distl O (2004) Genetic diversity in German draught horse breeds compared with a group of primitive, riding and wild horses by means of microsatellite DNA markers. Anim Genet 35: 270–277. doi: 10.1111/j.1365-2052.2004.01166.x

23. .Achmann R, Curik I, Dovc P, Kavar T, Bodo I, et al. (2004) Microsatellite diversity, population subdivision and gene flow in the Lipizzan horse. Anim Genet 35: 285–292. doi: 10.1111/j.1365-2052.2004.01157.x

24. .Bjornstad G, Gunby E, Roed KH (2000) Genetic structure of Norwegian horse breeds. J Anim Breed Genet 117: 307–317. doi: 10.1046/j.1439-0388.2000.00264.x

25. .Bömcke E, Gengler N, Cothran EG (2010) Genetic variability in the Skyros pony and its relationship with other Greek and foreign horse breeds. Genet and Mol Biol 34: 68–76. doi: 10.1590/s1415-47572010005000113

26. .Canon J, Checa ML, Carleos C, Vega-Pla JL, Vallejo M, et al. (2000) The genetic structure of Spanish Celtic horse breeds inferred from microsatellite data. Anim Genet 31: 39–48. doi: 10.1046/j.1365-2052.2000.00591.x

27. .Cothran EG, Canelon JL, Luis C, Conant E, Juras R (2011) Genetic analysis of the Venezuelan Criollo horse. Genet Mol Res 10: 2394–2403. doi: 10.4238/2011.october.7.1

28. .Cunningham EP, Dooley JJ, Splan RK, Bradley DG (2001) Microsatellite diversity, pedigree relatedness and the contributions of founder lineages to thoroughbred horses. Anim Genet 32: 360. doi: 10.1046/j.1365-2052.2001.00785.x

29. .Felicetti M, Lopes MS, Verini-Supplizi A, Machado Ada C, Silvestrelli M, et al. (2010) Genetic diversity in the Maremmano horse and its relationship with other European horse breeds. Anim Genet 41 Suppl 253–55. doi: 10.1111/j.1365-2052.2010.02102.x

30. .Glowatzki-Mullis ML, Muntwyler J, Pfister W, Marti E, Rieder S, et al. (2005) Genetic diversity among horse populations with a special focus on the Franches-Montagnes breed. Anim Genet 37: 33–39. doi: 10.1111/j.1365-2052.2005.01376.x

31. .Leroy G, Callede L, Verrier E, Meriaux JC, Ricard A, et al. (2009) Genetic diversity of a large set of horse breeds raised in France assessed by microsatellite polymorphism. Genet Sel Evol 41: 5. doi: 10.1186/1297-9686-41-5

32. .Marletta D, Tupac-Yupanqui I, Bordonaro S, Garcia D, Guastella AM, et al. (2006) Analysis of genetic diversity and the determination of relationships among western Mediterranean horse breeds using microsatellite markers. J Anim Breed Genet 123: 315–325. doi: 10.1111/j.1439-0388.2006.00603.x

33. .Shasavarani H, Rahimi-Mianji G (2010) Analysis of genetic diversity and estimation of inbreeding coefficient within Caspian horse population using microsatellite markers. Afr J Biotechnol 9: 293–299.

34. .Solis A, Jugo BM, Meriaux J-C, Iriondo M, Mazon LI, et al. (2005) Genetic diversity within and among four South European native horse breeds based on microsatellite DNA analysis: Implications for conservation. J Hered 96: 670–678. doi: 10.1093/jhered/esi123

35. .Tozaki T, Takezaki N, Hasegawa T, Ishida N, Kurosawa M, et al. (2003) Microsatellite variation in Japanese and Asian horses and their phylogenetic relationships using a European horse outgroup. J Hered 94: 374–380. doi: 10.1093/jhered/esg079

36. .Luis C, Juras R, Oom MM, Cothran EG (2007) Genetic diversity and relationships of Protuguese and other horse breeds based on protein and microsatellite loci variation. Anim Genet 38: 20–27. doi: 10.1111/j.1365-2052.2006.01545.x

37. .Weir BS, Hill WG (1980) Effect of mating structure on variation in linkage disequilibrium. Genetics 95: 477–488.

38. Nei M (1972) Genetic distance between populations. Amer Nat 106: 283–292. doi: 10.1086/282771

39. Evanno G, Regnaut S, Goudet J (2005) Detecting the number of clusters of individuals using the software STRUCTURE: a simulation study. Mol Ecol 14: 2611–2620. doi: 10.1111/j.1365-294x.2005.02553.x

40. .Wade CM, Giulotto E, Sigurdsson S, Zoli M, Gnerre S, et al. (2009) Genome sequence, comparative analysis, and population genetics of the domestic horse. Science 326: 865–867. doi: 10.1126/science.1178158

41. .McCue ME, Bannasch DL, Petersen JL, Gurr J, Bailey E, et al. (2012) A high density SNP array for the domestic horse and extant perissodactyla: utility for association mapping, genetic diversity, and phylogeny studies. PLOS Genet 8: e1002451. doi: 10.1371/journal.pgen.1002451

42. .Weatherby and Sons (1791) An Introduction to a General Stud Book. London, UK.

43. .Corbin LJ, Blott SC, Swinburne JE, Vaudin M, Bishop SC, et al. (2010) Linkage disequilibrium and historical effective population size in the Thoroughbred horse. Anim Genet 41 Suppl 28–15. doi: 10.1111/j.1365-2052.2010.02092.x

44. .Binns MM, Boehler DA, Bailey E, Lear TL, Cardwell JM, et al. (2012) Inbreeding in the Thoroughbred horse. Anim Genet 43: 340–342. doi: 10.1111/j.1365-2052.2011.02259.x

45. .Hendricks BL (2007) International Encyclopedia of Horse Breeds. Norman, OK: University of Oklahoma Pres. 486 p.

46. .Weatherley L (1978) Great horses of Britain. Hindhead: Spur Publications. viii, 269 p.

47. .Gates S (1979) A study of the home ranges of free-ranging Exmoor ponies. Mammal Rev 9: 3–18. doi: 10.1111/j.1365-2907.1979.tb00228.x

48. .Peilieu C (1984) Livestock Breeds of China. FAO: Animal Production and Health paper 46: 1–217.

49. .Hund A (2008) The Stallion's Mane: The Next Generation of Horses in Mongolia. Available: http://digitalcollections.sit.edu/cgi/viewcontent.cgi?article=1543&context=isp_collection. Accessed 2012 Dec 31.

50. .Ticklen M, editor (2006) Get to know the Finnhorse. Available:http://www.suomenhevonen.info/hippos/sh2007/pdf/SHjulkaisu_englanti_nettiin.pdf. Accessed 2012 Dec 31.

51. .Bowling AT, Ruvinsky A (2000) The genetics of the horse. Wallingford: CABI Pub. viii, 527 p.

52. .Lynghaug F (2009) The Official Horse Breeds Standards Guide: The Complete Guide to the Standards of All North American Equine Breed Associations. Stillwater, MN: Voyageur Press.

53. .Conant EK, Juras R, Cothran EG (2012) A microsatellite analysis of five Colonial Spanish horse populations of the southeastern United States. Anim Genet 43: 53–62. doi: 10.1111/j.1365-2052.2011.02210.x

54. .Lopes MS (2011) Molecular tools for the characterisation of the Lusitano horse. Angra do Heroísmo: University of Azores. 219 p.

55. .Tryon RC, Penedo MC, McCue ME, Valberg SJ, Mickelson JR, et al. (2009) Evaluation of allele frequencies of inherited disease genes in subgroups of American Quarter Horses. J Am Vet Med Assoc 234: 120–125. doi: 10.2460/javma.234.1.120

56. .Cothran EG, MacCluer JW, Weitkamp LR, Bailey E (1987) Genetic differentiation associated with gait within American standardbred horses. Anim Genet 18: 285–296. doi: 10.1111/j.1365-2052.1987.tb00772.x

57. .Florida Cracker Horse Association. Available:http://www.floridacrackerhorses.com/index.htm. Accessed 2012 Jun 5.

58. .Firouz L (1969) Conservation of a domestic breed. Biol Conserv 1: 1–2. doi: 10.1016/0006-3207(69)90117-7

59. .Bjornstad G, Nilsen NO, Roed KH (2003) Genetic relationship between Mongolian and Norwegian horses? Anim Genet 34: 55–58. doi: 10.1046/j.1365-2052.2003.00922.x

60. .Conrad DF, Jakobsson M, Coop G, Wen X, Wall JD, et al. (2006) A worldwide survey of haplotype variation and linkage disequilibrium in the human genome. Nat Genet 38: 1251–1260. doi: 10.1038/ng1911

61. .Li JZ, Absher DM, Tang H, Southwick AM, Casto AM, et al. (2008) Worldwide human relationships inferred from genome-wide patterns of variation. Science 319: 1100–1104. doi: 10.1126/science.1153717

62. .Mackay-Smith A (2000) Speed and the thoroughbred : the complete history. Lanham, MD: Derrydale Press. xxvii, 193 p.

63. .Bower MA, Campana MG, Whitten M, Edwards CJ, Jones H, et al. (2011) The cosmopolitan maternal heritage of the Thoroughbred racehorse breed shows a significant contribution from British and Irish native mares. Biol Letters 7: 316–320. doi: 10.1098/rsbl.2010.0800

64. .Silvestrelli M (1991) The Maremmano horse. Animal Genetic Resources Information (FAO/UNEP) 8: 74–83. doi: 10.1017/s1014233900003126

65. .Bower MA, Campana MG, Nisbet RER, Weller R, Whitten M, et al. (2012) Truth in the bones: Resolving the identify of the founding elite Thoroughbred racehorses. Archaeometry 54: 916–925. doi: 10.1111/j.1475-4754.2012.00666.x

66. .Andersson LS, Larhammar M, Memic F, Wootz H, Schwochow D, et al. (2012) Mutations in DMRT3 affect locomotion in horses and spinal circuit function in mice. Nature 488: 642–646. doi: 10.1038/nature11399

67. .Petersen JL, Mickelson JR, Rendahl AK, Valberg SJ, Andersson LS, et al. (2012) Genome-wide analysis reveals selection for important traits in domestic horse breeds. PLOS Genet. In press.

68. .Fletcher JL (1946) A study of the first fifty years of Tennessee walking horse breeding. J Hered 37: 369–373.

69. .Tennessee Walking Horse Breeders and Exhibitors Association. Available:http://www.twhbea.com/breed/history.php. Accessed 2012 Jun 5.

70. .MacCluer JW, Boyce AJ, Dyke B, Weitkamp LR, Pfennig DW, et al. (1983) Inbreeding and pedigree structure in Standardbred horses. J Hered 74: 394–399.

71. .Purcell S, Neale B, Tood-Brown K, Thomas L, Ferreira MAR, et al. (2007) PLINK: a toolset for whole-genome association and population-based linkage analysis. Amer J Hum Genet 81: 559–575. doi: 10.1086/519795

72. .Glaubitz JC (2004) CONVERT: A user-friendly program to reformat diploid genotypic data for commonly used population genetic software packages. Mol Ecol Notes 4: 309–310. doi: 10.1111/j.1471-8286.2004.00597.x

73. .Lischer H, Excoffier L (2012) An automated data conversion tool for connecting population genetics and genomics programs. Bioinformatics 28: 298–299. doi: 10.1093/bioinformatics/btr642

74. .Excoffier L, Laval G, Schneider S (2005) Arlequin (version 3.0): An integrated software package for population genetics data analysis. Evol Bioinform 1: 47–50.

75. .Belkhir KP, Borsa P, Chikhi L, Raufaste N, Bonhomme F (2001) GENTETIX, logiciel sous Windows TM pour la genetique des population. Universite de Montpellier II, Montpellier, France.

76. .Reynolds J, Weir BS, Cockerham CC (1983) Estimation of the coancestry coefficient: basis for a short-term genetic distance. Genetics 105: 767–779.

77. .Felsenstein J (1989) PHYLIP-Phylogeny Inference Package (version 3.2) Cladistics. 5: 164–166.

78. .Goloboff PA, Farris JS, Nixon K (2003) TNT: tree analysis using new technology. Syst Biol 54: 176–178.

79. .Falush D, Stephens M, Pritchard JK (2003) Inference of population structure using multilocus genotype data: linked loci and correlated allele frequencies. Genetics 164: 1567–1587.

80. .Pritchard JK, Stephens M, Donnelly P (2000) Inference of population structure using multilocus genotype data. Genetics 155: 945–959.

81. .Pritchard JK, Wen X, Falush D (2010) Documentation for Structure software: version 2.3.38.

82. .Jakobsson M, Rosenberg NA (2007) CLUMPP: a cluster matching and permutation program for dealing with label switching and multimodality in analysis of population structure. Bioinformatics 23: 1801–1806. doi: 10.1093/bioinformatics/btm233

83. .Rosenberg NA (2004) DISTRUCT: a program for the graphical display of population structure. Mol Ecol Notes 4: 137–138. doi: 10.1046/j.1471-8286.2003.00566.x

84. .Barrett JC, Fry B, Maller J, Daly MJ (2005) Haploview: analysis and visualization of LD and haplotype maps. Bioinformatics 21: 263–265. doi: 10.1093/bioinformatics/bth457

Chapter 12

Y-CHROMOSOME AND MTDNA GENETICS REVEAL SIGNIFICANT CONTRASTS IN AFFINITIES OF MODERN MIDDLE EASTERN POPULATIONS WITH EUROPEAN AND AFRICAN POPULATIONS

Danielle A. Badro[1], Bouchra Douaihy[1]., Marc Haber[1,2], Sonia C. Youhanna[1], Ange´lique Salloum[1], Michella Ghassibe-Sabbagh[1] , Brian Johnsrud[3], Georges Khazen[1], Elizabeth Matisoo-Smith[4], David F. Soria-Hernanz[2,5], R. Spencer Wells[5], Chris Tyler-Smith[6], Daniel E. Platt[7], Pierre A. Zalloua[1,8]

[1]The Lebanese American University, Chouran, Beirut, Lebanon

[2]Institut de Biologia Evolutiva (CSIC-UPF), Departament de Cie`ncies de la Salut i de la Vida, Universitat Pompeu Fabra, Barcelona, Spain

[3]Modern Thought and Literature, Stanford University, Stanford, California, United States of America

[4]Allan Wilson Centre for Molecular Ecology and Evolution, University of Otago, Dunedin, New Zealand, [5]The Genographic Project, National Geographic Society, Washington, DC, United States of America

[6]The Wellcome Trust Sanger Institute, Wellcome Trust Genome Campus, Hinxton, United Kingdom

[7]Computational Biology Centre, IBM TJ Watson Research Centre, Yorktown Heights, New York, United States of America

[8]Harvard School of Public Health, Boston, Massachusetts, United States of America

ABSTRACT

The Middle East was a funnel of human expansion out of Africa, a staging area for the Neolithic Agricultural Revolution, and the home to some of the earliest world empires. Post LGM expansions into the region and subsequent population movements created a striking genetic mosaic with distinct sex-based genetic differentiation. While prior studies have examined the mtDNA and Y-chromosome contrast in focal populations in the Middle East, none have undertaken a broad-spectrum survey including North and sub-Saharan Africa, Europe, and Middle Eastern populations. In this study 5,174 mtDNA and 4,658 Y-chromosome samples were investigated using PCA, MDS, mean-linkage clustering, AMOVA, and Fisher exact tests of F_{ST}'s, R_{ST}'s, and haplogroup frequencies. Geographic differentiation in affinities of Middle Eastern populations with Africa and Europe showed distinct contrasts between mtDNA and Y-chromosome data. Specifically, Lebanon's mtDNA shows a very strong association to Europe, while Yemen shows very strong affinity with Egypt and North and East Africa. Previous Y-chromosome results showed a Levantine coastal-inland contrast marked by J1 and J2, and a very strong North African component was evident throughout the Middle East. Neither of these patterns were observed in the mtDNA. While J2 has penetrated into Europe, the pattern of Y-chromosome diversity in Lebanon does not show the widespread affinities with Europe indicated by the mtDNA data. Lastly, while each population shows evidence of connections with expansions that now define the Middle East, Africa, and Europe, many of the populations in the Middle East show distinctive mtDNA and Y-haplogroup characteristics that indicate long standing settlement with relatively little impact from and movement into other populations.

INTRODUCTION

As a crossroad between Africa, Arabia, Asia, and Europe, the Levant has been a primary historical stepping stone in the first modern human expansions out of Africa and for later migrations into and out of Europe, Asia, and Africa [1]–[7]. As such, it has also become a land of remarkable human diversity. The earliest fossil and archaeological evidence of modern humans outside of the African continent are from the Levant, presumably indicating a migration via the northern route, and date to 125–95 kya [8], [9]. Additionally, genetic studies suggest that the initial peopling of Eurasia occurred through the northern Levantine (modern day Lebanon and Syria) route [10]–[12]. Two proposed routes chart the dispersal of anatomically modern humans out of the African continent: (1) a northern route, reaching west and central Asia through the Sinai Peninsula and the Levant, and (2) a southern route via the Bab el-Mandeb Strait and along the south Asian coast, ultimately reaching Australia [13]–[15].

While the out-of-Africa migrations have been major determining factors, other migratory events have strongly influenced genetic marker distributions throughout the Levant and the surrounding geographical areas. During the last glacial maximum (LGM, 26.5–19 kya), most of the Levant was an uninhabitable desert, with forested hills in Levantine Mediterranean coastal areas [7]. The genetics of the modern Levant were largely determined by subsequent repopulation (especially during the Neolithic agricultural revolution) and mass movements associated with empire building. Neolithic expansions in particular, beginning around 10 kya, induced gene flow between the Fertile Crescent and Europe, which shaped the genetic structure of both regions [16]–[22].

Most genetic studies of the Levant as a geographical area have focused exclusively on either Y-chromosome [23]–[27] or mitochondrial markers [28]. Further, contrasts between Y-chromosome and mtDNA data provide distinct insights into human expansions unavailable to somatic genome analyses [29]. While comparative analyses among the two marker types have been undertaken in the Middle East and Africa [30]–[34], none of these studies have explored the contrasting relationships of expansions throughout Europe, North Africa, the Levant and Arabian Peninsula after the LGM. Building on a previous study that reported phylogeographic characteristics of Y-chromosome markers in the Levantine region [25], we now compare and contrast Y and mtDNA phylogeographic distributions in the Levant and investigate the affinities of Middle Eastern populations with European and African populations.

MATERIALS AND METHODS

Ethics Statement

The samples were collected from donors after they had given their written informed consent to the project and to the data analysis, which was approved by the IRB of the Lebanese American University.

mtDNA Data

3,663 mtDNA records collected from the literature represented populations from Burkina-Faso[35], Cyprus [36], Egypt [37], Ethiopia [38], France [39], [40], Greece [36], Iraq [31], Jordan[30], Kenya [41], Libyan Sahara [42], Mali [35], Morocco [43], [44], Niger [35], Saudi Arabia[45], Slovakia [46], Tunisia [44], [47], and Yemen [38], [48]. In addition to this data, we added 1,511 new samples from Lebanon, Libya, Jordan, Palestine, and Syria. Samples were collected from unrelated blood donors from five countries. Surname repetitions were avoided and used as a criterion for absence of relatedness among volunteers, appropriate for Y-chromosome analysis. All demographic data were provided by self-assignment.

Given the broad cultural and genetic diversity in the region, terms such as "Middle East" may be problematical. Historically, the term evolved during the era of European Imperialism, and included all lands between Arabia and India, but came to include Turkey through Saudi Arabia, extending east through Afghanistan and Pakistan. In this report, "Middle Easterners" refers to Iraqis,

Jordanians, Lebanese, Palestinians, Saudis, Syrians, and Yemenis. The Greek data represent the Southeastern Europe region for mtDNA analyses, and is labelled "Southeastern Europe" in the rest of this report. France represents Western Europe mtDNA, and is labelled "Western Europe" throughout the rest of this report. All haplogroups were reduced to the most informative sets for the purpose of homogeneous representation and comparative analyses. mtDNA haplogroup frequencies are displayed inTable 1 and shown as pie charts in Figure 1.

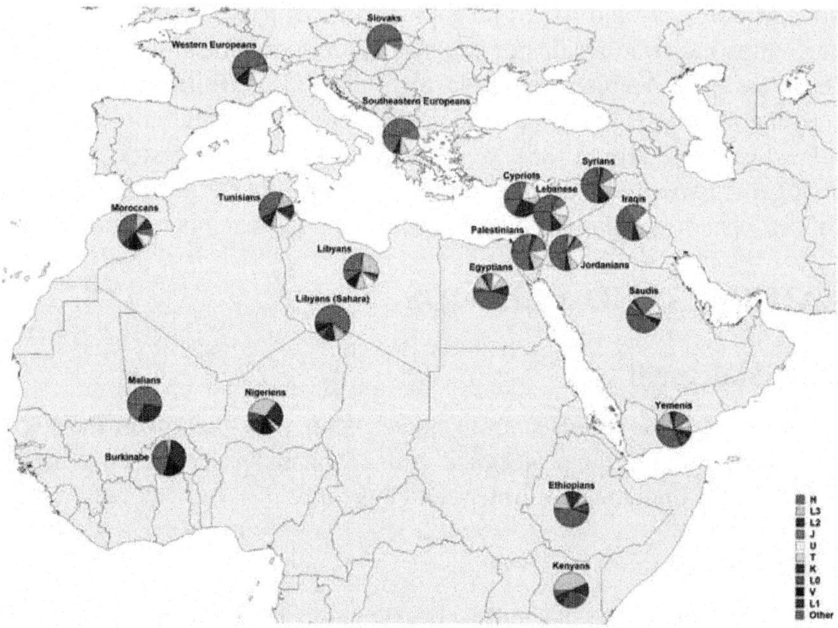

Figure 1: Geographic distribution of mtDNA haplogroups.

Frequencies distribution from the current study and from the published data [30], [31],[35]–[48] as reported in Table 1.

doi:10.1371/journal.pone.0054616.g001

Table 1: mtDNA Haplogroup frequencies of 1509 newly sequenced Levantine samples and 3665 samples collected from the literature

Country	n	L*	L3*	M*	D	N*	R*	R0	HV	H	U	References
Levant												
Lebanon	979	0.6	1.4	1.4	0.2	10.5	21.2	5.0	8.5	29.9	21.1	This study
Syria	234	3.4	2.6	0.9	0.0	11.1	25.6	2.6	9.0	24.4	20.5	This study
Palestine	120	5.0	2.5	3.3	0.0	8.3	27.5	5.8	8.3	25.0	14.2	This study
Cyprus	79	3.8	0.0	2.5	0.0	12.7	17.7	0.0	0.0	22.8	40.5	Irwin et al. [36]
Europe												
France	871	0.6	0.3	0.1	0.0	5.4	18.0	0.5	6.9	45.4	22.8	Dubut et al. [39] Richard et al. [40]
Greece	372	0.0	0.0	1.1	0.0	9.1	19.4	1.9	5.9	42.2	20.4	Irwin et al. [36]
Slovakia	200	1.0	0.0	1.0	0.0	10.0	20.0	0.0	4.0	47.0	17.0	Malyarchuk et al. [46]
Arabian Peninsula												
Jordan	290	4.1	5.9	1.4	0.3	10.7	17.9	3.4	6.9	26.2	23.1	This study Gonzales et al. [30]
Iraq	52	0.0	0.0	1.9	0.0	0.0	34.6	5.8	11.5	17.3	28.8	Al-Zahery et al. [31]
Saudi Arabia	539	6.5	3.5	7.1	0.0	12.2	28.0	18.2	0.7	8.7	15.0	Abu-Amero et al. [45]
Yemen	300	22.0	16.3	5.3	0.0	6.7	21.7	9.3	2.3	4.7	11.7	Kivisild et al. [38] Cerny et al. [48]
North Africa												
Egypt	278	8.3	12.2	6.8	0.0	10.4	21.6	17.27	6.1	0.0	13.7	Saunier et al. [37]
Libya	31	6.5	25.8	0.0	0.0	3.2	12.9	6.5	9.7	25.8	9.7	This study
Libyan Sahara	129	21.7	11.6	1.6	0.0	0.0	0.0	0.0	3.9	61.2	0.0	Ottoni et al. [42]
Morocco	137	18.2	10.9	7.3	0.0	2.2	10.2	3.6	7.3	23.4	16.8	Turchi et al. [44] Harich et al. [43]
Tunisia	160	20.6	14.4	1.9	0.0	0.6	7.5	0.0	7.5	31.3	16.3	Cherni et al. [47] Turchi et al. [44]
East sub Saharan												
Ethiopia	232	36.6	18.1	14.7	0.0	3.9	5.2	11.6	2.6	0.9	6.5	Kivisild et al. [38]
Kenya	84	47.6	44.0	4.8	0.0	0.0	1.2	1.2	0.0	0.0	1.2	Brandstätter et al. [41]
West sub Saharan												
Burkina-Faso	40	35.0	5.0	17.5	0.0	0.0	0.0	0.0	20.0	22.5	0.0	Pereira et al. [35]
Mali	21	28.6	0.0	19.0	0.0	0.0	0.0	0.0	0.0	52.4	0.0	Pereira et al. [35]
Niger	25	44.0	32.0	4.0	0.0	0.0	0.0	0.0	12.0	4.0	4.0	Pereira et al. [35]

doi:10.1371/journal.pone.0054616.t001

doi:10.1371/journal.pone.0054616.t001

Y-Chromosome Data

1,774 previously published Y-chromosome records were obtained from the literature representing populations from the Balkans [49], Burkina-Faso [50], Ethiopia [50], Italy [51], Kenya [50], Saudi Arabia [52], Slovakia [53], and Yemen [52]. In addition 2,884 previously published data from our laboratory representing populations from Cyprus, Egypt, Lebanon, Libya, Jordan, Morocco, Palestine, Syria and Tunisia were added to this study. The Italian Y-chromosome samples represent Western Europe in this study, and are labelled "Western European" through the rest of this report. The Balkan samples represent the Southeastern Europe region, and are labelled "Southeastern European" in the rest of this report. Haplogroups of the Saudi, Yemeni, and Slovak populations were not available, thus we have predicted those haplogroups using the populations haplotypes and the online haplogroup prediction tool [54], [55]:www.hprg.com/hapest5/hapest5a/hapest5.htm. We have also computed STR-predicted Y haplogroups across populations that had been SNP defined to ascertain STR-based haplogroup assignment accuracy, and identify geographically correlated trends in assignment error rates that may impact our conclusions.

mtDNA Sequencing and in Silico Prediction of Haplogroups

Total DNA was extracted from the peripheral leukocyte fraction of whole blood drawn in EDTA anticoagulant or cheek swab samples using a standard phenol/chloroform extraction procedure. The hypervariable region I (HVS-I) was amplified using primers designed by Maca-Meyer et al. [12]. Amplified HVS-I products were sequenced using a forward primer at position 15876 and a reverse primer at position 639 with ABI Big Dye Terminator v3.1 Cycle Sequencing kit (Applied Biosystems) and analysed on an Applied Biosystems 3130 xl Genetic Analyser.

Mutations in the HVS-I region were defined by aligning and comparing the sequences to the revised Cambridge Reference Sequence (rCRS) using the SeqScape software.

mtDNA haplogroups were predicted using the Genographic Project's online haplogroup prediction tool: nnhgtool.nationalgeographic.com.

mtDNA Genotyping of Samples

Haplogroup affiliations were confirmed using the Taqman approach with customized primers and probe sets to identify the SNPs listed (Applied Biosystems). Samples with incompatible prediction and Taqman results were excluded from the study. Mitochondrial nomenclature was assigned according to prior studies since established as standards [38],[56]–[59]. Data archiving was manually organized and edited.

Reduction to Most Informative Derived Set of Haplogroups

MtDNA Haplogroups reported in the literature were updated and reconciled to the 2009 phylogeny reported by van Oven et al. [59]. Construction of the most informative derived set was achieved by identifying the maximum level of resolution shared across all included studies. If some subhaplogroup markers were not typed in any given study, but no samples in that study resolved to a less-derived paragroup, then the most derived resolution was retained for the constructed most informative derived set.

Further, HVS-I regions reported by different sources varied. The range representing the largest common subset of HVS-I SNPs reported included 16090 through 16365.

Statistical Analyses

Fisher Exact Tests

Fisher exact tests were performed for haplogroup frequencies within populations. These tests were performed against a background of all populations, as well as among Middle Eastern populations (from Iraq, Jordan, Lebanon, Palestine, Saudi Arabia, Syria, and Yemen) only, with very low-power tests excluded.

PCA

Numbers of samples bearing mtDNA and Y-Chromosome reduced haplogroups within each population, and relative haplogroup frequencies within populations, were computed using R[60]. Principal Component Analysis was computed using prcomp in R [61]. Results were displayed with principal component contributions from each haplogroup using biplot. Agglomerative clustering with mean linkage (UPGMA) was applied to Euclidean distances computed between relative frequency vectors for each population using agnes and displayed in Figure 2 for mtDNA Haplogroups for Y-Haplogroups. These dendrograms should not be taken as population histories, but rather provide a repeatable description of population similarities also visible in the PCA.

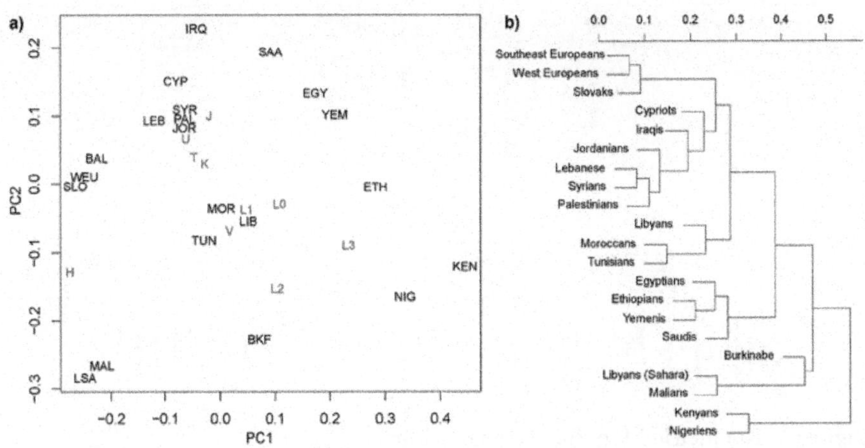

Figure 2: Populations comparison based on mtDNA haplogroups.

a) Principal Component Analysis of relative frequencies of haplogroups within populations, b) with mean-linkage (UPGMA) dendrogram determined from Euclidean distances.

doi:10.1371/journal.pone.0054616.g002

MDS

HVS-I SNPs were constructed against CRS [62] as revised rCRS [63], and the subrange common to all publications was selected. ARLEQUIN [64] was employed to compute F_{ST}'s[65], which were used as distances for non-metric MDS analysis [66], as implemented in isoMDS [61] in R. Agglomerative clustering with mean linkage was applied to the F_{ST}distances in the same way that they were applied to the Euclidean distances as described in the PCA section. An identical MDS and clustering were applied to Slatkin's R_{ST} distances [67]obtained from Y-chromosome samples. These results were displayed in Figures 3.

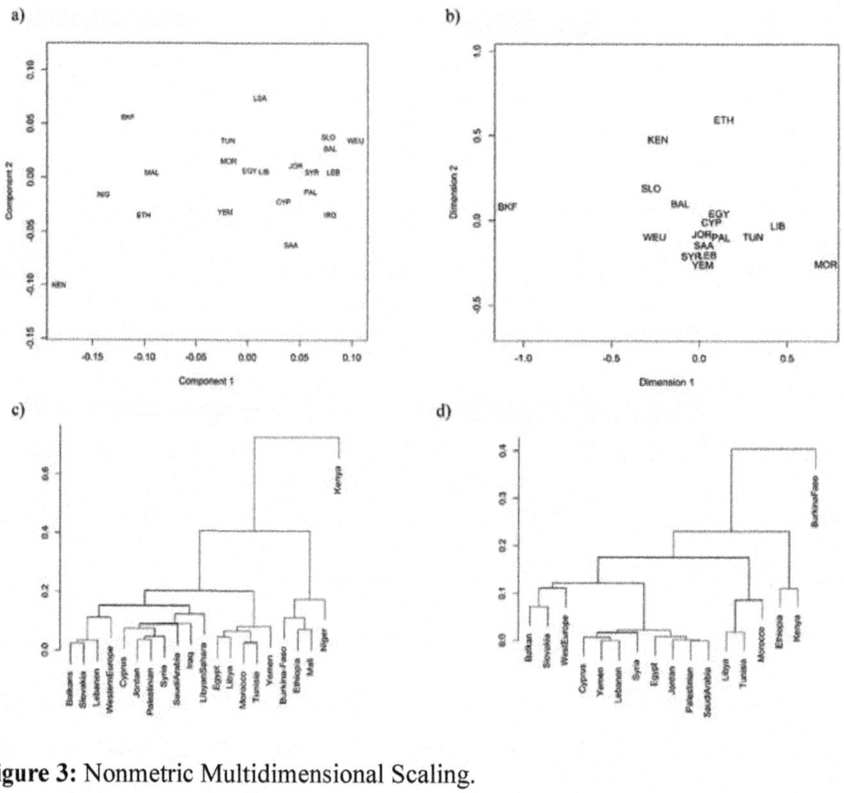

Figure 3: Nonmetric Multidimensional Scaling.

a) mtDNA F_{ST} and b) Y-STR R_{ST} distances with c) mtDNA F_{ST} and d) Y-STR R_{ST}mean-linkage dendrogram.

doi:10.1371/journal.pone.0054616.g003

Heatmap

Relative comparisons between mtDNA F_{ST} and Y-chromosome R_{ST} distances were constructed using a heatmap based on the normalized ratio of the Y-chromosome R_{ST}distance with respect to the total distance $(R_{ST}/(R_{ST}+F_{ST}))$ for mtDNA HVS-1 F_{ST} and Y-chromosome R_{ST} distances (Figure 4). The dendrograms are obtained using complete linkage hierarchical clustering with Euclidean distances between Y/(mtDNA HVS-I+Y) scores.

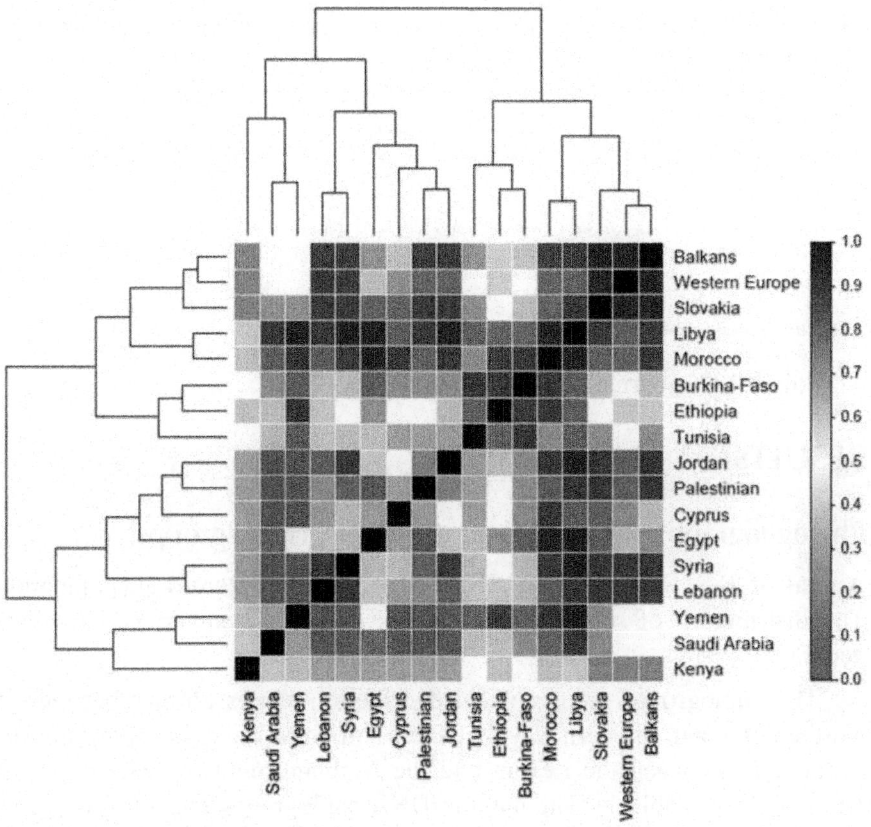

Figure 4: Heatmap of the Y(Y+mtDNA) distances.

The heatmap shows the normalized ratio of the Y F_{ST} distance with respect to the total distance (Y R_{ST}+mtDNA F_{ST} distances). The dendrograms are obtained using complete linkage hierarchical clustering with the Euclidean distance measure.

doi:10.1371/journal.pone.0054616.g004

AMOVA

The agglomerative clusters reflecting the results from the MDS analysis were used to identify groups of populations representing affinity of Middle Eastern populations with European and African populations observed in Y-chromosome and mtDNA genetics. AMOVA [68] was applied to the mtDNA and Y sets for each of the mtDNA and Y affinity sets, yielding a 2 by 2 measure of the differences between mtDNA and Y affinities, reported in Table 2.

Table 2: mtDNA vs. Y-chromosome AMOVA results contrasting mtDNA and Y dendrogram-based classifications

		mtDNA dendrogram based affinities				Y-Chromosome dendrogram based affinities			
		Variation	% Variation	F	p-value	Variation	% Variation	F	p-value
mtDNA AMOVA	Within Populations	2.43	97.15	0.029	$<10^{-4}$	2.43	97.59	0.024	$<10^{-4}$
	Within Groups	0.028	1.11	0.014	$<10^{-4}$	0.045	1.84	0.018	$<10^{-4}$
	Between Groups	0.043	1.74	0.017	3.6×10^{-3}	0.014	0.57	0.006	1.1×10^{-3}
Y-Chromosome AMOVA	Within Populations	6.56	87.05	0.129	$<10^{-4}$	6.03	89.04	0.11	$<10^{-4}$
	Within Groups	0.456	6.05	0.065	$<10^{-4}$	0.329	4.86	0.052	$<10^{-4}$
	Between Groups	0.519	6.89	0.068	6.3×10^{-3}	0.414	6.11	0.061	$<10^{-4}$

doi:10.1371/journal.pone.0054616.t002

doi:10.1371/journal.pone.0054616.t002

RESULTS

Phylogeographic Distribution of mtDNA Haplogroups

A total of 185 distinct HVS-I SNPs were identified across all populations. The distribution of mtDNA haplogroups shows systematic variation with geography.

The haplogroups' geographical distribution shows affinity between the Northern Levant (modern day Lebanon and Syria) and Europe with clear distinctions between the Levant and the Arabian Peninsula with regards to Africa (Fig. 1, Table 1). The main mtDNA haplogroups for both Europe and the Northern Levant are H and R*. The subhaplogroup H is more frequent in Europe (45%) than in the Levant (25%). Among the Levantine populations, only the Lebanese share Western Europe's overrepresentation of H.

Fisher exact tests were applied to determine when haplogroup frequency differences among populations over both Pan-Mediterranean tests, and regional Middle Eastern tests, were significant. They reveal patterns of significant over- and under- representation of haplogroups marking regional affinities.

In Lebanese, haplogroups H, HV, T, and K are over-represented, while Syrians are overrepresented in haplogroups T and K. Western Europeans show overrepresentation of haplogroups H, K.

By contrast, haplogroups J, R0, and M are significantly overrepresented in Saudis, and underrepresented in Western Europeans. Haplogroup J was also significantly overrepresented in Iraqis, among Palestinians, and Yemenis.

The African haplogroups L* and L3* are very rare (frequencies less than 1%) and underrepresented in Europe, noting that rarity reduces the power of these tests. In the Levant, Lebanese have the lowest frequency for these haplogroups with generally highly significant underrepresentation, The L haplogroups show rather broad penetration into Yemen, with most being significantly overrepresented, with Yemenis being the only population with an overrepresentation of L6. We have not found haplogroup L6 in our Lebanese (N=980), Syrian (N=234), and Jordanian samples (N=290). Further, they are absent from Abu-Amero et al's samples from Saudi Arabia, as well as other published results included in his study. A lower bound on the relative frequency of L6 in non-Yemeni Middle East of $f \geq 8.29 \cdot 10^{-4}$ would guarantee that at least one or more would be observed at least 95% of the time out of the 3614 samples collected across the data on which we are reporting. Assuming independent sampling following a binomial test, there is 5% or less chance of seeing zero L6 by chance with a relative frequency of f or higher, establishing $8.29 \cdot 10^{-4}$ as an upper bound to the relative frequency of L6 with a 95% confidence.

We note that subtypes HV0, HV1 and HV2 are generally too weakly represented among our populations to yield tests with adequate power. HV0 and HV1 show sufficient power when pooled into regions. We note that HV0 appears primarily in Europe (34 out of 42 HV0 samples are European, with $p<0.0001$), HV1 is primarily non-European (4 out of 60 were European, with 20 among the African sample, and 36 among Middle Eastern samples, $p<0.0001$). HV2 were very rare, with no significant p-values.

Some U subhaplogroups show regional localization, but none of them rose to sufficient frequency to make any significant contribution to the PCA. U3 appears most frequently in Jordan (Fisher›s test: p=2.39e-8), with representation throughout the Middle East. U4 (p=1.49e-7) and U5 (p=2.2e-16) appear to be more heavily European.

The two leading principal components displayed in Figure 2 capture 47.9% and 26.9% of the variance showing a well-defined separation between Mediterranean African populations and sub-Saharan populations (Fig 2a). There is a clear cluster of North African populations comprised of Libyans, Moroccans, and Tunisians. The Nile River marks another boundary of mtDNA differentiation within Africa, linking Egypt, Ethiopia and Kenya but also extending through to Yemen. Yemenis and Saudis both associate strongly with Egyptians, whereas the Jordanian, Lebanese, Palestinian, and Syrian populations clustered together. Thus, the Arabian Peninsula population clusters were relatively differentiated from the more northern Levantine populations.

Mitochondrial DNA Haplogroups showing significant contributions to the principal components include H, L3, L2, L0, V, L1, M, J, U, T, K, HV, and R0. The principal vectors for HV, T, K, J, and U point almost directly at the Levantine cluster (Fig 2a). H marks Western Europe and is a significant contributor to Libyan Sahara and Mali mtDNA diversity. L2 and L3 frequencies distinguish the populations of Kenya, Niger, Burkina Faso, Mali, Tunisia, and Libyan Sahara, with a decrease in frequencies of L haplotypes from Kenya through Saudi Arabia.

The dendrogram based on mtDNA haplogroup frequencies (Fig 2b) reveals the strongest differentiation across the Sahara, showing the northern populations differentiated from the southern ones (with Nigeria, Kenya, Mali, Libyan South Sahara, and Burkina-Faso). Egyptian, Yemeni, Saudi Arabian, and Ethiopian populations form a cluster that is distinct from the rest of North Africa, the remaining parts of the Middle East, and Europe. Among these, Libyans, Moroccans, and Tunisians, form a cluster.

UPGMA and PCA showed Yemenis and Saudis (two of the STR predicted Hg populations) closely associated, forming a clear outlier to clusters identifying more northerly Middle Eastern populations and Europe. Slovaks (the third predicted population) also formed a distinct outlier to all of these. Africans were partitioned into northern African populations and Sahel populations, and distinct from the other populations. Burkinabe formed a very distinct outlier to every other population.

MDS analyses were performed using mitochondrial HVS-I based F_{ST} and Y-chromosome STR R_{ST} data (Figure 3a & 3b). The F_{ST}'s computed with mtDNA HVS-I data and R_{ST}'s computed from Y-STRs of Ethiopian and Levantine populations tended to be less than 1/3, with $Nm>1$ [69], roughly corresponding to gene flow between populations over the course of time [5], [20], [38], [70], [71].

For the mtDNA HVS-I F_{ST} MDS analysis, the European populations formed a clear cluster very close to the Cypriots, Jordanians, Lebanese, Palestinians, and Syrians. Egyptian, Libyan, Moroccan, and Tunisian populations form a clear cluster. Significantly, Yemenis are on the far side of North Africans, distinct from the Levantine populations and the Libyan Sahara population stands significantly separated from the North African group. The sub-Saharan populations are clearly distinguished from the Mediterranean populations and show significant distances between them in comparison to the Mediterranean populations. The mtDNA HVS-I MDS and dendrogram show most of the

Levantine and Arabian Peninsula populations clustering together. Significantly, Yemenis do not seem to cluster with proximal African populations or with Saudis. The entire Levant population seems to cluster with Western Europeans, Southeastern Europeans, and Slovaks.

In contrast to mtDNA, the Y-STR-based MDS shows a tight cluster of Cypriots, Egyptians, Jordanians, Lebanese, Palestinians, Saudis, Syrians, and Yemenis, though Libyans, Tunisians, and Moroccans extend away from this cluster. The Southeastern Europeans, Slovaks, and Western Europeans lie in the opposite direction. The dendrogram shows a European cluster closer to the Levant/Arabian Peninsula cluster and the North African cluster acting as out-group to those.

In general, the MDS plots for mtDNA and Y-STRs show general agreement of European populations extending from the Levant in one direction and North Africans tending to extend in another direction. This places the Levant as a middle ground, either by averaging of in-migration, as a source feeding both North African populations and European populations, or both. The Y and mtDNA MDS plots differ in identifying affinities of Lebanese with Europeans and Yemenis with Egyptians.

Comparative Analyses of Paternal and Maternal Lineages in the Levant

The relative distance heatmap plot (Figure 4) shows proportion of genetic distances of mtDNA vs. Y. Red colors indicate greater distance of mtDNA vs. Y, while blue colors indicate greater distance of Y vs. mtDNA. Hierarchical clustering organizes the plot relating populations showing similar profiles of Y vs. mtDNA isolation. Most striking is that Saudis, Kenyans, and Yemenis cluster together away from Lebanese, Syrians, Palestinians, Cypriots and Jordanians in terms of showing relatively high differentiation of mtDNA vs. Y-chromosome genetics. Dendrograms provide a consistent description of the organization of data that may be easily compared with PCA or MDS plots. The application of mean-linkage dendrograms to Y STR data, mtDNA HVS-I data, and mtDNA haplogroup frequency data provides a consistent basis of comparison. Application of AMOVA to clustering results provides an independent test characterized by p-values and percent variances between vs. within groups. We are not inferring relationships of heritage among populations by application of mean-linkage clustering.

In order to preserve normalization, common subsets comprised of the 11 populations in common in both dendrograms were included. Each of the candidate partitions marking mtDNA affinities and Y affinities formed three groups. The groups representing mtDNA affinities were: (1) Southeastern Europeans, Lebanese, Slovaks, and Western Europeans vs. (2) Cypriots, Jordanians, Palestinians, Saudis, and Syrians, vs. (3) Egyptians and Yemenis. The groups representing Y affinities were: (1) Southeast Europeans, Slovaks, and Western Europeans vs. (2) Cypriots, Lebanese, Syrians, and Yemenis, vs. (3) Egyptians, Jordanians, Palestinians, and Saudis. These two affinity groupings were applied to both the Y and the mtDNA data, yielding results presented in Table 2. Both Y and mtDNA tend to cluster African, European, and Middle Eastern populations separately, and all combinations showed highly significant between-group vs. within-group variations. This reflects the dominating clustering distinguishing Africa, Europe, and the Middle East populations that mean-linkage clustering is picking up. Affinities of Lebanese and the Levantine populations with Europeans vs. Africans depend on comparisons of AMOVA variations within and between groups. Notably, the mtDNA affinity grouping increased AMOVA between-group variation of mtDNA HVS-I data by a factor of 3.05 compared to the result obtained applying the Y affinity grouping to the mtDNA HVS-I data, and decreased AMOVA within-group variation by a factor of 1.66. However, application of the Y affinity grouping reduced AMOVA between-group variation in Y STR data by a factor of 1.13 while reduced AMOVA within-group variations in the Y STR data by a factor of nearly 1.2 compared to the mtDNA affinity grouping. These factors are relatively neutral in contrasting Lebanese Y-chromosome affinity with Europe vs. North Africa, and actually place Lebanese Y-chromosome organization closer to Europeans than Africans. It is expected that mean-linkage clustering would minimize AMOVA within-groups variation, leading to larger AMOVA between-groups variation. Observation did not meet expectation. Instead, the AMOVA within groups' variations for the Y-chromosomes were reduced using the mtDNA clustering compared to Y clustering, suggesting reduced discrimination using the Y clustering for Y-STR genetics.

Limitations

Y Chromosome haplogroup frequency analyses are limited by a relatively high misclassification rate, with more than half of the populations showing more than 10% misclassification, and Ethiopia showing nearly 50%. Since PCA is a non-linear computation which folds in all populations, the apparent locations of any two populations may shift relative to each other when a third population is added or distorted.

Discussion

Here we present mitochondrial characteristics of a large group of newly typed samples from five populations (Lebanese, Libyans, Jordanians, Palestinians, and Syrians) and compare their geographical affinity, distribution, and frequency with those of Y-chromosome markers from populations across the broader region of Africa, Europe and the Arabian Peninsula.

The Y-chromosome results of the current study are in agreement with previous studies, suggesting a Middle Eastern gene pool with greater affinity to Africa. Maternal lineages of the Levantine populations studied here, however, reveal stronger European genetic affinities, while not showing Arabian peninsular influences.

The Contrast between the Two Lineages

Our results show a contrast of mtDNA affinities with previous Y-DNA results [25]. While our Y-DNA MDS and mean-linkage clustering showed a much greater proportion of East African and Near East Y-chromosomes in the Levant, evidence of much less mtDNA affinity, however, was found between the Levant and its southern neighbours.

European mtDNA affinity with the Levant was established in haplogroup frequency data through Fisher exact tests, PCA, and mean-linkage clustering based on Euclidean distances, and in HVS-I derived F_{ST} distances via MDS and mean-linkage cluster analysis. The mtDNA results are distinct from the Y-STR R_{ST}-based mean-linkage cluster analysis that showed closer affinity of the Levant populations with Cypriots, North Africans, and Yemenis, than to Europeans.

This cluster analysis suggests that the position of Lebanese relative to European Y-chromosome genetics represented in STR haplotype data is also much more ambiguous than suggested entirely by frequency analysis, revealing otherwise cryptic relationships between Lebanese's Y-STR structure and that of Europeans. Cluster analysis of Y-chromosome frequency based data shows similar partitioning of Europe, Africa, and Middle East, with the Levant much more strongly associated with the Middle East than Europe. As with mtDNA, African Y-chromosome haplogroup data also shows a clear partition between Northern populations and Sahel populations. Due to uncertainties in haplogroup inference from STRs, affinities of Yemenis with Ethiopians vs. Egyptians are uncertain, as are the relationships of Saudi Arabian haplogroups both similar to Yemenis or differentiated from Yemenis in affinity with African populations.

The Levant and Europe

Beyond the associations noted above, Lebanese show affinity with Europeans for mtDNA haplogroups H, HV, T, K, J, and U, all of which have been identified as markers of agricultural expansions from the Fertile Crescent into Europe [17].

Colonization of West Eurasia by modern humans is believed to have been a consequence of the Out-of-Africa dispersal and to have occurred via the Levant [6]. Indeed, migrating modern humans are believed to have settled near the Arabian Sea until climate changes allowed them to reach the Levant and then Europe [17], [72]–[74]. The LGM, followed by re-expansions from smaller LGM communities relatively isolated by widespread arid conditions, further impacted the coastal-inland contrast of Y-chromosome genetics [25], [75]. The significant overrepresentation of mtDNA haplogroup HV among Levantine populations compared to their southern neighbours has suggested these lineages were most likely derived from a single maternal Levantine source population [17].

Arabian genetic expansions: Arabia East Africa, and North Africa

From the 7th millennium B.C.E., empire expansions and trade, including the slave trade, heavily influenced genetic migration between Yemen and East Africa. Alternatively, known trade networks linking Egypt with Yemen included those for obsidian, and later through Aksum, spices, incense and other precious materials, as well as slaves [76]–[78]. It is particularly clear from prior mtDNA studies of this region that East African migration into Arab populations involved females to an extensive degree [79]. While Ethiopian and other East African populations may appear to be better candidates for the origins of modern Yemeni populations, our PCA and MDS analyses, and their associated mean-linkage clustering of Yemen›s mtDNA, show greater affinity between Yemenis, Egyptians and North Africans. They share in common haplogroups J, L0, L2, and N1. Comparison of mtDNA HVS-I F_{ST} distances also suggest that Yemen appears more similar to Egypt than Ethiopia.

Two haplogroups in this region show significant evidence of relative isolation. First, mtDNA patterns for haplogroup J reflect relatively moderate genetic outflow from Saudi Arabia, and haplogroup L6 is strongly localized within Yemen. Haplogroup J is evenly distributed throughout the Middle East, except in Saudi Arabia where it is significantly overrepresented.

It is likely the pattern of Hg J›s significant penetration, and the shared underrepresentation of Hg H, tips the balance for Yemenis› mtDNA affinity with Egyptians. Given the significant underrepresentation of Hg J in East

Africa, while not being significantly uncommon in Egypt, it is therefore plausible that Arabian female gene flow followed well established trade routes on the Red Sea with Egypt and North Africa while avoiding assimilation of Yemeni L6›s on the way.

The most striking feature of the heat map (Fig 4) is the relative isolation of mtDNA genetics of Yemenis and Saudis from the other populations in the Middle East in comparison to Y-chromosome variation. While Yemenis appear to share overrepresented haplogroups that characterize each of its neighbouring populations, none of the African populations have become dominated by Saudi Arabian J›s, nor have Middle Eastern populations been differentially dominated by the in-migration of African L›s the way Yemenis have.

The expansion of trade through the Red Sea and into the Indian Ocean basin starting in Classical times has provided the largest opportunities for genetic transfers from Africa into Yemen, being dominated by the Red Sea superpower: Egypt. The distribution of mtDNA haplogroup L6 provides a measure of the limited impact of genetic outflow from Yemen, and this flow seems to have been primarily unidirectional. This establishes the upper limits for Yemeni female-mediated gene flow during the Muslim Expansions, as well as identifying possible routes for the expansions.

Whether considering haplogroup composition revealed in Fisher tests, PCA, or F_{ST} based MDS analysis of HVS-I data, mtDNA shows a much stronger affinity between Levantine populations and Europeans compared with the rest of the Middle Eastern populations, or with North Africans. While Lebanese and Yemeni mtDNA epitomize very distinct affinities to different populations and regions well outside of the Middle East, Saudi Arabia seems to display strong local over-representation haplogroup J, while Yemen is even more localized in its L6. Further, these large-scale differences in affinity between mtDNA genetics appear in sharp contrast to regional affinities seen in their Y-chromosomal counterparts. While the mtDNA signal is sharp and clear in its affinities, the Y-chromosome results show somewhat more ambiguous associations in R_{ST} based analyses, with Lebanese showing less within-group variation when organized consistently with mtDNA and demonstrating associations closer to Europeans than Africans. This would suggest that while male migrants accompanied female migrants, especially to Europe, females did not always accompany male migrants, especially into North Africa. This leaves a more ambiguous signal for male compared to female migrations.

The historical and archaeological record reveals how trade and labour, colonization and settlement events, and military expansions all contributed to the immigration and displacement of individuals throughout these regions. As a distinct crossroad between geographic regions and civilizations, the Levant and the Near East harbour unique genetic affinities which are revealed most clearly through the comparison of Y-chromosome and mtDNA data.

Due to uncertainties in haplogroup inference from STRs, specific questions regarding affinities of Yemen with Ethiopia vs. Egypt are inaccessible, as are questions regarding the relationship of Saudi Arabian haplogroups both similar to Yemenis or differentiated from Yemenis in affinity with African populations.

ACKNOWLEDGMENTS

We thank the sample donors for taking part in this study. We would like to thank Professor Colin Renfrew for his insights, critical comments, and suggestions which helped us improve this manuscript significantly.

The Genographic Consortium includes: Janet S. Ziegle (Applied Biosystems, Foster City, California, United States); Li Jin & Shilin Li (Fudan University, Shanghai, China); Pandikumar Swamikrishnan (IBM, Somers, New York, United States); Asif Javed, Laxmi Parida & Ajay K. Royyuru (IBM, Yorktown Heights, New York, United States); Lluis Quintana-Murci (Institut Pasteur, Paris, France); R. John Mitchell (La Trobe University, Melbourne, Victoria, Australia); Syama Adhikarla, ArunKumar GaneshPrasad, Ramasamy Pitchappan & Arun Varatharajan Santhakumari (Madurai Kamaraj University, Madurai, Tamil Nadu, India); Angela Hobbs & Himla Soodyall (National Health Laboratory Service, Johannesburg, South Africa); Elena Balanovska & Oleg Balanovsky (Research Centre for Medical Genetics, Russian Academy of Medical Sciences, Moscow, Russia); Daniela R. Lacerda & Fabrício R. Santos (Universidade Federal de Minas Gerais, Belo Horizonte, Minas Gerais, Brazil); Pedro Paulo Vieira (Universidade Federal do Rio de Janeiro, Rio de Janeiro, Brazil); Jaume Bertranpetit, David Comas, Begoña Martínez-Cruz & Marta Melé (Universitat Pompeu Fabra, Barcelona, Spain); Christina J. Adler, Alan Cooper, Clio S. I. Der Sarkissian & Wolfgang Haak (University of Adelaide, South Australia, Australia); Matthew E. Kaplan & Nirav C. Merchant (University of Arizona, Tucson, Arizona, United States); Colin Renfrew (University of Cambridge, Cambridge, United Kingdom); Andrew C. Clarke & Elizabeth A. Matisoo-Smith (University of Otago, Dunedin, New Zealand); Matthew C. Dulik, Jill B. Gaieski, Amanda C. Owings, Theodore G. Schurr & Miguel G. Vilar (University of Pennsylvania, Philadelphia, Pennsylvania, United States).

AUTHOR CONTRIBUTIONS

Conceived and designed the experiments: PZ CTS RSW EMS. Performed the experiments: DB BD SY AS. Analyzed the data: DP MH DSH GK. Contributed reagents/materials/analysis tools: BJ MGS. Wrote the paper: PZ DP DB.

REFERENCES

1. Stringer CB, Grun R, Schwarcz HP, Goldberg P (1989) ESR dates for the hominid burial site of Es Skhul in Israel. Nature 338: 756–758. doi: 10.1038/338756a0
2. Bar-Yosef O (1992) The role of western Asia in modern human origins. Philosophical transactions of the Royal Society of London Series B, Biological sciences 337: 193–200. doi: 10.1098/rstb.1992.0097
3. Lahr MM, Foley RA (1998) Towards a theory of modern human origins: geography, demography, and diversity in recent human evolution. American journal of physical anthropology Suppl 27: 137–176. doi: 10.1002/(sici)1096-8644(1998)107:27+<137::aid-ajpa6>3.3.co;2-h
4. Tchernov E (1994) New comments on the biostratigraphy of the Middle and Upper Pleistocene of the southern Levant. In: Ben-Yosef O, Kra RS, editors. Late Quaternary Chronology and Paleoclimates of the Eastern Mediterranean: Radiocarbon. pp. 333–350.
5. Luis JR, Rowold DJ, Regueiro M, Caeiro B, Cinnioglu C, et al. (2004) The Levant versus the Horn of Africa: evidence for bidirectional corridors of human migrations. American journal of human genetics 74: 532–544. doi: 10.1086/382286
6. Olivieri A, Achilli A, Pala M, Battaglia V, Fornarino S, et al. (2006) The mtDNA legacy of the Levantine early Upper Palaeolithic in Africa. Science 314: 1767–1770. doi: 10.1126/science.1135566
7. Bar-Yosef O (1998) The Natufian Culture in the Levant, Threshold to the Origins of Agriculture. Evolutionary Anthropology 6: 159–177. doi: 10.1002/(sici)1520-6505(1998)6:5<159::aid-evan4>3.0.co;2-7
8. Valladas H, Reyss JL, Joron JL, Valladas G, Bar-Yosef O, et al. (1988) Thermoluminescence data of Mousterian Troto-Cro-Magnon remains from Israel and the origin of modern man. Nature 331: 614–616. doi: 10.1038/331614a0
9. Mercier N, Valladas H, Bar-Yosef O, Stringer CB, Joron JL (1993) Thermoluminescence dates for the Mousterian Burial Site of Es-Skhul, Mt. Carmel. Journal of Archaeological Science 20: 169–174. doi: 10.1006/jasc.1993.1012
10. Cann RL, Stoneking M, Wilson AC (1987) Mitochondrial DNA and human evolution. Nature 325: 31–36. doi: 10.1038/325031a0

11. Vigilant L, Stoneking M, Harpending H, Hawkes K, Wilson AC (1991) African populations and the evolution of human mitochondrial DNA. Science 253: 1503–1507. doi: 10.1126/science.1840702
12. Maca-Meyer N, Gonzalez AM, Larruga JM, Flores C, Cabrera VM (2001) Major genomic mitochondrial lineages delineate early human expansions. BMC genetics 2: 13. doi: 10.1186/1471-2156-2-13
13. Nei M, Roychoudhury AK (1993) Evolutionary Relationships of Human Populations on a Global Scale. Molecular biology and evolution 10: 927–943.
14. Cavalli-Sforza LL, Menozzi P, Piazza A (1994) The History and Geography of Human Genes. Princeton, NJ: Princeton University Press.
15. Foley RA, Lahr MM (1997) Mode 3 technologies and the evolution of modern humans. Cambridge Archaeological Journal 7: 3–36. doi: 10.1017/s0959774300001451
16. Cavalli-Sforza LL (1997) Genes, peoples, and languages. Proceedings of the National Academy of Sciences of the United States of America 94: 7719–7724. doi: 10.1073/pnas.94.15.7719
17. Richards M, Macaulay V, Hickey E, Vega E, Sykes B, et al. (2000) Tracing European founder lineages in the Near Eastern mtDNA pool. American journal of human genetics 67: 1251–1276. doi: 10.1016/s0002-9297(07)62954-1
18. Richards M, Macaulay V, Torroni A, Bandelt HJ (2002) In search of geographical patterns in European mitochondrial DNA. American journal of human genetics 71: 1168–1174. doi: 10.1086/342930
19. Simoni L, Calafell F, Pettener D, Bertranpetit J, Barbujani G (2000) Geographic patterns of mtDNA diversity in Europe. American journal of human genetics 66: 262–278. doi: 10.1086/302706
20. Semino O, Magri C, Benuzzi G, Lin AA, Al-Zahery N, et al. (2004) Origin, diffusion, and differentiation of Y-chromosome haplogroups E and J: inferences on the neolithization of Europe and later migratory events in the Mediterranean area. American journal of human genetics 74: 1023–1034. doi: 10.1086/386295
21. Plaza S, Calafell F, Helal A, Bouzerna N, Lefranc G, et al. (2003) Joining the pillars of Hercules: mtDNA sequences show multidirectional gene flow in the western Mediterranean. Annals of human genetics 67: 312–328. doi: 10.1046/j.1469-1809.2003.00039.x
22. Scozzari R, Cruciani F, Pangrazio A, Santolamazza P, Vona G, et al. (2001) Human Y-chromosome variation in the western Mediterranean

area: implications for the peopling of the region. Human immunology 62: 871–884. doi: 10.1016/s0198-8859(01)00286-5

23. Di Giacomo F, Luca F, Popa LO, Akar N, Anagnou N, et al. (2004) Y chromosomal haplogroup J as a signature of the post-neolithic colonization of Europe. Human genetics 115: 357–371. doi: 10.1007/s00439-004-1168-9
24. Flores C, Maca-Meyer N, Gonzalez AM, Oefner PJ, Shen P, et al. (2004) Reduced genetic structure of the Iberian peninsula revealed by Y-chromosome analysis: implications for population demography. European journal of human genetics : EJHG 12: 855–863. doi: 10.1038/sj.ejhg.5201225
25. El-Sibai M, Platt DE, Haber M, Xue Y, Youhanna SC, et al. (2009) Geographical structure of the Y-chromosomal genetic landscape of the Levant: a coastal-inland contrast. Annals of human genetics 73: 568–581. doi: 10.1111/j.1469-1809.2009.00538.x
26. Zalloua PA, Platt DE, El Sibai M, Khalife J, Makhoul N, et al. (2008) Identifying genetic traces of historical expansions: Phoenician footprints in the Mediterranean. American journal of human genetics 83: 633–642. doi: 10.1016/j.ajhg.2008.10.012
27. Cadenas AM, Zhivotovsky LA, Cavalli-Sforza LL, Underhill PA, Herrera RJ (2008) Y-chromosome diversity characterizes the Gulf of Oman. European journal of human genetics : EJHG 16: 374–386. doi: 10.1038/sj.ejhg.5201934
28. Rowold DJ, Luis JR, Terreros MC, Herrera RJ (2007) Mitochondrial DNA geneflow indicates preferred usage of the Levant Corridor over the Horn of Africa passageway. Journal of human genetics 52: 436–447. doi: 10.1007/s10038-007-0132-7
29. Underhill PA, Kivisild T (2007) Use of y chromosome and mitochondrial DNA population structure in tracing human migrations. Annual review of genetics 41: 539–564. doi: 10.1146/annurev.genet.41.110306.130407
30. Gonzalez AM, Karadsheh N, Maca-Meyer N, Flores C, Cabrera VM, et al. (2008) Mitochondrial DNA variation in Jordanians and their genetic relationship to other Middle East populations. Annals of human biology 35: 212–231. doi: 10.1080/03014460801946538
31. Al-Zahery N, Semino O, Benuzzi G, Magri C, Passarino G, et al. (2003) Y-chromosome and mtDNA polymorphisms in Iraq, a crossroad of the early human dispersal and of post-Neolithic migrations. Molecular phylogenetics and evolution 28: 458–472. doi: 10.1016/s1055-7903(03)00039-3

32. Pilkington MM, Wilder JA, Mendez FL, Cox MP, Woerner A, et al. (2008) Contrasting signatures of population growth for mitochondrial DNA and Y chromosomes among human populations in Africa. Molecular biology and evolution 25: 517–525. doi: 10.1093/molbev/msm279
33. Coelho M, Sequeira F, Luiselli D, Beleza S, Rocha J (2009) On the edge of Bantu expansions: mtDNA, Y chromosome and lactase persistence genetic variation in southwestern Angola. BMC evolutionary biology 9: 80. doi: 10.1186/1471-2148-9-80
34. Wood ET, Stover DA, Ehret C, Destro-Bisol G, Spedini G, et al. (2005) Contrasting patterns of Y chromosome and mtDNA variation in Africa: evidence for sex-biased demographic processes. European journal of human genetics : EJHG 13: 867–876. doi: 10.1038/sj.ejhg.5201408
35. Pereira L, Cerny V, Cerezo M, Silva NM, Hajek M, et al. (2010) Linking the sub-Saharan and West Eurasian gene pools: maternal and paternal heritage of the Tuareg nomads from the African Sahel. European journal of human genetics : EJHG 18: 915–923. doi: 10.1038/ejhg.2010.21
36. Irwin J, Saunier J, Strouss K, Paintner C, Diegoli T, et al. (2008) Mitochondrial control region sequences from northern Greece and Greek Cypriots. International journal of legal medicine 122: 87–89. doi: 10.1007/s00414-007-0173-7
37. Saunier JL, Irwin JA, Strouss KM, Ragab H, Sturk KA, et al. (2009) Mitochondrial control region sequences from an Egyptian population sample. Forensic science international Genetics 3: e97–103. doi: 10.1016/j.fsigen.2008.09.004
38. Kivisild T, Reidla M, Metspalu E, Rosa A, Brehm A, et al. (2004) Ethiopian mitochondrial DNA heritage: tracking gene flow across and around the gate of tears. American journal of human genetics 75: 752–770. doi: 10.1086/425161
39. Dubut V, Chollet L, Murail P, Cartault F, Beraud-Colomb E, et al. (2004) mtDNA polymorphisms in five French groups: importance of regional sampling. European journal of human genetics : EJHG 12: 293–300. doi: 10.1038/sj.ejhg.5201145
40. Richard C, Pennarun E, Kivisild T, Tambets K, Tolk HV, et al. (2007) An mtDNA perspective of French genetic variation. Annals of human biology 34: 68–79. doi: 10.1080/03014460601076098
41. Brandstätter A, Peterson CT, Irwin JA, Mpoke S, Koech DK, et al. (2004) Mitochondrial DNA control region sequences from Nairobi (Kenya): inferring phylogenetic parameters for the establishment of a forensic

database. International journal of legal medicine 118: 294–306. doi: 10.1007/s00414-004-0466-z

42. Ottoni C, Martinez-Labarga C, Loogvali EL, Pennarun E, Achilli A, et al. (2009) First genetic insight into Libyan Tuaregs: a maternal perspective. Annals of human genetics 73: 438–448. doi: 10.1111/j.1469-1809.2009.00526.x
43. Harich N, Costa M, Fernandes V, Kandil M, Pereira J, et al. (2010) The trans-Saharan slave trade - clues from interpolation analyses and high-resolution characterization of mitochondrial DNA lineages. BMC evolutionary biology 10: 138. doi: 10.1186/1471-2148-10-138
44. Turchi C, Buscemi L, Giacchino E, Onofri V, Fendt L, et al. (2009) Polymorphisms of mtDNA control region in Tunisian and Moroccan populations: an enrichment of forensic mtDNA databases with Northern Africa data. Forensic science international Genetics 3: 166–172. doi: 10.1016/j.fsigen.2009.01.014
45. Abu-Amero KK, Larruga JM, Cabrera VM, Gonzalez AM (2008) Mitochondrial DNA structure in the Arabian Peninsula. BMC evolutionary biology 8: 45. doi: 10.1186/1471-2148-8-45
46. Malyarchuk B, Grzybowski T, Derenko M, Perkova M, Vanecek T, et al. (2008) Mitochondrial DNA phylogeny in Eastern and Western Slavs. Molecular biology and evolution 25: 1651–1658. doi: 10.1093/molbev/msn114
47. Cherni L, Loueslati BY, Pereira L, Ennafaa H, Amorim A, et al. (2005) Female gene pools of Berber and Arab neighboring communities in central Tunisia: microstructure of mtDNA variation in North Africa. Human biology 77: 61–70. doi: 10.1353/hub.2005.0028
48. Cerny V, Mulligan CJ, Ridl J, Zaloudkova M, Edens CM, et al. (2008) Regional differences in the distribution of the sub-Saharan, West Eurasian, and South Asian mtDNA lineages in Yemen. American journal of physical anthropology 136: 128–137. doi: 10.1002/ajpa.20784
49. Bosch E, Calafell F, Gonzalez-Neira A, Flaiz C, Mateu E, et al. (2006) Paternal and maternal lineages in the Balkans show a homogeneous landscape over linguistic barriers, except for the isolated Aromuns. Annals of human genetics 70: 459–487. doi: 10.1111/j.1469-1809.2005.00251.x
50. de Filippo C, Barbieri C, Whitten M, Mpoloka SW, Gunnarsdottir ED, et al. (2011) Y-chromosomal variation in sub-Saharan Africa: insights into the history of Niger-Congo groups. Molecular biology and evolution 28: 1255–1269. doi: 10.1093/molbev/msq312

51. Ferri G, Ceccardi S, Lugaresi F, Bini C, Ingravallo F, et al. (2008) Male haplotypes and haplogroups differences between urban (Rimini) and rural area (Valmarecchia) in Romagna region (North Italy). Forensic science international 175: 250–255. doi: 10.1016/j.forsciint.2007.06.007
52. Alshamali F, Pereira L, Budowle B, Poloni ES, Currat M (2009) Local population structure in Arabian Peninsula revealed by Y-STR diversity. Hum Hered 68: 45–54. doi: 10.1159/000210448
53. Petrejcikova E, Sotak M, Bernasovska J, Bernasovsky I, Rebala K, et al. (2011) Allele frequencies and population data for 11 Y-chromosome STRs in samples from Eastern Slovakia. Forensic science international Genetics 5: e53–62. doi: 10.1016/j.fsigen.2010.08.003
54. Athey T (2005) Haplogroup Prediction from Y-STR Values Using an Allele-Frequency Approach. Journal of Genetic Genealogy 1: 1–7.
55. Athey T (2006) Haplogroup Prediction from Y-STR Values Using a Bayesian-Allele-Frequency Approach. Journal of Genetic Genealogy 2: 34–39.
56. Salas A, Richards M, De la Fe T, Lareu MV, Sobrino B, et al. (2002) The making of the African mtDNA landscape. American journal of human genetics 71: 1082–1111. doi: 10.1086/344348
57. Salas A, Richards M, Lareu MV, Scozzari R, Coppa A, et al. (2004) The African diaspora: mitochondrial DNA and the Atlantic slave trade. American journal of human genetics 74: 454–465. doi: 10.1086/382194
58. Trejaut JA, Kivisild T, Loo JH, Lee CL, He CL, et al. (2005) Traces of archaic mitochondrial lineages persist in Austronesian-speaking Formosan populations. PLoS biology 3: e247. doi: 10.1371/journal.pbio.0030247
59. van Oven M, Kayser M (2009) Updated comprehensive phylogenetic tree of global human mitochondrial DNA variation. Human mutation 30: E386–394. doi: 10.1002/humu.20921
60. R Development Core Team (2011) R: A Language and Environment for Statistical Computing. Vienna, Austria: R Foundation for Statistical Computing.
61. Venables WN, Ripley BD (2002) Modern Applied Statistics with S. New York, NY: Springer-Verlag.
62. Anderson S, Bankier AT, Barrell BG, de Bruijn MH, Coulson AR, et al. (1981) Sequence and organization of the human mitochondrial genome. Nature 290: 457–465. doi: 10.1038/290457a0

63. Andrews RM, Kubacka I, Chinnery PF, Lightowlers RN, Turnbull DM, et al. (1999) Reanalysis and revision of the Cambridge reference sequence for human mitochondrial DNA. Nature genetics 23: 147. doi: 10.1038/13779
64. Excoffier L, Lischer HE (2010) Arlequin suite ver 3.5: a new series of programs to perform population genetics analyses under Linux and Windows. Molecular ecology resources 10: 564–567. doi: 10.1111/j.1755-0998.2010.02847.x
65. Reynolds J, Weir BS, Cockerham CC (1983) Estimation of the coancestry coefficient: basis for a short-term genetic distance. Genetics 105: 767–779.
66. Cox TF, Cox MAA (2001) Multidimensional Scaling, Second Edition. New York, NY: Chapman and Hall.
67. Slatkin M (1995) A measure of population subdivision based on microsatellite allele frequencies. Genetics 139: 457–462.
68. Excoffier L, Smouse PE, Quattro JM (1992) Analysis of molecular variance inferred from metric distances among DNA haplotypes: application to human mitochondrial DNA restriction data. Genetics 131: 479–491.
69. Wright S (1951) The genetical structure of populations. Ann Eugenics 15: 323–354. doi: 10.1111/j.1469-1809.1949.tb02451.x
70. Semino O, Santachiara-Benerecetti AS, Falaschi F, Cavalli-Sforza LL, Underhill PA (2002) Ethiopians and Khoisan share the deepest clades of the human Y-chromosome phylogeny. American journal of human genetics 70: 265–268. doi: 10.1086/338306
71. Underhill PA, Shen P, Lin AA, Jin L, Passarino G, et al. (2000) Y chromosome sequence variation and the history of human populations. Nature genetics 26: 358–361.
72. Macaulay V, Hill C, Achilli A, Rengo C, Clarke D, et al. (2005) Single, rapid coastal settlement of Asia revealed by analysis of complete mitochondrial genomes. Science 308: 1034–1036. doi: 10.1126/science.1109792
73. Mellars P (2006) Why did modern human populations disperse from Africa ca. 60,000 years ago? A new model. Proceedings of the National Academy of Sciences of the United States of America 103: 9381–9386. doi: 10.1073/pnas.0510792103

74. Van Andel TH, Tzedakis PC (1996) Paleolithic landscapes of Europe and environs: 150,000–25,000 years ago: an overview. Quaternary Science Reviews 15: 481–500. doi: 10.1016/0277-3791(96)00028-5

75. Chiaroni J, King R, Underhill PA (2008) Correlation of annual precipitation with human Y-chromosome diversity and the emergence of Neolithic agricultural and pastoral economies in the Fertile Crescent. Antiquity 82: 281–289.

76. Fattovich R (1997) The Near East and Eastern Africa: Their Interaction. In: Vogel JO, editor. Encyclopedia of precolonial Africa. Walnut Creek: AltaMira Press. pp. 479–484.

77. Segal R (2001) Islam's Black Slaves: The Other Black Diaspora. New York: Farrar, Straus, and Giroux.

78. Lewis B (1990) Race and slavery in the Middle East: an historical enquiry. New York: Oxford University Press.

79. Richards M, Rengo C, Cruciani F, Gratrix F, Wilson JF, et al. (2003) Extensive female-mediated gene flow from sub-Saharan Africa into near eastern Arab populations. American journal of human genetics 72: 1058–1064. doi: 10.1086/374384

Chapter 13

IDENTIFICATION OF GENETIC VARIATION ON THE HORSE Y CHROMOSOME AND THE TRACING OF MALE FOUNDER LINEAGES IN MODERN BREEDS

Barbara Wallner, Claus Vogl, Priyank Shukla, Joerg P. Burgstaller, Thomas Druml, Gottfried Brem

Institute of Animal Breeding and Genetics, Department of Biomedical Sciences, University of Veterinary Medicine Vienna, Vienna, Austria

ABSTRACT

The paternally inherited Y chromosome displays the population genetic history of males. While modern domestic horses (*Equus caballus*) exhibit abundant diversity within maternally inherited mitochondrial DNA, no significant Y-chromosomal sequence diversity has been detected. We used high throughput sequencing technology to identify the first polymorphic Y-chromosomal markers useful for tracing paternal lines. The nucleotide variability of the modern horse Y chromosome is extremely low, resulting in six haplotypes (HT), all clearly distinct from the Przewalski horse (*E. przewalskii*). The most widespread HT1 is ancestral and the other five haplotypes apparently arose on the background of HT1 by mutation or gene conversion after domestication. Two haplotypes (HT2 and HT3) are widely distributed at high frequencies among modern European horse breeds. Using pedigree information, we trace the distribution of Y-haplotype diversity to particular founders. The mutation leading to HT3 occurred in the germline of the famous English Thoroughbred stallion "Eclipse" or his son or grandson and its prevalence demonstrates the influence of this popular paternal line on modern sport horse breeds. The pervasive introgression of Thoroughbred stallions during the last 200 years to refine autochthonous breeds has strongly affected the distribution of Y-chromosomal variation in modern horse breeds and has led to the replacement of autochthonous Y chromosomes. Only a few northern European breeds bear unique variants at high frequencies or fixed within but not shared among breeds. Our Y-chromosomal data complement the well established mtDNA lineages and document the male side of the genetic history of modern horse breeds and breeding practices.

INTRODUCTION

Mitochondrial DNA (mtDNA) and the paternally transmitted portion of the Y chromosome (NRY) are inherited uniparentally and do not recombine. They accumulate mutations, such as single nucleotide polymorphisms (SNPs), insertions and deletions (Indels) and structural rearrangements [1]–[3]. Genealogies inferred from the distribution of mutations on mtDNA and NRY haplotypes reflect gender-specific population genetic forces. Furthermore, gene conversion i.e. the transfer of a short sequence from the homologous, but non-recombining region on the X- to the Y-chromosome can also contribute to Y-chromosomal variation [4], [5]. mtDNA- and NRY-variant distributions have been widely analysed in humans and in a broad range of wild and domesticated animals to provide indications of the origin of species, domestication processes, the characterization of genetic diversity within and between populations and sex specific demographic behaviors [6]. Depending on the genetic regions analysed, uniparentally inherited markers can be used to trace individual founder lines or families [7]. In livestock, Y-chromosomal and mtDNA variation has been used to study the domestication and population structure of the domestic dog [8]–[10], pig [11], [12], sheep [13]and cattle [14].

Populations of the domestic horse (*E. caballus*) show high levels of mtDNA diversity with limited geographic structure [15]–[19]. Estimates of the coalescence times of mtDNA variants of extant horses far predate the date of domestication [18], [19]. Hence, multiple female lineages contributed to the domestic horse gene pool. Nuclear microsatellite data are also consistent with a scenario of high variation that predates domestication [20].

Nevertheless, genetic variability in the domestic horse represents a paradox: although horses have the largest diversity of maternal mtDNA among domestic species, no noteworthy sequence diversity can be detected on the NRY [21]–[23]. Whereas Y chromosomes of modern horses are clearly distinct from that of the Przewalski horse (*E. przewalskii*) [22] and ancestral Y-chromosomal diversity is found in prehistoric horses [24], the only polymorphism observed in modern horses is a microsatellite mutation found in autochthonous Chinese horse breeds [25]. Screening approaches for the identification of polymorphisms on the domestic horse NRY have proven unsuccessful [21]–[23]. The low NRY diversity in horses is assumed to be the result of the extremely low effective population size of males due to the specific mating behaviour and to several bottlenecks that occurred early in the domestication process[20], [21]. Moreover, Y-chromosomal variation might be further diminished in modern horse breeds by regulated breeding programmes and intensive horse-trading. Today's European horse breeds (e.g. the Lipizzan horse and the English Thoroughbred) are largely the result of centralized and organized breeding over the past 200 years. The breeding effort has mainly focused on stallions. Breeding programmes have been based on so-called "multiplier studs", which were founded at the end of the 18th century and were responsible for the production of stallions needed in rural regions. As a result of these early breeding programmes, modern European horse breeds show the consequences of several waves of introgression of imported stallions into local breeds. These include (a) the "Neapolitan" wave from the 15th to the 18thcentury, when the now extinct "Neapolitan horse" was introgressed; (b) the "Oriental" wave from the late 18th to the late 19th century, when "Original Arabian" stallions, imported from Syria to Egypt, were introgressed and (c) the "English" wave from the early 19th century to the present, when Thoroughbred stallions were introgressed [26]–[28]. In the early 19th century new breeding practices were introduced and rapidly supplanted existing practices. Inbreeding and line breeding concepts became popular and with the integration of private breeding into state breeding programmes the entire population of male horses became highly selected. The result of the modern breeding practices may have been the complete replacement of autochthonous Y-chromosomal variants by imported bloodlines.

In the Lipizzan breed, 89 different sire lines existed in the late 18th century, while only eight are present today [29]. Among these eight male lines, one can be attributed to the "Oriental" wave from the Original Arabian stallion "Siglavy" (born 1810, imported to Lipizza 1814). The other Lipizzan stallion lines derive from the earlier, less documented phase (the "Neapolitan" wave). The male gene pool is even more restricted in the English Thoroughbred, with only three paternal lines remaining [30]. All three lines can be attributed to the "Oriental" wave through the import of the three stallions Byerley Turk, Darley Arabian and Godolphin Arabian. Pedigrees in other European breeds are less well documented but reductions in male lines are similar and an influence of Neapolitan and Original Arabian horses can be observed or is presumed to have occurred. Pedigree information on northern European horse breeds also indicates reduced male diversity. Nevertheless, these breeds, notably the Icelandic horse, might be the only European breeds not to have been subjected to the recent introgression waves [30].

Due to the lack of polymorphic markers on the NRY, it is not possible to trace male-mediated population-genetic dynamics in modern horses. We now describe a systematic screen for horse Y-chromosomal variants in modern domestic horse breeds. As there is only scarce sequence information relating to the horse Y chromosome, we sequenced Y-chromosomal BAC clones to obtain reference sequences for our screen. Based on de novo assembled contigs, we amplified long-range PCR (LRP) products covering about 186 kb for targeted resequencing [31]. To identify variants, we selected (a) nine male horses from phenotypically, genetically and geographically highly distinct breeds, (b) eight Lipizzan stallions, each representing a classical founder line [26], and (c) one Przewalski horse (*E. przewalskii*). LRP products were pooled ("seq-pools") and sequenced with Illuminás Solexa technology. Based on the mutations identified, we describe the first Y-chromosomal haplotypes (HT) in domestic horses and their phylogenetic relationships (a detailed workflow is given in Fig. S1). Furthermore, we present the results of a screen for the distribution of the HTs among purebred modern horse breeds. Using pedigree data, we are able to trace the paternal roots of the extant males. Our data show the strong influence of influential founders, mainly from the Near East and a certain Thoroughbred line, on extant horse breeds.

MATERIALS AND METHODS

Ethics Statement

1) Blood Samples

Genomic DNA samples from the breed pool and the Przewalski horse isolated from blood were collected as part of routine diagnostics at the Institute of Animal Breeding and Genetics, University of Veterinary Medicine, Vienna, during the 1970's, 80′s, and 90′s. For the breed pool, information on breeds was available but the dates of collection, owners and horse identification were not recorded. The Przewalski horse was provided by Dr. Meltzer (Zoo Munich) and the Icelandic horse by B. Wallner.

The Lipizzan horse blood samples were collected before 1999 during an INCO Copernicus project (1996–2001). Permission for the scientific use of the samples was granted by all involved stud farms, which were partners in the project. A summary of the findings was published in "Der Lipizzaner im Spiegel der Wissenschaft" [26], edited by G. Brem, Publisher: Österreichische Akademie der Wissenschaften. We are not aware whether the blood samples were taken during routine diagnostics.

Samples were collected before the establishment in 2004 of the ethics commission of the University of Veterinary Medicine, Vienna.

2) Hair Root Samples

165 hair samples were recently taken and permission was granted by the private owners or breeding associations. The following breeding associations gave their consent: Associacion française du Lipizzan (France), Escola Portuguesa de Arte Equestre (Portugal), Fundación Real Escuela Andaluza del Arte Ecuestre (Spain), Lipizzanergestüt Piber and the Spanish Riding School (Austria).

450 genomic DNA samples isolated from hair roots derive from routine parentage testing from the archive of the company AGROBIOGEN GmbH, Hilgertshausen. DNA samples kindly provided by Agrobiogen are several years old.

All samples were anonymized.

454 Sequencing of BAC Clones

BAC DNA from 5 Y-chromosomal BAC clones (Fig. 1) [32], [33] was isolated as described at http://dga.jouy.inra.fr/grafra/BAC_DNA_midiprep.htm and treated with RNAseA. BAC DNA was purified with the Mammalian Genomic DNA Miniprep Kit (Sigma-Aldrich, cat. no G1N350-1KT). Parallel sequencing on the 454 Roche system was performed according to standard procedures at the Core Facility Molekularbiologie (Meduni Graz, AT). 50 ng of total BAC DNA were prepared with the Nextera™ DNA Sample Prep Kit (Epicentre, cat.no. NT09115) according to the manufacturer's instructions. DNA was incubated in the enzyme buffer for 5 minutes at 55°C and purified with the Qiagen MinElute PCR Purification Kit (Qiagen cat.no. 28004) and titanium adaptors were ligated to the fragments. After emulsion PCR (GS titanium emPCR reagents Lib-L Roche cat. no. 05618444001), bead recovery and enrichment sequencing was performed with the GS Titanium Sequencing Reagent XLR70 (Roche, cat. no. 05233526001) according to Roche standard 454 protocols.

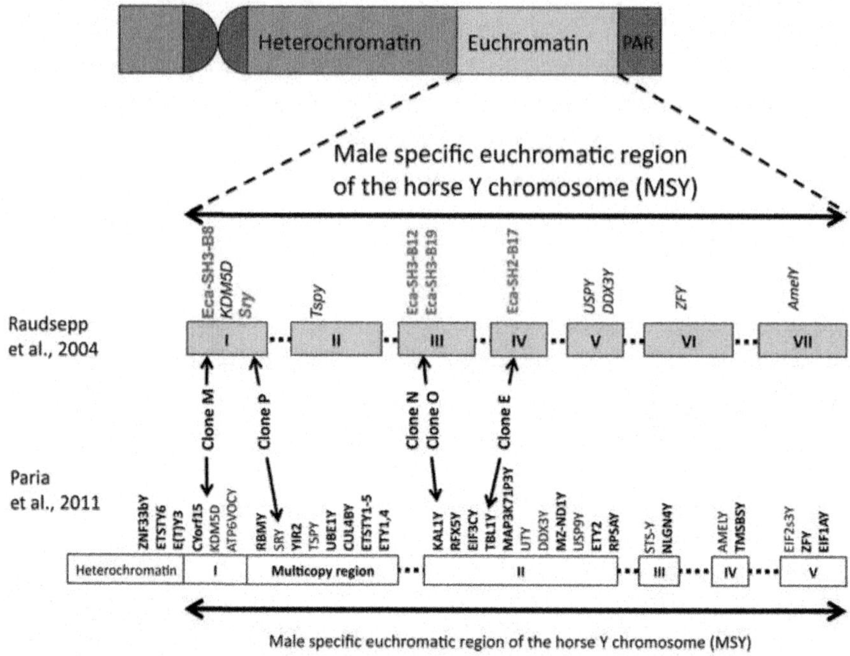

Figure 1: The localisation of the BAC-clones on the horse Y chromosome (ECAY).

The heterochromatic, male specific euchromatic (MSY) and the pseudoautosomal region (PAR) of the horse Y chromosome are shown at the top. The seven MSY contigs (I-VII) described in the study of Raudsepp et al., 2004 [33] are illustrated in the middle and indicated as boxes; gaps between contigs are spanned with dotted lines. The locations of some MSY genes are given (italics). The chromosomal position of each BAC clone used in this study is marked by an arrow. On the bottom the approximate positions of the BAC clones on the horse Y-chromosomal gene map after Paria et al. 2011 [40] are indicated.

doi:10.1371/journal.pone.0060015.g001

The raw sequence reads were filtered and trimmed to reject too short or bad quality sequences. No ambiguous base was allowed and the minimum sequence length exceeded 50 bases. Separate assemblies were performed for BAC clones M, P and E using Roche GS De Novo Assembler v2.3 to generate the contigs. Overlapping clones N and O were assembled together. Contig sequences were trimmed for BAC vector sequence and *E.coli* genome sequences. Assembled sequences longer than 5000 bp were selected for further analysis.

Analysis of Contig Sequences

Contig consensus sequences generated in each library were screened for homologies by manual blast search (blastN) against the complete nucleotide collection in Genbank and by BLAT (http://genome.ucsc.edu/cgi-bin/hgBlat) against the horse genome. Repetitive elements were identified with Repeat Masker (http://www.repeatmasker.org/).

Long–Range PCRs

Primers were designed with Primer3. Amplicon length ranged from 5.5 to 11.8 kilobase pairs (kb) and some contigs overlapped. PCR products were amplified with the Expand Long Template PCR System (Roche, cat. no. 11681842001) as described in the Kit and checked for male specificity by comparative amplification of blood genomic DNA from male and female horses (Fig. S2). Sequence specificity was controlled by restriction enzyme digests.

DNA Amplification for Sequence Analysis

To maximize diversity in the sample set, nine male purebred domestic horses from phenotypically different breeds and geographically distinct regions were selected for the first seq-pool ("breed"). The second pool ("lipp") contained 8 Lipizzan stallions, each representing a particular paternal founder line [26]. In the third pool ("prz"), only one Przewalski horse was sequenced. To detect all variants in a pooled sample [31], single LRP products were generated from each individual using high molecular-weight genomic DNA isolated from whole blood. The LRP products were visualized on a 0.8% agarose gel and the concentration of each product was measured with the Qubit ds DNA HS Assay (Invitrogen cat. no. Q32851). In case of multiple PCR products the Y chromosome specific amplicon was isolated from the agarosegel prior NGS library preparation. Y-specific LRP products were pooled equimolarily and cleaned with the High Pure PCR Product Purification Kit (Roche, cat. no. 11732668001), resulting in 2 μg clean LRP product per pool.

Illumina Library Preparation and Data Generation

The pools were fractionated to a size range of 300–700 bp using a Covaris sonicator and fragments were purified using the High Pure PCR Product Purification Kit. For high-throughput sequencing, libraries were prepared with NEBNext DNA Library Prep (NEB, cat. no. E6000S) using NEBNext Multiplex Oligos (NEB cat. no. E7335L) to index the three pools for multiplex sequencing according to the manufacturer's instructions. Indexed pools were submitted together on one lane of 76 bp paired-end sequencing at the CSF NGS

Unit (http://csf.ac.at/) using the Illumina GA II system [34]. The quality of the data was checked with FASTQC [35], sequences were aligned to the reference BAC sequence with BWA [36] and quality control (filtering) was performed with SAMtools [37]. After quality control and removal of duplicate reads, ~4500×, ~5000× and ~2500× mapped sequence coverage of the available Y reference sequence was obtained from Lipizzan, domestic and Przewalski samples, respectively. SNP calling (identifying positions that differed from the reference sequence) was performed with SAMtools and in-house python scripts.

Data Filtering to Identify Candidate Mutations

As there were nine individuals in the pool "breed", theoretically 11.1% of genomic DNA was contributed by each horse in the sequencing sample. Considering random errors and experimental bias at the various stages of library preparation, 8% was decided as a safe threshold for SNP calling. As only one individual was sequenced in the prz-pool, sites that differed from the reference in the same proportion in all three pools were assumed to be base-calling or alignment errors. For the second-class SNPs, we lowered the threshold for SNP calling to 6% minor allele frequency and/or allowed 3% in the prz-pool horse.

Verification of Candidate Polymorphic Sites by Sanger Sequencing and Sequence Analysis

For the filtered candidate mutations, we designed PCR primers by using Primer3 to amplify 240–760 bp fragments. PCR products were amplified from genomic DNA from each horse from the breed-, lipp-, and prz-pools and from a second Shetland pony (n=18). Amplicons were resequenced by conventional Sanger sequencing for forward and reverse strands at LGC genomics. See Figs. S3, S4, S5, S6 and S7 for the sequence alignment with primer binding sites. PCR products amplified from male and female DNA and the results of the Sanger sequencing for each polymorphic region. *E. przewalskii*-determining sites (positions 4086 and 4161 on contig YM23) and Eca-Y2B17 [22] and intron2 from AMELY [24], representing regions harbouring ancestral diversity, were sequenced for all horses in the pools. Sequences were aligned with the CLC workbench. Nucleotide diversity was calculated in R according to the formula of Nei [38].

Microsatellite Analysis

Individuals were characterized using five equine Y-chromosomal microsatellites described previously [23] . Genotypes were determined with MegaBACE 500 at the University of Veterinary Medicine, Vienna. Electropherograms were evaluated using MegaBACE Genetic Profiler v2.2 (GE Healthcare). Microsatellite variation was investigated for 100 domestic horses from various breeds, including the horses that were used for the seq-pools and three Przewalski horses.

Screening of Domestic Horses for Y Haplotypes

615 male horses, representing 58 mainly European horse breeds, were screened for their Y-chromosomal haplotypes. Genomic DNA samples isolated from hair roots (purified with nexttec™) derive from routine parentage testing. Genotyping was performed using the Sequenom MassARRAY iPLEX system (Sequenom, Germany) at the Department for Agrobiotechnology, IFA Tulln. A section of DNA containing the variant position was amplified from each individual by PCR, before a high-fidelity single-base primer extension reaction over the SNP being assayed was undertaken, using nucleotides of modified mass. The different alleles therefore produce oligonucleotides with mass differences that can be detected using highly accurate Matrix-Assisted Laser Desorption/Ionization Time-Of-Flight mass spectrometry [39]. The five SNP multiplex assay was designed using the Sequenom Assay Design 3.1 software, SNP genotyping was performed using the iPLEX® GOLD Complete Genotyping kit with SpectroCHIPs® II in the 384 format (Sequenom, Germany) in duplicate and female genomic DNA was included as a negative control. We followed the manufacturer's protocol with a single modification: to reduce unspecific primer extension, 5 ng sheared salmon sperm DNA (Invitrogen, Austria) per reaction was added to the PCR mastermix. Results were analysed with the Sequenom Typer 4.0 software (Sequenom, Germany) and the results validated by SANGER resequencing of ten (when available) alleles from each position.

The screening criteria for Shetland pony HT6 were (a) no results in Y-E17.1mut_SNP1 & SNP2, and (b) amplification of the 966 bp deletion and visualization on an Agarose gel.

Pedigree Analysis

For pedigree analysis, web-based databases were used: for Thoroughbreds the Pedigree Online Thoroughbred Database (availiable at: www.pedigreequery.com/) and the Galoppsieger database (www.galopp-sieger.de/); for Shagya

Arabians the Shagya Database (www.shagya-database.ch/hengste.php); for Icelandic horses the Stormhestar Database (www.stormhestar.de/german/default.asp) and for many other breeds the Pedigree Online All Breed Database (www.allbreedpedigree.com) or the Sporthorse Horse Show and Breed Database (www.sporthorse-data.com/breed.htm). All database informations were last accessed in October 2012.

Thirteen breeds (Appaloosa, Barb, Konik, Mangalarga Paulista, Mangalarga Marchador, New Forest, Paint, Russian Arabian, Saddlebred, Shire, Tinker horse, Wuerttemberg, Camargue) are not included in the pedigree analysis, as we had no males with a proven ancestry. For founder classification we call Arabian, so-called Oriental and Turkoman horses (i.e. horses from the middle East ranging from Turkmenistan to Egypt) that were imported to Europe in the 18th and 19th century "Original Arabians".

RESULTS

Identification of Y Chromosome Polymorphisms

To detect polymorphisms, we generated reference sequences from the horse Y chromosome by 454 sequencing of 5 Y-chromosomal BACs from certain regions on the chromosome (Fig. 1) [23], [32], [33], [40]. Sequence reads were de novo assembled and 11 contigs and an *Sry* containing BAC sequence (AC215855.2) were selected for designing 21 LRPs, covering a total of 186,122 bp. Y-chromosomal specificity was checked by comparative amplification of the LRPs on male and female DNA (Fig. S2).

We produced Y-chromosomal LRP products from 17 domestic and one Przwalski horse separately and pooled the products. LRPs of nine males from different breeds (pool-breed), of eight Lipizzan stallions (pool-lipp) and the Przewalski horse (pool-prz) were sequenced with the Illumina platform to high depth. Whereas the Przewalski horse showed 37 base substitutions and one 3051 bp deletion compared to the domestic horse samples, only 4 positions/regions promised to be polymorphic within the domestic horse pools. With relaxed criteria (see Materials and methods) another 19 second-class SNPs were added. For validation, we amplified the region spanning each polymorphic candidate from each domestic horse in the seq-pools and sequenced them by conventional Sanger sequencing. The four top candidate regions were confirmed but none of the 19 second-class SNPs were. The resequencing results and the male specificity of the confirmed candidates are shown in Figs. S3, S4, S5, S6 and S7. The variants detected in domestic horses were two SNPs, one single base deletion and a complex variant consisting of 20 SNPs and four indels. In addition we identified a 966 bp deletion in a Shetland pony and a transition in a second Przewalski horse during the screening procedures.

The two SNPs were found on discrete contigs (YE17 and YXX_24I23) but the indel polymorphism, the complex mutation and the 966 bp deletion are all located in close proximity or even overlapping in a restricted region on contig YE3. Screening for homologies to this region in the horse genome using BLAT revealed high similarities (>95%) with the homologous region on the horse X chromosome and the occurrence of a LINE element approximately 700 bp upstream of the region that is mutated in the domestic horse (Fig. S9). The alignment of X - and Y-sequences shows the need for careful selection of PCR amplification primers to amplify Y-chromosomal sequences. Furthermore, one can see from the alignment (Fig. S5) that the multiple variants on locus YE3 - Pos 1007–12040 conform to the X- rather than to the Y chromosome for a length of 300 bases. We thus assume that the complex variant on YE3 - Pos 1007–12040 comprising 25 SNPs and four indels is the result of a gene conversion event between the X and Y chromosome [5]. The deletion of a single T on locus YE3 - Pos 10594 is also found on the homologous X-chromosomal sequence (Fig. S4) and thus another putative gene conversion event.

Y-Chromosomal Diversity in Modern Horses - Haplotype Network and Estimation of Divergence Time

The polymorphisms result in six haplotypes in the domestic and two in the Przewalski horse (Tables S9, S10 and S11). The ancestral and the derived status in modern horses were determined by comparison to the Przewalski horse sequence. The haplotype network in Fig. 2gives the relationship between the HTs. *E. caballus* haplotypes are separated by only one mutational or gene conversion step, i.e. there is no deep bifurcation with many segregating mutations. Three haplotypes (HT1-3) occur at relatively high frequencies. HT1 is the ancestral haplotype when rooted with the Przewalski Y haplotype. HT2 is differentiated from HT1 by a single nucleotide mutation. The deletion that defines HT3 arose on the background of HT2. Clearly, HT1 is the most prominent haplotype in the evolution of modern horse breeds: of the five mutations observed in the panel, four arose on the background of HT1 and only one on the background of HT2. Thus, although the proportion of HT1 and HT2 in our sample is about equal, all except one variant arose on the background of the older and thus evolutionarily more important HT1. Lippold et al. found a Y-chromosomal HT in an ancient domestic horse that differed at three positions from the common domestic horse HT chromosome [24]. We observed none of these ancient bases when screening a set of extant horse breeds that represent all six haplotypes. The Y chromosome of the Przewalski horse forms a separate clade with a sequence divergence of 0.021% from *E. caballus*.

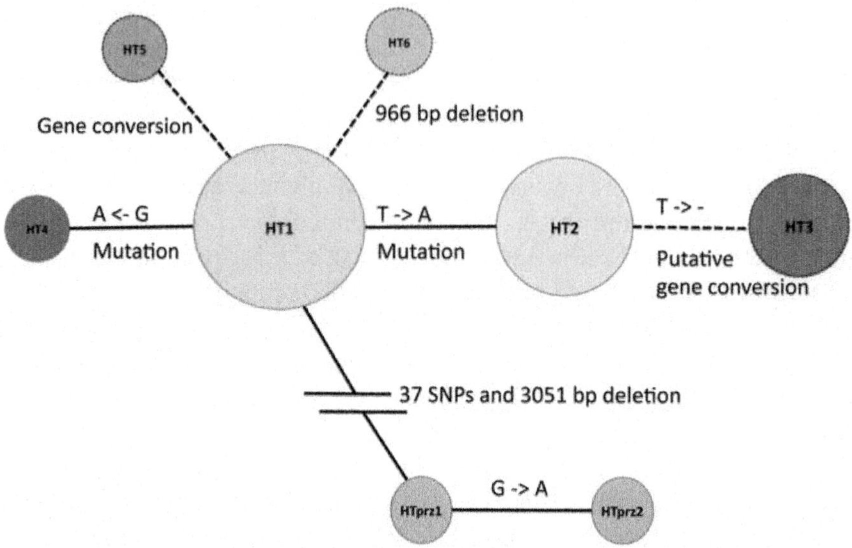

Figure 2: Haplotype network of the six modern and two Przewalski horse HTs.

Circles represent the haplotypes with the area proportional to the observed frequency in 20 male horses in the initial Y-chromosomal sequence analysis. HT1, n=7 (three Lipizzan, two Arabian, one Shetland pony, one Shire horse); HT2, n=5 (five Lipizzan); HT3, n=3 (one Thoroughbred, one Trakehner, one Quarter horse); HT 4 (one Icelandic horse), HT5 (one Norwegian Fjord horse), HT6 (one Shetland pony), HTPrz1 (one Przewalski horse), HTPrz2 (one Przewalski horse). A dashed line between the haplotypes indicates, that the polymorphism is located on the highly variable contig YE3, which was omitted when estimating divergence time and nucleotide diversity.

doi:10.1371/journal.pone.0060015.g002

For inference of population genetic diversity and divergence time estimates, we rely only on putative single nucleotide polymorphisms, i.e., we exclude the region affected by gene conversion (YE3). The rate of gene conversion is known to depend strongly on the genomic location, and is particularly high at translocation hotspots [41], [42]. The prediction of these hotspots seems complex [4], but the occurrence of a LINE element and the X/Y-sequence similarity indicate that the YE3 region is a candidate. Furthermore, we infer from pedigree data, that the deletion on YE3 leading to HT3 arose, presumably by a gene conversion event, very recently (see below). In addition, the occurrence of an independent 966 bp deletion in HT6 points to a general instability of the YE3 region. We thus base our estimates of diversity and dating solely on the observation of two SNPs (YXX_24I23 - Pos 25345 and YE17.1 - Pos 1277) per 170 731 bp of total length. We assume a rate of single nucleotide mutations per generation of about $1.1–5\times10^{-8}$ as found for humans [1], [43], [44] and expect the rate of mutations per generation in our dataset to be: $\mu\times$ (sequence length) $\approx$1,8 to 8×10^{-3}. Thus only 234 to 1064 meioses are theoretically necessary to give rise to the two SNPs over the region under investigation. We conclude that the present Y-chromosomal diversity in modern breeds most likely arose by mutations from HT1 after domestication, about 6000 years or 1000 generations ago [45]. We note, however, that the NRY is inherited as a single locus, so inferences based on the NRY are subject to large stochastic errors for parameters such as the overall tree depth.

Considering the frequencies of haplotypes with SNPs, pairwise nucleotide diversity [38] is $\pi=3.71\times10^{-6}$, an extremely low value compared to those observed in domestic pigs ($\pi=1.38\times10^{-3}$) [11], [12] and dogs ($\pi=2.91\times10^{-4}$) [46].

Phylogeography of Horse Y-chromosomal Haplotypes

We examined the distribution of the Y-chromosomal haplotypes in 615 males from 56, mainly European, domestic horse breeds. To avoid father-son pairs or paternal brothers in the dataset, only purebred horses with confirmed paternity were investigated. 91% of the male horses were found to carry one of the three major haplotypes. The ancestral HT1 is distributed across almost all breeds and the entire geographical region under investigation, underlining its importance (Fig. 3A). HT2 is also found at high frequencies across a broad range of breeds, although not in the northern European breeds and not in horses from the Iberian Peninsula. HT3 is almost fixed in the English Thoroughbred and is further distributed across many warm-blooded breeds. HT4-6 are only found in three local northern European breeds but in high frequencies in these breeds: HT4 in half the Icelandic horses and HT6 in 74% of the Shetland ponies while HT5 is fixed in Norwegian Fjord horses. We undertook genotyping of five Y-specific microsatellite markers [23] from a subset of 100 horses distributed over all HTs. Microsatellite analysis showed a uniform pattern of amplification over all domestic horses, distinct from the Przewalski horse at one locus, as in previous studies [23], [25].

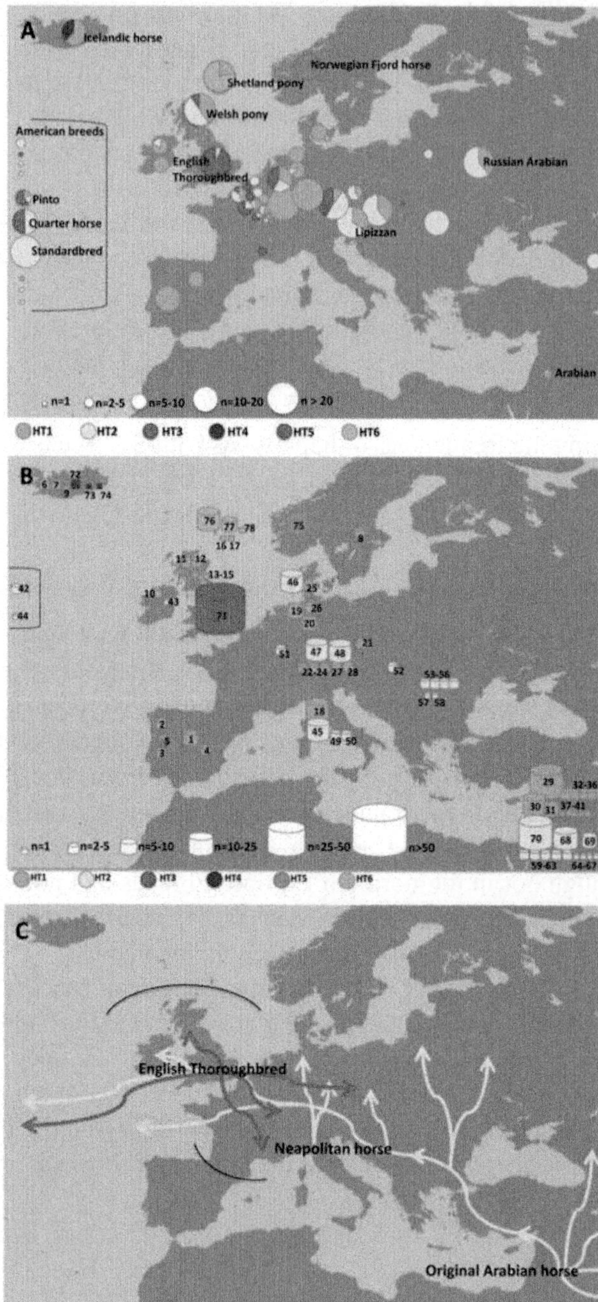

Figure 3: Geographic distribution and history of Y-haplotypes in modern horse breeds.

(a) Geographic distribution of Y-chromosomal haplotypes in a set of modern horse breeds. Only a few important breeds are specified, (b) Origin of modern domestic horse founders deduced from pedigree data. Each founder is represented by a drum with its size proportionally to the number of offspring in the dataset. The number in the drums serve as founder identifiers. Detailed information on founders (name, year of birth, breed, origin, information on import) is listed in Table 1. (c) Male introgression routes deduced from the pedigree and the distribution of HT2 and HT3 in our dataset. HT2 (yellow arrows) arrived from South-East at early times and has been spread during the Neapolitan and Oriental introgression waves, but did not reach Northern Europe and the Iberian peninsula. The English wave in red is well documented through pedigree data and the spread of HT3 (red arrows). Due to the ubiquitous occurrence of HT1, this haplotype is not considered. The black solid lines reflect the limits of the observation of HT2 and HT3.

doi:10.1371/journal.pone.0060015.g003

Incorporation of Pedigree Data to Uncover Founder Traces

We used the documentation of horse ancestry in studbooks and pedigree data from databases to trace the origin of the purebred horses. Sufficient pedigree information was available for 418 (67.97%) males from our dataset. Whereas the paternal ancestry is accurately documented back to the 18th century for old historic breeds such as the English Thoroughbred and the Lipizzan horse, data on most other modern breeds only extend to about 1900. Pedigree analysis revealed that the 418 males from 43 breeds originate from 78 paternal founders whose geographical/breed origin is shown in Fig. 3B. Founder specifications and haplotype information on renowned stallions are available in Table 1. No founder from the Iberian Peninsula, the British Isles (with the exception of a Connemara pony born in 1921, whose origin and possible Arabian influence is unclear) or northern Europe shows HT2. Instead, HT1 predominates, although autochthonous haplotypes (HT4, 5 and 6) are present at high frequencies in local breeds or are even fixed (e.g. in the Norwegian Fjord horse). In contrast, central, eastern and southern European as well as Original Arabian founders from the near East exclusively show HT1 and HT2 in about equal proportions.

Table 1: Founders contributing to modern horses

Founder identifier (Fig. 3B)	Founder name	Born	Inferred haplotype	Breed and/or geographical origin	Documented import	Descendants in dataset (n)	Modern breeds distributed
1	Favorito	1889	HT1	Spanish purebred/Spain		6	Pura Raza Española
2	Pereira	1917	HT1	Lusitano/Portugal		4	Lusitano
3	Nice	1915	HT1	Lusitano/Portugal		3	Lusitano
4	Descindido	1840	HT1	Spanish purebred/Spain		2	Pura Raza Española
5	Marialva	1905	HT1	Alter Real/Portugal		1	Lusitano
6	Brúnn frá Svaðastöðum	1900	HT1	Icelandic horse		5	Icelandic horse
7	Brúnn frá Árnanesi	1910	HT1	Icelandic horse		4	Icelandic horse
8	Hirgst fr Gulbrandsdalen	1846	HT1	Dole Gudbrandsdal/Norway		2	Swedish coldblood trotter
9	Kjarval frá Sauðárkróki	1981	HT1	Icelandic horse		1	Icelandic horse
10	Mountain Lad	1928	HT1	Connemara/Ireland		3	Connemara
11	Dalence	1896	HT1	Welsh/UK		3	Connemara pony
12	Charlie	1880	HT1	Welshmountain/UK		3	Riding Pony, Dartmoor Pony
13	Trotting Comet	1830	HT1	Dale or Exmoor/UK		1	Fell Pony
14	Broccoli	1855	HT1	Welsh/UK		1	Welsh-D
15	Prince Llewelly	1904	HT1	Welsh/UK		1	Connemara pony
16	Fairy Prince	1940	HT1	Shetlandpony/UK		1	Tigerpony
17	Pallieter de Bevelboekhof	1960	HT1	Shetlandpony/UK		1	Shetland pony
18	Conversano	1767	HT1	Neapolitan Horse/Italy		12	Lipizzan horse
19	Nemo	1885	HT1	Friesian/Netherlands		8	Baroque Pinto, Friesian, Kladruby
20	Old Flyer	1830	HT1	Brandenburg/Germany		8	Welsh-B, Welsh-D
21	Sacramoso Olomouc	1800	HT1	Kladruber/Czech Republik		4	Kladruby
22	Amor	1888	HT1	Pinzgauer horse/Austria		4	Noriker
23	Bravo 149	1877	HT1	Pinzgauer horse/Austria		4	Noriker
24	635 Vulkan	1887	HT1	Pinzgauer horse/Austria		4	Noriker
25	Zarif Sejer	1921	HT1	Frederiksborger/Denmark		3	Knabstrupper
26	Smoky	1955	HT1	Spotted horse/Denmark		3	Knabstrupper
27	80 Amsel 55	1866	HT1	Pinzgauer horse/Austria		3	Noriker
28	126 Optimus	1890	HT1	Pinzgauer horse/Austria		3	Noriker
29	El Bedavi	1830	HT1	Arabian	1833, Babolna	34	Haflinger
30	Siglavy	1810	HT1	Arabian	1814, Lipizza	11	Lipizzan horse
31	Kuhailan Haifi	1923	HT1	Arabian, desert bred	1931, Poland	6	Shagya Arabian, Arabian, Partbred Arabian, Riding pony
32	Ibrahim	1899	HT1	Arabian/Egypt	Poland/1910 GB	3	Riding pony, Welsh-A, Arabian
33	Khalil a Saklawi Jidran	1876	HT1	Arabian/Egypt	1910, GB, 1925 BRD	3	Arabian, AngloArabian, Shagya Arabian
34	Kohailan Adjuze	1876	HT1	Arabian, desert bred	1885 Babolna	2	Shagya Arabian
35	Kuhailan Zaid	1924	HT1	Arabian, desert bred	1931, Babolna	2	Shagya Arabian
36	Siglavy Bagdady	1895	HT1	Arabian, desert bred	1902, Babolna	2	Shagya Arabian
37	Dahoman	1846	HT1	Arabia, Syria	1852, Babolna	1	Shagya Arabian
38	Hadban	1891	HT1	Arabian, Egypt	1897, Babolna	1	Shagya Arabian
39	Ilderim	1894	HT1	Arabian	1901, Poland	1	Austrian Warmblood
40	Mahmoud Mirza	1851	HT1	Arabian, Iraq	1866, Babolna	1	Shagya Arabian
41	Bariq	1840	HT1	Arabian/Egypt	1880, GB	1	Riding pony
42	Traveler	1880	HT2	Quarter horse/U.S.		3	Quarter horse
43	Connemara Boy	1921	HT2	Connemara pony/Ireland		1	Connemara pony
44	Mahoma I	1900	HT2	Paso Fino/Colombia		1	Paso Fino
45	Neapolitano	1790	HT2	Neapolitan Horse/Italy		13	Lipizzan horse
46	Pluto	1765	HT2	Fredriksborger/Denmark		12	Lipizzan horse
47	Favory	1779	HT2	Kladruber/Czech Republik		10	Lipizzan horse
48	Maestoso	1773	HT2	Kladruber/Czech Republic		9	Lipizzan horse
49	Pepoli	1767	HT2	Orig. Italian/Italy		4	Kladruby
50	Tulipan	1880	HT2	Neapolitan Horse/Italy		3	Lipizzan horse
51	Trowert 1885	1883	HT2	German classic pony/Germany		2	German Shetland pony
52	Incitato	1802	HT2	Siebenbuerger stallion/ Hungary		2	Lipizzan horse
53	Goral	1808	HT2	Hucul/Romania		5	Hucul
54	Hroby	1894	HT2	Hucul/Romania		3	Hucul
55	Ousor	1929	HT2	Hucul/Romania		2	Hucul
56	Prislop	1932	HT2	Hucul/Romania		2	Hucul
57	Gurgul	1924	HT2	Hucul/Romania		1	Hucul
58	Polan	1929	HT2	Hucul/Romania		1	Hucul
59	Gazlan	1840	HT2	Arabian, desert bred	1852, Lipizza	4	Shagya Arabian
60	O Bajan	1881	HT2	Arabian	1886, Babolna	3	Shagya Arabian, Austrian Warmblood
61	Shagya	1830	HT2	Arabian	1836, Babolna	3	Shagya Arabian
62	Bairactar	1813	HT2	Arabian, desert bred	1817, Weil	2	Riding pony
63	Gook Yuchik	1890	HT2	Arabian	1895, BRD	2	Akhal Theke
64	Dahman Amir	1887	HT2	Arabian	1890 Szamrajówka	1	Partbred Arabian
65	Latif	1903	HT2	Arabian	1909, Pompadour	1	Partbred Arabian
66	Mersuch	1898	HT2	Arabian	1902, Babolna	1	Partbred Arabian
67	Soulakie	1894	HT2	Arabian	1899, Weil	1	Shagya Arabian
68	Byerley Turk	1684	HT2	Arabian	1689, England	8	Warmblood, English Thoroughbred, Hannoverian, Oldenburg, Quarter Horse, Sachsen Anhaltiner
69	Godolphin Arabian	1724	HT2	Arabian or Turkoman	1729, England	5	Pinto, Austrian Warmblood, Warmblood, Hungarian Warmblood
70	Darley Arabian, exluding descendants of Whalebone	1849	HT2	Arabian, Syria	1704, England	44	Standardbred, Oldenburger, Trakehner, Warmblood, Dutch Warmblood, Welsh-A, Welsh-B
71	Darley Arabian, via Whalebone	1807	HT3	Arabian, Syria		59	Bavarian Warmblood, Austrian Warmblood, English Thoroughbred, Hanoverian, Holstein, Oldenburg, Quarterhorse, Partbred Arabian, Riding pony, Rhinelander Horse
72	Hörður frá Geitaskarði	1915	HT4	Icelandic horse/Iceland		5	Icelandic horse
73	Hrottur frá Vilmundarstöðum	1946	HT4	Icelandic horse/Iceland		1	Icelandic horse
74	Íslands-Gráni frá Geitaskarði	1910	HT4	Icelandic horse/Iceland		1	Icelandic horse
75	Njål	1891	HT5	Norwegian Fjord/Norway		15	Norwegian Fjord horse
76	Jack	1817	HT6	Shetlandpony/U.K.		14	Shetland pony
77	John Bert	1880	HT6	Shetlandpony/U.K.		7	Shetland pony, Tigerpony
78	Prince of Thule	1872	HT6	Shetlandpony/U.K.		1	Shetland pony

Stallions with descendants in our dataset are listed, giving their origin, HT and their distribution in extant horse breeds (as estimated from our dataset).
doi:10.1371/journal.pone.0060015.t001

doi:10.1371/journal.pone.0060015.t001

At least 202 (48.3%) of the modern horses in our dataset obtained their Y chromosomes from Original Arabian founders. Among the HT2 ancestors are many influential stallions, including the famous founders of the English Thoroughbred: Darley Arabian, Byerley Turk and Godolphin Arabian.

The English Thoroughbred, which is best known for its use in horse racing, has a complete studbook since 1791 and all registered males can be traced back to one of three popular founders. Among them, the paternal line of Darley Arabian currently represents almost all male Thoroughbreds [47]. Pedigree analysis revealed that all HT3-carrying males can be traced to the single Thoroughbred "Whalebone", born 1807" a son of "Pot8os". Hence, the mutation leading from HT2 to HT3 must have occurred either in the germline of the famous racehorse "Eclipse" or in that of his son "Waxy" or grandson "Pot8os" (Fig. 4). The frequency of HT3 rose to 96.5% in the English Thoroughbred and to 41% in modern sport horse breeds in our dataset within 15–20 generations (Fig. 3A). Among 418 long pedigrees, we observed no obvious pedigree errors in the English Thoroughbred and Standardbreds, whereas 17 (4.06%) errors were found in Shetland ponies and Lipizzan, Warmblood and Welsh horses. This observation does not permit us to check the correctness of the studbook, which must be undertaken with maternal lines using mtDNA [48]–[51]. The low initial variation in the founders leads to a low resolution, making pedigree errors in the male line difficult to detect.

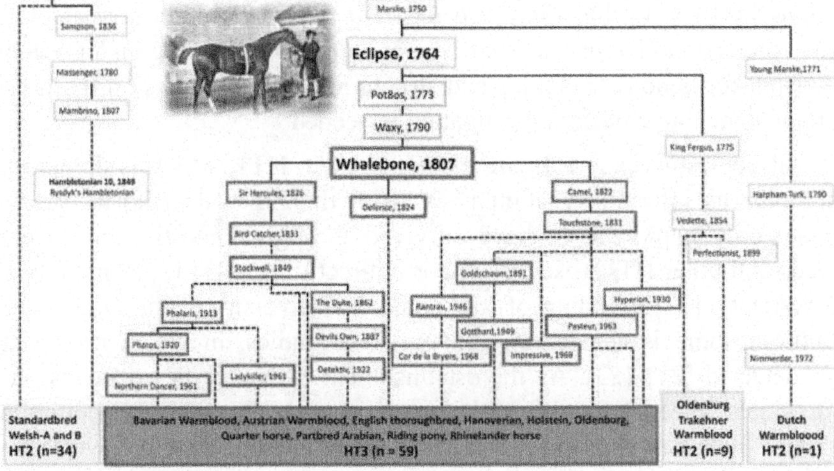

Figure 4: Pedigree of Darley Arabians progeny depicting the origin of HT3 from HT2.

Breeds of analysed males are listed on the bottom and the haplotypes of their ancestors are reconstructed (HT2-yellow, HT3-red, unknown-grey). Selected famous stallions are shown by name; dotted lines connect relatives where at least one ancestor is omitted. No descendants from "Pot8os" and "Waxy" were available apart from "Whalebone, 1807". The mutation leading to HT3 must have occurred either in the germline of stallion "Eclipse" [54] or in his son "Pot8os" or in his grandson "Waxy" and rose to very high frequency in the English Thoroughbred and many sport horse breeds through the progeny of the stallion "Whalebone".

doi:10.1371/journal.pone.0060015.g004

DISCUSSION

The analysis of Y-chromosomal and mtDNA markers offers an invaluable tool for the demographic characterization of populations. To date, the domestic horse was the only livestock species for which paternal lines could not be traced due to the lack of Y-chromosomal variability. We have identified five variants that presumably arose independently and that result in three major and three breed-specific HTs.

Of these five variants, only two arose by single basepair mutations. The other three events arose in a restricted region YE3 with extensive sequence similarity to a X-chromosomal region and consist of a gene conversion tract of about 300 basepairs, a single basepair deletion, which may also be caused by a gene conversion, and a deletion of about 900 bps. X to Y gene conversion strongly influences allelic diversity in specific human Y-chromosomal regions [4],[41]. In this study, we report the first observation of an X to Y gene conversion in a farm animal. The hyperpolymorphic region YE3 may be a hotspot for gene conversions and structural rearrangements on the horse Y chromosome, but a closer investigation is needed.

All domestic horse HTs are closely related. HT1, which is the ancestral haplotype as inferred by comparison with the Przewalski horse, seems to be the only haplotype to have survived through domestication to extant breeds. All other HTs arose directly or indirectly from HT1, presumably after domestication. The finding of low nucleotide diversity of the modern horse Y chromosome is consistent with previous studies, in which no variation was detected [21]–[23]. As the establishment of haplotypes depends on the individuals selected for the initial screen the significance of ascertainment bias has to be kept in mind [6], [52]. To overcome this problem, we screened microsatellite markers in a random sample over all haplotypes/regions and detected no variation. As microsatellites are highly mutable, the absence of significant microsatellite variation on the horse Y chromosome confirms the very recent origin of all haplotypes in our sample. Since we only used puredbred horses from a restricted region we note, that the global horse population may harbour more y-chromosomal variability.

The low diversity of the Y chromosome contrasts with the high diversity of mtDNA haplotypes observed in modern horses [15]–[19]. The difference is likely caused by the low effective population size of the horse Y chromosome due to a strong variation in male reproductive success. This may be due to the polygynous breeding patterns in wild horses and to a stronger bottleneck in male horses during domestication and might be further exacerbated by the intensive breeding practices in this species [20], [21].

The samples used in our study derive from purebred modern, mainly European, breeds, which have undergone intensive selection for particular traits during the past two centuries [30]. The refinement was mainly achieved through the disproportionate use of selected popular stallions and their descendants that were crossed to local mares. With the use of pedigree information, available in studbooks and open access databases, we inferred the impact of the upgrading process on the male horse population. We find that only a limited number of founders contribute to the extant horse haplotypes (Fig. 3B, Table 1).

Based on the pedigree information, we traced the effects of three introgression waves (the Neapolitan, the Oriental and the English waves) on NRY markers. The importance of the Thoroughbred in the English wave is clearly seen through the spread of HT3. In the "Original Arabians", the Neapolitan horse and the central and Eastern European founders, the proportion of HT2 is about 50%. In founders from northern Europe, i.e. Iceland, Norway and the British Isles, and the Iberian Peninsula the frequency of HT2 was very low. The distribution of HT2 is consistent with the movement of stallions from the Middle East to Central and Western Europe via the Neapolitan and Oriental waves (Fig. 3C, Fig. S10). The high proportion of HT2 in Central European horses may also be derived from earlier introgression of horses from the Middle East or even from ancient colonization [20].

Only three northern European horse breeds, Icelandic horses, Shetland ponies and Norwegian Fjord horses, were either not or were hardly subjected to these introgression waves and were therefore able to maintain autochthonous Y chromosome variants. These breeds have a comparatively isolated history. Due to the specific geographical and social structures in northern areas, locally adapted breeds were presumably of a higher value than imported animals from Central Europe and the Near East. In the case of Icelandic horses, the import of horses was restricted since 930 A.D. and has been prohibited since 1909 [30]. The three breeds are not necessarily closely related to one another. Icelandic and Norwegian Fjord horses branch from the root of modern breeds, when using 46244 autosomal loci [53]. As we only detected the ubiquitous HT1, little can be deduced about the ancestry of the Barb, the Iberian breeds, or the Swedish Coldblood Trotter.

The history of the domestic horse is marked by recurrent adjustment to changing socio-economic needs. Over the course of the last two centuries, when most modern breeds were established, the horse has been undergoing a transition from a working animal also used by the military use towards a domestic animal used in leisure and sports activities. This has largely been achieved through the use of a few, very popular, sires that have been extensively shifted among breeds. Due to the breeding practices during the last 200 years, "popular sires" and their sons have fathered an inordinate amount of offspring and replaced local Y chromosomes. Unique variants can only be found in a few northern European breeds. The restricted genetic diversity of the modern horse Y chromosome reflects what has survived through the species' dynamic history.

CONCLUSIONS

We describe the first polymorphic markers on the paternally inherited part of the Y chromosome of the domestic horse. We have used the new markers to investigate the paternal gene flow between horse populations and breeds. We document the influence of popular sires, particularly a single Thoroughbred stallion line, on many extant horse breeds. Our data on male lineages complement the information on the well established maternal mtDNA lineages and enhance the genetic documentation of the history and dynamics of modern horse breeds. The new polymorphic markers and haplotypes enable horse breeding practices to be monitored and verified. Although six haplotypes do not offer a high resolution, this study provides a first backbone phylogeny for deeper population genetic studies of the Y chromosome variation in horses.

DATA DEPOSITION

BAC sequences have been deposited in Genbank (http://www.ncbi.nlm.nih.gov/genbank) under accession numbers JX565700–JX565709, sequence alignments under accession numbers JX646942– JX647045. Illumina short read sequences generated in this study are available at the Sequence Read Archive (SRA) under the Accesssion number ERP001668 (http://www.ebi.ac.uk/ena/data/view/ERP001668). Y-chromosomal SNPs have been submitted to the NCBI dbSNP database: ss#711581504, ss#711581506, ss#711581507.

ACKNOWLEDGMENTS

We thank C Guelly at the Core Facility Molekularbiologie (Meduni Graz, AT) for 454 and A Sommer at the CSF NGS Unit Vienna for Illumina sequencing; A Ertl, M and V Dobretsberger and private horse owners for sample supply; N Kreutzmann, M Schwender and D Rigler for assistance in the lab. We are grateful to P Burger, K Traxler, T Leeb, S Mueller and S Macho-Maschler for their comments on the manuscript and and we thank G Tebb for editing.

AUTHOR CONTRIBUTIONS

Conceived and designed the experiments: BW GB. Performed the experiments: BW JB. Analyzed the data: BW PS CV TD. Contributed reagents/materials/ analysis tools: JB PS. Wrote the paper: BW CV TD GB.

REFERENCES

1. Xue Y, Wang Q, Long Q, Ng BL, Swerdlow H, et al. (2009) Human Y chromosome base-substitution mutation rate measured by direct sequencing in a deep-rooting pedigree. Curr Biol 19: 1453–1457. doi: 10.1016/j.cub.2009.07.032
2. Repping S, van Daalen SK, Brown LG, Korver CM, Lange J, et al. (2006) High mutation rates have driven extensive structural polymorphism among human Y chromosomes. Nat Genet 38: 463–467. doi: 10.1038/ ng1754
3. Jobling MA, Samara V, Pandya A, Fretwell N, Bernasconi B, et al. (1996) Recurrent duplication and deletion polymorphisms on the long arm of the Y chromosome in normal males. Hum Mol Genet 5: 1767–1775. doi: 10.1093/hmg/5.11.1767
4. Trombetta B, Cruciani F, Underhill PA, Sellitto D, Scozzari R (2009) Footprints of X-to-Y gene conversion in recent human evolution. Mol Biol Evol 27: 714–725. doi: 10.1093/molbev/msp231
5. Ellegren H (2011) Sex-chromosome evolution: recent progress and the influence of male and female heterogamety. Nat Rev Genet 12: 157–166. doi: 10.1038/nrg2948
6. Underhill PA, Kivisild T (2007) Use of y chromosome and mitochondrial DNA population structure in tracing human migrations. Annu Rev Genet 41: 539–564. doi: 10.1146/annurev.genet.41.110306.130407
7. King T, Jobling M (2009) What's in a name? Y chromosomes, surnames and the genetic genealogy revolution. Trends Genet 25: 351–360. doi: 10.1016/j.tig.2009.06.003

8. Ding ZL, Oskarsson M, Ardalan A, Angleby H, Dahlgren LG, Tepeli C, Kirkness E, Savolainen P, Zhang YP (2012) Origins of domestic dog in Southern East Asia is supported by analysis of Y-chromosome DNA. Heredity 108: 507–514. doi: 10.1038/hdy.2011.114
9. Brown SK, Pedersen NC, Jafarishorijeh S, Bannasch DL, Ahrens KD, et al. (2011) Phylogenetic distinctiveness of Middle Eastern and Southeast Asian village dog Y chromosomes illuminates dog origins. PLoS One 6: e28496. doi: 10.1371/journal.pone.0028496
10. Sundqvist A, Björnerfeldt S, Leonard J, Hailer F, Hedhammar A, et al. (2006) Unequal contribution of sexes in the origin of dog breeds. Genetics 172: 1121–1128. doi: 10.1534/genetics.105.042358
11. Ramirez O, Ojeda A, Tomas A, Gallardo D, Huang L, et al. (2009) Integrating Y-chromosome, mitochondrial, and autosomal data to analyze the origin of pig breeds. Mol Biol Evol 26: 2061–2072. doi: 10.1093/molbev/msp118
12. Cliffe KM, Day AE, Bagga M, Siggens K, Quilter CR, et al. (2010) Analysis of the non-recombining Y chromosome defines polymorphisms in domestic pig breeds: ancestral bases identified by comparative sequencing. Anim Genet 41: 619–629. doi: 10.1111/j.1365-2052.2010.02070.x
13. Meadows J, Hanotte O, Drögemüller C, Calvo J, Godfrey R, et al. (2006) Globally dispersed Y chromosomal haplotypes in wild and domestic sheep. Anim Genet 37: 444–453. doi: 10.1111/j.1365-2052.2006.01496.x
14. Edwards CJ, Ginja C, Kantanen J, Pérez-Pardal L, Tresset A, et al. (2011) Dual Origins of Dairy Cattle Farming – Evidence from a Comprehensive Survey of European Y-Chromosomal Variation. PLoS ONE 6(1): e15922. doi: 10.1371/journal.pone.0015922
15. Vila C, Leonard J, Gotherstrom A, Marklund S, Sandberg K, et al. (2001) Widespread origins of domestic horse lineages. Science 291: 474–477. doi: 10.1126/science.291.5503.474
16. Jansen T, Forster P, Levine M, Oelke H, Hurles M, et al. (2002) Mitochondrial DNA and the origins of the domestic horse. Proc Natl Acad Sci U S A 99: 10905–10910. doi: 10.1073/pnas.152330099
17. Cieslak M, Pruvost M, Benecke N, Hofreiter M, Morales A, et al. (2010) Origin and history of mitochondrial DNA lineages in domestic horses. PLoS One 5: e15311. doi: 10.1371/journal.pone.0015311
18. Lippold S, Matzke NJ, Reissmann M, Hofreiter M (2011) Whole mitochondrial genome sequencing of domestic horses reveals

incorporation of extensive wild horse diversity during domestication. BMC Evol Biol 11: 328. doi: 10.1186/1471-2148-11-328

19. Achilli A, Olivieri A, Soares P, Lancioni H, Hooshiar Kashani B, et al. (2012) Mitochondrial genomes from modern horses reveal the major haplogroups that underwent domestication. Proc Natl Acad Sci U S A 109: 2449–2454. doi: 10.1073/pnas.1111637109
20. Warmuth V, Eriksson A, Bower MA, Barker G, Barrett E, et al. (2012) Reconstructing the origin and spread of horse domestication in the Eurasian steppe. Proc Natl Acad Sci U S A 109: 8202–8206. doi: 10.1073/pnas.1111122109
21. Lindgren G, Backstroem N, Swinburne J, Hellborg L, Einarsson A, et al. (2004) Limited number of patrilines in horse domestication. Nat Genet 36: 335–336. doi: 10.1038/ng1326
22. Wallner B, Brem G, Mueller M, Achmann R (2003) Fixed nucleotide differences on the Y chromosome indicate clear divergence between Equus przewalskii and Equus caballus. Anim Genet 34: 453–456. doi: 10.1046/j.0268-9146.2003.01044.x
23. Wallner B, Piumi F, Brem G, Mueller M, Achmann R (2004) Isolation of Y chromosome-specific microsatellites in the horse and cross-species amplification in the genus Equus. J Hered 95: 158–164. doi: 10.1093/jhered/esh020
24. Lippold S, Knapp M, Kuznetsova T, Leonard JA, Benecke N, et al. (2011) Discovery of lost diversity of paternal horse lineages using ancient DNA. Nat Commun 2: 450. doi: 10.1038/ncomms1447
25. Ling Y, Ma Y, Guan W, Cheng Y, Wang Y, et al. (2010) Identification of y chromosome genetic variations in Chinese indigenous horse breeds. J Hered 101: 639–643. doi: 10.1093/jhered/esq047
26. Brem G (2011) Der Lipizzaner im Spiegel der Wissenschaft. Vienna: Austrian Academy of Sciences Press. 338 p.
27. Hamann H, Distl O (2008) Genetic variability in Hanoverian warmblood horses using pedigree analysis. J Anim Sci 86: 1503–1513. doi: 10.2527/jas.2007-0382
28. Druml T (2009) Functional Traits in Early Horse Breeds of Mongolia, India and China from the Perspective of Animal Breeding. In: B F, V S, R P, A S, editors. Horses in Asia: History, Trade and Culture. Vienna: Austrian Academy of Sciences. 9–16.
29. Grilz-Seger G and Druml T (2011) Lipizzaner Hengststämme. Graz: Vehling Medienservice und Verlag GmbH. 312 p.

30. Hendricks B (1995) International Encyclopedia of Horse Breeds. Norman: University of Oklahoma Press. 486 p.

31. Out AA, van Minderhout IJ, Goeman JJ, Ariyurek Y, Ossowski S, et al. (2009) Deep sequencing to reveal new variants in pooled DNA samples. Hum Mutat 30: 1703–1712. doi: 10.1002/humu.21122

32. Godard S, Schibler L, Oustry A, Cribiu E, Guerin G (1998) Construction of a horse BAC library and cytogenetical assignment of 20 type I and type II markers. Mamm Genome 9: 633–637. doi: 10.1007/s003359900835

33. Raudsepp T, Santani A, Wallner B, Kata S, Ren C, et al. (2004) A detailed physical map of the horse Y chromosome. Proc Natl Acad Sci U S A 101: 9321–9326. doi: 10.1073/pnas.0403011101

34. Quail MA, Kozarewa I, Smith F, Scally A, Stephens PJ, et al. (2008) A large genome center's improvements to the Illumina sequencing system. Nat Methods 5: 1005–1010. doi: 10.1038/nmeth.1270

35. Andrews S. (2010) FastQC: a quality control tool for high throughput sequence data. Available: www.bioinformatics.bbsrc.ac.uk/projects/fastqc/Accessed 2012 January.

36. Li H, Durbin R (2009) Fast and accurate short read alignment with Burrows-Wheeler transform. Bioinformatics 25: 1754–1760. doi: 10.1093/bioinformatics/btp324

37. Li H, Handsaker B, Wysoker A, Fennell T, Ruan J, et al. (2009) The Sequence Alignment/Map format and SAMtools. Bioinformatics 25: 2078–2079. doi: 10.1093/bioinformatics/btp352

38. Nei M (1987) Molecular evolutionary genetics. New York: Columbia University Press. 512 p.

39. Buggs RJ, Chamala S, Wu W, Gao L, May GD, et al. (2010) Characterization of duplicate gene evolution in the recent natural allopolyploid Tragopogon miscellus by next-generation sequencing and Sequenom iPLEX MassARRAY genotyping. Mol Ecol 19 Suppl 1132–146. doi: 10.1111/j.1365-294x.2009.04469.x

40. Paria N, Raudsepp T, Pearks Wilkerson AJ, O'Brien PC, Ferguson-Smith MA, et al. (2011) A gene catalogue of the euchromatic male-specific region of the horse y chromosome: comparison with human and other mammals. PLoS One 6: e21374. doi: 10.1371/journal.pone.0021374

41. Iwase M, Satta Y, Hirai H, Hirai Y, Takahata N (2010) Frequent gene conversion events between the X and Y homologous chromosomal regions in primates. BMC Evol Biol 10: 225. doi: 10.1186/1471-2148-10-225

42. Rosser ZH, Balaresque P, Jobling MA (2009) Gene conversion between the X chromosome and the male-specific region of the Y chromosome at a translocation hotspot. Am J Hum Genet 85: 130–134. doi: 10.1016/j.ajhg.2009.06.009
43. Conrad DF, Keebler JE, DePristo MA, Lindsay SJ, Zhang Y, et al. (2011) Variation in genome-wide mutation rates within and between human families. Nat Genet 43: 712–714. doi: 10.1038/ng.862
44. Roach JC, Glusman G, Smit AF, Huff CD, Hubley R, et al. (2010) Analysis of genetic inheritance in a family quartet by whole-genome sequencing. Science 328: 636–639. doi: 10.1126/science.1186802
45. Clutton-Brock J (1999) A Natural History of Domesticated Mammals. Cambridge, Massachusetts: Cambridge University Press. 248 p.
46. Natanaelsson C, Oskarsson M, Angleby H, Lundeberg J, Kirkness E, et al. (2006) Dog Y chromosomal DNA sequence: identification, sequencing and SNP discovery. BMC Genet 7: 45. doi: 10.1186/1471-2156-7-45
47. Cunningham E, Dooley J, Splan R, Bradley D (2001) Microsatellite diversity, pedigree relatedness and the contributions of founder lineages to thoroughbred horses. Anim Genet 32: 360–364. doi: 10.1046/j.1365-2052.2001.00785.x
48. Kavar T, Brem G, Habe F, Sölkner J, Dovc P (2002) History of Lipizzan horse maternal lines as revealed by mtDNA analysis. Genet Sel Evol 34: 635–648. doi: 10.1186/1297-9686-34-5-635
49. Bower MA, Campana MG, Whitten M, Edwards CJ, Jones H, et al. (2011) The cosmopolitan maternal heritage of the Thoroughbred racehorse breed shows a significant contribution from British and Irish native mares. Biol Lett 7: 316–320. doi: 10.1098/rsbl.2010.0800
50. Bower MA, Campana MG, Nisbet RE, Weller R, Whitten M, et al. (2012) Truth in the bones: resolving the identity of the founding elite thoroughbred racehorses. Archaeometry 54: 916–925. doi: 10.1111/j.1475-4754.2012.00666.x
51. Hill EW, Bradley DG, Al-Barody M, Ertugrul O, Splan RK, et al. (2002) History and integrity of thoroughbred dam lines revealed in equine mtDNA variation. Anim Genet 33: 287–294. doi: 10.1046/j.1365-2052.2002.00870.x
52. Lenstra JA, Groeneveld LF, Eding H, Kantanen J, Williams JL, et al. (2012) Molecular tools and analytical approaches for the characterization of farm animal genetic diversity. Anim Genet 43: 483–502. doi: 10.1111/j.1365-2052.2011.02309.x

53. McCue ME, Bannasch DL, Petersen JL, Gurr J, Bailey E, et al. (2012) A High Density SNP Array for the Domestic Horse and Extant Perissodactyla: Utility for Association Mapping, Genetic Diversity, and Phylogeny Studies. PLoS Genet 8(1): e1002451. doi: 10.1371/journal.pgen.1002451

54. *Eclipse At New Market With Groom* (British racehorses of the 18th century), by Stubbs George (1724–1806)*;* source: Wikimedia Commons, public domain, *copyright has expired.*

CITATION

CHAPTER 1

Mao, W. and Lee, J. (2008) A Combinatorial Analysis of Genetic Data for Crohn's Disease. Journal of Biomedical Science and Engineering, 1, 52-58. doi: 10.4236/jbise.2008.11008.

CHAPTER 2

Anil V Parwani M.D., Ph.D, Joseph Geradts M.D., Eric Caspers M.D., G Johan Offerhaus M.D., Charles J Yeo M.D, John L Cameron M.D., David S Klimstra M.D., Anirban Maitra M.D., Ralph H Hruban M.D. and Pedram Argani M.D., Immunohistochemical and Genetic Analysis of Non–Small Cell and Small Cell Gallbladder Carcinoma and Their Precursor Lesions, DOI: 10.1097/01.MP.0000062656.60581.

CHAPTER 3

Kuramoto T, Nakanishi S, Ochiai M, Nakagama H, Voigt B, Serikawa T (2012) Origins of Albino and Hooded Rats: Implications from Molecular Genetic Analysis across Modern Laboratory Rat Strains. PLoS ONE 7(8): e43059. doi:10.1371/journal.pone.0043059.

CHAPTER 4

Dominik T Schneider1, Susanne Zahn1, Sonja Sievers1,2, Katayoun Alemazkour1, Guido Reifenberger3, Otmar D Wiestler4,†, Gabriele Calaminus1, Ulrich Göbel1 and Elizabeth J Perlman, Molecular genetic analysis of central nervous system germ cell tumors with comparative genomic hybridization, doi:10.1038/modpathol.3800607.

CHAPTER 5

Sue S Yom, Asif Rashid, David I Rosenthal, Danielle D Elliott, Ehab Y Hanna, Randal S Weber and Adel K El-Naggar, Genetic analysis of sinonasal adenocarcinoma phenotypes: distinct alterations of histogenetic significance, doi:10.1038/modpathol.3800315.

CHAPTER 6

Sousa-Neves R, Rosas A (2010) An Analysis of Genetic Changes during the Divergence of Drosophila Species. PLoS ONE 5(5): e10485. doi:10.1371/ journal.pone.0010485.

CHAPTER 7

Huang Q, Li G, Husseneder C, Lei C (2013) Genetic Analysis of Population Structure and Reproductive Mode of the Termite Reticulitermes chinensis Snyder. PLoS ONE 8(7): e69070. doi:10.1371/journal.pone.0069070.

CHAPTER 8

Pan Y, Zhang H, Zhang D, Li J, Xiong H, Yu J, et al. (2015) Genetic Analysis of Cold Tolerance at the Germination and Booting Stages in Rice by Association Mapping. PLoS ONE 10(3): e0120590. doi:10.1371/journal.pone.0120590.

CHAPTER 9

Yates WR, Johnson C, McKee P, Cannon-Albright LA (2013) Genetic Analysis of Low BMI Phenotype in the Utah Population Database. PLoS ONE 8(12): e80287. doi:10.1371/journal.pone.0080287.

CHAPTER 10

Dong Zhang, Wenqian Kong, Jon Robertson, Valorie H Goff, Ethan Epps, Alexandra Kerr, Gabriel Mills, Jay Cromwell, Yelena Lugin, Christine Phillips and Andrew H Paterson, Genetic analysis of inflorescence and plant height components in sorghum (Panicoidae) and comparative genetics with rice (Oryzoidae), DOI: 10.1186/s12870-015-0477-6.

CHAPTER 11

Petersen JL, Mickelson JR, Cothran EG, Andersson LS, Axelsson J, Bailey E, et al. (2013) Genetic Diversity in the Modern Horse Illustrated from Genome-Wide SNP Data. PLoS ONE 8(1): e54997. doi:10.1371/journal.pone.0054997

CHAPTER 12

Badro DA, Douaihy B, Haber M, Youhanna SC, Salloum A, Ghassibe-Sabbagh M, et al. (2013) Y-Chromosome and mtDNA Genetics Reveal Significant Contrasts in Affinities of Modern Middle Eastern Populations with European and African Populations. PLoS ONE 8(1): e54616. doi:10.1371/journal.pone.0054616.

CHAPTER 13

Wallner B, Vogl C, Shukla P, Burgstaller JP, Druml T, Brem G (2013) Identification of Genetic Variation on the Horse Y Chromosome and the Tracing of Male Founder Lineages in Modern Breeds. PLoS ONE 8(4): e60015. doi:10.1371/journal.pone.0060015.

INDEX

M

N

O

P

Q

R

S

T

U

W